Series Editors: I. Appenzeller, Heidelberg, Germany
G. Börner, Garching, Germany
M. Harwit, Washington, DC, USA
R. Kippenhahn, Göttingen, Germany
J. Lequeux, Paris, France
P. A. Strittmatter, Tucson, AZ, USA
V. Trimble, College Park, MD, and Irvine, CA, USA

Springer
Berlin
Heidelberg
New York
Barcelona
Hong Kong
London
Milan
Paris
Singapore
Tokyo

ASTRONOMY AND ASTROPHYSICS LIBRARY

Series Editors: I. Appenzeller · G. Börner · M. Harwit · R. Kippenhahn
J. Lequeux · P. A. Strittmatter · V. Trimble

Theory of Orbits (2 volumes)
Volume 1: Integrable Systems and Non-perturbative Methods
Volume 2: Perturbative and Geometrical Methods
By D. Boccaletti and G. Pucacco

Galaxies and Cosmology
By F. Combes, P. Boissé, A. Mazure and A. Blanchard

The Solar System 2nd Edition By T. Encrenaz and J.-P. Bibring

The Physics and Dynamics of Planetary Nebulae By G. A. Gurzadyan

Astrophysical Concepts 2nd Edition By M. Harwit

Stellar Structure and Evolution By R. Kippenhahn and A. Weigert

Modern Astrometry By J. Kovalevsky

Astrophysical Formulae 3rd Edition (2 volumes)
Volume I: Radiation, Gas Processes and High Energy Astrophysics
Volume II: Space, Time, Matter and Cosmology
By K. R. Lang

Observational Astrophysics 2nd Edition
By P. Léna, F. Lebrun and F. Mignard

Galaxy Formation By M. S. Longair

General Relativity, Astrophysics, and Cosmology
By A. K. Raychaudhuri, S. Banerji and A. Banerjee

Tools of Radio Astronomy 3rd Edition
By K. Rohlfs and T. L. Wilson

Atoms in Strong Magnetic Fields
Quantum Mechanical Treatment and Applications
in Astrophysics and Quantum Chaos
By H. Ruder, G. Wunner, H. Herold and F. Geyer

The Stars By E. L. Schatzman and F. Praderie

Gravitational Lenses By P. Schneider, J. Ehlers and E. E. Falco

**Relativity in Astrometry, Celestial Mechanics
and Geodesy** By M. H. Soffel

The Sun An Introduction By M. Stix

Galactic and Extragalactic Radio Astronomy 2nd Edition
Editors: G. L. Verschuur and K. I. Kellermann

Reflecting Telescope Optics (2 volumes)
Volume I: Basic Design Theory and its Historical Development
Volume II: Manufacture, Testing, Alignment, Modern Techniques
By R. N. Wilson

P. Schneider   J. Ehlers
E. E. Falco

Gravitational Lenses

With 112 Figures

 Springer

Dr. Peter Schneider
Max-Planck-Institut für Astrophysik
Karl-Schwarzschild-Strasse 1
85748 Garching, Germany

Professor Dr. Jürgen Ehlers
Max-Planck-Institut für Gravitationsphysik
Albert-Einstein-Institut
Schlaatzweg 1
14473 Potsdam, Germany

Dr. Emilio E. Falco
Harvard-Smithsonian Center for Astrophysics
60 Garden Street
Cambridge, MA 02138, USA

Cover picture: When a source, e.g., a QSO, lies behind a foreground galaxy, its light boundle is affected by the individual stars of this galaxy. This microlensing effect, so far observed in at least one QSO (see Sect. 12.4), leads to a change in the flux we observe from the source, relative to an unlensed source. The flux magnification depends sensitively on the position of the source relative to the stars in the galaxy. Here we see the magnification as a function of the relative source position: red and yellow indicates high magnification, green and blue low magnification. The superimposed white curves are the caustics produced by the stars, projected into the source plane. The figure (taken from Wambsganss, Witt, and Schneider, *Astr. Astrophys., 258,* 591 (1992)) has been produced by combining the ray-shooting method (Sects. 10.6 and 11.2.5) with the parametric representation of caustics (Sect. 8.3.4). The parameters for the star field are $\kappa_* = 0.5, \gamma = \kappa_c = 0$, with all stars having the same mass. The shape of the acoustics is analyzed in Chap. 6.

Library of Congress Cataloging-in-Publication Data
Die Deutsche Bibliothek – CIP-Einheitsaufnahme

Schneider, Peter: Gravitational lenses / P. Schneider; J. Ehlers; E. E. Falco. – Study ed., 2. printing. –
Berlin; Heidelberg; New York; Barcelona; Hong Kong; London; Milan; Paris; Singapore; Tokyo: Springer, 1999
(Astronomy and astrophysics library)
ISBN 3-540-66506-4

1st Edition 1992
2nd Printing 1999

ISSN 0941-7834
ISBN 3-540-66506-4 Study Edition Springer-Verlag Berlin Heidelberg New York
ISBN 0-387-97070-3 First Edition Springer-Verlag Berlin Heidelberg New York

This work is subject to copyright. All rights are reserved, whether the whole or part of the material is concerned, specifically the rights of translation, reprinting, reuse of illustrations, recitation, broadcasting, reproduction on microfilm or in any other way, and storage in data banks. Duplication of this publication or parts thereof is permitted only under the provisions of the German Copyright Law of September 9, 1965, in its current version, and permission for use must always be obtained from Springer-Verlag. Violations are liable for prosecution under the German Copyright Law.

© Springer-Verlag Berlin Heidelberg 1999
© Springer-Verlag New York 1992
Printed in Germany

The use of general descriptive names, registered names, trademarks, etc. in this publication does not imply, even in the absence of a specific statement, that such names are exempt from the relevant protective laws and regulations and therefore free for general use.

Typesetting: Data conversion by Springer-Verlag
Cover design: *design & production* GmbH, Heidelberg

SPIN: 10744664   55/3144/XO - 5 4 3 2 1 0 - Printed on acid-free paper

Preface

The theory, observations, and applications of gravitational lensing constitute one of the most rapidly growing branches of extragalactic astrophysics. The deflection of light from very distant sources by intervening masses provides a unique possibility for the investigation of both background sources and lens mass distributions. Gravitational lensing manifests itself most distinctly through multiply imaged QSOs and the formation of highly elongated images of distant galaxies ('arcs') and spectacular ring-like images of extragalactic radio sources. But the effects of gravitational light deflection are not limited to these prominent image configurations; more subtle, since not directly observable, consequences of lensing are the, possibly strong, magnification of sources, which may permit observation of intrinsically fainter, or more distant, sources than would be visible without these natural telescopes. Such light deflection can also affect the source counts of QSOs and of other compact extragalactic sources, and can lead to flux variability of sources owing to propagation effects.

Trying to summarize the theory and observational status of gravitational lensing in a monograph turned out to be a bigger problem than any of the authors anticipated when we started this project at the end of 1987, encouraged by Martin Harwit, who originally approached us. The development in the field has been very rapid during the last four years, both through theory and through observation, and many sections have been rewritten several times, as the previous versions became out of date. Writing a book on such a timely field forced us to compromise between presenting the most recent results and concentrating on more fundamental topics. We have tried to separate these two strategies, in the hope that most of the sections will be useful by the time the remaining sections need an update because of new developments.

Since gravitational lensing encroaches upon an increasing number of fields in astrophysics, we have decided to present, in Chap. 2, the basic notions and facts of lensing life, with only a minimum of theory, and also to give a fairly detailed account of the current observational status. This chapter is meant to be of interest to those astrophysicists who may not intend to dig deeper into the theory. However, in the discussion of observed gravitational lens candidate systems we refer to results which are derived in later chapters, so that some details of this discussion can be fully understood only after reading the respective sections. The physics of gravitational light deflection is contained in Einstein's Theory of General Relativity. According to it, gravitational lens research investigates the shape, in particular the caustics, of our past light cone which, via the gravitational field equation, is related to the distribution of matter. We therefore thought it useful to derive the equations of gravitational lensing as far as possible from the basic laws of that theory; Chap. 3 deals with optics in arbitrarily curved spacetimes, and Chap. 4 concerns the "local" and "global" use of Einstein's field equation for lensing. Only by following these arguments in detail can one judge the validity and also recognize the shortcomings of the present state of gravitational lens theory. A more heuristic 'derivation' of the relevant lens equations is provided in Chap. 2 for readers not familiar with General Relativity.

Although not planned as a textbook, we have tried to make this book available to students of physics and astrophysics. Since many results of gravitational lens theory can be obtained with fairly simple mathematical tools, we consider lensing an ideal subject for courses or small projects on extragalactic astronomy and cosmology.

The selection of material is, of course, partly a matter of taste and strongly influenced by the expertise of the authors. Since we expect that lensing will continue to be an important field of research, we consider the presentation of methods at least as important as reporting results. A large fraction of this book is dedicated to statistical lensing, since we believe this to be the field where most progress is to be expected during the next several years. On the other hand, a few subjects such as the possible influence of lensing on the isotropy of the microwave background radiation are mentioned briefly only. Instead of modeling observed lens systems in detail, we describe the necessary methods and present reasonably simple models. We apologize to all who may feel that their favorite subject is not represented fairly.

During the preparation of this book, we have profited directly from many colleagues, too numerous all to be mentioned. M. Bartelmann, R.D. Blandford, H. Erdl, M. Freyberg, C.S. Kochanek and J. Wambsganss read and commented on the largest part of the manuscript; P. Haines, E.V. Linder, T. Schramm, and A. Weiss critically read selected chapters. Their advice has been most valuable to us, even if we did not always follow their suggestions; we are deeply indebted to them. Further, we thank the participants of the gravitational lensing conferences in Cambridge, MA, and Toulouse, for the lively and inspiring discussions, and those who have encouraged us to proceed with this project. Those colleagues who communicated to us unpublished work enabled us to include very recent results; we would like to thank them, as well as those who agreed to have their graphic material reproduced in this book.

Although the preparation of the manuscript was mainly done by the authors, it would never have been finished without P. Berkemeyer, who TEXed the difficult parts, in particular all tables and those sections where the density of equations is highest; she also provided invaluable help in preparing the bibliography. Many colleagues offered their expertise in TEX; in particular, M. Lottermoser provided invaluable assistance in the preparation of the index. Most of the figures in the text were prepared by G. Wimmersberger, who managed to produce fine drawings from sloppy sketches; to all, we are most grateful.

We thank the Max-Planck-Institut für Astrophysik, in particular its director R. Kippenhahn, for constant support, by letting us make use of all facilities, and financial support, without which this transatlantic collaboration would have been much more difficult.

EEF wishes to thank in particular his wife, J. Titilah, for her love and understanding, M. Geller and H. Huchra for their support, and I. Shapiro for his encouragement.

Finally we wish to thank Springer-Verlag for their cooperation and patience, waiting for the manuscript for about two years after the original deadline.

Garching *P. Schneider* and *J. Ehlers*
Cambridge *E.E. Falco*
April 1992

Contents

1. **Introduction** 1
   1.1 Historical remarks 1
      1.1.1 Before 1919 1
      1.1.2 The period 1919–1937 3
      1.1.3 The period 1963–1979 6
      1.1.4 Post-1979 9
   1.2 Outline of the book 11
   1.3 Remarks about notation 21

2. **Basic facts and the observational situation** 25
   2.1 The Schwarzschild lens 25
   2.2 The general lens 29
   2.3 The magnification factor 33
   2.4 Observing gravitational lens systems 41
      2.4.1 Expectations for point sources 42
      2.4.2 Expectations for extended sources ... 46
   2.5 Known gravitational lens systems 47
      2.5.1 Doubles 48
      2.5.2 Triples 60
      2.5.3 Quadruples 64
      2.5.4 Additional candidates 71
      2.5.5 Arcs 72
      2.5.6 Rings 77
      2.5.7 A rapidly growing list of candidates .. 84
      2.5.8 Speculations on other gravitational lens systems . 84
      2.5.9 Gravitational lenses and cosmology .. 89

3. **Optics in curved spacetime** 91
   3.1 The vacuum Maxwell equations 91
   3.2 Locally approximately plane waves 93
   3.3 Fermat's principle 100
   3.4 Geometry of ray bundles 104
      3.4.1 Ray systems and their connection vectors 104
      3.4.2 Optical scalars and their transport equations ... 106
   3.5 Distances based on light rays. Caustics ... 110
   3.6 Luminosity, flux and intensity 115

4.	**Derivation of the lens equation**	119
4.1	Einstein's gravitational field equation	119
4.2	Approximate metrics of isolated, slowly moving, non-compact matter distributions	121
4.3	Light deflection by quasistationary, isolated mass distributions	123
4.4	Summary of Friedmann–Lemaître cosmological models	127
4.5	Light propagation and redshift–distance relations in homogeneous and inhomogeneous model universes	132
	4.5.1 Flux conservation and the focusing theorem	132
	4.5.2 Redshift–distance relations	134
	4.5.3 The Dyer–Roeder equation	137
4.6	The lens mapping in cosmology	143
4.7	Wave optics in lens theory	150

5.	**Properties of the lens mapping**	157
5.1	Basic equations of the lens theory	157
5.2	Magnification and critical curves	161
5.3	Time delay and Fermat's principle	166
5.4	Two general theorems about gravitational lensing	172
	5.4.1 The case of a single lens plane	172
	5.4.2 Generalizations	176
	5.4.3 Necessary and sufficient conditions for multiple imaging	177
5.5	The topography of time delay (Fermat) surfaces	177

6.	**Lensing near critical points**	183
6.1	The lens mapping near ordinary images	184
6.2	Stable singularities of lens mappings	185
	6.2.1 Folds. Rules for truncating Taylor expansions	186
	6.2.2 Cusps	192
	6.2.3 Whitney's theorem. Singularities of generic lens maps	197
6.3	Stable singularities of one-parameter families of lens mappings; metamorphoses	198
	6.3.1 Umbilics	199
	6.3.2 Swallowtails	203
	6.3.3 Lips and beak-to-beaks	207
	6.3.4 Concluding remarks about singularities	211
6.4	Magnification of extended sources near folds	215

7.	**Wave optics in gravitational lensing**	217
7.1	Preliminaries; magnification of ordinary images	217
7.2	Magnification near isolated caustic points	220
7.3	Magnification near fold catastrophes	222

8. Simple lens models ... 229
8.1 Axially symmetric lenses ... 230
8.1.1 General properties ... 230
8.1.2 The Schwarzschild lens ... 239
8.1.3 Disks as lenses ... 240
8.1.4 The singular isothermal sphere ... 243
8.1.5 A family of lens models for galaxies ... 244
8.1.6 A uniform ring ... 247
8.2 Lenses with perturbed symmetry (Quadrupole lenses) .. 249
8.2.1 The perturbed Plummer model ... 252
8.2.2 The perturbed Schwarzschild lens ('Chang-Refsdal lens') ... 255
8.3 The two point-mass lens ... 261
8.3.1 Two equal point masses ... 261
8.3.2 Two point masses with arbitrary mass ratio ... 264
8.3.3 Two point masses with external shear ... 264
8.3.4 Generalization to N point masses ... 265
8.4 Lenses with elliptical symmetry ... 266
8.4.1 Elliptical isodensity curves ... 267
8.4.2 Elliptical isopotentials ... 268
8.4.3 A practical approach to (nearly) elliptical lenses . 271
8.5 Marginal lenses ... 274
8.6 Generic properties of "elliptical lenses" ... 277
8.6.1 Evolution of the caustic structure ... 277
8.6.2 Imaging properties ... 278

9. Multiple light deflection ... 281
9.1 The multiple lens-plane theory ... 282
9.1.1 The lens equation ... 282
9.1.2 The magnification matrix ... 285
9.1.3 Particular cases ... 287
9.2 Time delay and Fermat's principle ... 288
9.3 The generalized quadrupole lens ... 291

10. Numerical methods ... 295
10.1 Roots of one-dimensional equations ... 296
10.2 Images of extended sources ... 298
10.3 Interactive methods for model fitting ... 299
10.4 Grid search methods ... 300
10.5 Transport of images ... 302
10.6 Ray shooting ... 303
10.7 Constructing lens and source models from resolved images ... 307

11. Statistical gravitational lensing: General considerations ... 309
11.1 Cross-sections ... 310
11.1.1 Multiple image cross-sections ... 311
11.1.2 Magnification cross-sections ... 313
11.2 The random star field ... 320
11.2.1 Probability distribution for the deflection ... 322
11.2.2 Shear and magnification ... 328
11.2.3 Inclusion of external shear and smooth matter density ... 330
11.2.4 Correlated deflection probability ... 334
11.2.5 Spatial distribution of magnifications ... 337
11.3 Probabilities in a clumpy universe ... 344
11.4 Light propagation in inhomogeneous universes ... 348
11.4.1 Statistics for light rays ... 350
11.4.2 Statistics over sources ... 364
11.5 Maximum probabilities ... 366

12. Statistical gravitational lensing: Applications ... 371
12.1 Amplification bias and the luminosity function of QSOs ... 373
12.1.1 Amplification bias: Preliminary discussion ... 373
12.1.2 QSO source counts and their luminosity function ... 378
12.2 Statistics of multiply imaged sources ... 380
12.2.1 Statistics for point-mass lenses ... 381
12.2.2 Statistics for isothermal spheres ... 385
12.2.3 Modifications of the lens model: Symmetric lenses ... 395
12.2.4 Modification of the lens model: Asymmetric lenses ... 399
12.2.5 Lens surveys ... 401
12.3 QSO–galaxy associations ... 404
12.3.1 Observational challenges ... 404
12.3.2 Mathematical formulation of the lensing problem ... 407
12.3.3 Maximal overdensity ... 408
12.3.4 Lens models ... 411
12.3.5 Relation to observations ... 415
12.4 Microlensing: Astrophysical discussion ... 419
12.4.1 Lens-induced variability ... 421
12.4.2 Microlensing in 2237 + 0305 ... 425
12.4.3 Microlensing and broad emission lines of QSOs ... 429
12.4.4 Microlensing and the classification of AGNs ... 433

	12.5	The amplification bias: Detailed discussion	435
		12.5.1 Theoretical analysis	435
		12.5.2 Observational hints of amplification bias	444
		12.5.3 QSO–galaxy associations revisited	447
	12.6	Distortion of images	448
	12.7	Lensing of supernovae	453
	12.8	Further applications of statistical lensing	456
		12.8.1 Gravitational microlensing by the galactic halo	456
		12.8.2 Recurrence of γ-ray bursters	460
		12.8.3 Multiple imaging from an ensemble of galaxies, and the 'missing lens' problem	461
13.	**Gravitational lenses as astrophysical tools**		**467**
	13.1	Estimation of model parameters	468
		13.1.1 Invariance transformations	471
		13.1.2 Determination of lens mass and Hubble constant	473
		13.1.3 Application to the 0957 + 561 system	476
	13.2	Arcs in clusters of galaxies	483
		13.2.1 Introduction	483
		13.2.2 The nearly spherical lens	485
		13.2.3 Analysis of the observations; arcs as astronomical tools	492
		13.2.4 Statistics of arcs and arclets	498
	13.3	Additional applications	501
		13.3.1 The size of QSO absorption line systems	501
		13.3.2 Scanning of the source by caustics	504
		13.3.3 The parallax effect	508
		13.3.4 Cosmic strings	509
		13.3.5 Upper limits to the mass of some QSOs	511
		13.3.6 Gravitational lensing and superluminal motion	512
	13.4	Miscellaneous topics	513
		13.4.1 Lensing and the microwave background	513
		13.4.2 Light deflection in the Solar System	514
		13.4.3 Light deflection in strong fields	514
References			**517**
Index of Individual Objects			**545**
Subject Index			**547**

1. Introduction

1.1 Historical remarks

1.1.1 Before 1919

Nowadays, the behavior of light rays in a gravitational field must be described with Albert Einstein's Theory of General Relativity. Long before the creation of this theory, however, it was suspected that gravity influences the behavior of light. Already in the first edition of his *Opticks*, published in 1704, Sir Isaac Newton formulated as the first *Query*: "Do not Bodies act upon Light at a distance, and by their action bend its Rays; and is not this action (*caeteris paribus*) strongest at the least distance?" It took almost three generations until this problem was carried further. In 1783, the British geologist and astronomer John Michell communicated in a letter to Henry Cavendish that "... if the semi-diameter of a sphere of the same density with the Sun were to exceed that of the Sun in the proportion of 500 to 1, a body falling from an infinite height towards it, would have acquired at its surface a greater velocity than that of light, and consequently, supposing light to be attracted by the same force in proportion to its *vis inertiae*, with other bodies, all light emitted from such body would be made to return towards it, by its own proper gravity" [MI84.1]. He even proposed how one might discover such "black" bodies, by observing companion stars revolving around these invisible objects. Presumably stimulated by the correspondence with his friend Michell, Henry Cavendish calculated, around 1784, the deflection of light by a body, assuming the corpuscular theory of light and Newton's law of gravitation. He did not publish his calculation, but stated the result on "an isolated scrap" of paper (for details, see [WI88.1]).

Apparently independently of Michell, Peter Simon Laplace ([LA95.1]; for a translation, see Appendix A of [HA73.1]) noted in 1796 "that the attractive force of a heavenly body could be so large, that light could not flow out of it." Indeed, a test particle can leave the gravitational field of a spherical mass M of radius R, starting at the surface, if its initial velocity v_0 is larger than the escape velocity

$$v_e = \sqrt{\frac{2GM}{R}} \ . \tag{1.1}$$

This escape velocity increases with increasing compactness of the attracting object, and reaches the velocity of light c if the radius is smaller than

$$R_S \equiv \frac{2GM}{c^2} \approx 2.95 \frac{M}{M_\odot} \text{ km} \quad , \qquad (1.2)$$

where $M_\odot \approx 2.0 \times 10^{33}$ g is the mass of the Sun. Like Michell, Laplace concluded that a body of mass M, with radius smaller than R_S, is so compact that even light cannot escape from its surface: the object would appear completely black.

Michell and Laplace had in effect anticipated the possible existence of black holes, the non-rotating equilibrium states of which are represented by the first solution of the equations of GR ever found. A black hole has an event horizon (in the static case, a sphere of surface area $4\pi R_S^2$) out of which no signal can travel to the outside world; in particular, no light can escape from within the event horizon. The radius R_S is called the Schwarzschild radius, in honour of the discoverer of the spherically symmetric vacuum solution of Einstein's field equations.

In 1801, the Munich astronomer J. Soldner [SO04.1] published a paper entitled "Über die Ablenkung eines Lichtstrahls von seiner geradlinigen Bewegung durch die Attraktion eines Weltkörpers, an welchem er nahe vorbeigeht"[1], in which he investigated the error in the determination of the angular positions of stars due to the deflection of light. Motivated by Laplace's discussion, he computed the orbit (hyperbola) of a body with constant velocity c which passes near a spherical mass M with impact parameter r. As is derived in textbooks on classical mechanics, a particle starting with velocity v at large separation from the gravitating mass is deflected by the angle α, given by

$$\tan \frac{\alpha}{2} = \frac{GM}{v^2 r} \quad .$$

If we consider only very small deflection angles, the left-hand side of the preceding equation can be approximated by $\alpha/2$, so that

$$\alpha \simeq \frac{2GM}{v^2 r} \quad . \qquad (1.3)$$

One obtains the 'Newtonian' value for the deflection of light by setting $v = c$ in (1.3), in agreement with the result of Cavendish and Soldner. In particular, a light ray that grazes the surface of the Sun should be deflected by $0\rlap{.}''85$. Soldner concluded that "at the present status of practical astronomy one need not take into account the perturbation of light rays due to attracting heavenly bodies." Finally, he remarked: "I hope, nobody will find it disquieting that I treated a light ray as a massive body. That light rays

---

[1] Concerning the deflection of a light ray from its straight path due to the attraction of a heavenly body which it passes closely.

have all absolute properties of matter can be seen from the phenomenon of aberration, which is possible only if light rays are indeed material-like. – And, in addition, one cannot think of anything which exists and acts on our senses, which does not have the properties of matter." (Further comments on Soldner's paper can be found in [TR81.1].)

In 1911, Einstein, unaware of Soldner's work, obtained the same value for the deflection angle from the principle of equivalence and the assumption that the spatial metric is Euclidean, unaffected by gravity [EI11.1]. He noted that it would be desirable that astronomers take on the question of light deflection in gravitational fields. In fact, E. Freundlich of the Royal Observatory in Berlin got sufficiently interested in Einstein's ideas that he planned an expedition to a region of total eclipse of the Sun in the Russian Crimea. A few weeks after this expedition arrived, World War I broke out, and Freundlich was arrested by the Russians (and later exchanged for Russian officers captured by Germans).

Only with the full equations of General Relativity did Einstein obtain twice the Newtonian value [EI15.1]

$$\alpha = \frac{4GM}{c^2 r} = \frac{2R_S}{r} \ . \tag{1.4}$$

The prediction of this value and the measurements of the angle during a solar eclipse immediately following World War I were a great success for Einstein and made him famous. Although these first measurements were not very precise (containing errors of about 30% for the deflection angle), the Newtonian value could clearly be rejected.

[Not everybody, however, accepted this observational test of General Relativity, not for scientific reasons, but owing to the growing anti-Semitism in the early 1920s in Germany. P. Lenard [LE21.1], a Nobel laureate, in 1921 misused Soldner's paper to 'show' that the measurements of the light deflection at the limb of the Sun – which he claimed were compatible with the value $0''\!.85$ obtained from Newtonian considerations – can be explained without referring to Relativity. His polemic attack on "Herrn Einstein" is an infamous historic document; it shows the difficulties Einstein had to face in Germany (and, in particular, in Berlin).]

With the help of radio-interferometric methods, the Einstein deflection angle was verified to within 1% [FO76.1]; such precise measurements also impose strong constraints on possible deviations from General Relativity.

1.1.2 The period 1919–1937

The term 'lens' in the context of gravitational light deflection was first (ab)used by O.J. Lodge, who remarked that it is "not permissible to say that the solar gravitational field acts like a lens, for it has no focal length" [LO19.1] (in this subsection, we follow partly the presentation in [LI64.1]). Lodge discussed the radial dependence the refractive index must have to provide the same light deflection as gravity, arguing that an extended solar atmosphere "would have to vary with the inverse distance, which seems unlikely; but this is just the way the æther tension ought to vary in order to cause gravitation."

A.S. Eddington appears to have been the first to point out that multiple images can occur if two stars are sufficiently well aligned [ED20.1]. Besides the primary image, there should appear a dimmer, second image of the more distant star, on the opposite side of the nearer star as seen from Earth. He also noted that the fluxes of the images depend on the degree of alignment, but his calculation of the magnification factor was incorrect.

In 1924, O. Chwolson published a short note, entitled "Über eine mögliche Form fiktiver Doppelsterne" [CH24.1].[2] He considered a star in the foreground of much more distant stars. If the maximum deflection angle of light that can be caused by the foreground star is α_0, then all background stars with angular separation $\lesssim \alpha_0$ should produce a secondary image very close to the foreground star, and on the side opposite to the primary image. This secondary image, together with the foreground star, would form a fictitious double star, which would not be separable with a telescope, but its spectrum would consist of the superposition of two, probably completely different spectra. (Chwolson did not consider the change of the flux of the images due to light deflection; in the situation he considered, the secondary image would be highly demagnified.) As Chwolson remarked, if the background and foreground stars were perfectly aligned, a ring-shaped image of the background star centered on the foreground star would result (as remarked in [BA89.1], such rings should be termed 'Chwolson rings', but they are usually called 'Einstein rings' today; "the biggest cat gets all the milk"). Finally, Chwolson noted: "Whether the case of a fictitious double star considered here actually occurs, I cannot judge."

R.W. Mandl, a Czech electrical engineer, noted in a letter to Einstein that a star should be expected to act as a "gravitational lens" when light from another star passes it. On a visit to Einstein, he asked him "to publish the results of a little calculation, which I had made at his request" [EI36.1]. Einstein calculated the deflection, due to a star, of the light from a background star, and also found that the apparent luminosities of the images are changed by the gravitational light deflection; in particular, an image can be highly magnified if the observer, source, and deflector are sufficiently aligned. Einstein remarked at the end of his note that "there is no great chance of observing this phenomenon."

Einstein's conclusion is based on the fact that the angular separation of the two images of the background star would be too small to be resolved with optical telescopes then available. It took the unusual vision of Fritz Zwicky to suggest that the formation of multiple images of background objects through gravitational light deflection is not only observable, but that they actually should be discovered! In 1937, he published a paper

---

[2] Regarding a possible form of fictitious double stars.

entitled "Nebulae as Gravitational Lenses" [ZW37.1], from which we quote a few remarkable passages:

"Last summer Dr. V.K. Zworykin (to whom the same idea had been suggested by Mr. Mandl) mentioned to me the possibility of an image formation through the action of gravitational fields. As a consequence I made some calculations which show that extragalactic *nebulae* offer a much better chance than *stars* for the observation of gravitational lens effects."

The revolutionary character of this claim becomes clear if we remember that at that time it was believed that the masses of extragalactic nebulae (as galaxies were termed then) were about $10^9 M_\odot$. Zwicky, however, by applying the virial theorem to the Coma and Virgo clusters of galaxies, estimated the mass of these nebulae to be much higher, averaging about $4 \times 10^{11} M_\odot$.

He estimated the deflection angle caused by nebulae (an estimate too high by about a factor of 10), considered the formation of ring-shaped images, calculated the total flux magnification, and continued: "The discovery of images of nebulae which are formed through the gravitational fields of nearby nebulae would be of considerable interest for a number of reasons.

(1) It would furnish an additional test for the general theory of relativity.

(2) It would enable us to see nebulae at distances greater than those ordinarily reached by even the greatest telescopes. Any such *extension* of the known parts of the universe promises to throw very welcome new light on a number of cosmological problems.

(3) The problem of determining nebular masses at present has arrived at a stalemate ... Observations on the deflection of light around nebulae may provide the most direct determination of nebular masses."

In a second letter, Zwicky estimated the probability that the gravitational lens effect can be observed: "On the probability of detecting nebulae which act as gravitational lenses." [ZW37.2] He noted: "Provided that our present estimates of the masses of *cluster nebulae* are correct, the probability that nebulae which act as gravitational lenses will be found becomes practically a *certainty*." He then estimated that about 1/400 of the area of photographic plates is on average covered by nebular images; including the effect of gravitational focusing, he obtained the result that for "around one in about one hundred nebulae the ring-like image of a distant nebula should be expected, *provided* that the chosen nebula has an apparent angular radius smaller than the angles through which light is deflected on grazing the surface of this nebula."

Both of Zwicky's papers have proven remarkably prescient regarding the importance of gravitational lens effects. He pointed out that the magnification effect would allow astronomers to study objects at greater distance (as shown by the spectroscopy of luminous arcs, i.e., lensed galaxies at high redshift), that the magnification leads to a selection bias (because lensed sources will lead to an overestimate of the abundance of bright sources in samples – this effect is now called amplification bias, and is actually thought to occur), that lensing can serve as a method to determine galaxy masses, and that the probability to detect lensing is high. The reason he considered ring-like images only is that no compact extragalactic objects were known in 1937.

1.1.3 The period 1963–1979

The advent of radio astronomy in 1963 led to the discovery and identification of quasars [SC63.1], 'quasi-stellar radio sources', so-called because of their point-like appearance on optical plates (in this book, we will use the term quasar in its original sense, and talk of QSOs for quasi-stellar objects, without regard for their radio emission; hence, quasars are radio-loud QSOs). Owing to their point-like appearance, their prominent spectral features (strong, broad emission lines), their high redshift and thus large distance (with correspondingly high probability of intervening material), and their high luminosity, they are ideal sources for gravitational lensing.

At the same time, the topic of gravitational lensing which had been dormant since Zwicky's work began to be reconsidered in the astronomical literature. Yu.G. Klimov [KL63.1], S. Liebes [LI64.1] and S. Refsdal [RE64.1] independently reopened the subject. Klimov considered lensing of galaxies by galaxies, concluding that, for sufficient alignment, a ring-shaped image would occur and could be easily singled out from the general field of galaxies, whereas if the alignment were imperfect, multiple galaxy images would appear, which would be difficult to separate from double or multiple galaxies. Liebes considered lensing of stars on stars, of stars on globular clusters in our galaxy, and of stars on stars if both are members of the same globular cluster; he also considered the possibility that stars in our galaxy can lens stars in the Andromeda galaxy (this has recently been reconsidered [PA86.4] as a possible way to detect compact objects in the halo of our galaxy). In his paper, Liebes also estimated the probability that lensing events of the kind just mentioned can actually be observed, thereby emphasizing the importance of space-based observations.

Refsdal (and also Liebes) gave a full account of the properties of the point-mass GL,[3] and he also considered the time delay for the two images, due to the different light-travel-time along light rays corresponding to each image.[4] In particular, he argued that geometrical optics can be used safely in considering gravitational lensing effects (which was not generally accepted at that time; as Sjur Refsdal communicated to us, his thesis evaluation committee was fairly sceptical about this point, until his work was published in a scientific journal).

In a second paper [RE64.2], Refsdal considered potential applications of gravitational lensing. His abstract stated: "The gravitational lens effect is applied to a supernova lying far behind and close to the line of sight

---

[3] Hereafter, 'gravitational lens' will be represented by the acronym GL.
[4] In the same year, I. Shapiro pointed out the measurability of the retardation of light signals in the gravitational fields of massive bodies; this effect was measured with an accuracy of 10^{-3} [RE79.1].

through a distant galaxy. The light from the supernova may follow two different paths to the observer, and the difference Δt in the time of light travel for these two paths can amount to a couple of months or more, and may be measurable. It is shown that Hubble's parameter and the mass of the galaxy can be expressed by Δt, the red-shifts of the supernova and the galaxy, the luminosities of the supernova 'images' and the angle between them. The possibility of observing the phenomenon is discussed." Refsdal also pointed out the possibility that QSOs may be variable (which was not at all clear in 1964) and could then also be used for the determination of lens masses and the Hubble constant (hereafter H_0).

In two further papers ([RE66.1], [RE66.2]), Refsdal considered the possibility of testing cosmological theories by using the lens effect, and of determining the distances and masses of stars from lensing if a space observatory and a telescope on Earth could observe a lens event simultaneously; the telescope in space should be at a distance of 5% of one Astronomical Unit (henceforth AU) from Earth.

"In 1965, at the 119th meeting of the AAS, based on the astonishing similarity of the spectra of quasars and nuclei of Seyfert I galaxies, I proposed [BA65.1] that quasars are not novel superluminous constituents of the Universe, as was generally assumed, but are nuclei of Seyfert I galaxies intensified through the gravitational lens action of foreground galaxies. Moreover, the observed short-term brightness variations of quasars do not prove the extreme smallness of quasars but result from the scanning motion of the optical axis passing across areas of different brightness in the source." This quotation was selected from a paper by J.M. Barnothy [BA89.1] on the history of GLs. His idea was not received well by astronomers; in order to obtain the observed number of QSOs from the observed number of Seyfert I galaxies, he and his wife had to resort to fairly exotic cosmological models [BA68.1]. Although their ideas were widely rejected at that time (and still are), they have been revitalized in more recent years; today, the idea that source counts of QSOs can be affected by the magnification caused by GLs is considered a serious possibility, and it has even been suggested that one member of the active galactic nuclei (AGN) family, namely BL Lac objects, may in fact be magnified QSOs. These points will be discussed in more detail in the course of this book.

At the same time as the 'modern history' of gravitational lensing came the realization that light propagation through an inhomogeneous universe differs from that through a homogeneous one. Light propagating in an inhomogeneous universe is deflected due to gravitational fields, thus limiting the accuracy with which positions of distant sources can be determined [GU67.1]. The basic equations governing the change of cross-sections and shapes of ray bundles had already been set up (in connection with gravitational radiation theory) by R.K. Sachs [SA61.1]. They show how light bundles propagating through a region of space where the density is lower than the average density of the universe diverge relative to corresponding light bundles in a homogeneous universe of the same mean matter density

([ZE64.1], [DA65.1], [DA66.1], [GU67.2]). R. Kantowski [KA69.1] considered light propagation in the 'Swiss-Cheese' cosmological model (the only exact cosmological solution of Einstein's equations for a locally inhomogeneous universe; see Chap. 4).

Owing to the lack of any realistic exact solution of the equations of General Relativity that can describe an inhomogeneous universe, it was (and still is) assumed that it can be described on average as a Friedmann–Lemaître universe whose metric is locally perturbed by matter inhomogeneities (how this works is not obvious at all, due to the non-linearity of the Einstein equations – see Chap. 4 for a discussion). For the limiting case where a light bundle propagates far away from all matter inhomogeneities, C.C. Dyer and R.C. Roeder obtained a modified redshift-distance relation ([DY72.1], [DY73.1], [DY74.1]), which hereafter will be called the 'Dyer–Roeder prescription for light propagation' for an inhomogeneous universe.

For light bundles which do not stay far away from matter inhomogeneities, the deflecting action of such inhomogeneities must be taken into account. For a random distribution of clumps, the problem becomes statistical. Refsdal investigated "the propagation of light in universes with inhomogeneous mass distribution" [RE70.1], using the GL formalism. As we discuss in Sect. 11.4, lensing offers a most valuable tool for studying such cosmological problems.

N. Sanitt, in 1971, published a remarkable paper on the influence of gravitational lenses on QSOs [SA71.1]. To our knowledge, he was the first to consider extended masses (galaxies) as lenses. Sanitt studied the effect of lensing on the source counts of QSOs (amplification bias), concluding that "one cannot interpret all QSOs as gravitational lens images, unless there is a substantial amount of evolution with z in the sources of the lens images, and also a large fraction of the mass of the universe in the form of compact dark objects." He pointed out that the continuum region should be magnified more than the line-emission region, since the latter is thought to be larger, so that "the high luminosity tail of the lens images would all have their continuum enhanced relative to their line emission." Unfortunately, this paper seems to have escaped the attention of most of later research on the amplification bias.[5]

W.H. Press and J.E. Gunn discussed a method "for detecting a cosmologically significant density of condensed objects (subluminous stars, dead clusters of galaxies, black holes): If the Universe is filled with roughly a critical density of such objects, the probability is high that a distant point source will be gravitationally imaged into two roughly equal images. This conclusion is independent of the mass M of the objects, which only sets the scale of the image doubling, $\simeq 2.6 \times 10^{-6}(M/M_\odot)^{1/2}$ arcseconds" [PR73.1]. Very Long Baseline Interferometry (VLBI) could resolve double images caused by masses as small as $10^4 M_\odot$; Press and Gunn proposed a systematic search for

---

[5] Sanitt's office mate in Cambridge advised him to change subjects, as this lensing business at that time appeared too esoteric. This office mate was R.D. Blandford, one of the most active researchers in the field today, whereas N. Sanitt has left astrophysics.

such closely-spaced double images. They concluded from the known bright 'optical' QSOs with high redshift that a significant density of objects with mass larger than $10^{12} M_\odot$ "can probably be ruled out already, since pairs of QSOs with identical redshifts separated by a few seconds of arc would hardly have escaped the attention of observers"; this last statement has been proven partly wrong by history. In order to distinguish double images from double sources, Press and Gunn proposed to compare high-redshift sources with low-redshift ones; since the lensing probability increases with redshift, a redshift dependence for the fraction of double sources can be a sign of lensing. Furthermore, since compact radio sources are variable, and since the time delay for double images is small, provided the angular separation is also small, double images should vary nearly synchronously, in contrast to double sources, which can vary independently.

In a series of papers, R.R. Bourassa and coworkers ([BO73.1], [BO75.1], [BO76.1]) analyzed the lensing properties of a spheroidal mass distribution. These papers were the first to consider lenses whose mass distribution is not spherically symmetric. It appears that the first discussion of caustics in gravitational lensing (see Chap. 6) was presented in these papers. To calculate the deflection angle caused by a spheroidal mass distribution, the authors introduced a formalism wherein the sky is mapped onto the complex plane. They estimated the probability for a point source at high redshift to appear as multiple images, due to deflection produced by transparent spheroidal masses. A general expression for the time delay for multiple images was calculated by J.H. Cooke and R. Kantowski in 1975 [CO75.1]; the time delay was shown to be separable into two parts: the geometrical part, due to the different lengths of the light rays corresponding to different images, and the potential part, due to the different depths of the gravitational potential that the different light rays traverse.

K. Chang and S. Refsdal [CH79.1] pointed out that a single star (a 'microlens') in a lens galaxy can cause flux variations on time scales of a year. Although the image separation caused by a star is not observable, the corresponding change of magnification can be, thus providing a means for detecting compact objects in a lens galaxy. At the same time, the determination of the time delay for image pairs due to the 'macro' part of a GL can become significantly more difficult, as observed flux changes in different images cannot necessarily be attributed to intrinsic variability of the source. It was a happy coincidence that the work by Chang and Refsdal was finished at about the same time that the first gravitational lens system was discovered, so that their paper had an immediate application.

1.1.4 Post-1979

Until 1979, gravitational lensing was considered as a fairly esoteric field of research, and those working in that field did not always receive full respect from their fellow astrophysicists. That changed suddenly after D. Walsh, R.F. Carswell, and R.J. Weymann announced the detection of the first GL candidate [WA79.1]. The source 0957+561 was selected from a radio survey. It has two optical counterparts A and B, which turned out to be QSOs at a redshift of $z \sim 1.4$; their angular separation is about 6 arcseconds. The similarity of the spectra of A and B, and of the ratio of optical and radio fluxes, and the subsequent detection of a lensing galaxy at redshift $z_d \sim 0.36$ [ST80.1], [YO80.1] verified the lensing nature of this system. For an account of the history of the discovery of this first GL system, see [WA89.2].

Only one year later, a second GL candidate was discovered [WE80.1], apparently consisting of three images at first, one of which was much brighter than the other two. It was later discovered that this bright image is actually a pair of images ([HE80.1], [HE81.1]) with separation $\sim 0\rlap{.}''5$, whereas the separation relative to the other images is $\sim 2\rlap{.}''3$. A galaxy was found near the images quite some time later [SH86.1]. In Sect. 2.5, these two GL systems and several others discovered serendipitously are described in detail. The first GL to be discovered in a systematic search for gravitational lenses was 2016+112 [LA84.1].

The discovery of the first GL systems triggered an enormous outpour of publications, concerning both observational and theoretical work. It would be impossible to mention even just a small fraction of them here, but we would like particularly to point out the work of P. Young and his colleagues on the first two lens systems ([YO80.1], [YO81.2], [YO81.3], [YO81.4]) and Young's paper on microlensing in 0957+561 [YO81.1], which was not only the first paper to discuss the influence of an ensemble of stars in the lens galaxy, but also a milestone in the formulation of GL theory. His untimely death ended a most promising career.

A second type of GL systems was announced in an abstract for the 169th AAS meeting in Pasadena in 1987, by R. Lynds and V. Petrosian [LY86.1]: "We announce the existence of a hitherto unknown type of spatially extragalactic structure having, in the two most compelling known examples, the common properties: location in clusters of galaxies, narrow arc-like shape, enormous length, and situation of center of curvature toward both a cD galaxy and the apparent center of gravity of the cluster. The arcs are in excess of 100 kpc in length, have luminosities roughly comparable with those of giant E galaxies, and are distinctly bluer than E galaxies – especially so in one case." These so-called 'luminous arcs' are now thought to be highly distorted images of high-redshift galaxies, seen through a compact foreground cluster of galaxies. The reader is referred to [PE89.1] for a presentation of the discovery of arcs, and to Sect. 2.5.5 for an account of the present status of observations.

Finally, in a search for GL systems with radio telescopes, a third type of lensed images was found: radio sources with the shape of nearly full rings ([HE88.1], [LA89.1]). By the time of their discovery, the astronomical community was ready to accept the lensing hypothesis for these sources, although the observational evidence for a GL (at least for the first ring) was nearly entirely based on the image morphology. Nevertheless, it turns out that images of extended sources contain much more information than can be extracted from multiply imaged QSOs, and the impressive theoretical reconstruction of the first (so-called 'Einstein-') ring [KO89.1] erased all doubts about the lens nature of this source.

In 1983, the first international conference took place where lensing was a main topic ("Quasars and Gravitational Lensing", [SW83.1]). A workshop on lensing was held in Toronto in 1986 (reported in [WE86.1]), and two conferences devoted entirely to lensing were held in Cambridge, MA in 1988 [MO89.1] and Toulouse in 1989 [ME90.2]. A lensing conference in Hamburg took place in September 1991, just when the work on this book was completed. In addition, there is hardly any conference on AGNs or observational cosmology where lensing topics are not discussed. By now, lensing has become one of the most active fields of research in extragalactic astronomy.

There are a number of reviews of the gravitational lens phenomenon. Besides those contained in the proceedings mentioned above ([SW83.1], [MO89.1], [ME90.2]), we refer the reader to [BL87.3], [BU86.1], [NA86.2], [CA87.1], and [BL89.1]. In addition, a book on gravitational lensing has appeared in Russian [BL89.2].

1.2 Outline of the book

Chapter 2 contains the basic facts of gravitational lensing; we consider the simplest GL, namely, the point-mass lens, in some detail, to obtain typical length and angular scales of the separation of images expected in a GL system. We then consider the more general lensing situation, and the basic language for the description of lensing is introduced. We derive the deflection angle of a light ray passing through (or by) a mass distribution, and explain the concept of magnification due to distortion of light bundles by tidal gravitational forces. We then proceed to the notion of critical curves and caustics, which are essential for understanding GL systems. We compute the magnifications for the point-mass lens and discuss the effect of the source size on the magnification; this yields an upper limit to the size of a source if it is to be significantly magnified by a lens of given mass. We then develop the concept of the amplification bias which can affect the source counts of very distant sources, in particular QSOs. Since the relative positions of source, lens, and observer can change in time due to proper motions, the magnification can be time-dependent, which can lead to lens-induced variability. We show that lensing of stars by stars in our Galaxy is unlikely to be discovered. We next discuss the current state of observations, including a desription of strategies and obstacles, and in particular give a detailed account of the discovery, confirmation and models of observed GL systems. The latter constitute an ever-growing, no longer select club of varied sources.

Chapter 3 presents the propagation of light in arbitrary space-times. Starting from the covariant form of the vacuum Maxwell equations, we consider solutions which correspond to 'locally plane waves', using a WKB-type expansion for the electromagnetic field tensor. Light rays are described as null-geodesics; they propagate orthogonally to the wavefronts. We derive the transport equations for the amplitude and the polarization vector along light rays and obtain the law of conservation of photon number. Null-geodesics can be obtained from a variational principle; we formulate and prove a form of Fermat's principle, which states that light rays are those null-curves for which the arrival time on the timelike world line of the observer is stationary under small variations that keep the source fixed. We then consider the geometry of ray bundles, described by the optical scalars, and derive their transport equations; the resulting focusing equation is used later in the book to study the influence of matter inhomogeneities on the propagation of light bundles through the universe. We define two distance concepts which are intimately related to observations, the angular-diameter distance and the luminosity distance, and prove that they are related to each other in a simple fashion. In a final section, we show how angular diameter distance and redshift enter the flux-luminosity relations, and derive the dependence of surface brightness on redshift.

Chapter 4 begins with a brief review of the foundations of General Relativity, focussed on the role of the metric and Einstein's gravitational field equation. Then, the metrics of isolated systems with slowly moving sources and weak gravitational fields are treated in linear approximation. Fermat's principle is then used to derive the lens equation and the time delay for images for such "local" systems. To prepare for the cosmological generalization of these basic relations of lens theory, we next summarize the main facts about the homogeneous-isotropic Friedmann–Lemaître model universes. In the subsequent treatment of light propagation in models which are homogeneous and isotropic on average, we first use the photon number conservation law to obtain a condition for the mean magnification of fluxes in a clumpy universe, which is important for statistical applications of lens theory. Then we analyze the Dyer–Roeder approximation for the dependence of angular diameter distances on redshift and cosmological model parameters. We continue to set up the lens, or ray-trace and time delay equations in this cosmological setting and indicate their application to observational data. Finally, we show how the effect of a transparent gravitational lens on light can be obtained by wave optics, via Maxwell's equations in a curved spacetime.

Chapter 5 discusses general properties of the GL (or ray-trace) equation. This equation can be written in several ways; in particular, it is convenient to scale it to dimensionless quantities, using appropriate scales. The scaling directly leads to a characteristic surface mass density; if the surface mass density of a lens is larger than this characteristic value, the lens is capable of producing multiple images. By introducing the (two-dimensional) deflection potential, the ray-trace equation can be written as a gradient equation, in accordance with Fermat's principle. The magnification due to light deflection is a purely geometrical effect, since the surface brightness (or specific intensity) is not affected by deflections. Specifically, the distortion of a light bundle is described by a 2×2 matrix, whose determinant yields the magnification. From the scalar form of the ray-trace equation, we obtain a classification of images in a GL system. Using the geometry of wavefronts, we rederive the formula for the time delay of multiple images in a GL situation. We then formulate two general properties of gravitational light deflection: for transparent mass distributions and for fixed source and observer, there exists an odd number of light rays connecting these two, i.e., the observer sees an odd number of images of the source, and at least one of these images (that with the shortest light-travel-time) appears brighter than it would be if the deflecting mass were absent. Necessary and sufficient conditions for a lens to be able to produce multiple images are briefly discussed. We conclude this chapter with an investigation of the possible image topographies, which are explicitly discussed for the three-image and five-image case, and a general recipe is given for higher multiplicities. This classification of image topographies yields an intuitive geometrical insight into specific GL models.

Chapter 6 is devoted to the study of the local properties of lens mappings near points where that mapping is singular, i.e., not locally bijective. Images at such "critical" points are exceptionally bright and occur if sources lie on (or near) the caustic of an observer's past light cone. Knowing the caustics provides a good qualitative understanding of a lensing model, in particular since the number of images changes if, and only if, the source crosses a caustic. We first show that the stable singularities form smooth, so-called critical curves in the lens plane, while the corresponding caustics form curves with cusps, and we derive how the brightness of images increases if the source approaches a caustic. Subsequently, we consider unstable singularities at which discontinuous changes of the patterns of critical curves and caustics, so-called metamorphoses, occur if a parameter of the lens model varies continuously. We determine all those metamorphoses which are stable with respect to perturbations of one-parameter families of mappings. In all cases, we find the shapes of the caustics and the changes of the number of images which take place if a source crosses a caustic. This chapter may be considered as an application of, or elementary introduction to, catastrophe theory. We finally derive the magnification of an extended source near a fold singularity.

Chapter 7 provides the justification for using geometrical optics in GL theory. On the basis of results derived at the end of Chap. 4, we discuss light propagation in GL systems using wave optics, and demonstrate that in all cases of practical importance, geometrical optics provide a superb approximation, because of the absence of sufficiently compact sources in extragalactic systems.

Chapter 8 considers some simple mass distributions as GL models. The aim of this chapter is twofold: to introduce and analyze the most commonly used GL models in some detail, and to explain in depth the methods for investigating (parametrized) lens models. Axially symmetric GLs are fairly easy to investigate, and much is known about their properties; in particular, the ray-trace equation becomes one-dimensional in this case. For this type of lens, there is a natural distinction between two kinds of critical curves, 'radial' and 'tangential' critical circles, with the caustic corresponding to the latter ones degenerating to a single point. The nomenclature is then justified by investigating the shape of images close to the critical curves, which also provides further geometrical interpretation of the eigenvalues of the Jacobian matrix of the lens equation. Several general properties of axi-symmetric lenses are formulated and proven, e.g., the necessary and sufficient conditions for the existence of multiple images.

We next consider some specific axi-symmetric lens models: the point-mass lens, a uniform disk, the singular isothermal sphere, a family of matter distributions which are frequently used to model galaxy-type lenses, and a uniform ring. For all these GL models, we derive the lens equation, the Jacobian determinant (and thus the magnification of images), and classify the regions of the source plane according to the image multiplicities.

No real GL is axially symmetric; even a symmetric mass concentration would likely be part of a larger system of masses, like a galaxy in a cluster of galaxies. With this situation in mind, we consider next the so-called quadrupole lenses, corresponding to an axi-symmetric deflector embedded in a larger-scale mass distribution. The deflection angle caused by this larger-scale mass distribution is approximated by its lowest-order Taylor expansion, which corresponds to a quadrupole term (including shear and convergence). A geometric and an algebraic method are described for solving the corresponding lens equation, and the structure of the caustics is outlined. We then consider two quadrupole lenses in detail, one of which is the so-called Chang–Refsdal lens; in particular, we present a complete classification of its caustic structure.

We then describe briefly the two point-mass lens, which has been analyzed in detail in the literature, as well as a generalization to N point masses.

Lenses with elliptically-symmetric mass distributions are treated next; it turns out that the calculation of the deflection law for such GLs is fairly involved; thus, lenses with an elliptical deflection potential are frequently discussed in their stead. In particular, we consider the special case where an elliptical lens is just able to lead to image splitting of a background source (marginal lenses).

Chapter 9 analyzes the situation in which light is deflected more than once between the source and the observer. The corresponding ray-trace equation is derived geometrically, as in the case of a single deflector, and a recursion relation for the impact vectors in the lens planes is obtained. Similarly, the

Jacobian matrix of the lens equation is obtained iteratively. In contrast to the case of a single lens plane, the Jacobian matrix, in general, is no longer symmetric. The magnification is a fairly complicated function of the local values of shear and convergence in the individual lens planes, as shown explicitly for the special case of two deflectors. Two particular GL models are discussed briefly. The light-travel-time from the source to the observer is determined as a function of the impact vectors of the light ray in all lens planes. According to Fermat's principle, the physical light rays are those with stationary light-travel-time; we demonstrate that this condition indeed reproduces the multiple-deflection ray-trace equation. If the deflections in all lens planes, except one, vary slowly with impact parameter, they can be replaced by their lowest-order Taylor expansion, thus yielding a generalization of the quadrupole lens discussed in Chap. 8. In fact, the resulting lens equation shares many properties with that of a single lens plane.

Chapter 10 discusses several numerical tools useful for studying GLs. If the lens equation can be reduced to a one-dimensional form, its solution can be reduced to finding the roots of a single (non-linear) equation. Whereas no 'black box' routine is available for finding all the roots of an equation in a given interval, the problems occurring in GL theory suggest the use of a method which is quite reliable in finding all roots. There exists a fairly simple method for the construction of GL images of an extended source, making use of standard contour plot routines. An interactive procedure for fitting GL models to observed systems is presented. If the lens equation is to be inverted for a large number of source positions (as is needed for statistical studies), a grid search method is appropriate; we describe such a method and discuss its use. The grid search yields only approximate solutions of the lens equation; if higher accuracy is desired, the approximate roots can be used as a starting position for an iterative approximation, via Newton's method. We discuss the ray shooting method, which is used to determine the magnifications of (extended) sources seen through a lens. The main application of this method is found in microlensing (ML) studies, where the lens is assumed to consist of a large number of stars. In order to make this problem tractable, several techniques to reduce computing time are needed, such as a hierarchical tree code. Finally, we describe a method for building a lens model and reconstructing the source structure from an observed extended GL image; the method was applied successfully to the radio ring MG1131+0456.

Chapter 11 introduces the concepts and methods of statistical gravitational lensing. Since we readily detect only spectacular cases of lensing, such as multiple or highly distorted images, we are likely to miss cases where lensing is less prominent. Gravitational light deflection can have an impact on observations, even if it cannot be recognized ab initio. In particular, the flux from distant sources can be magnified. As in atomic and nuclear physics, the statistical measure of the lensing properties of a matter distribution can be described by a cross-section; we introduce this concept and provide a number of examples of lensing cross-sections, to be used later in specific applications. Since the lens mapping exhibits universal properties close to caustics, and since high magnifications for point sources occur only near caustics, it is possible to derive a general equation for the cross-section for high magnifications.

Lens models involving point-masses only admit certain scaling relations when a smooth surface mass density is added; two examples of such scaling are presented, for the Chang–Refsdal lens and for the random star field. The random star field is a natural lens model of the local lensing properties of a galaxy; we discuss the basic assumption of this model, statistical homogeneity. Statistical properties of random star fields can be obtained by using Markov's method, which we introduce and apply to the calculation of the deflection probability distribution and the shear distribution caused by a random star field. The probability distribution for the deflection angle depends on the size of the star field, so that the limit of an infinite star field does not exist. This is due to the long-range nature of gravity and very similar to the infrared divergence occurring in the scattering cross-section of charged particles in an infinite plasma ('Coulomb logarithm'). For small deflection angles, the probability distribution is a Gaussian, whereas it becomes a power-law for large deflections. The latter behavior is due to a single star near the light ray under consideration, whereas the Gaussian distribution arises from the combined action of many stars, and can also be derived from the central limit theorem. The deflection probability can be used to estimate the statistical properties of a microlensed image; in particular, we estimate the number of stars necessary in numerical ML simulations.

The probability distribution for the shear (or tidal force) is independent of the mass spectrum of the lenses, and depends only on the dimensionless surface mass density of stars, sometimes called 'optical depth'; in particular, the limit of an infinite star field exists. We then include a smooth surface mass density and an external shear, and derive the magnification probability for single light rays. Concentrating on high magnifications, it is possible to derive the probability for magnification of point sources, which we then compare with a simple, linear analysis. The deflections of two neighboring light rays is correlated, if their separation is not large compared to the mean separation of stars; we discuss the probability distribution for this correlated deflection.

Next, we describe in detail numerical ML studies, which yield the spatial distribution of magnifications as a function of the relative source position behind a random star field. Magnification patterns form the basis of synthetic light curves of microlensed sources, to be discussed in Chap. 12. The magnification patterns yield considerable insight into the ML process, for instance, the transition from linearity (where caustics produced by single stars can be clearly distinguished) to non-linearity (where the caustic structure is very complicated and is due to the combined effect of many stars, and the influence of an external shear and an underlying smooth surface mass density).

We then derive an equation for the lensing probabilities in a clumpy universe, where the lenses are distributed randomly. This equation relates lensing cross-sections to the 'self-consistent' probabilities, and forms the basis of the study of statistical lensing in Chap. 12.

Next, we take up the discussion of light propagation in an inhomogeneous universe. The problem is first transformed into a lensing problem, since the multiple deflection lens theory is suited to account explicitly for the deflection caused by inhomogeneities in a clumpy universe. Our model universe is cut into thin slices, each with matter distributed according to the parameters of the cosmological model. From the multiple lens plane equation, we obtain an alternative form of the Dyer–Roeder equation, which we show is equivalent to that derived in Chap. 4. Light rays in our model of the clumpy universe can then be studied and, depending on the nature of the lenses, analytical estimates for the statistical properties of light propagation can be obtained. We define an 'optical depth' in a universe filled with point masses and show that a linear consideration underestimates this optical depth. A comparison between extended lenses and point-mass lenses follows, and we also briefly mention several results from a study where collections of point masses represent extended lenses. We conclude Chap. 11 with a theorem regarding maximum probabilities, to be used later.

Chapter 12 treats the applications of statistical gravitational lensing. We start with a preliminary discussion of the amplification bias, i.e., the effect of magnification, which brings otherwise excessively faint sources into flux-limited samples, thus affecting the source counts. From a simple example, we obtain the basic properties of the amplification bias; in particular, it is stronger for steeper intrinsic luminosity functions of the sources under consideration. A brief discussion on QSO source counts is included, concerning both the difficulties in obtaining them, and selected results.

We then turn to one of the standard problems of statistical gravitational lensing, the estimation of the expected fraction of multiply-imaged QSOs in a flux-limited sample. As we will demonstrate, it is fairly difficult to obtain reliable results: the lensing population in the universe is not well known; in particular, the (cluster) environments of galaxies can strongly affect the multiple-image statistics. Furthermore, the amplification bias must be taken into account properly. Since QSOs are supposed to be sufficiently compact, so that stellar objects can significantly magnify them, the 'image splitter' (galaxy) and the potential 'magnifier' (compact object) are not necessarily coincident. We start with the simplest model, a universe filled with point-mass lenses. The resulting probability for a high-redshift source to be multiply imaged is readily obtained, and from observational limits of the fraction of multiple images, the cosmological density of compact objects in certain mass ranges can be constrained. We then use a simplified prescription of the galaxy distribution in the universe to obtain the fraction of multiply-imaged sources in flux-limited samples, and to study the influence of intracluster matter and the amplification bias on this fraction. Whereas the probability for any compact source at high redshift to be multiply imaged by the known galaxy population is less than about 1/200 (irrespective of angular separation), the fraction of lensed sources in flux-limited samples can be high. In particular, due to the amplification bias, the fraction of multiply-imaged sources among the apparently most luminous QSOs is expected to be significantly larger. This latter result seems to have been verified observationally, thus demonstrating the importance of the amplification bias. We briefly mention several modifications of the lensing population, including ML, different mass distributions of the galaxies, as well as the inclusion of a finite ellipticity of galaxy lenses. These modifications are important for studies of the properties of multiple image systems, e.g., the expected distribution of brightness ratios, the fraction of 5-image versus 3-image systems, and the distribution of angular separations. We describe some of the lens surveys currently underway, one of the radio and two of the optical variety. Their preliminary results are very encouraging: they have yielded candidate GL systems, and they can constrain the amount of lensing material in the universe.

The apparent association of high-redshift QSOs with nearby galaxies, which has been claimed to exist for nearly two decades (see [AR87.1] for an account of these observations), is investigated in the context of lensing. As we show, this effect, if present, cannot be explained by lensing, even if fairly extreme assumptions about the magnitude of the amplification bias are included. A recently found statistical association of foreground galaxies with high-redshift QSOs can be accounted for by lensing, if an amplification bias of the QSO population is taken into account. After a brief description of the observations, we derive the mathematical framework to study this effect, assuming that it is due to lensing. We show that a strict upper limit on the possible overdensity of galaxies around high-redshift QSOs can be derived, if the QSO population as a whole is not significantly affected by the amplification bias. For reasonable assumptions about the mass distribution in the associated galaxies, the observed overdensity is larger than this limit, thus providing strong evidence for the importance of the amplification bias on QSO counts. We estimate the expected overdensity for a few specific GL models, including ML; we predict that the overdensity should be larger if the flux threshold of the QSO sample is increased, i.e., if only more luminous QSOs are considered.

We then turn to an astrophysical discussion of ML. Due to changes of the relative position of source, lens and observer in a GL system, ML produces a time-varying magnification of lensed sources, and thus leads to lens-induced variability. We consider two situations, a single star and a random star field; the former can be considered a limiting case (for small optical depth) of the latter. Whereas a single star yields a fairly simple prediction for the expected light curve of the source, ML by an ensemble of stars can lead to a variety of light curves, as we show with examples. The variability of some QSOs is similar to that expected from ML, but it will be very difficult to glean clear evidence to show that ML is the cause of the fluctuating brightness. However, in multiple-image GL systems, ML can be detected, as intrinsic variations of the source must appear in all images (having accounted for the respective time delay). For the 2237+0305 GL system, clear evidence for ML has been observed, as we discuss next. Due to the importance of this GL system, we present a detailed study of ML for this case. The impact of ML on the profile of broad emission lines in QSOs is studied next; by comparing the profiles of the same emission line in different images of a multiply-imaged QSO, it may be possible to detect ML, and also to obtain information about the geometry of the broad-line region. Furthermore, a net shift of the central wavelength of broad emission lines can be caused by ML; this means that a certain amount of redshift difference is tolerable in multiple QSOs without necessarily rejecting the lensing hypothesis. The 2237+0305 system is the most likely candidate to be able to reveal the impact of ML on the broad emission lines, and there is preliminary evidence that a profile difference in different images has already been observed. Finally, we discuss whether ML can change the classification of an AGN, i.e., whether ML can turn the spectrum of a QSO into that of a BL Lac object.

We then offer a thorough discussion of amplification bias. We show that a finite source size is essential for a realistic study of the amplification bias caused by compact objects. For a universe filled with point masses (in a first approximation, the results do not depend on whether the point masses, e.g., stars, black holes, are distributed randomly throughout the universe, or concentrated in galaxies), we derive the magnification probability for extended sources, together with a useful analytic approximation. Then, for any given intrinsic luminosity function of the sources, the expected source counts can be obtained. These depend on the luminosity function, the size of the sources, and the mass spectrum of the lenses. In particular, if the mass spectrum of the lenses is a Salpeter mass function, the slope of the observed QSO counts for high fluxes can be obtained, but from a much steeper luminosity function. Such an occurrence would imply a strong amplification bias. We then discuss the observational hints for the amplification bias to occur in QSO samples. In fact, the statistical properties of recent lens surveys provide strong indications for a strong amplification bias for at least the apparently most luminous sources.

Apart from magnifying a source, gravitational light deflection affects the shape of images of extended sources. A sample of circular sources seen through an inhomogeneous universe would be seen as a distribution of images with finite ellipticity. Since real sources (e.g., galaxies) have an intrinsic distribution of ellipticity, intrinsic and lens-induced effects are not easily distinguished. We discuss a theoretical study of this effect, as well as some observational efforts to detect any statistical influence of lensing on image shapes. We also discuss briefly the effect of lensing on the distortion of intrinsically colinear sources (such as radio jets).

The basic difficulty in detecting statistical lensing is the lack of a population of bright, distant sources with well-known intrinsic properties. If we had a standard candle (i.e., a type of source with known intrinsic luminosity), the measurement of its flux and its distance would allow a determination of its magnification. The best standard candles known (and that are sufficiently bright) are supernovae. We discuss the probability for supernovae to be lensed, and the observational prospects. Although it will be difficult to obtain statistically significant results, this method may prove to be potentially useful in detecting a population of compact objects in the universe.

By observing a large number of stars ($\geq 10^6$) in a very nearby galaxy (e.g., the LMC), it may be possible to detect light variations due to lensing by compact objects in the halo of our galaxy (if they exist). We discuss the observational prospects of such a program. As for lensing of supernovae, the observational difficulties are considerable, but the effort will almost certainly be rewarded, and this observational strategy is the only one known to be able to probe for compact objects down to a mass of $\sim 10^{-8} M_\odot$.

We conclude Chapter 12 by discussing the possible impact of lensing on the observations of gamma-ray bursts, if they are at cosmological distance, and a potential solution to the 'missing lens problem', i.e., the problem that in many of the GL candidate systems, no lens is seen close to the images, thus implying, for a simple lensing scenario, that the lens must have a large mass-to-light ratio.

Chapter 13 discusses further applications of gravitational lensing. From an observed GL system one can estimate the lens parameters, such as the lens mass. We discuss the observables of a GL system, and point out the special role played by the time delay for pairs of images: this is the only dimensional observable. From the relation between observables and the lens equation, we derive the number of independent constraints for a lens model. There exist a number of invariance transformations for lens models, which leave all observables unchanged; therefore, successful lens models can at best be specified up to these transformations. We then discuss a highly simplified method for determining the lens mass and the Hubble constant, and apply it to the GL system 0957+561. The resulting estimates are fairly uncertain, as long as no use is made of the information obtained from the VLBI structure of the two images. By taking this information into account, as well as further refinements, the parameters of the lens models can be strongly constrained (of course, only up to the invariance transformations). We discuss the prospects for determining the Hubble constant from GL systems.

We then study lens models for luminous arcs, interpreted as images of an extended source situated near a cusp of a GL. Nearly-spherical lens models are investigated in detail; they allow a fairly simple geometrical construction of arc-like images. The observed arcs are then analyzed in view of the theoretical models, and the potential use of arcs for determining cluster masses, or as 'telescopes' for spectroscopy of distant, faint galaxies are mentioned. We illustrate these methods for one particular arc, which can be explained by a very simple GL model. Statistical investigations predict that for every spectacular, highly elongated arc, there must be numerous smaller ones, so-called arclets. They have been discovered recently, and their potential as astronomical tool is discussed.

Close QSO pairs, or multiply-imaged QSOs, provide a unique means to study the material associated with QSO absorption systems. From two GL systems, fairly strong constraints on the size of Lyα absorption clouds have been derived. A potential bias in QSO source counts might affect the statistics of metal absorption systems. An extended source moving across a caustic will show light variations that depend on its brightness profile. Hence, from an observed microlensing event, one might be able to reconstruct the (one-dimensional) brightness profile of sources. This method in principle allows microarcsecond resolution of sources. Observing the same QSO from two widely separated telescopes (with a separation of several, or up to 1000 Astronomical Units), microlensing in individual QSO images can be revealed; we discuss this parallax effect and some of its applications. A brief subsection discusses lensing by cosmic strings. For one QSO, the absence of lensing yields an upper limit of its mass. The possibility that the observed superluminal expansion of compact radio components in QSOs can be explained by lensing has been discussed in the literature; we present a number of arguments why this explanation is unlikely to work. Finally, we present a few topics which are usually not summarized under the name gravitational lensing: light deflection in strong gravitational fields (e.g., around neutron stars), light deflection in the Solar System and its impact on positional measurements of cosmic sources, and lensing of the microwave background.

1.3 Remarks about notation

The diverse roots of gravitational lensing, its young age, and its rapid development have prevented the creation of a standardized notation, a state of affairs that complicates reading the literature. We hope that our book will be of help in this respect.

Physical constants. The constant of gravity is $G \approx 6.67 \times 10^{-8}$ dyn cm^2 g^{-2}, the velocity of light in vacuo is $c \approx 3.0 \times 10^{10}$ cm s^{-1}.

Astronomical units. Astronomers measure distances in units of parsecs (pc), where 1 pc $\approx 3.09 \times 10^{18}$ cm ≈ 3.26 light years. Another convenient measure of distance is one Astronomical Unit (AU), 1 AU$\approx 1.5 \times 10^{13}$ cm. The mass of the Sun is $M_\odot \approx 1.99 \times 10^{33}$ g, and its radius is $R_\odot \approx 6.96 \times 10^{10}$ cm. Frequently used combinations of these parameters include the Schwarzschild radius of the Sun, $\frac{2GM_\odot}{c^2} \approx 2.95 \times 10^5$ cm, and the deflection angle for light rays at the solar limb, $\frac{4GM_\odot}{R_\odot c^2} \approx 8.48 \times 10^{-6} \approx 1''\!.75$.

Cosmology. Throughout this book, we consider a Friedmann–Lemaître (FL) cosmological model, mostly with zero pressure and zero cosmological constant. The basic notation is introduced in Chap. 4; in particular, H_0 denotes the Hubble constant, $\Omega = \varrho_0/\varrho_{\mathrm{cr}}$ is the cosmological density parameter, ϱ_0 the current mean mass density of the universe, and $\varrho_{\mathrm{cr}} = \frac{3H_0^2}{8\pi G}$ the critical (or closure) density. Frequently, we write the Hubble constant as $H_0 = h_{50} \times 50\,\mathrm{km\,s^{-1}\,Mpc^{-1}}$. The Hubble length is $\frac{c}{H_0} \simeq h_{50}^{-1}\,1.85 \times 10^{28}\,\mathrm{cm} \approx 6h_{50}^{-1}\,\mathrm{Gpc}$, and the Hubble time is $H_0^{-1} \approx h_{50}^{-1}\,1.96 \times 10^{10}$ years.

For an inhomogeneous Friedmann–Lemaître universe (introduced in Chap. 4), we denote the fraction of matter which is smoothly distributed by the smoothness parameter $\tilde{\alpha}$; the standard, smooth Friedmann–Lemaître universe is characterized by $\tilde{\alpha} = 1$.

Cosmological distances. We denote the angular-diameter distance of a source at redshift z_2 as seen from an observer at redshift z_1 by $D(z_1, z_2)$, and its scaled value by $r(z_1, z_2) = \frac{H_0}{c} D(z_1, z_2)$. In general, $r(z_1, z_2)$ depends on Ω and $\tilde{\alpha}$, and is determined as the solution of the initial value problem (4.52), (4.53). Further, we define $r(z) \equiv r(0, z)$. For $\tilde{\alpha} = 1$ (i.e., the smooth Friedmann–Lemaître universe), $r(z; \tilde{\alpha} = 1)$ is sometimes denoted by $r_1(z)$, which is explicitly given by (4.57). For a GL system with deflector redshift z_d and source redshift z_s, we usually write $r(z_\mathrm{d}) \equiv r_\mathrm{d}$, $r(z_\mathrm{s}) \equiv r_\mathrm{s}$, and $r(z_\mathrm{d}, z_\mathrm{s}) \equiv r_\mathrm{ds}$, in complete analogy with the definition of dimensional distances [e.g., $D_\mathrm{d} \equiv D(z_\mathrm{d}) \equiv D(0, z_\mathrm{d})$]. The luminosity distance $D_\mathrm{L}(z)$ to a source at redshift z is related to the angular-diameter distance $D(z)$ by $D_\mathrm{L}(z) = (1+z)^2 D(z)$.

Lensing notation. Throughout this book, we denote deflection angles (measured in radians) by $\hat{\boldsymbol{\alpha}}$. Physical vectors in the lens and source plane are usually denoted by $\boldsymbol{\xi}$ and $\boldsymbol{\eta}$, respectively. The surface mass density is denoted by Σ. All vectors in this book are, except for some of those in Chap. 3, two-dimensional. Usually, the quantities entering the lens (or ray-trace) equation are made dimensionless (see Sect. 5.1): one defines a length scale ξ_0 in the lens plane and defines $\mathbf{x} = \boldsymbol{\xi}/\xi_0$. The corresponding length scale in the source plane is $\eta_0 = \frac{D_\mathrm{s}}{D_\mathrm{d}}\xi_0$, and $\mathbf{y} = \boldsymbol{\eta}/\eta_0$. The scaled deflection angle is $\boldsymbol{\alpha} = \frac{D_\mathrm{d} D_\mathrm{ds}}{\xi_0 D_\mathrm{s}}\hat{\boldsymbol{\alpha}}$. For the particular choice $\xi_0 = D_\mathrm{d}$, $\mathbf{x}$ and $\mathbf{y}$ are angular coordinates in the lens and source plane, which are also denoted by $\boldsymbol{\theta}$ and $\boldsymbol{\beta}$, respectively. Another particular choice of ξ_0 is given by (2.6b), which we always use when dealing with point-mass lenses. The Schwarzschild radius is denoted by R_S.

The surface mass density is made dimensionless by defining $\kappa = \Sigma/\Sigma_{\mathrm{cr}}$, where Σ_{cr} is defined in (5.5). The deflection potential ψ is defined in terms of κ in (5.9), such that $\boldsymbol{\alpha} = \nabla\psi$, and the Fermat potential ϕ is given in (5.11). The shear caused by a lens is denoted by γ, and if a distinction is made between smooth matter and matter concentrated in stars, we denote the surface mass density of the former one by κ_c.

The change of observed flux from a lensed source is called magnification and denoted by μ (in the literature, the term 'amplification' instead of 'magnification' is frequently used; B. Burke convinced us, that 'magnification' is more appropriate, except for the term 'amplification bias'). In cases where a distinction is necessary between magnification of an image and the total magnification of a source, we sometimes denote the latter by μ_p and μ_e, for point sources and extended sources, respectively. The Jacobian matrix of the lens equation is denoted by the 2×2 matrix $A = \frac{\partial \mathbf{y}}{\partial \mathbf{x}}$, its determinant by D.

In Chaps. 11 and 12, we define various lensing cross-sections; there, we adopt the convention that $\hat{\sigma}$ denotes a cross-section in the source plane with dimensions of squared length, whereas $\sigma = \hat{\sigma}/\eta_0^2$ is the corresponding dimensionless cross-section. From cross-sections, we derive lensing probabilities; the term 'optical depth' will be introduced as a parameter in these probabilities, which, as we want to stress here, is not related to any absorption processes of photons.

Active galactic nuclei. We use the term active galactic nuclei (henceforth AGNs) for those extragalactic objects which show clear signs of nonthermal emission; in particular, these include Seyfert galaxies, QSOs, radio galaxies, and BL Lac objects. We use QSO to refer to both radio-loud (sometimes called quasars) and radio-quiet quasi-stellar objects (for an overview of AGNs, see [WE86.2]). The luminosity function of QSOs is introduced in Sect. 12.1.1, and its dimensionless form is denoted by Φ_Q.

Mathematical notation. ∇ denotes a two-dimensional gradient. The two-dimensional identity matrix is $\mathcal{I} = \text{diag}(1,1)$. $\text{H}(x)$ denotes the Heaviside step function, i.e., $\text{H}(x) = 1$ for $x \geq 0$, $\text{H}(x) = 0$ for $x < 0$. The derivative of $\text{H}(x)$ is Dirac's delta 'function' $\delta(x)$. i denotes the complex number $i = \sqrt{-1}$, * is complex conjugation, $\mathcal{R}e(z)$ and $\mathcal{I}m(z)$ are the real and imaginary parts of the complex number z, respectively. Relativistic notation, needed only in Chaps. 3 and 4, will be introduced at the beginning of Chap. 3.

Abbreviations. Only a few acronyms and abbreviations are used in this text; they include: GL (gravitational lens), FL (Friedmann–Lemaître), ML (microlensing), QSO, and AGN, as defined above.

Citations. We refer to papers in the format [ABij.k], where AB are the first two letters of the first author of a paper, ij are the two last digits of the year when the paper was published, and if two papers have the same combination ABij, a running number k distinguishes between them. Example: [SO04.1] corresponds to a paper by Soldner, published in 1804. (The ordering of papers by the running number k is completely arbitrary, and does not at all imply a chronology.) Although our ambition to produce a complete list of references about gravitational lensing was quite limited, we hope that the list given at the end of the book will be a useful source of information. In particular, we have not attempted to search through conference proceedings to find all contributions related to lensing: there are simply too many![6] A complete list of references on GL which appeared in refereed journals has been prepared by P. Véron; it can be obtained by request from him (Prof. P. Véron, Observatoire de Haute-Provence, 04870 Saint-Michel-L'Observatoire, France).

---

[6] As noted in [WA90.4], the number of papers cited in the *Astronomy and Astrophysics Abstracts* under the subject "Gravitational deflection" and "Gravitational lenses" in the periods 1974–78, 1979–83, 1984–88 is 36, 191, and 583, respectively.

2. Basic facts and the observational situation

Before presenting a detailed, critical derivation and investigation of GL theory, we offer an overview of the subject, using simple arguments and equations. This chapter is intended mainly for readers who prefer a quick look at the basic relations, e.g., between lens mass and angular separation between images, and who are not interested in the more detailed aspects of gravitational lens theory discussed in later chapters. In particular, a "guide for observers" is presented, concerning the search for, and the detection and verification of GL events. An overview of observed GL candidates is presented in Sect. 2.5.

2.1 The Schwarzschild lens

Einstein's deflection law. One of the early tests of General Relativity concerned light deflection by the Sun. General Relativity predicts that a light ray which passes by a spherical body of mass M at a minimum distance ξ, is deflected by the "Einstein angle"

$$\hat{\alpha} = \frac{4GM}{c^2 \xi} = \frac{2R_S}{\xi} \quad , \tag{2.1}$$

provided the impact parameter ξ is much larger than the corresponding Schwarzschild radius

$$R_S = \frac{2GM}{c^2} \quad . \tag{2.2}$$

To derive this result (e.g., [WE72.1],[MI73.1]) the linearized Schwarzschild metric suffices (see Chap. 4 below). The validity of (2.1) in the gravitational field of the Sun has been confirmed with radio-interferometric methods with an uncertainty of less than 1% ([FO75.1], [FO76.1]).

To demonstrate the effect of a deflecting mass, we show in Fig. 2.1 the simplest GL configuration. A "point mass"[1] M is located at a distance D_d

---

[1] According to General Relativity, a static spherical body with Schwarzschild radius R_S has a geometrical radius R larger than $(9/8)R_S$ while for a static black hole, $R = R_S$. Nevertheless, in lens theory the term "point mass" is used whenever one is concerned with light rays deflected with impact parameters $\xi \gg R_S$ by a spherical object, irrespective of the (unobservable) behavior of light rays with $\xi \sim R_S$; see the remark below (2.14).

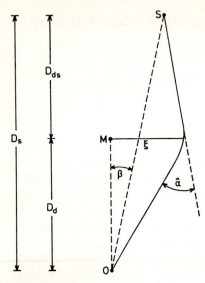

Fig. 2.1. The GL geometry for a point-mass lens M at a distance D_d from an observer O. A source S at a distance D_s from O has angular separation β from the lens. A light ray from S which passes the lens at distance ξ is deflected by $\hat{\alpha}$; the observer sees an image of the source at angular position $\theta = \xi/D_\mathrm{d}$

from the observer O. The source is at a distance D_s from the observer, and its true angular separation from the lens M is β, the separation which would be observed in the absence of lensing.[2] A light ray which passes the lens at a distance ξ is deflected by $\hat{\alpha}$, given by (2.1). The condition that this ray reach the observer is obtained solely from the geometry of Fig. 2.1, namely

$$\beta D_\mathrm{s} = \frac{D_\mathrm{s}}{D_\mathrm{d}} \xi - \frac{2R_\mathrm{S}}{\xi} D_\mathrm{ds} \quad . \tag{2.3}$$

Here, D_ds is the distance of the source from the lens. In the simple case with a Euclidean background metric considered here, $D_\mathrm{ds} = D_\mathrm{s} - D_\mathrm{d}$; however, since gravitational lensing occurs in the universe on large scales (among the hitherto detected GL events, most of the sources are QSOs with redshift typically larger than 1), one must use a cosmological model (see Sect. 4.5). There, "distance" does not have an unambiguous meaning, but several distances can be defined in analogy with Euclidean laws. The distances in (2.3) must then be interpreted as angular-diameter distances,[3] for which

---

[2] Due to the symmetry of the Schwarzschild lens, any light ray traveling from the source to the observer is confined to the plane spanned by source, lens, and observer. Therefore, instead of the two-dimensional separation vector of light rays from the lens (see Sect. 2.2), a one-dimensional treatment suffices here. Note that in the case $\beta = 0$, where the source, lens, and observer are colinear, no such plane is defined; this situation is discussed further below.

[3] to be more precise, as empty cone angular diameter distances, to be explained in Sect. 4.5 below

in general, $D_{ds} \neq D_s - D_d$. Denoting the angular separation between the deflecting mass and the deflected ray by

$$\theta = \xi/D_d \quad , \tag{2.4}$$

we obtain from (2.3) the lens equation,

$$\beta = \theta - 2R_S \frac{D_{ds}}{D_d D_s} \frac{1}{\theta} \quad ; \tag{2.5}$$

we allow β and θ to have either sign. This expression clearly shows that the problem under consideration involves a characteristic angle α_0 and a characteristic length, ξ_0, in the lens plane, given by

$$\alpha_0 = \sqrt{2R_S \frac{D_{ds}}{D_d D_s}} \quad , \tag{2.6a}$$

$$\xi_0 = \sqrt{2R_S \frac{D_d D_{ds}}{D_s}} = \sqrt{\frac{4GM}{c^2} \frac{D_d D_{ds}}{D_s}} = \alpha_0 D_d \quad . \tag{2.6b}$$

In addition, there is a characteristic length scale in the source plane, given by

$$\eta_0 = \sqrt{2R_S \frac{D_s D_{ds}}{D_d}} = \sqrt{\frac{4GM}{c^2} \frac{D_s D_{ds}}{D_d}} = \alpha_0 D_s \quad . \tag{2.6c}$$

The interpretation of these quantities will soon become clear. Writing (2.5) as $\beta = \theta - \alpha_0^2/\theta$ or

$$\theta^2 - \beta\theta - \alpha_0^2 = 0 \quad , \tag{2.7a}$$

we obtain

$$\theta_{1,2} = \frac{1}{2} \left(\beta \pm \sqrt{4\alpha_0^2 + \beta^2} \right) \quad . \tag{2.7b}$$

The angular separation between the images[4] is

$$\Delta\theta = \theta_1 - \theta_2 = \sqrt{4\alpha_0^2 + \beta^2} \geq 2\alpha_0 \quad , \tag{2.8a}$$

and the "true" angular separation between the source and the deflector is related to the image positions by

$$\theta_1 + \theta_2 = \beta \quad . \tag{2.8b}$$

Thus, the ray-trace equation always has two solutions of opposite sign. This means that the source has an image on each side of the lens. We show below

---

[4] According to the terminology of optics, the deflector does not act as a "lens", which produces "images"; in general, the rays are not focused at the observer. However, it has become customary to (ab)use the terms lens and image in the context of gravitational "lens" theory.

that the two images are of comparable brightness only if β, and hence $|\theta|$, is of order α_0. Therefore, the typical angular separation of the two images is close to its minimum $2\alpha_0$, and the corresponding typical impact parameter is ξ_0.

Typical length and angular scales. In practice, two cases of lensing situations are of importance. First, if both lens and source are at cosmological distances, the factors $D_\mathrm{d} D_\mathrm{ds}/D_\mathrm{s}$ and $D_\mathrm{d} D_\mathrm{s}/D_\mathrm{ds}$ will be fair fractions η and η', respectively, of the Hubble length c/H_0:

$$\frac{D_\mathrm{d} D_\mathrm{ds}}{D_\mathrm{s}} = \eta \frac{c}{H_0} \;, \tag{2.9a}$$

$$\frac{D_\mathrm{d} D_\mathrm{s}}{D_\mathrm{ds}} = \eta' \frac{c}{H_0} \;, \tag{2.9b}$$

where H_0 is the Hubble constant (see Sect. 4.4). Measuring H_0 in units of $50 \text{ km s}^{-1} \text{ Mpc}^{-1}$, $H_0 = h_{50} \cdot 50 \text{ km s}^{-1} \text{ Mpc}^{-1}$, the Hubble length is

$$\frac{c}{H_0} \approx h_{50}^{-1} \, 1.8 \times 10^{28} \text{cm} \approx 5.8 \times 10^9 \, h_{50}^{-1} \text{ pc} \;. \tag{2.10}$$

The angular scale α_0 is then

$$\alpha_0 \approx 1.2 \times 10^{-6} \sqrt{\frac{M}{M_\odot}} \sqrt{\frac{h_{50}}{\eta'}} \text{ arcsec} \;, \tag{2.11}$$

where $M_\odot = 1.989 \times 10^{33}$ g is the Sun's mass, and $4GM_\odot/c^2 = 2R_{\mathrm{S}\odot} \approx 5.9$ km. The corresponding length scale is

$$\xi_0 \approx 10^{17} \sqrt{\frac{M}{M_\odot}} \sqrt{\frac{\eta}{h_{50}}} \text{ cm} \approx 0.03 \sqrt{\frac{M}{M_\odot}} \sqrt{\frac{\eta}{h_{50}}} \text{ pc} \;. \tag{2.12}$$

The second case that is frequently discussed is $D_\mathrm{d} \ll D_\mathrm{ds} \approx D_\mathrm{s}$; then

$$\alpha_0 \approx 3 \times 10^{-3} \sqrt{\frac{M}{M_\odot}} \left(\frac{D_\mathrm{d}}{1 \mathrm{kpc}}\right)^{-1/2} \text{ arcsec} \tag{2.13}$$

and

$$\xi_0 \approx 4 \times 10^{13} \sqrt{\frac{M}{M_\odot}} \left(\frac{D_\mathrm{d}}{1 \mathrm{kpc}}\right)^{1/2} \text{ cm} \;. \tag{2.14}$$

For the nearest observed gravitational lens system, the lens is a spiral galaxy at redshift $z_\mathrm{d} \approx 0.039$; if we take $M = 10^{12} M_\odot$, we get $\xi_0 \approx 7 h_{50}^{-1/2}$ kpc. As mentioned above, the images are of comparable brightness only if β is sufficiently small that $|\theta| \sim \alpha_0$. This means that, with respect to lensing, any spherically symmetric matter distribution can be treated as a point mass if its radius is less than, say, $\xi_0/3$, since the exterior of a spherically symmetric mass distribution is always described by the Schwarzschild metric.

"Einstein rings". A special situation arises if source, lens and observer are colinear, i.e., $\beta = 0$. In this case, there is no preferred plane for the light rays to propagate, but the whole configuration is rotationally symmetric about the line-of-sight to the lens. For $\beta = 0$, the solutions (2.7) become $\theta_{1,2} = \pm\alpha_0$; hence, due to symmetry, the whole ring of angular radius $\theta = \alpha_0$ is a solution to the lens equation. A point source exactly behind the lens would thus appear, according to ray optics, as a circle around the lens. However, in this idealized case, infinitely many light rays with exactly equal path length intersect at the observer, and therefore diffraction and interference invalidate geometrical optics. In contrast, any real source is extended and not strictly monochromatic. Then, (i) those parts of a wave train emitted from a source point S located not on, but near the axis OM, which reach the observer from directions corresponding to the two geometrical images of S, arrive there separated in time, if the time delay exceeds the coherence time; and (ii) waves from different source points are incoherent [the time delay will be given in (4.78) below, and extended sources are considered, in this context, in Sect. 7.2]. Therefore, unresolved, extended sources, nearly aligned with deflector and observer, produce nearly the same ring images which geometrical optics predicts for the idealized case.

Ring-shaped images of point sources can only occur in lensing situations with axially symmetric matter distributions; they arise solely from symmetry. On the other hand, extended sources can have ring-shaped images even if the lens is not perfectly symmetric, as we describe in Sect. 2.5.6. Such images are frequently called "Einstein rings".

2.2 The general lens

The description of a mass distribution as a point mass (Schwarzschild lens) is only rarely sufficient for GL considerations. Even if the deflector is a star, its gravitational field is distorted in most realistic cases, either because the star is part of a galaxy which provides a tidal gravitational field, or a disturbance is due to galaxies lying near the line-of-sight to the source which, therefore, introduces an additional distortion. Hence, the Schwarzschild lens is an idealization; however, it is extremely useful, not only because of its simple properties, but also because such a simple model provides relations, e.g., between lens mass, the distances to lens and source, and angular separation between the images of a lensed source, which are of the same order-of-magnitude as those for more realistic lenses. Nevertheless, we should warn the reader from the outset: without a good intuitive understanding of a lens situation, relations based on a Schwarzschild lens should be considered with care. In the course of this book, this remark is explained in considerable detail; we only give one example here: whereas a Schwarzschild lens (and other matter distributions with spherical symmetry) can produce ring-shaped images of arbitrarily small sources, a general lens does not have this property.

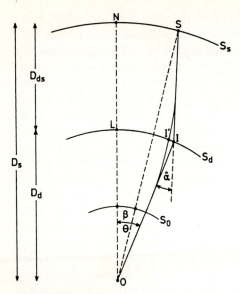

Fig. 2.2. A general GL system; the 'center' of the lens is at L, and the line through L and the observer O is the 'optical axis'. Relative to that, the source S has an undisturbed angular position β. A light ray $SI'O$ from the source is deflected by an angle $\hat{\alpha}$, so that an image of the source is observed at position θ. Due to the smallness of all angles present, we can replace the real light ray by its approximation SIO, and the source and lens spheres by their tangent planes. — In figures such as this one, observer and source may be interchanged; in equations such as (2.15), the D's have to be substituted by means of (4.55); we demonstrate this explicitly in Sect. 5.3.

Description of a general lensing situation. In Fig. 2.2 we show a typical lensing situation. Consider the source sphere $\mathbf{S_s}$, i.e., a sphere with radius D_s, centered on the observer O, or, in the cosmological situation, the set of sources with redshift z_s, and, correspondingly, the deflector sphere $\mathbf{S_d}$ with radius D_d, i.e., the distance to the center of the lens L.[5] Throughout this book, we assume that the lens has a velocity relative to a comoving observer which is much smaller than the velocity of light.[6] We call the straight line through O and L the optical axis, which serves as a reference line; it intersects the source sphere at N. In addition, consider the observer sphere $\mathbf{S_o}$ which is the apparent "sky" of the observer. On $\mathbf{S_o}$, the source would have angular position β if the light rays from the source S were not influenced by the gravitational field of the deflector. However, since the lens does bend light rays, the straight line SO is no longer a physical ray path. Rather, there are light rays which connect source and observer but which are curved near $\mathbf{S_d}$. One such ray $SI'O$ is drawn, together with its

---

[5] In a cosmological setting, geometrical terms (sphere, radius, straight line) refer to the 3-space of constant curvature of the background Friedmann-Lemaître model; see Sect. 4.6.

[6] More precisely, we treat lens systems as being static; when a relative velocity of either source, lens, or observer is involved, we consider a series of 'static' configurations with different values of the relative source position.

approximation *SIO*, consisting of the two asymptotes of the real ray. The angle $\hat{\alpha}$ between the two asymptotes *SI* and *IO* is the deflection angle caused by the matter distribution *L*. The observer will thus see the source at the position θ on his sphere $\mathbf{S_o}$.

Lens plane and source plane. In all cases of astrophysical interest, the deflection angles are very small. For all the observed lens cases, angular image separations are below 30″. As we saw in the preceding section, the typical impact parameter of light rays for a Schwarzschild lens (2.6b) is much larger than the Schwarzschild radius (2.2); hence, the fields under consideration are weak in the regions of interest. Therefore, only a small cone around the optical axis *OL* needs to be considered in general. Within such a small cone, the three spheres introduced above can be replaced by the corresponding tangent planes. Those corresponding to $\mathbf{S_s}$ and $\mathbf{S_d}$ will be called *source plane* and *lens plane*, respectively, in the following. The separation of the light ray from the optical axis, *LI*, will be described by the two-dimensional vector $\boldsymbol{\xi}$ in the lens plane – if $\hat{\alpha}$ is small, the distinction between *I* and *I'* is unnecessary, and the angles θ and $\hat{\alpha}$ will be described as (angular) vectors in the tangent plane to $\mathbf{S_o}$.

Ray-trace (or lens) equation. From the geometry of Fig. 2.2, we can easily derive a relation between the source position (described by the unlensed position angle β) and the positions of the images $\theta = \boldsymbol{\xi}/D_\mathrm{d}$ of the source:

$$\boldsymbol{\beta} = \boldsymbol{\theta} - \frac{D_\mathrm{ds}}{D_\mathrm{s}} \hat{\boldsymbol{\alpha}}(\boldsymbol{\xi}) \quad , \tag{2.15a}$$

or, in terms of the distance $\eta = D_\mathrm{s}\beta$ from the source to the optical axis,

$$\boldsymbol{\eta} = \frac{D_\mathrm{s}}{D_\mathrm{d}} \boldsymbol{\xi} - D_\mathrm{ds} \hat{\boldsymbol{\alpha}}(\boldsymbol{\xi}) \quad . \tag{2.15b}$$

Given the matter distribution of the lens and the position η of the source, (2.15a,b) may have more than one solution $\boldsymbol{\xi}$. This means that the same source can be seen at several positions in the sky. We shall show later that *any transparent mass distribution with finite total mass and with a weak gravitational field produces an odd number of images* [BU81.1], except for special source positions (see Sect. 2.3 for details). In our discussion of the Schwarzschild lens we did not consider rays with impact parameters comparable to the Schwarzschild radius for which the linear approximation fails (for this case, see, e.g., [MI73.1]). Within the lens approximation, a Schwarzschild lens always produces two images. If the lens is taken to be a black hole, the above theorem does not apply since a black hole captures rays incident with small impact parameters. For normal stars, the same is true for rays with an impact parameter smaller than the stellar radius. If, on the other hand, the lens is assumed to be a transparent ball of matter, the theorem does apply.

We see that the ray-trace equation (2.15) allows us to determine directly the "true" (unperturbed) source position β, if an image position θ is given, and if the deflection law $\hat{\boldsymbol{\alpha}}(\boldsymbol{\xi})$ is known. However, the typical problem in GL theory is to invert the lens equation, i.e., to find all the images of a source for a given matter distribution or to find, for given image positions, a suitable matter distribution. These problems can only be solved numerically for general deflectors. Numerical methods are described in Chap. 10. In Chap. 8, we investigate a number of "simple" lens models for which analytical results were obtained, or which can be analyzed with relatively simple numerical procedures.

The deflection angle. As will be shown in Sect. 4.3, for geometrically-thin lenses the deflection angles of several point masses simply add. Then, one can decompose a general matter distribution into small parcels of mass m_i and write the deflection angle for such a lens as

$$\hat{\boldsymbol{\alpha}}(\boldsymbol{\xi}) = \sum_i \frac{4G\, m_i}{c^2} \frac{\boldsymbol{\xi} - \boldsymbol{\xi}_i}{|\boldsymbol{\xi} - \boldsymbol{\xi}_i|^2} \quad , \tag{2.16}$$

where $\boldsymbol{\xi}$ describes the position of the light ray in the lens plane, and $\boldsymbol{\xi}_i$ that of the mass m_i.

We can take the continuum limit in (2.16), and replace the sum by an integral. This is most conveniently done by defining $dm = \Sigma(\boldsymbol{\xi}) d^2\xi$, where $d^2\xi$ is the surface element of the lens plane, and $\Sigma(\boldsymbol{\xi})$ is the surface mass density at position $\boldsymbol{\xi}$ which results if the volume mass distribution of the deflector is projected onto the lens plane. We then find

$$\hat{\boldsymbol{\alpha}}(\boldsymbol{\xi}) = \frac{4G}{c^2} \int_{\mathbb{R}^2} d^2\xi'\, \Sigma(\boldsymbol{\xi}') \frac{\boldsymbol{\xi} - \boldsymbol{\xi}'}{|\boldsymbol{\xi} - \boldsymbol{\xi}'|^2} \quad , \tag{2.17}$$

and the integral extends over the whole lens plane. Let us again point out the conditions which must be fulfilled for (2.17) to be valid: the gravitational fields under consideration must be weak, hence the deflection angle must be small. In addition, the matter distribution of the lens must be nearly stationary, i.e., the velocity of the matter in the lens must be much smaller than c. All these conditions are very well satisfied for astrophysical situations relevant to gravitational lensing. However, (2.17) cannot be used, for instance, to study the propagation of light rays in the vicinity of a black hole, but we do not treat such cases in this book (for photon propagation in the vicinity of black holes, see [ME63.1], [BA73.1]).

In the next section we consider some simple properties of GLs, always using the Schwarzschild lens as an example. Although it may be atypical in many respects, its mathematical description is sufficiently simple to serve as an illustration for several properties of the lens equation. More general lenses will be treated in detail later.

2.3 The magnification factor

Solid angles of images. Light deflection in a gravitational field not only changes the direction of a light ray, but also the cross-section of a bundle of rays. Since gravitational light deflection is not connected with emission or absorption, the *specific intensity* I_ν is constant along a ray, if measured by observers with no frequency shift relative to each other [this will be shown below, see (3.82)]. Moreover, gravitational light deflection by a localized, nearly static deflector does not introduce an additional frequency shift, besides the cosmological one, between source and observer; therefore, the *surface brightness* I for an image is identical to that of the source in the absence of the lens (both per unit frequency, and as an integral over frequency). The *flux* of an image of an infinitesimal source is the product of its surface brightness and the solid angle $\Delta\omega$ it subtends on the sky. Since the former quantity is unchanged during light deflection, the ratio of the flux of a sufficiently small image to that of its corresponding source in the absence of the lens, is given by

$$\mu = \frac{\Delta\omega}{(\Delta\omega)_0} \quad , \tag{2.18}$$

where 0-subscripts denote undeflected quantities.

We illustrate a typical GL arrangement in Fig. 2.3, which represents schematically the distortion of a light bundle. For an extended source, the flux magnification for corresponding images is obtained as the integral of the magnification (2.18) over the source, weighted by the surface brightness of the source.

Consider now an infinitesimal source at $\boldsymbol{\beta}$ that subtends a solid angle $(\Delta\omega)_0$ on the source sphere and – in the absence of lensing – also on the sphere-of-vision of the observer (image sphere). Let $\boldsymbol{\theta}$ be the angular position of an image with solid angle $\Delta\omega$. The relation of the two solid angles is determined by the area-distortion of the lens mapping (2.15) and given by

$$\frac{(\Delta\omega)_0}{\Delta\omega} = \left|\det\frac{\partial\boldsymbol{\beta}}{\partial\boldsymbol{\theta}}\right| = \frac{A_s}{A_I}\left(\frac{D_d}{D_s}\right)^2 \quad ; \tag{2.19}$$

i.e., the area distortion caused by the deflection is given by the determinant of the Jacobian matrix of the lens mapping $\boldsymbol{\theta} \to \boldsymbol{\beta}$. From (2.18), the magnification factor is

$$\mu = \left|\det\frac{\partial\boldsymbol{\beta}}{\partial\boldsymbol{\theta}}\right|^{-1} \quad . \tag{2.20}$$

The magnification μ is thus the ratio of the flux of an image to the flux of the unlensed source. If a source is mapped into several images, the ratios of the respective magnification factors are equal to the flux ratios of the images.

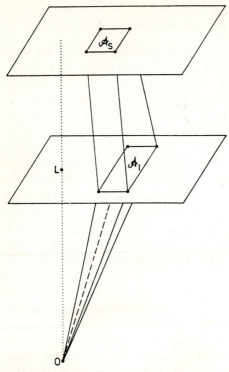

Fig. 2.3. The distortion of the solid angle subtended by a source. The source spans an area $\mathcal{A}_s$ and thus subtends a solid angle $(\Delta\omega)_0 = \mathcal{A}_s/D_s^2$ at the observer O in the absence of lensing. If lensing takes place, the solid angle of the image is $\Delta\omega = \mathcal{A}_I/D_d^2$, in general different from $(\Delta\omega)_0$. Since the surface brightness of the source is unchanged by the light deflection, the apparent brightness of the source is magnified in proportion to the solid angle $\Delta\omega$

Parity of images. The magnification factor is a function of $\boldsymbol{\theta}$, determined by the deflection law $\boldsymbol{\beta}(\boldsymbol{\theta})$. The determinant of the Jacobian matrix may have either sign; the images of the source for which the determinant is positive (negative) are said to possess *positive (negative) parity*. A source can also have images for which the determinant vanishes. Formally, one could assign the parity 0 to such 'critical' images; the lens mapping close to such points behaves in a particular way, as we discuss in detail in Chap. 6. Unless otherwise mentioned, such critical images are excluded here. As will be shown in Chap. 5, the parity of an image determines its circulation (orientation) relative to that of the unperturbed image. To illustrate this point, consider a source which consists of a "core" and a curved "jet". In the images of positive parity, the direction of curvature is preserved, whereas in negative parity images it is reversed.

Critical curves and caustics. Regions in the lens plane where the Jacobian determinant has opposite sign are separated by curves on which the latter vanishes; these curves are termed *critical curves*. On these curves, the magnification factor diverges – see (2.20). However, this divergence does not mean that the image of a source is actually infinitely bright, because two additional facts, hitherto neglected, must be taken into account. First, real sources are extended; for such sources, the magnification is the weighted mean of (2.20) over the source. This always leads to finite magnifications. Second, even if a source were point-like, the magnification would still not be infinite: so far, we have treated the propagation of light in the approximation of geometrical optics. This approximation is, quite generally, sufficient for GL studies. As will be shown in Chap. 7, sources must be unrealistically small for interference effects to be important. However, even for a hypothetical point source, the maximum of the magnification factor for images on the critical curve is finite, due to wave effects.

The critical curves of a deflection mapping are of great importance for a qualitative understanding of its properties. Consider the images of the critical curves under the lens mapping (2.15); these curves in the source plane are called *caustics*. Given the positions of observer and lens, the number of images in general varies with the source position. We show in Chap. 6 that *the number of images changes by two if, and only if, the source crosses a caustic*. Depending on the direction of crossing, two images with opposite parity merge into one on the critical curve and then disappear, or vice versa. Shortly before their fusion (or after their creation) the images are very bright, since they are in the vicinity of the corresponding critical curve in the lens plane.

If the locations of the caustics are known, it is easy to determine the dependence of the number of images on the position of the source, just by using the aforementioned property. The only additional information required is that *for any transparent matter distribution with finite mass, the number of images of a point source is one if the source is sufficiently misaligned with the lens* (see Sect. 5.4); again, transparency of the lens excludes mass distributions which have points of infinite surface density (e.g., a black hole). In addition, from these arguments one can infer the validity of the odd-number theorem mentioned in the last section. Note, however, that the imaging of extended sources is much more complicated, as will become clear from the examples given in Chap. 8.

Magnification theorem. A second general property of gravitational light deflection should be mentioned here. It is established in Chap. 5 that *among the images of any point source produced by an arbitrary transparent matter distribution, there is always one with positive parity which is at least as bright as the unlensed source* ([SC84.2], [BL86.1]), i.e., at least one image has $\mu \geq 1$.

One may wonder how this is possible – at first sight it may seem that GLs violate the law of conservation of energy. Imagine a source surrounded by equidistant observers. If lenses are brought into the space between source and observers, each observer appears to receive more energy per unit time than before, although the luminosity of the source has not been changed.

This apparent contradiction will be resolved in Chap. 4, when we discuss the cosmological models that are necessary in GL theory. Briefly, insertion of deflecting matter in the sphere spanned by the observers changes spacetime; this change has to be necessarily taken into account when discussing flux conservation (see Sect. 4.5.1).

Magnification factor for a Schwarzschild lens. We now calculate μ for the Schwarzschild lens. Let us define the normalized angles

$$\tilde{\theta} = \theta/\alpha_0 \;, \tag{2.21a}$$

$$\tilde{\beta} = \beta/\alpha_0 \;. \tag{2.21b}$$

Consider an infinitesimal source at $\tilde{\beta}$ with (normalized) solid angle $\Delta\tilde{\beta}\,\tilde{\beta}\,\Delta\varphi$, where $\tilde{\beta}\,\Delta\varphi$ is the size of the source perpendicular to the plane of Fig. 2.1. This source will have two images at positions $\tilde{\theta}_+$, $\tilde{\theta}_-$:

$$\tilde{\theta}_\pm = \frac{1}{2}\left(\tilde{\beta} \pm \sqrt{4 + \tilde{\beta}^2}\right) \;, \tag{2.22}$$

with (normalized) solid angle $\Delta\tilde{\theta}_\pm\,\tilde{\theta}_\pm\,\Delta\varphi$; due to the symmetry of the lens, $\Delta\varphi$ is unchanged. The absolute values of the magnification factors of the images, according to (2.20) are

$$\mu_\pm = \left|\frac{\Delta\tilde{\theta}_i \cdot \tilde{\theta}_i}{\Delta\tilde{\beta} \cdot \tilde{\beta}}\right| \;. \tag{2.23}$$

From (2.22), after a little algebra,

$$\mu_\pm = \frac{1}{4}\left[\frac{\tilde{\beta}}{\sqrt{\tilde{\beta}^2 + 4}} + \frac{\sqrt{\tilde{\beta}^2 + 4}}{\tilde{\beta}} \pm 2\right] \;. \tag{2.24}$$

Here, $\tilde{\beta} \geq 0$, and the indices $\pm$ indicate the parities of the images.

It is interesting to note that the magnification factor diverges for $\tilde{\beta} \to 0$. As shown before, for $\tilde{\beta} = 0$ the image of the source is a ring; we expected that in such a case the magnification was large. The ring with radius $\tilde{\theta} = 1$ is thus the critical curve of the Schwarzschild lens, and the caustic in the source plane degenerates to the point $\tilde{\beta} = 0$.

The two images are of quite different brightness if $\tilde{\beta}$ is not near 0. The flux ratio

$$\frac{\mu_+}{\mu_-} = \left[\frac{\sqrt{\tilde{\beta}^2 + 4} + \tilde{\beta}}{\sqrt{\tilde{\beta}^2 + 4} - \tilde{\beta}}\right]^2 \tag{2.25}$$

is shown, together with other relevant quantities, in Fig. 2.4. Thus, the flux ratio and the image separation $\Delta\theta$ determine the Einstein angular radius α_0 and the angular distance β of the "true" position of the source from the deflector. The limiting behavior for the magnification factor is:

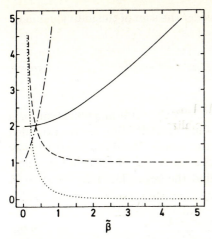

Fig. 2.4. The angular separation of the two images produced by a Schwarzschild lens, in units of α_0 (solid line), the magnification of the primary (μ_+, dashed line) and the absolute value of that of the secondary image (μ_-, dotted line), and the absolute value of the ratio μ_+/μ_- of these magnifications (dashed-dotted line), as a function of the scaled angular coordinate $\tilde{\beta}$ of the source

$$\text{for } \tilde{\beta} \to 0 : \begin{cases} \mu_+ = \dfrac{1}{2\tilde{\beta}} + \dfrac{1}{2} + \mathcal{O}(\tilde{\beta}) \;, \\ \mu_- = \dfrac{1}{2\tilde{\beta}} - \dfrac{1}{2} + \mathcal{O}(\tilde{\beta}) \;, \\ \mu_+/\mu_- = 1 + \tilde{\beta} + \mathcal{O}(\tilde{\beta}^2) \;; \end{cases}$$

$$\text{for } \tilde{\beta} \to \infty : \begin{cases} \mu_+ = 1 + \tilde{\beta}^{-4} + \mathcal{O}(\tilde{\beta}^{-6}) \;, \\ \mu_- = \tilde{\beta}^{-4} + \mathcal{O}(\tilde{\beta}^{-6}) \;, \\ \mu_+/\mu_- = \tilde{\beta}^4 + \mathcal{O}(\tilde{\beta}^0) \;. \end{cases}$$
(2.26)

As we discuss below, we are not likely to find a Schwarzschild lens that produces multiple images of a source which can be resolved: either the angular separation and brightness ratio of the images are far below the resolution and sensitivity of current telescopes, or the lens is sufficiently massive and too extended to be considered as a point mass. The importance of the Schwarzschild lens model, in addition to the properties discussed in Sect. 2.2, is due to the fact that a compact mass can significantly magnify sources. We show in Sect. 12.5 that compact objects can influence source counts, since they cause a selection bias (called amplification bias): if a source is intrinsically below the threshold of a flux-limited sample, but sufficiently magnified by gravitational light deflection, it will be included in the sample.

Magnification of extended sources. Consider the sum of the individual magnifications (2.24),

$$\mu_p \equiv \mu_+ + \mu_- = \frac{\tilde{\beta}^2 + 2}{\tilde{\beta}\sqrt{\tilde{\beta}^2 + 4}}, \qquad (2.27)$$

where the p-subscript indicates that (2.27) is the total magnification for a point source. To obtain the corresponding magnification for an extended source, (2.27) must be integrated over the source. We do so later; here we only consider the special case where a circular source of uniform brightness and angular radius $\vartheta \alpha_0$ is exactly behind the lens. The magnification in this case is

$$\mu = \frac{1}{\pi \vartheta^2} \int_0^\vartheta 2\pi \tilde{\beta} \, d\tilde{\beta} \, \mu_p(\tilde{\beta}) = \frac{\sqrt{\vartheta^2 + 4}}{\vartheta}. \qquad (2.28)$$

This result can also be obtained using geometry, based on the original definition (2.18) of the magnification factor. The source under consideration, in the absence of lensing, subtends a solid angle $(\Delta \omega)_0 = \alpha_0^2 \pi \vartheta^2$. Its image under the lens action is a ring with inner and outer angular radii $\frac{\alpha_0}{2}\left(\sqrt{\vartheta^2 + 4} - \vartheta\right)$ and $\frac{\alpha_0}{2}\left(\sqrt{\vartheta^2 + 4} + \vartheta\right)$, respectively [see (2.8)], and with solid angle $\Delta \omega = \pi \vartheta \sqrt{\vartheta^2 + 4}\, \alpha_0^2$. Inserting this into (2.18) reproduces (2.28). Since the magnification of a source is largest when it is colinear with lens and observer, (2.28) also provides a relation between the maximum magnification factor and the size of the source. Significant magnification is possible if the (normalized) angular radius ϑ of the source is less than about 1, i.e., if the true angular radius is less than α_0, which corresponds to a length scale

$$\eta_0 = \alpha_0 D_s = \sqrt{\frac{4GM}{c^2} \frac{D_{ds} D_s}{D_d}}. \qquad (2.29)$$

Hence, if the spatial extent of the source is less than η_0, significant magnification may occur.

Estimate of physical scales. To convert (2.29) to a physically meaningful scale, we consider the case $D_d \ll D_{ds} \approx D_s$, which is of importance for magnification of distant sources by compact objects in nearby galaxies or in our own galaxy (see Sect. 12.4). Then,

$$\eta_0 \approx 2.7\, h_{50}^{-1} \left(\frac{M}{M_\odot}\right)^{1/2} \frac{D_s}{c/H_0} \left(\frac{D_d}{1\,\mathrm{Mpc}}\right)^{-1/2} \mathrm{pc}, \qquad (2.30)$$

i.e., sources smaller than about 1 pc can be affected by stellar masses, provided the lens is sufficiently nearby. To illustrate the significance of the value of η_0 further, we introduce our "standard source", a hypothetical source with the following properties: it has a high redshift, $z_s \sim 2$, and consists of three components of spatial extents 10^{15} cm (R1), 10^{18} cm (R2) and 10^{22} cm (R3), respectively. This "standard source" includes the sizes thought to describe different emission regions of QSOs. R1 corresponds to the continuum-emitting region, R2 to the broad-line emission region and R3 to the host galaxy. Whenever numerical estimates are given, we refer to the "standard source", since gravitational lensing is most important for QSOs and AGNs. If a compact object of mass M at a distance of ~ 10 Mpc (the distance of nearby galaxies) acts on our "standard source", R1 can be magnified significantly, even if M is as small as $\sim 10^{-5} M_\odot$; for R2, masses of about $10 M_\odot$ are needed, whereas component R3 can only be magnified significantly if $M \gtrsim 10^{8.5} M_\odot$.

Amplification bias and lens-induced variability. It is important to see that even Jupiter-like objects can influence the apparent brightness of the continuum region of QSOs quite significantly. The observability of this effect, however, depends on the probability of occurrence of such magnification events, which must be investigated by statistical analysis, as shown in Chap. 12. One of the important consequences of gravitational magnification is its ability to cause apparent variability of sources, due to changes in the relative position of the source with respect to the lens as a consequence of transverse motion. The resulting light curves depend on the lens mass, transverse velocity, distances of the source and the lens from the observer, and the size of the source. In particular, the light curves induced by general GLs are quite different from those induced by Schwarzschild lenses, because in general the caustic curves in the source plane do not degenerate to a single point (see especially Chap. 8). We postpone a detailed discussion of light curves to Chap. 12. Here, only a rough estimate of the time-scale of variability will be given. Consider a lens of mass M that approaches the line-of-sight to a distant source with transverse velocity $v_\perp$. The variability time-scale is roughly the time for the lens to cross a region of radius ξ_0 (2.6b). Consider first the case where both source and lens are at cosmological distances [compare, e.g., (2.9–11)]. Then, the variability time-scale is

$$t_{\text{var}} \approx 100 \left(\frac{M}{M_\odot}\right)^{1/2} \left(\frac{\eta}{h_{50}}\right)^{1/2} \left(\frac{v_\perp}{300 \text{ km/s}}\right)^{-1} \text{ yrs} . \qquad (2.31)$$

When $D_{\text{d}} \ll D_{\text{ds}} \approx D_{\text{s}}$, we have

$$t_{\text{var}} \approx 5 \times 10^{-2} \left(\frac{M}{M_\odot}\right)^{1/2} \left(\frac{D_{\text{d}}}{1 \text{ kpc}}\right)^{1/2} \left(\frac{v_\perp}{300 \text{ km/s}}\right)^{-1} \text{ yrs} . \qquad (2.32)$$

The last expression shows that stars in our Galaxy could cause rapid variability of sources; however, the "optical depth" for lensing[7] by stars in our

---

[7] The term "optical depth" is used in GL theory in a fairly loose sense. In general, it describes the magnitude of lensing probabilities – see Chap. 11. Here, optical depth means the fraction of the sky covered by the Einstein circles of all lenses between us and sources at a specified distance.

own Galaxy is very small (see below), so events between Milky Way stars occur rarely. It was recently proposed [PA86.4] that such variability could, in principle, be studied by monitoring a large number of individual stars in a nearby galaxy. From a study of a very large number of stars, one could detect the presence, and determine the properties of compact clumps of dark matter in our own Galaxy. In fact, two observational projects are currently underway to search for MACHOs (massive compact halo objects) using this method. In Chap. 12, we discuss further applications of lens-induced variability. Once again, the estimates (2.31) and (2.32) only give orders of magnitude; the variability time-scales depend sensitively on the detailed nature of the lens event and can be significantly shorter than the estimates we have given.

Lensing within our Galaxy. We now show that lensing is unimportant for observations of galactic sources, due to the small probability that a star in our Galaxy is sufficiently aligned with a source and the Earth to produce an appreciable distortion. For a source at distance D_s, consider the probability that its flux is changed markedly. For such an event to occur, a star at distance λ must lie within $x\xi_0(\lambda)$ from the line-of-sight to the source, where ξ_0 is given by (2.6b), x is a dimensionless number, and the magnification μ_p is a function of x only [see (2.27)]. For example, for $x = 1$, $\mu_p = 1.34$. For simplicity, we assume that all the mass in our Galaxy is in stars with density $n_* M \simeq 0.12\, M_\odot/\text{pc}^3$ [AL73.1]. It is straightforward to show that the volume described by the condition that the transverse distance ξ of a star be less than $x\xi_0(\lambda)$ is an ellipsoid [NE86.1]. The expected number of stars in this volume is

$$N_{\exp} = n_* \int_0^{D_s} d\lambda\, \pi x^2 \xi_0^2 = \pi n_* x^2 \int_0^{D_s} d\lambda \frac{4GM}{c^2} \frac{\lambda(D_s - \lambda)}{D_s} .$$

We see that $N_{\exp}$ depends only on the product $n_* M$, i.e., on the mass density in stars, but not on the mass of individual stars. This is due to the specific aspect of lensing we discuss; the probability for a source to have multiple images with separation greater than $\Delta\theta$ *does* depend the mass spectrum of the lenses. The integral above is easily evaluated:

$$N_{\exp} = 1.26 \times 10^{-7} x^2 \left(\frac{D_s}{1\text{kpc}}\right)^2 . \qquad (2.33)$$

We see that the probability of finding a star in the line-of-sight to a compact source in our Galaxy is negligibly small (remember that $\mu_p \simeq x^{-1}$ for $\mu_p \gg 1$) and therefore Galactic lens systems are not likely to occur. For a discussion of the probability of detection of lensing of *extended* Galactic objects, see [SA85.1].

2.4 Observing gravitational lens systems

A GL does not change the surface brightness of a source; it also preserves the spectrum of deflected radiation. As we discuss below, in practice this is an approximate statement, but let us suppose for the moment that it were true. Then, it should be straightforward to discover a GL system (a set of objects that comprises a foreground GL and the images it produces of background sources): one could search for at least two objects separated by, say, between one arcsecond and one arcminute on the sky, whose spectra are identical. Such an apparently uncomplicated project is, in fact, a challenge for astronomers, as we will explain in the course of this section.

Objects very far away from the Earth, such as QSOs and distant galaxies, are the most promising targets to search for GL systems. Features in favor of QSOs are their high intrinsic luminosities, their simple optical morphology, and the prominent emission lines that can be found in their spectra. Less favorable features of QSOs are their stellar appearance, which prevents their classification as QSOs without spectra, their large redshifts, which result in a faint appearance, and their relatively low population density (per square degree, brighter than $m_V = 20$, there are $\simeq 20$ QSOs [MA84.1], and for galactic latitudes greater than $\simeq 20°$, there are $> 10^4$ stars [AL73.1]). The tireless forward march of astronomical observations, aided by serendipity, initially did favor QSOs: the first GL systems found and confirmed consist of GL images of QSOs that are formed by intervening galaxies. Galaxies are not as luminous as QSOs, they are extended, and the lines in their spectra are much weaker than for QSOs, or they can be detected only as 'breaks' in continuum spectra. Extracting information from the spectra of distant galaxies can therefore be extremely difficult. However, their population density is orders of magnitude higher than that of QSOs. Thus, galaxies are being vindicated: several GL systems were found within the past few years, thanks to the very strong distortion of the images of distant galaxies due to the lens action of intervening clusters of galaxies.

In the following two subsections, we set the stage first for point sources (QSOs) and then for extended sources (galaxies). For each type of source, we describe the kind of GL systems one might expect to find, and we discuss the methods of confirmation of putative GL systems. We also discuss potential problems that await the unwary observer. In the next section, we describe the known GL systems, in order of increasing number of component images. Although a non-singular mass distribution produces an odd number of images of a point source (Sect. 5.3), in effect, even numbers of images are also found, because of demagnification and confusion effects acting on 'odd' images. Therefore, a progression is found, beginning with two images of point sources. Beyond four images (the current observed maximum), we find ourselves in the area of extended arcs and rings, a field where intense work and abundant new findings may come to dominate GL work in terms of numbers, and of popularity.

2.4.1 Expectations for point sources

Let us now consider QSOs simply as point sources that can be multiply imaged. For the remainder of this subsection, when we refer to a 'GL system', it should be understood that we are assuming a point source. In spite of their very high intrinsic luminosities, QSOs are very distant, and thus faint. Since the probability for a QSO to be affected by a lens increases with its distance away from Earth (there is simply more mass near the line-of-sight), the farthest QSOs are the most promising candidates for multiple imaging. Therefore, faintness represents the first observational obstacle against discovery. The second observational obstacle is the expected angular separation of images. Galaxies are thought to be the most common massive mass concentrations that are also likely to be sufficiently compact to yield multiple images of a distant source. The requirement of compactness means roughly that a mass M must lie fully within its own Einstein circle of radius ξ_0 – see (2.6b). We will see that the first seven or so years of the GL era were in fact dominated by lenses that turned out to be single, or cluster galaxies. Clusters of galaxies, obviously much more massive than a single galaxy, are sufficiently compact to yield multiple images [NA84.3] when they are very rich. From (2.11), we see that the typical angular scale of image separation for even a massive galaxy ($M \sim 10^{12} M_\odot$) is only of order one arcsecond. Thus, one expects multiple images to occur at angular separations near the resolution limit of optical telescopes. The fact that, in the known GL systems, the image separations are as large as 7 arcseconds may be due to either the collusion of deflection effects due to a cluster with those due to one of the galaxies in the cluster (for instance a massive cD galaxy) or to unknown dark lenses.

We have seen that finding GL candidates is far from trivial; in addition, the fraction of QSOs among stellar sources at about 17th magnitude is about 10^{-4} – the rest are stars. Thus, we can easily understand why it took so long before the first lens system was discovered in 1979 [WA79.1]. Since radio-interferometric observations yield much higher angular resolution than optical observations, it may be fruitful to search for GL candidates in samples of compact radio sources; in fact, many of the known GL candidates were first identified through radio observations. Of course, compact radio sources can be accompanied by complicated extended emission, and it may be difficult to distinguish intrinsic features from lensing effects. Furthermore, only $\sim 10\%$ of all QSOs are radio-loud. At any rate, radio observations have been shown to be very useful for generating lists of GL candidates (Sect. 12.2.5).

Every extragalactic point source is affected, albeit generally weakly, by the GL effect of inhomogeneities along the line-of-sight to each object. Such weak effects must be taken into account, for example, when one studies population statistics of QSOs. Such statistical effects will be discussed further in Sect. 2.5.8.

Confirmation. Suppose we have found at least two images separated by a few arcseconds on the sky, in either a radio map or an optical frame. Individual images may differ in apparent luminosity, due to the magnification caused by the GL. Due to anisotropic differential deflection of light bundles, called shear, radio contours need not have the same shape or size. To attempt to confirm an occurrence of the GL effect, we must measure spectra. This is one of the main burdens for GL searches with radio surveys: although the angular resolution is better for radio-interferometric than for optical observations, a GL candidate must also be observed at optical wavelengths, since only the optical part of the spectrum contains enough information to distinguish distinct QSOs from multiple images of a single QSO – although, as we will see later on, even a high-quality optical spectrum may not differentiate between images or separate objects.

Let us assume that we can measure the optical spectrum of each image, and that each turns out to be a QSO. The most basic useful test is a comparison of their redshifts. If they are different, the GL hypothesis becomes improbable. If they are equal, the system can be tentatively accepted as a GL candidate. But the question of whether the redshifts are 'equal' can be answered only with high signal-to-noise-ratio spectra (of faint objects); for instance, if only broad, noisy lines are observed, it may be difficult to determine the redshift to better than $\sim 200 \,\mathrm{km/s}$ accuracy. However, if the quality of the spectra suffices, if narrow lines can be observed, and if the redshift difference can be constrained to be less than, say, $50 \,\mathrm{km/s}$, the system is an excellent GL candidate. We must also check the strengths of the lines, which should be similar for distinct GL images, but not necessarily identical, as we discuss below. If a potential lens, such as a galaxy, is found near the images, the GL hypothesis is significantly more likely. Such a best-case scenario, however, is not the most common one.

If the images correspond to a QSO, and the latter is variable (a particularly common occurrence for radio-loud QSOs), the fluctuations in each image should be correlated: since the light-travel-time along different ray paths varies from image to image, the observed light curves have the same shape, but translated in time. We calculate the time delay for image pairs in Sects. 4.6 and 5.3; an order-of-magnitude estimate for image separation $\Delta\theta$ is about $(\Delta\theta/5'')^2$ years.[8] If we observe for a time longer than the time delay, the correlated nature of the light curves should become apparent for images of a single QSO. One word of caution is that stars, crossing the beams that form images due to a lens such as a galaxy, may mask intrinsic fluctuations. The effect, called *microlensing* (hereafter ML), cannot be neglected, as we discuss below.

To summarize the preceding discussion, let us say that, to classify a set of observed objects as a GL system, we must have:

---

[8] This estimate is based on the fact that a typical QSO has a redshift of $z_s \sim 2$, for which the most probable lens redshift is $z_d \sim 0.5$, and that there is no special geometrical alignment.

1. at least two images close together on the sky;
2. flux ratios in different spectral bands that are the same for all images;
3. redshifts that are the same for all images;
4. line flux-ratios and shapes that are similar for different images;
5. a possible lens in the vicinity of the images, but at redshift smaller than that of the images;
6. temporal variations in different images that are correlated.

Pitfalls. The list of necessary properties for a GL system was established in the idealized world of a theorist. In general, not all the stated conditions will be satisfied. We discuss next the modifications to the list in the real world.

Condition (1.) is of course essential, because GL systems are identified by closely spaced images, and all candidates satisfy condition (1.). Property (2.) is a quick check as to whether the images can originate from a single source; e.g., a blue and a red object near each other would probably be classified as a chance projection of a star with a QSO (or another star) and thus be rejected. In an analogous fashion, for radio observations, the images should share a common spectral index. However, the colors need not be identical in a real situation. The light that makes up different images has pierced the lens at different positions, and in general has sampled different values of its mass density. Therefore, one image can show stronger absorption, and can be more reddened by the lens. The radio spectral index should be free of this effect, so radio observations can help greatly.

A strong argument for lensing is condition (3.): if several images correspond to a single QSO, they should all share a common redshift. More precisely, if the redshifts of the images can be measured from narrow emission lines, they should agree to within the measurement errors. Such an agreement is the most sensitive check for lensing, if the measurements are sufficiently accurate.

Condition (4.) on the agreement of the spectra is fairly weak, for several reasons. First, reddening and absorption in the lensing galaxy can change the underlying continuum. Second, the spectrum of the source can actually be changed by lensing. The spectrum of an ideal point source is unaltered by gravitational light deflection. However, the spectrum of a real (extended) source is an integral of the spectrum over differential patches covering the source, weighted by the point source magnification factor. If the emission spectrum varies across the source, and if the point source magnification factor is not constant across the source, the spectrum of the extended source will be modified by lensing. A typical situation where this problem can arise occurs when a star in the lens comes close to an image; the star will lead to small-scale variations of the point source magnification, and can thus modify the broad-emission line spectrum of the QSO. In particular, the equivalent width of the lines can be affected, if the continuum source is magnified and the broad-emission line region is not. We will discuss this and related effects in detail in later sections; the study of such ML effects has become a very active field of study.

ML can produce measurable differences in the spectra of the images of a QSO. It may even affect the central wavelength of the broad emission lines, thus changing the observed redshift. Only narrow lines can be reasonably expected not to be modified by lensing. QSOs are variable, and their spectra can also vary over times as short as a few days, but certainly over times of the order of a year, which is comparable to the time delay expected from galaxy-type lenses. We observe the spectrum of the source at two different times of emission, and therefore, until the time delay is known, the spectra of different images cannot be compared properly.

We have seen that condition (4.) is at times weak. In extreme cases, such as when one image shows a strong line that is missing in the other image, we would at first reject the GL hypothesis. Image spectra should be similar, but we cannot give a quantitative measure of such similarity, because the spectra of QSOs with very different redshifts can have very similar appearances. In fact, presently, there is no quantitative measure that could be used to define 'similarity of QSO spectra'.

Condition (5.) appears to be a natural requirement for a GL system. However, there are systems that are regarded as very good GL candidates, for which no lens has been observed. We describe some of these in detail in Sect. 2.5; for a few cases, very stringent upper limits on the luminosity of a lens have been established. This observational difficulty is not well understood; one explanation is that the lenses in these cases have very large mass-to-light ratios, much larger than for normal galaxies. We discuss such possibilities later on in more detail. We just note here that the absence of an observed lens does not necessarily lead to rejection of GL candidacy.

Condition (6.) is a natural requirement for sources with intrinsic variability, but the history of observations of lens systems is rather disappointing in this respect. Since no generally accepted value for the time delay is available for 0957+561 (the best- and most-observed case – but see end of Sect. 13.1), we see that its measurement is much more complicated than previously thought. Although the QSO in this system varies only weakly, we might have expected that one could determine the time delay after a few years of observations. Unfortunately, ML can cause variations in one or more images, with no correlation, but with amplitudes similar to those of intrinsic fluctuations. As mentioned in the previous subsection, a measurement of the time delay would be of great interest, as it is the only dimensional observable in a lensing situation.

The discussion above has shown that it is difficult to confirm the GL hypothesis. There is not a single case for which all the conditions listed above are satisfied. A few of the known systems have only some of the properties we listed, but they are still regarded as GL systems, simply because other explanations appear to be even less likely. In Sect. 2.5, we discuss specific examples.

2.4.2 Expectations for extended sources

Let us now consider extended sources, that is, in general, distant (optical) galaxies or extended radio sources. For the remainder of this subsection, when we refer to a 'GL system', it should be understood that we assume an extended source. Galaxies are even fainter than QSOs, and they suffer from the difficulties of spectroscopy that we have mentioned. However, the larger density of galaxies (compared to that of QSOs) and their extension, which means that they are easily distinguishable from stars and that they cover larger areas than point sources, now seem to be significant factors in their favor. In addition, the very spectacular nature of the phenomena of arcs and rings should increase their a priori probability of discovery.

The requirements of lens compactness that were described for point sources still apply to extended ones. Of course, the statistics and morphologies we expect for galaxies will differ from those for QSOs. For example, an extended source with intrinsic size of a few arcseconds, when magnified by typical factors of 5 to 10, will span at least tens of arcseconds. For such highly magnified images of extended sources to occur, the lens and the images must have at least a comparable angular size; thus, clusters of galaxies are the most natural lensing objects.

It remains true that finding GL candidates is not trivial. In addition to high resolution, extended images demand even more sensitivity from telescopes. However, the number of detectable galaxies per magnitude and per unit sky area is several orders of magnitude larger than that for QSOs. More exotic occurrences of extended images (see Sect. 2.5) depend on the lower-probability alignment of critical, high-magnification regions of the source plane with the source.

As for point sources, weak deflections must affect the background of extended sources at some level. For instance, most clusters of galaxies produce only statistically detectable lensing, e.g., in terms of the number of elongated and tangentially aligned (single) images of background galaxies. We discuss statistical lensing in Sect. 2.5.8.

Confirmation. The basic requirement that applied to point sources is weaker for extended sources: the presence of a single arc (or that of a single ring!) suffices to establish the possibility of lensing. Several alternative explanations have been proposed to account for elongated arcs, for example star formation induced by a shock wave in the cluster [BE87.1], or light echoes of a previously bright object at the center of such a cluster. All these interpretations physically associate elongated features with the lensing cluster, and therefore they assign the redshift of the cluster to the features. However, if the arcs are formed by an intervening cluster [PA87.1], their redshifts should be considerably larger than that of such a cluster. Therefore, once again redshifts provide a crucial test for confirmation of the GL effect. Unfortunately, this test is more difficult to carry out, because of the lack of sharp features in the spectra of galaxies.

If we have an arc, or a nearly full ring, and if the background galaxy has a variable subcomponent, for example if a supernova appears, and is imaged more than once along the arc, the ensuing temporal variations in separate images are correlated. A supernova is ideal in terms of having a recognizable light curve, if not in terms of frequency of occurrence.

The list of properties that was summarized in Sect. 2.4.1 is similar for extended images, save for a relaxation of some of the requirements: multiple distinct images may no longer be necessary. To summarize, then, to classify a set of extended objects as a GL system, the set must have:

1. at least one arc or ring;
2. a possible lens in the vicinity of the images, at redshift smaller than that of any image;
3. luminosity ratios in different spectral bands that are the same in all images;
4. redshifts that are the same in all images;
5. line flux-ratios and shapes that are similar for different images;
6. temporal variations in different images that are correlated.

Pitfalls. The same pitfalls found for point sources also affect extended sources. Condition (1.) stipulates just one arc, because single arcs are quite possible (Sect. 13.2). Since just one such object is required, instead of two, and arcs are extended and unusually-shaped, this condition is less stringent than for point sources. However, arcs are diffuse and faint, and their discovery and analysis are still a significant technical challenge. Condition (2.) is a similar demand to that posed by point images, except that it may be visually simpler to identify a cluster of galaxies, but it is a heavier burden to measure the redshifts of a large fraction of its galaxies. Conditions (3.) through (6.) apply if the extended source has more then one image; they are more restrictive than for point sources, because they require difficult spectral measurements, and because the extensions of the images complicate the comparison of colors, redshifts and spectral lines. Finally, condition (6.) is quite analogous to the corresponding one for point sources.

2.5 Known gravitational lens systems

The first GL candidate was detected in 1979 [WA79.1]; since then, the number of additional candidates has grown at an increasing rate. Although some of these candidates are now generally accepted as bonafide GL occurrences, others remain questionable or difficult to interpret, for instance, when a lens cannot be found. In the remainder of this section, we discuss several of the known GL systems and candidates. The list we have chosen is necessarily incomplete at any given time, and it may not do full justice to the latest additions, because of the continuing efforts of observers. However, we have attempted to give a flavor of the observational state of the art, by discussing all those cases that demonstrate the joy and the difficulties of observations.

The main two types of GL systems we discuss are the now 'classical' lensing candidates, with point sources, and with arcs and rings. We also describe speculative GL systems, that reveal the problems of interpretation encountered by observers. We finally discuss the less spectacular, but no less important effect of statistical lensing, as it applies to unusual and time-variable AGNs and to high-redshift sources near foreground galaxies.

In Table 2.1 we have listed the currently accepted systems of gravitational lensing.

2.5.1 Doubles

Doubles are the simplest configurations of images, and historically they were the first type to be discovered, sixty years after Eddington's measurements of the deflection of starlight by the Sun. Because of their relatively simple morphology, it might seem that such systems make 'clean' conclusions possible, but history has proven otherwise.

- **0957+561: a venerable first**

Discovery. The first GL system was discovered as part of a project of identification of optical counterparts to a survey of radio sources, which was carried out at 966 MHz at Jodrell Bank. Some of these sources were either too extended or confused for an accurate determination of their radio positions, so more than one candidate could be found within the error box of the corresponding radio position. However, the typical positional accuracy was 5 to 10 arcseconds, and the identifications were considered to be 80% reliable.

The suggested optical counterpart to one of the sources in the Jodrell Bank survey was a pair of blue stellar objects, 0957+561 A and B, separated by about 6 arcseconds, with A located almost due North of B. A and B overlap on the Palomar Observatory Sky Survey (a standard photographic sky atlas which covers the sky at declinations above $-24°$), but they are easily distinguishable and have comparable brightness. The mean of the positions of A and B was 17 arcseconds away from the estimated radio position, so the identification was tentative.

Walsh, Carswell & Weymann [WA79.1] undertook the first spectral investigation of 0957+561 A and B. They found that these images have very similar spectra. Their redshifts, estimated on the basis of the strong emission lines CIV $\lambda1549$ and CIII] $\lambda1909$, were both $z \approx 1.405$. The continua of the objects appeared to have very similar shapes below about 5300Å, with A brighter than B by 0.35 mag. For longer wavelengths, B appeared to be brighter than A, so in other words, B was redder than A.

In both images, low-ionization absorption line systems were visible. Their redshifts were both $z \approx 1.390$, with a velocity difference of only 45 km/s, compatible with zero difference. Since the absorption lines were narrow, their redshifts could be measured with higher accuracy than those

Table 2.1. Observed Properties of GL systems[1]

System	No. of images	Lens redshift	Image redshift	Maximum separation (arcsec)
QSO images				
0957+561	2	0.36	1.41	6.1
0142−100	2	0.49	2.72	2.2
0023+171	3	?	0.946	5.9
2016+112	3	1.01	3.27	3.8
0414+053	4	?	?	3.0
1115+080	4	?	1.72	2.3
1413+117	4	?	2.55	1.1
2237+0305	4	0.039	1.69	1.8
Arcs[2]				
Abell 370		0.374	0.725	
Abell 545		0.154	?	
Abell 963		0.206	0.77	
Abell 2390		0.231	0.913	
Abell 2218		0.171	0.702	
Cl0024+16		0.391	?	
Cl0302+17		0.42	?	
Cl0500−24		0.316	0.913	
Cl2244−02		0.331	2.237	
Rings[3]				
MG1131+0456		?	?	2.2
0218+357		?	?	0.3
MG1549+3047		0.11	?	1.8
MG1654+1346		0.25	1.75	2.1
1830−211		?	?	1.0

[1] Only those systems that have achieved acceptance by at least some measure of consensus (as of the Hamburg GL meeting of September 1991) are listed.

[2] The maximum "separation" that could be assigned to arcs is the diameter of the corresponding Einstein ring, which would measure about one arcminute for lensing clusters at redshifts between 0.2 and 0.4.

[3] The maximum "separation" of images was taken to be the diameter in the case of ring systems.

based on the broad emission lines. For these lines, the velocity difference was estimated to be 120±150 km/s. Hence, both absorption and emission redshift differences were compatible with zero.

From their spectral measurements, Walsh and his coworkers concluded that it was very unlikely that A and B were separate QSOs. They therefore considered the possibility that the two objects were gravitationally lensed images of the same QSO, referring to the theoretical work of Sanitt [SA71.1]. From the magnitude difference they estimated the magnification of the A image to be a factor of about 4. The estimated lens mass was of the order of $z_d 10^{13} M_\odot$, where z_d is the redshift of the lens. The difference in color of the two images was explained by differential reddening along the two light rays in the lensing galaxy; from the fact that the B image is more reddened than the A image, it was predicted that the lensing galaxy lies closer to image B.

The lens. The first question to arise regarded the nature of the lensing object. Given the angular separation between A and B, it was suspected that a massive galaxy could cause the image splitting. It was natural to look for a galaxy near the two images of the QSO, as history shows, since two observations were performed with different telescopes within 12 days of each other ([ST80.1], [YO80.1]). Both observations revealed a galaxy, labeled G1, very near image B, as expected from the reddening of B compared to A. The center of G1 is about 1 arcsecond almost exactly to the North of B (see Fig. 2.5). The radial profile of G1 was well fit by a King profile, with a core radius of about 0".24 [ST80.1], and an ellipticity of about 0.13 [YO80.1].

As a significant complication, G1 seems to be a member of a rich cluster of galaxies [YO81.3]. The redshift of G1 had been determined to be about 0.39 in [YO80.1], but it was later lowered to 0.36 in [YO81.3], where a list of 180 objects was presented, of which 146 were galaxies. The redshifts of the great majority of the 146 galaxies were not known, save for two of them, whose redshifts were determined to be about 0.54 and 0.36 [YO81.3]. G1 is the brightest galaxy of the cluster; it is important to note that it does not lie on the straight line connecting A and B, but is offset from it. This means that the lens cannot be represented as a single, circularly-symmetric galaxy.

Early radio observations. 0957+561 was selected as a radio source, and its identification as a double QSO occurred just before the completion of the Very Large Array (VLA), a radio interferometer near Socorro, New Mexico, consisting of 27 antennas with a maximum separation between antennas of 40 km. With this instrument, radio maps of sources became possible with a resolution higher than that of optical images. 0957+561 was mapped soon after its identification by several groups ([PO79.1], [RO79.1]). The 6-cm maps showed four components, two of which coincided with the optical positions. In fact, the radio structure of this source was found to be "unusually complex, and unlike any other single radio source" [PO79.1]. The

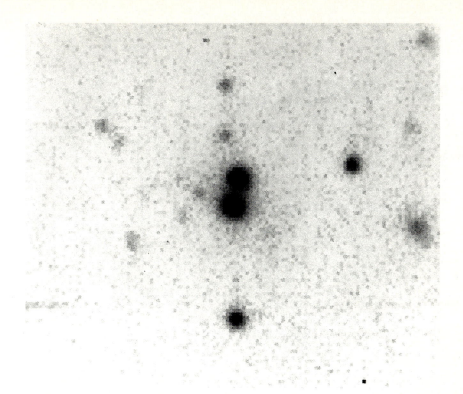

Fig. 2.5. Optical view of 0957+561 (courtesy of R. Schild)

flux ratio B/A was about 0.8, nearly the same value as for optical fluxes. At long wavelengths, component B was not detected, which implied an upper limit on the flux ratio of 0.5 at 408 MHz. However, at this wavelength the resolution becomes worse, and the flux measured for the A component was contaminated by the other two components. There was a slight indication that 0957+561 is variable at $\lambda = 6$ cm. We show a later radio map from [RO85.1] in Fig. 2.6. The two additional components seen by [PO79.1] are the component labeled E in Fig. 2.6, and the sum of components C and D. The complex radio structure of the source, in particular the fact that the A image seems to be associated with components C, D, and E was taken by [RO79.1] as an argument against the lens interpretation of this source. To summarize their discussion: why does A have a jet but B doesn't? Such an argument shows how much the thinking at the time was restricted to axi-symmetric matter distributions; only after the lens was detected did it become clear that 0957+561 cannot be accounted for by a axi-symmetric lens.

Further evidence for a GL came from the fact that the radio components A and B are compact, whereas C, D, and E are resolved. The properties of this system were tested further by two groups ([PO79.2], [GO80.1]), with Very Long Baseline Interferometry (VLBI). It was thus found that A and

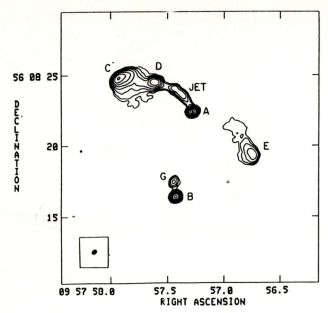

Fig. 2.6. VLA 6-cm map of 0957+561 (courtesy of D. Roberts)

B are unresolved even on a scale of 20 milliarcseconds, and that their separation and flux ratio are the same as those for their optical counterparts. The data excluded the presence of any additional compact components.

Early models. As we stated above, the lensing hypothesis requires a matter distribution in the deflector which is more complicated than a simple point mass lens. This was clearly recognized by Young and his collaborators [YO80.1]. This paper, which also presented the detection of the host cluster for G1, was the first to present a model for the lens. The matter distribution of the galaxy G1 was represented by a spherically-symmetric King model [KI66.1]; their best fit yielded a velocity dispersion of 425 km/s. In [YO80.1], image B was assumed to be double, with components B1 and B2. The third image B2 (a transparent, non-singular GL always produces an odd number of images, [BU81.1]) could be brought very close to B1, so that B1 and B2 would be unresolvable at optical wavelengths. However, since the calculated magnifications of B1 and B2 were comparable, this model was in conflict with the VLBI results. In addition, a single symmetric lens could not produce images that were not colinear with G1 on the sky.

In a second model, a 'cluster' mass distribution was added to that of G1 [YO80.1]. Since the cluster center did not coincide with that of G1, the deflection law was no longer circularly symmetric, and thus could produce images not colinear with G1. The contribution to the deflection of the cluster allowed a reduction of the velocity dispersion for G1. Thus, reasonable model parameters were easily found, to account for the image positions rel-

ative to G1, with a 'missing' image B2 forming an unresolvable double with
B1. The dynamic range of the VLBI observations [PO79.2] available at the
time was too small to rule out the double nature of B. A second possibility
was that B2 was faint and so near the center of brightness of G1, as to be
effectively undetectable.

Later observations. Spectroscopic measurements of G1 yield a redshift of
$z_d = 0.36$ [YO81.3]. Using the galaxies in the observed field, the cluster
center was estimated to lie to the west of G1, which is important to note
as this position yields the direction of the shear perturbation caused by the
cluster (see Chap. 5). The velocity difference of the absorption system at
$z = 1.391$ was determined to be less than $\sim 20\,\mathrm{km/s}$, and the difference
between the emission line redshifts, determined from the same broad line in
both components, was estimated to be less than about $100\,\mathrm{km/s}$. However,
the redshift estimates inferred from different lines in the same object dif-
fered by larger amounts ([WE79.1], [WI80.1]). All the equivalent widths of
the emission lines agreed within the measurement errors. The fact that B is
redder than A, as was originally thought, could be interpreted as contamination from G1, after it became clear that G1 was so close to B. A further
absorption line system at $z \approx 1.125$ was discovered [YO81.4], with equal
strength and redshift in both images.

The complex radio structure, which was one of the main difficulties
for simple lens models, was further investigated with the VLA ([GR80.1],
[GR80.2]). A jet originating at A and emanating towards D was clearly
resolved, and an elongated structure close to B was found not to be a jet,
but an additional component whose position is compatible with the optical
position of the galaxy G1. Thus, since the jet of the A image has no coun-
terimage, stringent constraints on lens models could be set, as the source,
assumed to be of the 'core-jet' type, must be positioned such that the compact component lies inside the caustic (so it is multiply imaged), but the
jet lies outside the caustic. In a later paper [RO85.1], an additional weak
radio source was found $0\overset{''}{.}35$ north of B; since it is very difficult to find a
model for this GL system with a third image sufficiently weak and so close
to B1, this component was interpreted as a weak counterpart of the jet in
A, a small portion of which is thus multiply imaged.

More information about the structure of the images was obtained from
further VLBI observations. Since the compact radio flux of 0957+561 is
weak, these observations are difficult.

It was found that both components have a compact radio structure that
comprises a core and a jet [PO81.1]. The flux ratio of these compact components was $B/A \approx 2/3$. The direction of the jet relative to the core is
nearly the same for each component, a property that turns out to yield a
very stringent constraint on the lens models. Utilizing a new, very sensitive
VLBI system, the core-jet structure in each image was confirmed, and a
third compact component was found, with a flux about 3% of that of the
compact core in A [GO83.1]. The position of this new component (termed

G', after the VLA component close to G1 was termed G) is the same as that of G1 and G, within the error with which the optical and VLA position can be determined. The compact flux of G' is smaller by a factor of $\simeq 6$ than the flux of G. The question was, and still remains, whether G' is the third image of the QSO, or a compact component in G1, or a mixture of both.

Imaging of the VLBI components of A and B can yield very sensitive constraints on lens models. One of them was already mentioned, the near-parallel orientations of the VLBI jets. In addition, the size of VLBI components is special, because the optical continuum emitting region can be sufficiently compact so that ML can influence the magnification of these components. On the other hand, the resolution of the VLA yields a size over which the mass distribution of the lens can vary. The VLBI scale is just between these two scales; therefore, it can yield information about the local properties of the lens. The image distortion can be described by a 2×2 matrix, whose elements depend on the local parameters of the lens at the position of the image. Since the source structure is not known, this matrix cannot be determined from observations. However, the product of this 'distortion' matrix for image A and the inverse of the analogous matrix at image B yields the (linear) transformation between images A and B ([GO84.1], see Sect. 13.1). This product, the relative magnification matrix, can be determined observationally from VLBI observations, such as those that yield the brightness models shown in Fig. 2.7. In particular, one finds that this matrix has a negative determinant which shows that the parities of images A and B are different [GO88.2], as predicted by all lens models.

Additional observations at ultraviolet [GO82.1] and infrared [SO80.1] wavelengths confirmed the positions of the images and the ratio of bright-

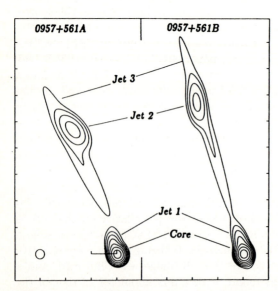

Fig. 2.7. 13-cm brightness model for 0957+561 based on VLBI observations [GO88.2].

nesses found at other wavelengths. Finally, unpublished X-ray observations with the Einstein observatory yielded positions consistent with those at other wavelengths, but the weakness of the QSO in X rays will require high sensitivity, such as that of the ROSAT satellite, to determine accurate fluxes for A and B.

Lens models. The observations described above permit estimation of parameters of models for the lens in 0957+561. Several such models have been published; they consist of two components that describe G1 and the cluster. In the models proposed in [YO81.3], [NA84.1] and [GR85.1], both lens components are represented as ellipsoidal mass distributions, each with four free parameters (velocity dispersion, core size, ellipticity, and orientation of the major axis). In addition, the center of the cluster had to be fixed, because the observations impose weak constraints on its coordinates.

The model for the lens that was proposed in [YO81.3] accounted well for all the constraints known at that early time: the optical positions and brightness ratio for A and B. The parameters for the galaxy and the cluster were reasonable, so the lens hypothesis appeared very natural after this work.

Reasonable models that accounted for the optical, VLA and early VLBI observations of 0957+561 were proposed in [NA84.1] and [GR85.1]. Again, G1 and its host cluster were represented as ellipsoidal mass distributions, and required about ten parameters. Thus, more than half of the parameters were fixed at arbitrary values, and the models were not unique. The orientation of the VLBI jets relative to their corresponding core in each image imposed a significant constraint for model fitting in both cases. After the relative magnification matrix for the A and B images was measured using VLBI observations [GO88.2], a new model was described in [FA88.1], and discussed in detail in [FA91.2]. The new model accounts for all the observations mentioned above, and in particular for the relative magnification matrix for A and B, which is not affected by ML. The main characteristic of the new model is that the effect of the cluster is represented by its shear (Sect. 5.2), and therefore the number of parameters in the model is reduced to the point that none need be fixed arbitrarily. The reduction in the number of parameters also allowed the first complete study of uniqueness of GL models. It was shown in [FA91.2] that the observations available allow an indeterminate amount of dark matter in the lens. Additional observations can constrain the dark matter present in both G1 and its host cluster, and should yield constraints on H_0 (see the following subsection).

It should be noted that the last three models we have discussed predict a third image near the nucleus of G1, with flux less than 3% of that of B, and yield single images of the extended radio lobes found with the VLA [RO85.1], and thus satisfy all the available observational constraints.

The modeling results can be considered to be an indirect argument in favor of lensing, because the newer observations have resulted in fewer, instead of more, free parameters, and physically reasonable parameters remain appropriate.

Time delay. The observations described above strongly argue for the interpretation of 0957+561 as a GL system. Very strong additional evidence in favor of this interpretation would be the detection of correlated variability of the two images, separated by a time delay. In addition to the verification of the GL hypothesis, the measurement of the time delay can be used, in principle, to determine the Hubble constant, as was first noticed by Refsdal [RE66.1]. We discuss this possibility in some detail in Sect. 13.1. It is therefore natural that after the identification of this system, several groups have looked for variability of the images.

The time delay could have been observed only if it were smaller than about ten years. Simple estimates yield a value which is considerably smaller than this. More elaborate estimates [DY80.1], based on a two-component lens (galaxy plus cluster) gave a range of time delays from 0.03 years to 1.7 years. These two-component models were constrained by requiring that the flux ratio $B/A \approx 3/4$, and that the third image B2 lies within $0\overset{''}{.}42$ of B1. These constraints could be met if the center of the cluster was roughly to the north of G1, and as we show in Sect. 13.1, in this case the time delay is extremely sensitive to the exact position of the cluster center.

In contrast to the models in [DY80.1], the lens models discussed here predict that the third image is near the nucleus of G1, and is demagnified sufficiently to satisfy the observed absence of a third image. The detailed model in [YO81.3] predicts a time delay of nearly six years (all the estimates for the time delay cited in this section are for $H_0 = 60\,\mathrm{km\,s^{-1}\,Mpc^{-1}}$); however, as was pointed out in [NA84.1], this result appears to stem from a sign error in the potential time delay, and the correct expected values are in the vicinity of one year.[9]

The time delay can only be measured if the QSO is intrinsically variable.[10] 0957+561 is variable at optical wavelengths: both the fluxes of the images and their spectra change in time [MI81.1]. Since the amplitude of the variations was found to be at least 1 mag [LL81.1], there was much optimism that the time delay could be accurately determined within a few years. B seemed to have a larger amplitude of variations than A. Long-time variability was verified with archival plates [KE82.1]. By 1982, no correlation between the light curves of the images had been observed, which led to the conclusion that the time delay must be larger than about 2.5 years.

---

[9] Moreover, it was claimed in [YO81.3] that the light-travel-time of the unseen image B2 is shorter than that of B1, which is also incorrect (as can be seen directly from the scalar formulation of lens theory which will be introduced in Chap. 5; in such a formulation, images A, B1, and B2 correspond to a minimum, saddle point, and maximum of the time delay surface, respectively).

[10] A different method for determining the time delay is the monitoring of its VLBI structure [VA82.1]. If the separation of a compact VLBI component from its corresponding core was measured as a function of time in both images, one could extrapolate backwards to the time the component was expelled from the core; the difference in this time for the two images is then an estimate of the time delay. Unfortunately, such experiments are still fraught with the problems of following components with the low density of time coverage allowed by VLBI observations.

By that time, however, it had become clear that variations of the flux of the images can also be caused by ML, due to a single star moving across the light beams ([CH79.1], [GO81.1], [CH84.2]). Since the density of stars at the position of image B was expected to be larger than that for image A, the enhanced variations of B could be due to ML, which, of course, would not produce corresponding variations in the A image. The simple picture of a single star in the beam may be an oversimplification, as was pointed out in the pioneering paper by Young [YO81.1].[11] Therefore the question arose as to how we can determine the time delay from observations of variability, if one (or both) images are affected by ML. There is no clear recipe for such a determination; the best one can hope for is that a clear feature is present in both light curves.

The first claim of a measured time delay was published in 1984 [FL84.1], based on monitoring with a fairly small telescope. The estimate of about 1.5 years agreed well with expectations; however, the data were too noisy for this value to become widely accepted. In fact, in the same year another light curve of 0957+561 was published [SC84.1], in which no clear indication of pronounced variability was detected. In 1986 a cross-correlation analysis of the two light curves yielded a value of 1.03 years for the time delay [SC86.1]. However, the peak in the cross-correlation function was not very strong, which prompted doubts about the reliability of the estimate. At the time of this writing, the latest estimates for the time delay were published in [VA89.1] and [SC90.5], with a value of about 415 days.[12] This result appears to depend critically on a single feature that occurred in each light curve separated by an interval equal to the delay estimate. We find this result, obtained after nearly 10 years of observations, should still be considered with some scepticism [FA90.1]. In fact, the determination of the time delay turned out to be much more complicated than it was anticipated at the time 0957+561 was discovered. However, continued observations are warranted, because of the possible payoff regarding estimates of H_0: according to [FA91.2], if both the time delay $\Delta \tau_{BA}$ for A and B and the velocity dispersion σ_v of G1 are measured with uncertainties of 10% or less, H_0 can be estimated to be

$$H_0 = (90 \pm 20) \left(\frac{\sigma_v}{390 \, \text{km/s}} \right) \left(\frac{\Delta \tau_{BA}}{1 \, \text{yr}} \right)^{-1} \text{km s}^{-1} \, \text{Mpc}^{-1} \ .$$

A 20% uncertainty is a significant improvement over 'classical' estimates of H_0.

---

[11] This paper was not only the first to study the statistical properties of randomly-distributed stars between us and a compact source, but it also set the standard for notation in lens theory, introduced the concepts of convergence and shear, and took the first significant step toward a scalar formulation of lens theory.

[12] A recent reanalysis of the data published in [VA89.1], using the construction of an 'optimal underlying signal' (intrinsic light curve) has obtained a value of 536^{+14}_{-12} days for the time delay [PR91.1]. This value is in much better accord with the recent determination of the time delay from radio observations [RO91.1].

Discussion. We have presented overwhelming evidence that 0957+561 is a GL system. Going back to the list of properties a GL system should have (Sect. 2.4.1), we see that nearly all are satisfied in this case; only the final verification by cross-correlation of the light curves is missing, but this is mainly due to the fact that the QSO is not strongly variable, and ML may influence the observations considerably.

The lack of a third image is not considered a serious problem, as such an image can be sufficiently demagnified to be unobservable. We will discuss this matter in the next section in somewhat more detail, but it is largely believed at the moment that the third image is hidden somewhere near the nucleus of G1. It remains to be seen whether the third compact VLBI component G' [GO83.1] is in fact the third image, or the nucleus of G1, itself a radio emitter. We note again that this compact source is so weak that its observation is at the limit of current capabilities.

The first known GL system has shown that nature is more complicated than theorists want it to be. Circularly-symmetric lens models are destined to fail, as are models that neglect the influence of the cluster. The matter distribution in the lens is quite complicated, but simple models for it account for the image configuration, the VLBI structure, and the absence of a third image. Thus, indirectly, 0957+561 can be used to determine galaxy and cluster masses at a high redshift. As pointed out in [GR85.1], the mass-to-light ratio of the cluster at the position of G1 must be extremely large, which confirms expectations for dense clusters of galaxies. The lens models and the observations still leave us considerable freedom, so estimates of H_0 will remain only a possibility until additional measurements become available [FA91.2]; for further discussion of this topic, see Sect. 13.1.

- **2345+007: two images without a lens?**

In a spectroscopic search for QSOs, a pair of QSO components, A and B, separated by 7″.3 was found [WE82.1]. They have quite different brightnesses, the A image being 1.7 mag brighter than the B image. One strong and two weak emission lines are seen in both spectra, and their wavelengths and profiles are indistinguishable within observational accuracy. The redshift of the QSOs is about 2.15, and the redshift difference has been narrowed down to 44 ± 40 km/s [ST90.1]. The equivalent widths of the emission lines are different in both images, and vary from line to line [ST90.1]. The continuum colors are nearly the same, with a slight tendency of the B image to be redder [TY86.2]. The agreement in the redshifts of the two images and in the line ratios suggests that this may be a GL system. The large image separation of the two QSO images makes this an ideal candidate to investigate separately the absorption line spectra of the components; as will be discussed in Sect. 13.3.1, this has led to interesting constraints on the size of absorbing clouds [FO84.1].

No indication of a lens has been found in deep images of the system [TY86.2], up to stringent limits. In addition, there are significant constraints on the brightness of a lens, assuming it is superposed on one of the QSO images. There are several galaxies in the field, but their concentration is not significantly high to consider them a cluster. Both QSO images show a strong absorption system at a redshift of 1.5, with two components separated by 1000 km/s, a typical velocity dispersion in a cluster of galaxies. If the splitting is due to a single isolated galaxy, it must have an extremely large velocity dispersion and mass-to-light ratio. Five lines of FeII are identified in image B, with no counterpart in image A [ST90.1], suggesting that the lensing galaxy is closer to image B than image A. Embedding a galaxy lens in a cluster would ease these constraints. It has been suggested that this system can be produced by two lenses at different redshifts [SU84.1]. However, there is no observational evidence for a lens anywhere near the images. Stringent limits on the brightness of any additional image of the QSO are also available.

The discovery that the B image can be resolved into two components [NI88.1], as was suspected earlier [SO84.1], supports the GL hypothesis. Since the nature of the system remains unclear, we included it under the category of doubles. The separation of the two subcomponents is about $0''.36$, and the direction of their separation is nearly, but not quite, coincident with the direction of a line that joins A and B. The two subcomponents of B have comparable brightness, and one of them shows an elongation, whose nature is unclear. The colors of the images are not compatible with the hypothesis that one of the subcomponents is the lens galaxy, although there may be an underlying galaxy in the image. Its contribution to the flux of the image cannot be larger than about 10%; otherwise, the B image would be too red. The double nature of the B component has not been confirmed, although the elongation of B perpendicular to the line separating the A and B component has [WE91.1].

There are too few observational constraints to attempt to find reasonably unique parameters for a lens model that accounts for the properties of 2345+007. The double nature of the B image, however, is a hint that lensing is taking place, if not a clear sign. The large separation of the images makes this a highly interesting case to search for compact dark matter.

- **1635+267**

A slitless spectroscopic survey yielded a QSO pair with an angular separation of $3''.8$ [SR78.1]; later observations showed that the two QSO images have nearly indistinguishable redshifts [DJ84.1], $z \approx 1.961$. The redshift difference, expressed as a relative velocity is smaller than 150 km/s. Such a low value, taken together with the similarity of the spectra, suggests that this is a GL system. Deep CCD images, however, revealed no lens near the two images. The spectrum of the brighter component A has an absorption feature that was interpreted in [DJ84.1] as MgII at $z \approx 1.118$ and as a possible indication of a lens at that redshift.

Higher-quality spectroscopic observations of 1635+267 permitted a comparison of the spectra of the images in greater detail [TU88.1]. These confirmed that the spectra are remarkably similar, since the line ratios, widths and shapes agree in both images. As noted earlier, there is as yet no quantitative measure for 'similarity of QSO spectra', but a comparison of the spectrum of 1635+267A with that of a QSO at nearly the same redshift yields clear differences in line profiles and strength ratios, much larger than the differences in the spectra of the QSO pair. Although it is difficult to evaluate the significance of this test, it nevertheless supports the GL hypothesis for this system.

The subtraction of the continuum of B, scaled by the relative magnification (≈ 2.83) of the images, from the continuum of A, revealed a residual flux, which was interpreted in [TU88.1] as a 20th magnitude galaxy at a redshift of 0.57. This then could be a sign of a lens galaxy superposed on the A image. Before spectra of higher quality become available, this conclusion must be considered tentative. The same conclusion about the nature of the lens as in 2345+007 applies: if the lens is not near one of the images, it must have a very large mass-to-light ratio.

- **UM673**

A systematic search for multiple images in a sample of highly luminous QSOs yielded a pair of QSOs, labeled UM673 (0142−100) [SU87.3]. The images are $2\rlap{.}''2$ apart; they differ in brightness by about two magnitudes. The redshift common to both images is $z = 2.719$, within the measurement errors. Emission and absorption lines appear quite identical [SU87.3]. A galaxy was found between the two images, and very near to the line joining the images. It is about $0\rlap{.}''8$ away from the weaker image, and its redshift is $z = 0.493$. As argued in [SU87.3], it is unlikely that the brightness difference between the images is due to ML, as this would affect the line-to-continuum ratio differently in each image, which is not supported by the observations [SU88.1]. Therefore, it is likely that the A image is highly magnified by the deflection field of the galaxy. Although the observations provide insufficient constraints to model the lens, it is clear that this system can be accounted for by the effect of a single elliptical galaxy. A preliminary model yields a lens mass of about $2 \times 10^{11} M_\odot$ and a time delay of a few weeks. The source seems to be a radio emitter, but no high-resolution data are yet available.

2.5.2 Triples

- **2016+112: a triple radio source**

Discovery. The first GL system found by a systematic search for lenses at radio frequencies (see Sect. 12.2.5 for a description of the survey) was 2016+112 [LA84.1]. It was initially identified as a very good lensing candidate with three distinct components. Two of them, A and B, had identical spectral indices, significantly steeper than that of the third one, C. A schematic diagram of the 2016+112 system is shown in Fig. 2.8. The flux of

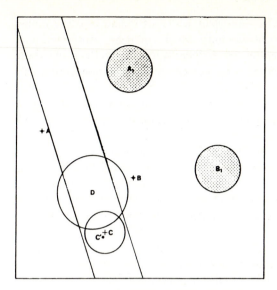

Fig. 2.8. Schematic diagram for 2016+112 (courtesy of D. Schneider)

C was nearly three times that of A or B, which were nearly equally bright. The optical counterparts of A and B, two star-like optical sources, had very similar spectra; in particular, they shared the same redshift, $z \approx 3.27$, within the observational uncertainty of 40 km/s. The continuum of B was redder than that of A, and the continuum flux ratio A/B was about 1.3. In contrast, the flux ratio of the emission lines was 1.64, significantly different from the flux ratio at radio wavelengths. One peculiarity noted in [LA84.1] was that the emission lines are the narrowest ever observed for an object at such a high redshift.

The spectrum of the optical counterpart of the radio component C could not be measured with high accuracy [LA84.1], but its color, morphology, and apparent magnitude suggested that it might be an elliptical galaxy at a redshift of about 0.8. Certainly, C cannot be a third image of the background QSO, as was already clear from its different radio spectral properties, as well as from the fact that in contrast to A and B, C is resolved in radio maps. It was suspected that C might contribute to the image splitting, but the configuration of the images made it necessary to invoke additional lens components. The field around the system is crowded, so it cannot be excluded that a cluster of galaxies contributes significantly to the GL effect.

Additional observations. A subsequent paper [SC85.2] described the discovery of an additional component of 2016+112 (object D in Fig. 2.8). This new object appeared to be a faint and extremely red galaxy, positioned in the midst of the three radio sources. On the assumption that D is the brightest member of a cluster of galaxies, its redshift was estimated to be $z \approx 1$. In addition, it was found that the B image had dimmed by 0.3 mag, but

A had remained constant. The equivalent width of B had nearly doubled in the 9 months between the observations, which could be interpreted as a constant line flux with a decrease in the continuum flux. Precisely these types of variations are expected to be caused by ML, so monitoring this system is quite warranted.

Finally, a third image was discovered, by imaging the system through a filter centered on the Lyα emission line [SC86.2]. This new image, labeled C' in Fig. 2.8, was nearly superposed on the radio component C. The optical source (C+C') is about 1.4 mag fainter than B, and the optical data are consistent with the interpretation that most of the optical flux from C is indeed due to the third image of the QSO. Then, roughly 6% of the radio flux of C can be attributed to the QSO, while the rest is attributed to the galaxy C. If this interpretation is correct, VLBI observations should be able to detect an additional compact component in the radio source C.

The spectrum of the galaxy D confirmed the prediction for the redshift, $z \approx 1.01$. D cannot help being an important lens component. Its spectrum closely resembles that of a brightest cluster galaxy, but there is no indication of significant clustering of fainter galaxies around D. Imaging with a Lyα-filter [SC85.2] revealed two additional sources in the field, A_1 and B_1 in Fig. 2.8. They are resolved, and thus are not additional images of the QSO; this is also clear from the fact that there is no radio emission connected to them. Since they appeared in the narrow-band filter, it can be speculated that they are at the same redshift as the QSO; it is unclear, however, whether they are two separate sources or images of the same object. This question may be clarified by spectroscopy with higher sensitivity.

Interpretation. As opposed to the lens systems described above, 2016+112 poses a problem for reasonable lens models. In particular, the extreme misalignment of the three images A, B, and C' precludes a simple three-image lensing geometry. The detection of the third image C' invalidates the early lens models [NA84.2]. Lens models that predict an image at the position of C' are complicated [NA87.1], with two lensing galaxies (D and C) and, in addition, a cluster of galaxies. The ellipticities of the galaxies in such models are also rather extreme. These models predict five images of the source, with one very faint image hidden in the core of galaxy D, and two images very close to C. In some of these models, the two images near C are faint, in apparent agreement with observations. However, as mentioned above, whenever two images are very close together, they are usually highly magnified. In order to avoid this, the models appear to be unnaturally fine-tuned. It may well be that this GL system will require a model with lenses at different redshifts [BL87.3].

In addition to the difficulty in obtaining a simple lens model for this system, there are other puzzles, e.g., the different ratios of magnifications in images A and B in the radio and in the spectral line flux. One way to solve the puzzle is to assume variability of the source spectrum, which, in connection with the time delay (expected to be of the order of one year), can lead to different flux ratios in different spectral bands. A second possibility is ML in at least one image. ML might also account for the dimming of only image B. A third possibility is differential magnification, where parts of the source are more magnified than others. This can occur if the scale on which the magnification varies in the source plane is comparable with the size of the emitting regions; the effect can thus be different for radiation of different wavelengths.

2016+112 is generally considered to be a GL system, although it is not well understood at the moment. The main argument in favor of the lensing nature is the small redshift difference between the two images A and B, the observation of a third image, and the detection of two galaxies very close to the images. The small separation of galaxy D from image B essentially requires the occurrence of multiple images. An improved understanding of this lens system should be derived from VLBI observations, which would improve our knowledge of C' [HE90.2]. The relative orientation of compact VLBI structures in images A and B may lead, as for 0957+561, to tighter constraints for lens models.

The example of 2016+112 shows that the interpretation of a given system is not straightforward, and that a typical GL system may be difficult to observe, due to the faintness of the images and of potential lenses.

- **0023+171: a puzzling triple**

The radio search for GL systems described in Sect. 12.2.5 yielded a composite radio source with multiple optical counterparts [HE87.1]. A radio map revealed three components, two of which, A and B, were resolved. Both of these were brighter than the third unresolved component, C. A and B appeared to be opposed radio jets. The separation between A and B was $3''\!.0$, and that between A and C was $5''\!.9$. Optical counterparts were found for A and C. The optical counterpart of C was brighter than A's, the opposite as for the corresponding radio components. Spectra of the optical components showed that they are at the same redshift ($z \approx 0.946$), and the flux ratios in two emission lines agreed in both components. The optical counterpart of C shows a possible elongation towards A; if the latter were interpreted as an additional source, it would be an extended, very weak object about $1''\!.0$ away from the compact source C. Since the components were fairly weak, a more detailed comparison of the spectra was not possible.

If we consider only the radio morphology, the data suggest three radio sources: an extended double and an unresolved single source. These components are strong emitters, with a luminosity typical of radio galaxies. On the other hand, the luminosities of their optical counterparts are smaller than those of typical radio galaxies. The similarity between the two optical components is remarkable, as can be seen by comparing the redshifts, line ratios, and especially the ratios of optical and radio fluxes. However, the optical continuum and line flux ratios are different; this could be understood under the GL hypothesis by invoking ML. The small optical extension of the C image could be interpreted as a galaxy, but one that would be much too faint to be a promising lens candidate on its own. No other sign of a lens is observed, and the observational limits on any additional object in the field correspond to a mass-to-light ratio in excess of a few hundred. The nature of the 0023+171 system must be labeled unclear at the time of this writing.

2.5.3 Quadruples

• **PG1115+080: initially a triple**

Discovery. In the course of high resolution studies of bright QSOs, selected from the Bright Quasar Survey [SC83.1], Weymann and his coworkers [WE80.1] discovered that the source PG1115+080, although point-like on the Palomar Observatory Sky Survey, reveals two other stellar objects about 2.5 mag fainter within 3 arcseconds of the QSO itself. The redshift of the QSO is 1.722, which, together with its visual magnitude of 15.8, makes it one of the most luminous QSOs known. The spectra of the companion stellar objects are indistinguishable within the accuracy of observations, which led [WE80.1] to hypothesise that this system of QSOs was the second example of a GL system.

The optical appearance of PG1115+080 is displayed in Fig. 2.9, from [HE86.1]; it consists of three objects, A being brightest, B faintest, with about 10% of the flux of A, and C, with about 18% of the flux of A. C lies farther away from A than B. The separation of C from A is about $2''.3$, which renders this system difficult to observe. In the discovery paper [WE80.1], it was noted that B is slightly redder than A and C. Despite that, and the fact that the image configuration looks unusual, this object was

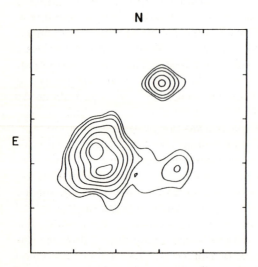

Fig. 2.9. Optical view of PG1115+080 (courtesy of J.N. Heasley)

considered a good GL candidate, because the redshifts of the emission lines and their equivalent widths agreed for all three components. Unfortunately, the source is radio-quiet.

The double nature of image A. Additional imaging of PG1115+080 with good seeing revealed that image A is not star-like, but elongated [HE80.1]. In particular, the data are consistent with A being a double image, with a separation between components of about $0\rlap{.}''5$. This possibility was confirmed by speckle interferometry [HE81.1], which yields a separation of $0\rlap{.}''54$. The two subcomponents of A are of approximately equal brightness and unresolved.

Models for this lensing system [YO81.2] failed to reproduce the three-image configuration well, but a five-image configuration was much easier to produce with a single galaxy acting as a lens. In particular, the two bright images (the subcomponents of A) were naturally found to be highly magnified. As we will see in Chap. 6, this is a generic property of GLs: close pairs of images are highly magnified, because the source is close to a caustic. The model proposed in [YO81.2] is a five-image configuration, where one image is near the center of the lensing galaxy and is strongly demagnified. Similar models were presented in [NA82.1]. The main difference between the models in [YO81.2] and in [NA82.1] is that the former required an edge-on spiral galaxy with a bulge and a disk, whereas the latter managed to reproduce the image geometry with a normal elliptical galaxy. Note that there is considerable freedom in these models, as the position of the galaxy is not constrained by observations.

The lens galaxy. The brightness of PG1115+080, together with the small separation for the images, makes it much more difficult to find the lens in PG1115+080 than in 0957+561. The models mentioned above predict that the galaxy should be inside the triangle defined by the images (counting A as one). In addition, it was found by several authors ([WE80.1], [YO81.2]) that the B component is redder than the other two; since B is the faintest image, its color would be more affected by the presence of a red galaxy than the other two images. However, early observations (e.g., [HE80.1]) failed to detect any sign of a lens galaxy, and thus yielded strong constraints on its magnitude.

The situation changed after it was reported that one of the A subcomponents is itself elongated [FO85.1]. Several interpretations of this elongation could be given, including a possible galaxy superposed on this image, a chance projection with a foreground object, or a further image of the background QSO. Then, it was claimed that the lens galaxy lies just between the two A images [HE86.1]. However, both of these lens positions would complicate the explanations for the image configuration. Once again, the models account well for the observations if the lens galaxy lies within the approximate triangle defined by the images.

In two independent observations, another object was found near the images ([SH86.1], [CH87.1]). Although the positions given for this additional object, which is interpreted as the lens galaxy, differ slightly in detail, the two groups of observers agree that its position is within the area where the models predict it to be. It is not yet clear whether the objects seen in [FO85.1] and [HE86.1] are real; one should bear in mind that these observations are at the current limit of resolution of ground-based telescopes.

Further comments. Because the angular separation of the images is much smaller than that of the double QSO0957+561, and since PG1115+080 is radio-quiet, no comparable high-resolution map of this object can be obtained; in particular, one should not expect to encounter such well-constrained models for this system as for 0957+561. Nevertheless, since the spectra of the components agree very well, and the redshifts have been shown to differ by less than 100 km/s, and since the image configuration is typical for an elliptical matter distribution [BL87.2] – particularly the fact that the two close images are much brighter than the other ones – this has led to a general consensus to accept the GL hypothesis for PG1115+080. The relatively small image separation is of the order expected from a moderately massive, isolated galaxy, in contrast to the double QSO, where a cluster assists the lensing galaxy in splitting the images.

In several papers, PG1115+080 was reported to be variable (e.g., [FO85.1], [VA86.1]), although, as the QSO is radio-quiet, no large intrinsic variations are expected. However, a measurement of the time delays for the images could constrain the lens models significantly. The delays are expected to be much smaller than for 0957+561, on the order of a month for the delay between B, C and A, and of the order of a day for the two A components, as can be seen from (6.21). In fact, the brightness ratio of the two A components has been claimed to vary [FO85.1]. But again, ML can account for the brightness variations of the images; in particular, the double image, which lies close to a critical curve of the lens, can be easily affected by the granularity of the matter distribution in the deflector. Unfortunately, monitoring the brightness of these two images will be extremely difficult, due to their proximity to each other.

- **2237+0305, the "Einstein cross": a unique case**

Discovery. The spiral galaxy 2237+0305 was first identified in the Center for Astrophysics Redshift Survey of galaxies. It was found to have a broad emission line, which on further study turned out to belong to a QSO at redshift $z \approx 1.695$ [HU85.1]. CCD images of the QSO showed that it shines through the nucleus of the spiral galaxy, whose redshift is about $z \approx 0.039$. These redshifts by themselves strongly suggest the GL hypothesis, since the surface mass density of this galaxy can be expected to be sufficient to affect the light from the background QSO. However, the redshifts can also be considered to support the hypothesis that the origin of the redshift of QSOs is non-cosmological – for a review, see [AR87.1]. This was pointed out by G. Burbidge [BU85.1], who argued that the QSO could be the nucleus of that spiral galaxy, i.e., a high-redshift QSO would be physically associated with a nearby galaxy. If this was true, then indeed the cosmological nature of QSO redshifts would be in trouble.[13] However, as will be shown below, the lens nature of this system has been convincingly demonstrated, to the relief of the majority of astronomers.

The QSO optical continuum slope is much steeper than usual, a feature interpreted as reddening by dust in the galaxy. The rotational velocity of the galaxy, although not well defined from the measurements, and its luminosity are typical of early-type luminous spirals. No evidence for multiple images was found with a resolution of 2″. The QSO is radio-quiet, as we must expect in general.

2237+0305 calls for a few statistical considerations. There are about 15000 galaxies in the CfA Redshift Survey (about 7000 at the time of discovery). The surface density of QSOs with magnitude brighter than 17 is about $0.05/\deg^2$. From the average slit area used in the observations of these galaxies one can estimate that the probability of finding a QSO with the given brightness in the spectrum of a galaxy is about 10^{-3}; if we require in addition that the QSO be within $0\rlap{.}''3$ from the center of a galaxy, the probability of finding such a system in the sample falls below 10^{-5}. Although a posteriori statistics can be misleading, they hint that 2237+0305 is a remarkable object. To increase these probabilities it was proposed in [HU85.1] that the QSO is highly magnified by the GL action of the galaxy. In fact, from usual mass-to-light ratio arguments, it was estimated that the surface mass density in the center of the galaxy is a significant fraction of the critical surface mass density [see eq. (5.5)], so that a certain amount of magnification necessarily occurs. Assuming a magnification of about 40, the above probabilities increase to 0.03 and 2×10^{-4}, respectively, due to the artificially higher apparent density of fainter QSOs.

Resolving the images. Additional observations of 2237+0305 resolved the QSO images [TY85.2]. Besides the nucleus, four point sources were found within the inner few arcseconds of the nucleus of the galaxy ([SC88.1], [YE88.1]). The largest angular separation between these images is $1\rlap{.}''8$. The images are all bluer than the galaxy, but by varying amounts. However, on a two-color diagram, the images can be shown to lie on an extinction curve, thus strongly supporting the interpretation that the different colors are produced by differential reddening in the galaxy. Isophotes for this system are shown in Fig. 2.10; the plot shows the relative location of the four images, as well as the nature of the galaxy, which has a prominent bar. Surface photometry shows that the galaxy has a bulge in addition to its

---

[13] In fact, the clear association of a high-redshift QSO with a low-redshift galaxy, as in 2237+0305, was that kind of object the 'non-cosmological redshift community' had awaited for nearly two decades.

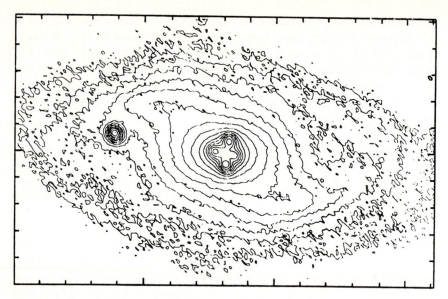

Fig. 2.10. Optical isophotes for 2237+0305 (courtesy of H. Yee)

disk; these are well represented by a de Vaucouleurs law and an exponential disk, respectively.

Spatially resolved spectroscopy ([DE88.1], [AD89.1]) removed the remaining doubts about the lens nature of this system. The CIII] line has been seen in all four components with the same redshift and, within the observational uncertainty, the same line-width. Imaging with a narrow-band filter centered on this emission line determined the positions of the images much more accurately than wide-band imaging; it was thus found that images A and D are closer together than the other images.

Models and interpretation. 2237+0305 provides a unique opportunity to observe and resolve a lens, thanks to its small redshift. Observers were able to map the light distribution of the galaxy in great detail to help constrain lens models. If we assume a constant mass-to-light ratio, a map of the galaxy provides the surface mass density distribution of the deflector, up to an overall constant. One can then estimate whether a mass model proportional to the light distribution can account for the image properties ([SC88.1]; the reader interested in modeling lens systems is referred to this paper as an instructive example). Given the observed light distribution, the lens configuration has three free parameters, the two unknown coordinates of the source and the overall mass-to-light ratio of the galaxy. One then defines a quantity which measures the deviation of the predicted image configuration from the observed one; this deviation can also include the brightness ratios for the images (see Sect. 10.3). Minimizing this quantity then yields the best-fit lens model for this system. In practice, the approach is somewhat more cumbersome, as the seeing smooths the intrinsic light distribution of the galaxy; such an effect must be accounted for. The best model in [SC88.1] satisfactorily approximates the observed image configuration, as well as the flux ratios for all four images. A fifth image is produced, which is highly demagnified and therefore hidden from our view, in the center of the galaxy.

If a source is exactly aligned with the center of a symmetric lens which is sufficiently strong to produce multiple images, a ring image (Einstein ring) of this source will occur. When the source is slightly displaced from exact alignment, the ring breaks up into two highly elongated images. On the other hand, if the symmetry of the lens is broken, the ring splits up into four images, with opposing images having comparable magnification. Such a 'broken symmetry' model is the best-fit model in [SC88.1]. The data on which this model is based support the near mirror-symmetric arrangement of the images. To obtain this, the source must be well aligned with the center of the galaxy. Using slightly different and higher quality data [YE88.1], a somewhat different model can be constructed [KE88.1] which places the source somewhat further away from the center of the galaxy. In particular, the A and D images are closer together in this model. Both models predict a time delay for image pairs of the order of 1 day, and a total magnification between 10 and 15.

It is clear that more detailed observations will allow refinements of the extant GL models. For 2237+0305, the observational constraints on the image configuration and the light distribution of the galaxy are better than for all the other lens systems (with the possible exception of 0957+561, where the VLBI data place considerable constraints on the models). One can then ask how unique these models are. This question is partly answered in [KE88.1], where in addition to the deflecting galaxy the allowance was made for a black hole in the center of this galaxy. It turns out that even for a large mass of such a black hole, acceptable fits for this lens system can be obtained. Thus, the models for this system are far from unique. Therefore, model fitting should not be regarded as a possibility to obtain detailed information about the matter distribution in the deflector, but more as a hint of whether an observed image system can plausibly be regarded as being due to lensing. Only with more refined observations, such as a measurement of the velocity dispersion in the bulge of the lens, can we expect better constraints on model parameters.

2237+0305 deserves the label 'unique', as it is highly unlikely that another system will ever be found where the lens is as close to Earth, and thus can be studied in such great detail ([SC88.1], [KE88.1]). The proximity of the lens has an additional consequence concerning ML, in that the effective transverse velocity of source and lens is expected to be much higher than for all the other lens systems, where the lens is roughly halfway between Earth and the source. Such a high transverse velocity should therefore reduce the timescale for ML events. In addition, due to the small time delay between the images, variations of the flux ratios between different images are unlikely to be due to an intrinsic variability of the source, since radio-quiet QSOs usually do not vary over time scales as short as one day. Hence, this system is ideally suited for investigating the effects of ML ([WA90.1], [WA90.2], [KA89.2]). We describe ML studies of this source in Sect. 12.4. We also want to point out that the 2237+0305 GL system provides a unique opportunity to study the interstellar matter in the spiral galaxy with absorption spectra of the QSO images [HI90.1].

- **H1413+117: the 'clover leaf'**

Surprisingly, another system with a striking pattern of four images was recently found [MA88.1], and was labeled a 'clover leaf'. The discovery was the result of the survey [SU90.1] that also yielded the UM673 system (Sect. 2.5.1). The system was also resolved at 3.6 cm with the VLA [KA90.2]. Figure 2.10 shows an optical view of the four images of H1413+117; they are separated by about 1 arcsec.

Since its discovery, high-resolution spectra have confirmed the very strong similarity of the four images [AN90.1], which thus share the redshift

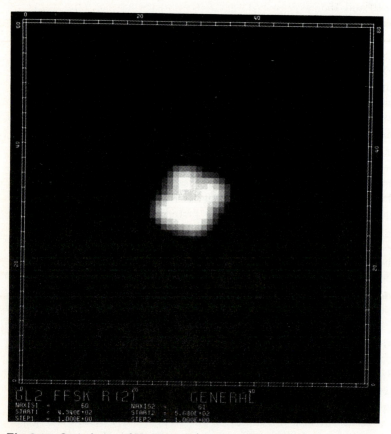

Fig. 2.11. Optical view of H1413+117 (courtesy of P. Magain)

$z \approx 2.55$. The high redshift of the QSO in this case lessens the problem of the absence of a lens, because it may itself have a relatively high redshift, and therefore may easily escape detection. The only hints about a possible lens may be the absorption-line systems at $z \approx 1.44$ and $z \approx 1.66$ [MA88.1]. Models proposed in [KA90.2] that adopt $z \approx 1.44$ account reasonably for the properties of the images.

Two tantalizing properties of H1413+117 make it remarkable, if not as unique as 2237+0305. First, the QSO itself belongs to the broad absorption-line (BAL) class, and at $m_v \approx 17$, it is one of the brightest such QSOs. Of course, a possibly substantial magnification factor of the QSO lessens the significance of the latter property. But, as pointed out in [KA90.2], membership in the BAL class is a novelty among lensed QSOs. Second, again according to the models of [KA90.2], and according to expectations due to the small image separations, the time delays for image pairs are between a few days and a month, so H1413+117 is another important target for ML studies. However, the lever arm of a lens at $z \approx 1.5$ is much smaller than for 2237+0305, so the longer time-scales of variation may complicate observations, to say nothing of the tightness of the image configuration. Recently, the spectral differences in one of the images has led to the suspicion that in fact ML does affect the spectrum of the images [AN90.2].

2.5.4 Additional candidates

Although, as we describe below, recent fruitful observations now encompass more complicated morphologies, systems with multiple images of point sources are still being found, as attested by UM425 (1120+019) [ME89.1], and by the triple radio source MG0414+0534 [HE89.1].

Several systematic searches for GL systems have been announced. In the first one, still underway, radio sources from the MIT-Green Bank (MG) survey were mapped with the VLA (Sect. 12.2.5). Three "filters" were applied: first, one regarding brightness and proximity to the galactic plane; second, one regarding the VLA morphology; and third, one regarding optical morphology and spectra. Only those candidates that survived all filters began to be considered as confirmed candidates [BU90.1]. Fortunately, the collaboration in this case comprises a sufficient number of astronomers!

At least three systematic searches by direct optical imaging have been, or still are being performed, with yields of a few percent. Surdej [SU90.1] reviewed the state of all of these. Two other searches fall under the classification of investigation of statistical effects: an automated search for gravitationally lensed QSOs [WE88.1], which is actually the descendant of a search for QSO-galaxy associations that is meant to discern the amplification bias (Chap. 12), and a search for magnified, and possibly multiply imaged, 3CR radio galaxies [HA86.1].

The work necessary to carry out the surveys we have described is monumental, but the results expected are worth the investment. Beyond the handfuls of good lensing candidates they unveil, surveys are increasing our knowledge of QSOs significantly.

2.5.5 Arcs

So far, we have considered the imaging of point sources. For the first several years of the GL era, these were regarded as the most important sources in the study of GL systems. However, it is also natural to expect images of extended sources such as galaxies to occur. That is the subject to which we now turn.

The essential theoretical understanding of GL models in the case of point sources was presented in Sects. 2.1–2.3. However, a detailed description of the theory and models behind the formation of arcs is relegated to Chap. 13, because these objects were discovered quite recently, and the development of theories to understand them is truly contemporaneous to our writing. We describe three examples below; a list of arcs can be found in [FO90.1].

- **Abell 370: a spectacular first**

Discovery. The Abell 370 and Cl2244−02 systems were the first two arc systems found [LY86.1]. Abell 370 is a rich, well-studied cluster of galaxies at redshift $z \approx 0.37$, the highest in the Abell catalog. Relatively recent pictures were published in 1973, and the first mention of "filamentary" structures appears to have been published in 1981 [LY89.1]. Figure 2.12 shows a CCD image of Abell 370. In 1976, during tests of a new video camera at Kitt Peak National Observatory, a program to obtain multicolor images of clusters was started [LY89.1]. Images of one of the clusters, Abell 2218, appeared to have an interesting structure: Lynds and Petrosian described "a subjective impression of patterned circularity." Out of 29 observed clusters, Abell 370 and Cl2244−02 revealed even better-delineated "arclike features", but the quality of the data did not permit a detailed analysis. One property was clear: the arcs were distinctly bluer than typical E galaxies in the clusters [LY89.1]. The observing program proceeded, but did not yield improved pictures of the arcs until 1985, when the clusters were included as incidental targets for other observing programs. Thus, observing conditions were far from ideal in the majority of the observations. However, the presence of these new astronomical objects was undeniable, and they were announced in 1986 [LY86.1]. These arcs were independently discovered by the "Toulouse group", and published in 1987 [SO87.1].

At the time when arcs were first noticed, GLs were still a mere theoretical curiosity. The advance of technology, as usual in astronomy, played a significant role in helping the discovery.

As can be seen from Fig. 2.12, the large arc in the Abell 370 system is a crescent. Its length is $\simeq 21''$, its mean thickness $\simeq 2''$, and its mean radius of curvature $\simeq 15''$ [LY89.1]. The figure also shows that the arc is neither uniformly bright, nor uniformly circular; it appears knotty and has clear bends, particularly at the eastern end. There are also variations in its thickness. Such a complicated morphology provides constraints on the GL models as we discuss in the following subsection.

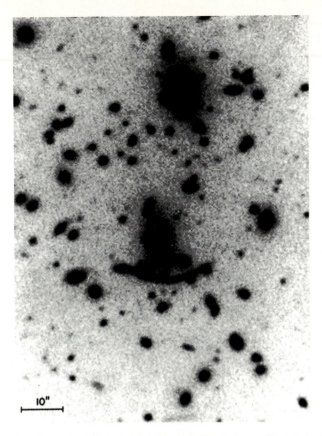

Fig. 2.12. Optical view of Abell 370; North is up, East is to the left (courtesy of V. Petrosian)

We see that an arc like that in Abell 370, if due to a GL, is in an imaging regime that is totally different from that of the QSO images we discussed in previous sections. Not only are the images in a different regime, so is the lens. Figure 2.12 shows that a handful of members of the cluster are within a few arcseconds from the arc. The first published spectroscopy of two of the member galaxies by the "Toulouse group" [SO87.1] placed them at the redshift of the cluster. The attempts at explaining the new structure first concentrated on local effects, such as star bursts induced by cooling flows [SO87.2], bow shocks [BE87.1], or light-echo models ([BR89.1], [KA89.1]). The GL hypothesis was first mentioned, but also discounted, due to the absence of a counterarc in [SO87.1] (see Chap. 13). This is the analog of the initial thoughts against lensing in 0957+561, when circular symmetry was the guiding concept.

The appeal of the new structures seems to have focused intense observational efforts on Abell 370, for in 1987, the "Toulouse group" published extensive photometry and spectroscopy of the cluster [ME88.1], and in 1988,

73

the redshift of those parts of the large arc without superimposed galaxies was measured, and found to be $z \approx 0.724$ [SO88.1] (see also [MI88.1] for earlier spectroscopy of A370 and Cl2244−02). The quality of the spectrum was sufficient to dispel any doubts, and the GL hypothesis had won once again. The spectrum of the arc was also found to be similar to that of a spiral galaxy. Further photometry [FO88.1] yielded an integrated magnitude for the arc of $m_B \approx 22.4$, which corresponds to only 7% of the sky brightness, and illustrates the difficulties of these measurements. The spectral type of the arc was further confirmed by infrared measurements in the infrared 2-μm band [AR90.1]. The confirmed blue color of the large arc, its reassuring spectral similarity to a spiral galaxy, and the fact that high-redshift clusters ($z > 0.1$) present a strong blue-galaxy excess, were all in favor of the GL hypothesis.

Additional evidence in favor of the GL hypothesis was provided by the identification of several faint, elongated blue objects near the cluster core [FO88.1]. In particular, arclet A5 appears to have redshift $z \approx 1.305$ [SO90.1]. Since the redshifts of at least two of the blue objects are thus larger than that of the lens cluster ($z_d \approx 0.374$; [SO88.1]), these measurements excluded the in situ explanations for the arcs.

Models. The first published models for the lens that forms the large arc in Abell 370 ([SO87.2], [HA87.1], [PE89.1]) accounted in a simple way for the general shape of the arc. For instance, the model described in [SO87.2] consisted of nine point-masses, to represent an overall cluster potential, not centered on any galaxy, and to represent the unavoidable effects of the galaxies located near the arc. Since a simple, constant mass-to-light ratio was assumed, it was comforting to see that such a simple model accounted for the overall shape of the arc. The models discussed in [PE89.1] also indicated that dark matter associated with the brightest galaxies near the large arc would need to be present in unheard-of amounts that would result in a visible counterarc. Thus, dark matter that is not associated with luminous galaxies remained a valid requirement. The main conclusion from these first efforts was that off-center, dark matter in the cluster was necessary to explain the observations, with a mass-to-light ratio of > 100 in solar units.

Since then, several improvements to the early models have been proposed. The first modification was the abandonment of the assumption of circular symmetry for the potential of the cluster (e.g., [KO89.3], [HA89.2], [GR89.1]). The later models account for the detailed shape of the large arc ([ME90.1], [BE90.1]), and they show again that the constraints available permit a wide variety of models, including for example 'bimodal' types, where two non-concentric elliptical potentials are assigned to the cluster [ME90.1]. Some of the theoretical aspects of modeling lenses where extended sources overlap various caustic shapes are discussed in Chap. 13.

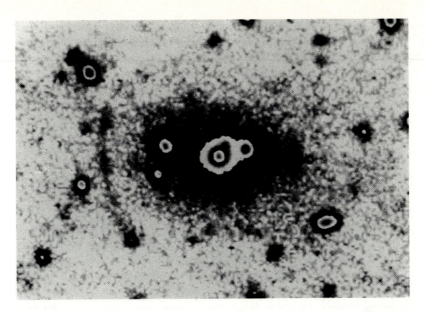

Fig. 2.13. Optical view of Abell 963 [LA88.1]. North is up, and East to the left. (coutesy of R. Lavery)

- **Abell 963**

Abell 963 is a rich cluster of galaxies at redshift $z \sim 0.2$. Its core is dominated by a cD galaxy, and thanks to observations performed in 1987 [LA88.1], the cluster became the first GL candidate to reveal a possible counterarc as a companion to a main arc. We show an optical view of the system in Fig. 2.13. The separation of the two arcs is $\sim 30''$; they extend over 15–25 arcseconds.

Based on intuition derived from circularly symmetric lenses, one might expect to find counterarcs nearly always. However, once circular symmetry is abandoned, the absence of counterarcs is no longer surprising. Abell 963 represents a return to early expectations regarding counterarcs, but a full confirmation of the GL hypothesis awaits measurements of the redshifts of the arcs. As for Abell 370, the arcs in Abell 963 are very blue, and very faint (approximate B magnitudes derived in [LA88.1] are ~ 22–24). Thus, once again, confirmation by measurement of redshifts remains a significant technical challenge.

Given the lack of measured arc redshifts, the only models available, discussed in [LA88.1] and [KO89.3] remain tentative. However, by placing an extended source at redshift $z = 1$, and representing the lens as a point mass at the position of the cD galaxy and a uniform sheet of dark matter, the arcs could be easily accounted for. Unfortunately, the amount and center of such dark matter are not well defined, again due to the lack of a source redshift.

Recently, results on the spectroscopy of the two arcs were published [EL91.1]. From a strong emission line in the northern arc, its redshift is determined to be $z = 0.771$. The fainter southern arc shows no obvious spectral feature. The near-constant blue color along both arcs support the idea that both are due to gravitational lensing of a background object. The optical and infrared color is consistent with that of a spiral galaxy at this redshift, which undergoes strong star formation.

- **Abell 2390**

Abell 2390, a rich cluster of galaxies at redshift $z \approx 0.23$, was selected in a search for arcs on the basis of its large X-ray luminosity [FO90.1] (see Sect. 2.5.7). The brightness of the core of Abell 2390 is dominated by a bright radio galaxy. Figure 2.14 shows an optical view of the cluster, and of the unusual, large arc, which appears to be nearly straight over its length of $\sim 15''$, with radius of curvature greater than $\sim 5'$. The arc is not uniformly bright: it appears to have two breaks, roughly symmetric about its center.

The spectral and photometric properties of this new arc are what may now be called typical, i.e., blue and faint. In addition, Pello and coworkers managed to obtain high-quality spectra of the arc [PE90.1], and to determine a possible redshift of $z \approx 0.9$, based on a single line. In addition, the part of the arc where contamination by a nearby galaxy is minimized shows a spectrum that is consistently similar to that of a spiral galaxy at $z \sim 0.9$.

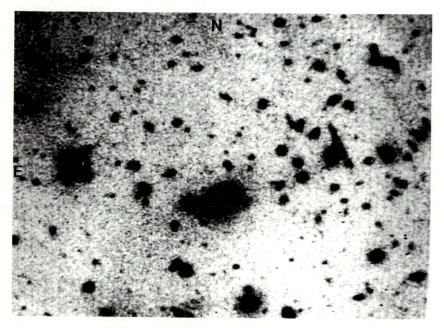

Fig. 2.14. Optical view of Abell 2390; North is up, and East to the left (courtesy of G. Soucail)

The observations we have mentioned are quite recent, and can be expected to be improved, to test further the GL hypothesis. The first models, proposed in [ME90.1], that accounted well for the shape of the arc, and in particular its breaks, require a 'bimodal' potential, i.e., two components offset from each other. One such component was centered on the bright radio galaxy, and the other was not associated with any luminous object. The lens model is successful because it yields a cusp-type caustic on the source plane (see Chap. 6), which, for the appropriate alignment with the source (assumed to have an elliptical shape), produces three images that nearly merge. Such special geometrical coincidences are not surprising, given all the successful models for various GL systems. However, the statistics of such occurrences will need to be considered carefully as the number of systems grows. "Straight arcs" can be easily produced by placing the source near a beak-to-beak singularity – see Sect. 6.3.3. This is also in agreement with new observations reported in [PE91.1]; there, the redshift of the arc was confirmed. The arc shows two breaks which lie symmetrically to the position of a galaxy close to the arc. This observational result can be understood as follows: the critical curve produced by the cluster is distorted and 'detours' around this nearby galaxy. Hence, the critical curve crosses the arcs at two positions, yielding a configuration of the caustics close to a beak-to-beak metamorphosis. The same kind of model seems to apply for the straight arc in Cl0500−24 ([WA89.1], [GI89.1]). In addition, several arclets have been found near the main arc, with orientation very similar to the main arc. This also would be in accord with the interpretation given above. One of these faint objects shows the same redshift and spectrum as the main arc. A velocity gradient along the arc is observed, leading to the interpretation that the source is a spiral galaxy seen edge-on, so that the velocity gradient could be interpreted as a rotational velocity of the galaxy.

We also refer the reader to [PE88.1] for the observation of several arcs in the cluster A2218.

2.5.6 Rings

- **MG1131+0456: a nearly full Einstein ring**

Discovery. The radio search for lenses described in Sect. 12.2.5 yielded a ring-like radio source [HE88.1]. A 6-cm VLA map of this source is displayed in Fig. 2.15; in the map, an elliptical ring can be seen, with major and minor axes measuring $2\rlap{.}''2$ and $1\rlap{.}''6$, respectively. In addition, there are four compact components (A1,A2,B,C), where A1 and A2 are clearly separated in the 2-cm map. There is practically no flux at the center of this ring, and its edges are well-defined. Optical imaging of this source revealed a faint ($m_R \approx 22$) extended object. The major and minor axes of an elliptical Gaussian that fits the optical source measured $2\rlap{.}''5$ and $2\rlap{.}''1$, respectively. However, the position angle of the major axis of the optical source was different from that of the radio ring. A spectrum of the optical source revealed

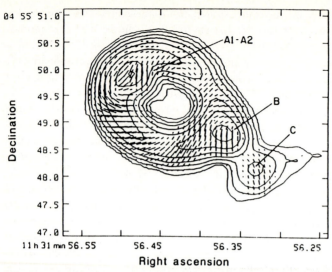

Fig. 2.15. VLA 6-cm map of MG1131+0456 (courtesy of J. Hewitt)

a continuum without any emission lines; therefore, no redshift could be determined. On the other hand, the lack of emission lines excludes galactic objects such as HII regions, supernova remnants or planetary nebulae. Due to its high galactic latitude, it is likely to be an extragalactic source, but it is not a known infrared source. The radio spectral index of MG1131+0456 and its optical and infrared properties are consistent with the interpretation that this source is a radio galaxy. However, no radio galaxy with comparable radio morphology had been observed before the discovery of MG1131+0456.

The unusual radio morphology led to the suspicion that MG1131+0456 is a lensed radio galaxy. If the symmetry of a lens is perturbed, the singular point on the symmetry axis in the source plane breaks up into a caustic figure with (at least) four cusps. An extended source covering part of this caustic curve can have a highly elongated and curved image. If the alignment is sufficiently good and the source size comparable with, or larger than, the size of the caustic curve, a complete ring can be produced. For less colinear or smaller sources, a broken ring can occur. The 2-cm observations of MG1131+0456 show a broken ring structure, whereas the 6-cm data show a complete ring; this may be due to a larger source size at longer wavelengths, or to the different angular resolution of the observations. A source point outside the caustic figure would have three images, whereas source points inside the caustic have five images; however, if the core of the lens is sufficiently compact, the number of images would effectively be reduced by one. In particular, the two (main) images of a source point just outside the caustic would be seen on opposite sides of the lens. The compact components A and B may be images of a compact source component which lies just outside the caustic. If Faraday rotation in the lens is negligible, image points belonging to the same source points should have the same direction of linear polarization. It is noteworthy, therefore, that the linear polarization vectors of the two compact components A and B are nearly parallel.

Lens model. A simple lens model was proposed in [HE88.1], to explain the observed properties of the MG1131+0456 system, in particular the elliptical shape of the ring, which is produced by an elliptical lens, and the arrangement of the compact components. Of course, there is considerable freedom for modeling this system, because we have no redshift measurements. One might wonder whether a system with so many unknowns can be modeled reliably. Nevertheless, in this case, we know that the source is extended; thus, one does not have a few point images to constrain a lens model, but a two-dimensional brightness distribution and thus potentially much more information. Such a possibility was recognized in [KO89.1], where a reconstruction technique for extended images was developed. We describe this method in Sect. 10.7; the idea is to use a parametrized GL model, and a minimizing procedure is applied simultaneously to the lens parameters and to the source structure.

For MG1131+0456, a six-parameter lens model was used to attempt to account for the observations. The parameters were: two coordinates for the center of the lens, the ellipticity, the position angle of the major axis, the angular core radius, and the 'strength' of the lens (e.g., the velocity dispersion). One might think that, with six parameters, one could fit nearly everything, but this is far from true. In particular, four of the six parameters are purely geometrical, and as it turns out, the core radius is unimportant and can be set to zero with no significant degradation of the resulting model fit. The best model consists of a set of these six parameters and the source structure. The fitting was performed using the 2-cm data, as high angular resolution is more important for this method than a high signal-to-noise ratio. Figure 2.16a shows a 2-cm map together with the critical curves for the best lens model. The corresponding source model and caustics are shown in Fig. 2.16c. Figure 2.16b shows the reconstruction of the image from the source model and the corresponding lens model; we see that the reconstructed image is very similar to the original map. The source model has the appearance of a normal 'core-jet' source, a common type of extragalactic radio source. The lensing potential has a moderate ellipticity of $\epsilon \approx 0.14$; due to the unknown redshifts of the source and the lens, the velocity dispersion of the galaxy cannot be determined, but for $z_d = 0.5$ and $z_s = 2.0$, it amounts to about 250 km/s, a 'reasonable' value. Except for the core radius, the model parameters are well determined, since even small variations lead to considerably worse fits.

Once the lens model is specified, it can be checked with additional data, such as 6-cm maps. Such a map is shown in Fig. 2.17, along with the source model and the reconstructed image. The agreement between the observed map and the reconstruction is quite satisfactory, save for the feature near the center of the lens. However, this spurious feature is probably due to the

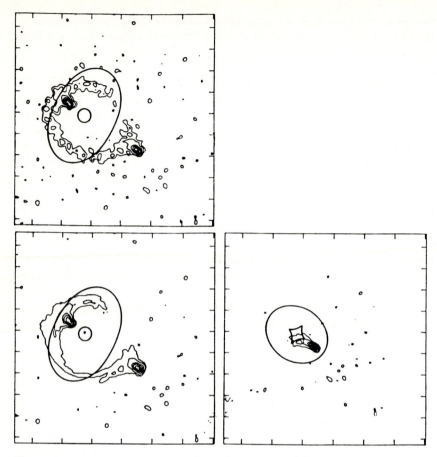

Fig. 2.16a–c. 2-cm map and model for MG1131+0456. (**a**) – upper left – shows the observed brightness distribution, together with the critical curves obtained for the best lens model. (**b**) – lower left – shows the reconstruction of the image, as obtained from the best-fit model of the lens and the source, the latter being shown in (**c**) – right panel – together with the corresponding caustics (courtesy of C. Kochanek)

finite resolution of the map and will vanish if higher resolution data become available. Apart from this problem, it is quite gratifying that a simple lens model that accounts for the 2-cm data can also reproduce the 6-cm image.

There is an additional and completely independent check of the lens model. Note that, since the deflection law is fixed, one has implicitly identified points on the image which stem from the same point of the source. If there is no differential Faraday rotation of the linear polarization vector in the lens, images that map to the same source point should have the same direction of linear polarization. Figure 2.18 depicts the polarization map at 6-cm, the corresponding polarized source model, and the reconstructed-image polarization map. Comparing the original map with the reconstructed image, one cannot help but be impressed by the high level of agreement between observations and predictions at λ 6 cm. We

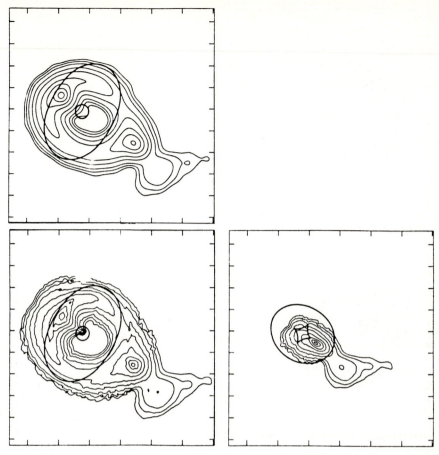

Fig. 2.17. 6-cm map and model for MG1131+0456, as in Fig. 2.16 (courtesy of C. Kochanek)

stress again that there is no freedom in this reconstruction. Therefore, we consider MG1131+0456 a very strong GL candidate, although the observations have not revealed any information about the lens, or the redshift of the source.

Recently, using the New Technology Telescope (NTT) of ESO, Hammer and Le Fevre [HA91.1] have obtained deep broad band images in several spectral bands. The optical image looks more extended than an elliptical galaxy. They subtracted the brightness profile of an elliptical galaxy from the image; the residual light distribution reveals a ring-like feature which has a gap at the same position as in the 2 cm map. These observations, if confirmed, have thus detected the first optical ring. Two breaks in the spectrum have been preliminarily interpreted in [HA91.1] as corresponding to redshifts 0.85 and 1.13, which may hint at the redshifts for the lens and the source, both being elliptical galaxies.

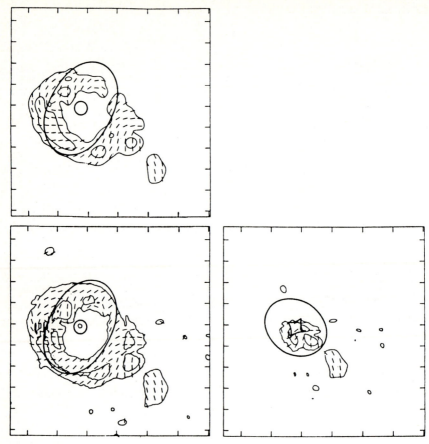

Fig. 2.18. 6-cm calculated polarization map for MG1131+0456, as in Fig. 2.16 (courtesy of C. Kochanek)

- **MG1654+1346: a second ring**

Unexpectedly, rings have made a second appearance in the GL universe, with the new candidate MG1654+1346 [LA89.1], shown in Fig. 2.19.

The second ring specimen has one important resemblance to MG1131+0456: its diameter is $\sim 2''$, but it is also significantly different from the first one, because the ring is accompanied by two optical sources, one a QSO (at the position labeled Q in Fig. 2.19), with redshift $z \approx 1.7$, the other a galaxy with $z \approx 0.25$ (at the position labeled G in Fig. 2.19) [LA89.1]. Although no detailed fitting was performed in [LA89.1], it was shown that a GL interpretation was plausible, especially since the two compact components are on opposite sides of the ring. Models are still under study [KO90.3]; preliminary results indicate that the morphology of all objects can be accounted for by assuming an elliptical potential associated with G, similar to that for MG1131+0456, and by assuming that the ring is an image of a radio lobe of

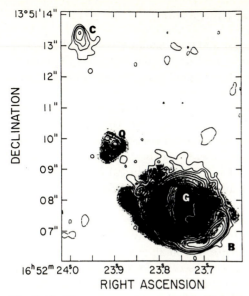

Fig. 2.19. The contours show a 8-GHz VLA map of MG1654+1346, which is superimposed on an R-band optical image of the quasar Q and the galaxy G (from [LA90.1])

emission that emanates from the core of Q. Higher-resolution maps in this case will be able to constrain the lens models particularly well. The more recent observational results have permitted a lens model for the ring, using an elliptical mass distribution [LA90.1]; this model accounts for the overall shape of the ring, its thickness, its orientation, as well as for the location of the compact components. Also, the polarization data of the system strongly supports the lensing hypothesis [LA90.1]. Due to the relative simplicity of this lens system, the mass of the lensing galaxy contained in the ring image can be estimated fairly well. Unfortunately, as pointed out in [KO90.3], since the object that is multiply imaged seems to be a radio lobe, it is not expected to vary significantly, and estimating H_0 will not be possible using MG1654+1346. We can only hope that another ring system will be found that combines both known systems [KO90.3].

In fact, very recently a new candidate system for an Einstein ring has been found [JA91.2], which has been suspected previously to be a gravitational lens system ([RA88.1], [SU90.2]). This ring source, PKS1830−211, shows two compact components on an elliptical ring. The separation of these two compact components is about 1 arcsecond, i.e., smaller than the size of the other two rings. Most noticible, this source is one of the strongest flat-spectrum radio sources in the sky. It lies close to the galactic plane, which renders optical identification difficult; a faint optical counterpart has been reported. Several strong arguments are given in [JA91.2] for this source and the corresponding lens to be extragalactic. Its brightness, together with the fact that the compact components are probably variable, make this an ideal target for further studies.

2.5.7 A rapidly growing list of candidates

Since arcs are one of the more recent and spectacular GL types discovered, and their properties are even more complicated than those of QSO images, their study is only just beginning. Direct optical imaging surveys are underway (for a review, see [FO90.1]), and are rapidly increasing the number of candidates. Early candidates are Cl0024+16 and Abell 2218. More recent additions are Cl0500−24, Abell 1689, Abell 545 and Abell 1525 [FO90.1]. Several of these clusters are known, strong X-ray sources [KE90.1], and their study as GL systems is particularly interesting, because of the possibility of combining studies of the mass distribution at optical and X-ray wavelengths. The improved sensitivity and resolution of ROSAT will doubtless contribute significantly to such studies [KE90.1].

Surveys of the radio type, such as the MG survey mentioned in Sect. 12.2.5, have yielded two ring systems, and more may yet be found. It may even be that ring images predominate [KO90.2], mainly due to the large size of MG sources compared to the cross-section for multiple imaging, but they may not be resolvable. However, the majority of the rings may be just below the resolution limits of the VLA. Optical surveys are proceeding, and their results are beginning to be made public. The early successes should be only the beginning of our database of GLs.

2.5.8 Speculations on other gravitational lens systems

The confirmation of any system as a GL depends on the quality of the observations; for some of the systems described earlier the observational evidence in favor of lensing is so strong that it can hardly be disputed. In other cases, such as 1635+267, there is some evidence which points towards the lens nature of this system, but the acceptance of the GL interpretation depends, at least in part, on personal views. We outlined the difficulties of confirming the lens nature of an observed system of multiple images in the previous section. As long as no quantitative measure for the similarity of QSO spectra is available, multiple images not accompanied by a possible deflector will be the subject of disputes with respect to their lensing nature.

It was after the detailed investigation of the suspected lens system 1146+111 (described below) that it became clear that 'similar' QSO images do not necessarily imply a common origin. As is true for galaxies, QSOs are clustered in space, which implies clustering in redshift. A pair of QSO images with the same redshift may be a physical pair of QSOs. That was the 'sobering' conclusion from 1146+111.

Besides the splitting of a background source into distinct images, the fluxes of such images can be magnified, compared to the flux of the unlensed source, which can lead to the amplification bias or, in cases where the magnification is due to an ensemble of microlenses (e.g., stars in a foreground galaxy), time-dependent magnification can occur due to the change of relative position of lens and source in time. Both of these effects will be discussed in more detail in later chapters; here we mention briefly just two aspects of these effects and discuss possible lensing candidates that are relevant to these considerations.

High-redshift radio galaxies. The 3CR sample of radio sources (a flux-limited sample at low radio frequencies containing several hundred sources) has been nearly completely identified optically; some of the sources, however, have fairly weak optical counterparts, and there are just a few 3CR sources left for which no redshift is yet available. The high-redshift radio galaxies in this sample have the largest known redshifts, and as a rule, they are very luminous. These properties led to the suspicion that some of these high-redshift radio galaxies are gravitationally magnified [HA86.3], as was discussed as early as 1976 by N. Sanitt [SA76.1]. In a series of recent papers, some of these radio galaxies were optically imaged with very good seeing (e.g., [LE87.1], [LE88.1], [LE88.2], [LE88.3], [LE90.1], [HA90.1]). These observations lend support to the magnification hypothesis for at least some radio galaxies, i.e., that the amplification bias affects the measureable properties of the sample.

The radio galaxy 3C324 [LE87.1] has a redshift of $z = 1.206$. Broadband imaging reveals that 3C324 has multiple components. There is an additional line system in the spectrum at a redshift of 0.84. Narrow-band imaging revealed separate components at the two different redshifts. The central object was interpreted as a spiral galaxy at a redshift of ≈ 0.85, and the outer ones as multiple components of the radio galaxy, with angular separations of about $2''$. Since the luminosity of the galaxy at $z \approx 0.85$ is about ten times that of our Galaxy, it is unavoidable for it to create multiple images of the background source; the minimum mass required is only $1.5 \times 10^{11} M_\odot$. The observed ellipticity of the foreground galaxy suggests a very asymmetric lens model. Although detailed modeling is not justified, due to the sparseness of the observational constraints, a qualitative comparison of the observed image geometry (which contains additional components) with theoretical lens calculations may indicate that 3C324 is indeed a GL system. From the image splitting and the position of the foreground galaxy, it was estimated that the total magnification for the radio galaxy is about two magnitudes [LE87.1].

The radio galaxy 3C238 (with redshift $z = 1.405$) has a very elongated optical shape and can be resolved into three components. The two outer components appear stellar, whereas the central component seems to be resolved. In addition, the two outer components, with a separation of about $3''.5$, have the same color, much redder than the third component. These properties, and the fact that 3C238 lies in the direction of a distant Abell Cluster (Abell 949, with a redshift of about 0.14), suggest that the two outer components are images of a compact source, while the central component is the lens responsible for the image splitting, with the assistance of the galaxy cluster. If the central component is a galaxy at the redshift of the cluster, it must have a mass-to-light ratio of the order of 100 (in solar units), smaller than the corresponding constraints on the lensing objects in 2345+007 and 1635+267. If, however, the central component is a galaxy at higher redshift, its required mass-to-light ratio decreases. Clearly, a spectrum of this central component is needed before the GL nature of this system can be confirmed.

In a number of other high-redshift 3CR radio galaxies, a foreground galaxy was detected very near the optical counterpart of the radio source [HA90.1]. For reasonable values of the mass-to-light ratios for these foreground galaxies, significant magnification of the radio galaxy is unavoidable. In other cases, the multiple optical structure of the radio galaxy agrees very well with the radio structure, which may also be taken as evidence for possible lensing. Higher-resolution imaging, which is becoming possible with new telescopes, is needed to verify the GL nature in these cases. However, the high success rate for finding foreground galaxies in the vicinity of 3CR radio galaxies suggests that the amplification bias is strongly affecting this sample of high-redshift radio galaxies.

1146+111: a dubious candidate. In a paper titled "Will cosmic strings be discovered using the Space Telescope?" B. Paczyński [PA86.3] discussed a few QSO pairs with angular separation of a few arcminutes and very similar redshifts. The pair with the best agreement in redshift was 1146+111, with components labeled B and C, which had been reported earlier to lie within a large concentration of QSOs; subsequent observations [AR80.1] determined their redshifts: $z \approx 1.01$ for both objects. The separation of the QSOs is 157 arcseconds, nearly two orders of magnitude larger than for known lens systems. Image splittings of this order may be produced by cosmic strings (see Sect. 13.3.4), if their mass per unit length is sufficiently large. Lensing by a straight cosmic string would cause no magnification of the images, so the nearly equal brightness of B and C is suggestive of such an interpretation.

Following this suggestion, 1146+111 was investigated in detail [TU86.1]. It was found that the redshifts of the two images are within 300 km/s of each other, the difference being statistically indistinguishable from zero; the spectra agreed closely in their significant features. Imaging showed a possible rich cluster of galaxies between the two QSOs and an unusual number of low redshift galaxies in the field. The spectral similarity and the nearly identical redshifts supported the lens hypothesis for this system.

Unfortunately, additional spectroscopic observations of 1146+111 yielded only evidence against the GL hypothesis ([HU86.1], [SH86.2]). In particular, a significant difference in the Balmer lines was found: they are at least 7 times weaker in the C image than in the B image. These spectral differences, together with a detailed investigation of the colors of other objects in the field [TY86.1], led to the conclusion that 1146+111 is probably not a GL system.

Three different potential kinds of lenses have been proposed for this system: a $10^{15} M_\odot$ mass black hole [PA86.5] (see also [AR86.1]), a cosmic string [GO86.1] and a cluster of galaxies ([OS86.2], [CR86.1]). If the lens were a supermassive black hole, the other QSOs nearby in the field would be multiply imaged, and other background objects, such as high-redshift galaxies, should show clear signs of shape distortions [BL87.1]. The same is true for a cosmic string, which would also create multiple images of other nearby sources. A cluster of galaxies of the required mass and compactness to lead to such an image splitting would almost certainly reveal its presence in the microwave spectrum, due to the Sunyaev-Zeldovich effect [OS86.2]. Radio observations of the 1146+111 field ([ST86.1], [LA86.1]) did not detect this effect; a temperature jump would be expected across a moving cosmic string, but none was found. Thus, all three proposed types of lenses would produce effects that are ruled out by observations. Although the arguments can in principle be circumvented, a lens model for 1146+111 would require rather implausible parameters [BL87.1]. Hence, although the GL nature of this system cannot be ruled out (e.g., the spectral differences can be due to variability of the source, combined with the time delay for the images, expected to be of the order of 10^3–10^4 years), 1146+111 is the weakest GL candidate, in spite of many spectral similarities.

In [BA86.1], it was pointed out that the clustering of QSOs makes it not unlikely to detect pairs of QSOs with (nearly) the same redshift, and that the system 1146+111 can equally well be interpreted as a physical pair of objects. A similar argument was given in [PH86.1]. Moreover, it was argued in [BA86.1] that other lens candidates (2345+007, 1635+267) may be pairs of QSOs in clusters. The proposed statistical arguments are important, but they neglect the fact that very large portions of the spectra of the QSO pairs are 'very similar'. Nevertheless, 1146+111 has raised significant doubts about the lensing nature for systems without a luminous lens.

Scepticism about the true nature of lens candidates has been refuelled recently with the discovery of close QSO pairs with (nearly) identical redshifts. For instance, the members of the QSO pair 1343+164 [CR88.1] at redshift $z \approx 2.03$ are separated by $9\rlap{.}''5$, but substantial spectral differences preclude the GL hypothesis in this case. Even more tempting is the QSO pair 1145−071 [DJ87.1], where two QSOs at $z = 1.35$ are separated by only $4\rlap{.}''2$, and except for the CIV emission lines, their spectra are very similar, as are their redshifts. In fact, if one neglects the CIV lines, their redshifts are estimated to agree to within 100 km/s. However, this system cannot be a GL candidate, as only one of the objects is radio-loud, with a ratio of radio fluxes in excess of 500, about 200 times the optical flux ratio. We thus see that not every QSO pair with (nearly) identical redshifts need be part of a GL system.

Lens-induced variability. The light curves generated when the line-of-sight to a compact source sweeps through a dense field of stars (due to relative motion) were calculated in a number of papers (see, e.g., [WA90.4] and references therein); these theoretical light curves, which are further discussed in Sect. 12.4, can be compared to observed light curves to see whether they are 'similar'. However, this endeavour may appear easier than it really is, for several reasons, the most important being the fact that for only a few sources has there been accurate monitoring of optical light curves. Furthermore, it is difficult to quantify the 'similarity' of light curves. Finally, even if quantitative tests for 'similarity' are developed, it would still not be convincing if observed and theoretical light curves were similar; intrinsic variability cannot be dismissed, and it is extremely difficult, if not impossible, to rule out this last possibility.

There are in fact several well-documented light curves which appear similar to synthesized ML light curves, such as those for 3C345 and 1156+295. Other sources display an outburst behavior, which can be interpreted as a source crossing a caustic; such an explanation has been postulated for the eruptive BL Lac object 0846+51 [NO86.1], which is also known to be near a foreground galaxy [ST89.1]. We shall not elaborate on this here (but see Sect. 12.4) as no case of lens-induced variability has yet been convincingly confirmed, except in the images of the GL system 2237+0305, although we consider it possible (even likely) that the light curves of some AGNs are affected by ML.

BL Lac objects as lensed QSOs. Since the continuum radiation emitted by BL Lac objects is quite similar to that of QSOs, but, by definition, BL Lac objects have no strong emission lines, it has been suggested that at least some BL Lac objects may be gravitationally lensed QSOs, where the continuum source is highly magnified, so as to outshine the line emission, which is possible because the line-emitting region is assumed to be much larger than the region emitting the optical continuum [OS85.1]. In fact, some objects turn from a QSO into a BL Lac object if their luminosity increases. If the ML hypothesis applies, some BL Lac objects should have a foreground galaxy superposed on them. Since the redshifts of BL Lac objects are difficult to measure (emission lines are virtually absent), it is difficult to distinguish between a foreground galaxy and a possible host galaxy for these objects. We discuss these issues further in Sect. 12.4.4.

These few remarks should suffice to demonstrate that lensing is not necessarily always apparent, only the 'spectacular' effects of observing the same source at different positions on the sky, and highly distorted images of extended sources, are easily identified as being due to gravitational light deflection.

Statistical lensing. Unseen compact objects and clusters affect the appearance of distant sources. As we discussed in Sect. 2.3, the source counts of AGN may be affected by the amplification bias. Only a small fraction of magnified sources will be split into multiple images (at least if the amplification bias is due to ML); most of them will just be single (or unresolved multiple) images. Magnified but single images of extended distant sources form a class of objects for which there are hints of magnification that can be deduced only from the statistics of source counts and morphologies. When a sufficient number of objects is distorted by a cluster of galaxies ([TY90.1],

[GR90.1], [NE90.1]), such hints can be turned into constraints on the distribution of light and dark matter in clusters. One striking example, the technique used by Tyson [TY90.2] to derive the mass distribution, for example for cluster Cl1409+52, involved measuring the statistical alignment and elongation properties of blue galaxies (those favored by the spectacular arcs) within the projected area of the cluster. The measurements were conducted automatically by pattern-recognition software, and processing yielded the surface mass density of all matter in the cluster. No surprises were found in the preliminary results, i.e., large mass-to-light ratios were required, but the studies must be extended to yield improved statistics. Since the density of blue galaxies found in extremely deep exposures is high (to ~ 29 mag, there are 2×10^5 galaxies per square degree per magnitude [TY90.2]), very sensitive surveys promise to yield new measurements of the density of dark matter in clusters of galaxies.

The properties of populations of lens systems and their effect on other background populations can only be obtained with survey techniques. For example, if we know the fraction of QSOs at a certain redshift which have multiple images with a separation larger than a given threshold, we can deduce the number density of galaxies able to produce the required splitting. The statistical distribution of the image splitting in a well-defined lens sample can yield the mass spectrum of the galaxies. If the fraction of multiply-imaged QSOs is larger for brighter sources than for fainter ones, we would suspect that the amplification bias plays a crucial role in the number counts of QSOs, at least for the apparently most luminous ones. We investigate these questions in much more detail in Chap. 12. Here we only wish to argue that well-defined and unbiased samples of objects that are affected by a population of GLs are potentially very useful in observational cosmology studies.

2.5.9 Gravitational lenses and cosmology

The foregoing examples of lens systems reveal a multitude of different image configurations, observational constraints, and theoretical understanding of these systems. It seems that every new lens candidate provides a new puzzle to the theorist. We have also seen that observations of lens systems are extremely difficult and at the forefront of modern observing techniques (just to mention one example, consider the spectroscopy of the four images of 2237+0305 which have a maximum separation of $1.''8$). None of the above systems satisfies all the criteria listed in Sect. 2.4; nevertheless, they all are regarded as very likely lens systems. One of the strong arguments in favor of lensing is the relative ease with which these systems can be modeled. In fact, these models are usually better than one can expect, as they reproduce the flux ratios for image pairs, when one should expect that at least in some systems ML might influence the brightness of one (or more) images.

The known lens systems have yielded no cosmological surprises. Fitting lens models to observations yields values for the ellipticity of the lens, its mass and core radius, but such models are far from uniqueness, as we have seen, except for 0957+561, for which significant limits have been derived, and for MG1131+0456, whose extended image contains much more information about the lens than in systems with a multiply-imaged point source. Model fitting usually proceeds by starting with a 'reasonable' parametrized family of deflectors and adjusting these parameters to fit the observations. Hence, in a certain way, we adjust our prejudices about 'reasonable' lens models to observed lens and image properties. The early hopes of determining the Hubble constant by measuring time delays for image pairs in lens systems have not yet been satisfied, but additional observations may [FA91.2]. The possible accuracy of these measurements ($\sim 20\%$) represents a significant improvement over the 'classical' methods. We discuss further the remaining pitfalls for such measurements in Sect. 13.1. In spite of the expected difficulties, the potential payoff remains so high that GLs should remain high-priority targets for observations.

The known GLs have confirmed that our basic understanding of the universe is a generally self-consistent framework. In particular, lensing is yet another strong argument in favor of the cosmological origin of QSO redshifts. In addition, lensing probes the mass distributions of lenses in a new fashion, different from other methods used to measure masses of extragalactic celestial objects. More than a decade after the first lens was discovered, we can say that the systems we have described fit well with the rest of our standard extragalactic picture.

The prominent arcs found superposed on so many clusters of galaxies are a strong indicator of dark matter in those clusters. Refinements of the initial models and methods now appear capable of achieving the expected potential of GLs as cosmological tools. The key question regarding the accuracy of these methods for measurements of the ratio of dark to luminous matter remains to be answered. However, accuracies of order $\sim 20\%$ might not be too far off, especially when X-ray studies of the matter distribution in clusters are combined with GL studies [KE90.1].

3. Optics in curved spacetime

Notation. In this and the next chapter, we use standard tensor notation. Greek indices refer to spacetime, as in $x^\alpha = (x^0, x^i)$, whereas Roman indices label spatial coordinates and components; $x^0 = ct$. Round and square brackets indicate symmetrization and alternation, respectively, as in $A_{(\alpha\beta)} = \frac{1}{2}(A_{\alpha\beta} + A_{\beta\alpha})$, $B_{[\alpha\beta]} = \frac{1}{2}(B_{\alpha\beta} - B_{\beta\alpha})$. Throughout, we use the Einstein summation convention. The spacetime metric $g_{\alpha\beta}$ is taken to have signature -2, or $(+ - - -)$; the special-relativistic Minkowski metric in orthonormal coordinates is written $\eta_{\alpha\beta} = \mathrm{diag}(1, -1, -1, -1)$. The determinant of $g_{\alpha\beta}$ is denoted by g. Spacetime is always assumed to be time-oriented so that we may speak, e.g, of future-directed vectors, future (half-) lightcones. Covariant derivatives with respect to the Levi-Civita connection $\Gamma^\alpha_{\beta\gamma}$ of $g_{\alpha\beta}$ are indicated by semicolons as in $g_{\alpha\beta;\gamma} = 0$, partial derivatives by commas. The Riemann-, Ricci-, scalar-, and Einstein curvature tensors are defined by $2A_{\alpha;[\beta\gamma]} = A_\delta R^\delta{}_{\alpha\beta\gamma}$, $R_{\alpha\beta} = R^\gamma{}_{\alpha\gamma\beta}$, $R = R^\alpha{}_\alpha$, $G_{\alpha\beta} = R_{\alpha\beta} - \frac{1}{2}R g_{\alpha\beta}$. The Weyl conformal curvature tensor is defined in (3.62). More special notations will be explained where they are used for the first time. We employ electromagnetic units and dimensions according to the system of Heaviside–Lorentz, specialized in Chap. 3 to $c = 1$; see, e.g., [JA75.1], appendix.

Basic references which may be consulted in connection with Chaps. 3 and 4 include [WE72.1], [MI73.1], [ST84.1], [WA84.2].

3.1 The vacuum Maxwell equations

Form of the equations, conformal invariance. In spacetime language, an electromagnetic field is represented as a real, skew symmetric, 2-covariant tensor field $F_{\alpha\beta}$; $F_{\beta\alpha} = -F_{\alpha\beta}$. The relation of $F_{\alpha\beta}$ to the familiar electric and magnetic fields is as follows. Let (x^α) be a coordinate system which is orthonormal at some spacetime point P, and denote by E_a and B_a the cartesian components of **E** and **B** at P, respectively, which would be measured by an observer for whom (x^α) is a local rest system. Then one has, at P,

$$F_{oa} = E_a \ , \qquad F_{ab} = -B_c \ , \tag{3.1}$$

where (a, b, c) form a cyclic permutation of $(1,2,3)$.

Maxwell's equations in a matter-free region, generalized to an arbitrary spacetime M with metric $g_{\alpha\beta}$, are

$$F_{[\alpha\beta;\gamma]} = \frac{1}{3}\left(F_{\alpha\beta;\gamma} + F_{\beta\gamma;\alpha} + F_{\gamma\alpha;\beta}\right) = 0 \quad , \tag{3.2a}$$

$$F^{\alpha\beta}{}_{;\beta} = 0 \quad . \tag{3.2b}$$

Both equations can also be written in terms of ordinary rather than covariant partial derivatives,

$$F_{[\alpha\beta,\gamma]} = 0 \quad , \tag{3.2a'}$$

$$\left(\sqrt{-g}\, F^{\alpha\beta}\right)_{,\beta} = \left(\sqrt{-g}\, g^{\alpha\gamma} g^{\beta\delta} F_{\gamma\delta}\right)_{,\beta} = 0 \quad . \tag{3.2b'}$$

These equations show that the metric acts on the electromagnetic field through its conformal structure only: if $g_{\alpha\beta}$ is replaced by $\tilde{g}_{\alpha\beta} = \Omega^2 g_{\alpha\beta}$, where the scalar field Ω is an arbitrary positive conformal factor, then $\tilde{g}^{\alpha\beta} = \Omega^{-2} g^{\alpha\beta}$ and $\sqrt{-\tilde{g}} = \Omega^4 \sqrt{-g}$; hence, $\sqrt{-g}\, g^{\alpha\gamma} g^{\beta\delta}$ remains unchanged.

Energy-momentum and its "covariant conservation". The distribution of energy and momentum of an electromagnetic field is accounted for by the tensor field

$$T^{\alpha\beta} = F^{\alpha\gamma} F_{\gamma}{}^{\beta} + \frac{1}{4} g^{\alpha\beta} F_{\gamma\delta} F^{\gamma\delta} \quad , \tag{3.3}$$

which is symmetric and trace-free,

$$T^{\alpha\beta} = T^{\beta\alpha} \quad , \qquad T^{\alpha}{}_{\alpha} = 0 \quad , \tag{3.4}$$

and obeys the divergence law[1]

$$T^{\alpha\beta}{}_{;\beta} = 0 \quad , \tag{3.5}$$

which follows from (3.2). The physical meaning of $T^{\alpha\beta}$ can be expressed by equations similar to (3.1). With respect to an orthonormal frame, one has:

$$T^{00} = \frac{1}{2}\left(\mathbf{E}^2 + \mathbf{B}^2\right) = \text{energy density} \quad , \tag{3.6a}$$

$$\left(T^{01}, T^{02}, T^{03}\right) = \mathbf{E} \times \mathbf{B} \tag{3.6b}$$
$$= \text{energy flux density (Poynting vector)}$$
$$= \text{momentum density} \quad ,$$

$$T^{ab} = -E_a E_b - B_a B_b + \frac{1}{2}\delta_{ab}\left(\mathbf{E}^2 + \mathbf{B}^2\right) \tag{3.6c}$$
$$= \text{Maxwell's stress tensor} \quad .$$

In flat spacetime, i.e., in Special Relativity, (3.5) expresses the local conservation of energy and momentum of a free electromagnetic field. In curved

---

[1] Note that, in contrast to $F_{\alpha\beta}$, $T^{\alpha\beta}$ is not conformally invariant; if $g_{\alpha\beta} \to \Omega^2 g_{\alpha\beta}$, then $T^{\alpha\beta} \to \Omega^{-6} T^{\alpha\beta}$. Equation (3.5) is preserved.

spacetime, electromagnetic energy and momentum are not conserved because of the influence of the gravitational field whose "somewhat elusive" energy and momentum [PE66.1] are not unambiguously localizable according to General Relativity. Equation (3.5), which is usually (though misleadingly) called a "covariant conservation law", takes the place of the special-relativistic conservation law in the presence of gravitational fields. We shall see below how (3.5) is related, in the short-wave approximation, to a true conservation law which can be interpreted as the conservation of "photon number".

3.2 Locally approximately plane waves

The short-wave (WKB) approximation. In curved spacetimes, Maxwell's equations cannot be solved explicitly, except in cases of high symmetry; in particular, plane waves do not exist. In many cases of interest, however, especially in GL theory, one is interested in electromagnetic waves which appear, relative to the observers of interest, as nearly plane and monochromatic on a scale large compared with a typical wavelength, but very small compared with the typical radius of curvature of spacetime. Such "locally plane" waves are represented by approximate solutions of Maxwell's equations of the form

$$F_{\alpha\beta} \sim \mathcal{R}e\left\{e^{\frac{i}{\varepsilon}S}\left(A_{\alpha\beta} + \frac{\varepsilon}{i}B_{\alpha\beta}\right) + \mathcal{O}\left(\varepsilon^2\right)\right\} \quad . \tag{3.7}$$

Here, S denotes a real scalar field, $A_{\alpha\beta}$ and $B_{\alpha\beta}$ denote skew-symmetric, complex tensor fields, and $\mathcal{R}e$ indicates that the real part of the following expression is to be taken. The parameter ε is introduced for book-keeping; S, $A_{\alpha\beta}$ and $B_{\alpha\beta}$ are supposed to be independent of ε. By choosing ε sufficiently small, one can make the *phase* $\varepsilon^{-1}S$ change as rapidly as one wants, compared with the *amplitude* $A_{\alpha\beta} + \frac{\varepsilon}{i}B_{\alpha\beta}$. During the calculation ε serves to identify orders of magnitude; at the end, one can absorb ε^{-1} into S and ε into $B_{\alpha\beta}$, or simply put $\varepsilon = 1$.

Expressions such as the right-hand side of (3.7) do, of course, not converge if ε tends to zero, but under rather general assumptions, they can be shown to provide asymptotic approximations to actual solutions, at least in domains where the gradient of S does not vanish (see, e.g., [CO62.1], [MA81.1], [AR89.1]).

For an observer with proper time τ, world line $x^\alpha(\tau)$ and 4-velocity $u^\alpha = \frac{dx^\alpha}{d\tau}$, the *circular frequency* ω of the wave (3.7) with $\varepsilon = 1$ is defined as

$$\omega = -\frac{dS}{d\tau} = -S_{,\alpha}u^\alpha = k_\alpha u^\alpha \quad , \tag{3.8}$$

and

$$k_\alpha = -S_{,\alpha} \tag{3.9}$$

is called the *wave vector* or *frequency vector*.

In the next two subsections, we review the laws imposed on S and $A_{\alpha\beta}$ by requiring the complex field {...} in (3.7) to solve Maxwell's equations asymptotically for $\varepsilon \to 0$. In preparation, we insert the ansatz (3.7) into (3.2) and equate to zero the terms of order ε^{-1} and ε^0, respectively, obtaining the following relations:

$$\text{order } \varepsilon^{-1}: \quad A_{[\alpha\beta}k_{\gamma]} = 0 \quad, \qquad A_{\alpha\beta}k^\beta = 0 \quad, \tag{3.10}$$

$$\text{order } \varepsilon^0: \quad A_{[\alpha\beta;\gamma]} = B_{[\alpha\beta}k_{\gamma]} \quad, \qquad A_\alpha{}^\beta{}_{;\beta} = B_{\alpha\beta}k^\beta \quad. \tag{3.11}$$

Eikonal equation, light rays. Multiplying the first of (3.10) by k^γ, and using the second one, we find $A_{\alpha\beta}k_\gamma k^\gamma = 0$. Thus, if $A_{\alpha\beta}$ is assumed to vanish at most on hypersurfaces, it follows that

$$k_\alpha k^\alpha = 0 \quad, \tag{3.12a}$$

i.e., the wave vector $k^\alpha = -g^{\alpha\beta}S_{,\beta}$ is a null (light-like) vector, and the phase must obey the *eikonal equation*

$$g^{\alpha\beta}S_{,\alpha}S_{,\beta} = 0 \quad. \tag{3.12b}$$

It generalizes the time-dependent eikonal equation of classical optics (see, e.g., [BO80.1], Chap. 3). The hypersurfaces of constant phase or *wavefronts* are thus everywhere tangent to the local light cone.

Since[2] $\mathcal{Re}\left\{e^{\frac{i}{\varepsilon}S}(A_{\alpha\beta}+\ldots)\right\} = \mathcal{Re}\left\{e^{\frac{-i}{\varepsilon}S}(A^*_{\alpha\beta}+\ldots)\right\}$ we may require, without loss of generality, that k^α be future-directed. Then, $\omega > 0$ for any observer.

The integral curves $x^\alpha(v)$ of the vector field k^α, defined by

$$\frac{dx^\alpha}{dv} = k^\alpha = -g^{\alpha\beta}S_{,\beta} \quad, \tag{3.13}$$

are called *light rays*. Because $k^\alpha S_{,\alpha} = 0$, they are contained in the hypersurfaces $S = \text{const}$. (At the same time, k^α is orthogonal to these hypersurfaces. One should remember that the spacetime metric is indefinite and therefore admits self-orthogonal, non-zero vectors.)

Propagation of light rays. Differentiating the tangent vector k^α of a light ray covariantly along the ray and remembering that k_α is a gradient, whence $k_{\alpha;\beta} = k_{\beta;\alpha}$, we infer from (3.12)

$$k^\beta k^\alpha{}_{;\beta} = k^\beta g^{\alpha\gamma}k_{\gamma;\beta} = k^\beta g^{\alpha\gamma}k_{\beta;\gamma}$$
$$= \frac{1}{2}g^{\alpha\gamma}\left(k^\beta k_\beta\right)_{;\gamma} = 0 \quad.$$

---

[2] Here and throughout, the asterisk * denotes complex conjugation.

The result

$$k^\alpha{}_{;\beta} k^\beta = 0 \tag{3.14a}$$

or, written out more explicitly by means of (3.13) and the formula for covariant derivatives in terms of Christoffel symbols $\Gamma^\alpha{}_{\beta\gamma}$,

$$\frac{d^2 x^\alpha}{dv^2} + \Gamma^\alpha{}_{\beta\gamma} \frac{dx^\beta}{dv} \frac{dx^\gamma}{dv} = 0 \quad , \tag{3.14b}$$

means that *light rays are null geodesics*. The parameter v introduced in (3.13) is called an affine parameter since a reparametrization $v \to \tilde{v}$ of a ray preserves the form (3.14b) of the geodesic equation if, and only if, it is an affine transformation, $\tilde{v} = av + b$. Equation (3.13) shows that v has the physical dimension of a squared length. This is in agreement with the following remark: if S obeys the eikonal equation with respect to the metric $g_{\alpha\beta}$, it also does so with respect to the conformally rescaled metric $\tilde{g}_{\alpha\beta} = \Omega^2 g_{\alpha\beta}$, since $\tilde{g}^{\alpha\beta} = \Omega^{-2} g^{\alpha\beta}$. Thus, from (3.13), $\tilde{v} = \int \Omega^2 dv$ is an affine parameter of a ray with respect to the rescaled metric.

Often, it is useful to regard light rays as world lines of particles with light-like four-momenta $\hbar k^\alpha$. We shall see later that this classical "photon" picture is compatible with Maxwell's theory as far as statements concerning energy-momentum transport by locally plane waves are concerned.

Any phase function S obeying the eikonal equation (3.12b) determines a family of light rays. In a region of spacetime where $S_{,\alpha}$ does not vanish, the rays do not intersect; through each event of such a region passes exactly one light ray of the family. Intersections of rays occur along world lines of point sources and at self-intersections of wavefronts; excluded from such regions are also caustics which will be defined in Sect. 3.5 and further considered in Chap. 6. At and near such points, the geometrical optics approximation is not valid, and the mathematical description of wavefronts in terms of smooth functions S on spacetime breaks down there. A systematic treatment of such wavefront singularities will not be attempted in this book (see, e.g., [AR89.1] and [FR83.1], and the references cited therein).

It is easily verified that light rays correspond to those solutions of Hamilton's equations

$$\frac{dx^\alpha}{dv} = \frac{\partial H}{\partial k_\alpha} \quad , \quad \frac{\partial k_\alpha}{\partial v} = -\frac{\partial H}{\partial x^\alpha} \tag{3.15a}$$

for which the value of the Hamiltonian

$$H = \frac{1}{2} g^{\alpha\beta}(x^\gamma) k_\alpha k_\beta \tag{3.15b}$$

vanishes.

According to the theory of partial differential equations, the eikonal equation (3.12b) determines the characteristics (wavefronts) of Maxwell's equations (3.2), and the solution curves $[x^\alpha(v), k_\alpha(v)]$ of (3.15) in (x^α, k_α)-phase space are the bicharacteristics associated with the characteristics

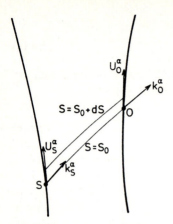

Fig. 3.1. Spacetime diagram frequency shift. Two consecutive light rays connect the worldlines of source and observer. The phases are constant on the rays, which establishes the relation (3.16a) between the proper time intervals $d\tau_S$, $d\tau_O$

[CO62.1]. Not only can rays be found from S via (3.13), but conversely, solutions S of (3.12b) can be constructed from ray-systems obeying (3.15) (sometimes called the method of ray-tracing; see, e.g., [AR89.1]).

Frequency shift. The transport law (3.14a) for the wave vector k^α permits the calculation of the frequency shift between a point source at an event S with 4-velocity u_S^α and an observer at O with 4-velocity u_O^α (see Fig. 3.1). According to the definition (3.8) of ω and the fact that the phase change dS is the same at the source and the observer, one has

$$\frac{\omega_O}{\omega_S} = \frac{(k_\alpha u^\alpha)_O}{(k_\beta u^\beta)_S} = \frac{d\tau_S}{d\tau_O} =: \frac{1}{1+z} \quad , \tag{3.16a}$$

where k_O^α arises from k_S^α, by parallel transport along the ray connecting S to O. A rescaling of S and k_α does not change the ratio (3.16a), so the frequency shift is independent of frequency or wavelength. (The separation of the shift into a "Doppler" and a "gravitational" part is merely conventional and has no intrinsic meaning in General Relativity. The metric field between S and O determines, at the same time, the gravitational field there and the relative velocity between the source and the observer.)

In a spacetime whose metric is conformally stationary – i.e., which admits a coordinate system (t, x^a) with respect to which the components $g_{\alpha\beta}$ of the gravitational potential are independent of the time coordinate t except for an overall scale factor Ω^2 which may depend on all four coordinates, $g_{\alpha\beta}(t, x^a) = [\Omega(t, x^a)]^2 h_{\alpha\beta}(x^a)$ – the eikonal equation (3.12b) obviously has solutions of the form $S = -\tilde{\omega}t + \tilde{S}(x^a)$ with a positive constant $\tilde{\omega}$. Such a phase function represents radiation due to "comoving" sources ($x^a = $ const.). According to (3.8), the proper circular frequency of such radiation with respect to a comoving observer O is

$$\omega_O = \frac{\partial S}{\partial t}\left(\frac{dt}{d\tau}\right)_O = \frac{\tilde{\omega}}{\Omega(t, x_O^a)\sqrt{h_{tt}(O)}} \quad ,$$

where $h_{tt}(O)$ denotes the value of the t,t-component of $h_{\alpha\beta}$ at the position x_O^a of the observer O. Since an analogous formula holds for the source, the frequency ratio is given under these conditions by

$$\frac{\omega_O}{\omega_S} = \frac{\Omega(t_S, x_S^a)}{\Omega(t_O, x_O^a)}\sqrt{\frac{h_{tt}(S)}{h_{tt}(O)}} \quad . \tag{3.16b}$$

If there happen to be several rays connecting a comoving source to a comoving observer, a situation of prime interest to GL theory, the frequency ratios of two "images" of a source can differ only if the corresponding Ω-ratios in (3.16b) differ, due to a change of the scale factor between the times at which those rays were emitted which arrive at the observer simultaneously. If Ω does not change appreciably during this "time delay", as is the case in actual GL systems, the frequency ratios for different images will be equal, since according to (3.16b), only the events of emission and observation determine those ratios, not the ray-paths. If source and observer are not comoving, the frequency shift has to be corrected by local Doppler factors, which are the same for all images in all relevant cases.

The transport of the amplitude and the polarization vector. Let us continue studying the implications of (3.10) and (3.11). If we multiply the first equation of (3.10) by a vector P^γ obeying $k_\gamma P^\gamma = 1$ – there are many such vectors, of course – put $A_\alpha = P^\beta A_{\beta\alpha}$, and take into account the second of (3.10), we obtain an expression for $A_{\alpha\beta}$ in terms of the ray vector k_α and a complex amplitude A_α orthogonal ("transverse") to k^α:

$$A_{\alpha\beta} = 2k_{[\alpha}A_{\beta]} \quad , \qquad A_\alpha k^\alpha = 0 \quad . \tag{3.17}$$

With this result and the foregoing statements, the consequences of the lowest-order WKB equations (3.10) have been exhausted.

Next, we insert $A_{\alpha\beta}$ from the last equation into the first of (3.11). This gives

$$2A_{[\alpha;\beta}k_{\gamma]} = B_{[\alpha\beta}k_{\gamma]} \quad ,$$

an equation of the same form as the first of (3.10), this time for $2A_{[\alpha;\beta]} - B_{\alpha\beta}$ instead of $A_{\alpha\beta}$. Thus, as in the earlier case, we conclude that there exists a vector B_α such that

$$B_{\alpha\beta} = 2\left(A_{[\alpha;\beta]} + B_{[\alpha}k_{\beta]}\right) \quad . \tag{3.18}$$

Putting $A_{\alpha\beta}$ and $B_{\alpha\beta}$ from (3.17) and (3.18), respectively, into the remaining second equation of (3.11), we obtain

$$2A_{\alpha;\beta}k^\beta + A_\alpha k^\beta{}_{;\beta} = k_\alpha\left(B_\beta k^\beta - A^\beta{}_{;\beta}\right) \quad .$$

Now, it is clear from (3.17) that A_α can be replaced by $A_\alpha + fk_\alpha$, for any arbitrary function f. That function can be chosen so as to make the right-hand side of the last equation vanish. Then

$$A_{\alpha;\beta}k^\beta = -\frac{1}{2}A_\alpha k^\beta{}_{;\beta} \quad . \tag{3.19}$$

This equation can be restricted to each ray; it determines how the amplitude A_α is transported along each ray.[3]

Equations (3.17) show that A_α can be changed into $A_\alpha + fk_\alpha$ without affecting $A_{\alpha\beta}$. Under this change, the real scalar $A_\alpha^* A^\alpha$ is invariant. Since it is non-positive – A_α being spacelike or null because of the second equation of (3.17) – we may define a nonnegative *scalar amplitude* a for our field as $a := (-A_\alpha^* A^\alpha)^{1/2}$, and split A_α into a and a complex "unit" vector,

$$A_\alpha = aV_\alpha \quad , \quad a \geq 0 \quad , \quad V_\alpha^* V^\alpha = -1 \quad , \quad V_\alpha k^\alpha = 0 \quad . \tag{3.20}$$

Using this decomposition, we can replace (3.19) equivalently by the two simpler transport laws

$$\dot{V}_\alpha := V_{\alpha;\beta}k^\beta = 0 \tag{3.21a}$$

and

$$\dot{a} = -\frac{1}{2}a\, k^\alpha{}_{;\alpha} \quad . \tag{3.21b}$$

The (complex) *polarization vector* V_α is seen to be parallel-transported along the rays. Since k^α is also parallel on rays, the "transversality" condition $V_\alpha k^\alpha = 0$ is preserved if it holds at one event on a ray.

Summary. The result of the WKB analysis may be summarized as follows [EH67.1]. Any phase function S satisfying the eikonal equation (3.12b), together with a nonnegative scalar amplitude a transported along the rays of S according to (3.21b) and a complex, parallel-propagated, transverse polarization vector V_α, determines (in lowest WKB approximation) a locally plane wave

$$F_{\alpha\beta} = 2a\mathcal{R}e\left\{e^{iS}k_{[\alpha}V_{\beta]}\right\} \quad , \tag{3.22}$$

and any locally plane wave has this structure.[4] The scalar amplitude a and the wave vector $k_\alpha = -S_{,\alpha}$ are uniquely determined by the field strength $F_{\alpha\beta}$, whereas S and V_α are so determined only up to the (gauge) transformations

$$S \to S + s \quad , \quad V_\alpha \to e^{-is}V_\alpha + bk_\alpha \quad , \tag{3.23}$$

where s denotes an arbitrary real constant and b denotes a complex scalar field, constant on rays but otherwise arbitrary. Given an observer (world

---

[3] Equation (3.19) is conformally invariant, the splitting $A_\alpha = aV_\alpha$ and (3.20), (3.21) are not.

[4] In the lowest WKB approximation, (3.22) may be written as $F_{\alpha\beta} = \mathcal{R}e\left(2A_{[\beta;\alpha]}\right)$, where $A_\alpha := iaV_\alpha e^{iS}$ is a complex vector potential in Lorentz gauge.

line) with four velocity u^α, one can use the gauge freedom to make the amplitude V_α spatial for the observer, i.e., set $V_\alpha u^\alpha = 0$. V^α is then orthogonal to the wave normal as "seen" by the observer.

In flat spacetime, one can choose a, k_α and V_α to be constant, with $k_\alpha k^\alpha = k_\alpha V^\alpha = 0$ and $S = -k_\alpha x^\alpha$; this gives the familiar exact plane, monochromatic waves.

The linear, homogeneous ordinary differential equation (3.21b) shows that, if the amplitude a vanishes at one event on a ray, it vanishes all along that ray. Consequently, in the lowest order WKB approximation (in the "limit of vanishing wavelength"), wave packets can have sharp boundaries, i.e., cast sharp shadows. This no longer holds if higher-order amplitudes such as B_α in (3.18) are included. The latter are transported along the rays too, but according to inhomogeneous transport equations whose source terms contain derivatives of the lowest-order variables S, a, V_α. We shall not need these refinements, however.

The preceding results imply that a vacuum gravitational field acts on radiation like a non-dispersive medium. For gravitational lensing, this implies that the spectra, directions and degrees of polarization – of point sources or resolved extended sources – of different images can differ only when their respective ray bundles have been affected differently by intervening matter (e.g., Faraday rotation, absorption).

Energy-momentum transport, photons. The energy-momentum-stress tensor of a locally plane wave is easily obtained from its definition (3.3) and the field (3.22):

$$T^{\alpha\beta} = \frac{a^2}{2} k^\alpha k^\beta \left(1 - \mathcal{Re}\left\{e^{2iS} V_\gamma V^\gamma\right\}\right) \quad .$$

Although this expression was obtained from an approximate solution of Maxwell's equations, it obeys the "conservation law" (3.5) exactly if a, S, k_α and V_α satisfy the WKB laws (3.12b, 9, 20, 21). The term $\mathcal{Re}\{\ldots\}$ represents rapid oscillations transverse to the phase hypersurfaces; averaging over a wavelength provides the "effective" energy tensor

$$T^{\alpha\beta}_{\text{eff}} = \frac{a^2}{2} k^\alpha k^\beta \quad . \tag{3.24}$$

It is suggestive to write

$$T^{\alpha\beta}_{\text{eff}} = N^\alpha P^\beta \quad , \tag{3.25a}$$

to interpret $P^\alpha := \hbar k^\alpha$ as the 4-momentum of a photon, and

$$N^\alpha := \frac{a^2}{2\hbar^2} P^\alpha \tag{3.25b}$$

as the photon number 4-current density. The law

$$N^\alpha{}_{;\alpha} = 0 \quad , \tag{3.26}$$

equivalent to (3.21b), then implies (via Gauss' integral theorem) that *the number of photons in a ray bundle is conserved and observer-independent*, as required by a classical photon picture, whereas (3.14a) says that *the 4-momentum of each photon is transported parallely along the photon's world line*. (The possibility of such a particle-interpretation within classical electrodynamics rests on the fact that, for a Maxwellian wave packet, the "actions" energy/frequency and momentum/wavenumber are observer-independent, conserved, adiabatic invariants [EH59.1].) One can treat an incoherent radiation field as a photon gas whose state is given by a distribution function on phase space. This description is particularly useful to account for emission, absorption, and scattering processes (see, e.g., [EH73.1], [LI66.1]).

3.3 Fermat's principle

Light rays (in vacuo) can be characterized either as the null solutions of the geodesic equation (3.14b) or as the solutions of Hamilton's equations (3.15). Alternatively, one may use Lagrange's equation with the Lagrangian

$$\mathcal{L}\left(x^{\alpha}, \dot{x}^{\beta}\right) = \frac{1}{2} g_{\alpha\beta}(x^{\gamma}) \dot{x}^{\alpha} \dot{x}^{\beta} \tag{3.27}$$

and, again, pick out those solutions along which $\mathcal{L}$ vanishes. This amounts to saying that light rays are those extremals of the action principle

$$\delta \left\{ \frac{1}{2} \int g_{\alpha\beta} \dot{x}^{\alpha} \dot{x}^{\beta} dv \right\} = 0 \tag{3.28}$$

along which $\mathcal{L}$ equals zero. (If one drops the conditions $H = 0$, $\mathcal{L} = 0$ respectively, one obtains timelike and spacelike geodesics.)

The intuitively most appealing and, for GL theory, most useful way to characterize light rays, however, is given by the following theorem.

Theorem (Fermat's principle). [5] Let S be an event ("source") and ℓ a timelike world line ("observer") in a spacetime $(M, g_{\alpha\beta})$. Then a smooth null curve γ from S to ℓ is a light ray (null geodesic) if, and only if, its arrival time τ on ℓ is stationary under first-order variations of γ within the set of smooth null curves from S to ℓ (see Fig. 3.2),

$$\delta\tau = 0 \quad . \tag{3.29}$$

Note that (i) this version of Fermat's principle does not refer to the "time" a light ray needs to travel from the source to the observer – which in general has no intrinsic meaning in General Relativity – but states a

---

[5] This relativistic version of Fermat's priciple is due to Kovner [KO90.1], its proof was given by Perlick [PE90.2] who quotes earlier results. Fermat formulated his principle in 1661.

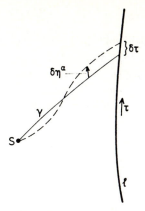

Fig. 3.2. Geometry of Fermat's principle. See the theorem above (3.29)

stationarity property of the (invariant) *time of arrival* at the observer who, in contrast to the assumptions made in classical optics, may be moving relative to the source, in a time-dependent (metric) optical field; (ii) no preferred parameters on ℓ or γ enter the theorem; on ℓ one may use proper time τ or any monotonic function of it; and (iii) the assertion is conformally invariant.

One may also interpret the theorem using the opposite time-orientation: ℓ is then to be considered as the worldline of a source, and one asks for light rays emitted by that source which arrive simultaneously at an observer.

Proof: The necessity of this condition is easily established. Indeed, let ℓ be given by $\xi^\alpha(\tau)$ with 4-velocity $u^\alpha = \frac{d\xi^\alpha}{d\tau}$, and let $\eta^\alpha(v, \varepsilon)$ describe a family of null curves $\gamma(\varepsilon)$ from S to ℓ, so that $0 \leq v \leq 1$, $|\varepsilon| < \varepsilon_0$ (for some small positive ε_0), $\eta^\alpha(0, \varepsilon) = x^\alpha(S)$, $\eta^\alpha(1, \varepsilon) = \xi^\alpha(\tau(\varepsilon))$, and, for all v and ε, $\dot\eta_\alpha \dot\eta^\alpha = 0$. Here and during the following proof we use the dot to indicate covariant differentiation with respect to v on $\gamma(\varepsilon)$. Then, by the standard integration by parts of variational calculus, and with the abbreviations $A =$ arrival event, $\delta := \frac{d}{d\varepsilon}()_{\varepsilon=0}$ and $\int \ldots \equiv \int_0^1 dv \ldots$, one has

$$0 = \delta \int \frac{1}{2} \dot\eta_\alpha \dot\eta^\alpha = (\dot\eta_\alpha u^\alpha)_A \delta\tau - \int \ddot\eta_\alpha \delta\eta^\alpha \quad . \tag{3.30}$$

Now, if $\gamma(0)$ is geodesic and, without loss of generality, affinely parametrized, we have $\ddot\eta^\alpha = 0$; since $\dot\eta_\alpha$ is lightlike and u^α timelike and both are future directed, $(\dot\eta_\alpha u^\alpha)_A$ is positive. Therefore, $\delta\tau = 0$.

Sufficiency is trickier to prove. Let us first find out the conditions which a variational vector field $\delta\eta^\alpha$ on $\gamma(0)$ must satisfy. Varying $\dot\eta_\alpha \dot\eta^\alpha = 0$ gives

$$\dot\eta_\alpha (\delta\eta^\alpha)^{\cdot} = 0 \quad . \tag{3.31a}$$

In addition, of course, we must have

$$\delta\eta^\alpha(0) = 0 \quad , \qquad \delta\eta^\alpha(1) \text{ is parallel to } u_A^\alpha \quad . \tag{3.31b}$$

Let us assume for a moment that, given a vector field $\delta\eta^\alpha$ on a null curve $\gamma(0)$ with the properties (3.31), there exists a family $\{\gamma(\varepsilon)\}$ of varied curves having that $\delta\eta^\alpha$ as its "derivative". Then, we can prove sufficiency rather simply: we take an arbitrary smooth vector field w^α on $\gamma(0)$ which vanishes at the end points, $w^\alpha(0) = w^\alpha(1) = 0$, and we parallel-transport u_A^α "backwards" on $\gamma(0)$, obtaining $u^\alpha(v)$. Then

$$\delta\eta^\alpha(v) := w^\alpha(v) - u^\alpha(v) \int_0^v \frac{\dot{w}_\beta(\bar{v})\dot{\eta}^\beta(\bar{v})}{u_\gamma(\bar{v})\dot{\eta}^\gamma(\bar{v})} d\bar{v}$$

satisfies (3.31). It is therefore, according to our preliminary assumption, a "possible" variation. If $\delta\tau = u_\alpha(1)\delta\eta^\alpha(1) = 0$ for all such variations, then $\delta\eta^\alpha(1) = 0$ (due to the argument given above), i.e.

$$\int_0^1 \dot{w}_\beta \frac{\dot{\eta}^\beta}{u_\gamma\dot{\eta}^\gamma} dv = -\int_0^1 w_\beta \left(\frac{\dot{\eta}^\beta}{u_\gamma\dot{\eta}^\gamma}\right)^{\cdot} dv = 0 \quad .$$

Since w_β is arbitrary, this implies constancy of $\dot{\eta}^\alpha/u_\gamma\dot{\eta}^\gamma$ on $\gamma(0)$, i.e., $\gamma(0)$ has a covariantly constant direction, whence it is a geodesic. So far, so good. It remains to show that, given a vector field $\delta\eta^\alpha$ on $\gamma(0)$ obeying (3.31), there actually exists a corresponding varied family $\{\gamma(\varepsilon)\}$ of null curves. To do that, we assume for simplicity that a neighborhood of the null curve $\gamma(0)$ can be covered by a coordinate system x^α such that, (i) the x^0-curves are future directed timelike, and (ii) the timelike curve ℓ is given by $x^a = \xi^a = $ const.[6] To construct $\eta^\alpha(v, \varepsilon)$, we take

$$\eta^a(v, \varepsilon) = \eta^a(v) + \varepsilon\delta\eta^a(v) \tag{3.32}$$

and define $\eta^0(v, \varepsilon)$ as the solution of the differential equation

$$g_{\alpha\beta}(\eta^a, \eta^0) \dot{\eta}^\alpha \dot{\eta}^\beta = 0 \tag{3.33}$$

with the initial value $\eta^0(0, \varepsilon) = x^0(S)$; in (3.33) the η^a of equation (3.32) are to be inserted so that (3.33) becomes a first order ordinary differential equation $\dot{\eta}^0 = F(\eta^0, v, \varepsilon)$. The solvability of equation (3.33) for $\dot{\eta}^0$ and the smoothness of F are guaranteed by our assumptions (i) and (ii) made above. Inspection shows that the functions $\eta^\alpha(v, \varepsilon)$ constructed in this way do define a smooth family of null curves $\gamma(\varepsilon)$ containing the given curve $\gamma(0)$, connecting S with ℓ. By construction, the "a-components" of the variational vector field $\bar{\delta}\eta^\alpha$ of this family equal the given ones, $\delta\eta^a$. Since, in addition, $\bar{\delta}\eta^0(0) = \delta\eta^0(0) = 0$ and both $\delta\eta^\alpha$ and $\bar{\delta}\eta^\alpha$ obey (3.31a), it also follows that $\bar{\delta}\eta^0 = \delta\eta^0$. This remark concludes the proof.

---

[6] For a proof without this simplifying assumption, see [PE90.2].

We remark without proof that the arrival time τ of a light ray γ is not only a stationary value, but in fact a strict local minimum, if there is no point conjugate to S on γ (including A).[7] If there is such a point before γ reaches A, τ corresponds to a saddle point (see [PE90.2]). Under the assumptions of gravitational lensing considered in Chap. 5 there always exists a light ray which arrives at ℓ first; that ray is free of conjugate points and minimizes τ (which is not trivial since it refers to non-geodesic null curves). If there is more than one light ray or "image", the arrival time functional also has saddle points.

Fermat's principle in conformally stationary spacetimes. A special case of Fermat's theorem is of particular interest in connection with the standard approximation so far used in GL theory; it concerns conformally stationary spacetimes, i.e., spacetimes whose physical metric $\widetilde{ds}^2$ is conformal to a stationary (time-independent) metric ds^2:

$$\widetilde{ds}^2 = \Omega^2 ds^2 \quad , \quad \Omega > 0 \quad , \tag{3.34}$$

$$ds^2 = e^{2U}\left(dt - w_i dx^i\right)^2 - e^{-2U} d\ell^2 \quad , \tag{3.35}$$

$$d\ell^2 = \gamma_{ij} dx^i dx^j \quad . \tag{3.36}$$

In these equations, U, w_i, γ_{ij} denote functions of the spatial coordinates x^i only, and $d\ell^2$ is a spatial Riemannian (positive definite) metric. The conformal factor Ω may depend on all four coordinates. Since curves which are lightlike or lightlike geodesics with respect to $\widetilde{ds}^2$ have these properties also with respect to ds^2, one may apply Fermat's theorem to ds^2 to find the light rays of $\widetilde{ds}^2$. On a future-directed null curve, $ds^2 = 0$ gives

$$dt = w_i dx^i + e^{-2U} d\ell \quad . \tag{3.37}$$

If, at emission, $t = 0$ (without loss of generality) and if we consider a stationary observer, i.e., one with worldline $x^i = \text{const.}$, parametrized by t, the arrival time of a null curve whose spatial projection $\tilde{\gamma}$ is given by functions $x^i(\lambda)$, is

$$t = \int_{\tilde{\gamma}} \left(w_i dx^i + e^{-2U} d\ell\right) \quad . \tag{3.38}$$

Thus under these assumptions, Fermat's principle reduces to

$$\delta \int_{\tilde{\gamma}} \left(w_i dx^i + e^{-2U} d\ell\right) = 0 \quad , \tag{3.39}$$

where the spatial paths $\tilde{\gamma}$ are to be varied with fixed endpoints ([PH62.1], [BR73.1]). This version of Fermat's principle is formally identical with the classical one if

---

[7] Conjugate points are defined at the end of Sect. 3.5.

$$n = \mathrm{e}^{-2U} + w_i \frac{dx^i}{d\ell} \tag{3.40}$$

is considered as a (position and direction-dependent) effective index of refraction, and $d\ell$ is viewed as the geometrical arc length. In the even more special (conformally) static case, $w_i = 0$, the vacuum behaves like an isotropic, non-dispersive medium with index

$$n = \mathrm{e}^{-2U} \; . \tag{3.41}$$

The law (3.39) can also be stated as follows: the spatial paths of light rays are geodesics with respect to the Finsler metric $w_i \, dx^i + \mathrm{e}^{-2U} \, d\ell$, which is Riemannian if $w_i = 0$. Note that in the static case, $\mathrm{e}^{-2U} \, d\ell$ is *not* the spatial part of the physical arc length, $\mathrm{e}^{-U} \, d\ell$. This difference is the origin of "the factor 2" in the light deflection law. These results are useful not only in GL theory, but whenever one is concerned with geometrical optics in stationary or static gravitational fields, e.g., near the rotating Earth. Applications of Fermat's principle will be given in later sections of this book.

3.4 Geometry of ray bundles

3.4.1 Ray systems and their connection vectors

Ray systems. We now study the system of rays associated with a fixed phase function S. Let each ray be parametrized by an affine parameter v, and let the rays be labeled by three parameters y^a. (For example, y^1 could be the phase value of the ray, y^2, y^3 could be angles distinguishing "photons" travelling in different directions.) The ray system is then given by

$$x^\alpha = f^\alpha(v, y^a) \; , \tag{3.42}$$

where

$$k^\alpha = \frac{\partial f^\alpha}{\partial v} = -g^{\alpha\beta} S_{,\beta} \tag{3.43}$$

is the frequency vector. The parametrization is not unique; changes of the form

$$y^a = g^a\left(\tilde{y}^b\right) \; , \quad v = \tilde{v} + h\left(\tilde{y}^b\right) \; ,$$
$$x^\alpha = f^\alpha\left(\tilde{v} + h\left(\tilde{y}^b\right), g^a\left(\tilde{y}^b\right)\right) = \tilde{f}^\alpha\left(\tilde{v}, \tilde{y}^b\right) \tag{3.44}$$

are allowed.

Connection vectors. Consider the ray γ given by particular values y^a. The rays infinitesimally near γ are given by $y^a + \delta y^a$, and the vectors

$$\delta x^\alpha = \frac{\partial f^\alpha}{\partial y^a} \delta y^a \tag{3.45}$$

connect γ with its neighbors. A reparametrization (3.44) implies the change

$$\begin{aligned}
\widetilde{\delta x}^\alpha &= \frac{\partial \tilde{f}^\alpha}{\partial \tilde{y}^b} \delta \tilde{y}^b = \frac{\partial f^\alpha}{\partial y^a} \frac{\partial g^a}{\partial \tilde{y}^b} \delta \tilde{y}^b + \frac{\partial f^\alpha}{\partial v} \frac{\partial h}{\partial \tilde{y}^b} \delta \tilde{y}^b \\
&= \frac{\partial f^\alpha}{\partial y^a} \delta y^a + k^\alpha \delta h \quad , \\
\widetilde{\delta x}^\alpha &= \delta x^\alpha + k^\alpha\, \delta h
\end{aligned} \tag{3.46}$$

of connection vectors. Note that δx^α and $\widetilde{\delta x}^\alpha$ connect the same nearby rays (see Fig. 3.3).

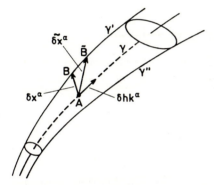

Fig. 3.3. A bundle of light rays. The vectors δx^α, $\widetilde{\delta x}^\alpha$ connect the central ray with tangent k^α to one of its neighbors

Beam properties which are valid for all observers. Two rays connected by δx^α belong to the same phase if, and only if,

$$k_\alpha \delta x^\alpha = 0 \quad , \tag{3.47}$$

a parametrization-independent property. Let us consider two rays γ_1, γ_2 close to the "central" ray γ of a narrow ray bundle. If $\gamma, \gamma_1, \gamma_2$ all have the same phase, then according to (3.46) and (3.47), the inner product

$$g_{\alpha\beta} \widetilde{\delta x_1}^\alpha \widetilde{\delta x_2}^\beta = g_{\alpha\beta} \delta x_1^\alpha \delta x_2^\beta \tag{3.48}$$

at an event on γ is uniquely determined by $\gamma, \gamma_1, \gamma_2$, independent of the parametrization used.

The relations (3.47) and (3.48) can be interpreted as follows. Suppose a (point) observer O with 4-velocity u^α is hit by the ray γ at an event P.

Then, according to (3.46), a parametrization can be chosen such that, at P, all vectors δx^α connecting γ to its neighbors satisfy

$$u_\alpha \delta x^\alpha = 0 \quad , \tag{3.49}$$

i.e., they connect events which are simultaneous for O. Then, (3.47) is equivalent to the 3-vector equation $\mathbf{k} \cdot \delta \mathbf{x} = 0$ in the 3-space of O. Thus, the intersection of the phase hypersurface $S = $ const. containing γ with the instantaneous 3-space of the observer O at P, the wave surface "seen" by O, is orthogonal to the ray direction $\mathbf{k}$; it consists of precisely those rays ("photons") close to γ which arrive at the observer simultaneously with γ. This holds for all observers at P, irrespective of their velocities; in particular the invariance of (3.47) under reparametrizations implies that two neighboring photons reach all or none of these observers simultaneously [SA61.1]. We shall henceforth call such a collection of photons a *beam*. Equation (3.48) can then be interpreted as follows: the size and shape of a cross-section of an infinitesimal beam at an event P is the same for all observers at P [KE32.1].

3.4.2 Optical scalars and their transport equations

Transport law for connection vectors. In view of the preceding results it is natural to ask how the sizes, shapes and orientations of cross-sections of a beam change along the central ray γ. To find the answer we note that, as a consequence of (3.43) and (3.45), we have for constant δy^a:

$$\frac{\partial k^\alpha}{\partial y^b} \delta y^b = \frac{\partial^2 f^\alpha}{\partial y^b \partial v} \delta y^b = \frac{\partial}{\partial v} \delta x^\alpha \quad .$$

Since k^α may also be considered as depending on x^β rather than on (v, y^a), we may rewrite this as

$$\frac{\partial}{\partial v} \delta x^\alpha = k^\alpha{}_{,\beta} \delta x^\beta \quad .$$

Adding the same Christoffel symbol terms on both sides, we obtain the covariant transport equation

$$(\delta x^\alpha)^{\cdot} := \frac{D}{dv} \delta x^\alpha = k^\alpha{}_{;\beta} \delta x^\beta \tag{3.50}$$

for connection vectors of a ray bundle [which is invariant under reparametrizations (3.46) because of (3.14a)]. It implies that (3.47) is preserved along a ray γ; i.e., those rays which form a beam at one event P of the central ray γ, do so also at other events on γ.

Optical scalars. Owing to the observer-independence of the beam properties derived above, we may, in investigating the consequences of the transport law (3.50), choose observers whose 4-velocities u^α are parallel along our central ray γ. It is also convenient to choose, at each event of γ, a pair $E^\alpha_{(i)}$ of unit vectors, orthogonal to each other as well as to k^α and u^α, and to put

$$\varepsilon^\alpha := E^\alpha_{(1)} + iE^\alpha_{(2)} \quad ; \tag{3.51}$$

the $E^\alpha_{(i)}$ may be visualized as spanning a screen in the three-space of the observer, orthogonal to the ray direction. The $E^\alpha_{(i)}$ and thus ε^α are also to be taken as parallel on γ. Finally, we choose the affine parameter on γ such that $k_\alpha u^\alpha = 1$ everywhere on γ. We then have the algebraic relations

$$k_\alpha k^\alpha = k_\alpha E^\alpha_{(i)} = u_\alpha E^\alpha_{(i)} = k_\alpha \varepsilon^\alpha = u_\alpha \varepsilon^\alpha = E^{(1)}_\alpha E^\alpha_{(2)} = \varepsilon_\alpha \varepsilon^\alpha = 0 \quad ,$$
$$k_\alpha u^\alpha = u_\alpha u^\alpha = E^{(i)}_\alpha E^\alpha_{(i)} = \frac{1}{2}\varepsilon^*_\alpha \varepsilon^\alpha = 1 \quad , \tag{3.52}$$

and at each event of γ, the tensor

$$P^\alpha{}_\beta = \delta^\alpha{}_\beta + k^\alpha k_\beta - k^\alpha u_\beta - u^\alpha k_\beta = \frac{1}{2}\left(\varepsilon^\alpha \varepsilon^*_\beta + \varepsilon^{*\alpha}\varepsilon_\beta\right) \tag{3.53}$$

projects orthogonally onto the space orthogonal to u^α and k^α, spanned by $E^\alpha_{(1)}$ and $E^\alpha_{(2)}$. Using these tools, we re-express $k_{\alpha;\beta} = k_{\rho;\mu}\delta^\rho{}_\alpha\delta^\mu{}_\beta$, using the first, and then the second of (3.53) together with the null and geodesic properties of k_α:

$$\begin{aligned}k_{\alpha;\beta} &= k_{\rho;\mu}\left(P^\rho{}_\alpha - k^\rho k_\alpha + k^\rho u_\alpha + u^\rho k_\alpha\right)\left(P^\mu{}_\beta - k^\mu k_\beta + k^\mu u_\beta + u^\mu k_\beta\right) \\ &= k_{\rho;\mu}P^\rho{}_\alpha P^\mu{}_\beta + P_{(\alpha}k_{\beta)} \\ &= \frac{1}{2}\mathcal{R}e\left\{\varepsilon_\alpha\varepsilon_\beta k_{\rho;\mu}\varepsilon^{*\rho}\varepsilon^{*\mu} + \varepsilon_{(\alpha}\varepsilon^*_{\beta)}k_{\rho;\mu}\varepsilon^{*\rho}\varepsilon^\mu\right\} + P_{(\alpha}k_{\beta)} \\ &= \sigma_{\alpha\beta} + \theta P_{\alpha\beta} + P_{(\alpha}k_{\beta)} \quad ,\end{aligned} \tag{3.54}$$

where P_α is of no interest and θ, σ, the *optical scalars*, and $\sigma_{\alpha\beta}$ are defined by

$$\theta = \frac{1}{2}k^\alpha{}_{;\alpha} \quad , \qquad \sigma = \frac{1}{2}k_{\alpha;\beta}\varepsilon^{*\alpha}\varepsilon^{*\beta} \quad , \qquad \sigma_{\alpha\beta} = \mathcal{R}e(\sigma\varepsilon_\alpha\varepsilon_\beta) \quad . \tag{3.55}$$

Geometrical interpretation of optical scalars. The rate of change of the distance ℓ between two neighboring rays of a beam now follows from (3.50) as

$$(\ell^2)^{\cdot} = \left(g_{\alpha\beta}\delta x^\alpha \delta x^\beta\right)^{\cdot} = 2k_{\alpha;\beta}\delta x^\alpha \delta x^\beta \quad ,$$

or, if we write $\delta x^\alpha = \ell e^\alpha$ and use (3.54),

$$\frac{d\ell}{\ell\, dv} = \theta + \sigma_{\alpha\beta}e^\alpha e^\beta \quad . \tag{3.56}$$

The first term on the right-hand side of this equation corresponds to an isotropic expansion or contraction of a beam, the second one describes a distortion or "shear" whereby a spherical cross-section is deformed into an elliptical one. The *shear tensor* $\sigma_{\alpha\beta}$ is trace-free, and its magnitude is easily computed by means of (3.52) and (3.55) to be

$$\frac{1}{2}\sigma_{\alpha\beta}\sigma^{\alpha\beta} = |\sigma|^2 , \tag{3.57a}$$

which in turn can be expressed in terms of k_α by means of (3.54) and the orthogonality and trace relations, with the result

$$|\sigma|^2 = \frac{1}{2}k_{\alpha;\beta}k^{\alpha;\beta} - \frac{1}{4}(k^\alpha{}_{;\alpha})^2 . \tag{3.57b}$$

Since the (essentially 2-dimensional) shear tensor is symmetric and trace-free, it has two eigenvalues with opposite signs, given, according to (3.57a), by $\pm|\sigma|$. Thus, we can infer from (3.56) the extremal values (with respect to the direction e^α) of the rate of stretching:

$$\max \frac{d\ell}{\ell\, dv} = \theta + |\sigma| , \qquad \min \frac{d\ell}{\ell\, dv} = \theta - |\sigma| .$$

Consequently, 2θ measures the relative rate of change, with respect to v, of the area A of the cross-section of an infinitesimal beam of light rays,

$$k^\alpha{}_{;\alpha} = 2\theta = \frac{dA}{A\, dv} , \tag{3.58}$$

and

$$|\sigma| = \frac{1}{2}\left(\max \frac{d\ell}{\ell\, dv} - \min \frac{d\ell}{\ell\, dv}\right) \tag{3.59}$$

measures its rate of distortion, or *shear*. The phase of σ determines, according to (3.56), the orientation of the principal axes of the shear tensor relative to the reference basis $E^\alpha_{(i)}$ used to form ε^α.

Equation (3.58) allows a simple geometrical interpretation of the transport law (3.21b) for the amplitude a of a locally plane wave: it follows from these two equations that $a\sqrt{A}$ is constant for any thin beam of rays. This is nothing other than a version of the law (3.26) of the conservation of photon number.

Transport of optical scalars. With the representations (3.55) of θ and σ, one can obtain transport equations for them. From Ricci's identity,

$$k_{\beta;\gamma\delta} - k_{\beta;\delta\gamma} = R_{\alpha\beta\gamma\delta}k^\alpha , \tag{3.60}$$

which characterizes the curvature tensor $R_{\alpha\beta\gamma\delta}$, we obtain

$$\dot{\theta} = \frac{1}{2}k^\alpha{}_{;\alpha\beta}k^\beta = \frac{1}{2}k^\alpha{}_{;\beta\alpha}k^\beta - \frac{1}{2}R_{\alpha\beta}k^\alpha k^\beta$$

$$= -\frac{1}{2}k^\alpha{}_{;\beta}\, k^\beta{}_{;\alpha} - \frac{1}{2}R_{\alpha\beta}k^\alpha k^\beta ;$$

and, remembering (3.55) and (3.57b),

$$\dot{\theta} + \theta^2 + |\sigma|^2 = -\frac{1}{2} R_{\alpha\beta} k^\alpha k^\beta \quad . \tag{3.61}$$

A similar derivation for σ, based on (3.55) and the parallelism of ε^α on γ (used here for the first and only time), yields

$$\dot{\sigma} + 2\theta\sigma = -\frac{1}{2} R_{\alpha\beta\gamma\delta} \, \varepsilon^{*\alpha} k^\beta \varepsilon^{*\gamma} k^\delta \quad .$$

At this point, it is useful to employ the decomposition of the curvature tensor into its Weyl and Ricci parts,

$$R_{\alpha\beta\gamma\delta} = C_{\alpha\beta\gamma\delta} + g_{\alpha[\gamma} R_{\delta]\beta} - g_{\beta[\gamma} R_{\delta]\alpha} - \frac{1}{3} R \, g_{\alpha[\gamma} g_{\delta]\beta} \quad . \tag{3.62}$$

The orthogonality relations (3.52) show that only the Weyl part contributes to the transport of σ,

$$\dot{\sigma} + 2\theta\sigma = -\frac{1}{2} C_{\alpha\beta\gamma\delta} \, \varepsilon^{*\alpha} k^\beta \varepsilon^{*\gamma} k^\delta \quad . \tag{3.63}$$

The vanishing of $C_{\alpha\beta\gamma\delta}$ is necessary and sufficient for the spacetime to be (locally) conformally flat [SC21.1]. Except for a few special cases – see [KR80.1], Sect. 22.2 – spacetimes satisfying Einstein's gravitational field equation (4.1) with a physically reasonable source are *not* conformally flat. Therefore, (3.63) shows that *in generic spacetimes, almost all lightbeams exhibit some shear.* For an interesting cosmological implication of this fact, see [KR66.1]. The same equation also shows that, in general, even in vacuo the principal axes of shear are not parallel along a ray, in contrast to the situation in classical geometrical optics.

The non-linear system (3.61 & 63) of coupled transport equations for the optical scalars θ and σ [SA61.1] is fundamental for the analysis of light propagation in general, curved spacetimes. (In contrast to its individual terms, the system is conformally invariant, as are null geodesics.) We can summarize its meaning as follows [PE66.1]. The effect of (the trace-free part of) Ricci-curvature on a beam is to introduce *anastigmatic positive focusing*[8], like a magnifying lens, whereas the effect of conformal curvature is to produce *purely astigmatic focusing*.

The focusing equation. Using (3.58), we may transform (3.61) into a slightly simpler equation for $\sqrt{A}$,

$$\ddot{\sqrt{A}} = -\left(|\sigma|^2 + \frac{1}{2} R_{\alpha\beta} k^\alpha k^\beta\right) \sqrt{A} \quad , \tag{3.64}$$

which will be used in Chap. 4 to discuss the influence of matter on light propagation. Hence, shearing always contributes to the focusing of a beam.

---

[8] because of the field equation (4.1) and the positivity of energy density

3.5 Distances based on light rays. Caustics

Luminosity distance and angular-diameter distance. In cosmology, "distance" is not a simple, directly measurable quantity. Instead, several theory-dependent distance concepts have been defined, whose values can be determined only indirectly. Two kinds of distances which are intimately linked with the propagation of radiation are defined as follows. Consider a thin beam of light rays emanating from a source-event S and reaching an observation-event O and its neighborhood (see Fig. 3.4). We know from the previous subsection that the area dA_O of the cross-section of the beam at O is well-defined independently of an assignment of a 4-velocity at O. On the other hand, in order to measure the size of the beam at S in terms of a solid angle $d\Omega_S$, it is necessary to take into account a 4-velocity U_S^α at S, and to measure $d\Omega_S$ in the tangent 3-space orthogonal to U_S^α. The *corrected luminosity distance* of the source-at-S-with-4-velocity-U_S^α from the observation-event O is defined as

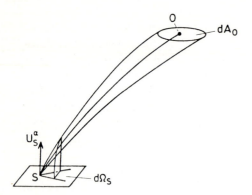

Fig. 3.4. A light beam from the source event S to the observation event O

$$D(U_S^\alpha, O) := \left(\frac{dA_O}{d\Omega_S}\right)^{1/2} . \tag{3.65}$$

The name will be motivated in the next subsection. Similarly, interchanging the roles of source and observer, one defines (see Fig. 3.5a)

$$D(U_O^\rho, S) := \left(\frac{dA_S}{d\Omega_O}\right)^{1/2} , \tag{3.66}$$

the *distance from apparent solid-angular size* of S from (O, U_O^ρ).

The dependence of this distance on the 4-velocity of the observer, given the events S and O, is due to the phenomenon of *aberration*. In fact, one infers from Fig. 3.5b that, if k^α is any vector tangent to the ray SO at O

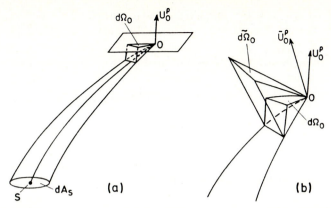

Fig. 3.5. (a) A beam of light rays from a surface element dA_S of the source at S with vertex at the observer O, of size $d\Omega_O$, used to define the angular diameter distance of S from (O, U_O^ρ). (b) The solid angles $d\Omega$, $\widetilde{d\Omega}$ of the beam reaching O depend on the four-velocities U_O^ρ, $\widetilde{U_O^\rho}$; "aberration", see text

and U^ρ, $\tilde{U}^\rho$ are two 4-velocities at O with corresponding solid angles $d\Omega$, $\widetilde{d\Omega}$, then $\frac{d\Omega}{\widetilde{d\Omega}} = \frac{(k_\rho U^\rho)^2}{(k_\rho \tilde{U}^\rho)^2}$. An analogous formula holds for solid angles at S, of course. Needless to say, the "d's" (dA_O etc.) in (3.65) and (3.66) are supposed to indicate that the limits of infinitely thin beams are to be taken.

If two events S, O are connected by a light ray and 4-velocities U_S^α, U_O^α are given, both distances (3.65) and (3.66) are defined. In any spacetime, they are related by the *reciprocity-relation* [ET33.1]

$$D(U_S^\alpha, O) = (1+z)D(U_O^\rho, S) \quad ; \tag{3.67}$$

here, z denotes the redshift of the source as seen by the observer, $1 + z = \omega_S/\omega_O$.

Next, we show how these distances can in principle be computed, and prove the remarkable law (3.67).

Geodesic deviation, Jacobi vectors and distances. Near an event S, the 2-parametric set of light rays starting at S generates a conical hypersurface, the future light (half-)cone $\mathcal{C}_S^+$. Let any smooth, one-parametric subfamily of these rays be given by $x^\alpha = f^\alpha(v, y)$, where y labels the rays and v is an affine parameter on each ray. We put

$$k^\alpha = \frac{\partial f^\alpha}{\partial v} \quad , \quad Y^\alpha = \frac{\partial f^\alpha}{\partial y} \quad , \tag{3.68}$$

so that $\delta x^\alpha = \delta y \, Y^\alpha$ "connects nearby rays", as in (3.45). The same reasoning which led from (3.43) to (3.50) [where the second of (3.43) was not used] gives

$$\frac{DY^\alpha}{dv} = \frac{Dk^\alpha}{dy} , \qquad (3.69)$$

where $\frac{D}{dv}$ and $\frac{D}{dy}$ denote covariant differentiation along the curves $y = $ const. and $v = $ const., respectively. Since the curves $y = $ const. are affinely parametrized geodesics and $k_\alpha k^\alpha = 0$, (3.69) implies

$$\frac{d}{dv}(k_\alpha Y^\alpha) = k_\alpha \frac{DY^\alpha}{dv} = k_\alpha \frac{Dk^\alpha}{dy} = \frac{1}{2}\frac{d}{dy}(k_\alpha k^\alpha) = 0 \; ; \qquad (3.70)$$

hence, $k_\alpha Y^\alpha$ is constant on each ray. At the vertex S, where we may put $v = 0$, all rays intersect, so that there $Y^\alpha = 0$. Thus,

$$k_\alpha Y^\alpha = 0 \qquad (3.71)$$

everywhere on $\mathcal{C}_S^+$. As long as neighboring rays do not intersect again or touch each other to first order, the connection vectors at an event P of a ray span a two-dimensional spacelike plane orthogonal to the ray direction k^α. Since the vectors $\lambda Y^\alpha + \mu k^\alpha$ form the tangent space to $\mathcal{C}_S^+$ at P, (3.71) shows that k^α is the normal to the light cone at P, which in turn implies that $\mathcal{C}_S^+$ is a lightlike hypersurface. If neighboring rays form envelopes or intersect, this is no longer true; then, $\mathcal{C}_S^+$ develops *caustics*, which will be considered further at the end of this section.

So far, the scales of the affine parameters on the various rays have not been restricted (apart from smoothness). Let us henceforth require that, at S, $U_\alpha k^\alpha = U_\alpha \frac{dx^\alpha}{dv} = \omega_S$ for some source 4-velocity U^α and for all ray tangents, with a fixed circular frequency ω_S. Then, for all parameter values y and for $v = 0$, $U_\alpha \frac{Dk^\alpha}{dy} = 0$ and, by (3.69), $U_\alpha \frac{DY^\alpha}{dv} = 0$. Thus, remembering (3.70), we have, writing $\frac{DY^\alpha}{dv} = \dot{Y}^\alpha$, that

$$\dot{Y}_S^\alpha \text{ is orthogonal to } k^\alpha \text{ and } U_S^\alpha \; . \qquad (3.72)$$

Henceforth, we consider a ray γ from S to an event O and a narrow beam containing γ. Since $U_\alpha dx^\alpha = \omega_S dv = d\ell$ is the (infinitesimal) distance of a point on γ from S in the source's rest frame, the vector $\frac{1}{\omega_S}\dot{Y}_S^\alpha \delta y = \frac{DY^\alpha}{d\ell}\delta y = \delta\theta_S^\alpha$ at S, which by (3.72) is contained in the source's rest space and tangent to the wave surface, connects γ with a neighboring ray γ'; its magnitude is the angular separation of these rays (see Fig. 3.6).

To obtain a transport equation for Y^α along the ray γ, we differentiate (3.69) and use Ricci's identity and the geodesic law, $\frac{Dk^\alpha}{dv} = 0$:

$$\frac{D^2 Y^\alpha}{dv^2} = \frac{D^2 k^\alpha}{dv\, dy} = \frac{D^2 k^\alpha}{dy\, dv} + R^\alpha{}_{\beta\gamma\delta} k^\beta k^\gamma Y^\delta \; ,$$

i.e.,

$$\ddot{Y}^\alpha = R^\alpha{}_{\beta\gamma\delta} k^\beta k^\gamma Y^\delta \; . \qquad (3.73)$$

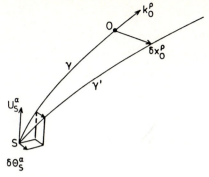

Fig. 3.6. The deviation of neighboring light rays from source S to observer O; see text

This linear, ordinary differential equation for the propagation of a deviation vector Y^α, or *Jacobi vector*, on a ray γ, is called the *geodesic deviation equation* or Jacobi equation. It is fundamental for the investigation of focusing properties of bundles of geodesics. The linearity of (3.73) implies that the value of a solution at the end point O depends linearly on the initial values at S; thus, for $Y_S^\alpha = 0$ one has

$$Y_O^\rho = J^\rho{}_\alpha(O,S) \frac{1}{\omega_S} \dot{Y}_S^\alpha \tag{3.74a}$$

or, equivalently,

$$\delta x_O^\rho = J^\rho{}_\alpha(O,S) \, \delta\theta_S^\alpha \,. \tag{3.74b}$$

Here and in the following argument, it is useful to denote indices referring to tensors at S by letters $\alpha, \ldots$ from the beginning of the Greek alphabet, and those referring to O by $\rho, \ldots$ The "Jacobi map" $J^\rho{}_\alpha(O,S)$ depends on both events O and S; it maps vectors at S into vectors at O, as the notation is intended to indicate. $J^\rho{}_\alpha(O,S)$ depends functionally on the curvature tensor along the ray SO and can be computed iteratively from it (see, e.g., [SY64.1]).[9]

If the angular separation vector $\delta\theta_S^\alpha$ ranges over the two-dimensional space corresponding to the solid angle $d\Omega_S$ of a beam, its image δx_O^ρ according to (3.74b) and (3.72), sweeps out a two-dimensional, spacelike area at O whose size dA_O measures the cross-section of the beam for any observer at O, as was shown in Sect. 3.4 in connection with (3.48). One can refer $J^\rho{}_\alpha$ to orthonormal bases in the spacelike planes at S and O spanned by $\delta\theta_S^\alpha$ and δx_O^ρ, respectively. Let J denote the resulting 2×2 matrix. A look at (3.74b) and Fig. 3.6 then shows that the corrected luminosity distance is given by

---

[9] Let v_{OS} denote the affine distance of O from S. Then the linear map $(\omega_S v_{OS})^{-1} J^\rho{}_\alpha = H^\rho{}_\alpha$ is dimensionless and depends on O and S only; it is the derivative of the exponential map, frequently used in differential geometry.

$$D(U_S^\alpha, O) = \sqrt{|\det J|} \ . \tag{3.75}$$

A similar formula holds, of course, for $D(U_O^\rho, S)$; the roles of S and O must be simply interchanged. In this case one has to use the map $J^\alpha{}_\rho(S,O)$,

$$Z_S^\alpha = J^\alpha{}_\rho(S,O) \frac{1}{\omega_O} \dot{Z}_O^\rho \ , \tag{3.76a}$$

$$\delta x_S^\alpha = J^\alpha{}_\rho(S,O) \, \delta\theta_O^\rho \ . \tag{3.76b}$$

According to the last equation, a linear segment of length $\delta\ell_S$ in the source subtends an angle $\delta\theta_O$ at the observer, where

$$\frac{\delta\ell_S}{\delta\theta_O} = \sqrt{J_{\alpha\rho} J^\alpha{}_\sigma E^\rho E^\sigma} \ ,$$

if that segment is seen at O in the direction of the unit vector **E**. Since beams in general exhibit shear – recall (3.56), (3.59) – $\delta\theta_O$ depends not only on $\delta\ell_S$, but also on **E**. The term "angular diameter distance" for $D(U_O^\rho, S)$ is thus somewhat misleading, since it suggests that this quantity generally relates one-dimensional angles at O to linear distances at S, which is not the case. Keeping this in mind, we shall nevertheless use the expression "angular diameter distance".

In order to link the two maps $J^\rho{}_\alpha(O,S)$ and $J^\alpha{}_\rho(S,O)$, and thus the two distances, to each other, one must find a relation between Jacobi vector fields Y^α which vanish at S and those, Z^α (say), which vanish at O. Any two Jacobi vectors Y^α, Z^α obey

$$\left(Z_\alpha \dot{Y}^\alpha - Y_\alpha \dot{Z}^\alpha \right)^{\cdot} = Z_\alpha \ddot{Y}^\alpha - Y_\alpha \ddot{Z}^\alpha = 0 \ ,$$

due to the symmetry $R_{\alpha\beta\gamma\delta} = R_{\gamma\delta\alpha\beta}$ of the curvature tensor. Hence, $Z_\alpha \dot{Y}^\alpha - Y_\alpha \dot{Z}^\alpha = \text{const.}$ Therefore, if $Y_S^\alpha = 0$ and $Z_O^\rho = 0$, then $\left(Z_\alpha \dot{Y}^\alpha \right)_S = -\left(Y_\rho \dot{Z}^\rho \right)_O$ or, by (3.74a) and (3.76),

$$\dot{Y}_\alpha^S J^\alpha{}_\rho \frac{c\dot{Z}_O^\rho}{\omega_O} = -\dot{Z}_\rho^O J^\rho{}_\alpha \frac{c\dot{Y}_S^\alpha}{\omega_S} \ .$$

Since the initial values $\dot{Y}_S^\alpha$, $\dot{Z}_O^\rho$ are arbitrary, it follows that

$$\frac{1}{\omega_O} g_{\alpha\beta} J^\beta{}_\rho = -\frac{1}{\omega_S} g_{\rho\sigma} J^\sigma{}_\alpha \ . \tag{3.77}$$

(Here, $g_{\alpha\beta}$ is the metric at S, $g_{\rho\sigma}$ that at O.) With respect to orthonormal bases, the 2×2 matrices representing $\frac{1}{\omega_O} J^\alpha{}_\rho$ and $\frac{-1}{\omega_S} J^\rho{}_\alpha$ are, therefore, transposes of each other. This fact, combined with formula (3.75) and its analog for $D(U_O^\rho, S)$, establishes (3.67).

The two distances $D(U_S^\alpha, O)$ and $D(U_O^\rho, S)$, while positive for O close to S and initially monotonically increasing if O "recedes" from S, may after a while decrease and even (both) vanish at some $O \neq S$, due to focusing of light beams. When this happens, the points S and O are said to be *conjugate* with respect to the geodesic connecting them. It follows from the preceding consideration that S and O are conjugate if, and only if, there exists a non-identically vanishing Jacobi field which vanishes at S and O. Conjugacy is a conformally invariant relation.

Caustics. The future (half) light "cone" $\mathcal{C}_S^+$ of an event S, generated by future-directed null geodesics originating at S, is a smooth, cone-like hypersurface near S. "Later", however, it in general develops several sheets which may touch or intersect each other. (This, after all, is why General Relativity predicts gravitational lensing.) The set of points conjugate to S on the various null geodesics starting at S is called the *caustic* of $\mathcal{C}_S^+$; a similar definition applies to the past (half) cones of the source and the observer, $\mathcal{C}_S^-$ and $\mathcal{C}_O^-$, respectively. In general, the caustic of $\mathcal{C}_S^+$ consists of several two-dimensional surfaces bounded by one-dimensional edges and/or vertices. When a beam originating at S passes through, or touches the caustic at P, its cross-section degenerates into a line segment or a point there. In the first case, the conjugate point P is said to be simple, in the second case, it is said to have multiplicity two. The multiplicity of a point O conjugate to S equals the co-rank of the Jacobi map $J^\rho{}_\alpha(O,S)$. Visualizing the behaviour of a beam one recognizes that the orientation of its cross-section changes in passing through a simple conjugate point, but remains unchanged in passing through a "double" conjugate point.

The structure of singularities (caustics, self-intersections) of generic wavefronts – in particular, light "cones", wavefronts due to point sources – in arbitrary spacetimes has been discussed and illustrated in detail in [FR83.1].

3.6 Luminosity, flux and intensity

The radiation emitted by a point-source with world line ℓ_S travels on the null geodesics starting at events on ℓ_S. The corresponding phase-hypersurfaces are the null (half-) cones with vertices on ℓ_S.

The number of photons emitted during the proper time interval $d\tau_S$ into the solid angle $d\Omega_S$ with energy in the range $\hbar d\omega_S$ equals

$$d\tau_S \, d\Omega_S \, d\omega_S \, \frac{L_{\omega_S}}{4\pi\omega_S} \quad ,$$

according to the definition of the specific luminosity L_ω, provided the source radiates isotropically. These photons form a thin ray bundle. According to the photon number conservation law derived in Sect. 3.2, these same photons pass through an area dA_O orthogonal to the ray direction in an observer's 3-space in proper time $d\tau_O$ with energies in the range $\hbar d\omega_O$; their number is

$$d\tau_O \, dA_O \, d\omega_O \, \frac{S_{\omega_O}}{\omega_O} \quad,$$

where S_ω denotes the specific flux measured by the observer. Equating the two expressions, and using the redshift relation (3.16a) and the definition (3.65) of the corrected luminosity distance, now abbreviated as D'_L,

$$S_\omega = \frac{L_{(1+z)\omega}}{4\pi(1+z) \, D'^2_L} \quad . \tag{3.78}$$

Integration over all frequencies gives the (total) flux

$$S = \frac{L}{4\pi(1+z)^2 \, D'^2_L} =: \frac{L}{4\pi \, D^2_L} \tag{3.79}$$

in terms of the bolometric luminosity L. This last equation motivates the name of D'_L, viz. (redshift-) corrected luminosity distance; $D_L = \left(\frac{L}{4\pi S}\right)^{1/2}$ is called the ("uncorrected") luminosity distance. Writing now D_A for the angular diameter distance and using (3.67), we note

$$D_L = (1+z)D'_L = (1+z)^2 D_A \quad . \tag{3.80}$$

For an *extended source*, considered as an assembly of incoherently radiating point sources, one can combine (3.78) with the definition (3.66) of the angular diameter distance D_A and (3.80) to obtain the formula

$$\frac{dS_\omega}{d\Omega_O} = \frac{dL_{(1+z)\omega}}{4\pi dA_S} \frac{1}{(1+z)^3} \quad .$$

Recalling the definition of the specific intensity I_ω as

$$I_\omega = \frac{\text{radiative energy}}{\text{area} \times \text{time} \times \text{solid angle} \times \text{frequency range}} \quad ,$$

we rewrite the preceding equation as

$$I_\omega(O) = \frac{I_{(1+z)\omega}(S)}{(1+z)^3} \quad ; \tag{3.81}$$

it relates the specific intensity at the observer, $I_\omega(O)$, to that at the source, $I_{(1+z)\omega}(S)$. The last result may also be expressed by saying that *in any non-interacting radiation field, the ratio I_ω/ω^3 is observer-independent and constant on each ray.*

Integration on ω gives

$$I(O) = \frac{I(S)}{(1+z)^4} \quad , \tag{3.82}$$

the relativistic generalization of the law of constancy of the surface brightness. The remarkable simplicity and generality of the laws (3.81), (3.82) is due to Etherington's reciprocity relation (3.67).

We emphasize that (3.78–82) follow from the WKB-approximated, covariant Maxwell equations in an arbitrary spacetime; the photon-picture was used only for ease of expression.

Generalizations of the foregoing equations which take into account absorption and emission along the ray can be found, e.g., in [EL71.1]. At places where the WKB approximation fails, e.g., at focal points, the foregoing equations have to be replaced by wave-optical refinements; see Sect. 4.7.

4. Derivation of the lens equation

As discussed in Chap. 2, the gravitational lens phenomenon is important mainly for distant, extragalactic sources ($z \gtrsim 1$) and deflectors ($z \gtrsim 0.05$). The corresponding theory must, therefore, account for (i) the bending and distortion of light bundles by nearly isolated, bound masses, and (ii) the imbedding of sources, deflectors, and observers into their cosmological environment, how this affects the propagation of light, and, in particular, the relations between observables of sources and lenses. The first three sections of this chapter are devoted to the first topic. The fourth section contains a summary of those aspects of cosmology which are relevant for lens theory, together with some background material. On this basis, the second topic is treated in Sects. 4.5 and 4.6. Finally, in Sect. 4.7 we consider some aspects of wave optics, in preparation for Chap. 7.

4.1 Einstein's gravitational field equation

The metric in General Relativity (GR). According to the general theory of relativity, the metric of spacetime plays a double role. On the one hand, it represents the *metric* in the original sense: it defines, via its field of null cones, the local distinction between space and time ("causal structure"), and it determines spatial and temporal distances and local inertial frames. On the other hand, it also determines, through its connection (Christoffel symbols, covariant derivative operator), the geodesics, interpreted as the world lines of freely falling test particles and light rays, i.e., it serves as the *gravitational potential*. By assigning this double role to $g_{\alpha\beta}$, the identity ("equivalence") of spacetime geometry, inertia and gravity is built into the theory. This dual role implies that the metric must be considered not as given once and for all as in special relativity, but as a dynamical field coupled to matter.

The field equation. In Chap. 3 we used the metric in its first, kinematical role only, as if it were a given, external field. Now we take into account its second role by adopting *Einstein's gravitational field equation*

$$G^{\alpha\beta} := R^{\alpha\beta} - \frac{1}{2} R g^{\alpha\beta} = \frac{8\pi G}{c^4} T^{\alpha\beta} , \qquad (4.1)$$

which relates the Einstein tensor $G^{\alpha\beta}$, respectively the Ricci tensor $R^{\alpha\beta}$, to the stress-energy-momentum tensor $T^{\alpha\beta}$ of matter and non-gravitational fields. $R = R^\alpha_\alpha = g^{\alpha\beta} R_{\alpha\beta}$ denotes the trace of $R^{\alpha\beta}$, G denotes Newton's constant of gravity, and c the speed of light in vacuo. The Ricci tensor is obtained by contraction from the Riemann–Christoffel curvature tensor $R^\alpha{}_{\beta\gamma\delta}$,

$$R_{\alpha\beta} = R^\gamma{}_{\alpha\gamma\beta} \;, \tag{4.2}$$

and that in turn is defined in terms of the connection coefficients (Christoffel symbols), which do *not* form a tensor, by

$$R^\alpha{}_{\beta\gamma\delta} = -2 \left(\Gamma^\alpha_{\beta[\gamma,\delta]} + \Gamma^\epsilon_{\beta[\gamma} \Gamma^\alpha_{\delta]\epsilon} \right) \;. \tag{4.3}$$

Finally, the connection coefficients are given in terms of the metric by the equations

$$\Gamma^\alpha_{\beta\gamma} = \frac{1}{2} g^{\alpha\delta} (g_{\delta\beta,\gamma} + g_{\delta\gamma,\beta} - g_{\beta\gamma,\delta}) \;. \tag{4.4}$$

Here, and in similar equations, a comma followed by an index denotes partial differentiation with respect to the coordinate labeled by that index.

In analogy with Newton's theory of gravity, one may interpret $g_{\alpha\beta}, \Gamma^\alpha_{\beta\gamma}, R^\alpha{}_{\beta\gamma\delta}$ as (the components of) the gravitational potential, field strength and field gradient (tidal field), respectively. Then, $G^{\alpha\beta}$ is analogous to the result of applying the Laplace-operator to the potential, and (4.1) takes the place of Poisson's equation. (These correspondences can be made rigorous by formalising the Newtonian limit of GR; see, e.g., [EH81.1], [EH86.2], [LO88.1].)

Energy tensors. The field equation (4.1) attains a physical meaning only if the matter tensor $T^{\alpha\beta}$ is specified. In vacuo, $T^{\alpha\beta} = 0$ or, if one wishes to take into account a "cosmological term", perhaps derivable from quantum field theory as an effective energy tensor of the vacuum, $T^{\alpha\beta}_\Lambda = \frac{\Lambda c^4}{8\pi G} g^{\alpha\beta}$. The interpretation of $T^{\alpha\beta}$ is generally, as in the special case of the electromagnetic field – compare (3.6) – as follows: in any local inertial frame, T^{00} represents the energy density, the spatial vector cT^{0i} represents the energy flux density which equals $c^2 \times$ momentum density (Planck's law), and $-T^{ij}$ represents the (spatial) stress tensor which equals the momentum flux density (these quantities refer to matter, whose energy-momentum is considered as localizable in GR, not to the gravitational field).

For most astrophysical purposes, one idealises bulk matter as a *perfect fluid*, for which

$$T^{\alpha\beta} = \left(\varrho c^2 + p \right) U^\alpha U^\beta - p g^{\alpha\beta} \;, \tag{4.5}$$

where ϱ denotes the mass density (which includes the mass-equivalents of short range interaction and thermal energies) and p the pressure, both measured by a comoving observer, and U^α is the 4-velocity, normalized to one,

$$g_{\alpha\beta}U^\alpha U^\beta = 1 \quad. \tag{4.6}$$

Another important example of an energy tensor is that of an electromagnetic field given in (3.3) and, in leading WKB-approximation for a locally plane wave, in (3.24).

Three important differences between Newton's and Einstein's theory are worth recalling:

(a) the operations leading from the potential to the field, (4.4), and from the latter to the tidal field, (4.3), are both nonlinear in GR, in contrast to the analogous equations in Newton's theory;

(b) the time-evolution of the metric is governed by hyperbolic equations, i.e., changes of the gravitational field (gravitational waves) propagate with the (finite) speed c, unlike Newtonian forces;

(c) since the Einstein tensor obeys the contracted Bianchi identity $G^{\alpha\beta}{}_{;\beta} = 0$, the field equation (4.1) implies the divergence law (3.5) which, although it is not a conservation law [see the comments following (3.6)], restricts the motion of material sources and leads, at least for simple models of matter and in combination with approximation methods, to equations of motion for those sources.

4.2 Approximate metrics of isolated, slowly moving, non-compact matter distributions

The linearized field equation and its retarded solution. Suppose a metric $g_{\alpha\beta}$ differs little from the flat, Minkowskian metric $\eta_{\alpha\beta} = \text{diag}(1,-1,-1,-1)$ in orthonormal coordinates $x^0 = ct$, $\mathbf{x} = (x^i)$. Then, one can write

$$g_{\alpha\beta} = \left(1 - \frac{1}{2}h\right)\eta_{\alpha\beta} + h_{\alpha\beta} \quad,$$
$$h := \eta^{\alpha\beta}h_{\alpha\beta} \quad, \quad |h_{\alpha\beta}| \ll 1 \quad. \tag{4.7}$$

In linear approximation with respect to the metric deviation components $h_{\alpha\beta}$, one can, without loss of generality, choose the coordinates such that the coordinate gauge condition

$$h^{\alpha\beta}{}_{,\beta} = 0 \tag{4.8}$$

is satisfied.[1] Then the gravitational field equation (4.1), linearized in $h_{\alpha\beta}$, reads

---

[1] Note that, in this approximation, the indices of $h_{\alpha\beta}$ may be raised by means of the background metric $\eta^{\alpha\beta}$ rather than with $g^{\alpha\beta}$.

$$\left(\Delta - \frac{1}{c^2}\frac{\partial^2}{\partial t^2}\right)h^{\alpha\beta} = \frac{16\pi G}{c^4}T^{\alpha\beta} \quad . \tag{4.9}$$

Its solution for an isolated source without incoming gravitational radiation is the retarded one,

$$h^{\alpha\beta}(t,\mathbf{x}) = \frac{-4G}{c^4}\int\frac{T^{\alpha\beta}\left(t - \frac{|\mathbf{y}|}{c}, \mathbf{x}+\mathbf{y}\right)}{|\mathbf{y}|}d^3y \quad , \tag{4.10}$$

provided a possible Λ-term is negligible.

Specialisation to slowly moving, perfect fluid sources. We now assume a matter tensor of the form (4.5) and assume that, (a) matter moves slowly with respect to the coordinate system x^α, i.e., $v^i := \frac{dx^i}{dt}$ obeys $|\mathbf{v}| \ll c$, and (b) $|p| \ll \varrho c^2$. Then, (4.5) implies

$$T^{00} \approx \varrho c^2 \quad , \quad T^{0i} \approx c\varrho v^i \quad , \quad T^{ij} \approx \varrho v^i v^j + p\delta^{ij} \quad , \tag{4.11}$$

where $\approx$ indicates that terms of relative order $\frac{v^2}{c^2}, \frac{p}{\varrho c^2}$ have been neglected. If we now introduce the retarded potentials

$$U(t,\mathbf{x}) := -G\int\frac{\varrho\left(t - \frac{|\mathbf{y}|}{c}, \mathbf{x}+\mathbf{y}\right)}{|\mathbf{y}|}d^3y \quad ; \tag{4.12a}$$

$$\mathbf{V}(t,\mathbf{x}) := -G\int\frac{(\varrho\mathbf{v})\left(t - \frac{|\mathbf{y}|}{c}, \mathbf{x}+\mathbf{y}\right)}{|\mathbf{y}|}d^3y \quad , \tag{4.12b}$$

we obtain from (4.7) and (4.10–12) the metric

$$ds^2 = g_{\alpha\beta}dx^\alpha dx^\beta$$
$$\approx \left(1 + \frac{2U}{c^2}\right)c^2dt^2 - 8c\,dt\frac{\mathbf{V}\cdot d\mathbf{x}}{c^3} - \left(1 - \frac{2U}{c^2}\right)d\mathbf{x}^2 \quad . \tag{4.13}$$

(In this approximation, the stresses T^{ij} do not affect the metric.) In the near zone of a system of slowly moving bodies, the retardation in (4.12) can be neglected; there,

$$U(t,\mathbf{x}) \approx -G\int\frac{\varrho(t,\mathbf{x}+\mathbf{y})}{|\mathbf{y}|}d^3y \quad , \tag{4.14a}$$

$$\mathbf{V}(t,\mathbf{x}) \approx -G\int\frac{(\varrho\mathbf{v})(t,\mathbf{x}+\mathbf{y})}{|\mathbf{y}|}d^3y \quad . \tag{4.14b}$$

The "post-Minkowskian" metric (4.13) satisfies the weak-field condition $|h_{\alpha\beta}| \ll 1$ if, and only if, in addition to assumptions (a) and (b) above, the Newtonian potential U of the mass distribution ϱ obeys

$$|U| \ll c^2 \quad ; \tag{4.15a}$$

then

$$\left|\frac{\mathbf{V}}{c^3}\right| \lesssim \left|\frac{\mathbf{v}}{c}\right| \cdot \left|\frac{U}{c^2}\right| \ll 1 \quad . \tag{4.15b}$$

(For spherical bodies, the first inequality implies $\frac{2GM}{c^2} = R_S \ll R$; hence, compact objects as black holes and neutron stars have to be excluded.)

Since the gravitational vector potential $\mathbf{V}$ is smaller than U/c^2 by one order in v/c, (4.13) shows that, in the near zone, the metric at one instant t is completely determined in lowest order by the density at the same instant, just like the potential in Newton's theory.

Equations of motion. In order to obtain an approximate solution of the field equation, the gauge condition (4.8) also has to be solved approximately. If one insists on (4.8) exactly, then (4.9) requires the special-relativistic conservation law $T^{\alpha\beta}{}_{,\beta} = 0$ for matter, i.e., matter would have to move without "feeling" the gravitational field. To resolve this apparent paradox, one has to consider the linear approximation leading to the metric (4.13) as the first step of an iteration procedure. Within such a scheme, it turns out to be correct in lowest order to compute the motion of matter by means of the Newton-Euler equations of motion, and to use the metric (4.13) to determine, in lowest ("first post-Minkowskian") order, the behavior of clocks, test particles and light rays (see, e.g., the review [DA87.1] and the references therein). This method is used, in particular, in gravitational lens theory, to determine the deflection of light rays by local concentrations of matter.

4.3 Light deflection by quasistationary, isolated mass distributions

Effective index of refraction. We now assume that during the time a light ray interacts essentially with a local distribution of matter, the configuration of that matter does not change significantly. Then we may consider the metric (4.13) as stationary and apply the version of Fermat's principle given in (3.39) and (3.40). This leads to

$$n = 1 - \frac{2U}{c^2} + \frac{4}{c^3}\mathbf{V}\cdot\mathbf{e} \tag{4.16}$$

as the effective index of refraction of the gravitational field, where $dl = |d\mathbf{x}|$ denotes the Euclidean arc length, and $\mathbf{e} := \frac{d\mathbf{x}}{dl}$ the unit tangent vector of a ray.[2]

---

[2] Note that the vacuum gravitational field acts as a non-dispersive medium; gravitational "lenses" are achromatic.

The equation for the spatial light paths now follows as the Euler-Lagrange equation of the variational principle $\delta \int n \, dl = 0$. Using ordinary vector notation, we obtain

$$\frac{d\mathbf{e}}{dl} = \frac{-2}{c^2}\nabla_\perp U + \frac{4}{c^3}\mathbf{e}\times(\nabla\times\mathbf{V}) \ . \tag{4.17}$$

Here, $\nabla_\perp U := \nabla U - \mathbf{e}(\mathbf{e}\cdot\nabla U)$ denotes the projection of ∇U onto the plane orthogonal to the direction $\mathbf{e}$ of the ray. This equation can also be derived, though with more work, from the geodesic law (3.14b).

The first, Coulomb-type contribution corresponds to an attraction towards the deflecting mass. The second part is due to the gravitomagnetic field produced, according to GR, by moving matter (mass currents); it is related to the famous, but so far not directly observed "dragging of inertial frames" by moving matter, particularly by rotating bodies (see, e.g., [IB83.1] for calculations concerning a rotating "lens"). Owing to its smallness, even compared to the already small first term, the gravitomagnetic term appears to be unobservable, and we therefore neglect it after (4.19).

The deflection angle. Defining the deflection "angle" $\hat{\boldsymbol{\alpha}}$ as the difference of the initial and final ray direction,

$$\hat{\boldsymbol{\alpha}} := \mathbf{e}_{\text{in}} - \mathbf{e}_{\text{out}} \ , \tag{4.18}$$

we obtain from (4.17)

$$\hat{\boldsymbol{\alpha}} = \frac{2}{c^2}\int \nabla_\perp U \, dl - \frac{4}{c^3}\int \mathbf{e}\times(\nabla\times\mathbf{V})dl \ . \tag{4.19}$$

Since, in general, the second order, nonlinear differential equation (4.17) for the ray $\mathbf{x}(l)$ cannot be solved explicitly, (4.19) would seem to be of little use since one has to integrate along the ray. Under most realistic conditions, however, the deflection is very small and occurs essentially in a region whose size is a few times $d + \delta$, where d is the diameter of the deflecting mass and δ is the smallest distance of the ray from that mass. In the special case of a point mass, $U(\mathbf{x}) = -\frac{GM}{|\mathbf{x}|}$, integration over the unperturbed ray, $\mathbf{x}(l) = \boldsymbol{\xi} + l\mathbf{e}$ (with the impact vector $\boldsymbol{\xi}$ orthogonal to the tangent vector $\mathbf{e} = \mathbf{e}_{\text{in}}$) leads to the "Einstein angle"

$$\hat{\boldsymbol{\alpha}} = \frac{4GM}{c^2}\frac{\boldsymbol{\xi}}{|\boldsymbol{\xi}|^2} \ . \tag{4.20}$$

Let us now assume not only that the total deflection angle $\hat{\boldsymbol{\alpha}}$ is very small, but that the extent L of the deflecting mass in the direction of the incoming ray is so small that the value of the transverse gravitational field strength $\nabla_\perp U$ on the actual ray deviates but little from that on the unperturbed (straight) ray, i.e., that the maximal deviation $\Delta s_{\text{max}} \sim \hat{\alpha} L$ of the ray is small compared to the length scale on which the field changes,

$$|\Delta s_{\text{max}} \nabla_\perp \nabla_\perp U| \ll |\nabla_\perp U| \ . \tag{4.21}$$

Then, one can again integrate in (4.19) over the unperturbed ray, $\mathbf{e} \equiv \mathbf{e}_{in}$, and the deflection angle due to such a *geometrically-thin lens* is equal to the sum of the Einstein angles (4.20) due to the mass elements of the lens. Accordingly, all mass elements in an infinitesimal cylindrical tube parallel to $\mathbf{e}$ have the same impact vector. We may therefore project all mass elements of the lens onto a plane orthogonal to $\mathbf{e}$, passing through some (largely arbitrary) "center" of the lens, and characterize the deflector by the resulting surface mass density $\Sigma(\boldsymbol{\xi})$. Thus,

$$\hat{\boldsymbol{\alpha}}(\boldsymbol{\xi}) = \frac{4G}{c^2} \int_{\mathbb{R}^2} \frac{(\boldsymbol{\xi} - \boldsymbol{\xi}')\Sigma(\boldsymbol{\xi}')}{|\boldsymbol{\xi} - \boldsymbol{\xi}'|^2} d^2\xi' \quad , \tag{4.22}$$

where the integral now is over the *lens plane* just introduced, and $\boldsymbol{\xi}$ is a two-dimensional vector in that plane. This expression for the deflection angle, which was given in Sect. 2.2 already, is one of the basic formulae of gravitational lens theory.

Arrival time and Fermat potential. Instead of first using Fermat's principle to derive the photon path equation (4.17) and then to obtain the deflection law (4.22) from the latter, we now follow a different method, which is of interest mainly because of its generalization to cosmology, to be given in Sect. 4.6 below. This time, we start from Fermat's principle in its original form, i.e., the principle of stationary arrival time. We assume the (point) source, the deflecting mass distribution and the observer to be parts of an isolated system and to be at rest with respect to the coordinate system $(t, \mathbf{x})$ we are using. Again, we assume the lens to be geometrically thin and the deflection to be small. Motivated by these assumptions, we stipulate that the actual ray path can be approximated by combining its incoming and outgoing asymptotes (compare Fig. 2.2). The principle of stationary arrival time then serves to single out, from among all kinematically possible broken ray paths, the best approximations to actual paths.

Let a light signal be emitted at the source S at time $t = 0$, proceed rectilinearly to the deflection point I near the lens, and then go straight on to the observation point O. According to the metric (4.13) (in which we again neglect $\mathbf{V}$), it will arrive at time

$$t = c^{-1} \int \left(1 - \frac{2U}{c^2}\right) dl = c^{-1}l - 2c^{-3} \int U \, dl \quad , \tag{4.23}$$

where l is the Euclidean length of the path SIO, and the "potential term" is to be integrated along that path. (The reality of that "Shapiro time delay" has been established in the Solar System to an accuracy of about 10^{-3} [RE79.1].) From Fig. 2.2, we read off

$$\begin{aligned} l &= \sqrt{(\boldsymbol{\xi} - \boldsymbol{\eta})^2 + D_{ds}^2} + \sqrt{\boldsymbol{\xi}^2 + D_d^2} \\ &\approx D_{ds} + D_d + \frac{1}{2D_{ds}}(\boldsymbol{\xi} - \boldsymbol{\eta})^2 + \frac{1}{2D_d}\boldsymbol{\xi}^2 \quad , \end{aligned} \tag{4.24}$$

where the distances D_d, D_{ds}, and $D_s = D_d + D_{ds}$ refer to the Euclidean background metric.

To determine the potential term, we first compute the integral of the potential U of a point mass at the origin of the lens plane along a straight ray from a source S at η to an image point I at $\boldsymbol{\xi}$. The result is

$$\int_S^I U\, dl = GM \left[\ln \frac{|\boldsymbol{\xi}|}{2D_{ds}} + \frac{\boldsymbol{\xi} \cdot (\boldsymbol{\eta} - \boldsymbol{\xi})}{|\boldsymbol{\xi}| D_{ds}} + \mathcal{O}\left(\left(\frac{\boldsymbol{\eta} - \boldsymbol{\xi}}{D_{ds}}\right)^2\right) \right] \quad . \quad (4.25a)$$

Under the conditions of lensing, we may approximate this by

$$\int_S^I U\, dl = GM \ln \frac{|\boldsymbol{\xi}|}{2D_{ds}} \quad ; \quad (4.25b)$$

the dependence on η is negligible. If D' denotes a distance small compared to D_{ds} but large compared to the relevant impact parameters $|\boldsymbol{\xi}|$, the last expression can be decomposed as

$$GM \left(\ln \frac{|\boldsymbol{\xi}|}{2D'} + \ln \frac{D'}{D_{ds}} \right) \quad , \quad (4.25c)$$

in which the first part is due to the ray contained in a slab of thickness D' above the lens plane, while the second part is due to the ray outside this slab. Only the first part, which arises in a neighborhood of size D' of the deflector, depends on $\boldsymbol{\xi}$.

Adding to the foregoing result an analogous term for a ray from I to the observer O, decomposing it as in (4.25c) and using the linearity of U in the mass distribution, we get for the potential time delay the expression

$$\frac{-2}{c^3} \int U\, dl = \frac{-4G}{c^3} \int d^2\xi'\, \Sigma(\boldsymbol{\xi}')\, \ln\left(\frac{|\boldsymbol{\xi} - \boldsymbol{\xi}'|}{D'} \right) + \text{const.} \quad . \quad (4.26)$$

Of course, D' may be replaced by an arbitrary length scale ξ_0 since the value of the constant is of no interest. For the extension of the present consideration to cosmology in Sect. 4.6 it is important, however, that the first part of the right-hand member of (4.26) describes a "local" effect which arises in a neighborhood of the lens.

Adding the geometrical and potential contributions to the arrival time and subtracting the – of course purely geometrical – arrival time for an unlensed ray from S to O, we obtain for the *time delay* of a kinematically possible ray relative to the undeflected ray,

$$c\, \Delta t = \hat{\phi}(\boldsymbol{\xi}, \boldsymbol{\eta}) + \text{const.} \quad , \quad (4.27a)$$

where the *Fermat potential* $\hat{\phi}$ is given by

$$\hat{\phi}(\boldsymbol{\xi}, \boldsymbol{\eta}) = \frac{D_d D_s}{2 D_{ds}} \left(\frac{\boldsymbol{\xi}}{D_d} - \frac{\boldsymbol{\eta}}{D_s} \right)^2 - \hat{\psi}(\boldsymbol{\xi}) \quad , \quad (4.27b)$$

the *deflection potential* is

$$\hat{\psi}(\boldsymbol{\xi}) = \frac{4G}{c^2} \int d^2\xi' \, \Sigma(\boldsymbol{\xi}') \ln\left(\frac{|\boldsymbol{\xi} - \boldsymbol{\xi}'|}{\xi_0}\right) \quad, \tag{4.27c}$$

and const. is independent of $\boldsymbol{\xi}$ and $\boldsymbol{\eta}$.

Fermat's principle now asserts that the actual light paths – under our approximations and assumptions – are those for which, given $\boldsymbol{\eta}$, the arrival time t or, what amounts to the same, the arrival time delay is stationary with respect to variation of the deflection point I, i.e., of $\boldsymbol{\xi}$. Forming $\frac{\partial(\Delta t)}{\partial \boldsymbol{\xi}} = 0$, we obtain the so-called *lens mapping*

$$\boldsymbol{\eta} = \frac{D_s}{D_d}\boldsymbol{\xi} - D_{ds}\hat{\boldsymbol{\alpha}}(\boldsymbol{\xi}) \quad, \tag{4.28}$$

as stated in Sect. 2.2 already. It relates source and image positions, for a given deflecting mass. The deflection angle $\hat{\boldsymbol{\alpha}} = \nabla\hat{\psi}$ is seen to agree with the previous result (4.22).

In terms of the Fermat potential, the lens equation can be abbreviated as

$$\nabla_{\boldsymbol{\xi}}\hat{\phi}(\boldsymbol{\xi}, \boldsymbol{\eta}) = 0 \quad, \tag{4.29}$$

and the *arrival time difference* (or *time delay*) for two images $\boldsymbol{\xi}^{(1)}, \boldsymbol{\xi}^{(2)}$ of a source at position $\boldsymbol{\eta}$ can be expressed as

$$c(t_1 - t_2) = \hat{\phi}\left(\boldsymbol{\xi}^{(1)}, \boldsymbol{\eta}\right) - \hat{\phi}\left(\boldsymbol{\xi}^{(2)}, \boldsymbol{\eta}\right) \quad. \tag{4.30}$$

$(t_1 - t_2)$ in (4.30) is the difference of the coordinate times at which the light rays arrive at the observer. To obtain the corresponding proper time difference, $(t_1 - t_2)$ must be multiplied by the time dilation factor $1 + U_o/c^2$ where U_o is the value of the gravitational potential at the observer according to (4.13). This is practically irrelevant in applications of gravitational lensing, however, since on Earth, $U_o/c^2 \approx 10^{-9}$. In Sect. 4.6 the results of this section will be generalized to the cosmological context.

4.4 Summary of Friedmann–Lemaître cosmological models

As a theoretical framework for the interpretation of cosmological observations, we accept the homogeneous and isotropic, general-relativistic big bang models of the universe, augmented by perturbations to account for inhomogeneities. The assumptions underlying these *standard models*, to be reviewed below, are to be considered as working hypotheses which so far have been fairly successful in coordinating and suggesting observations.

The Robertson–Walker metrics. One way of setting up these models proceeds as follows. One assumes that, on a sufficiently large averaging scale – 500 Mpc, say – there exists a mean motion of matter and radiation in the universe with respect to which all (averaged) observable properties are *isotropic*, i.e., independent of direction. In addition, one assumes that all fundamental observers – i.e., all imagined observers who follow the mean motion – will experience the same history of the universe, the same observable properties, provided they set their clocks suitably; i.e., the universe is *observer-homogeneous*. Formulating these two assumptions about the mean motion and the light rays mathematically, H.P. Robertson and A.G. Walker have shown ([RO35.1], [WA35.1]) – without assuming any a priori geometric structure of spacetime – that such an idealized universe admits a metric of the type

$$ds^2 = c^2 dt^2 - R^2(t) d\sigma^2 \ , \tag{4.31}$$

where

$$d\sigma^2 = \frac{d\mathbf{x}^2}{\left(1 + \frac{k}{4}\mathbf{x}^2\right)^2} \tag{4.32}$$

is the metric of the 3-dimensional simply-connected Riemannian space of constant curvature $k = +1, -1$, or 0, such that the mean motion is represented by the geodesics $\mathbf{x} = $ const. and the light rays are given by the lightlike geodesics. Such a metric is called a *Robertson–Walker* (RW) *metric*.[3] In it, t represents a *cosmic time*, and $R(t)$ denotes a positive and otherwise arbitrary *scale function*. The cosmic time t coincides for each fundamental observer $\mathbf{x} = $ const. with the proper time, and for neighboring observers, equal values of t correspond to events which are simultaneous in Einstein's sense. Thus, $R^2(t) d\sigma^2$ is the spatial metric of the universe at time t; it depends on time only through a conformal scale factor.

Red shift. Equations (3.16) and (4.31) imply that electromagnetic radiation emitted at time t_S from a source S with wavelength λ_S and received by an observer O at time t_0, exhibits a cosmic redshift

$$z := \frac{\lambda_0 - \lambda_S}{\lambda_S} = \frac{R(t_0)}{R(t_S)} - 1 \ , \tag{4.33a}$$

where λ_0 is the wavelength seen by the observer. For sources close to the observer, $z \ll 1$, $c(t_0 - t_S) = D$ is the distance, and, by (4.31), $v = \left(\frac{\dot{R}}{R}\right)_0 D =$

---

[3] If one requires the model to be only *locally* isotropic with respect to all fundamental observers, one obtains the same metrics (4.31), (4.32), but the three-spaces need not be simply connected. Instead, for each value of k, various topologies, particularly compact ones, are possible. Examples are flat, toroidal models, "small universes" [EL86.1]. The field equations (4.34), (4.35) are not affected by the choice of the topology.

$H_0 D$ is the radial velocity of the source relative to the observer. Taylor expansion of $R(t)$ at t_0 then gives the original, simple version

$$z \approx \frac{v}{c} \approx \frac{H_0}{c} D \quad , \quad H_0 = \left(\frac{\dot{R}}{R}\right)_0 \quad , \quad z \ll 1 \qquad (4.33b)$$

of Hubble's law, where H_0 denotes Hubble's constant.

Friedmann–Lemaître models. Only now, having fully exploited isotropy and homogeneity, one imposes the field equation (4.1) on the metric (4.31), (4.32). It then turns out that the matter tensor necessarily has the form (4.5) corresponding to a perfect fluid, with a 4-velocity tangent to the timelines $\mathbf{x} = $ const., and with a density $\varrho(t)$ and pressure $p(t)$ depending on time only; moreover, ϱ and p must be related to R by Friedmann's equation[4]

$$\left(\frac{\dot{R}}{R}\right)^2 = \frac{8\pi G}{3}\varrho - \frac{kc^2}{R^2} + \frac{1}{3}\Lambda c^2 \qquad (4.34)$$

and the equation for adiabatic expansion,

$$\left(\varrho c^2 R^3\right)^{\cdot} + p\left(R^3\right)^{\cdot} = 0 \quad , \qquad (4.35)$$

first derived from (4.1) in this context by G. Lemaître. Λ denotes the cosmological constant or, if one prefers, the vacuum energy constant (mentioned in Sect. 4.1). Cosmological models having an RW metric (4.31–32), and obeying the field equations (4.34) and (4.35) are called *Friedmann–Lemaître models* (FL models).

From (4.34), the FL model singled out by $\Lambda = 0$, $k = 0$, today has a density

$$\varrho_{\mathrm{cr}} := \frac{3H_0^2}{8\pi G} \quad , \qquad (4.36)$$

called the *critical* (or *closure*) *density*.

Matter model. For an overall description of the history of a FL-model of the universe, it suffices to represent the averaged cosmic matter as a non-interacting mixture of "cold", non-relativistic matter, frequently called *dust*, and "hot", relativistic matter, in short called *radiation*. For dust with mass density ϱ_d, $p_\mathrm{d} = 0$; for radiation with mass density ϱ_r, $p_\mathrm{r} = \frac{1}{3}\varrho_\mathrm{r} c^2$. For both of these components, (4.35) holds; therefore, $\varrho_\mathrm{d} \propto R^{-3}$ and $\varrho_\mathrm{r} \propto R^{-4}$.

Parameters. A (dust + radiation) FL model, together with the present epoch t_0, is uniquely determined by the values of four parameters: $H_0 := \left(\dot{R}/R\right)_0$, the *Hubble constant*;

---

[4] We note that this equation is also valid in Newtonian cosmology [HE59.1], where $-k$ means the sign of the total energy.

$$\Omega_d := \frac{\varrho_{d0}}{\varrho_{cr}} \quad ; \quad \Omega_r := \frac{\varrho_{r0}}{\varrho_{cr}} \quad ; \quad q_0 := -\left(\frac{R\ddot{R}}{(\dot{R})^2}\right)_0 , \qquad (4.37)$$

the *density parameter of dust*, the *density parameter of radiation*; and the *deceleration parameter*, respectively. Instead of q_0, one can use $\Omega_v := \frac{\Lambda c^2}{3H_0^2} = \frac{1}{2}\Omega_d + \Omega_r - q_0$, the *vacuum density parameter*. The index 0 always indicates values at the present epoch. The parameters Ω_d, Ω_r, Ω_v and q_0 are dimensionless and determine the *shape* of the universe, whereas H_0 fixes the *scales* of time, length and mass.

Specializing (4.34) to the present epoch $t = t_0$, we see that the (Gaussian) space curvature today, k/R_0^2, is related to the total density parameter $\Omega := \Omega_d + \Omega_r + \Omega_v$ by

$$\frac{k}{R_0^2}\frac{c^2}{H_0^2} = \Omega - 1 . \qquad (4.38)$$

Observations indicate that ([BO88.1], [KO90.5])

$$H_0 = h_{50} \times 50 \text{ km s}^{-1} \text{ Mpc}^{-1} , \quad 1 \lesssim h_{50} \lesssim 2 , \qquad (4.39a)$$
$$0.1 \lesssim \Omega_d \lesssim 2 , \qquad (4.39b)$$
$$1.6 \times 10^{-5} \lesssim \Omega_r \lesssim 6.4 \times 10^{-5} , \quad \Omega_r \lesssim 2 \times 10^{-4}\Omega_d , \qquad (4.39c)$$
$$-1 \lesssim q_0 \lesssim 2.5 . \qquad (4.39d)$$

These data, combined with QSO redshifts, imply that realistic FL models begin with a big bang when $R(t)$, followed into the past, tends to zero, so that the metric (4.31) degenerates, and the densities ϱ_d, ϱ_r as well as the curvature become infinite [BO88.4].

For such models, the *age t_0 of the universe*,

$$t_0 = H_0^{-1} \int_0^1 \{\Omega_r(1-x^2) + \Omega_d x(1-x) - \Omega_v x^2(1-x^2) + x^2\}^{-1/2} x \, dx , \qquad (4.40)$$

is another observable; it is estimated to have a value in the range ([BO88.1], [KO90.5])

$$12 \times 10^9 \text{ years} \lesssim t_0 \lesssim 20 \times 10^9 \text{ years} . \qquad (4.41)$$

Only some time after the decoupling of matter and radiation, which occurred at about the time when $\varrho_r \approx \varrho_d$, $z \approx 10^3$, did galaxies and clusters begin to form (for $z \lesssim 10$); in this matter-dominated era, the pressure is negligible in equation (4.35), so that the expansion is governed by the Friedmann equation

$$\dot{a}^2 = H_0^2 \left(\frac{\Omega_d}{a} + \Omega_v a^2 - \frac{kc^2}{H_0^2 R_0^2}\right) \qquad (4.42a)$$

for the dimensionless expansion factor

$$a := \frac{R}{R_0} = \frac{1}{1+z} \quad .$$

Mainly because of the poorly understood evolution of galaxies, observational data do not yet permit a decision between a locally spherical ($k = +1$), a locally Euclidean ($k = 0$) and a locally Lobatschevskian ($k = -1$, saddle-like) cosmic space; in particular, they do not allow one of the three terms on the right-hand side of (4.42a) to be singled out as dominant over the other two. Especially, the present fractional contribution $\left(\frac{\Omega_v}{\Omega_d + \Omega_r + \Omega_v}\right)_0$ of the *vacuum density* to the total mass density is not well known, though ϱ_v is exceedingly small compared to expectations based on particle physics, especially in view of changes of vacua during symmetry-breaking phase transitions in the very early universe.

Owing to this uncertainty, the fact that $\Omega_v = \Lambda = 0$ is at least compatible with observations and current prejudice, *we henceforth put the cosmological constant equal to zero.* Then, (4.42a) simplifies to

$$\dot{a}^2 = H_0^2 \left\{ \Omega \left(a^{-1} - 1 \right) + 1 \right\} \quad , \tag{4.42b}$$

with $\Omega \equiv \Omega_d$. The presently *observed distribution* of visible, i.e., photon-emitting *matter* is far from isotropic and homogeneous. Galaxies tend to form groups and clusters, clusters often form superclusters, and the overall galaxy distribution pattern appears to be sponge-like. Only on scales above some hundreds of Megaparsecs, the distribution may be isotropic and homogeneous, as indicated by counts of radio sources.

The strongest support for the assumed large-scale isotropy and the hot initial state of the universe comes from the very high degree of temperature isotropy, $\left|\frac{\Delta T}{T}\right| \lesssim 3 \times 10^{-5}$ on the scale of arcminutes, and the thermal distribution, with $T \approx 2.7$ K, of the *microwave background radiation* [MA90.1], supposedly a relic of the hot equilibrium state of the cosmic "ylem". It even appears that this degree of isotropy requires, in contrast to visible matter, a rather smooth distribution of *dark matter*, in agreement with mass estimates based on rotation curves of galaxies, the distribution of kinetic and potential gravitational energies in clusters (virial theorem), and estimates based on counts of galaxies with photometrically determined redshifts. Further evidence for an initially hot, on average homogeneous FL-model is provided by the linearity of the *redshift-magnitude relation* (Hubble law, see below) for bright galaxies, and the good agreement between the observed *abundances of light elements* (D, ^{3}He, ^{4}He, ^{7}Li) and those calculated on the basis of the standard model [OL90.1].

A difficult, unduly neglected, fundamental task for theoretical cosmologists is to clarify the relation between the homogeneous and isotropic "background models" of the universe and more detailed, fine-grained models. Even in the absence of explicit fine-grained models, one would like to know how, and under which conditions, one could in principle extract a background model from an inhomogeneous one such that (i) both obey (approximately?) Einstein's equation, in spite of the fact that one has to "average", or "smooth out" the solution of a *non-linear* equation, and (ii) observational determinations of "global" quantities like H_0, Ω do, in fact, correspond to that mathematical *averaging* or *fitting procedure* (for discussions of that problem, see, e.g., [EL88.1]). This problem is, in particular, relevant to gravitational lens theory. For now, we have to proceed on the basis of working hypotheses; see Sect. 4.5 for examples.

Without entering a discussion, we only mention that some shortcomings of the standard model (horizon, homogeneity, monopole problem) related to the problem of explaining, rather than postulating, initial conditions for the universe, may be resolvable by *inflation models*.

4.5 Light propagation and redshift-distance relations in homogeneous and inhomogeneous model universes

In contrast to times, angular separations and flux densities, distances are not directly observable in astronomy and, in particular, in cosmology, as we mentioned in Sect. 3.5 already. Nevertheless, several distance-like quantities are indispensable theoretical intermediaries. Of particular importance to us are affine parameters on light rays, luminosity distances and angular diameter distances as functions of redshift.

4.5.1 Flux conservation and the focusing theorem

Flux conservation. Before turning to these relations, we draw an important consequence of equation (3.79) which gives the flux in terms of the luminosity, redshift and "distance". Imagine an arbitrarily inhomogeneous universe filled with transparent matter. Let a point source at an event S have luminosity L. Consider a set of observers with redshift z whose receiving areas cover a 2-dimensional cross-section of the future light cone $\mathcal{C}_S^+$ of S, i.e., which fill out a wavefront (which may have self-intersections). Then, according to (3.79) and the definition (3.65) of the corrected luminosity distance,

$$\int_{z=\text{const}} S \, dA = \frac{L}{(1+z)^2} \quad , \tag{4.43a}$$

i.e., the total flux is independent of the distortion of the system of rays between emission and reception which may be caused by intervening matter. This *flux conservation law*, which is simply a restatement of the conservation of photon number, is important for resolving an apparent paradox: According to the magnification theorem stated in Sect. 2.3, any point source has an image which is at least as bright and, as will be shown in Sect. 5.4, in general brighter than it would appear in the absence of a lens, at a given redshift. The apparent contradiction with the conservation of the total flux disappears upon the realization that deflecting masses not only increase the flux density values, but change the geometry such as to decrease the corresponding receiving areas, leaving the total flux unchanged. (Note that this compensation follows from spacetime geometry and photon kinematics; the field equation does not enter into the argument.)

Equation (4.43a) expresses a powerful result which, however, can be applied only under special conditions [WE76.1]. Consider an ensemble of sources, all at the same redshift z and with the same luminosity L. If we lived in a smooth Friedmann–Lemaître universe (without local inhomogeneities), we would observe all these sources to have the same flux, S_{FL}. In an inhomogeneous universe, the fluxes of these sources are different, due to light propagation effects. However, (4.43a) states that the flux of each of these sources, averaged over all directions of emission, is the same as it would be in a smoothed-out Friedmann–Lemaître universe, *provided* the area of a surface $z = $ const. on the future light cone of a source in the inhomogeneous universe is equal to that of the corresponding spherical wave front in the Friedmann–Lemaître model. Suppose now many such sources are observed in various directions, and assume that the light bundles which reach the observation event may be taken to have randomly distributed directions. Then, in an inhomogeneous, on-average homogeneous and isotropic, transparent universe, the average flux $\langle S \rangle$ from sources at z with luminosity L, measured by a "typical" observer, equals the corresponding flux S_{FL},

$$\langle S \rangle = S_{\text{FL}} = \frac{L}{4\pi \left(\bar{D}_{\text{L}}(z)\right)^2} \quad , \tag{4.43b}$$

where $\bar{D}_{\text{L}}(z)$ denotes the luminosity distance in the smooth Friedmann–Lemaître model, fitted to the inhomogeneous universe according to the area condition mentioned above. Equation (4.43b) imposes an important consistency condition on probability distributions for fluxes in clumpy universe models.

If some areas on the sky, which may depend on the depth of the sample considered, are practically opaque, (4.43b) requires corrections, as discussed in the paper quoted above. For example, it has been suggested that high-redshift QSO samples are affected by dust absorption [HE88.2].

Since the flux-conservation law has given rise to misunderstandings in the literature, we would like to stress that it is in general impossible to compare "mean fluxes of sources" in two different spacetimes (see also [SC89.3]). For example, comparing the mean flux of a source in a Schwarzschild metric, as measured by a set of observers situated at the same radial coordinate r, with the flux of a source in a flat spacetime, measured by observers at the same distance D [AV88.1], is a useless exercise: there is no unique way to relate r and D to each other, since, as we have seen, "distance" does not have a unique meaning in curved spacetimes. Of course, one can always define a relation between r and D such that the mean flux in the former situation equals the one in the latter geometry, but this is merely a matter of defining distances in the Schwarzschild metric. Actually, (4.43a) immediately provides the recipe for how one would define the radius in the Schwarzschild metric so as to make the mean flux equal to that in the Newtonian situation, as the left-hand side of (4.43a) equals $A \langle S \rangle$; therefore, the area of the sphere of observers of constant radial coordiante has to be chosen to be $4\pi r^2$, so that in this case the usual radial coordinate satisfies the above property – as found in [AV88.1].

The reason why we can make the comparison (4.43b) is that we hypothesized the homogeneous and the clumpy universe to have the same large-scale structure; in particular, spheres of the same redshift in both universes are assumed to have the same surface area. *Only because of this* can we make a reasonable comparison between the "fluxes of sources at constant redshift" in these two situations

Focusing theorem. Next, we turn to the focusing equation (3.64) for light beams. According to the field equation (4.1), the Ricci term in that equation is equal to $\frac{4\pi G}{c^4} T_{\alpha\beta} k^\alpha k^\beta$. For a perfect fluid matter tensor (4.5), this expression reduces to $\frac{4\pi G}{c^2} \varrho (U^\alpha k_\alpha)^2$, a non-negative quantity. [This seems to hold more generally for all kinds of matter. Note also that the energy-momentum of the beam itself, given by (3.24), does not contribute to the Ricci term since $k_\alpha k^\alpha = 0$. Also, a cosmological term would not affect the focussing law, for the same reason.] Thus, equation (3.64) implies the inequality $\left(\sqrt{A}\right)^{\cdot\cdot} \leq 0$. Integrating both members of this inequality, starting at the observation event O where $\left(\sqrt{A}\right)^{\cdot} = \sqrt{\Omega_O}$ (if the affine parameter v on the ray is chosen to agree with the Euclidean distance[5] close to O), one obtains $A(v) \leq v^2 \Omega_O$ as long as $A(v) \geq 0$. If there were neither shear nor matter in the beam, (3.64) would give $A(v) = v^2 \Omega_O$. We thus obtain the *focusing theorem.* If there is any shear or matter along a beam connecting a source to an observer, the angular diameter distance of that source from the observer is smaller than that which would occur if that source was seen through an empty, shear-free cone, provided the affine parameter distance is the same and the beam has not gone through a caustic.

The last condition is essential, since after a transition through a caustic the inequality $A(v) \leq v^2 \Omega_O$ need not hold any more. As the derivation shows, an increase of shear or matter density along the beam decreases the angular diameter distance and, consequently, increases the observable flux, given z.

4.5.2 Redshift-distance relations

Differential relation between affine parameter and redshift along light rays in arbitrary cosmological models. One would like to express the angular diameter distance D defined in (3.66) as a function of redshift; for, as we

---

[5] For notational convenience, we here scale v such that it becomes a length.

shall see in the next section, these distances occur in the cosmological lens mapping, and such a function $D(z)$ is also needed to convert equations (3.78) and (3.79) (and similar relations for number counts of galaxies) into flux-redshift (or number-redshift) laws. To obtain such a relation requires a metric. Metrics describing an inhomogeneous universe resembling the real one are not known. Using a RW-metric would be too crude and, in particular, would exclude gravitational lensing. Faced with this dilemma, one tries to set up a relation $D(z)$ which holds approximately in inhomogeneous universes which are homogeneous and isotropic on a large scale. As a preparation, we here derive an exact, generally valid differential equation relating $\frac{dz}{dv}$ to the differential motion of matter along a light ray.

Differentiating the definition $\omega = ck^\alpha U_\alpha$ of the circular frequency of an electromagnetic wave along a light ray with frequency vector k^α and taking into account (3.14a), we obtain $\frac{d\omega}{dv} = cU_{\alpha;\beta} k^\alpha k^\beta$.[6] We now split $k^\alpha = \frac{\omega}{c}(U^\alpha + e^\alpha)$ into a part parallel to the 4-velocity U^α of matter along the ray, and a part orthogonal to U^α. Then, the unit vector e^α represents the direction of the ray for an observer comoving with the matter. Also, we decompose the gradient of the 4-velocity,

$$cU_{\alpha;\beta} = \hat{\omega}_{\alpha\beta} + \theta_{\alpha\beta} + \dot{U}_\alpha U_\beta \quad , \tag{4.44}$$

into the local rate-of-rotation $\hat{\omega}_{\alpha\beta} = \hat{\omega}_{[\alpha\beta]}$,[7] the rate of deformation $\theta_{\alpha\beta} = \theta_{(\alpha\beta)}$, and a term containing the 4-acceleration; both $\hat{\omega}_{\alpha\beta}$ and $\theta_{\alpha\beta}$ are spatial with respect to the fluid, $\hat{\omega}_{\alpha\beta} U^\beta = \theta_{\alpha\beta} U^\beta = 0$. [The decomposition (4.44) is the 4-dimensional analogue of the classical Helmholtz decomposition of the fluid velocity gradient; see, e.g., [HA73.1], Sect. 4.1.] Then,

$$\frac{c^2}{\omega^2} \frac{d\omega}{dv} = \theta_{\alpha\beta} e^\alpha e^\beta + \dot{U}_\alpha e^\alpha \quad .$$

Since $\omega/\omega_0 = 1 + z$, where ω_0 refers to the observer as before, we can rewrite the last equation as

$$\frac{c^2}{\omega_0} \frac{dz}{dv} = (1+z)^2 \left(\theta_{\alpha\beta} e^\alpha e^\beta + \dot{U}_\alpha e^\alpha \right) \quad .$$

We now substitute for v the dimensionless affine parameter $w := \frac{H_0 \omega_0}{c^2} v$, scaled such that, at the observer, $\frac{c}{H_0} dw$ equals the element $d\ell$ of proper length, as is seen from $\omega_0 = c(U_\alpha k^\alpha)_0 = c \left(\frac{U_\alpha dx^\alpha}{dv} \right)_0 = c \left(\frac{d\ell}{dv} \right)_0$. Then,

$$H_0 \frac{dz}{dw} = (1+z)^2 \left(\theta_{\alpha\beta} e^\alpha e^\beta + \dot{U}_\alpha e^\alpha \right) \quad . \tag{4.45}$$

This formula [EH61.1] says that the change of redshift along a ray intercepted by fluid world lines is due partly to the relative motion of neighbor-

---

[6] In contrast to the previous subsection, here we use the original affine parameter v defined in (3.13), which has the dimension of a squared length.
[7] To distinguish the rate-of-rotation from the circular frequency ω, we mark the former by a ˆ

ing particles – Doppler shift – and partly to the "acceleration of gravity" in the ray direction – gravitational shift. Such a decomposition of the redshift is possible only relative to a fluid flow or "observer field", of course. — If matter is modeled as dust, $p = 0$, then $\dot{U}^\alpha = 0$ in consequence of $(\varrho U^\alpha U^\beta)_{;\beta} = 0$.

$\frac{dz}{dw}$ **in RW and "on-average-RW" models.** In an RW-model, matter is freely falling and undergoes an isotropic expansion,

$$\dot{U}_\alpha = 0 \quad \text{and} \quad \theta_{\alpha\beta} = \frac{\dot{R}}{R}(U_\alpha U_\beta - g_{\alpha\beta}) \quad ,$$

as can be formally verified using $U^\alpha = \delta_0^\alpha$ and the metric (4.31). Then, (4.45) simplifies to

$$\frac{dz}{dw} = (1+z)^3 \frac{\dot{R}}{\dot{R}_0} \quad . \tag{4.46}$$

Let us now imagine a universe which is populated by randomly distributed matter inhomogeneities and resembles an RW model on a large scale. Then it is not unreasonable to assume that, at least along the majority of light rays which traverse a large part of such a universe, the average rate of deformation of matter in the direction of a ray, $\langle \theta_{\alpha\beta} e^\alpha e^\beta \rangle$, equals the average RW expansion rate $\dot{R}/R$, and the component of the acceleration of matter in that direction, if any, averages to zero, $\langle \dot{U}_\alpha e^\alpha \rangle = 0$. (One may, indeed, consider these statements as part of a – not yet available – *definition* of an inhomogeneous, "on average RW" universe model.) Under these assumptions, the general equation (4.45) again leads to the RW relation (4.46) on average. This relation also holds in a Swiss–Cheese model [KA69.1].

In an expanding RW model, $\dot{R} > 0$, and we may consider $\dot{R}$ as a function of R or, finally, as a function of $z = R_0/R - 1 = a^{-1} - 1$. Thus, an expansion function $R(t)$ specifies $\dot{R}[z]$ and, via (4.46), the function $z(w)$.

$w(z)$ **in FL dust models.** Taking into account the Friedmann equation (4.42b), we obtain from (4.46)

$$w = \int_0^z \frac{d\zeta}{(1+\zeta)^3 \sqrt{\Omega\zeta + 1}} \quad . \tag{4.47a}$$

(The integral is elementary; after some manipulation one gets

if $\Omega = 1$: $w = \frac{2}{5}\left[1 - (\Omega z + 1)^{-5/2}\right]$,

if $\Omega > 1$: $w = \frac{3\Omega^2}{4(\Omega-1)^{5/2}} \arctan \frac{\sqrt{\Omega-1}(1+\sqrt{\Omega z + 1})}{\Omega - 1 + \sqrt{\Omega z + 1}}$

$+ \frac{3}{4(\Omega-1)^2}\left(\frac{(\Omega z + \frac{5}{3}\Omega - \frac{2}{3})\sqrt{\Omega z + 1}}{(z+1)^2} - \frac{5}{3}\Omega + \frac{2}{3}\right)$;

if $\Omega < 1$, $\sqrt{\Omega - 1}$ in the above expression has to be replaced by $\sqrt{1 - \Omega}$, and arctan by arctanh.)

This equation enables us to parametrize light rays in terms of the observable z rather than in terms of the unobservable w.

Proper distance. For some applications, we also need the relation between the increment of redshift, dz, and the corresponding increment of radial proper distance, dr_{prop}. On a past-directed light ray, we have by (4.31)

$$dr_{\text{prop}} = -c\, dt = -\frac{c\, dR}{\dot{R}} = \frac{c\, R_0\, dz}{(1+z)^2 \dot{R}} = \frac{c\, dz}{(1+z)\, H} \quad,$$

where we used that $R_0 = R(1+z)$ and $H = \dot{R}/R$. If we express H in terms of z via (4.42b), we get the desired relation,

$$dr_{\text{prop}} = \frac{c\, dz}{H_0\, (1+z)^2 \sqrt{1+\Omega z}} = \frac{c}{H_0}(1+z)\, dw \quad; \qquad (4.47b)$$

the last relation is obvious from (4.47a).

Angular diameter distances in FL and "on-average FL" dust models. Any RW spacetime is conformally flat; its conformal curvature tensor $C^\alpha{}_{\beta\gamma\delta}$ is zero. Consequently, the shear in the focusing equation (3.64) vanishes, too. [This follows from (3.63) with the initial condition that, at the vertex of any light cone, $\sigma \to 0$; it can also be deduced directly from the RW metric (4.31).]

Evaluating the Ricci term by means of the field equation as above, and remembering that, with our choice of the dimensionless affine parameter, $U_\alpha k^\alpha = \frac{c}{H_0}(1+z)$, we obtain the focusing equation

$$\frac{d^2}{dw^2} \sqrt{A} + \frac{4\pi G}{H_0^2}(1+z)^2 \varrho \sqrt{A} = 0 \quad.$$

Eliminating w in favor of z by means of (4.47a), using the conservation law (4.35), specialized to $p = 0$, $\varrho = \varrho_0(1+z)^3$ and the definition (3.66) of the angular diameter distance D, we get the following differential equation for $D(z)$:

$$(z+1)(\Omega z + 1)\frac{d^2 D}{dz^2} + \left(\frac{7}{2}\Omega z + \frac{\Omega}{2} + 3\right)\frac{dD}{dz} + \frac{3}{2}\Omega D = 0 \quad. \qquad (4.48)$$

Before giving its general solution, we consider arguments leading to a generalized distance-redshift equation.

4.5.3 The Dyer–Roeder equation

Swiss–Cheese and similar models. We want to generalize (4.48), which holds exactly in FL dust models, to partly clumpy, "on-average-FL" models. No rigorous derivation of such a generalization is possible at present, for (i) no such cosmological models are known, (ii) no well-defined averaging procedure for nonlinear equations like those which occur in this context has been developed, and (iii) in an inhomogeneous universe, the angular diameter distance depends not only on redshift, but also on the direction, i.e., the distribution of matter around the line-of-sight to a source.

One may object to (i), pointing out that the Swiss–Cheese models are exact, inhomogeneous models [KA69.1]. They are constructed by removing matter from non-overlapping, otherwise arbitrary comoving balls, and then filling the holes by (static or non-static) spherically symmetric matter distributions. However, the fitting conditions between the inner Schwarzschild metric and the outer Friedmann–Lemaître metric at the boundaries of the balls require the total mass of the filling to be equal to that of the matter which was removed. In the region outside these balls, the gravitational potential remains unchanged – as in the corresponding Newtonian models – and each condensation is surrounded by an underdense halo. The restriction to strictly spherical "clumps" in these models is quite unrealistic, though. Nevertheless, lensing occurs in these models, and one may use them for illustration and for testing approximation methods. (Studies of light propagation in the models are found in, e.g., [KA69.1], [DY81.2], [DY88.1], [WA90.3].)

A similar model is obtained by describing the matter inside each hole of a Swiss–Cheese universe by another Friedmann–Lemaître model, the expansion rate of which may differ from that of the overall universe ([NO82.1], [NO82.2], [NO83.1], [HA85.1], [HA85.2], [SA87.1]). If the expansion rate of the interiour Friedmann–Lemaître models is different from the overall cosmic expansion rate, the metric of this model is not stationary; therefore, light passing through such *vacuoles* experience redshift effects; for a recent discussion, see [NO90.1].

"Derivation" of the Dyer–Roeder equation. The best one can do in this situation is to proceed on the basis of a "plausible" working hypothesis. To do that, we first recall that the appearance of Ω in the coefficients of the two derivatives in (4.48) is due solely to the transformation from the affine parameter w to the redshift z via (4.47a), and we have already argued that this transformation, based on (4.46), remains trustworthy under the more general conditions envisaged. Secondly, we recall that the Ω in front of the undifferentiated D is due to the Ricci, or matter term in the focusing equation (3.64). We now assume, following Dyer and Roeder [DY73.1] that a mass-fraction $\tilde{\alpha}$ (*smoothness parameter*) of matter in the universe is smoothly distributed, i.e., not bound in galaxies, while a fraction $1 - \tilde{\alpha}$ is bound. For bundles of light rays not passing through bound "clumps", we therefore replace the Ω in the third term of (4.48) by $\tilde{\alpha}\Omega$. Note that the mass conservation law for dust, $\varrho_{,\alpha}U^\alpha + \varrho U^\alpha{}_{;\alpha} = 0$, requires the volume expansion rate $U^\alpha{}_{;\alpha}$ to remain unchanged if ϱ is changed into $\tilde{\alpha}\varrho$ with a constant $\tilde{\alpha}$, in agreement with using the mean expansion rate in computing the coefficients of the derivatives in (4.48). Thirdly, we have to consider the shear in (3.64). If we keep it in the equation and substitute $\tilde{\alpha}\Omega$ for Ω, we obtain as a generalization of (4.48):

$$(z+1)(\Omega z+1)\frac{d^2 D}{dz^2} + \left(\frac{7}{2}\Omega z + \frac{\Omega}{2} + 3\right)\frac{dD}{dz}$$
$$+ \left(\frac{3}{2}\tilde{\alpha}\Omega + \frac{|\sigma|^2}{(1+z)^5}\right)D = 0 \quad . \tag{4.49}$$

The complicating term is the last one, containing the shear σ. The transport equation (3.63) for σ, now written in terms of the dimensionless affine parameter w, can be rewritten under our present assumptions as

$$\frac{d}{dw}(D^2 \sigma) = -\frac{1}{2}D^2 T \quad ,$$

where

$$T := \left(\frac{c}{H}\right)^2 C_{\alpha\beta\gamma\delta}\, \varepsilon^{*\alpha} k^\beta \varepsilon^{*\gamma} k^\delta \tag{4.50}$$

is a dimensionless measure of the gravitational tidal field, here to be considered as a function on the light ray. Thus, by (4.47a), the (dimensionless) shear is

$$\sigma(z) = -\frac{1}{2}\int_0^z T\left(\frac{D(\zeta)}{D(z)}\right)^2 (1+\zeta)^{-3}(\Omega\zeta+1)^{-1/2} d\zeta \quad . \tag{4.51}$$

An estimate of the shear term thus requires a model of the distribution of "clumps" around the light beam considered and a statistical evaluation of the integral, since light propagation in an inhomogeneous universe is an intrinsically statistical problem. Therefore, one cannot expect to obtain a unique distance-redshift relation, but the angular diameter distance along a light bundle depends on the distribution of clumps around the beam; the distances of sources at constant redshift will follow a probability distribution. For Swiss–Cheese models, such a probability distribution has been determined in [DY88.1].

Loosely speaking, one can single out two limiting cases: a light bundle propagates "far from all clumps" and is thus only "weakly" affected by the fine-grained structure of the gravitational field, or its path gets close to one or more clumps, in which case the explicit lens action of these clumps must be taken into account. It is clear, however, that this distinction does not cover the intermediate cases: one cannot divide light rays into those making "strong" and "weak" lensing. However, it makes sense to consider the limiting case of a light bundle which propagates "very far away from all clumps" and thus is not affected by shear at all (we shall denote such a limiting situation as *light propagating through an empty cone*). For these idealized light bundles, the shear term in (4.49) vanishes; then, that equation simplifies to the (generalized) *Dyer–Roeder equation* ([DY72.1], [DY73.1], [DY74.1], [DY81.2])

$$(z+1)(\Omega z+1)\frac{d^2 D}{dz^2} + \left(\frac{7}{2}\Omega z + \frac{\Omega}{2} + 3\right)\frac{dD}{dz} + \frac{3}{2}\tilde{\alpha}\Omega D = 0 \quad . \quad (4.52)$$

The appropriate solution of this equation does define an angular-diameter distance as a function of redshift; its usefulness is due to the fact that it yields the largest possible distance (for given redshift) for light bundles which have not passed through a caustic; it thus provides a strong constraint on the probability distribution for distances. In fact, for universes not "too much filled by clumps", this distribution will have a strong peak very close to, but at smaller values than the Dyer–Roeder distance. Owing to that strong peak, it is justified to use these Dyer–Roeder distances for those light rays which are not strongly affected by lensing. On the other hand, if the universe is not very clumpy, the Dyer–Roeder distances and those of the smooth FL universe are not too different.

Studies of the influence of shear on the propagation of light bundles through a clumpy universe have been performed in [FU89.2], [KA90.3], [WA90.5] and [WA90.3], by integrating the optical scalar equations in such a clumpy universe. We will show later (Chap. 11) that gravitational lens theory provides a powerful tool to study these effects, and in Sect. 11.4 we present some results of pertinent investigations.

General properties of solutions to the Dyer–Roeder equation. For each pair of values of the parameters $\Omega, \tilde{\alpha}$ ($0 \leq \Omega, 0 \leq \tilde{\alpha} \leq 1$), any solution of (4.52) can be expressed as a linear combination of two linearly independent ones. Since z, Ω, $\tilde{\alpha}$ are dimensionless, it is convenient to write $D = \frac{c}{H_0} r$ and work with r, the "dimensionless Dyer–Roeder distance".

Let $r(z_1, z_2)$ denote the angular diameter distance of a source at z_2 from a fictitious observer at z_1, the real observer being at $z = 0$, of course. Then, $r(z_1, z_2)$ is the value at z_2 of that solution $r(z_1, z)$ which satisfies the initial conditions

$$r(z_1, z_1) = 0 \quad , \quad \frac{d}{dz}r(z_1, z)\bigg|_{z=z_1} = \frac{\text{sign}(z_2 - z_1)}{(z_1 + 1)^2 \sqrt{\Omega z_1 + 1}} \quad . \quad (4.53)$$

[To obtain the second initial condition, apply the Hubble law (4.33b) to a fictitious observer at z_1. For her, the Hubble constant $\dot{R}/R = \dot{a}/a$ was, by (4.42b),

$$H_1 = H_0(z_1 + 1)\sqrt{\Omega z_1 + 1} \quad ,$$

and the redshift from a source at $z + dz$ was

$$\delta z = \frac{\lambda(z_1)}{\lambda(z_1 + dz)} - 1 = \frac{\lambda(z_1)}{\lambda_0}\frac{\lambda_0}{\lambda(z_1 + dz_1)} - 1 = \frac{dz}{1 + z_1} \quad .$$

Thus, her Hubble law was $\delta z = \frac{H_1}{c} dD = \frac{H_1}{H_0} dr$, which gives the required $\frac{dr}{dz}$.]

Let, in particular, $r(z) := r(0, z)$ for $z > 0$, i.e., $r(0) = 0$, $\frac{dr}{dz}(0) = 1$. It can then easily be verified that the function[8] $r(z)r(z_1)(z_1 + 1)$
$\times \int_{z_1}^{z} \frac{d\zeta}{r^2(\zeta)(1+\zeta)^3 \sqrt{\Omega\zeta + 1}}$ obeys (4.52) and satisfies the initial conditions (4.53). Therefore,

$$r(z_1, z_2) = (1 + z_1)r(z_1)r(z_2) \left| \int_{z_1}^{z_2} \frac{dz}{r^2(z)(z + 1)^3 \sqrt{\Omega z + 1}} \right| . \quad (4.54)$$

This equation shows that

$$\frac{r(z_1, z_2)}{1 + z_1} = \frac{r(z_2, z_1)}{1 + z_2} , \quad (4.55)$$

which proves that *Etherington's reciprocity law remains valid exactly in the Dyer–Roeder approximation.*[9] It therefore suffices to consider, from now on, the case $z_2 \geq z_1$. Note that, for $z_1 < z_2$, $\frac{c}{H_0} r(z_1, z_2)$ is the angular-diameter distance of a source at z_2 from an observer at z_1, whereas $\frac{c}{H_0} r(z_2, z_1)$ corresponds to the "corrected luminosity distance" introduced in Sect. 3.5.

An increase in clumpiness $1 - \tilde{\alpha}$ for fixed overall density Ω leads to an increase of $r(z_1, z_2)$, as follows most easily from the focusing equation in the form preceding (4.48). This is intuitively clear, since less matter in the beam decreases the focusing; an additional proof will be given in Sect. 11.4.1.

Before listing particular explicit solutions to (4.52), we mention in passing that if Ω is different from 0 and 1, the change $z \to x = \frac{1+\Omega z}{1-\Omega}$, $r(z) = r[x]$ of variable transforms the Dyer–Roeder equation into the hypergeometric equation

$$x(1-x)\frac{d^2r}{dx^2} + [c - (1 + a + b)x]\frac{dr}{dx} - abr = 0 ,$$

with $a = \frac{5}{4} + \beta$, $b = \frac{5}{4} - \beta$, $c = \frac{1}{2}$, $\beta := \frac{1}{4}\sqrt{25 - 24\tilde{\alpha}}$. This fact is not very useful in practice, however, since the evaluation of the hypergeometric function $F(a, b, c; x(z))$ takes much longer than integrating the Dyer–Roeder equation numerically.

Explicitly solvable cases are listed below.

Explicit solutions for dust models. (1) $\Omega = 1$. This concerns the Einstein–de Sitter background model with a locally flat 3-space and arbitrary amount of clumpiness, presumably the most realistic case among the explicitly integrable cases. Then, with β as above,

$$r(z) = \frac{(1+z)^\beta - (1+z)^{-\beta}}{2\beta(1+z)^{5/4}} ,$$

$$r(z_1, z_2) = \frac{1}{2\beta} \left[\frac{(1+z_2)^{\beta - 5/4}}{(1+z_1)^{\beta + 1/4}} - \frac{(1+z_1)^{\beta - 1/4}}{(1+z_2)^{\beta + 5/4}} \right] . \quad (4.56)$$

---

[8] We use Liouville's formula, to obtain by "variation of constants" a second solution of (4.52) from a given one; see, e.g., [AR85.1], Sect. 8.6.
[9] Equation (4.55) holds exactly, not only in the Dyer–Roeder approximation, as follows immediately from (3.67) and the multiplicativity of redshifts along a light ray, (3.16a).

The angular diameter distance $r(z)$ has a maximum at

$$z_m = \left(\frac{5+4\beta}{5-4\beta}\right)^{\frac{1}{2\beta}} - 1 \quad .$$

If all matter is smoothly distributed, $\tilde{\alpha} = 1$, $z_m = 1.25$; z_m increases with increasing clumpiness and tends to infinity if $\tilde{\alpha} \to 0$. For $\tilde{\alpha} = 2/3$, $z_m \approx 1.52$ [DA66.1].

(2) $\tilde{\alpha} = 1$. This concerns the unperturbed models, having no clumps. Then,

$$r(z) = \frac{2}{\Omega^2(1+z)^2}\left[\Omega z - (2-\Omega)\left(\sqrt{\Omega z+1}-1\right)\right] \quad ,$$
$$r(z_1, z_2) = \frac{2}{\Omega^2}(1+z_1)\left[R_1(z_2)R_2(z_1) - R_1(z_1)R_2(z_2)\right] \quad ,$$

(4.57a)

where

$$R_1(z) = \frac{\Omega z - \Omega + 2}{(1+z)^2} \quad , \quad R_2(z) = \frac{\sqrt{\Omega z+1}}{(z+1)^2} \quad . \tag{4.57b}$$

(3) $\tilde{\alpha} = \frac{2}{3}$. In these models, one third of the matter is bound. Then,

$$r(z) = \frac{2}{\Omega^2(z+1)^2}\left[\frac{1}{3}\Omega z\sqrt{\Omega z+1} - \left(\frac{2}{3}-\Omega\right)\left(\sqrt{\Omega z+1}-1\right)\right] \quad ,$$

(4.58a)

$$r(z_1, z_2) = \frac{2(z_1+1)}{3\Omega^2}\left[R_1(z_1)R_2(z_2) - R_2(z_1)R_1(z_2)\right] \quad ,$$

where

$$R_1(z) = \frac{1}{(z+1)^2} \quad , \quad R_2(z) = \frac{\sqrt{\Omega z+1}\,(\Omega z + 3\Omega - 2)}{(z+1)^2} \quad . \tag{4.58b}$$

(4) $\tilde{\alpha} = 0$. This is the extreme opposite of case 2; all matter is bound in clumps. In this case, the Dyer–Roeder equation is effectively of first order and has the solution [compare (4.47a)]

$$r(z) = w(z) \quad ,$$
$$r(z_1, z_2) = (1+z_1)(w(z_2) - w(z_1)) \quad . \tag{4.59}$$

(5) $\Omega = 0$. This corresponds to Milne's special-relativistic, kinematic model with a hyperbolic 3-space (see, e.g., [RI79.1]). Then,

$$r(z) = \frac{1}{2}\left(1 - \frac{1}{(z+1)^2}\right) \quad ,$$
$$r(z_1, z_2) = \frac{z_1+1}{2}\left(\frac{1}{(z_1+1)^2} - \frac{1}{(z_2+1)^2}\right) \quad . \tag{4.60}$$

Using the preceding formulae for $D(z)$, one can convert (3.78) and (3.79) into *flux-redshift* or *magnitude-redshift* relations which contain the parameters H_0, Ω, $\tilde{\alpha}$ specifying the cosmological model. In particular, (4.57a) leads to the magnitude-redshift law for dust models without cosmological term [MA58.1]. Note that these relations are valid (at best) for unlensed sources; the influence of lensing (amplification bias) will be considered in Sect. 12.5.

4.6 The lens mapping in cosmology

In Sect. 4.3 the basic equations of lens theory were derived under the assumptions that source, lens, and observer may be considered as being at rest relative to each other and form parts of an isolated system in an asymptotically flat spacetime. Of course, these assumptions do not hold for realistic lensing situations. It turns out, however, that the relevant equations require very little modification to become valid in the cosmological context, granted some plausible assumptions. This will now be shown.

Assumptions. We assume that the universe can be modeled, on a very large scale and in the matter-dominated, late phase ($z \lesssim 100$, say), by a FL dust model as reviewed in Sect. 4.4. To account very roughly for the inhomogeneity of matter, we introduce the smoothness parameter $\tilde{\alpha}$ and adopt the "scenario" of an "on-average FL universe", as in the previous section. In agreement with this picture, we assume that in a typical lensing situation, the light rays from the source to the neighborhood of the deflector and, after deflection, those from that neighborhood to the observer, form beams which are (nearly) unaffected by gravitational tidal fields and subject only to the focusing of the smooth, $\tilde{\alpha}\Omega$-part of matter. (Of course, shear *is* produced during the deflection, close to the lensing mass.)

To determine, under these assumptions, the lens mapping and the arrival time differences for images, we follow essentially the strategy used in the last subsection of Sect. 4.3, using Fermat's principle. As "kinematically possible" light rays we again take piecewise smooth world lines, consisting of a null geodesic of the RW metric (4.31) from a fixed emission event S to some deflection event I near the lens, and another such null geodesic from I to a fixed observation event O on the observer's world line. Among these possible rays there is exactly one which is smooth, the "unperturbed" ray from S to the observer. Projected into the comoving 3-space Σ_k of constant curvature $k = 1, 0$, or -1, introduced in connection with (4.31) and (4.32), the events S, I, O and the rays give a geodesic triangle $\hat{S}$, $\hat{I}$, $\hat{O}$ in Σ_k; see Fig. 4.1. As in the earlier treatment, we split the time delay Δt of the perturbed rays SIO relative to the unperturbed ray SO into a geometrical part Δt_{geom}, due to the difference of the lengths of the paths $\hat{S}\hat{I}\hat{O}$ and $\hat{S}\hat{O}$, and a potential part Δt_{pot}, due to the retardation of the deflected rays SIO

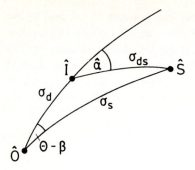

Fig. 4.1. The geodesic triangle $\hat{S}\hat{I}\hat{O}$ which arises by projecting the rays from the emission event S to the deflection event I, from the latter to the observation event O, and the unperturbed ray from S to O, respectively, into the three-space Σ_k of constant curvature. Only the difference angle $\theta - \beta$ is shown; the individual angles θ, β refer to an optical axis whose choice is irrelevant here

caused by the gravitational field of the lens near I. We now carry out the calculations in three steps.

Geometrical time delay. To obtain the *geometric time delay* Δt_{geom}, we write the RW metric (4.31) as

$$ds^2 = R^2[\eta]\{d\eta^2 - d\sigma^2\} \quad , \tag{4.61}$$

where $\eta := c \int \frac{dt}{R(t)}$ denotes the so-called conformal time and $R[\eta] = R(t)$ is the scale function in terms of η.

Let the emission "time" be $\eta = 0$ (by choice of the η-origin). According to (4.61) and because of $ds^2 = 0$ along rays, the geometric "time" delay is given by

$$\Delta \eta_{\text{geom}} = \sigma_{\text{ds}} + \sigma_{\text{d}} - \sigma_{\text{s}} \quad , \tag{4.62}$$

where the σ's denote distances measured by means of the metric $d\sigma^2$ given in (4.32). Since the time delay is very small compared to the Hubble time H_0^{-1}, we may put

$$c\Delta t_{\text{geom}} = R_0 \Delta \eta_{\text{geom}} \quad . \tag{4.63}$$

To evaluate $\Delta \eta_{\text{geom}}$, we consider first the case of a locally spherical 3-space, $k = 1$. Then, the law of cosines of spherical trigonometry gives

$$\cos \sigma_{\text{s}} = \cos \sigma_{\text{ds}} \cos \sigma_{\text{d}} - \sin \sigma_{\text{ds}} \sin \sigma_{\text{d}} \cos \hat{\alpha}$$

or, with

$$\cos \hat{\alpha} = 1 - 2\sin^2\left(\frac{\hat{\alpha}}{2}\right) \quad ,$$

$\cos \sigma_s = \cos(\sigma_{ds} + \sigma_d) + 2 \sin \sigma_{ds} \sin \sigma_d \sin^2\left(\frac{\hat{\alpha}}{2}\right)$. Using again the cosine addition law we obtain, remembering (4.62),

$$\sin\left(\frac{1}{2}\Delta\eta_{geom}\right) = \frac{\sin \sigma_{ds} \sin \sigma_d}{\sin \frac{1}{2}(\sigma_{ds} + \sigma_d + \sigma_s)} \sin^2\left(\frac{\hat{\alpha}}{2}\right) \quad .$$

If we take $\hat{O}$ and $\hat{S}$ as fixed, but $\hat{I}$ as variable and approaching, as a function of $\hat{\alpha}$, the point $\hat{I}_0$ on the unperturbed path, we infer from the last equation that, for $\hat{\alpha} \to 0$, $\Delta\eta_{geom} \sim \hat{\alpha}^2$, and thus, since $\sigma_s \approx \sigma_{ds} + \sigma_d$,

$$\Delta\eta_{geom} = \frac{\sin \sigma_{ds} \sin \sigma_d}{2 \sin \sigma_s} \hat{\alpha}^2 + \mathcal{O}\left(\hat{\alpha}^4\right) \quad . \tag{4.64}$$

Next, we relate the σ-distances to (unperturbed) angular diameter distances. Consider a light beam such as was used to define the angular diameter distance D_{ds} of a source at S from I, i.e., (practically) from the lens. If its diameter at S is d and it subtends an angle δ at I, its projection from spacetime into the 3-space Σ_1 has diameter d/R_s at $\hat{S}$, by (4.61), and it subtends the same angle δ at $\hat{I}$ since the projection preserves angles. Therefore, $D_{ds} = d/\delta = R_s \sin \sigma_{ds}$. The first of these two equations is just the definition of D_{ds}, the second one states the relation which holds on a unit sphere between the length of an arc and the angle it subtends at a distance σ_{ds}. Similarly one gets $D_d = R_d \sin \sigma_d$ and $D_s = R_s \sin \sigma_s$. Further, Fig. 4.1 shows that $(\boldsymbol{\theta} - \boldsymbol{\beta}) \sin \sigma_s = \hat{\boldsymbol{\alpha}} \sin \sigma_{ds}$. Putting these equations into (4.63) and (4.64) and recalling that $R_0/R_d = 1 + z_d$, leads to the result,

$$c\Delta t_{geom} = (1 + z_d) \frac{D_d D_s}{2 D_{ds}} (\boldsymbol{\theta} - \boldsymbol{\beta})^2 \quad . \tag{4.65}$$

This formula also holds if $k = 0$ or $k = -1$. In the Euclidean case, $k = 0$, it follows by repeating the preceding considerations with plane rather than spherical geometry, and in the case of hyperbolic geometry, $k = -1$, the analogous calculation holds with hyperbolic functions instead of trigonometric ones.

Potential time delay. To obtain the *potential time delay*, we can use the result (4.26). As was emphasized below that equation, the $\boldsymbol{\xi}$-dependent part of the potential time delay arises locally when a ray traverses the neighborhood of the lens. Since the corresponding time-differences between rays are "carried from the lens to the observer" along those light rays, the cosmological potential time delay is obtained from the local one, (4.26), by "redshifting" it. Thus, we may write

$$c\Delta t_{pot} = -(1 + z_d) \hat{\psi}(\boldsymbol{\xi}) + \text{const.} \quad , \tag{4.66}$$

where the constant is the same for all rays from the source plane to the observer.

Total time delay. We now collect the results in order to obtain the *total time delay*, up to an irrelevant constant. From (4.65) and (4.66) we get

$$c\Delta t = (1 + z_{\rm d}) \left\{ \frac{D_{\rm d} D_{\rm s}}{2 D_{\rm ds}} (\boldsymbol{\theta} - \boldsymbol{\beta})^2 - \hat{\psi}(\boldsymbol{\xi}) \right\} + \text{const.} \quad . \tag{4.67}$$

To put this result into a more transparent form, we first employ the results about angular diameter distances obtained in Sect. 4.5. We insert $D = \frac{c}{H_0} r$ and apply (4.54) with $z_1 \to z_{\rm d}$, $z_2 \to z_{\rm s}$, to get the useful equation

$$(1 + z_{\rm d}) \frac{D_{\rm d} D_{\rm s}}{D_{\rm ds}} = \frac{c}{H_0} \left[\chi(z_{\rm d}) - \chi(z_{\rm s}) \right]^{-1} \quad , \tag{4.68a}$$

where

$$\chi(z; \Omega, \tilde{\alpha}) := \int_z^\infty \frac{d\zeta}{r^2(\zeta; \Omega, \tilde{\alpha}) \, (\zeta + 1)^3 \sqrt{\Omega \zeta + 1}} \quad ; \tag{4.68b}$$

the dimensionless Dyer–Roeder distance $r(z; \Omega, \tilde{\alpha})$ was defined in Sect. 4.5. [Note that the preceding equations give $\eta' = \left[(1 + z_{\rm d}) \, (\chi(z_{\rm d}) - \chi(z_{\rm s})) \right]^{-1}$ for the coefficient introduced in (2.9b).]

Secondly, we split the deflection potential $\hat{\psi}$ of (4.27c) into a scale factor and a dimensionless function $\tilde{\psi}$, using the angular variable $\boldsymbol{\theta} = \boldsymbol{\xi}/D_{\rm d}$ and putting $\xi_0 = D_{\rm d}$, as follows:

$$\hat{\psi}(D_{\rm d} \boldsymbol{\theta}) = 2 R_{\rm S} \tilde{\psi} [\boldsymbol{\theta}] \quad , \tag{4.69}$$

$$\tilde{\psi} [\boldsymbol{\theta}] := \int \tilde{\Sigma} [\boldsymbol{\theta}'] \ln |\boldsymbol{\theta} - \boldsymbol{\theta}'| \, d^2 \theta' \quad . \tag{4.70}$$

Here,

$$\tilde{\Sigma} [\boldsymbol{\theta}] := \frac{\Sigma(D_{\rm d} \boldsymbol{\theta})}{M} D_{\rm d}^2 \tag{4.71}$$

represents the *fraction* of the total mass, M, of the lens contained in the solid angle $d^2 \theta'$, as seen by the observer, and $R_{\rm S} = \frac{2GM}{c^2}$.

Using these results and definitions, we rewrite (4.67) in the form

$$c\Delta t = \hat{\phi}(\boldsymbol{\theta}, \boldsymbol{\beta}) + \text{const.} \quad , \tag{4.72}$$

where the *cosmological Fermat potential* $\hat{\phi}$ is given by

$$\hat{\phi}(\boldsymbol{\theta}, \boldsymbol{\beta}) = \frac{c}{2 H_0} \frac{(\boldsymbol{\theta} - \boldsymbol{\beta})^2}{\chi(z_{\rm d}) - \chi(z_{\rm s})} - 2 R_{\rm S} (1 + z_{\rm d}) \, \tilde{\psi} [\boldsymbol{\theta}] \quad . \tag{4.73}$$

We see that the time delay Δt depends on the dimensionless parameters $z_{\rm d}$, $z_{\rm s}$, Ω, $\tilde{\alpha}$, the dimensionless relative mass distribution per solid angle, $\tilde{\Sigma}$, of the deflector, the position angles $\boldsymbol{\theta}$, $\boldsymbol{\beta}$ of images and source with respect to an (arbitrarily chosen) optical axis, and the two time scales H_0^{-1}, $c^{-1} R_{\rm S}$.

[For $z \ll 1$, we have $z \approx c^{-1} H_0 D$ and $\chi(z) \approx z^{-1} \approx \frac{c}{H_0 D}$, hence the cosmological Fermat potential (4.73) then reduces to the local one given in (4.27), written in terms of angles. Thus, (4.73) does indeed generalize (4.27).]

The cosmological lens mapping. Fermat's principle now provides the cosmological lens mapping, $\boldsymbol{\theta} \mapsto \boldsymbol{\beta}$, via $\frac{\partial \hat{\phi}}{\partial \boldsymbol{\theta}} = 0$:

$$\boldsymbol{\beta} = \boldsymbol{\theta} - \frac{2R_S}{cH_0^{-1}} (1 + z_d)[\chi(z_d) - \chi(z_s)] \frac{\partial \tilde{\psi}}{\partial \boldsymbol{\theta}} , \qquad (4.74a)$$

which may also be written in terms of $\boldsymbol{\xi} = D_d \boldsymbol{\theta}$, $\boldsymbol{\eta} = D_s \boldsymbol{\beta}$, $\hat{\boldsymbol{\alpha}}(\boldsymbol{\xi}) = \frac{2R_S}{D_d} \frac{\partial \tilde{\psi}}{\partial \boldsymbol{\theta}}$ as

$$\boldsymbol{\eta} = \frac{D_s}{D_d} \boldsymbol{\xi} - D_{ds} \hat{\boldsymbol{\alpha}}(\boldsymbol{\xi}) . \qquad (4.74b)$$

This equation is identical with (2.15) and (4.28). Now, however, it refers to the cosmological context and is no longer restricted to $z \ll 1$.

For actual (as opposed to kinematically possible) rays, the unobservable $\boldsymbol{\beta}$ can be eliminated from Δt by means of (4.74a), which gives $\boldsymbol{\theta} - \boldsymbol{\beta}$ in terms of $\boldsymbol{\theta}$. For two images at positions $\boldsymbol{\theta}_i$, $\boldsymbol{\theta}_j$ with separation $\boldsymbol{\theta}_{ij} := \boldsymbol{\theta}_i - \boldsymbol{\theta}_j$ and time delay Δt_{ij}, one obtains from (4.72–74):

$$\boldsymbol{\theta}_{ij} = \frac{2R_S}{cH_0^{-1}} (1 + z_d)[\chi(z_d) - \chi(z_s)] \left(\frac{\partial \tilde{\psi}}{\partial \boldsymbol{\theta}} [\boldsymbol{\theta}_i] - \frac{\partial \tilde{\psi}}{\partial \boldsymbol{\theta}} [\boldsymbol{\theta}_j] \right) \qquad (4.75)$$

and

$$\Delta t_{ij} = \frac{2R_S}{c} (1 + z_d) \left\{ \boldsymbol{\theta}_{ij} \frac{1}{2} \left(\frac{\partial \tilde{\psi}}{\partial \boldsymbol{\theta}} [\boldsymbol{\theta}_i] + \frac{\partial \tilde{\psi}}{\partial \boldsymbol{\theta}} [\boldsymbol{\theta}_j] \right) \right.$$
$$\left. - \left(\tilde{\psi}[\boldsymbol{\theta}_i] - \tilde{\psi}[\boldsymbol{\theta}_j] \right) \right\} . \qquad (4.76)$$

For any model of the relative mass distribution $\Sigma[\boldsymbol{\theta}]$ of the lens, the first formula relates the dimensionless parameters Ω, $\tilde{\alpha}$, z_d, z_s, $\frac{2R_S}{cH_0^{-1}}$ to the (in principle observable) angles $\boldsymbol{\theta}_i$, $\boldsymbol{\theta}_j$. The second one shows that Δt_{ij} is related directly to the Schwarzschild radius R_S, i.e., the mass, of the deflector without intervention of Ω, $\tilde{\alpha}$, z_s or H_0. Only a combination of (4.75) and (4.76) can, in principle, be used to determine both R_S and H_0, given the cosmological parameters Ω, $\tilde{\alpha}$, the redshifts z_d, z_s, the angles $\boldsymbol{\theta}_i$ and a deflector mass distribution model. The foregoing equations also indicate the difficulty of determining such a model, i.e., a function $\tilde{\psi}[\boldsymbol{\theta}]$, from a rather small number of data. Clearly, images of extended sources provide more information to restrict the deflector models than point sources. Note that the cosmological parameters Ω and $\tilde{\alpha}$ enter the equations (4.75), (4.76) only through the function χ; this function provides the link between lensing models and the underlying cosmology.

In order to illustrate possible applications of (4.75) and (4.76), we consider the case of a point source lensed by a point mass [RE64.1]. Then, by (4.70), $\tilde{\psi}[\boldsymbol{\theta}] = \ln|\boldsymbol{\theta}|$, and since in this case the relevant light rays are contained in a plane, we can use numerical, signed angles θ_i instead of vectors (as in Fig. 2.1) in evaluating (4.75) and (4.76). We obtain

$$|\theta_1 \theta_2| = \frac{2R_S}{cH_0^{-1}} (1+z_d)[\chi(z_d) - \chi(z_s)] \qquad (4.77)$$

and

$$\Delta t_{12} = \frac{2R_S}{c} (1+z_d) \left(\frac{\theta_2^2 - \theta_1^2}{2|\theta_1 \theta_2|} + \ln\left|\frac{\theta_2}{\theta_1}\right| \right) \qquad (4.78a)$$

Alternatively, by (2.25), denoting by ν the ratio of absolute values of the magnifications of the images,

$$\Delta t_{12} = \frac{R_S}{c} (1+z_d) \left(\nu^{1/2} - \nu^{-1/2} + \ln \nu \right) \qquad (4.78b)$$

Hence, observation of $\theta_1, \theta_2, z_d, z_s$ in one such case gives but one relation between $\frac{2R_S}{cH_0^{-1}}$, Ω, $\tilde{\alpha}$. Observation of $\theta_1, \theta_2, z_d, z_s$ and Δt in one such case would provide, besides the relation between $\frac{2R_S}{cH_0^{-1}}$, Ω, $\tilde{\alpha}$ already mentioned, the value of R_S. Observation of 3 (suitable) such cases would provide H_0, Ω, $\tilde{\alpha}$ and the masses of the lenses. This highly idealized consideration is, in practice, simplified if one has some rough information on Ω and $\tilde{\alpha}$, since χ depends only weakly on Ω and $\tilde{\alpha}$, particularly for redshifts $\lesssim 0.5$. On the other hand – and this is unfortunately more significant – the analysis becomes much more complicated for extended lenses whose mass distribution has to be determined too.

The function χ which occurs in (4.68) is displayed for a few values of Ω and $\tilde{\alpha}$ in the following equations and in Fig. 4.2:[10]

(i) $\Omega = 1$:

$$\chi(z; 1, \tilde{\alpha}) = \frac{2\beta}{(1+z)^{2\beta} - 1} \quad ; \quad \beta = \frac{1}{4}\sqrt{25 - 24\tilde{\alpha}} \quad ,$$

(ii) $\Omega = 0$:

$$\chi(z; 0, \tilde{\alpha}) = \frac{2}{(1+z)^2 - 1} \quad , \qquad (4.79)$$

(iii) $\tilde{\alpha} = 1$:

$$\chi(z; \Omega, 1) = \frac{\Omega}{2} \left[\frac{1}{\sqrt{1+\Omega z} - 1} + \frac{\Omega - 1}{\sqrt{1+\Omega + \Omega z} - 1} \right] \quad .$$

---

[10] We are grateful to M. Bartelmann for preparing this figure for us.

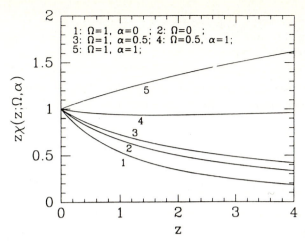

Fig. 4.2. The graph of the function $z\chi(z;\Omega,\tilde{\alpha})$ for several values of Ω and $\tilde{\alpha}$

Frequency shift. The derivation of the cosmological Fermat potential given above amounts to the assumption that in a neighborhood of the relevant light rays, the spacetime metric can be approximated by

$$ds^2 = R^2\left[\eta\right]\left\{\left(1+\frac{2U}{c^2}\right)d\eta^2 - \left(1-\frac{2U}{c^2}\right)d\sigma^2\right\} \tag{4.80}$$

where U is the time-independent gravitational potential of the deflector. The expression (4.80) combines the FL metric (4.61) with the "local" metric (4.13).

The metric (4.80) is conformally static, hence the formula (3.16b) for the frequency ratio applies. Since the gravitational potential of the deflector practically vanishes at the source and at the observer, the quoted formula reduces to the standard cosmological redshift law (4.33a), and since the scale function R does not change appreciably on the scale of the time delay Δt, it follows that all images exhibit practically the same redshift. The result that the potential does not affect the redshift may be said to be due to the fact that the photons do not change their energies by passing through the conservative field represented by U.

Magnification factor. In deriving the Fermat potential and the lens mapping, we have used Dyer–Roeder distances referring to empty cones, see (4.68a,b). This means, in particular, that the magnification factor μ, given by (2.18) and applied to (4.74a), has to be interpreted in the cosmological context as follows: μ is the ratio of the flux S actually observed in an image of a source and the flux S_0 which the same source would produce if seen, at the same redshift, through an empty cone, without deflection. Thus, we may write

$$S = \frac{L}{4\pi D_L^2(z)}\mu \quad, \tag{4.81}$$

where

$$D_{\text{L}}(z) := \frac{c}{H_0}(1+z)^2 \, r(z, \Omega, \tilde{\alpha})$$

denotes the luminosity distance related to an empty cone. This D_{L} differs from the one, $\bar{D}_{\text{L}}$, which occurs in the average-flux law (4.43b); in our present notation,

$$\bar{D}_{\text{L}}(z) = \frac{c}{H_0}(1+z)^2 \, r(z; \Omega, 1) \quad .$$

Thus, (4.43b) may be expressed in the equivalent form

$$\langle \mu \rangle = \left(\frac{r(z; \Omega, \tilde{\alpha})}{r(z; \Omega, 1)} \right)^2 \quad , \tag{4.82}$$

which shows how the angular average $\langle \mu \rangle$ depends on $z, \Omega, \tilde{\alpha}$.

4.7 Wave optics in lens theory

In the preceding chapter we used Maxwell's equations only to derive the basic laws of geometrical optics, which we employed in this chapter to set up the lens equation. Indeed, as will be shown in Chap. 7, in nearly all cases of practical importance for lens theory, geometrical optics suffices. However, in this section we go back to Maxwell's equations and derive a formula which will be used as a starting point in Chap. 7 and which provides a basis for investigating coherence properties of lensed radiation fields.

Assumptions and strategy. We consider again, as in the previous section, radiation travelling from a source past a thin lens to an observer, all three assumed to participate in the mean motion of a Friedmann–Lemaître universe, perturbed by a deflecting mass. We wish to determine the electromagnetic field $F_{\alpha\beta}$ at the observer to within an approximation which takes into account the wave properties of light in leading order, as required in particular if the observer is situated close to a caustic of the source's future light cone. (Since conjugacy of events on a light ray is a symmetric relation, this happens if, and only if, the source is close to a caustic of the observer's past light cone.) To do this, we make essential use of the fact that *all Robertson–Walker spacetimes are locally conformally flat* (see, e.g., [HA73.1], Chap. 5), so that instead of (4.80), we may write the perturbed metric as

$$ds^2 = \Psi^2 \widetilde{ds}^2 = \Psi^2 \left\{ \left(1 + \frac{2U}{c^2}\right) c^2 dT^2 - \left(1 - \frac{2U}{c^2}\right) d\mathbf{X}^2 \right\} \quad , \tag{4.83}$$

where the conformal factor Ψ depends on the curvature index k of the background Friedmann–Lemaître model and on the coordinates T, $\mathbf{X}$, which are nearly orthonormal for $\widetilde{ds}^2$ except near the lens, where the gravitational potential U differs appreciably from zero. (The domain of spacetime where (4.83) holds is sufficiently large to contain the light bundles of interest, as can be verified by means of the details given in [HA73.1].)

Since Maxwell's equations are conformally invariant, we can and will work with the static metric $\widetilde{ds}^2$ of (4.83). Let us, then, denote as E the lens plane in $\mathbf{X}$-space, and by E' a plane between the observer's position $\hat{O}$ and E, parallel to E; see Fig. 4.3. We assume that E' can be chosen such that, (a) E' is close to E in the sense that the (Euclidean) distances d' and d of E' and E from $\hat{O}$ obey $d - d' \ll d$, (b) between the source and E', there are no caustics, and (c) between E' and $\hat{O}$, the gravitational potential U of the lens is negligible, so that there the metric $\widetilde{ds}^2$ may be taken to be flat. (For actual lensing situations, these assumptions seem reasonable. The effect of the lens on the light happens essentially near the lens.)

Because of assumption (b), the field $F_{\alpha\beta}$ on E' can be well approximated by geometrical optics, as in (3.22). To compute the field at the position $\hat{O}$ of the observer in terms of its boundary values on E', we apply *Kirchhoff's formula*, known from diffraction theory, to (the cartesian components of) $F_{\alpha\beta}$. This is possible due to assumption (c) and the fact that, in flat spacetime, Maxwell's equations imply the ordinary *wave equation*, $\Box F_{\alpha\beta} = 0$ [to obtain it, apply the operation $(\ldots)^{;\gamma}$ to (3.2a) and use (3.2b)].

Since the metric $\widetilde{ds}^2$ is static, it is possible to Fourier-transform the real field $F_{\alpha\beta}$ with respect to the "time" T,

$$F_{\alpha\beta}(\mathbf{X}, T) = \int_{-\infty}^{\infty} \tilde{F}_{\alpha\beta}(\mathbf{X}, \Omega) e^{-i\Omega T} d\Omega \quad . \tag{4.84}$$

It is convenient to consider first a monochromatic, positive frequency

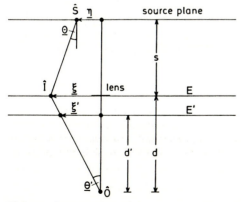

Fig. 4.3. Source plane, lens plane E, intermediate plane E' and observer position $\hat{O}$ in Euclidian $\mathbf{X}$-space as explained in the text

component $\tilde{F}_{\alpha\beta}(\mathbf{X})$, suppressing the argument Ω and the factor $\mathrm{e}^{-i\Omega T}$. Then, $\tilde{F}_{\alpha\beta}$ obeys, between E' and $\hat{O}$, the flat-space *Helmholtz equation* $\Delta \tilde{F}_{\alpha\beta} + \left(\frac{\Omega}{c}\right)^2 \tilde{F}_{\alpha\beta} = 0$, and consequently the field at $\hat{O}$ can be represented by the *Kirchhoff integral* ([BO80.1], Sect. 8.3.2)

$$\tilde{F}_{\alpha\beta}(\hat{O}) = \frac{1}{4\pi} \int_{E'} d^2 \xi' \left\{ \tilde{F}_{\alpha\beta} \frac{\partial}{\partial n} \left(\frac{\mathrm{e}^{i\frac{\Omega}{c}d'}}{d'} \right) - \frac{\mathrm{e}^{i\frac{\Omega}{c}d'}}{d'} \frac{\partial}{\partial n} \tilde{F}_{\alpha\beta} \right\} \quad . \tag{4.85}$$

Here, we denote the coordinates in E' by $\boldsymbol{\xi}'$, $\mathbf{n}$ denotes the normal to E' pointing towards $\hat{O}$, and $d'(\boldsymbol{\xi}')$ is the distance of $\hat{O}$ from a running point $\boldsymbol{\xi}'$ on E' (the "screen" in diffraction theory). We also consider, for the time being, a point source located at $\hat{S}$.

The following calculations closely parallel those of ordinary diffraction theory. This is possible due to three circumstances: (i) Maxwell's equations are conformally invariant, (ii) Robertson–Walker metrics are conformally flat, and (iii) the effect of a lens on an electromagnetic field is essentially restricted to a (cosmologically) small region.

The spatial part $\tilde{S}$ of the phase of the monochromatic field under discussion must obey the eikonal equation $(\nabla \tilde{S})^2 = \left(\frac{n\Omega}{c}\right)^2$, with the refractive index (4.16) due to the lens' field. Its change along the ray paths is given by $\frac{\Omega}{c} \int n\, dl$. Its values on E' can thus be expressed in terms of the Fermat potential as follows:

$$\tilde{S}(\boldsymbol{\xi}', \boldsymbol{\eta}) = \frac{\Omega}{c} \left(\hat{\phi}(\boldsymbol{\xi}, \boldsymbol{\eta}) - d'(\boldsymbol{\xi}') \right) + \alpha(\boldsymbol{\eta}) \quad . \tag{4.86}$$

Here, $\boldsymbol{\xi}$ and $\boldsymbol{\eta}$ denote position vectors in the lens and source planes, respectively, as in Fig. 4.3 and in the previous section; α is the $\boldsymbol{\xi}$-independent part of the phase.[11]

The complex amplitude A_α of our field [defined as in (3.17)] at $\boldsymbol{\xi}'$ is determined by the emission mechanism of the source and the transport law stated below (3.59). It is practically the same as at $\boldsymbol{\xi}$; see Fig. 4.3. [Recall assumption (a) and the fact that we are concerned with small-angle scattering, in which case the lens affects the amplitude, due to *differential* deflection near the lens, only on the long journey from E' to $\hat{O}$, as is obvious from (2.20).] Therefore, we may write

$$A_\alpha(\boldsymbol{\xi}') = \frac{\mathrm{e}^{-i\alpha} C_\alpha(\boldsymbol{\xi})}{\Omega^2 s} \quad ,$$

if s denotes the distance of $\hat{S}$ from $\hat{I}$. $C_\alpha(\boldsymbol{\xi})$ determines both the intensity and the polarisation of the source's radiation in the direction specified by $\boldsymbol{\xi}$. The factor $\Omega^{-2} \mathrm{e}^{-i\alpha}$ has been included for convenience; it simplifies the

---

[11] Since $\hat{\phi}$ has been defined as the time delay relative to the unperturbed ray, $\alpha(\boldsymbol{\eta})$ has to be introduced here to subtract the contribution due to the unperturbed ray.

following equations (we temporarily suppress the argument η since we are dealing with a fixed point source).

Combining phase and amplitude, we obtain for the *field on E'*,

$$\tilde{F}_{\alpha\beta}(\boldsymbol{\xi}') = \frac{2}{\Omega^2 s} k_{[\alpha} C_{\beta]}(\boldsymbol{\xi}) \, e^{i\frac{\Omega}{c}(\hat{\phi}(\boldsymbol{\xi})-d')} \quad . \tag{4.87}$$

If this is inserted into Kirchhoff's formula (4.85) and if derivatives of slowly changing functions are neglected compared to those of the rapidly changing phases, there results

$$\tilde{F}_{\alpha\beta}(\hat{O}) = (2\pi i c\Omega)^{-1} \int_{E'} d^2\xi' \, \frac{k_{[\alpha} C_{\beta]}(\boldsymbol{\xi}')}{(sd')(\boldsymbol{\xi}')} \, e^{i\frac{\Omega}{c}\hat{\phi}(\boldsymbol{\xi})}(\cos\theta + \cos\theta') \tag{4.88}$$

(the angles θ, θ' are shown in Fig. 4.3). The factor $e^{i\frac{\Omega}{c}\hat{\phi}(\boldsymbol{\xi})}$ in the integrand accounts for the retardations with which the partial waves (in the sense of Huyghens principle) arrive at $\hat{O}$. Since the phase factor oscillates very rapidly, the integrand contributes to the integral appreciably only for $\boldsymbol{\xi}$-values close to the "geometrical optics values" where $\frac{\partial}{\partial \boldsymbol{\xi}}\hat{\phi} = 0$ and θ, $\theta' \ll 1$ (stationary phase method, cf. [BO80.1], App. III). Moreover, because of (a), we may take $\boldsymbol{\xi}' = \boldsymbol{\xi}$ in (4.88), which then simplifies to

$$\tilde{F}_{\alpha\beta}(\hat{O}) = (i\pi c\Omega)^{-1} \frac{k_{[\alpha} C_{\beta]}}{sd} \int_E d^2\xi \, e^{i\frac{\Omega}{c}\hat{\phi}(\boldsymbol{\xi})} \quad . \tag{4.89}$$

The values of k_α, s, d, C_α in this formula are those corresponding to the optical axis. This equation is a tensorial version of the *Fresnel–Kirchhoff diffraction integral*, ([BO80.1], Sect. 8.3.3) applied to the case of interest here.[12]

We now express the flat-space frequency Ω by the physical frequency ω at the observer. If $\Psi(O)$ is the value of the conformal factor of (4.83) at the observation event O, then $\Omega = \Psi(O)\omega$. If we also replace the "T-time Fermat potential" by the "physical" one and call that again $\hat{\phi}$, and if we write $k_\alpha = -\frac{\omega}{c} n_\alpha$ for the wave covector at the observer, we finally obtain, reintroducing the variables ω and η, allowing for an extended source, but neglecting the very small variation of n_α as well as the $\boldsymbol{\xi}$-dependence of C_α,

$$\tilde{F}_{\alpha\beta}(\hat{O},\omega) = 2 A n_{[\alpha} \int_{\text{source}} C_{\beta]}(\omega,\boldsymbol{\eta}) \, V(\omega,\boldsymbol{\eta}) \, d^2\eta \quad , \tag{4.90}$$

$$V(\omega,\boldsymbol{\eta}) := \int_{\mathbf{R}^2} d^2\xi \, e^{i\frac{\omega}{c}\hat{\phi}(\boldsymbol{\xi},\boldsymbol{\eta})} \quad . \tag{4.91}$$

The complex constant A depends on the positions of source plane, lens plane and observer, but is independent of ω, $\boldsymbol{\eta}$ and the properties of the

---

[12] It is possible to prove an exact, tensorial formula of this type for a general, curved spacetime, provided the "screen" E is contained in a normal neighborhood of the "observation event" O [EL72.1]. However, that has not been evaluated under the conditions of gravitational lensing, where it presumably reduces practically to (4.89).

lens. In an orthonormal rest frame at the observer, $n_\alpha = (1, \mathbf{e})$, where $\mathbf{e}$ is the spatial unit vector pointing towards the source. The complex vector C_α characterizes the emission processes and may be taken to be spatial and orthogonal to $\mathbf{e}$ at the observer, $C_\alpha = (0, \mathbf{C})$, $\mathbf{C} \cdot \mathbf{e} = 0$. The complex function $V(\omega, \boldsymbol{\eta})$ contains the effect of the deflecting mass on the electromagnetic field.

A transparent, deflecting mass acts on the light like a "phase object" in the image formation process in a microscope. The part of the phase factor in (4.91) which is due to the gravitational potential of the lens corresponds to the "transmission function" of the object (cf. [BO80.1], Sect. 8.6.3). Therefore, we call V the *transmission factor* of a lensing configuration.

According to (3.1), (4.84) and (4.90), the electric field at the observer is the Fourier transform of

$$\tilde{\mathbf{E}}(\omega) = A' \int_{\text{source}} V(\omega, \boldsymbol{\eta}) \, \mathbf{C}(\omega, \boldsymbol{\eta}) \, d^2\eta \quad (A' = A\Psi(O)) \quad ; \qquad (4.92)$$

i.e.

$$\mathbf{E}(t) = \int_{-\infty}^{+\infty} \tilde{\mathbf{E}}(\omega) \, e^{-i\omega t} \, d\omega \quad , \qquad (4.93)$$

and the magnetic field is $\mathbf{B} = \mathbf{E} \times \mathbf{e}$. Since $\mathbf{E}(t)$ is real, $\tilde{\mathbf{E}}(-\omega) = \tilde{\mathbf{E}}^*(\omega)$; thus, all the information about the electromagnetic field is contained in $\tilde{\mathbf{E}}(\omega)$ for $\omega > 0$ or, equivalently, in the complex, positive frequency part of the field,

$$\mathbf{E}^+(t) = \int_0^\infty \tilde{\mathbf{E}}(\omega) \, e^{-i\omega t} \, d\omega \quad (\mathbf{E} = 2\mathcal{R}e\mathbf{E}^+) \quad . \qquad (4.94)$$

Given the configuration of observer, lens and source plane, this field or *complex signal* depends, in turn, on the emission process of the source, i.e., on $\mathbf{C}(\omega, \boldsymbol{\eta})$, (4.92). Due to incomplete information about, as well as quantum properties of, this process the latter is described as a stochastic process, represented by an ensemble $\{\mathbf{C}(\omega, \boldsymbol{\eta})\}$ of functions, and observables of the radiation field, such as Stokes polarization parameters, are given by ensemble averages indicated by brackets, $\langle...\rangle$.

For a *quasistationary* source – the case of main interest in the context of lensing – the specific flux S_ω at the observer can be expressed as the ensemble average

$$S_\omega = A'' \left\langle \left| \int_{\text{source}} d^2\eta \, V(\omega, \boldsymbol{\eta}) \, \mathbf{C}(\omega, \boldsymbol{\eta}) \right|^2 \right\rangle \quad . \qquad (4.95a)$$

In this formula, A'' is a geometrical constant such as A in (4.90) (for details, see [BO80.1], Sect. 10.2, and the references quoted there). Writing (4.95a) in the form

$$S_\omega = A'' \int d^2\eta \, d^2\eta' \, V^*(\omega,\boldsymbol{\eta}) \, V(\omega,\boldsymbol{\eta}) \, \langle \mathbf{C}^*(\omega,\boldsymbol{\eta}) \, \mathbf{C}(\omega,\boldsymbol{\eta}') \rangle \qquad (4.95b)$$

and assuming different points of the source to radiate *incoherently*,

$$\langle \mathbf{C}^*(\omega,\boldsymbol{\eta}) \, \mathbf{C}(\omega,\boldsymbol{\eta}') \rangle \propto I_\omega(\boldsymbol{\eta}) \, \delta(\boldsymbol{\eta}-\boldsymbol{\eta}') \quad , \qquad (4.96)$$

we finally get the specific flux in terms of the specific surface brightness $I_\omega(\boldsymbol{\eta})$ of the source,

$$S_\omega = A''' \int d^2\eta \, |V(\omega,\boldsymbol{\eta})|^2 \, I_\omega(\boldsymbol{\eta}) \quad , \qquad (4.97)$$

with still another constant A''', which will be determined in Chap. 7, where some consequences of (4.97) will be derived.

5. Properties of the lens mapping

In this chapter we describe general properties of the lens mapping. In Sect. 5.1 the basic equations of GL theory are presented and briefly discussed. A general description of the lens mapping in the infinitesimal neighborhood of a point is given in Sect. 5.2. It leads to the concepts of magnification and critical curves and to a classification of ordinary images. In Sect. 5.3 the light-travel-time along ray bundles that correspond to GL images is determined by means of the geometry of wavefronts; this again leads to the connection between Fermat's principle and the lens mapping. The difference in light-travel-time for different images of a source is an observable; this time delay is one of the central quantities in gravitational lensing. Two general theorems are proved in Sect. 5.4, and a classification of 3-image and 5-image topographies is presented in Sect. 5.5.

5.1 Basic equations of the lens theory

The ray-trace (or lens) equation. In Chap. 4, we derived a relation between the position of a source and the impact vector in the lens plane of those light rays which connect source and observer. Equivalently, we found a relation between the angular position of an unlensed source and the position of its images if the light rays emanating from the source are perturbed by a gravitational field. If the "optical axis" is defined to be the straight line connecting the observer and some reference point in the lens plane (e.g., the center of mass), and if the coordinate frames in the source and lens plane are chosen to have their origin on the point of intersection of the optical axis with the corresponding plane, this relation reads

$$\boldsymbol{\eta} = \frac{D_\mathrm{s}}{D_\mathrm{d}} \boldsymbol{\xi} - D_\mathrm{ds}\, \hat{\boldsymbol{\alpha}}(\boldsymbol{\xi}) \quad , \tag{5.1}$$

where $\boldsymbol{\eta}$ is the source position, $\boldsymbol{\xi}$ the impact vector in the lens plane, and D_d, D_s, D_ds are the angular diameter distances from us to the lens, from us to the source, and from the lens to the source, respectively. This equation agrees with (2.15b); it holds if the lens is embedded in an almost flat spacetime as well as in a nearly Robertson–Walker spacetime.

The deflection angle. It was shown above that in most cases of astrophysical interest, the deflection angle $\hat{\boldsymbol{\alpha}}(\boldsymbol{\xi})$ can be obtained by projecting the volume mass density of the deflector onto the lens plane, which results in a surface mass density $\Sigma(\boldsymbol{\xi})$; then, the deflection angle is

$$\hat{\boldsymbol{\alpha}}(\boldsymbol{\xi}) = \int_{\mathbb{R}^2} d^2\xi' \frac{4G\Sigma(\boldsymbol{\xi}')}{c^2} \frac{\boldsymbol{\xi} - \boldsymbol{\xi}'}{|\boldsymbol{\xi} - \boldsymbol{\xi}'|^2} \quad , \tag{5.2}$$

where the integral extends over the lens plane. The deflection angle is thus simply a superposition of Einstein angles for mass elements $dm = \Sigma(\boldsymbol{\xi}')d^2\xi'$.

The lens equations (5.1) and (5.2) describe a mapping $\boldsymbol{\xi} \mapsto \boldsymbol{\eta}$ from the lens plane to the source plane; for any mass distribution $\Sigma(\boldsymbol{\xi})$, this mapping is obtained straightforwardly. The problem of GL theory is the inversion of (5.1), i.e., to determine the image positions $\boldsymbol{\xi}$ for a given source position $\boldsymbol{\eta}$. Since the mapping $\boldsymbol{\xi} \mapsto \boldsymbol{\eta}$ is nonlinear, the inversion has been carried out analytically only for very simple matter distributions (see Chap. 8).

Scaling to dimensionless quantities. It is useful to rewrite equations (5.1) and (5.2) in dimensionless form. Let us define a length scale ξ_0 in the lens plane and a corresponding length scale $\eta_0 = \xi_0 D_s/D_d$ in the source plane. Then, we define the dimensionless vectors

$$\mathbf{x} = \frac{\boldsymbol{\xi}}{\xi_0} \quad ; \quad \mathbf{y} = \frac{\boldsymbol{\eta}}{\eta_0} \quad , \tag{5.3}$$

as well as the dimensionless surface mass density

$$\kappa(\mathbf{x}) = \frac{\Sigma(\xi_0 \mathbf{x})}{\Sigma_{\mathrm{cr}}} \quad , \tag{5.4}$$

where the *critical surface mass density* Σ_{cr} is

$$\begin{aligned} \Sigma_{\mathrm{cr}} &= \frac{c^2 D_s}{4\pi G D_d D_{ds}} \\ &= \frac{cH_0}{4\pi G}\left[(1+z_d)\left[\chi(z_d)-\chi(z_s)\right]r^2(z_d;\Omega,\tilde{\alpha})\right]^{-1} \quad . \end{aligned} \tag{5.5}$$

The last expression is obtained with (4.68a) and the definition of the dimensionless Dyer–Roeder distance r, given in Sect. 4.5. The physical significance of this critical surface mass density will become clear below. With these definitions, the lens equations (5.1) and (5.2) are rewritten as

$$\mathbf{y} = \mathbf{x} - \boldsymbol{\alpha}(\mathbf{x}) \quad , \tag{5.6}$$

where

$$\boldsymbol{\alpha}(\mathbf{x}) = \frac{1}{\pi}\int_{\mathbb{R}^2} d^2x' \, \kappa(\mathbf{x}')\frac{\mathbf{x}-\mathbf{x}'}{|\mathbf{x}-\mathbf{x}'|^2} = \frac{D_d D_{ds}}{\xi_0 D_s}\hat{\boldsymbol{\alpha}}(\xi_0\mathbf{x}) \tag{5.7}$$

will be referred to as the scaled deflection angle. The length scale ξ_0 is, at this point, arbitrary; in the consideration of a specific lens model, it can be chosen such that the ray-trace equation (5.6) is simplest. For example, for a point-mass lens, the length (2.6b) is a natural length scale. For $\xi_0 = D_d$, $\mathbf{x}$ and $\mathbf{y}$ are the angular positions of the image and the unlensed source relative to the optical axis, respectively, as can be seen from (5.3).

The deflection potential. The identity $\nabla \ln |\mathbf{x}| = \frac{\mathbf{x}}{|\mathbf{x}^2|}$ shows that the deflection angle is a gradient with respect to $\mathbf{x}$,

$$\boldsymbol{\alpha} = \nabla \psi \; , \tag{5.8}$$

where

$$\psi(\mathbf{x}) = \frac{1}{\pi} \int_{\mathbb{R}^2} d^2 x' \, \kappa(\mathbf{x}') \ln |\mathbf{x} - \mathbf{x}'| \tag{5.9}$$

is the logarithmic potential associated with the surface density $\kappa(\mathbf{x})$. Thus, the mapping $\mathbf{x} \mapsto \mathbf{y}$ is a gradient mapping,

$$\mathbf{y} = \nabla \left(\frac{1}{2} \mathbf{x}^2 - \psi(\mathbf{x}) \right) \; , \tag{5.10}$$

which can also be expressed in terms of the scalar function

$$\phi(\mathbf{x}, \mathbf{y}) = \frac{1}{2} (\mathbf{x} - \mathbf{y})^2 - \psi(\mathbf{x}) \tag{5.11}$$

as [SC84.2]

$$\nabla \phi(\mathbf{x}, \mathbf{y}) = 0 \; . \tag{5.12}$$

The functions ψ and ϕ are dimensionless counterparts of the functions $\hat{\psi}$ and $\hat{\phi}$ as defined in Chap. 4,

$$\hat{\psi} = \frac{D_s \xi_0^2}{D_d D_{ds}} \psi \quad ; \quad \hat{\phi} = (1 + z_d) \frac{D_s \xi_0^2}{D_d D_{ds}} \phi \; .$$

As was shown in Chap. 4, the function ϕ is closely related to the light-travel-time of light rays: equation (5.12) is the formulation of *Fermat's principle* in GL theory ([SC85.3], [BL86.1]). The *Fermat potential* ϕ renders the calculation of the time delay for image pairs of a source particularly simple. In addition, the formulation (5.12) turns out to be very useful for general considerations of the lens mapping. It will be used below to prove the two general theorems, mentioned in Chap. 4 in a general context, in the framework of gravitational lensing, and it forms the basis for the analysis of the singularities of lens mappings to be given in Chap. 6.

The relation (5.9) giving ψ in terms of κ can be inverted, via the identity $\Delta \ln |\mathbf{x}| = 2\pi \delta^2(\mathbf{x})$, as

$$\Delta \psi = 2\kappa \quad ; \tag{5.13}$$

here, δ^2 and Δ denote the two-dimensional delta-function and Laplacian with respect to $\mathbf{x}$, respectively.

Complex scattering function. Sometimes, it is useful to represent the two-dimensional vectors $\mathbf{x}$, $\mathbf{y}$, $\boldsymbol{\alpha}$ by complex numbers, e.g., $x_c = x_1 + i x_2$. With this notation,

$$y_c = x_c - I_c^*(x_c) \quad , \tag{5.14a}$$

where

$$I_c(x_c) = \frac{1}{\pi} \int_{\mathbb{C}} \kappa(x_c') \frac{1}{x_c - x_c'} d^2 x' \quad , \tag{5.14b}$$

and, as before, the asterisk denotes complex conjugation. Complex formulation is used when the calculation of the deflection angle from a given mass distribution is non-trivial, but when complex integration theory can lead to analytic results; then it is natural to start with complex quantities. The only case treated in the literature ([BO73.1], [BO75.1]) for which a complex formulation was found to be superior, is that of matter distributions with elliptical symmetry; recently, however, a way was found to calculate the deflection angle for elliptical matter distributions with the vector formulation, using a very elegant method [SC89.4], as we discuss in Sect. 8.4. Another case where the complex formulation is of advantage will be discussed in Sect. 8.3.4.

Mathematical properties of the surface density, the potentials, and the deflection angle. So far, the mathematical properties of the surface density have not been specified. Of course, we assume that κ is non-negative, $\kappa \geq 0$. For general considerations, we assume the dimensionless surface mass density to be smooth, although in specific models we permit point masses, i.e., densities with δ-singularities, and discontinuities. The integral representations (5.7), (5.9) of $\boldsymbol{\alpha}$ and ψ hold if κ decreases at infinity stronger than $|\mathbf{x}|^{-2}$; that also ensures that the lens has a finite total mass M. If κ is smooth and decreases stronger than $|\mathbf{x}|^{-2}$, the deflection potential ψ is smooth, obeys Poisson's equation (5.13), and for $|\mathbf{x}| \to \infty$, it has the asymptotic behavior

$$\psi(\mathbf{x}) = \frac{M}{\pi \xi_0^2 \Sigma_{\mathrm{cr}}} \ln |\mathbf{x}| + \mathcal{O}\left(\frac{1}{|\mathbf{x}|}\right) \quad , \tag{5.15a}$$

$$\boldsymbol{\alpha}(\mathbf{x}) = \frac{M}{\pi \xi_0^2 \Sigma_{\mathrm{cr}}} \frac{\mathbf{x}}{|\mathbf{x}|^2} + \mathcal{O}\left(\frac{1}{|\mathbf{x}|^2}\right) \quad , \tag{5.15b}$$

$$\alpha_{ij} = \frac{M}{\pi \xi_0^2 \Sigma_{\rm cr}} \frac{\delta_{ij} |\mathbf{x}|^2 - 2 x_i x_j}{|\mathbf{x}|^4} + \mathcal{O}\left(\frac{1}{|\mathbf{x}|^3}\right) \quad . \tag{5.15c}$$

Given such a κ, the solution ψ of (5.13) is determined uniquely up to an additive constant, if it increases at infinity not faster than $\ln |\mathbf{x}|$. Also, two functions $\psi, \tilde\psi$ determine the same lens mapping (5.6) via (5.8), if, and only if, they differ by a constant. For these reasons, we henceforth consider ψ, and consequently also ϕ, as defined up to an additive constant only, to be fixed in each case by convenience.

Equations (5.11) and (5.15a) show that, for any fixed $\mathbf{y}$, the Fermat potential $\phi(\mathbf{x}, \mathbf{y})$ increases quadratically at infinity. Therefore, it has a minimum, at which (5.12) holds. Thus, *for each source position, there exists at least one image*; in other words, any lens mapping is surjective.

5.2 Magnification and critical curves

Definition of the magnification. Gravitational light deflection can not only lead to multiple imaging but, because of the differential deflection across a light bundle, the deflection also affects the properties of the image(s) of a source. In particular, the flux of images is influenced since the cross-sectional area of a light bundle is distorted by the deflection. Since photon number is conserved, the flux of an image is determined by this area distortion. Consider an infinitesimal source with surface brightness I_ν (where ν is the observed frequency), which, in the absence of gravitational light deflection, subtends a solid angle $d\omega^*$ on the sky. The (monochromatic) flux from this source is

$$S_\nu^* = I_\nu \, d\omega^* \quad . \tag{5.16}$$

If the light bundle undergoes a deflection, the solid angle $d\omega$ of the image will differ from $d\omega^*$. Since the light bending changes neither ν nor I_ν, the observed flux of the image is

$$S_\nu = I_\nu \, d\omega \quad ; \tag{5.17}$$

hence, the light deflection leads to a change of the flux of the observed image by a factor

$$|\mu| = S_\nu / S_\nu^* = d\omega / d\omega^* \quad , \tag{5.18}$$

which is independent of the frequency of deflected radiation.

Magnification and image distortion. As we indicated in the preceding section, the dimensionless quantities $\mathbf{x}$ and $\mathbf{y}$ are equal to the angular positions $\boldsymbol{\theta}$ and $\boldsymbol{\beta}$, respectively, if $\xi_0 = D_{\rm d}$ in (5.3). Hence, the ratio (5.18) of solid angles is given by

$$\frac{d\omega}{d\omega^*} = \frac{d^2 x}{d^2 y} \quad . \tag{5.19}$$

It is thus seen that the magnification factor $|\mu|$ is obtained from the Jacobian determinant, which describes the area distortion of the lens mapping (5.6). Let us define the Jacobian matrix for (5.6),

$$A(\mathbf{x}) = \frac{\partial \mathbf{y}}{\partial \mathbf{x}} \quad , \quad A_{ij} = \frac{\partial y_i}{\partial x_j} \quad , \tag{5.20}$$

and the *magnification factor*

$$\mu(\mathbf{x}) = \frac{1}{\det A(\mathbf{x})} \quad . \tag{5.21}$$

The image at $\mathbf{x}$ of an infinitesimally small source is thus brightened or dimmed by a factor $|\mu(\mathbf{x})|$. The magnification factor μ can be positive or negative; the corresponding images are said to have *positive or negative parity*. For certain values of $\mathbf{x}$, $\det A$ can vanish and μ diverges: these points are called *critical points*. Such divergence indicates that near critical points the approximation of geometrical optics fails, and wave optics must be applied. However, as will be shown in Chap. 7, wave optics do not lead to any considerable correction to the magnification of real (i.e., extended) sources, which is obtained by averaging (5.21) over the source, weighted by the surface brightness.

Form of the magnification matrix: convergence and shear. Equations (5.10) and (5.11), together with (5.20), imply

$$A_{ij} = \phi_{ij} = \delta_{ij} - \psi_{ij} \quad , \tag{5.22}$$

where we have denoted partial derivatives of the scalar functions ϕ and ψ with respect to x_i by subscripts. This shows that A is symmetric. Using (5.13) we can write the Jacobian matrix as

$$A = \begin{pmatrix} 1 - \kappa - \gamma_1 & -\gamma_2 \\ -\gamma_2 & 1 - \kappa + \gamma_1 \end{pmatrix} \quad , \tag{5.23}$$

where

$$\gamma_1 = \frac{1}{2} (\psi_{11} - \psi_{22}) \quad ; \quad \gamma_2 = \psi_{12} = \psi_{21} \quad . \tag{5.24}$$

From (5.23), we read off the orthogonal invariants of A, viz. the determinant

$$\det A = (1 - \kappa)^2 - \gamma^2 \quad , \tag{5.25}$$

the trace

$$\operatorname{tr} A = 2(1 - \kappa) \tag{5.26}$$

and the eigenvalues

$$a_{1,2} = 1 - \kappa \mp \gamma \quad . \tag{5.27}$$

In these equations, γ is

$$\gamma = \sqrt{\gamma_1^2 + \gamma_2^2} \ . \tag{5.28}$$

The determinant and the eigenvalues consist of two terms: the first, "local" term depends only on the surface mass density κ within the beam and is called *convergence* or *Ricci focusing*, whereas the second term depends on the mass distribution outside of the beam and is called the *shear*. Note that these contributions to the magnification result from the corresponding terms in the optical focusing equation (3.64), applied to a ray bundle converging towards the observer.

Classification of ordinary images. For fixed $\mathbf{y}$, the Fermat potential $\phi(\mathbf{x}, \mathbf{y})$ defines a (two-dimensional) arrival-time surface. Ordinary, i.e., non-critical images of the source occur at those points $\mathbf{x}$, where $\nabla\phi$ vanishes and the Jacobi matrix $\phi_{ij} = A_{ij}$ of the lens equation does not degenerate. Thus they are located at local extrema and saddle points of the arrival time surface. Minima of ϕ are characterized by positive definiteness of A, maxima by negative definiteness, and saddle points by eigenvalues of opposite parity. Thus, the following *types of non-critical images* can occur:

Type I: minimum of ϕ,

$\det A > 0$; $\operatorname{tr} A > 0$; i.e.,

$\gamma < 1 - \kappa \leq 1, \ a_i > 0, \ \mu \geq \dfrac{1}{1 - \gamma^2} \geq 1$

Type II: saddle point of ϕ, (5.29)

$\det A < 0$; i.e., $(1 - \kappa)^2 < \gamma^2, \ a_2 > 0 > a_1$

Type III: maximum of ϕ,

$\det A > 0$; $\operatorname{tr} A < 0$ i.e., $(1 - \kappa)^2 > \gamma^2, \ \kappa > 1, \ a_i < 0$.

Light beams from an infinitesimal source which converge to a point at an observer have not passed through the caustic of the past light cone $\mathcal{C}_O^-$ of the observation event O, if they give rise to an image of type I. They are magnified ($\mu > 1$) except in the very special case when $\kappa = \gamma = 0$, $\mu = 1$, as shown by the relevant inequality in (5.29). Light bundles corresponding to images of type II have gone once through the caustic (i.e., those points where the cross-sectional area of a light bundle vanishes, see Sect. 3.5), those of type III have crossed or touched the caustic twice, provided conjugate points are counted with multiplicities (defined at the end of Sect. 3.5).[1]

---

[1] The arrival time surface can have local maxima, in contrast to the arrival time functional used in the proof of Fermat's priciple in Sect. 3.3. In fact, if *all* null curves from a source to an observer are permitted, one can always increase the arrival time by making a detour, e.g., near the observer. This is excluded in the definition of ϕ, which requires straight paths between source and lens plane, and between lens plane and observer. This argument is illustrated when one considers the multiple lens plane situation, described in Chap. 9. There, one can always formally introduce an additional lens plane which carries no mass distribution; hence, the gravitational delay in that plane vanishes identically. Bending a light ray in that plane thus always leads to an increase of its travel time.

Note also that the latter light bundles pass through a region of overcritical density. To prove these statements, consider as fixed the observer position, the deflecting matter distribution (described by its surface mass density Σ in the lens plane) and the location where light bundles pass through the lens plane, and regard the source's distance as variable. (Note that, within the thin lens approximation under discussion here, the spatial projection of the caustic of $\mathcal{C}_O^-$ may be identified with the union of the "caustics" of the lens mappings for $D_d < D_s < \infty$.) From the rescaling which led to (5.9) one infers that the ψ_{ij} are proportional to (D_{ds}/D_s), and thus, by (5.13) and (5.24), κ and γ also are proportional to this ratio. Equation (5.27) then implies:

$$a_1 = 1 - c_1 \frac{D_{ds}}{D_s} \quad , \quad a_2 = 1 - c_2 \frac{D_{ds}}{D_s} \quad , \quad c_1 \geq 0 \quad , \quad |c_2| \leq c_1 \quad ;$$

c_2 may be positive, negative, or zero. For small values of (D_{ds}/D_s), the eigenvalues obey $a_2 \geq a_1 > 0$. Since a transition through the caustic occurs exactly if one of the eigenvalues becomes zero, the preceding statements, combined with (5.29), establish our claims. It also follows that each eigenvalue can become zero at most once. Therefore, any light bundle going through a single geometrically-thin matter distribution can cross the caustic at most twice. For critical images, $\det A = 0$, i.e., $\gamma = |1 - \kappa|$.

Orientation and handedness of images. Consider a source at $\mathbf{y}$ and an image at $\mathbf{x}$, not on a critical curve. Let $\mathbf{Y}$ be a displacement vector at the source, mapped onto the displacement vector $\mathbf{X}$ at $\mathbf{x}$, where

$$\mathbf{Y} = A\mathbf{X} \quad . \tag{5.30}$$

Then,

$$\mathbf{X} \cdot \mathbf{Y} = \sum_{i,j=1}^{2} A_{ij} X_i X_j \quad ,$$

which means that, for images of type I, the position angle of the image vector differs by no more than $\pi/2$ from that of the source vector, whereas for images of type III they differ by more than $\pi/2$.

Let us consider two displacement vectors at $\mathbf{y}$, $\mathbf{Y}$ and $\mathbf{Z}$, and the corresponding image vectors $\mathbf{X}$ and $\mathbf{W}$. We define the handedness of two vectors $\mathbf{Y}$ and $\mathbf{Z}$ as the sign of

$$\mathbf{Y} \times \mathbf{Z} \equiv Y_1 Z_2 - Y_2 Z_1 = \det \begin{pmatrix} Y_1 & Z_1 \\ Y_2 & Z_2 \end{pmatrix}$$

With (5.30) we find

$$\mathbf{X} \times \mathbf{W} = \frac{1}{\det A} \mathbf{Y} \times \mathbf{Z} \quad .$$

Hence, for images of type I and III (positive parity images) the handedness is preserved, whereas for images of type II (negative parity) it is reversed. This illustrates the notion of parity. The last formula also restates (5.21) since $|\mathbf{X} \times \mathbf{Y}|$ is the area spanned by $\mathbf{X}$ and $\mathbf{Y}$, so that this last expression explicitly demonstrates that the magnification is the area distortion of the lens mapping.

Shape of ordinary images. To obtain the shapes of the images, consider a small circular source with radius R at $\mathbf{y}$, bounded by a curve described by

$$\mathbf{c}(t) = \mathbf{y} + R(\cos t, \sin t) \ .$$

The corresponding boundary curve of the image is

$$\mathbf{d}(t) = \mathbf{x} + A^{-1} R(\cos t, \sin t) \ .$$

Using equation (5.23), it is easily seen that the image curve is an ellipse centered on $\mathbf{x}$ with semi-axes parallel to the principal axes of A, with magnitudes

$$R \Lambda_\pm = \frac{R}{|1 - \kappa \mp \gamma|} \ , \tag{5.31}$$

which are functions solely of the invariants of A, and with position angles $\varphi_\pm$ for the axes, where

$$\tan \varphi_\pm = \frac{\gamma_1}{\gamma_2} \mp \sqrt{\left(\frac{\gamma_1}{\gamma_2}\right)^2 + 1} \ . \tag{5.32}$$

$\Lambda_\pm$ are the factors by which the image is stretched in the two eigendirections of A; they are the inverses of the eigenvalues of A. The area of the image is larger than the area of the source by a factor $|\Lambda_+ \Lambda_-| = |\det A|^{-1} = |\mu|$, as expected. The ellipse degenerates into a circle if, and only if, either the shear vanishes, $\gamma_1 = \gamma_2 = 0$, or $\operatorname{tr} A = 0$, i.e., $\kappa = 1$.

The considerations above are valid only if $\det A \neq 0$. At critical curves, i.e., those curves where $\det A = 0$, the lens mapping exhibits quite different properties which are explored in detail in the next chapter. Here, we only note that critical curves play a special role for the number of images a source produces. Since that number depends on the position of the source, a change of the source position can change the number of images. This can happen only if the source crosses a caustic, since at other points the lens mapping is locally invertible and therefore, no images can appear or disappear. In the following chapter, we show that the number of images changes by two if, and only if, the source crosses a caustic in the source plane, i.e., the image of a curve where $\det A = 0$ under the mapping (5.6).

5.3 Time delay and Fermat's principle

The time delay for GL images is the only dimensional observable. All other observables, image separations, redshifts, brightness ratios, alignment of extended images, are dimensionless.[2] The importance of this fact is illustrated by the following example: consider two lensing geometries which are identical except that, for one of them, all distances are doubled. An observer in each case will see the same angular separations and flux ratios of images. However, the time delay for any image pair is also doubled in the second case. Hence, a measurement of the time delay allows, at least in principle, to determine the overall length scale for a GL system. We will return to this point in Sect. 13.1. In this section, we rederive, with a different method, some of the results of Chap. 4.

Physical origin of the time delay. If a lens produces two (or more) images of a single source, the light-travel-times along the different light paths will, in general, be different, which may be ascribed to two effects which contribute to the light-travel-time [CO75.1]. First, the deflection causes a light ray to be bent, and since curved rays are geometrically longer than straight rays, there is a *geometrical time delay*. Second, the light traverses the gravitational field of the deflector. From the weak-field metric (4.13), the coordinate time interval dt for a ray to travel the Euclidian length dl is

$$c\, dt \approx (1 - 2U)dl \quad , \tag{5.33}$$

where U denotes the Newtonian gravitational potential. Thus, there is also a *potential time delay*, given by the integral of $-2U$ along the ray.

Wavefronts. The splitting of the time delay into a geometrical and a potential part as outlined above is useful for physical insight, but it is not the simplest way to derive an expression for the time delay in GL theory. More transparent is the consideration of the geometry of wavefronts ([RE64.1], [CH76.1], [KA83.1]). Consider a point source which emits a flash of radiation. A short time Δt later, all observers on a sphere with radius $c\Delta t$ can see the flash. These observers trace the location of the wavefront which is connected to the flash. The wavefront thus characterizes the locus of all points with equal light-travel-time from the source. As is known from wave optics, light rays *in vacuo* are perpendicular to the wavefronts; this property is also valid in General Relativity, as shown in Chap. 3. Wavefronts are spherical until they become distorted by a refracting medium. In this respect, a gravitational field acts like such a medium that delays light rays, according to (5.33), see also (4.16).

---

[2] The brightnesses of the individual images do not yield additional information because the intrinsic luminosity of the source is not known, due to the lack of an independent luminosity indicator.

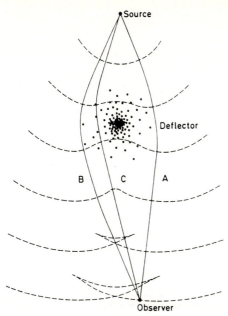

Fig. 5.1. The wavefronts in a typical GL system. Close to the source the wavefronts are nearly spherical, but as they approach the deflector, they become deformed and can self-intersect. Every passage of the wavefront past the observer corresponds to an image of the source in the direction of the light rays, perpendicular to the wavefront. Note that in this example, the light travel time is shortest along ray A and longest along ray C

A typical GL arrangement with corresponding wavefronts is shown in Fig. 5.1. The wavefronts, spherical close to the source, become distorted by traversing the gravitational field of the deflector. If the field is sufficiently strong, the wavefronts develop edges and intersect themselves, as indicated in the figure. An observer located behind such folded fronts will "see" different sheets of the same front passing him one after another. Since the directions of propagation of the various sheets during the crossings are different, the source is seen in different directions, given by the normals of the wavefront sheets. The time delay for pairs of images is then the time elapsed between two crossings of the wavefront. This geometrical picture is well suited for calculating the time delay. For the following argument, we must generalize the lens equation to allow for a variable observer position.

Ray-trace equation explicitly containing the observer position. So far, the coordinate system used for lensing calculations has been chosen such that the optical axis connects the observer and the "center" of the lens. Now, consider a description where the observer is not on the optical axis, but a distance ζ away from it, as shown in Fig. 5.2. The optical axis, as defined hitherto, is the line through the observer O and the "center" of the lens L, whereas we now consider the line through O' and L as the reference line. The observer is at distance ζ from O' in the observer plane. The source S,

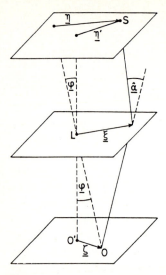

Fig. 5.2. Gravitational lensing geometry for an arbitrary choice of the reference line $O'L$. The observer at point O is at position ζ with respect to the optical axis, and the source at position η'. If the line OL is taken as the optical axis, the old ray-trace equation applies, with the source at position η with respect to this reference line. To obtain a relation between ζ, η' and η, one considers the angle φ between the two reference lines, as described in the text

which is at η from the old optical axis, is at η' from the new reference line. The old and new reference lines intersect at L, forming angle φ. From the figure, we see that the ray-trace equation

$$\eta = \frac{D_s}{D_d}\xi - D_{ds}\hat{\alpha}(\xi) \tag{5.34}$$

is still valid. A relation between ζ and η' is obtained by noting that

$$\varphi = \frac{\eta - \eta'}{D_{ds}} = \frac{\zeta}{D(z_d, 0)} \quad . \tag{5.35}$$

Eliminating η from (5.34) by means of (5.35), and using (4.55), i.e., $D(z_d, 0) = (1 + z_d)D_d$, yields

$$\eta' + \frac{D_{ds}}{D_d(1+z_d)}\zeta = \frac{D_s}{D_d}\xi - D_{ds}\hat{\alpha}(\xi) \quad . \tag{5.36}$$

Thus, only the combination $\eta = \eta' + \frac{D_{ds}}{D_d(1+z_d)}\zeta$ of source and observer positions enters the lens equation. As in Sect. 5.1, we can convert the lens equation (5.36) to dimensionless form by choosing a length scale ξ_0 in the lens plane and by applying the definitions (5.3); then, the dimensionless lens equation is identical to (5.6). The only difference is that now the expression for the dimensionless quantity

$$\mathbf{y} = \frac{1}{\eta_0}\left[\boldsymbol{\eta}' + \frac{D_{ds}}{D_d(1+z_d)}\boldsymbol{\zeta}\right] \tag{5.37}$$

contains both, the source and observer position explicitly. In particular, we may choose the optical axis such that $\boldsymbol{\eta}' = 0$; then **y** is equal to the dimensionless observer position; we will do so in the following consideration.

Derivation of the time delay. Let us now calculate the time delay for two images of a source. Consider the case where the observer at position $\boldsymbol{\zeta}$ sees two images of a fixed source, at $\boldsymbol{\xi}^{(1)}$ and $\boldsymbol{\xi}^{(2)}$. If the position of the observer changes, both image positions will also change. In particular, if the observer moves along a curve $\boldsymbol{\zeta}(\lambda)$, there will be two image curves $\boldsymbol{\xi}^{(i)}(\lambda)$ such that, for any λ, the lens equation is satisfied for $i = 1, 2$, as long as $\boldsymbol{\zeta}(\lambda)$ does not cross a caustic curve. In Fig. 5.3, we have drawn, for two observer positions which differ by $d\boldsymbol{\zeta}$, the light rays for the two images under consideration, together with one wavefront for each image, chosen such that they intersect one observer position simultaneously. These two fronts intersect the second observer position at different times. The sum of these two delays can be readily obtained from the figure:

$$d(c\Delta t) = \boldsymbol{\vartheta} \cdot d\boldsymbol{\zeta} \; , \tag{5.38}$$

where $\boldsymbol{\vartheta}$ is the angular separation of the images. If we integrate this expression along a curve $\boldsymbol{\zeta}(\lambda)$, the time delay becomes

$$c\,\Delta t(\boldsymbol{\zeta}) = \int_{\boldsymbol{\zeta}_0}^{\boldsymbol{\zeta}} \boldsymbol{\vartheta}(\boldsymbol{\zeta}') \cdot d\boldsymbol{\zeta}' \; + \; c\,\Delta t(\boldsymbol{\zeta}_0) \; , \tag{5.39}$$

where $\boldsymbol{\zeta}_0$ is an arbitrary reference point [KA83.1]. With equations (5.10) and (5.11), we can eliminate the arbitrary reference point $\boldsymbol{\zeta}_0$ and the line

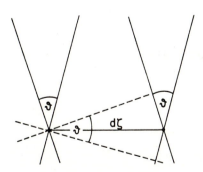

Fig. 5.3. Light rays (solid lines) and wavefronts (dashed lines) from two images of a source, which have angular separation ϑ. The wavefronts are chosen such that they intersect the observer to the left at the same time. An observer at a distance $d\boldsymbol{\zeta}$ will see these two wavefronts at different times, where the time delay follows from the geometry of the figure and is given by (5.38)

integral [SC85.3]. We will see that expression (5.39) contains both the geometrical and the potential time delay [BO83.1].

Relation to the Fermat potential $\phi(\mathbf{x}, \mathbf{y})$. The angular separation between the two images $\boldsymbol{\xi}^{(1)}(\lambda)$ and $\boldsymbol{\xi}^{(2)}(\lambda)$ is

$$\vartheta = \frac{1}{D_\mathrm{d}} \left(\boldsymbol{\xi}^{(2)} - \boldsymbol{\xi}^{(1)} \right) = \frac{\xi_0}{D_\mathrm{d}} \left(\mathbf{x}^{(2)} - \mathbf{x}^{(1)} \right) \quad , \tag{5.40}$$

and the dimensionless observer displacement $d\mathbf{y}$ that corresponds to $d\boldsymbol{\zeta}$ is obtained from

$$d\boldsymbol{\zeta} = (1 + z_\mathrm{d}) \xi_0 \frac{D_\mathrm{s}}{D_\mathrm{ds}} d\mathbf{y} \quad ; \tag{5.41}$$

hence, (5.39) becomes

$$c\Delta t(\mathbf{y}) = \xi_0^2 \frac{D_\mathrm{s}}{D_\mathrm{d} D_\mathrm{ds}} (1 + z_\mathrm{d}) \int_{\mathbf{y}_0}^{\mathbf{y}} d\mathbf{y}' \cdot \left[\mathbf{x}^{(2)}(\mathbf{y}') - \mathbf{x}^{(1)}(\mathbf{y}') \right] \\ + c\Delta t(\mathbf{y}_0) \quad . \tag{5.42}$$

The integral in (5.42) is readily evaluated: $\mathbf{x} \cdot d\mathbf{y} = d(\mathbf{x} \cdot \mathbf{y}) - \mathbf{y} \cdot d\mathbf{x}$. From the lens equation in the form $\mathbf{y} = \mathbf{x} - \nabla \psi(\mathbf{x})$,

$$\mathbf{x} \cdot d\mathbf{y} = d(\mathbf{x} \cdot \mathbf{y}) - \mathbf{x} \cdot d\mathbf{x} + d\psi(\mathbf{x}) = d \left[\frac{\mathbf{y}^2}{2} - \phi(\mathbf{x}, \mathbf{y}) \right] \quad ,$$

where, in the last step, (5.11) was used. Then,

$$\int_{\mathbf{y}_0}^{\mathbf{y}} d\mathbf{y}' \cdot \left[\mathbf{x}^{(2)}(\mathbf{y}') - \mathbf{x}^{(1)}(\mathbf{y}') \right] = \left[\phi(\mathbf{x}^{(1)}, \mathbf{y}) - \phi(\mathbf{x}^{(2)}, \mathbf{y}) \right] \\ - \left[\phi(\mathbf{x}_0^{(1)}, \mathbf{y}_0) - \phi(\mathbf{x}_0^{(2)}, \mathbf{y}_0) \right] , \tag{5.43}$$

where $\mathbf{x}_0^{(i)}$, $i = 1, 2$, are the image positions for the observer at $\mathbf{y}_0$. Suppose now that the two images fuse if the observer moves to $\mathbf{y}_0$, which can happen if $\mathbf{y}_0$ is on a caustic of the source's light cone (compare the discussion in Sect. 5.2 and in the next chapter). Then, $\mathbf{x}_0^{(1)} = \mathbf{x}_0^{(2)}$, $\Delta t(\mathbf{y}_0) = 0$, and the final result reads

$$c\Delta t(\mathbf{y}) = \xi_0^2 \frac{D_\mathrm{s}}{D_\mathrm{d} D_\mathrm{ds}} (1 + z_\mathrm{d}) \left[\phi\left(\mathbf{x}^{(1)}, \mathbf{y}\right) - \phi\left(\mathbf{x}^{(2)}, \mathbf{y}\right) \right] \quad . \tag{5.44}$$

Note that this expression does not contain $\mathbf{y}_0$ at all. Therefore, the formula (5.44) must be valid independently of any point $\mathbf{y}_0$ for which the images fuse. By combining suitable pairs of images which fuse at suitable critical curves, one can easily see that the time delay is independent of a particular point $\mathbf{y}_0$, due to the additivity of time delays.

Interpretation of ϕ as light travel time along rays. The last equation suggests that the two terms on the right-hand side represent individually the light-travel-times of the rays, even for kinematically possible rays. This is, in fact, true, as was shown in Chap. 4. Thus, the value of the function

$$T(\mathbf{x},\mathbf{y}) = \frac{\xi_0^2}{c} \frac{D_\mathrm{s}}{D_\mathrm{d} D_\mathrm{ds}} (1+z_\mathrm{d}) \phi(\mathbf{x},\mathbf{y})$$
$$= \frac{\xi_0^2}{c} \frac{D_\mathrm{s}}{D_\mathrm{d} D_\mathrm{ds}} (1+z_\mathrm{d}) \left(\frac{(\mathbf{x}-\mathbf{y})^2}{2} - \psi(\mathbf{x}) \right) \quad (5.45)$$

is, except for an additive constant[3], the travel time from the source to the observer for a light ray that crosses the lens plane at $\mathbf{x}$, relative to a ray which crosses the lens plane at $\mathbf{x} = \mathbf{y}$ and does not experience any gravitational potential. From the fact that only $\mathbf{y}$ enters this expression, we can now return to the original reference system where the observer is on the optical axis, and $\boldsymbol{\eta} = \eta_0 \mathbf{y}$ is the source position. Note that (5.45) is independent of ξ_0, since the Fermat potential ϕ scales like ξ_0^{-2}. The first term of ϕ in (5.45) describes the deviation of the light ray from a straight line, for which $\mathbf{x} = \mathbf{y}$. In fact, in agreement with [CO75.1] and our discussion in Chap. 4, the first term of (5.45) describes the increase of light-travel-time relative to an unbent ray, for a ray which originates at $\mathbf{y}$ and hits the lens plane at $\mathbf{x}$. Similarly, the second term is the time delay a ray experiences as it traverses the deflection potential $\psi(\mathbf{x})$. Hence, the splitting of the time delay into geometrical and potential terms emerges naturally from the wavefront method described above.

The interpretation of the Fermat potential ϕ immediately shows why the lens equation can be written in the form (5.12): it is just the formulation of Fermat's principle in GL theory: Of all the possible ray paths, parametrized by the impact vector $\mathbf{x}$ in the lens plane, physical rays correspond to those paths for which the light-travel-time is stationary.

The simplicity of the derivation for the time delay given above and the final result (5.44) show the usefulness of the wavefront method and the formulation of lens theory in terms of ϕ, for both intuition and for analytic work. For later applications, we rewrite (5.44) in a somewhat different form by making the choice $\xi_0 = D_\mathrm{d}$; then, $\mathbf{x} = \boldsymbol{\theta}$ and $\mathbf{y} = \boldsymbol{\beta}$ are the angular position of the image and the undisturbed angular position of the source, respectively. Then

$$c\Delta t(\boldsymbol{\beta}) = \frac{D_\mathrm{d} D_\mathrm{s}}{D_\mathrm{ds}} (1+z_\mathrm{d}) \left[\phi\left(\boldsymbol{\theta}^{(1)},\boldsymbol{\beta}\right) - \phi\left(\boldsymbol{\theta}^{(2)},\boldsymbol{\beta}\right) \right] \quad . \quad (5.46)$$

---

[3] The physical origin of this constant has been discussed in connection with (4.25); it is due to the long-range nature of the potential.

5.4 Two general theorems about gravitational lensing

In Sect. 2.3, we already mentioned two general theorems about the image systems produced by gravitational light deflection. Here, we consider these theorems in some detail. First, we state and prove them for the case of a single geometrically-thin deflecting mass which has been treated in the preceding sections of this chapter. Afterwards, we outline arguments indicating that the conclusions of these theorems remain valid under more general assumptions.

5.4.1 The case of a single lens plane

We consider a single, geometrically-thin GL whose surface mass density $\kappa(\mathbf{x})$ is smooth and decreases faster than $|\mathbf{x}|^{-2}$ for $|\mathbf{x}| \to \infty$. Then, the lens has a finite total mass, and the deflection angle $\boldsymbol{\alpha}(\mathbf{x})$ is continuous, tends to zero if the impact parameter $|\mathbf{x}|$ tends to infinity, and is therefore bounded; let

$$|\boldsymbol{\alpha}| \leq a \tag{5.47}$$

(these assumptions again exclude stars and black holes). We denote the number of ordinary images of type I of a source $\mathbf{y}$ as n_I, use n_II, n_III analogously, and put $n_\mathrm{I} + n_\mathrm{II} + n_\mathrm{III} = n$.

Theorem 1. Under the assumptions just stated, and if $\mathbf{y}$ denotes a source position not situated on a caustic, (a) $n_\mathrm{I} \geq 1$, (b) $n < \infty$, (c) $n_\mathrm{I} + n_\mathrm{III} = 1 + n_\mathrm{II}$, (d) for $|\mathbf{y}|$ sufficiently large, $n = n_\mathrm{I} = 1$. Hence, the total number $n = 1 + 2n_\mathrm{II}$ of images is odd, the number of even-parity images exceeds that of odd parity images by one [BU81.1], $n_\mathrm{II} \geq n_\mathrm{III}$, and $n > 1$ if and only if $n_\mathrm{II} \geq 1$.

Theorem 2. Under the same assumptions, that image of a source which arrives first at the observer is of type I and appears brighter than, or equally bright as the source would appear in the absence of the lens [SC84.2].

Index theorem. To prove the first of these theorems, we introduce the notion of the index of a continuously differentiable vector field $\mathbf{X}(\mathbf{x})$ on $\mathbb{R}^2$ at a stationary point, i.e., a point where $\mathbf{X} = 0$. By writing $\mathbf{X} = |\mathbf{X}|(\cos\varphi, \sin\varphi)$, a continuously differentiable function $\varphi(\mathbf{x})$ is defined on the open set of non-stationary points of the vector field.

Let $\mathbf{c}(\lambda)$ be an oriented, closed curve enclosing exactly one stationary point and not passing through such a point. Since the angle $\varphi(\lambda)$ which describes the direction of $\mathbf{X}(\lambda)$ along that curve returns to its original value after one loop, the value of the integral

$$\frac{1}{2\pi}\oint_{\mathbf{c}} d\varphi = \frac{1}{2\pi}\oint_{\mathbf{c}} d\lambda \frac{d\varphi}{d\lambda} \tag{5.48}$$

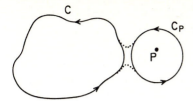

Fig. 5.4. If a closed curve **c** is deformed, the value of the integral (5.48) changes only if **c** moves across a stationary point **P** of **X**. The figure illustrates that if c_P is "added" to **c**, the integral changes by the index of **X** at **P**. Thus, by "adding" curves each enclosing a single stationary point, Poincaré's index theorem results

is an integer which cannot jump if **c** is deformed without meeting stationary points. This integer is, therefore, independent of **c** and called the index of **X** at the stationary point considered. An obvious geometrical argument (see Fig. 5.4) shows that if a curve **c** does not pass through stationary points of **X**, *the value of the integral* (5.48) *equals the sum of the indices of* **X** *at all stationary points enclosed by* **c**, provided these points are isolated (Poincaré). It is equally obvious that if a vector field on $\mathbb{R}^2$ is asymptotically constant at infinity, the sum of its indices vanishes, and if it is asymptotically radial, the sum of its indices is one.

Indices of ordinary images. Consider the vector field $\mathbf{X} = \nabla\phi(\mathbf{x},\mathbf{y})$, where again **y** is treated as a parameter. The stationary points of this vector field which correspond to ordinary images are either minima, maxima, or saddle points of the potential ϕ. To find the indices of these points, we expand the potential ϕ, for a fixed source position **y**, about a stationary point $\mathbf{x}_0$, i.e., where $\nabla\phi(\mathbf{x}_0,\mathbf{y}) = 0$ (see Fig. 5.5). One can choose a coordinate system such that $\mathbf{x}_0 = 0$ and that the Jacobian matrix A (5.20) is diagonal at $\mathbf{x}_0$. Then, up to second order in **x**,

$$\phi(\mathbf{x},\mathbf{y}) = \phi_0 + \frac{1}{2}\phi_{11}x_1^2 + \frac{1}{2}\phi_{22}x_2^2 \quad , \tag{5.49}$$

where the indices of ϕ denote partial derivatives, all taken at the origin $\mathbf{x} = 0$, and $\phi_0 = \phi(0,\mathbf{y})$. If ϕ_{11} and ϕ_{22} are both positive (negative), the point $\mathbf{x}_0$ is a minimum (maximum). If $\phi_{11}\phi_{22} < 0$, $\mathbf{x}_0$ is a saddle point. Hence, for small values of $|\mathbf{x}|$, the vector field is given by

$$\mathbf{X} = \begin{pmatrix} \phi_{11}x_1 \\ \phi_{22}x_2 \end{pmatrix} \; .$$

We choose the curve **c** to be an ellipse, sufficiently small to exclude any other stationary points. Let

$$\mathbf{c}(\vartheta) = \begin{pmatrix} \epsilon_1 \cos\vartheta \\ \epsilon_2 \sin\vartheta \end{pmatrix} \; ,$$

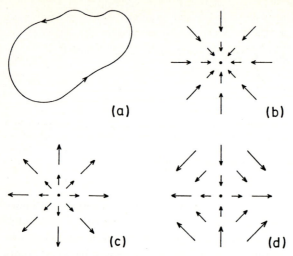

Fig. 5.5a–d. The index of stationary point. (a) shows an oriented closed curve. (b) shows the typical behavior of the vector field $\mathbf{X} = \nabla\phi$ near a maximum of ϕ. (c) the same for a minimum, (d) for a saddle point. From this figure, the index of stationary points can be obtained by inspection

where $\epsilon_1, \epsilon_2 > 0$; then, at the point $\mathbf{c}(\vartheta)$, the angle φ of the vector field is given by

$$\tan\varphi = \frac{\phi_{22}\epsilon_2 \sin\vartheta}{\phi_{11}\epsilon_1 \cos\vartheta} \ . \tag{5.50}$$

If the stationary point is a minimum ($\phi_{11}, \phi_{22} > 0$), we choose $\epsilon_1 = \epsilon\phi_{22}$, $\epsilon_2 = \epsilon\phi_{11}$. At a maximum, we take $\epsilon_1 = -\epsilon\phi_{22}$, $\epsilon_2 = -\epsilon\phi_{11}$, and at a saddle point, for which we assume without loss of generality that $\phi_{11} > 0$, $\phi_{22} < 0$, we take $\epsilon_1 = -\epsilon\phi_{22}$, $\epsilon_2 = \epsilon\phi_{11}$. Then, at an extremum, we have from (5.50) that $\varphi = \vartheta$, whereas at a saddle point, $\varphi = -\vartheta$. Evaluating the integral (5.48), we find that *the index of an extremum is* $+1$, *and that of a saddle point is* -1. We also infer from (5.49) that stationary points of $\mathbf{X}$ corresponding to ordinary images are isolated, i.e., in a suitable neighborhood of such a point there is no other image.

Proof of theorem 1. (a) $\phi(\mathbf{x}, \mathbf{y})$ has an absolute minimum, and that is an ordinary image of type I.

(b) For fixed $\mathbf{y}$, the lens equation $\mathbf{y} = \mathbf{x} - \boldsymbol{\alpha}(\mathbf{x})$ and the boundedness (5.47) of $|\boldsymbol{\alpha}|$ imply that the images $\mathbf{x}$ of $\mathbf{y}$ are contained in the disc $|\mathbf{x}| \leq |\mathbf{y}| + a$. Since the images are ordinary ones (by assumption), they are isolated, i.e., they cannot accumulate. Hence, $n < \infty$.

(c) The n images of $\mathbf{y}$ are contained in a disc with radius $R < |\mathbf{y}| + a$. If the radius of the circle $|\mathbf{x}| = R$ is large enough, the vector field $\mathbf{X} = \mathbf{x} - \mathbf{y} - \boldsymbol{\alpha}(\mathbf{x})$ is nearly radial on the circle since $|\boldsymbol{\alpha}|$ is bounded. The index theorem then implies, as remarked above, that $n_{\mathrm{I}} + n_{\mathrm{III}} - n_{\mathrm{II}} = 1$.

(d) For $|\mathbf{x}| \to \infty$, $A \to \mathcal{I}$, hence both det A and tr A are positive outside some circle of radius R. Images outside of that circle are thus of type I. Now, the lens equation shows that images $\mathbf{x}$ of $\mathbf{y}$ obey $|\mathbf{x}| \geq |\mathbf{y}| - |\boldsymbol{\alpha}(\mathbf{x})| \geq |\mathbf{y}| - a$, so if $|\mathbf{y}| > a + R$, $|\mathbf{x}| > R$, hence all images of such a $\mathbf{y}$ are of type I, i.e., $n_{\text{II}} = n_{\text{III}} = 0$. Then (c) gives $n_{\text{I}} = 1$.

Proof of theorem 2. According to the proof of theorem 1 (a), there is an image arriving first, and that is of type I; therefore, $\mu \geq 1$, by (5.29). The relevant inequalities show that $\mu = 1$ is extremely unlikely, because it requires that both $\kappa = 0$ and $\gamma = 0$ at the minimum, i.e., $A = \mathcal{I}$ - a single "point" in the 3-space of symmetric matrices.

Additional remarks. Inspection of the proof shows that theorem 1 does not depend on the inequality $\Delta\psi \geq 0$; moreover, the strong fall-off condition stated before (5.47) can be replaced by (5.47) and $\lim_{|\mathbf{x}| \to \infty} A = \mathcal{I}$. The theorem may thus be generalized to all gradient maps which are asymptotic to the identity map at infinity. Theorem 2, however, *does* depend on the positivity of mass, as one would expect.

The odd number theorem has led to the question of why, in most observed cases of gravitational lensing, either two or four images are observed. We discussed this point in Sect. 2.5; here we simply note that the theorem does not say anything about the brightness ratios of the images. It may well be, and is the case for many GL systems, that different images have vastly different brightnesses, which limits the usefulness of our theorem.[4] However, for individual lens models, the odd number theorem provides a (necessary, but not sufficient) check of whether one has found all images. The main consequence of the magnification theorem is that no source can be demagnified relative to the case where the lens is absent, i.e., relative to the case where one observes the source through an empty cone.

The proofs given above also show how the theorems must be modified to allow for point masses in the lens. Suppose that we replaced every point mass by an extended, transparent matter distribution. This would lead to local maxima of ϕ. Each point mass just turns such a local maximum into a logarithmic spike. Hence, the number of positive-parity images plus the number of point masses is greater by one than the number of negative-parity images. The magnification theorem is not influenced by point masses, since they only introduce additional spikes in ϕ; the fact that there has to be a minimum is unchanged by point masses. Hence, theorem 2 is valid in general.

---

[4] The absence of a third (or fifth) observable image in GL systems is usually explained by the large demagnification of that image near the center of the lensing galaxy [NA86.1].

5.4.2 Generalizations

We now imagine a general model of the universe filled with inhomogeneously distributed, transparent matter. The following reasoning indicates that an "odd number theorem" generalizing theorem 1 may be valid even then:

Consider a source which emits a flash of radiation. The corresponding wavefront will be spherical close to the source, but due to the inhomogeneity of the universe, it will in general be deformed farther away from the source (see Fig. 5.1). But no matter how strong the deformation, it remains a closed, connected two-dimensional surface homotopic to a 2-sphere. For such a surface, one can always define the inside and the outside: a point not on the surface is inside of it if a curve connecting it to the source crosses the front an even number of times (or not at all). Correspondingly, a point is outside the closed surface if a curve connecting it to the source crosses the surface an odd number of times. Now as the wavefront expands, an observer who initially was outside the front may be crossed by it and is then inside the front. If the front is strongly deformed, it may intersect itself and cross the observer several times, and in each crossing the state inside/outside with respect to the front is changed. If one waits long enough, the front may have expanded so as to enclose both the source and the observer. Then, in this final stage, the observer is inside the front. Since he or she was initially outside, the front must have passed by an odd number of times. Because each crossing corresponds to an image of the source, this proves that the observer has seen an odd number of images.

This demonstration, while intuitively convincing, tacitly uses several properties the validity of which requires assumptions which have not been formulated explicitly. It presupposes that, (a) spacetime is diffeomorphic to a product $\mathbb{R} \times \mathcal{S}$ with spacelike "leaves" $t \times \mathcal{S}$ such that the worldline of the observer is represented as a point in $\mathcal{S}$, so that one can describe the evolution of the wavefront in a single 3-space $\mathcal{S}$; (b) the emission event is not separated from the worldline of the observer by an event horizon (in which case the observer would never see an image); (c) all sheets of the wavefront ultimately cross the observer; (d) there exist but finitely many such sheets; and (e) the latter do not "propagate round the universe" (as could happen in a spatially closed universe) to reach the observer again.

Assumptions leading to a general odd number theorem have been formulated [MC85.1], but their physical significance remains, in our opinion, somewhat obscure.

As for the *magnification theorem* (theorem 2), one may argue as follows: there will be a light ray arriving first at the observer (see Fig. 5.1), namely that corresponding to the first crossing of the wavefront. That ray does not cross a caustic (see Fig. 5.6). The focusing theorem (Sect. 4.5.1) then ensures that the corresponding light beam is magnified ($\mu \geq 1$).

This reasoning would be a proof if it were indeed true that there always exists a light ray not crossing a caustic and arriving first, and while examples indicate that this is so, to our knowledge it has not been established in general.

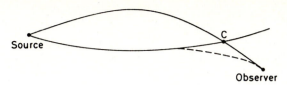

Fig. 5.6. The shortest ray path between a source and an observer does not pass through a caustic C. Any ray crossing a caustic can be shortened (dashed line)

The preceding analysis indicates that both general theorems admit generalizations to spacetime models which do not differ too much from Friedmann–Lemaître universes, provided source and observer are not too far apart, and light rays travelling round the universe can be separated from the lensing phenomenon of interest.

5.4.3 Necessary and sufficient conditions for multiple imaging

A matter distribution at distance D_d, described by its surface mass density Σ, may or may not be sufficiently strong to cause multiple images of sources at distance $D_s > D_d$. Here, we point out some criteria which can be used to decide this issue (see also [SU86.1]):

(a) An isolated transparent lens can produce multiple images if, and only if, there is a point $\mathbf{x}$ with $\det A(\mathbf{x}) < 0$. (b) A sufficient (but not necessary) condition for possible multiple images is that there exists a point $\mathbf{x}$ such that $\kappa(\mathbf{x}) > 1$.

These two statements are easily proved. (a) If $\det A(\mathbf{x}) > 0$ for all $\mathbf{x}$, the lens mapping is globally invertible and can thus not cause multiple imaging. On the other hand, if at $\mathbf{x}_0$, $\det A(\mathbf{x}_0) < 0$, a source at $\mathbf{y}_0 = \mathbf{y}(\mathbf{x}_0)$ has an image at $\mathbf{x}_0$ which is of type II. Then, from Theorem 1, there must be at least two additional images of positive parity.

(b) If at $\mathbf{x}_0$, $\kappa(\mathbf{x}_0) > 1$, a source at $\mathbf{y}_0 = \mathbf{y}(\mathbf{x}_0)$ has an image at $\mathbf{x}_0$, which according to (5.29) is a maximum. Again, from Theorem 1, there must be at least two additional images of that source.

In Sect. 8.2 we shall see that $\kappa > 1$ is not a necessary condition for multiple images; nevertheless, the sufficiency of this condition shows the relevance of the scale Σ_{cr}, (5.5), of the surface mass density. In Sect. 8.1, much more explicit conditions for the occurrence of multiple images from lenses with axial symmetry [i.e., $\kappa(\mathbf{x}) = \kappa(|\mathbf{x}|)$] will be obtained.

5.5 The topography of time delay (Fermat) surfaces

Elementary topographies and three-image geometries. The saddle points of the Fermat potential $\phi(\mathbf{x}, \mathbf{y})$ can be used to investigate the topography of GL images. This is not only useful to classify the possible image geometries,

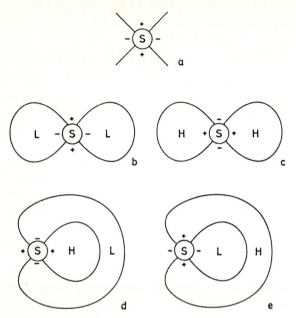

Fig. 5.7. (a)The local structure of a saddle point. In two directions that intersect at a vertex, as shown, the function is locally constant, and the '+' and '−' signs indicate the regions where the function is larger (smaller) than at the saddle point itself. (b)–(e) The four possible topographies to close the isochrones of a saddle point. Only (b) and (d) correspond to possible three-image geometries, as there the saddle point can be embedded in the asymptotic paraboloid without additional stationary points

but it also provides further geometrical insight into gravitational lensing [BL86.1]. To see why saddle points play a central role, we return to the expansion (5.49) of the Fermat potential. The equation

$$x_1 = \pm \sqrt{-\frac{\phi_{22}}{\phi_{11}} x_2} \qquad (5.51)$$

describes the directions of curves of constant $\phi = \phi_0$ at the saddle. Hence, locally a saddle point can be characterized, as in Fig. 5.7a, by two straight lines; the "+" and "−" signs indicate that in the corresponding region near a saddle point the value of ϕ is larger (smaller) than ϕ_0. The lines of constant ϕ, which we call *isochrones* because of their relation to the time delay (5.45), are of course curved. In particular, since ϕ behaves like a paraboloid for large values of **x**, they cannot go to infinity. Since these lines of constant ϕ cannot end, they have to be closed curves. They cannot intersect except at additional saddle points, and they cannot touch except at critical images which we here exclude, assuming **y** not to lie on a caustic. There are four different ways to close these curves, shown in Fig. 5.7b–e. The location of maxima (H) and minima (L) are indicated in the figures. The shape of the curves in Fig. 5.7b,c is that of a *lemniscate*, and Fig. 5.7d,e show curves with the shape of a *limaçon*. These two topographies are the building blocks for the investigation of possible image topographies.

Let us consider the case where a GL produces three images. From the last section, we know that exactly one of these corresponds to a saddle point of ϕ. It is not possible for the curves of Fig. 5.7c,e to correspond to the critical isochrones in this case, since outside of these curves, the value of ϕ is lower than at the saddle point S. Since ϕ behaves like $\mathbf{x}^2/2$ for large $|\mathbf{x}|$, this would imply that there is another minimum and another saddle point outside the critical isochrone. Since we wanted to consider a three-image geometry, this is not possible. Hence, the only possible topography for the three-image geometry is given by the curves in Fig. 5.7b,d.

If the image geometry is as in Fig. 5.7d, one can directly infer the time-ordering of the images in the sense that, if the source varies, the variation will first be seen in the image labeled L (since there the light-travel-time is a minimum), then at S and finally at H. On the other hand, if the geometry is described by Fig. 5.7b, the variation will be seen first at either L1 or L2, depending on whether $\phi(L1) < \phi(L2)$ or $\phi(L1) > \phi(L2)$, and S will vary last.

Five-image geometries. Next, we consider five-image geometries; in this case, there are exactly two saddle points. Each of these is connected with a critical isochrone, which can take either of the forms of Fig. 5.7b–e. In general, $\phi(S1) \neq \phi(S2)$, as we assume in the following, i.e., we neglect the degenerate case $\phi(S1) = \phi(S2)$. One of the two saddle points will be on the outer critical isochrone, in the sense that the second saddle point S2 will be inside the isochrone of S1. Now S1 can either be of the topography in Fig. 5.7b or Fig. 5.7d, the other two being excluded for the same reason that they were excluded before, i.e., there is no saddle point outside the isochrone connected to S1. Inside either of the two closed loops of the lemniscate in Fig. 5.7b, one can now locate S2, which can be of the form of either Fig. 5.7b or d. Since the two loops are equivalent, we therefore obtain the two possible image topographies shown in Fig. 5.8a,b, where the lemniscate is the outer critical isochrone. If S1 is connected to the limaçon of Fig. 5.7d, then S2 can be either in the inner lobe or in the outer lobe. In the inner lobe, S2 must be of the form of Fig. 5.7c or Fig. 5.7e, whereas in the outer lobe it must be of the form of Fig. 5.7b or d. Thus, we have the four possible configurations shown in Fig. 5.8c–f, where the limaçon is the outer critical isochrone. There are thus six different image topographies for the five-image GL configurations.

As in the three-image case, we can now consider the time-ordering of the images. The details will be left to the reader; we just consider one of the six cases, Fig. 5.8d. From that figure, $\phi(S1) > \phi(S2)$. There are three different possibilities in that case: if $\phi(H2) < \phi(S1)$, the ordering is (L,S2,H2,S1,H1). In the case that $\phi(S1) < \phi(H2) < \phi(H1)$, one obtains (L,S2,S1,H2,H1), and if $\phi(H1) < \phi(H2)$, then (L,S2,S1,H1,H2). Similar arguments lead to the time-ordering in the other cases.

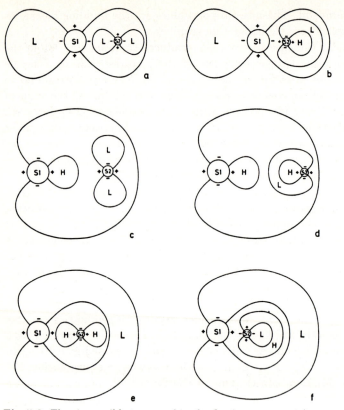

Fig. 5.8. The six possible topographies for five-image geometries

Finally, we present an example for the structure of the isochrones in a real GL model. In Fig. 5.9a, the isochrones are plotted for the case where there is no deflector; hence, they are just concentric circles, and only one image exists, corresponding to a minimum. If we consider an elliptical GL, whose strength can be scaled (e.g., by an overall factor for the surface mass density), the isochrones evolve as shown in Fig. 5.9b–d. If the lens is weak, they are just deformed, but their topography is conserved; hence, one is still left with one image. If the lens is sufficiently strong (Fig. 5.9c), a saddle point develops and a three-image topography is obtained. If we increase the strength of the lens further, the minimum splits into two minima and an additional saddle point, which gives a five-image topography. Again, this figure shows the utility of the Fermat potential ϕ to obtain insight in GL studies; we will use these geometrical pictures at several other points in this book, to explain lensing behavior qualitatively.

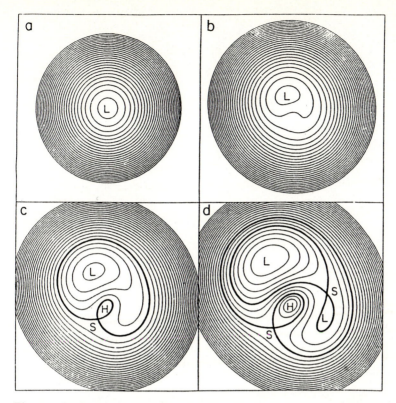

Fig. 5.9. Isochrones for an elliptical potential, with fixed source position. In (a), the 'strength' of the lens is set to zero, so the isochrones correspond to the geometrical time delay only. In (b), the deflection potential is weak, leading to a distortion of the isochrones. Increasing the strength of the lens, multiple images can be formed; (c) and (d) correspond to a three-image and five-image geometry, respectively (from [BL86.1])

6. Lensing near critical points

The lens equation, derived in Chap. 4 and discussed in some detail in the previous chapter, defines a surjective mapping $f : \mathbb{R}^2 \to \mathbb{R}^2$, $\mathbf{x} \mapsto \mathbf{y}$ from the lens plane to the source plane. We assume here that it is differentiable as often as is needed. If $\mathbf{y}^{(0)} = f\left(\mathbf{x}^{(0)}\right)$ is the image of $\mathbf{x}^{(0)}$ and if the Jacobian $D = \det A$ of the derivative $A = \frac{\partial \mathbf{y}}{\partial \mathbf{x}} = \frac{\partial(y_1,y_2)}{\partial(x_1,x_2)}$ of f does not vanish at $\mathbf{x}^{(0)}$, there exist neighborhoods of $\mathbf{x}^{(0)}$ and $\mathbf{y}^{(0)}$ on which f is bijective, i.e., the lens mapping is locally invertible. For an infinitesimal displacement $d\mathbf{y}$ of the source, the corresponding image position in the lens plane changes by $d\mathbf{x} = A^{-1}\left(\mathbf{x}^{(0)}\right) d\mathbf{y}$.

To avoid confusion we shall henceforth use the terms "image" and "source" in their *physical* meanings, so that images correspond to $\mathbf{x}$-values, sources to $\mathbf{y}$-values, though in mathematical terminology (used in the previous paragraph) $\mathbf{y}$-values would be called images, and $\mathbf{x}$-values pre-images of the lens mapping.

A lens mapping may be locally invertible at some images and not at other images of a given source. If a source position $\mathbf{y}$ does not lie on a caustic, the lens mapping is locally invertible at all of its images; this then also holds for source positions close to $\mathbf{y}$. Thus, displacements of the source $\mathbf{y}$ can lead to changes of the number of its images only if $\mathbf{y}$ crosses a caustic, and distinct images of a source can merge only at a critical point. It turns out that the number of images of a source does, in fact, change by two whenever the source crosses a caustic curve.[1]

In this chapter we study lens mappings of the type (5.10) locally near arbitrary points. In Sect. 6.1 we reconsider briefly the mappings at ordinary images. In the following sections we analyse the mappings near critical images, paying particular attention to (i) the shapes of the caustics, (ii) the changes of the number and the brightness of the images which occur if the source crosses a caustic, and (iii) the stability of the qualitative properties of the mapping under slight changes of the deflection potential.

---

[1] We can define the "inside" (or "positive" side) and "outside" (or "negative" side) of a caustic such that the number of images of a source on the inside of a caustic is larger by two than on the outside. In the special case where the source crosses two caustic curves at the point of their intersection, the number of images changes by four or not at all, depending on the direction (inside $\to$ outside, or vice versa) of the crossing relative to both individual caustics; see Fig. 6.8b below.

Mathematically, Sects. 6.2 and 6.3 provide simple examples of *singularities* of gradient maps of the plane into itself. (A differentially mapping is said to be singular at a point, if its Jacobian determinant vanishes there; such a point is then termed critical.) The study of singularities of differentiable maps is a rather young and active branch of analysis, which is also called *catastrophe theory*. "Catastrophes are violent sudden changes representing discontinuous responses of systems to smooth changes in the external conditions" (Arnold, *loc. cit.* below). Examples are changes of the number of images caused by continuous changes of the source position or of lens parameters in gravitational lens theory, the onset of instabilities in dynamics, phase transitions in thermodynamics, and collapse processes in the theory of galaxy formation. For details on catastrophe theory, we refer the reader to the elementary survey [AR84.2], the introduction [SA82.1] for scientists, and the more demanding mathematical treatments [AR85.2], [BR75.1], [GO73.1], [PO78.1], and [GI81.1]. Singularities in gravitational lensing have been discussed in [SC86.4], [SC86.5] [BL86.1] and [KO87.1].

Finally, in Sect. 6.4 we consider the magnification of an extended source near a fold caustic.

6.1 The lens mapping near ordinary images

Let us consider a parametrized lensing model characterized by the Fermat potential

$$\phi(\mathbf{x}, \mathbf{y}; \mathbf{p}) = \frac{1}{2}(\mathbf{x} - \mathbf{y})^2 - \psi(\mathbf{x}; \mathbf{p}) \quad , \tag{6.1}$$

where $\mathbf{p} = (p_\mu)$ denotes the N parameters of the deflection potential ψ. Examples are given in Chap. 8; parameters can be the core radii of galaxies, their ellipticities, the separation between lens components, or the source redshift, to name just a few.

For a given source position $\mathbf{y}$ and specified parameter values $\mathbf{p}$, the corresponding image positions $\mathbf{x}^{(\alpha)}$ are the critical points of the function $\mathbf{x} \mapsto \phi(\mathbf{x}, \mathbf{y}; \mathbf{p})$, i.e., the solutions of $\nabla \phi(\mathbf{x}, \mathbf{y}; \mathbf{p}) = 0$. As remarked in Sect. 5.1, for any lens model and source position there is at least one image if the deflection potential ψ is due to a finite mass distribution.

In catastrophe theory, the $N + 2$ variables y_1, y_2, p_μ are called *control parameters*, whereas x_1, x_2 are called *state variables*. Catastrophe theory investigates the local relationship between the "states" $\mathbf{x}^{(\alpha)}$ and the control parameters $\mathbf{y}$, $\mathbf{p}$.

Suppose that $\mathbf{x}^{(0)}$ is an ordinary image (sometimes called a "Morse point" of the lens mapping) of a source at $\mathbf{y}^{(0)}$ for parameter values $\mathbf{p}^{(0)}$, i.e., that

$$\phi_i^{(0)} = 0 \quad , \tag{6.2}$$

$$D^{(0)} = \det A^{(0)} = \det\left(\delta_{ij} - \psi_{ij}^{(0)}\right) = \det \phi_{ij}^{(0)} \neq 0 \ . \tag{6.3}$$

($D^{(0)}$ denotes the value of D at $\mathbf{x}^{(0)}$, $\mathbf{p}^{(0)}$; we shall use a similar notation throughout this chapter. As in Chap. 5, subscripts on scalar functions denote partial derivatives with respect to x_i.) Then, there exist neighborhoods U of $\mathbf{p}^{(0)}$ and V of $\mathbf{x}^{(0)}$ such that, for all parameter values $\mathbf{p}$ in U, the images $\mathbf{x}$ in V are again ordinary, simply by continuity of D. Moreover, since $\det A$ and $\operatorname{tr} A$ depend continuously on $\mathbf{x}$ and $\mathbf{p}$, the type of the image $\mathbf{x}$ as defined in Sect. 5.2 does not change under small changes of $\mathbf{x}$ and $\mathbf{p}$. In other words, for any lens mapping, the ordinary images as well as the images of types I, II and III [see (5.29)] each form open sets, and the property of an image of being ordinary or being of a particular type is *stable*, i.e., is preserved if the given lens mapping f is imbedded in an arbitrary, smooth, N-parameter family of maps and the parameters $\mathbf{p}$ are varied near the value $\mathbf{p}^{(0)}$ corresponding to the given f.

6.2 Stable singularities of lens mappings

Critical images of a lens mapping are characterized by the equation

$$D = \det \phi_{ij} = 0 \ ; \tag{6.4}$$

under the assumptions made at the end of Sect. 5.1, they form a compact subset of the image plane. The corresponding source positions form the caustic set. The latter is not only compact, but has measure zero (Sard's theorem; see, e.g., [ST64.1]), whereas the critical set may have a positive measure.[2]

At a critical image the Jacobian matrix $A = (\phi_{ij})$ may have rank 1 or 0. We first consider those critical images with $\operatorname{rank}(A) = 1$, i.e., where exactly one eigenvalue of A vanishes and where, in addition, $\nabla D \neq 0$. The critical images with these two properties form smooth curves without end points, as follows from the implicit function theorem applied to the equation $D = 0$ and the fact that the rank of a continuous, matrix-valued function is at least as large in a neighborhood of a point $\mathbf{x}^{(0)}$ as at that point.

Suppose then that, at $\left(\mathbf{x}^{(0)}, \mathbf{p}^{(0)}\right)$, $A^{(0)}$ has rank 1 and $\nabla D^{(0)} \neq 0$. We can then introduce cartesian coordinates such that $\mathbf{x}^{(0)}$ and its source $\mathbf{y}^{(0)}$ are at the origins of the image and source planes, respectively, the x- and y-axes are parallel, and the symmetric matrix $A^{(0)}$ is diagonal, with $A_{11}^{(0)} \neq 0$, $A_{22}^{(0)} = 0$, so that the x_2-direction at $\mathbf{x}^{(0)}$ is annihilated by $A^{(0)}$. A second preferred direction at $\mathbf{x}^{(0)}$ is defined by the critical curve given by (6.4).

---

[2] This happens, e.g., in the case of a spherically symmetric lens, where the central part is a uniform disk of critical surface mass density – see section 8.1.3 below.

Since
$$D = \phi_{11}\phi_{22} - \phi_{12}^2 \tag{6.5}$$

and $A_{ik} = \phi_{ik}$, at $\mathbf{x}^{(0)}$ the normal vector ∇D has components $D_i^{(0)} = \phi_{11}^{(0)}\phi_{22i}^{(0)}$. Hence, a tangent vector to the critical curve at $\mathbf{x}^{(0)}$ is obtained by applying a $\frac{\pi}{2}$-rotation $R\left(\frac{\pi}{2}\right)$ to ∇D,

$$\mathbf{T}^{(0)} = R\left(\frac{\pi}{2}\right)\nabla D^{(0)} = \phi_{11}^{(0)}\left(-\phi_{222}^{(0)}, \phi_{122}^{(0)}\right) . \tag{6.6}$$

Thus, there are two qualitatively different possibilities: Either the derivative of the restriction of the lens mapping to the critical curve does not vanish at $\mathbf{x}^{(0)}$,

$$A^{(0)} \cdot \mathbf{T}^{(0)} \neq 0 \quad , \quad \text{i.e.,} \quad \phi_{222}^{(0)} \neq 0 \quad , \tag{6.7}$$

or it vanishes,

$$A^{(0)} \cdot \mathbf{T}^{(0)} = 0 \quad , \quad \text{i.e.,} \quad \phi_{222}^{(0)} = 0 \quad . \tag{6.8}$$

We consider these two cases in turn. Note that, if $\mathbf{c}(\lambda)$ parameterizes a critical curve in the lens plane, the corresponding caustic is $\boldsymbol{\gamma}(\lambda) = \mathbf{y}(\mathbf{c}(\lambda))$, and the tangent vector at $\boldsymbol{\gamma}(\lambda)$ is $A(\lambda)\mathbf{T}(\lambda)$. Hence, if (6.7) applies, the caustic has a well-defined tangent vector at the point under consideration and is therefore smooth, whereas this is not necessarily the case if (6.8) applies.

6.2.1 Folds. Rules for truncating Taylor expansions

If at some point $\mathbf{x}^{(0)}$

$$D^{(0)} = 0 \quad , \quad A^{(0)}R\left(\frac{\pi}{2}\right)\nabla D^{(0)} \neq 0 \quad , \tag{6.9}$$

then these relations also hold at all points near $\mathbf{x}^{(0)}$ where D vanishes. Thus, singular images of this kind form curves without end points.

Local expansion of the Fermat potential. To study a lens mapping near an image where (6.9) is valid, we choose coordinates as in the last section, so that

$$\phi_1^{(0)} = \phi_2^{(0)} = \phi_{12}^{(0)} = \phi_{22}^{(0)} = 0 \quad , \quad \phi_{11}^{(0)} \neq 0 \neq \phi_{222}^{(0)} \quad . \tag{6.10}$$

Hence, the Taylor expansion of the Fermat potential ϕ at $\mathbf{x}^{(0)}$, $\mathbf{y}^{(0)}$ reads

$$\begin{aligned}\phi = \phi^{(0)} &+ \frac{1}{2}\mathbf{y}^2 - \mathbf{x}\cdot\mathbf{y} + \frac{1}{2}\phi_{11}^{(0)}x_1^2 + \frac{1}{6}\phi_{111}^{(0)}x_1^3 \\ &+ \frac{1}{2}\phi_{112}^{(0)}x_1^2 x_2 + \frac{1}{2}\phi_{122}^{(0)}x_1 x_2^2 + \frac{1}{6}\phi_{222}^{(0)}x_2^3 \\ &+ x_1 P_3 + R_4 \quad ,\end{aligned} \tag{6.11}$$

where P_i denotes a homogeneous polynomial of degree i and R_4 denotes a 4th order remainder term. Therefore, the lens mapping f, (5.12), reads (at fixed $\mathbf{p}^{(0)}$):

$$y_1 = \phi_{11}^{(0)} x_1 + \phi_{112}^{(0)} x_1 x_2 + \frac{1}{2}\phi_{122}^{(0)} x_2^2 + x_1 P_1 + R_3 \;,$$
$$y_2 = \frac{1}{2}\phi_{112}^{(0)} x_1^2 + \phi_{122}^{(0)} x_1 x_2 + \frac{1}{2}\phi_{222}^{(0)} x_2^2 + R_3 \;.$$
(6.12)

These equations exhibit the difficulty one must face to deduce general properties of the lens mapping near singular points: one has to prove that small remainder terms can legitimately be neglected, i.e., that they do not affect the properties of interest. To do this rigorously requires the exclusion of some rare, exceptional maps and the introduction of curvilinear coordinates in the image and source planes, respectively, which are adapted to the mapping in question; in terms of these, the mappings near singularities can be expressed in simple, polynomial normal forms. It is not our purpose here to carry out such a laborious procedure; we refer the interested reader to the references quoted. Instead, we proceed as follows: We note that omitting the terms $x_1 P_1 + R_3$ and R_3 in (6.12), respectively, introduces only arbitrarily small relative errors into the mapping $\mathbf{x} \mapsto \mathbf{y}$ near the origin. Therefore, we approximate the general mapping (6.12) by the representative example

$$y_1 = \phi_{11}^{(0)} x_1 + \frac{1}{2}\phi_{122}^{(0)} x_2^2 + \phi_{112}^{(0)} x_1 x_2 \;,$$
$$y_2 = \frac{1}{2}\phi_{112}^{(0)} x_1^2 + \phi_{122}^{(0)} x_1 x_2 + \frac{1}{2}\phi_{222}^{(0)} x_2^2 \;,$$
(6.13)

and study the latter.[3] Its properties derived below hold for all maps that it represents, i.e., those which obey the conditions (6.9).

Before continuing our study of the map (6.13), we give some rules for truncating Taylor expansions which we shall use in the remainder of this chapter.

Rules for truncating Taylor expansions. As a first step, we identify terms in ϕ which, in the case in question, do not vanish [such as $\frac{1}{2}\phi_{11}^{(0)} x_1^2$ and $\frac{1}{6}\phi_{222}^{(0)} x_2^3$ in (6.11)]; let us call them *significant* terms. Not every monomial $x_1^p x_2^q$ which is of higher degree than a significant term may be neglected, in spite of the fact that each contributes arbitrarily little to ϕ for $\mathbf{x} \to 0$; for we need not only ϕ itself, but also its first derivatives (to find the lens mapping) and its second derivatives (to find D and, with its help, the critical points). In (6.11), for example, the term involving $x_1^2 x_2$ is negligible compared to x_1^2, but its derivative with respect to x_2, x_1^2, dominates that of the significant term, which vanishes. Therefore, in a second step, we *pro-*

---

[3] We note that (6.13) is not the canonical form of the potential around a fold, as usually given in textbooks on catastrophe theory. The reason for this is that we confine ourselves to linear coordinate transformations. The canonical form can be obtained from (6.13) by introducing coordinates which depend nonlinearly on the ones we use.

visionally neglect those terms which, together with their first and second derivatives, are of higher degree than a corresponding expression due to a significant term. However, these rules are not sufficient to guarantee that the result concerning the approximate form of the critical curves and caustics is valid; for it may happen that in the course of the calculation of the solutions of $D = 0$ and the subsequent computation of the caustics, some terms originating in significant terms cancel so that, in the corresponding equations, a term which was provisionally neglected turns out, in fact, not to be negligible. (A case where such a cancellation occurs will turn up in Sect. 6.3.2.) Therefore, having determined the caustics, one has to check, in a third step, whether such a cancellation occurred. If so, one has to reintroduce the leading, provisionally neglected, terms and repeat the calculation until the neglected terms turn out to be truly negligible throughout the entire calculation. Fortunately, in the cases which follow, this procedure quickly leads to a consistent approximation; we shall henceforth not mention this checking procedure. — Note that some terms may be neither significant nor negligible, e.g., $\phi^{(0)}_{122}$ in (6.11). — It may happen that a term which does not appear to be negligible according to the preceding rules, turns out to be negligible eventually. This is the case, for example, with all terms of higher than third degree in (6.46) and (6.49) below. We then drop these terms without further ado. — Whereas the foregoing rules suffice for our purposes, it is clear that a general theory of singularities calls for more sophisticated concepts and methods such as those explained, e.g., in [BR75.1], [GO73.1].

After this slightly complicated, but unavoidable digression we return to the case under discussion.

Fold singularity. The Jacobian matrix of the map (6.13) is

$$A = \begin{pmatrix} \phi^{(0)}_{11} + \phi^{(0)}_{112}x_2 & \phi^{(0)}_{112}x_1 + \phi^{(0)}_{122}x_2 \\ \phi^{(0)}_{112}x_1 + \phi^{(0)}_{122}x_2 & \phi^{(0)}_{122}x_1 + \phi^{(0)}_{222}x_2 \end{pmatrix} . \tag{6.14}$$

Hence, the tangent of the *critical curve* at the origin is obtained via $D = \det A = 0$, as the line

$$\phi^{(0)}_{122}x_1 + \phi^{(0)}_{222}x_2 = 0 . \tag{6.15}$$

Eliminating x_2 from (6.13) by means of (6.15) gives as the leading approximation to the *caustic* near the origin of the source plane, the parabola

$$Q(\mathbf{y}) := 2\left(\phi^{(0)}_{11}\right)^2 \phi^{(0)}_{222} y_2 - \left(\phi^{(0)}_{112}\phi^{(0)}_{222} - \left[\phi^{(0)}_{122}\right]^2\right) y_1^2 = 0 . \tag{6.16}$$

Since, according to the second of the two conditions (6.9), the derivative of the lens mapping in the direction of the critical curve does not vanish, that curve is mapped bijectively onto the caustic.

Next we study the inversion of $\mathbf{x} \mapsto \mathbf{y}$. Solving the first of (6.13) for x_1, inserting the result into the second of those equations and neglecting terms containing $y_1 x_2^2$, x_2^3 and x_2^4 as negligible compared to the x_2^2-term [with coefficient $\left(\phi_{11}^{(0)}\right)^2 \phi_{222}^{(0)}$, which is nonzero according to our assumptions] gives a quadratic equation for x_2 which has real solutions if, and only if, the left-hand side of (6.16), $Q(\mathbf{y})$, is non-negative. Close to the origin, therefore, a source on the negative side of the caustic (i.e., where $Q < 0$) has no image, one on the caustic has exactly one image (on the critical curve), and one on the positive side has exactly two images,[4] situated on opposite sides of the critical curve, given in leading order by

$$x_1 = \frac{\phi_{222}^{(0)} y_1 - \phi_{122}^{(0)} y_2}{\phi_{11}^{(0)} \phi_{222}^{(0)}},$$

$$x_2 = \frac{-\phi_{122}^{(0)} y_1 \pm \sqrt{Q(\mathbf{y})}}{\phi_{11}^{(0)} \phi_{222}^{(0)}}.$$

(6.17)

Since D changes sign at the critical curve, the two images have opposite parities.

The prototype of a mapping exhibiting a singularity of the type under consideration is the parallel projection of a sphere onto a plane, see Fig. 6.1. In this case, the great circle parallel to the plane represents the critical curve. For an obvious intuitive reason the singular curves consisting of points obeying the conditions (6.9) are called *folds*.

If we denote the "vertical" separation of a source point from the caustic as Δy_2 and the corresponding separations of its images from the critical curve as Δx_2 we obtain, by (6.16) and (6.17),

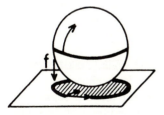

Fig. 6.1. The projection f of a sphere onto a plane exhibits a fold-type singularity

---

[4] Quite generally, the number of images for a source on a caustic is the mean of the numbers of images on both sides; for later use, we also note that the number of images of a source on a cusp is the same as on the outside of the cusp. In addition, we want to note that, since our present considerations are local, referring to the domain of validity of a Taylor expansion, the image numbers considered in the following and listed in the figure captions, count only those images which are located near that critical point which is mapped onto the caustic under consideration. There may, of course, be additional images located elsewhere.

$$\Delta y_2 = \frac{Q(\mathbf{y})}{2\left(\phi_{11}^{(0)}\right)^2 \phi_{222}^{(0)}} \qquad (6.18a)$$

and, for sources on the positive side of the caustic,

$$\Delta x_2 = \pm\sqrt{\frac{2\Delta y_2}{\phi_{222}^{(0)}}} \ . \qquad (6.18b)$$

The magnifications of these images, obtained from the Jacobian of the map $\mathbf{y} \mapsto \mathbf{x}$ given in (6.17), are

$$|\mu| = \frac{1}{\left|\phi_{11}^{(0)}\right|\sqrt{2\phi_{222}^{(0)}\Delta y_2}} \ . \qquad (6.19a)$$

Thus, the total magnification μ_p of a point source close to a caustic is

$$\mu_p = \frac{1}{\left|\phi_{11}^{(0)}\right|} \cdot \sqrt{\frac{2}{\phi_{222}^{(0)}\Delta y_2}} + \mu_0 \ , \qquad (6.19b)$$

where μ_0 is the magnification of other images of the source that are far away from the point $\mathbf{x}^{(0)}$ under consideration, which are not accounted for by our local analysis.

On approaching the caustic from the positive side, a source produces images coming closer and closer together, and nearer to the critical curve, thereby brightening. When the source crosses the caustic, the two images fuse and disappear. At the moment they fuse, their magnification becomes (formally) infinite. This infinite magnification poses no physical problem: for any finite source size, the magnification remains finite (see Sect. 6.4), and if the source were point-like, wave optics would have to be used to describe lensing very close to caustics (see Chap. 7); also in this case the magnification remains finite.

Physical units. So far, we have not specified the length scale ξ_0 used in defining the dimensionless quantities $\mathbf{x}$ and $\mathbf{y}$ [see (5.3)]. If we choose $\xi_0 = D_d$, $\mathbf{x}$ and $\mathbf{y}$ become angular separation vectors. Thus, the angular separation $\Delta\theta$ between the two images equals $2|\Delta x_2|$, (6.18b). Combining (6.18b) and (6.19a), the unobservable source position Δy_2 can be eliminated:

$$|\mu| = \frac{2}{\left|\phi_{11}^{(0)}\right|\left|\phi_{222}^{(0)}\right|} \frac{1}{\Delta\theta} \ ; \qquad (6.20)$$

thus, the image magnification is inversely proportional to the angular separation of the images.

From (5.46), we can obtain the time delay between the two nearly critical images, using (6.11) and the solutions x_1, x_2:

$$c\,\Delta t = \frac{D_\mathrm{d} D_\mathrm{s}}{D_\mathrm{ds}}(1+z_\mathrm{d})\frac{4}{3}|\Delta y_2|\sqrt{\frac{2\Delta y_2}{\phi_{222}^{(0)}}}\;,$$

which can be combined with (6.19a) and (6.18b) to yield

$$\frac{(c\,\Delta t)\,|\mu|}{(\Delta\theta)^2} = \frac{1}{6\left|\phi_{11}^{(0)}\right|}\frac{D_\mathrm{d}D_\mathrm{s}}{D_\mathrm{ds}}(1+z_\mathrm{d})\;. \qquad (6.21a)$$

From this equation we can estimate the expected time delay. The only parameter from the lens model is $\left|\phi_{11}^{(0)}\right| = \left|1-\psi_{11}^{(0)}\right|$, which can be directly related to the local dimensionless surface mass density κ by $\left|\phi_{11}^{(0)}\right| = |\mathrm{tr}\,A| = 2|1-\kappa|$, according to (5.26) and the assumed fact that $\phi_{22}^{(0)} = 0$. Then,

$$\Delta t = \frac{r(z_\mathrm{d})\,r(z_\mathrm{s})}{r(z_\mathrm{d},z_\mathrm{s})}(1+z_\mathrm{d})\left(\frac{\Delta\theta}{1''}\right)^2 \frac{1}{|\mu|}\frac{1}{|1-\kappa|}\,h_{50}^{-1}\,1.2\times 10^6\,\mathrm{sec}.\qquad (6.21b)$$

Note that the explicit dependence of the first factor in (6.21b) on the redshifts is best described by (4.68a) and the corresponding Fig. 4.2. This equation has been used in the discussion of PG1115+08, where a pair of nearly critical images is observed (see Sect. 2.5.3).

Coordinate-free expressions. It is convenient to investigate the local behavior of the lens equation in an appropriate coordinate system. However, the results can also be expressed without referring to specific coordinates. One example is equation (6.21b) which does not contain any coordinate-dependent term.

According to (6.14), we have

$$(\nabla\det A)^{(0)} = -\mathrm{tr}\,A\begin{pmatrix}\phi_{122}^{(0)}\\ \phi_{222}^{(0)}\end{pmatrix}\;.$$

Since $\nabla\det A$ is perpendicular to the critical curve, the direction of the critical curve is that of the vector $R(\pi/2)\nabla\det A$. Furthermore, we note that

$$R\left(-\frac{\pi}{2}\right)A\,R\left(\frac{\pi}{2}\right) = \begin{pmatrix}0 & 0\\ 0 & \mathrm{tr}\,A\end{pmatrix}\;,$$

so that

$$\left[R\left(-\frac{\pi}{2}\right)A\,R\left(\frac{\pi}{2}\right)\nabla(\det A)\right]\cdot\mathbf{y} = -(\mathrm{tr}\,A)^2\phi_{222}\,y_2\;,$$

where we have neglected the small difference between y_2 and Δy_2.

Since the left-hand side of this equation is coordinate-invariant, so is the right-hand side. Thus, a coordinate-free expression of the magnification reads

$$|\mu| = \left\{ 2 \left[R\left(-\frac{\pi}{2}\right) A\, R\left(\frac{\pi}{2}\right) \nabla(\det A) \right] \cdot \mathbf{y} \right\}^{-1/2} , \qquad (6.22)$$

This expression agrees with that of [KA89.3].

Stability of folds. Finally we show that a fold is stable against small perturbations of the lens model. Indeed, if

$$D\left(\mathbf{x}^{(0)}; \mathbf{p}^{(0)}\right) = 0 , \quad D_2\left(\mathbf{x}^{(0)}; \mathbf{p}^{(0)}\right) \neq 0 ,$$

then the implicit function theorem says that the solutions of $D = 0$ near $\mathbf{x}^{(0)}$, $\mathbf{p}^{(0)}$ are given by a function $x_2 = x_2(x_1, \mathbf{p})$, and clearly, close to $\mathbf{x}^{(0)}$, $\mathbf{p}^{(0)}$ on this curve, the characteristic properties (6.9) of a fold continue to hold, because of continuity.

Convexity of caustics. It is of some interest to note that if a critical curve, or a part of it, is situated outside the mass distribution of the lens or if κ is locally constant, Poisson's equation holds there, and since $\phi_{ijk} = -\psi_{ijk}$, we get by differentiation, $\phi^{(0)}_{112} = -\phi^{(0)}_{222}$. Equation (6.16) then shows that, viewed from the positive, 2-image side, the corresponding caustic looks convex. This property, which has been suspected earlier [GR89.2], will be demonstrated explicitly in several examples further below (e.g., the two point-mass lens, Sect. 8.3, and the caustics of the random star field, Sect. 11.2).

6.2.2 Cusps

Expansion of the Fermat potential. We now study the second kind of critical image considered in the introduction to this section, i.e., we assume – recall (6.8) –

$$D^{(0)} = 0 , \quad \operatorname{tr} A^{(0)} \neq 0 , \quad \nabla D^{(0)} \neq 0 , \quad A^{(0)} R\left(\frac{\pi}{2}\right) \nabla D^{(0)} = 0 . \quad (6.23)$$

The last condition means that, at $\mathbf{x}^{(0)} (= 0)$, the tangent vector to the critical curve passing through $\mathbf{x}^{(0)}$ is contained in the kernel of the Jacobian matrix A (the eigenvector of A with vanishing eigenvalue). Under these conditions, we can introduce coordinates such that

$$\begin{gathered} \phi^{(0)}_1 = \phi^{(0)}_2 = \phi^{(0)}_{12} = \phi^{(0)}_{22} = \phi^{(0)}_{222} = 0 , \\ \phi^{(0)}_{11} \neq 0 \neq \phi^{(0)}_{122} . \end{gathered} \qquad (6.24)$$

The Fermat potential near $\mathbf{x}^{(0)}$ then reads

$$\phi = \phi^{(0)} + \frac{1}{2}\mathbf{y}^2 - \mathbf{x}\cdot\mathbf{y} + \frac{1}{2}\phi_{11}^{(0)}x_1^2 + \frac{1}{6}\phi_{111}^{(0)}x_1^3$$
$$+ \frac{1}{2}\phi_{112}^{(0)}x_1^2 x_2 + \frac{1}{2}\phi_{122}^{(0)}x_1 x_2^2 + \frac{1}{24}\phi_{2222}^{(0)}x_2^4 \quad (6.25)$$
$$+ x_1 P_3 + R_5 \quad,$$

and hence the lens mapping is given by

$$y_1 = \phi_{11}^{(0)}x_1 + \frac{1}{2}\phi_{111}^{(0)}x_1^2 + \phi_{112}^{(0)}x_1 x_2 + \frac{1}{2}\phi_{122}^{(0)}x_2^2 + P_3 + R_4 \quad,$$
$$y_2 = \frac{1}{2}\phi_{112}^{(0)}x_1^2 + \phi_{122}^{(0)}x_1 x_2 + \frac{1}{6}\phi_{2222}^{(0)}x_2^3 + x_1 P_2 + R_4 \quad. \tag{6.26}$$

As in the previous subsection, we restrict attention to the representative mapping

$$y_1 = cx_1 - \frac{1}{2}bx_2^2 + dx_1 x_2 \quad,$$
$$y_2 = \frac{1}{2}dx_1^2 - bx_1 x_2 - ax_2^3 \quad; \tag{6.27}$$

we have used the abbreviations

$$a = \frac{-1}{6}\phi_{2222}^{(0)}, \quad b = -\phi_{122}^{(0)}, \quad c = \phi_{11}^{(0)}, \quad d = \phi_{112}^{(0)} \quad. \tag{6.28}$$

Critical curve and caustic. Equation (6.27) leads to

$$A = \begin{pmatrix} c + dx_2 & dx_1 - bx_2 \\ dx_1 - bx_2 & -bx_1 - 3ax_2^2 \end{pmatrix} \tag{6.29}$$

and

$$D = -bcx_1 - (3ac + b^2)x_2^2 - d(dx_1^2 - bx_1 x_2 + 3ax_2^3) \quad. \tag{6.30}$$

Thus, all terms in D involving d turn out to be negligible near the origin although this was not obvious before, and we obtain as the critical curve the parabola

$$x_1 = -\frac{3ac + b^2}{bc}x_2^2 \quad. \tag{6.31}$$

Its image under the mapping (6.27) is given asymptotically for $x_2 \to 0$ by

$$y_1 = -\frac{3}{2b}(2ac + b^2)x_2^2 \quad,$$
$$y_2 = \frac{1}{c}(2ac + b^2)x_2^3 \quad, \tag{6.32}$$

provided

$$2ac + b^2 \neq 0 \quad . \tag{6.33}$$

(The case where $2ac + b^2 = 0$ will be treated in Sect. 6.3.2.) Then, the caustic is a semicubic parabola

$$y_1^3 = -\frac{27c^2 (2ac + b^2)}{8b^3} y_2^2 \quad . \tag{6.34}$$

The geometrical significance of the inequality (6.33) is related to the behavior of the vector $AR\left(\frac{\pi}{2}\right) \nabla D$ [which occurs in (6.7) and (6.8)] along the critical curve. Since $R\left(\frac{\pi}{2}\right) \nabla D = (-D_2, D_1)$, that vector can be computed by means of (6.30–32); the result is, if we use x_2 as a parameter on the critical curve,

$$\left[AR\left(\frac{\pi}{2}\right) \nabla D\right]_{\text{critical curve}} = 3 \left(2ac + b^2\right) \left(cx_2, -bx_2^2\right) \quad .$$

This equation shows that the inequality (6.33) holds in addition to (6.23) if, and only if, the vector $AR\left(\frac{\pi}{2}\right) \nabla D$ along the critical curve through $\mathbf{x}^{(0)}$ does not vanish in a neighborhood of $\mathbf{x}^{(0)}$ except at $\mathbf{x}^{(0)}$; in fact, we see that

$$2ac + b^2 \neq 0 \iff \left(\frac{\partial}{\partial x_2} AR\left(\frac{\pi}{2}\right) \nabla D\right)^{(0)} \neq 0 \quad . \tag{6.35}$$

This inequality means that the angle β between the tangent $\mathbf{T} = R\left(\frac{\pi}{2}\right) \nabla D$ of the critical curve and the kernel of A has a simple zero at $\mathbf{x}^{(0)}$.[5] If this inequality holds in addition to (6.23), the point $\mathbf{x}^{(0)}$ separates two folds contained in a smooth curve through $\mathbf{x}^{(0)}$. The corresponding caustic (6.34) has a cusp at $\mathbf{y}^{(0)} = 0$. Therefore, critical images characterized by the conditions (6.23) and (6.35) are also called *cusps*.

In Fig. 6.2 we have plotted the critical curve and the caustic near a cusp singularity. The fact that a smooth curve $\mathbf{c}(\lambda)$ can have an image $\gamma(\lambda)$ which is not smooth can be understood from the remark made after (6.8).

Image multiplicity. Let us now study the inversion of the lens mapping (6.27) near a cusp-type singularity. Equations (6.27) are equivalent to

---

[5] The function β just introduced is defined along critical curves. The roots of β can be used to classify certain types of singularities, the so-called cuspuids. Folds and cusps are the lowest-order cuspuids, corresponding to no or a simple root of β, respectively. If β has a second order root, a swallowtail results (see Sect. 6.3.2 below), and for higher-order roots, butterflies, wigwams, etc. occur.

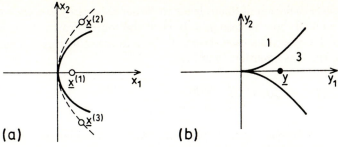

Fig. 6.2. The critical curve (**a**) and the corresponding caustic (**b**) near a cusp singularity, shown for the case $b > 0$, $c > 0$, $2ac + b^2 < 0$. In (b), the numbers of images for sources in the interior and the exterior of the caustic are indicated. The three images $\mathbf{x}^{(i)}$ of $\mathbf{y}$ approach the origin if $\mathbf{y}$ approaches the cusp, as described in the text

$$x_1 = \frac{y_1}{c} + \frac{b}{2c} x_2^2 \;,$$
$$x_2^3 + \frac{2by_1}{2ac + b^2} x_2 + \frac{2cy_2}{2ac + b^2} = 0 \;. \tag{6.36}$$

The second of these equations is of third order in x_2 and can be solved analytically. This solution can then be inserted in the first equation to yield x_1. The number of solutions of the second equation of (6.36) depends on the discriminant Δ,

$$\Delta = \frac{1}{(2ac + b^2)^2} \left[c^2 y_2^2 + \frac{8}{27} \frac{b^3 y_1^3}{(2ac + b^2)} \right] \;;$$

if $\Delta > 0$, only one real solution exists, whereas for $\Delta < 0$ there are three real solutions of (6.36). For $\Delta \to 0$, two of these three solutions merge. It is easily seen that $\Delta = 0$ corresponds to the caustic (6.34), which thus separates the regions of one and three solutions in the source plane: inside the cusp, a source has three images, whereas outside it has one. The standard example of a projection map exhibiting a cusp is shown in Fig. 6.3.

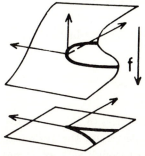

Fig. 6.3. The standard example of a projection of a surface onto a plane which illustrates the occurrence of a cusp

Since the general solution of (6.36) leads to rather messy expressions, we will concentrate on the special case $y_2 = 0$ when the source lies on the axis of symmetry of the cusp; this case shows all the important features. Then the solutions of the lens equation (6.36) are

$$\mathbf{x}^{(1)} = \left(\frac{y_1}{c}, 0\right) \quad ; \quad \mathbf{x}^{(2,3)} = \left(\frac{2ay_1}{2ac + b^2}, \pm\sqrt{-\frac{2by_1}{(2ac + b^2)}}\right) \quad , \quad (6.37)$$

where the last two solutions are real only if the expression under the square root is positive, i.e., if the source is inside the cusp. The magnifications of the images (6.37) are

$$\mu^{(1)} = -\frac{1}{by_1} \quad ; \quad \mu^{(2,3)} = \frac{1}{2by_1} \quad ,$$

i.e., the "single" image $\mathbf{x}^{(1)}$ has opposite parity to the "pair" of images $\mathbf{x}^{(2,3)}$. If y_1 approaches the cusp from the inside, the three images will tend to $\mathbf{x} = 0$ in such a way that the single image moves on a line perpendicular to the critical curve, and the pair of images approach the critical curve tangentially. For $y_1 = 0$ all three images fuse, their magnifications being (formally) infinite. If y_1 is outside the cusp, only one image remains. Note that this single image has a high magnification if $|y_1|$ is small, in contrast to the fold, where highly magnified images always occur in pairs. We state without proof that, in fact, for a source inside a cusp, the sum of the magnifications of the two images with the same parity is the same, up to a sign, as that of the third image with opposite parity [SC91.4].

Stability of cusps. Let a family of lens mappings have a cusp singularity at $\mathbf{x}^{(0)}$ for the parameter value $\mathbf{p}^{(0)}$. Then the last of conditions (6.23) may be written as

$$\chi_i\left(\mathbf{x}^{(0)}; \mathbf{p}^{(0)}\right) = 0 \quad (i = 1, 2) \quad , \qquad (6.38)$$

where $AR\left(\frac{\pi}{2}\right) \nabla D = (\chi_1, \chi_2)$. Explicitly,

$$\chi_1 = \phi_{12} D_1 - \phi_{11} D_2 \quad ,$$
$$\chi_2 = \phi_{22} D_1 - \phi_{12} D_2 \quad .$$

The Jacobian determinant of the pair of equations (6.38), evaluated at $\mathbf{x}^{(0)}$, $\mathbf{p}^{(0)}$, is

$$\left(\frac{\partial(\chi_1, \chi_2)}{\partial(x_1, x_2)}\right)^{(0)} = -3b^2 c^2 \left(2ac + b^2\right) \quad .$$

This expression does not vanish because of the inequalities (6.23) and (6.33). Therefore, again because of the implicit function theorem, the system of equations $\chi_i(\mathbf{x}; \mathbf{p}) = 0$ has a unique solution $\mathbf{x}(\mathbf{p})$ through $\mathbf{x}^{(0)}$, $\mathbf{p}^{(0)}$, and at each point $\mathbf{x}(\mathbf{p})$, conditions analogous to (6.23) and (6.33) hold, which characterize cusps. Thus, having a cusp is a stable property of a lens mapping.

6.2.3 Whitney's theorem. Singularities of generic lens maps

Stable and unstable singularities of lens maps. We have seen that folds and cusps are stable, i.e., preserved under small perturbations of the map. Folds are characterized by *one* equation, $D = 0$, and an inequality; cusps by *two* equations, $A^{(0)} R\left(\frac{\pi}{2}\right) \nabla D^0 = 0$ (which immediately implies $D^{(0)} = 0$ for $\nabla D^{(0)} \neq 0$) and an inequality. Suppose now that a lens map has a singularity at $\mathbf{x}^{(0)}$ which is neither a cusp nor contained in a fold. Then, according to the foregoing discussion, one of the following three sets of conditions (which are not mutually exclusive) must hold:

$$A^{(0)} = 0 \; , \tag{6.39a}$$

$$A^{(0)} \cdot \mathbf{T}^{(0)} = 0 \;, \quad (\mathbf{T} \cdot \nabla (A \cdot \mathbf{T}))^{(0)} = 0 \; , \tag{6.39b}$$

$$D^{(0)} = 0 \;, \quad \nabla D^{(0)} = 0 \; , \tag{6.39c}$$

where

$$\mathbf{T} = R\left(\frac{\pi}{2}\right) \nabla D \; .$$

In each case, $\mathbf{x}^{(0)}$ satisfies at least *three* equations $\chi_j^{(0)} = 0$, where the functions χ_j are polynomials in derivatives of ψ (of at most fourth order). Each of the three equations defines a subset S_j – generically a curve – in the image plane, and $\mathbf{x}^{(0)}$ is contained in the intersection of these three sets. Since in perturbing a map one can independently change the derivatives of ψ at $\mathbf{x}^{(0)}$, and therefore deform the subsets S_j differently, intuition suggests that, given a singular image $\mathbf{x}^{(0)}$ of one of the types listed above, one should be able to perturb the deflection potential ψ in such a way that the sets S_j no longer intersect in some neighborhood of $\mathbf{x}^{(0)}$, since, in general, three plane curves do not have a common point of intersection. This is, in fact, true; thus *the only stable singularities of lens maps* (and, more generally, of smooth maps from the plane $\mathbb{R}^2$ into itself) *are folds and cusps*. This theorem was proved by Hassler Whitney in 1955 [WH55.1].

Singularities of generic lens maps. According to Whitney's theorem, a lens map which has singularities other than folds and cusps can be slightly changed such that these other singularities disappear and only folds and cusps remain. The set of lens maps having unstable singularities is "small" in the space of all lens maps; more precisely, it is closed and nowhere dense in that space (with respect to a suitable topology). Generally, a statement referring to members of some topological space of maps is said to hold *generically*, or *for generic mappings*, if it is valid except on some "small" (in the above sense) subset; if that statement happens not to hold for a particular member of the space, it will nevertheless hold for nearly all of its neighbors ("perturbations").

Generically, the critical set of a lens map consists of finitely many, or possibly none, closed (i.e., compact without boundary), smooth, non-intersecting curves. The caustics corresponding to the critical curves are closed, piecewise-smooth curves which may have finitely many cusps; they may intersect themselves and each other.

6.3 Stable singularities of one-parameter families of lens mappings; metamorphoses

Folds and cusps are the only stable singularities of lens mappings; generically, no other singularities occur. Simple counting of equations and variables x_1, x_2; $p_1,\ldots,p_N$, as in the last subsection, suggests that those critical images which are determined by at most *three* independent equations can be stable in *one-parameter families* of lens mappings, and no others can occur in such families. This turns out to be true.

In a one-parameter family of lens mappings, the critical curves and caustics will depend on the parameter p. Qualitative changes of their patterns, called *metamorphoses*, will however take place only for particular values $p^{(0)}$ and at special points $\mathbf{x}^{(0)}$, $\mathbf{y}^{(0)}$. These are the singularities which occupy us in this section. The 3-dimensional control space associated with a 1-parameter family of lens mappings is foliated by source planes $p = $ const. in which y_1, y_2 are cartesian coordinates. In such a space, the unions of the caustics will in general form surfaces ("large caustics") with one-dimensional, sharp edges consisting of cusps. Points at which metamorphoses occur are singular points of these surfaces; the latter may self-intersect and bifurcate.

An important example of a one-parameter family of lens mappings is obtained as follows. Consider as fixed an observer position O and a lens plane L carrying a mass distribution, and view the ratio $p = D_{\mathrm{ds}}/D_{\mathrm{s}}$ as variable, $0 < p < 1$, as for the classification of ordinary images discussed in Sect. 5.2. Then, as remarked previously, the large caustic in the control space spanned by y_1, y_2, p may be identified, in the approximation of lens theory, with the spatial projection of that part of the caustic of the past light cone $\mathcal{C}_O^-$ which is caused by that mass distribution, i.e., which is generated by light rays traversing L in a neighborhood of the deflector. (An analogous remark applies to $\mathcal{C}_S^+$, of course.) This example shows how the three kinds of "caustics" used in lens theory – caustics of one mapping in one $\mathbf{y}$-plane, large caustics of 1-parameter families of mappings in $\mathbf{y}$, p-space, and caustics of light cones – are related to each other.

In catastrophe theory, singularities are considered as equivalent if the (large) caustic sets to which they belong can be transformed into each other by a diffeomorphism of the control space. In the case of lens mappings, only those diffeomorphisms which preserve the foliation by source planes are permitted; hence a finer classification of singularities results.

6.3.1 Umbilics

We first consider those critical images $\mathbf{x}^{(0)}$ which obey (6.39a),

$$A^{(0)} = 0 \quad . \tag{6.40}$$

The corresponding light beams pass through the lens plane at points where, by (5.26) and (5.4), the surface mass density has the critical value defined in (5.5). Equation (6.40) implies $D^{(0)} = 0$ and, since D is bilinear in the second derivatives of ϕ, also $\nabla D^{(0)} = 0$. Let Δ denote the determinant of the symmetric tensor D_{ik} (the Hessian of D). Since the equation $\Delta^{(0)} = 0$ imposes restrictions on the third derivatives of ϕ, and is thus in general independent of the three equations contained in the matrix equation (6.40), we consequently assume that either

$$\Delta^{(0)} > 0 \tag{6.41a}$$

or

$$\Delta^{(0)} < 0 \quad , \tag{6.41b}$$

where

$$\Delta^{(0)} = \left(6\phi_{111}\phi_{112}\phi_{122}\phi_{222} - 4\phi_{111}\phi_{122}^3 \right. \\ \left. - 4\phi_{112}^3\phi_{222} + 3\phi_{112}^2\phi_{122}^2 - \phi_{111}^2\phi_{222}^2 \right)^{(0)} \quad . \tag{6.42}$$

For reasons which will become clear below, critical images defined by (6.40) and (6.41a) are called *elliptic umbilics*, those defined by (6.40) and (6.41b), *hyperbolic umbilics*.

For both kinds of umbilics, $\phi_i^{(0)} = \phi_{ij}^{(0)} = 0$. The Taylor expansion of the Fermat potential therefore starts, apart from an irrelevant constant, with the cubic form

$$\frac{1}{6}\phi_{ijk}^{(0)} x_i x_j x_k \quad . \tag{6.43}$$

To keep the calculations simple, it is useful to employ non-orthogonal linear coordinates in the image and source planes, on the basis of the following statement:

Lemma: The real, binary cubic form (6.43) can be transformed, by an invertible, real linear transformation, into the form

$$\frac{1}{3}\bar{x}_1^3 - \bar{x}_1\bar{x}_2^2 \tag{6.44}$$

if (6.41a) holds, and into the form

$$\frac{1}{3}(\bar{x}_1^3 + \bar{x}_2^3) \tag{6.45}$$

if (6.41b) holds. (The cubic has four coefficients, as does a linear transformation. A suitable ansatz and comparison of terms gives the desired results.)

Elliptic umbilics. According to the preceding lemma, the leading, third order terms of the unperturbed Fermat potential near an elliptic umbilic can be written in suitable linear coordinates in the form (6.44). Making use of suitable coordinate translations $y_i \to y_i + p\alpha_i$, $x_i \to x_i + p\beta_i$, we can remove terms of the form px_i, px_1^2, px_1x_2 from the perturbed part of ϕ. Consequently, we may take as the representative potential

$$\phi = \phi^{(0)} + \frac{1}{2}\mathbf{y}^2 - \mathbf{x} \cdot \mathbf{y} + \frac{1}{3}x_1^3 - x_1x_2^2 + 2px_2^2 \quad . \tag{6.46}$$

(We have suppressed the bars on top of the x_i to simplify the notation.) The (representative) lens mapping thus reads

$$\begin{aligned} y_1 &= x_1^2 - x_2^2 \;, \\ y_2 &= -2x_1x_2 + 4px_2 \;, \end{aligned} \tag{6.47}$$

whence

$$-\frac{1}{4}D = (x_1 - p)^2 + x_2^2 - p^2 \quad .$$

Consequently, the critical curves are ellipses for $p \neq 0$ (not necessarily circles since we are using non-cartesian coordinates) which, for $p = 0$, degenerate into the singular point $\mathbf{x}^{(0)} = (0,0)$. A parameter representation of the critical ellipses is

$$\begin{aligned} x_1 &= p(1 + \cos\varphi) \;, \\ x_2 &= p\sin\varphi \;. \end{aligned} \tag{6.48a}$$

Insertion of these expression into (6.47) gives the caustics:

$$\begin{aligned} y_1 &= 2p^2\cos\varphi(1 + \cos\varphi) \;, \\ y_2 &= 2p^2\sin\varphi(1 - \cos\varphi) \;. \end{aligned} \tag{6.48b}$$

The tangent vector $\frac{d\mathbf{y}}{d\varphi}$ vanishes at $\varphi = 0, \frac{2}{3}\pi, \frac{4}{3}\pi$. At these values the caustic has cusps; their positions according to (6.48b) are shown in Fig. 6.4.

Thus, the following picture emerges: Near the singular point $\mathbf{x}^{(0)} = 0$, $p = 0$, the union of the critical curves is an elliptic cone with the singular

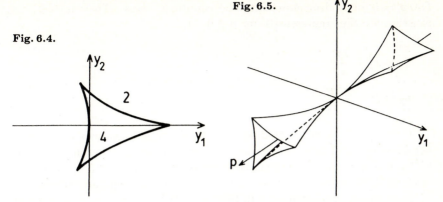

Fig. 6.4. The caustic of an elliptic umbilic near the metamorphosis. We also indicate the number of images for sources in their respective regions of the source plane

Fig. 6.5. The large caustic of the elliptic umbilic in the $(\mathbf{y}, p)$-control space

point as its vertex. The caustic has the shape of the surface of a pyramid whose three curved edges, also called cusp ridges, merge smoothly at the singular point $\mathbf{y} = 0$, $p = 0$ in control space; see Fig. 6.5.

The image multiplicities are now easily obtained. According to Fig. 6.4, it suffices to consider the images of sources situated on the y_1-axis. Then, according to (6.47), either $x_2 = 0$ and $x_1 = \pm\sqrt{y_1}$, or $x_1 = 2p$ and $x_2 = \pm\sqrt{4p^2 - y_1}$. This results in the numbers of images indicated in Fig. 6.4.

Hyperbolic umbilics. In this case, we obtain with the help of (6.45) instead of (6.46), by analogous arguments,

$$\phi = \phi^{(0)} + \frac{1}{2}\mathbf{y}^2 - \mathbf{x} \cdot \mathbf{y} + \frac{1}{3}\left(x_1^3 + x_2^3\right) + 2px_1x_2 \quad . \tag{6.49}$$

Accordingly, the lens mapping is

$$\begin{aligned} y_1 &= x_1^2 + 2px_2 \quad , \\ y_2 &= x_2^2 + 2px_1 \quad . \end{aligned} \tag{6.50}$$

Therefore,

$$\frac{1}{4}D = x_1x_2 - p^2 \quad , \tag{6.51}$$

so that, for $p \neq 0$, the critical curves are hyperbolae given by

$$x_2 = \frac{p^2}{x_1} \quad . \tag{6.52}$$

For $p = 0$, they degenerate into the coordinate axes. Their images, the caustics, are thus represented, for $p \neq 0$, by

$$y_1 = x_1^2 + \frac{2p^3}{x_1},$$

$$y_2 = 2px_1 + \frac{p^4}{x_1^2}. \tag{6.53}$$

It is seen from (6.50) that for $p = 0$, the caustic consists of the non-negative halves of the y_1, y_2 coordinate axes; it has a corner at the source $\mathbf{y}^{(0)} = 0$ corresponding to the critical image $\mathbf{x}^{(0)} = 0$. For $p \neq 0$, a consideration of (6.53) shows that the caustic consists of two separate curves. One of them, given by $x_1 \operatorname{sign}(p) > 0$, has a cusp which, for $p \to 0$, approaches the origin along the ray $y_1 = y_2 > 0$; the other one is a smooth curve – see Fig. 6.6. For $p \to 0$, the two curves approach each other and become identical at $p = 0$. One can visualize the large caustic in $(\mathbf{y}, p)$-space as a pair of boat-bow shaped surfaces intersecting each other in a curve with a corner at the point where the metamorphosis takes place, see Fig. 6.7.

To obtain the numbers of images for various regions it suffices, according to Fig. 6.6, to consider sources situated on the diagonal $y_1 = y_2$. Equations (6.50) can easily be solved for $\mathbf{y}$ on this diagonal; the resulting image numbers are indicated in Fig. 6.6.

Stability of umbilics. The foregoing descriptions of the critical and caustic surfaces in $(\mathbf{x}, p)$ and $(\mathbf{y}, p)$-space, respectively, confirm that both kinds of umbilics considered here are unstable as singularities of individual lens mappings. However, they are *stable as singularities of* generic, one-parameter *families* of such mappings. To prove this, let us assume that, according to (6.40), the Fermat potential $\phi(\mathbf{x}, \mathbf{y}; p)$ of such a family obeys

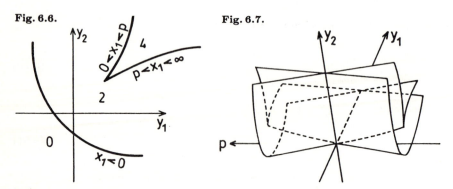

Fig. 6.6. The caustic of a hyperbolic umbilic near the metamorphosis. As before, we indicate the number of images

Fig. 6.7. The large caustic of the hyperbolic umbilic in the $(\mathbf{y}, p)$-control space

$$\phi_{ik}\left(\mathbf{x}^{(0)}, p^{(0)}\right) = 0 \ .$$

One can easily verify that the Jacobian $\frac{\partial(\phi_{11}, \phi_{22}, \phi_{12})}{\partial(x_1, x_2, p)}$ equals, at $\mathbf{x}^{(0)}$, $p^{(0)}$, the expression

$$-\frac{1}{2}\phi'_{11}D_{22} - \frac{1}{2}\phi'_{22}D_{11} + \phi'_{11}D_{12} \ ,$$

where the prime indicates differentiation with respect to the parameter p. It vanishes only if the ϕ'_{ij}-values are contained in a plane, i.e., a set of measure zero of the space $\mathbb{R}^3$ of all possible ϕ'_{ij}-values. For generic families, we thus have

$$\frac{\partial(\phi_{11}, \phi_{22}, \phi_{12})}{\partial(x_1, x_2, p)}\left(\mathbf{x}^{(0)}, p^{(0)}\right) \neq 0 \ .$$

The implicit function theorem shows that if the family $\phi(\mathbf{x}, \mathbf{y}; p)$ is embedded in any larger family $\phi(\mathbf{x}, \mathbf{y}; p, q)$ which reduces to the former one for $q = 0$, then the equations $\phi_{ik}(\mathbf{x}; p, q) = 0$ can be solved uniquely for $\mathbf{x}$, p near $\mathbf{x}^{(0)}$, $p^{(0)}$, $q = 0$, and the conditions analogous to (6.40), (6.41) continue to hold at $\mathbf{x}; p, q$. Thus, for small values $q \neq 0$ the family given by $\phi(\mathbf{x}, \mathbf{y}; p, q)$ again has an umbilic for $(\mathbf{x}, \mathbf{y}, p)$ near $\left(\mathbf{x}^{(0)}, \mathbf{y}^{(0)}, p^{(0)}\right)$, as we needed to show.

6.3.2 Swallowtails

Expansion of the Jacobian. We next turn to those critical images which satisfy the conditions (6.39b) or (6.39c), and at which $A^{(0)}$ has rank 1. Then, we can again choose coordinates such that $\mathbf{x}^{(0)} = 0$, $\mathbf{y}^{(0)} = 0$, $\phi_{22}^{(0)} = \phi_{12}^{(0)} = 0$, $\phi_{11}^{(0)} \neq 0$, as in Sect. 6.2, and $A^{(0)} R\left(\frac{\pi}{2}\right) \nabla D^{(0)} = 0$. These conditions imply

$$\begin{aligned}
D^{(0)} &= 0 \ , \\
D_1^{(0)} &= \phi_{122}^{(0)} \phi_{11}^{(0)} \ , \quad D_2^{(0)} = 0 \ , \quad \phi_{222}^{(0)} = 0 \ , \\
D_{11}^{(0)} &= \phi_{1122}^{(0)} \phi_{11}^{(0)} + 2\phi_{111}^{(0)} \phi_{122}^{(0)} - 2\left(\phi_{112}^{(0)}\right)^2 \ , \\
D_{12}^{(0)} &= \phi_{1222}^{(0)} \phi_{11}^{(0)} - \phi_{122}^{(0)} \phi_{112}^{(0)} \ , \\
D_{22}^{(0)} &= \phi_{2222}^{(0)} \phi_{11}^{(0)} - 2\left(\phi_{122}^{(0)}\right)^2 \ .
\end{aligned} \qquad (6.54)$$

Thus, we have

$$\nabla D = \begin{pmatrix} D_1^{(0)} + D_{11}^{(0)} x_1 + D_{12}^{(0)} x_2 \\ D_{12}^{(0)} x_1 + D_{22}^{(0)} x_2 \end{pmatrix} + R_2 \ ,$$

so that

$$\mathbf{T} = R\left(\frac{\pi}{2}\right)\nabla D = \begin{pmatrix} -D_{12}^{(0)}x_1 - D_{22}^{(0)}x_2 \\ D_1^{(0)} + D_{11}^{(0)}x_1 + D_{12}^{(0)}x_2 \end{pmatrix} + R_2 \qquad (6.55)$$

and

$$A = \begin{pmatrix} \phi_{11}^{(0)} + \phi_{111}^{(0)}x_1 + \phi_{112}^{(0)}x_2 & \phi_{112}^{(0)}x_1 + \phi_{122}^{(0)}x_2 \\ \phi_{112}^{(0)}x_1 + \phi_{122}^{(0)}x_2 & \phi_{122}^{(0)}x_1 \end{pmatrix} + R_2 \qquad (6.56)$$

and, therefore,

$$A\mathbf{T} = \begin{pmatrix} \left(D_1^{(0)}\phi_{112}^{(0)} - D_{12}^{(0)}\phi_{11}^{(0)}\right)x_1 + \left(D_1^{(0)}\phi_{122}^{(0)} - D_{22}^{(0)}\phi_{11}^{(0)}\right)x_2 \\ D_1^{(0)}\phi_{122}^{(0)}x_1 \end{pmatrix} + R_2 \,. \qquad (6.57)$$

On the critical curve, the vector field $\mathbf{T}$ is tangent to that curve, thus its image $A\mathbf{T}$ under the lens mapping is tangent to the caustic.

Swallowtail: Expansion of the Fermat potential. Let us now assume, in addition, that $\nabla D^{(0)} \neq 0$, i.e., $\phi_{122}^{(0)} \neq 0$, but that $\left(\mathbf{T} \cdot \nabla (A \cdot \mathbf{T})^{(0)}\right) = 0$, so that we are dealing with the case of (6.39b). Since at the origin, $\mathbf{T}$ points in the x_2-direction, the derivative of $A\mathbf{T}$ in that direction at the origin equals, by (6.57), $\left(D_1^{(0)}\phi_{122}^{(0)} - D_{22}^{(0)}\phi_{11}^{(0)}, 0\right)$. Substituting from (6.54) we obtain:

$$(\mathbf{T} \cdot \nabla (A \cdot \mathbf{T}))^{(0)} = 0 \iff 3\left(\phi_{122}^{(0)}\right)^2 = \phi_{11}^{(0)}\phi_{2222}^{(0)} \,. \qquad (6.58)$$

Since, under our present assumptions, $\phi_{11}^{(0)}$ and $\phi_{122}^{(0)}$ do not vanish,

$$\phi_{2222}^{(0)} \neq 0 \,. \qquad (6.59)$$

The unperturbed Fermat potential therefore has the form

$$\phi = \phi^{(0)} + \frac{1}{2}\mathbf{y}^2 - \mathbf{x} \cdot \mathbf{y} + \frac{bc}{2}x_1^2 + \frac{d}{2}x_1^2 x_2$$
$$+ \frac{b}{2}x_1 x_2^2 + \frac{f}{6}x_1 x_2^3 + \frac{b}{8c}x_2^4 + \frac{e}{5}x_2^5 + x_1^2 P_2 + x_1 P_4 + R_6 \,,$$

where we have written

$$b = \phi_{122}^{(0)}, \quad bc = \phi_{11}^{(0)}, \quad d = \phi_{112}^{(0)}, \quad e = \frac{1}{24}\phi_{22222}^{(0)}, \quad f = \phi_{1222}^{(0)}, \quad (6.60)$$

and we eliminated $\phi_{2222}^{(0)} = 3b/c$ by means of (6.58).

To determine the leading part of the perturbation of the potential proportional to p, we make use of the fact that terms linear in x_i and terms proportional to $x_1 x_2$ and x_2^2 in that potential can be removed by suitable coordinate-translations $y_i \to y_i + p\alpha_i$, $x_i \to x_i + p\beta_i$ in the unperturbed potential. Furthermore, to preserve the relation (6.58) also for the perturbed potential, we carry out a suitable rotation proportional to p. If, after these simplifications, we drop some terms which, together with their first and second derivatives, are negligible compared to other terms for $(x_i, p) \to (0, 0, 0)$, and for simplicity assume $d \neq 0$, we arrive at the following representative potential:

$$\phi = \phi^{(6)} + \frac{1}{2}\mathbf{y}^2 - \mathbf{x} \cdot \mathbf{y} + \frac{b}{2}cx_1^2 + \frac{1}{2}dx_1^2 x_2 + \frac{1}{2}bx_1 x_2^2 \\ + \frac{b}{8c}x_2^4 + \frac{f}{6}x_1 x_2^3 + \frac{1}{5}ex_2^5 + px_2^3 \; . \tag{6.61}$$

According to our present assumptions, b, c and d are non-zero, whereas the values of e and f so far are not subject to any conditions. We have kept the terms $\frac{1}{6}fx_1 x_2^3$ and $\frac{1}{5}ex_2^5$ although they are small compared to $\frac{b}{2}x_1 x_2^2$ and $\frac{b}{8c}x_2^4$, respectively, as are their derivatives. The reason for this apparently unnecessary complication will become clear soon.

Equation (6.61) leads to the lens mapping

$$y_1 = b\left(cx_1 + \frac{1}{2}x_2^2\right) + dx_1 x_2 + \frac{f}{6}x_2^3 \; , \\ y_2 = bx_1 x_2 + \frac{1}{2}dx_1^2 + \frac{f}{2}x_1 x_2^2 + \frac{b}{2c}x_2^3 + ex_2^4 + 3px_2^2 \; . \tag{6.62}$$

Critical curve and caustic. From (6.62) we obtain the asymptotic form of the equation $D = 0$ for the one-parameter family of critical curves, as

$$x_1 = -\frac{6p}{b}x_2 - \frac{1}{2c}x_2^2 + \left(\frac{3f}{2bc} - \frac{4e}{b} - \frac{2d}{bc^2}\right)x_2^3 \; . \tag{6.63}$$

Inserting this result into (6.62), we get the equations describing the caustics,

$$y_1 = -cx_2\left(6p + gx_2^2\right) \; , \\ y_2 = -3x_2^2\left(p + \frac{g}{4}x_2^2\right) \; , \tag{6.64}$$

where

$$g = \frac{5d}{2c^2} + 4e - \frac{5f}{3c} \; . \tag{6.65}$$

At this stage we see that the two "significant" b-terms of ϕ have cancelled out; they do not appear in the last two equations. Instead, the terms involving e and f, although negligible compared to the b-terms in ϕ, contribute to the last terms in (6.64) and thus cannot be neglected. Since we only

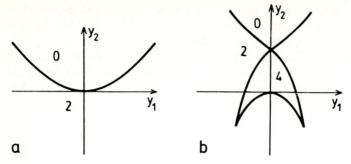

Fig. 6.8. The caustic of a swallowtail before and at the metamorphosis ($p \leq 0$, (**a**)) and thereafter ($p > 0$, (**b**)), again with image numbers

wish to treat critical images obeying exactly three independent equations, we assume $g \neq 0$ or, equivalently in view of (6.58) and (6.61),

$$\phi^{(0)}_{11}\phi^{(0)}_{22222} + 15\phi^{(0)}_{112}\phi^{(0)}_{122} - 10\phi^{(0)}_{1222} \neq 0 \ . \tag{6.66}$$

A somewhat laborious calculation reveals the invariant geometrical meaning of this inequality; it is that the angle β between the tangent of the critical curve and the direction of the kernel of A has a strict local extremum at the origin which is the critical point under discussion.[6]

Granted this condition, we obtain from (6.64):

$$\begin{aligned} \frac{dy_1}{dx_2} &= -3c\left(2p + gx_2^2\right) \ , \\ \frac{dy_2}{dx_2} &= -3x_2\left(2p + gx_2^2\right) \ . \end{aligned} \tag{6.67}$$

The tangent vector to the caustic (with respect to the parameter x_2) vanishes at $x_2 = \pm\sqrt{\frac{-2p}{g}}$ if $\frac{p}{g} \leq 0$ and does not vanish near $x_2 = 0$ if $\frac{p}{g} > 0$. In the case $\frac{p}{g} < 0$, the caustic has cusps for $x_2 = \pm\sqrt{\frac{-2p}{g}}$; in the case $\frac{p}{g} \geq 0$ the caustic is smooth. Equations (6.64) show that for $\frac{p}{g} < 0$, the caustic intersects itself for $x_2 = \pm\sqrt{\frac{-6p}{g}}$, i.e., at $\mathbf{y} = \left(0, -9\frac{p^2}{g}\right)$. The shape of the caustic for $\frac{p}{g} < 0$ shown in Fig. 6.8b motivated the name *swallowtail* for the type of singularity under discussion. The large caustic is indicated in Fig. 6.9.

Finally, we consider the number of images of a source situated near a swallowtail. For this purpose, the first of the lens equations (6.62) is solved for x_1 and inserted into the second, to yield an equation of fourth degree for x_2. Since the number of images changes only if the source position crosses

---

[6] We would like to thank H. Erdl for pointing out an error in the preliminary version of this subsection

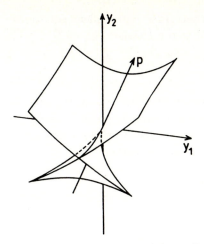

Fig. 6.9. The large caustic of the swallowtail in the $(\mathbf{y}, p)$-control space

a caustic, and since the line $y_1 = 0$ passes through all regions of the source plane separated by caustics, it suffices to consider that equation for $y_1 = 0$, where it reads

$$x_2^4 + 12\frac{p}{g}x_2^2 - \frac{4}{g}y_2 = 0 \quad, \tag{6.68a}$$

with solutions

$$x_2 = \pm\sqrt{\frac{-6p}{g} \pm \sqrt{\frac{36p^2}{g^2} + \frac{4}{g}y_2}} \tag{6.68b}$$

(the two signs $\pm$ are independent). If we take, without loss of generality, $g < 0$, the distribution of the number of images is as indicated in Fig. 6.8.

The *stability* of swallowtails in one-parameter families of lens mappings can be established in the same manner as in preceding cases: one can show that the equations

$$D^{(0)} = 0\,, \quad A^{(0)} \cdot \mathbf{T}^{(0)} = 0\,, \quad (\mathbf{T} \cdot \nabla (A \cdot \mathbf{T}))^{(0)} = 0 \tag{6.69a}$$

and inequalities

$$\mathrm{tr}A^{(0)} \neq 0\,, \quad \nabla D^{(0)} \neq 0\,, \quad \left((\mathbf{T} \cdot \nabla)^2 (A \cdot \mathbf{T})\right)^{(0)} \neq 0 \tag{6.69b}$$

which characterize a swallowtail, define implicitly $\mathbf{x}$ and p as functions of q in any perturbed family of the given family.

6.3.3 Lips and beak-to-beaks

We finally turn to those critical images at which $\mathrm{tr}A^{(0)} \neq 0$ and (6.39c) hold, i.e., $D^{(0)} = 0$ and $\nabla D^{(0)} = 0$. Since in this section we want to consider singularities obeying no more than three independent equations, we require, as in the case of umbilics, that the Hessian D_{ik} of D not be degenerate at $\mathbf{x}^{(0)}$, i.e.,

$$\Delta^{(0)} \neq 0 \ . \tag{6.70}$$

Under these conditions, we can again choose coordinates as in the preceding subsection, and since now $\phi_{122}^{(0)} = 0$, the equations (6.54–57) simplify somewhat. By means of arguments strictly analogous to those which led to (6.60) we now obtain the representative Fermat potential

$$\begin{aligned}\phi = \phi^{(0)} &+ \frac{1}{2}\mathbf{y}^2 - \mathbf{x} \cdot \mathbf{y} + \frac{c}{2}x_1^2 + \frac{d}{2}x_1^2 x_2 \\ &+ \frac{a}{2}x_1^2 x_2^2 + \frac{e}{3}x_1 x_2^3 + \frac{b}{12}x_2^4 - \frac{1}{2}px_2^2 \ ,\end{aligned} \tag{6.71}$$

where we have written

$$\begin{aligned}c = \phi_{11}^{(0)}, \quad d = \phi_{112}^{(0)}, \quad a &= \frac{1}{2}\phi_{1122}^{(0)}, \\ e = \frac{1}{2}\phi_{1222}^{(0)}, \quad b &= \frac{1}{2}\phi_{2222}^{(0)} \ ,\end{aligned} \tag{6.72}$$

and p again denotes the perturbation parameter. The lens equation implied by (6.71) is

$$\begin{aligned}y_1 &= cx_1 + dx_1 x_2 + ax_1 x_2^2 + \frac{e}{3}x_2^3 \ , \\ y_2 &= -px_2 + \frac{d}{2}x_1^2 + ax_1^2 x_2 + ex_1 x_2^2 + \frac{b}{3}x_2^3 \ ;\end{aligned} \tag{6.73}$$

the Jacobian is, up to terms of higher order,

$$D = -pc + 2pdx_2 + (ac - d^2) x_1^2 + 2ecx_1 x_2 + bcx_2^2 \ ; \tag{6.74}$$

and the inequality (6.70) reads

$$b(ac - d^2) \neq ce^2 \ . \tag{6.75}$$

By means of a p-independent rotation and a translation of the coordinates linear in p, one can eliminate the terms proportional to $x_1 x_2$ and x_2^2, respectively; this changes the form of (6.73). To avoid excessive calculations, we consider the special case $d = e = 0$ which exhibits all the important features. Then $D = 0$ reduces to

$$ax_1^2 + bx_2^2 = p \ , \tag{6.76}$$

and we must have $ab \neq 0$.

Lips. Consider the case $\Delta^{(0)} > 0$, i.e., in our special case $ab > 0$. If $ap < 0$, (6.76) has no (real) solution, for $p = 0$ the only solution is the point $\mathbf{x}^{(0)} = 0$, and for $ap > 0$ the critical curves are ellipses,

$$x_1 = \sqrt{\frac{p}{a}} \cos\varphi \ ,$$
$$x_2 = \sqrt{\frac{p}{b}} \sin\varphi \ . \tag{6.77}$$

Inserting this into the lens equation (6.73) one finds for the caustics, in leading order for $p \to 0$:

$$y_1 = c\sqrt{\frac{p}{a}} \cos\varphi \ ,$$
$$y_2 = \frac{-2}{3} p \sqrt{\frac{p}{b}} \sin^3\varphi \ . \tag{6.78}$$

The caustics are symmetric with respect to both reflections $y_1 \to -y_1$, $y_2 \to -y_2$; moreover $|y_1| \leq c\sqrt{\frac{p}{a}}$, $|y_2| \leq \frac{2}{3}p\sqrt{\frac{p}{b}}$. The tangent vector $\frac{d\mathbf{y}}{d\varphi}$ vanishes for $\varphi = 0$ and $\varphi = \pi$. Thus, the caustics are lip-shaped as shown in Fig. 6.10, with cusps on the y_1-axis; for $p \to 0$, they contract toward the singular point $\mathbf{y}^{(0)} = 0$ and then disappear.

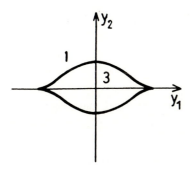

Fig. 6.10. The lips caustic after its creation, again with image numbers

To recognize the image multiplicities, we eliminate x_1 from the lens equation and ask without loss of generality for images of sources on the y_1-axis. They are given by

$$x_2 \left(x_2^2 + \frac{3}{b} \left[\frac{a}{c^2} y_1^2 - p \right] \right) = 0 \ .$$

Thus, there are three images for sources inside the closed caustic ($y_1^2 < \frac{c^2}{a}p$), two for sources on the caustic, and there is just one for sources outside the caustic.

Lips occur generically in one-parameter families of lens models. Consider, for example, a non-singular, asymmetric matter distribution, and let the model parameter p be the distance between source and lens. If this distance is sufficiently small, so that κ is small everywhere, $\det A > 0$ for all $\mathbf{x}$:

the lens is too weak to focus light bundles efficiently. Increasing p, there will be a value $p = p^{(0)}$ where at one point $\mathbf{x}$ the determinant vanishes, whereas at neighboring points it is still positive, so that det A has a minimum (this corresponds to the case $p = 0$ here). As p increases slightly, a lips caustic occurs. Thus, the onset of a caustic of a light cone generically corresponds to a metamorphosis in which lips suddenly emerge "out of nothing". For example, the transition from Fig. 5.9b to c occurs through a lips singularity; further examples will be given in Chap. 8 below. — Lips are also essential for the pancake model of structure formation in the universe [ZE70.1].

Beak-to-beaks. Suppose now that $\Delta^{(0)} < 0$, i.e., in our special case $ab < 0$. Then (6.76) has solutions for all values of p. If $ap > 0$, the critical curves are the hyperbolae

$$x_1 = \pm\sqrt{\frac{p}{a}}\cosh\varphi ,$$
$$x_2 = \sqrt{-\frac{p}{b}}\sinh\varphi ;$$

(6.79)

this yields the caustics

$$y_1 = \pm c\sqrt{\frac{p}{a}}\cosh\varphi ,$$
$$y_2 = \frac{-2}{3}p\sqrt{-\frac{p}{b}}\sinh^3\varphi .$$

(6.80)

Similar equations apply for $ap < 0$. For the unperturbed mapping $p = 0$, the critical curves are represented by $x_2 = \pm\sqrt{-\frac{a}{b}}\,x_1$, the caustics by $y_2 = \pm\frac{2a}{3c^3}\sqrt{-\frac{a}{b}}\,y_1^3$.

Whereas the critical curves are hyperbolae in both cases $ap > 0$ and $ap < 0$, the shapes of the caustics are quite different in these two cases, see Fig. 6.11: for $ap < 0$, there are two smooth caustic curves, convex if viewed from the y_1-axis, with minimum separation $2p\sqrt{\frac{p}{b}}$. A source between these curves has three images, one above or below has one image only. For $ap > 0$, one has two caustics with one cusp each on the y_1-axis, their separation

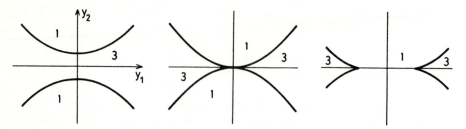

Fig. 6.11. The beak-to-beak caustic, before $(ap < 0)$, at $(p = 0)$ and after $(ap > 0)$ the metamorphosis, with image numbers

being $2c\sqrt{\frac{p}{a}}$. A source to the right (left) of the right (left) caustic has three images, one between the caustics has one image. For $p = 0$, when the metamorphosis occurs, the critical lines intersect at $\mathbf{x}^{(0)} = 0$; these two straight lines form the asymptotes of the hyperbolae belonging to $p \neq 0$. The corresponding caustics touch tangentially. The shapes of the caustics for $ap > 0$ suggested the name for this metamorphosis, *beak-to-beak*.

The *stability* of lips and beak-to-beak singularities under small perturbations of one-parameter families of lens mappings again follows straightforwardly: The Jacobian of the three equations $D = 0$, $D_1 = 0$, $D_2 = 0$ with respect to the control parameters x_1, x_2, p, evaluated at $\mathbf{x}^{(0)} = 0$, $p^{(0)} = 0$, is obviously equal to $\left(\frac{\partial D}{\partial p}\right)^{(0)} \cdot \Delta^{(0)}$, and thus does not vanish. As before, this implies stability via the implicit function theorem.

Lips and beak-to-beak singularities are not listed as "elementary catastrophes" in books on catastrophe theory, in contrast to folds, cusps, swallowtails and umbilics. The reason for this was given in the last paragraph of the introduction to this section, namely, that here two control parameters have a preferred status, the coordinates y_1, y_2 in the source plane. Indeed, if one permits arbitrary foliations of the control space, one can obtain lips and beak-to-beak metamorphoses easily from cusp ridges associated with any of the standard singularities (see Figs. 6.12a,b).

6.3.4 Concluding remarks about singularities

In this chapter we have treated those singularities of lens mappings which are stable as properties of individual mappings – folds and cusps – or as metamorphoses of one-parameter families of such mappings – umbilics, swal-

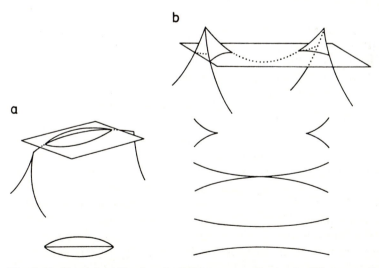

Fig. 6.12. Lips (**a**) and beak-to-beak (**b**) metamorphoses arise if large caustics are intersected by suitable families of planes near cusp ridges

lowtails, lips and beak-to-beak singularities. The method was to expand the Fermat potential ϕ, the resulting lens mapping and the determinant D, up to leading terms in the image coordinates x_1, x_2 and the perturbation parameter p. The goal was to exhibit the local structures of these singularities, and to derive conditions for their occurrence, expressed both in invariant form and in suitably adapted coordinates. Owing to its geometrical nature, gravitational lensing provides an ideal starting point for studying Catastrophe Theory.

The central mathematical concepts needed to analyze singularities of maps are stability and genericity. Their rigorous definitions are delicate and have not been given, but we hope to have made their meaning and role plausible. Another concept whose introduction we have avoided by simply terminating Taylor expansions, is that of a "finitely determined" function, i.e., a smooth function which, in a certain precise sense, is equivalent to one of its finite Taylor polynomials.

The classification of singularities is predicted on the vanishing of some, and the non-vanishing of other, invariants of a map constructed from ϕ and its derivatives such as $D = \det(\phi_{ik})$, $\operatorname{tr} A = \phi_{ii}$, $\Delta = \det(D_{ik})$. Generically, a singularity characterized by r independent equations for s control parameters gives rise to an $(s-r)$ dimensional (large) caustic in control space ($r \leq s$). Gravitational lensing provides us with two natural control parameters, the coordinates of the source position. A singularity of a generic lens model, or family thereof, is stable with respect to arbitrary small perturbations of that model (or family). It is important to note that the perturbations one considers are arbitrary. For example, as we will see in Chap. 8, rotationally symmetric lenses have caustics which consist of just a single point, whereas the corresponding critical curve is a circle. This is due to the fact that the tangent vector of such a ("tangential") critical curve is everywhere an eigenvector of the derivative of the lens mapping belonging to the eigenvalue 0. Any perturbation of the lens model will transform the isolated caustic point to a caustic curve, unless the perturbation itself is symmetric about the same axis. Thus, if one confined the consideration to a class of symmetric perturbations one would conclude that caustic points are generic. It is therefore important to consider general perturbations with no special symmetry.

We have listed all stable singularities with up to three control parameters. The fold and the cusp are characterized by one and two equations, respectively, and are thus generic properties of a lens model, since each lens model necessarily has two control parameters (the source coordinates). The catastrophes with three control parameters – elliptic and hyperbolic umbilic, swallowtail, lips and beak-to-beak – are generic properties of one-parameter families of lens models. They necessarily occur if the caustic topology is different for different values of the lens model parameter. In particular, we have seen that cusps are created (or destroyed) if the lens parameter changes through a critical value which corresponds to the "catastrophes" listed above. Thus, if one finds all catastrophes with three control parameters in a generic one-parameter family of lens models, one has obtained a complete classification of its caustic structure. Since the caustic structure provides a good qualitative understanding of the lensing behavior, it is important to be able to identify all of its singularities. Higher-order singularities can be treated similarly, but they seem less important for lens theory.

We have summarized the conditions for the occurrence of catastrophes (stable singularities) with up to three control parameters in Table 6.1. These conditions are arranged in the form of a tree, with more and more special conditions for higher-order catastrophes. We thus call this a "catastrophe tree".

Even number of cusps. The foregoing treatment of all stable metamorphoses in one-parameter families of lens models allows us now to prove that *all lenses have an even number of cusps*.[6] To show this, we recall the properties of metamorphoses with respect to the formation (or destruction) of cusps: beak-to-beaks, lips, and swallowtails mark a transition from zero to two cusps, elliptic umbilics from three to three cusps, and hyperbolic umbilics exchange a cusp between two caustic lines; thus, umbilics leave the number of cusps invariant. Therefore, these metamorphoses change the number of cusps by ±2 or not at all.

Next, consider a, say, three-parameter family of lens models, and let one of these parameters describe the "strength" of the lens [e.g., the distance of the source from the lens, which scales κ via (5.4) and (5.5)]. First, we assume that the lens is transparent; then, for small "strength", the lens forms no caustic at all, and each lens model in the three parameter family of models can be connected by a curve $\mathbf{p}(\lambda)$ with such a model without caustics. The metamorphoses treated in this section form two-dimensional surfaces in the three-dimensional parameter space, so the curve $\mathbf{p}(\lambda)$ can cut through one or more of these surfaces, each time changing the number of cusps by zero or two (the first surface cut by the curve corresponds to lips, since they are the only metamorphoses generating caustics "from nothing"). Higher-order catastrophes form at most curves in the three-dimensional parameter space, and the parameter curve $\mathbf{p}(\lambda)$ can always be chosen such that it avoids these higher-order catastrophes. This argument can, of course, be easily generalized to N-parameter families of lens models, thus proving the statement made above for transparent lenses.

This argument can now be generalized to the case that the lens contains point masses. If we again consider the distance between lens and source as one parameter describing the family of lens models, then for small separation, each such point mass lens acts nearly "on its own", since the shear and convergence from the other components of the lens are very weak. Thus, each point mass lens will form a caustic corresponding to that of a Chang–Refsdal lens (see Sect. 8.2.2), which has four cusps. With the foregoing

---

[6] this was suggested to us by A. Weiss

Table 6.1. Catastrophe tree. n_c denotes the minimal number of control parameters required for the occurrence of a catastrophe. $D = \det A$, $G = \nabla D$ is the gradient of $\det A$, $T = R(\pi/2) G$ a vector tangent to the critical line, and $\Delta = \det(D_{ij})$. All boxed inequalities correspond to a singularity, which is characterized by that inequality and the equations and inequalities preceding it on the tree. The left side of the table ($G \neq 0$) shows the so-called cuspoids, the first of which are fold, cusp, and swallowtail. Umbilics ($A = 0$) are shown in the right part of the table.

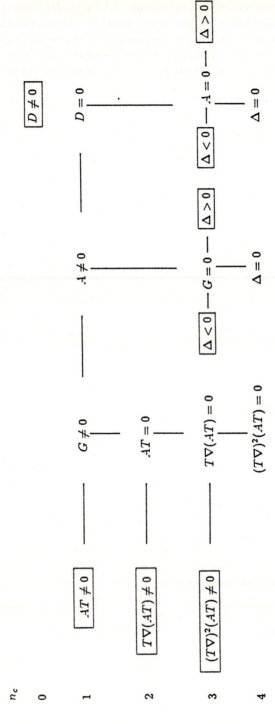

214

argument, every model in the family of lens models can be connected with a curve to one model in which only Chang–Refsdal caustics are present, which proves the statement made above.

6.4 Magnification of extended sources near folds

In this final section, we consider the magnification of an extended, circular source near a fold, since this situation is of some astrophysical interest; as we shall see in Sect. 12.4, the lightcurves of sources undergoing microlensing will partly be described by these sources crossing a fold caustic.

The magnification μ_e of an extended source with surface brightness profile $I(\mathbf{y})$ is given generally as

$$\mu_e = \frac{\int d^2y\, I(\mathbf{y})\, \mu_p(\mathbf{y})}{\int d^2y\, I(\mathbf{y})} \quad , \tag{6.81}$$

where $\mu_p(\mathbf{y})$ is the magnification of a point source at position $\mathbf{y}$.

We consider now the situation wherein an extended source is near a fold, and assume that the radius of curvature of the caustic is much larger than the size of the source. Then, the caustic can be considered locally as being straight, and we can choose coordinates such that the caustic is described by the line $y_1 = y_c$ and that the positive side of the caustic is given by $y_1 > y_c$. From (6.19b), we find

$$\mu_p(\mathbf{y}) = \sqrt{\frac{g}{(y_1 - y_c)}}\, H(y_1 - y_c) + \mu_0 \quad , \tag{6.82}$$

where μ_0 accounts for the magnification of all images which are not near the critical curve corresponding to the caustic under consideration; in particular, this magnification is assumed to depend much more weakly on $\mathbf{y}$ than the magnification of the critical images, and is therefore set to a constant μ_0. For a specific lens model, the constant of proportionality g can be calculated from (6.19b) in terms of the derivatives of the Fermat potential.

Let us assume that the surface brightness profile of the source is circularly symmetric, and that its center is at $\mathbf{y} = 0$, so that y_c measures the separation of the source's center from the caustic. We parametrize the surface brightness by

$$I(\mathbf{y}) = I_0\, f\left(\frac{y}{R}\right) \quad , \tag{6.83}$$

where $y = |\mathbf{y}|$, and R measures a typical radius of the source. Combining the last two equations, we find from (6.81):

$$\mu_e(y_c) = \mu_0 + \sqrt{\frac{g}{R}}\, \zeta\left(\frac{y_c}{R}\right) \quad , \tag{6.84a}$$

where the function

$$\zeta(w) = \frac{\int_w^\infty \frac{ds}{\sqrt{s-w}} \int_0^\infty dt\, f\left(\sqrt{s^2+t^2}\right)}{\pi \int_0^\infty ds\, s\, f(s)} \tag{6.84b}$$

depends on the shape of the brightness distribution only, but not on its scale R. Hence, the magnification of the images close to the critical curve factorizes; for fixed separation of the source from the caustic, in units of the source's radius R, i.e., fixed y_c/R, this magnification behaves as $R^{-1/2}$. In particular, the maximum magnification of a source shows an $R^{-1/2}$ behavior.

In Fig. 6.13, we have plotted the function $\zeta(w)$ for a source with uniform brightness [i.e., $f(x) = H(1-x)$] and a source with Gaussian brightness profile, $f(x) = e^{-x^2}$; analytic expressions for $\zeta(w)$ in these cases can be found in [SC87.3] (the case of limb-darkened brightness profiles of the form

$$f(x) = \left[1 - c\left(1 - \sqrt{1-x^2}\right)\right] H(1-x) \quad ,$$

has been considered in [SC87.6]). The non-smoothness of ζ at $w = 1$ is due to the discontinuous brightness profile of the uniform disc. Note that ζ for the uniform disc is much steeper than for the Gaussian source, since the latter has a smoother brightness profile. Both functions agree for $w \lesssim -3$, where they quickly approach the point source behavior, $\zeta = 1/\sqrt{-w}$.

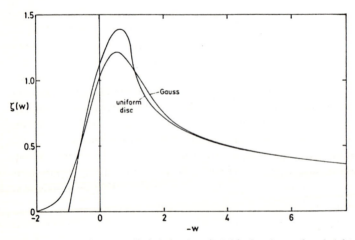

Fig. 6.13. The function $\zeta(w)$ defined in (6.84b), for the surface brightness profiles of a uniform disc and a Gaussian source

7. Wave optics in gravitational lensing

So far we have treated (and will do so in the rest of this book) the propagation of light for GL systems in the approximation of geometrical optics, ignoring the wave properties of light. In this chapter we investigate the conditions under which the latter may become noticeable. It will be shown that in (nearly) all cases of practical importance the geometrical optics treatment is an excellent approximation, at least as long as one considers those sources most likely to be lensed, namely QSOs and galaxies.

Nevertheless, there are good reasons to investigate the lensing behavior in the framework of wave optics. First, in a gravitational lensing situation, the observer receives two (or more) light bundles from the same source, and thus their light is, at least in principle, mutually coherent ([MA82.1], [SC85.1]). Second, in the approximation of geometrical optics the magnification of a point source becomes infinite as it approaches a caustic. This, of course, is an artifact of the geometrical optics approximation, and the wave character of light must be taken into account in those cases [OH83.1]. One can understand the need for a wave-optics treatment in that case also by recalling that the two images close to a critical curve in the lens plane have a very small mutual time delay (cf. Chap. 6); if this time delay is less than the coherence time of the light, one expects interference to occur. It usually suffices to work in the framework of geometrical optics because astrophysical sources are in general too large for wave effects to be important. Using the fomulae obtained in Sect. 4.7, we will derive the conditions for the validity of the geometrical optics approximation.

Wave optics effects at caustics have been treated in the context of scintillating radio sources [GO87.1], affected by interstellar electron density fluctuations. In this situation, wave optics effects are, in fact, important.

7.1 Preliminaries; magnification of ordinary images

Transmission factor. The specific flux of the radiation emitted by a quasistationary source, deflected by a geometrically thin lens, has been shown in Sect. 4.7 to be given in terms of the specific surface brightness by

$$S_\omega = A \int d^2y \ |V(\omega, \mathbf{y})|^2 I_\omega(\mathbf{y}) \quad ; \tag{7.1}$$

here,

$$V = \int_{\mathbb{R}^2} d^2x \, e^{if\phi(\mathbf{x},\mathbf{y})} \quad , \tag{7.2}$$

is the transmission factor which was introduced in Sect. 4.7. We now use dimensionless variables $\mathbf{x}$, $\mathbf{y}$ and the dimensionless Fermat potential ϕ, as in Chap. 5. The very large numerical constant f is then

$$f = \frac{\omega}{c} \frac{D_s}{D_d D_{ds}} (1 + z_d) \xi_0^2 = \frac{\omega}{H_0} \left(\frac{\xi_0}{D_d} \right)^2 [\chi(z_d) - \chi(z_s)]^{-1} \quad ; \tag{7.3}$$

the last expression results from (4.68a).

In the trivial case of no lens, $\phi(\mathbf{x}, \mathbf{y}) = \frac{1}{2}(\mathbf{x} - \mathbf{y})^2$, the transmission factor is

$$V = \int dx_1 \, dx_2 \, e^{if(x_1^2 + x_2^2)/2} = \frac{2}{f} \left(\int_{-\infty}^{\infty} e^{i\xi^2} d\xi \right)^2 = \frac{2\pi i}{f} \quad , \tag{7.4}$$

independent of the source position (use, e.g., (3.691.1) of [GR80.1]).

In the general case, the integral (7.2) can be evaluated asymptotically by the method of the stationary phase, since f is very large indeed. Since the phase is stationary at the images of $\mathbf{y}$ obtained from geometrical optics, V is a sum of terms corresponding to these images. Near an ordinary image, we have in Cartesian coordinates, chosen such that they diagonalize the Jacobian matrix and have their origin at that image, $\phi(\mathbf{x}, \mathbf{y}) = \phi^{(0)} + \frac{1}{2} \left(\phi_{11}^{(0)} x_1^2 + \phi_{22}^{(0)} x_2^2 \right)$. Therefore, evaluating the integral as above, with due attention to the signs of $\phi_{11}^{(0)}$, $\phi_{22}^{(0)}$, we obtain as the contribution of one image to V:

$$\frac{2\pi i}{f} \left| \det A^{(0)} \right|^{-1/2} e^{i \left(f\phi^{(0)} - n\pi/2 \right)} \quad . \tag{7.5}$$

The integer n depends on the type of the image [see (5.29)]: $n = 0, 1, 2$ for images of type I, II, III, respectively. It thus equals the number of focal points (caustics) traversed between source and observer by the ray corresponding to that image and is called the *Morse index* of that ray (see [AR89.1] for an account of the role of this concept). Summing over the terms (7.5), we obtain for the case that the source is not located on a caustic, labeling the images with an index l,

$$V = \frac{2\pi i}{f} \sum_l \left| \det A_l^{(0)} \right|^{-1/2} e^{i \left(f\phi_l^{(0)} - n_l \pi/2 \right)} \quad . \tag{7.6}$$

Magnification: General expression. Using (7.1) to compare the specific fluxes of a source with and without a lens and employing the result (7.4), we get the following general expression for the "specific" magnification $|\mu_\omega| = S_\omega / S_\omega^0$:

$$|\mu_\omega| = \frac{f^2}{4\pi^2} \frac{\int d^2y \, |V(\omega,\mathbf{y})|^2 \, I_\omega(\mathbf{y})}{\int d^2y \, I_\omega(\mathbf{y})} \tag{7.7a}$$

for an extended source, and

$$|\mu_\omega| = \frac{f^2}{4\pi^2} |V(\omega)|^2 \tag{7.7b}$$

for a point source (we now have omitted the $\mathbf{y}$).

Magnification for point source not located on a caustic. Next, we apply (7.7b) to the case where only ordinary images occur. Equation (7.6) gives

$$|\mu_\omega| = \sum_l \left|\det A_l^{(0)}\right|^{-1} + 2 \sum_{l<m} \sqrt{\left|\det A_l^{(0)}\right|^{-1} \left|\det A_m^{(0)}\right|^{-1}}$$
$$\times \cos\left[f\left(\phi_l^{(0)} - \phi_m^{(0)}\right) - \frac{\pi}{2}(n_l - n_m)\right] \quad .$$

The last term accounts for interference between the partial waves associated with the various images. If we recall that $f\left(\phi_l^{(0)} - \phi_m^{(0)}\right) = \omega \Delta t_{lm}$ is the time delay (or "phase difference") of the l-th relative the m-th "geometrical image", we can rewrite the argument of the cosine as $\omega \Delta t_{lm} - (n_l - n_m)\frac{\pi}{2}$. The very sensitive frequency dependence claimed by this formula is not measurable. Averaging over a narrow frequency band $\Delta \omega$ around $\bar{\omega}$, determined either by atomic properties as in the case of line radiation or by the observer's receiver, or both, gives, if $\tau_c \sim \frac{1}{\Delta \omega}$ denotes the coherence time,

$$|\bar{\mu}_{\bar{\omega}}| \approx \sum_l \left|\det A_l^{(0)}\right|^{-1} \quad \text{if} \quad \Delta t_{lm} \gg \tau_c \quad ,$$

$$|\bar{\mu}_{\bar{\omega}}| \approx \sum_l \left|\det A_l^{(0)}\right|^{-1}$$
$$+ 2\sum_{l<m} \frac{\cos\left[\bar{\omega} \Delta t_{lm} - \frac{\pi}{2}(n_l - n_m)\right]}{\sqrt{\left|\det A_l^{(0)}\right| \left|\det A_m^{(0)}\right|}} \quad \text{if} \quad \Delta t_{lm} \lesssim \tau_c \quad , \tag{7.8}$$

with obvious modifications if only some Δt_{lm} are small. (The exact dependence of the interference term on the Δt_{lm} depends, of course, on the shape of the line.) This result is as expected, except perhaps for the appearance of the phase differences $\frac{\pi}{2}(n_l - n_m)$, due to the different transitions of the partial waves through caustics. If two images approach a critical curve, they have opposite parities; hence, with suitable labeling, $(n_l - n_m) = 1$ and $\cos\left(\bar{\omega} \Delta t_{lm} - \frac{\pi}{2}\right) = \sin(\bar{\omega} \Delta t_{lm})$. For the case of well-separated images, the first line of (7.8) reproduces the expression for the magnification which until now was derived from the area change of the lens mapping. Now, it has been deduced more fundamentally from wave optics.

7.2 Magnification near isolated caustic points

The investigation of wave effects in gravitational lensing has mainly focused on axially-symmetric lenses ([OH74.1], [BL75.1], [ST76.1], [ST78.1], [BE79.1], [DE86.1]). Although such symmetric lenses do not occur in nature, we briefly outline a method to derive the magnification near point singularities.

Schwarzschild lens. Let us start with the Schwarzschild lens, for which we can obtain an analytic expression for the magnification of a point source, in the framework of wave optics. As will be shown in Sect. 8.1.2, the Fermat potential for a Schwarzschild lens is $\phi = (\mathbf{x} - \mathbf{y})^2/2 - \ln|\mathbf{x}|$, which after insertion into (7.2) yields

$$V = \int_0^\infty x\, dx\, e^{if(x^2/2 - \ln x)} \int_0^{2\pi} d\varphi\, e^{-ifxy\cos\varphi} \quad,$$

where we have chosen the source to be on the positive y_1-axis. The φ-integration can be performed by using (3.715.18) of [GR80.3]; then, V becomes

$$V = 2\pi \int_0^\infty dx\, x^{1-if}\, e^{ifx^2/2}\, J_0(fxy) \quad, \tag{7.9}$$

where $J_0(x)$ is the ordinary Bessel function, and we have dropped an unimportant phase factor. The remaining integral can be analyzed using (6.631.1) of [GR80.3], which yields

$$V = \frac{2\pi}{f}\, \Gamma\left(1 - i\frac{f}{2}\right)\, e^{\pi f/4}\, M\left(1 - i\frac{f}{2},\, 1,\, i\frac{fy^2}{2}\right) \quad, \tag{7.10}$$

where $\Gamma(x)$ is the gamma function and $M(a, b, z)$ is the confluent hypergeometric function. Using (6.1.31) of [AB65.1] we find for the magnification:

$$\mu = \frac{\pi f/2}{\sinh(\pi f/2)}\, e^{\pi f/2}\, \left|M\left(1 - i\frac{f}{2},\, 1,\, i\frac{fy^2}{2}\right)\right|^2 \quad. \tag{7.11}$$

This expression can only be used for small values of y since for larger values, the time delay for the two images[1] becomes larger than the coherence time of the light, in which case (7.7b) no longer holds, as discussed earlier. The analysis of this final equation is somewhat cumbersome, as can be seen from the treatment in [DE86.1]. We only remark here that the maximum magnification is achieved for $y = 0$, where

$$\mu = \mu_{\max} = \frac{\pi f}{1 - e^{-\pi f}} \quad, \tag{7.12}$$

in agreement with [DE86.1]; for more details, the reader is referred to that paper.

Estimate of f. Before treating a more general situation, let us estimate typical values for f. Since we used the Einstein radius of the lens of mass M as a length scale, we find

---

[1] recall, from Sect. 5.3, that the time delay for the two images of a point mass lens is roughly y times the light crossing time over the Schwarzschild radius

$$\pi f \approx 4 \times 10^5 \frac{M}{M_\odot} \frac{\nu}{1\,\text{GHz}} \quad . \tag{7.13}$$

Hence, even for radio frequencies, f is a large number. Therefore, the expression in (7.12) for most situations can be simplified to $\mu_{\max} = \pi f$; the maximum magnification of a point source in Schwarzschild geometry thus depends only on the mass of the lens and the frequency of the radiation observed.

General symmetric lens. A point singularity is due to the symmetry of the lens and occurs when a critical curve with radius $x = x_t$ is mapped onto the origin $\mathbf{y} = 0$ of the source plane, as we will show in Sect. 8.1. From the one-dimensional ray-trace equation, $y = x - \psi'(x)$ one finds that at the critical radius, $\psi'(x_t) = x_t$. Expanding the deflection potential ψ up to second order about x_t, we obtain (up to terms independent of x)

$$\psi(x) = \frac{1}{2}\psi''(x_t)x^2 + \left(1 - \psi''(x_t)\right) x\, x_t \quad .$$

If we place the source on the positive y_1 axis, the Fermat potential becomes

$$\phi = b\left(\frac{x^2}{2} - x\, x_t\right) - x\, y \cos\varphi \quad , \tag{7.14}$$

where, again, polar coordinates were introduced in the lens plane, and $b = 1 - \psi''(x_t)$. By taking the gradient of ϕ, we find that it predicts two images close to the critical curve at $x_\pm = y/b \pm x_t$.

Inserting (7.14) into (7.2),

$$V = \int_0^{2\pi} d\varphi \int_0^\infty dx\, x\, e^{if[b(x^2/2 - x x_t) - x y \cos\varphi]} \quad .$$

The integrand in the x integral is a rapidly oscillating function; the main contribution to the integral will come from values of x close to the stationary phase point. This stationary point is at $x_0 = (y/b)\cos\varphi + x_t$, which depends on φ. If we transform to the new variable $z = x - x_0$, the Fermat potential becomes $\phi = b\left(z^2 - x_0^2\right)/2$, where we have dropped constant terms that affect only the phase of V. Hence,

$$V = \int_0^{2\pi} d\varphi\, e^{-ifbx_0^2/2} \int_{-x_0}^\infty dz\,(z + x_0)\, e^{ifbz^2/2} \quad . \tag{7.15}$$

Since the main contribution to the z-integral will come from the region near $z = 0$, we can extend the lower integration limit to $-\infty$. Then, the term proportional to z vanishes by symmetry, and x_0 is independent of z. Using (3.691.1) of [GR80.3],

$$V = \sqrt{\frac{2\pi}{fb}} \int_0^{2\pi} d\varphi\, x_0\, e^{-ifbx_0^2/2} \quad ,$$

where, again, a phase factor was dropped. For $y \ll x_t$, we can expand x_0^2 in the exponential as $x_0^2 \approx x_t^2 + 2x_t(y/b)\cos\varphi$ and consider the x_0 factor independent of φ. Using this approximation, from (3.715.18) of [GR80.3] we find that

$$V = \sqrt{\frac{2\pi}{fb}}\, 2\pi x_t J_0(fx_t y) \quad .$$

Finally, from (7.7b) we obtain for the magnification

$$\mu = \frac{2\pi x_t^2 f}{b}\, J_0^2(fx_t y) \quad . \tag{7.16}$$

Since $J_0(z)$ has its maximum at $z = 0$, for which it is unity, the maximum magnification is

$$\mu_{\max} = \frac{2\pi x_t^2 f}{b} \ . \tag{7.17}$$

This result can be compared with the result given above for the Schwarzschild lens by noting that in this case, $\psi''(x_t) = -1$, i.e., $b = 2$, and $x_t = 1$; hence, (7.17) agrees with the previous expression. Note that $x_t^2 f$ is independent of the length scale ξ_0, and since b is of order unity, $\mu_{\max}$ is estimated to be of the same order as in (7.13).

Extended sources. The magnification of a point source is a rapidly oscillating function of the source position, as can be seen from the asymptotic behavior of the Bessel function,

$$J_0(z) \approx \sqrt{2/(\pi z)}\cos(z - \pi/4) \quad \text{for} \quad |z| \to \infty \ ;$$

thus, the period of the oscillation is

$$\Delta y = \frac{2\pi}{f x_t} \ . \tag{7.18}$$

The magnification of sources smaller than Δy will show oscillations (fringes); much larger sources will smear out the fringes and will not produce an observable effect. Such extended sources will have a magnification of $\langle \mu \rangle = 2/(by)$, as can be obtained from phase-averaging (7.16), which agrees with the result from geometrical optics. A source of dimensionless size r will average over $N = r/\Delta y$ oscillations of the Bessel function and will produce fringes of relative amplitude $\Delta\mu/\langle\mu\rangle = 1/N$. Taking the Schwarzschild lens as an example, one can see that observable fringes require extremely compact sources. If the source and the lens are at cosmological distances, the angular diameter δ of a source must satisfy

$$\delta \lesssim 10^{-17} \frac{\lambda'}{1\,\text{cm}} \sqrt{h_{50}\eta'} \left(\frac{M}{M_\odot}\right)^{-1/2} \tag{7.19}$$

to produce observable fringes, where λ' is the wavelength of the radiation at the lens, and η' was defined in (2.9b). For example, for a solar mass lens and a radio source, the source radius must be smaller than about 10^{10}cm. Except for pulsars, there are no known astrophysical sources which satisfy the constraint (7.19).

We can estimate the range of y for which the result (7.16) is valid. As mentioned before, we must require that the dominant contribution to the analytic signal V come from a region of the lens plane within which the variation of the light-travel-time is less than the coherence time of the light. This condition can be approximated by the statement that the two images obtained from ray optics have a time delay smaller than the coherence time. Using (7.14) and (5.45), the condition becomes

$$f y x_t < \frac{\omega}{\Delta\omega} \ .$$

7.3 Magnification near fold catastrophes

Point sources. As we mentioned in the introduction to this chapter, wave effects are expected to play a role if the source is near a caustic. Since fold catastrophes occur most generally, we now discuss the interference of light near them. We use the same notation as in Sect. 6.2, i.e., the coordinates in the lens and source plane are denoted by **x** and **y**, respectively, and their origin and orientation are chosen such that $\mathbf{x} = 0$ is a point on a critical curve which corresponds to the point $\mathbf{y} = 0$ on the caustic, and the Jacobian matrix of the lens equation is diagonal. From (6.11) the transmission factor is

$$V = \int_{-\infty}^{\infty} dx_2 \int_{-\infty}^{\infty} dx_1 \, e^{if\left[\phi_{11} x_1^2/2 - x_1 y_1 - x_2 y_2 + \phi_{222} x_2^3/6\right]} \quad , \tag{7.20}$$

where we have kept only the leading terms.[2] The integral over x_1 can be performed by substituting $z = x_1 - x_{10}$, where $x_{10} = y_1/\phi_{11}$ is the point of stationary phase in the x_1 integral. Using again (3.691.1) of [GR80.3], we obtain

$$V = \sqrt{\frac{2\pi}{f|\phi_{11}|}} \int_{-\infty}^{\infty} dx_2 \, e^{-if\left(x_2 y_2 - \phi_{222} x_2^3/6\right)} \quad . \tag{7.21}$$

From (10.4.32) of [AB65.1], we see that the remaining integral leads to an Airy function. Insertion of the resulting expression into (7.7b) yields for the magnification[3]

$$\mu = \frac{\mu_{\max}}{Q^2} \left[\text{Ai}\left(\frac{y_2}{Y_0}\right)\right]^2 \quad , \tag{7.22}$$

where we have assumed that $\phi_{222} < 0$ without loss of generality, $Q \approx 0.5357$ is the maximum value of the Airy function Ai(x) (which occurs for $x = -1.0188$),

$$\mu_{\max} = \frac{2^{5/3} \pi f^{1/3}}{|\phi_{11}||\phi_{222}|^{2/3}} Q^2 \tag{7.23}$$

is the maximum magnification a point source can have near a caustic, and

$$Y_0 = \left(\frac{|\phi_{222}|}{2f^2}\right)^{1/3} \tag{7.24}$$

is the natural length scale for the interference pattern. When both the source and the lens are at cosmological distances, one can obtain crude order-of-magnitude estimates for Y_0 and $\mu_{\max}$, by assuming that the length scale ξ_0 is to be identified with the Einstein radius of the lens, and that the parameters $|\phi_{222}|$ and ϕ_{11} are of order unity; in this case, one finds for the angular extent δ corresponding to Y_0

$$\delta \sim \left(\frac{\lambda}{10^{22} \text{cm}}\right)^{2/3} \left(\frac{M}{M_\odot}\right)^{-1/6} \sim \left(\frac{H_0 \lambda}{c}\right)^{1/2} \left(\frac{\lambda}{R_S}\right)^{1/6} \quad , \tag{7.25}$$

---

[2] Here, we need only the asymptotic behavior of ϕ itself, not of its derivatives, near the fold; therefore, the cubic terms in (6.11) may be neglected, except for the x_2^3 term.
[3] The reason for y_2 to enter the expression for the magnification is that y_1 is a 'trivial' coordinate for fold catastrophes; compare Sect. 6.2.

and for the maximum magnification

$$\mu_{\max} \sim \left(\frac{M}{M_\odot}\right)^{1/3} \left(\frac{\lambda}{10^6 \text{cm}}\right)^{-1/3}, \qquad (7.26)$$

where λ is the wavelength of the radiation. We notice that the magnification can reach very large values, even for radio frequencies. But again, the source must be unrealistically compact for fringes to be observable.

To investigate (7.22) further, we note that the Airy function falls off nearly exponentially for positive arguments, whereas it oscillates for negative arguments. Note that positive values of y_2 correspond to that side of the caustic where the geometrical optics consideration predicts no images. In wave optics, there will be a region of size $\sim Y_0$ on the "negative" side of the caustic where some radiation from a source will be seen. On the positive side of the caustic there will be fringes; more specifically, the behavior of $\text{Ai}(x)$ for $-x \gg 1$ is $\text{Ai}(-x) \approx \pi^{-1/2} x^{-1/4} \sin\left(\frac{2}{3} x^{3/2} + \frac{\pi}{4}\right)$. From this formula, one finds the phase-averaged magnification

$$\langle \mu \rangle = \frac{1}{|\phi_{11}|} \sqrt{\frac{2}{y_2 \phi_{222}}}, \qquad (7.27)$$

which agrees with the value (6.19b). In Fig. 7.1, we plot $\mu/\mu_{\max}$ as a function of y_2/Y_0, which clearly shows the fringe structure of the magnification for negative y_2.

Extended sources. To see how an extended source averages out the fringes, we have computed the magnification of an extended source (with a parabolic brightness profile) with radius rY_0, whose center is at a distance Y from the

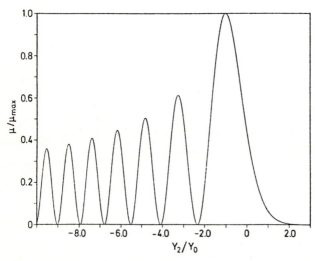

Fig. 7.1. The magnification of a point source in units of the maximum magnification (7.23), $\mu/\mu_{\max}$, as a function of the normalized distance y_2/Y_0 from the caustic

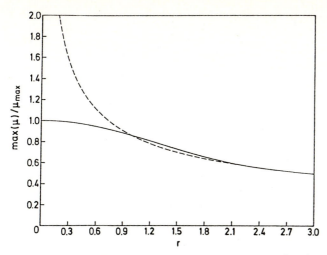

Fig. 7.2. The maximum magnification in units of $\mu_{\max}$ (7.23) of an extended source of radius r with parabolic brightness profile, as a function of the source radius. The solid curve results from wave-optics, whereas the dashed curve is the ray-optics approximation. For $r \gtrsim 2$, these two curves essentially coincide

caustic. We assumed that the individual source elements radiate incoherently, so the magnification μ_e is obtained simply by a brightness-weighted average over the source, using (7.1). Figure 7.2 shows the ratio of the maximum magnification and $\mu_{\max}$ [see (7.23)] as a function of the normalized source radius r; also plotted is the corresponding quantity as derived from geometrical optics. One sees that, for $r \gtrsim 2$, the two curves nearly coincide, but that they strongly disagree for $r \lesssim 1$, where geometrical optics predicts a much larger maximum magnification.

Figure 7.3 shows the magnification of an extended source as a function of its position Y, for three different source sizes; also shown are the corresponding results from geometrical optics. The magnification oscillates around the smooth curve predicted by geometrical optics with an amplitude that depends strongly on the source size. For $r \gtrsim 2$, the fringes have nearly disappeared, which shows that in order to lead to observable fringes, the spatial extent of a source should not be much larger than Y_0.

Finally, we estimate the values of y_2 for which (7.22) is valid, by equating the time delay between the two critical images with the coherence time $\tau_c \approx 1/\Delta\omega$. This yields

$$\frac{y_2}{Y_0} \approx \left(\frac{3}{4}\frac{\omega}{\Delta\omega}\right)^{2/3} ; \qquad (7.28)$$

for smaller values of y_2, the finite coherence time of the light does not influence the results presented in this section.

We now recognize that (6.19b) is not a good approximation for strictly monochromatic point sources; but it gives practically always correct results

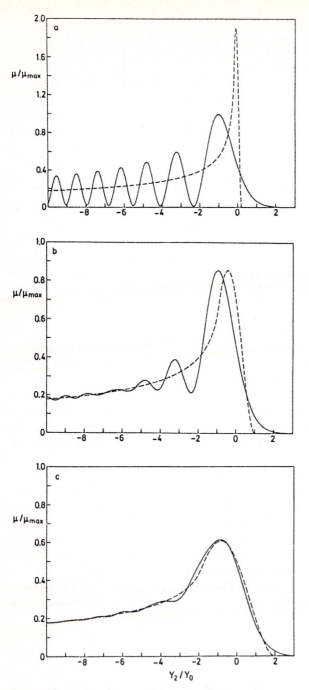

Fig. 7.3a–c. The magnification of a source with dimensionless radius r, normalized by $\mu_{\max}$ (7.23), as a function of the scaled distance y_2/Y_0 of the source center from the caustic, for (**a**) $r = 0.2$, (**b**) $r = 1.0$, and (**c**) $r = 2.0$. The dashed line in each figure is the corresponding expression from geometrical optics. The fringe amplitudes become smaller with increasing source size

if (i) either it is integrated over an even slightly extended source [where (7.25) indicates the angular size necessary], or (ii) if used for not strictly monochromatic sources, or (iii) if the distance from the caustic is larger than the value indicated by (7.28).

Summarizing this chapter, we have seen that, for sources near caustics, wave effects can in principle lead to interference patterns similar to the one for diffraction at a half plane. It turns out, however, that for this effect to be observable, very compact sources are necessary, such as pulsars in distant galaxies. Since no extragalactic source of sufficient compactness is known (no pulsar outside our galaxy has yet been observed), it is unlikely that these wave effects will become observable in the near future; thus we may safely proceed to use geometrical optics.

8. Simple lens models

The principal aim of GL theory is to determine which pairs, consisting of a deflecting mass distribution and a source, can lead to a given (e.g., observed) configuration of images. To approach the answer, one first chooses some simple, plausible mass distributions and studies the inversions of the corresponding lens mappings, i.e., one searches for all points ξ which solve the equation for a given source position η. As mentioned before, this is a nontrivial task in general, that has been carried out analytically only for special mass distributions of the lens. In general, one has to rely on numerical calculations; several methods are described in Chap. 10. Here, we concentrate on the simplest models which can be treated analytically. An understanding of these simple models is essential for the application of numerical methods, partly because the number of images is not known a priori for a given source position. It is therefore dangerous to use a numerical routine, e.g., for root searching, as a black box without useful estimates of the possible results.

We can understand many features of the GL mapping from a study of a few simple lens geometries. In addition, we consider the action of an ensemble of lenses in the chapters on statistical gravitational lensing (Chaps. 11 and 12). There, it is often assumed that the individual lenses can be treated separately, and in order to obtain useful results, the individual lenses are modeled as one of the simple types we describe below.

The problem of finding the solutions of the ray-trace equation simplifies considerably if symmetry allows reduction to a one-dimensional equation. In Sect. 8.1 we treat matter distributions with axial symmetry.[1] It will turn out that such lenses have some very special properties which are due only to their symmetry. Nevertheless, owing to their relative simplicity, symmetric lens models are frequently considered in the literature, e.g., for a first simple model of an observed lens case, or as elementary components in statistical gravitational lensing (see Chaps. 11 and 12). In particular, the

---

[1] An axially-symmetric mass distribution has a volume mass density that is invariant with respect to rotations about the optical axis; in particular, any spherical mass distribution is axi-symmetric. The corresponding surface mass density is then invariant with respect to rotation about its center of mass, i.e., lines of constant surface mass density are concentric circles. We use the expressions "axially symmetric" and "circularly symmetric" synonymously, and often just use "symmetric", if no ambiguity can occur.

formalism developed in Chap. 5 can be applied easily to symmetric lenses, and much is known about their general properties. Lenses with perturbed symmetry are treated in Sect. 8.2; they are obtained by superposition of the gravitational field of a symmetric mass distribution and a quadrupole perturbation. These types of models find application in situations where a symmetric deflector is associated with a larger matter distribution, whose gravitational field can be represented as a quadrupole distortion (e.g., for a star as a member of a galaxy). In Sect. 8.3 we investigate the lens action of two point masses. It turns out that such a system is just simple enough to be accessible to a detailed analytic treatment. GLs with elliptical symmetry are considered in Sect. 8.4, and Sect. 8.5 discusses deflectors which are just capable of producing multiple images. Finally, in Sect. 8.6 we discuss some generic properties of a fairly large class of lenses, including, in particular, the elliptical lenses and quadrupole lenses.

8.1 Axially symmetric lenses

8.1.1 General properties

The deflection angle. We begin with a class of matter distributions with a circularly-symmetric surface mass density, $\Sigma(\boldsymbol{\xi}) = \Sigma(|\boldsymbol{\xi}|)$. Then, the ray-trace equation reduces to a one-dimensional form, since all light rays from the (point) source to the observer must lie in the plane spanned by the center of the lens, the source, and the observer (i.e., this plane is a totally geodesic submanifold of the spacetime considered;[2] if the source, observer, and lens center are colinear, rays are not restricted to a single plane, and ring images can be formed, as will be described below). This can be seen explicitly by considering the scaled deflection angle – see (5.7)

$$\boldsymbol{\alpha}(\mathbf{x}) = \frac{1}{\pi} \int_{\mathbf{R}^2} d^2x' \, \kappa(\mathbf{x}') \frac{\mathbf{x} - \mathbf{x}'}{|\mathbf{x} - \mathbf{x}'|^2} \quad . \tag{8.1}$$

By symmetry, we can restrict the impact vector $\mathbf{x}$ to the positive x_1-axis in the lens plane, $\mathbf{x} = (x, 0)$, $x \geq 0$. Converting to polar coordinates, $\mathbf{x}' = x'(\cos\varphi, \sin\varphi)$, we have for a symmetric matter distribution $\kappa(\mathbf{x}') = \kappa(x')$. Using $d^2x' = x' \, dx' \, d\varphi$, we find

$$\alpha_1(x) = \frac{1}{\pi} \int_0^\infty x' dx' \kappa(x') \int_0^{2\pi} d\varphi \frac{x - x' \cos\varphi}{x^2 + x'^2 - 2xx' \cos\varphi} \quad , \tag{8.2a}$$

$$\alpha_2(x) = \frac{1}{\pi} \int_0^\infty x' dx' \kappa(x') \int_0^{2\pi} d\varphi \frac{-x' \sin\varphi}{x^2 + x'^2 - 2xx' \cos\varphi} \quad . \tag{8.2b}$$

---

[2] As before, we assume that there is no other source of gravity, other than the lens under consideration.

By symmetry, the second component (8.2b) vanishes; hence, $\boldsymbol{\alpha}$ is parallel to $\mathbf{x}$. Using the lens equation (5.6) we find that the source position vector $\mathbf{y}$ must also be parallel to $\mathbf{x}$, which proves the statement made above. For the first component, (8.2a), we note that the inner integral vanishes if $x' > x$ and equals $2\pi/x$ if $x' < x$. Thus, the matter within the disc of radius x around the center of mass contributes to the deflection at the point $\mathbf{x}$ as if it were located at that center, and the matter outside does not contribute, in a similar manner to gravitational forces of spherical mass distributions in three dimensions. From (8.2a),

$$\alpha(x) \equiv \alpha_1(x) = \frac{1}{x} 2 \int_0^x x' \, dx' \, \kappa(x') \equiv \frac{m(x)}{x} \quad , \tag{8.3}$$

where the last equality defines the dimensionless mass $m(x)$ within a circle of radius x.

Comparison of equations (5.2) and (5.7) yields the relation between the scaled deflection angle $\boldsymbol{\alpha}$ and the true deflection $\hat{\boldsymbol{\alpha}}$,

$$\hat{\boldsymbol{\alpha}}(\boldsymbol{\xi}) = \frac{\xi_0 D_\text{s}}{D_\text{d} D_\text{ds}} \boldsymbol{\alpha}(\boldsymbol{\xi}/\xi_0) \quad , \tag{8.4}$$

where ξ_0 is an arbitrary length scale in the lens plane. With (5.4), (5.5) and (8.3) we then find that, for a circularly-symmetric mass distribution,

$$\hat{\alpha}(\xi) = \frac{1}{\xi} \frac{4G}{c^2} 2\pi \int_0^\xi \xi' \, d\xi' \, \Sigma(\xi') \equiv \frac{4GM(\xi)}{c^2 \xi} \quad , \tag{8.5}$$

where $M(\xi)$ is the mass enclosed by the circle of radius ξ. The deflection due to a geometrically-thin symmetric mass distribution at a point ξ is thus the Einstein angle for the mass $M(\xi)$ enclosed by the circle with radius ξ.

Hence, the lens equation for circularly-symmetric matter distributions, $\kappa = \kappa(|\mathbf{x}|)$, is

$$y = x - \alpha(x) = x - \frac{m(x)}{x} \quad , \tag{8.6}$$

where now the range of x is taken to be the whole real axis, and $m(x) \equiv m(|x|)$. Owing to symmetry, we can restrict our attention to source positions $y \geq 0$. Since $m(x) \geq 0$, any positive solution x of (8.6) must have $x \geq y$, and any negative one must obey $\frac{m(x)}{-x} > y$.

The deflection potential ψ and Fermat potential ϕ. The deflection potential, generally defined by (5.9), for the case under consideration, is (for the time being, $x \geq 0$)

$$\psi(x) = \frac{1}{\pi} \int_0^\infty dx' \, x' \, \kappa(x') \int_0^{2\pi} d\varphi \, \ln\sqrt{x^2 + x'^2 - 2xx' \cos\varphi} \quad .$$

These integrals can be calculated using equation (4.224.14) of [GR80.3]:

$$\psi(x) = 2\ln x \int_0^x x' \, dx' \, \kappa(x') + 2 \int_x^\infty x' \, dx' \, \kappa(x') \ln x' \quad . \tag{8.7}$$

Since ψ is determined only up to an additive constant, we can add the term

$$-2 \int_0^\infty x' \, dx' \, \kappa(x') \ln x'$$

to (8.7), which then becomes[3]

$$\psi(x) = 2 \int_0^x x' \, dx' \, \kappa(x') \ln\left(\frac{x}{x'}\right) \quad . \tag{8.8}$$

Using

$$\frac{d}{dx} \int_a^x dx' \, F(x, x') = F(x, x) + \int_a^x dx' \, \frac{\partial F(x, x')}{\partial x} \quad ,$$

it is easily seen that $\alpha(x) = \frac{d\psi(x)}{dx}$. The Fermat potential $\phi(x, y)$ is

$$\phi(x, y) = \frac{1}{2}(x - y)^2 - \psi(x) \quad , \tag{8.9}$$

and the lens equation is equivalent to

$$\frac{\partial \phi}{\partial x} = 0 \quad . \tag{8.10}$$

The Jacobian matrix A. Let us now consider the Jacobian matrix A. First, we write the deflection angle at a point $\mathbf{x} = (x_1, x_2)$ as

$$\boldsymbol{\alpha}(\mathbf{x}) = \frac{m(x)}{x^2} \mathbf{x} \quad , \tag{8.11}$$

where $x = |\mathbf{x}|$. Differentiation then leads to

$$A = \mathcal{I} - \frac{m(x)}{x^4} \begin{pmatrix} x_2^2 - x_1^2 & -2x_1 x_2 \\ -2x_1 x_2 & x_1^2 - x_2^2 \end{pmatrix}$$

$$- \frac{dm(x)}{dx} \frac{1}{x^3} \begin{pmatrix} x_1^2 & x_1 x_2 \\ x_1 x_2 & x_2^2 \end{pmatrix} \quad , \tag{8.12}$$

where $\mathcal{I}$ is the two-dimensional identity matrix, and from (8.3),

$$\frac{dm}{dx} = 2x\kappa(x) \quad . \tag{8.13}$$

We can check from (8.12) that (5.26) is satisfied. From (5.24), we obtain the components of the shear:

---

[3] The results about $\alpha(x)$ given in (8.1–5) hold provided the surface density κ decreases faster than $|x|^{-1}$. Then, (8.8) represents a potential for α, even in cases when the integrals in (5.9) and in the calculation which led to (8.8) diverge, as in the case $\kappa \propto x^{-3/2}$. The weaker fall-off $- |x|^{-1-\varepsilon}$ rather than $|x|^{-2-\varepsilon}$ – can be permitted here since the "distant masses" do not contribute to α, as explained below (8.2).

$$\gamma_1 = \frac{1}{2}\left(x_2^2 - x_1^2\right)\left(\frac{2m}{x^4} - \frac{m'}{x^3}\right) , \tag{8.14a}$$

$$\gamma_2 = x_1 x_2 \left(\frac{m'}{x^3} - \frac{2m}{x^4}\right) , \tag{8.14b}$$

where $m' = \frac{dm}{dx}$. From these relations, with (5.28),

$$\gamma^2 = \left(\frac{m}{x^2} - \kappa\right)^2 , \tag{8.15}$$

and the determinant of A is evaluated from (5.25):

$$\det A = \left(1 - \frac{m}{x^2}\right)\left(1 + \frac{m}{x^2} - 2\kappa\right) . \tag{8.16}$$

This result can also be obtained directly from the original definition of A, equation (5.20), which reduces in the symmetric case considered here to

$$\begin{aligned}\det A &= \frac{y}{x}\frac{dy}{dx} = \left(1 - \frac{m}{x^2}\right)\left[1 - \frac{d}{dx}\left(\frac{m}{x}\right)\right] \\ &= \left(1 - \frac{\alpha(x)}{x}\right)\left(1 - \frac{d}{dx}\alpha(x)\right) ,\end{aligned} \tag{8.17}$$

in agreement with (8.16).

Tangential and radial critical circles. In the axisymmetric case, the critical curves in the lens plane are circles with radii determined by $\det A(x) = 0$.[4] As can be seen from (8.17), there are two kinds of critical circles: those where $m/x^2 = 1$, called *tangential critical curves* for reasons which will become clear in the following, and those where $\frac{d(m/x)}{dx} = 1$, called *radial critical curves*. As can be seen from the lens equation (8.6), tangential critical curves are mapped onto the point $y = 0$; hence, whenever a symmetric lens has a tangential critical curve, there exists a caustic which degenerates to a single point. This property is solely due to the symmetry; any perturbation thereon will remove the degeneracy.

At a critical point, the matrix A has an eigenvector belonging to the eigenvalue zero. Consider now a critical point $x_1 \equiv x$ on the x_1-axis, i.e., for $x_2 = 0$; there,

$$A = \mathcal{I} - \frac{m(x)}{x^2}\begin{pmatrix}-1 & 0 \\ 0 & 1\end{pmatrix} - \frac{m'}{x}\begin{pmatrix}1 & 0 \\ 0 & 0\end{pmatrix} . \tag{8.18}$$

The vector $\mathbf{X}_t = (0,1)$ is tangent to the critical curve at the point under consideration, whereas, for $x_1 \neq 0$, $\mathbf{X}_r = (1,0)$ is normal to the critical

---

[4] The degenerate case, where the point $\mathbf{x} = 0$ is critical, occurs if, and only if, $\kappa(0) = 1$, since in this case, $m(x) = x^2 + \mathcal{O}(x^3)$, and, from (8.16), $\det A = 0$. Actually, if $\mathbf{x} = 0$ is critical, both eigenvalues vanish – also a consequence of symmetry. Note that generally, $A(0) = [1 - \kappa(0)]\mathcal{I}$.

curve. Now, we can easily see that $\mathbf{X_t}$ is an eigenvector with zero eigenvalue if the curve under consideration is a tangential critical curve, and $\mathbf{X_r}$ is an eigenvector with eigenvalue zero for a radial critical curve, which explains the notation used above. Note that the radial critical curves give rise to fold catastrophes (see Sect. 6.2) whereas the point singularity at $\mathbf{y} = 0$ is unstable (Sect. 6.3.4).

Occasionally it happens that at a tangential critical circle, $\kappa = 1$ or, equivalently, $A = 0$; we then call the circle *degenerate*. If $\kappa \neq 1$ at a critical circle, A has rank 1 there. Let $\hat{P}$ denote the source position corresponding to a point $\hat{I}$ on a critical circle of an axially-symmetric deflection mapping, and let γ be a light ray in spacetime containing the events P, I, O whose spatial projections are, respectively, $\hat{P}$, $\hat{I}$, and the observer position O. Then, the geometrical meaning of the magnification matrix A implies: *P is conjugate to O on γ*, and *the multiplicity of P is equal to the corank of A at $\hat{I}$*. In other words, P is a simple (degenerate) conjugate point if rank$(A) = 1$ (0) there. In the first case (P simple), a light beam with vertex at O is astigmatic and traverses a focal *line* containing P, in the second case (P degenerate), such a beam is anastigmatic; it reconverges at the focal *point* P. These remarks apply, of course, also to non-symmetric lenses.

Image distortion near a critical curve. Now, consider a point $\mathbf{x_c} = (x_c, 0)$ very close to the critical curve; if at the critical curve $m/x^2 = 1$, then at $\mathbf{x_c}$, $m/x^2 = 1 - \delta$ with $|\delta| \ll 1$. The Jacobian matrix A at $\mathbf{x_c}$ is then

$$A \approx \begin{pmatrix} 2 - m'/x_c & 0 \\ 0 & \delta \end{pmatrix} ,$$

where we have neglected $\delta \ll 1$ in the first diagonal element. Consider an ellipse surrounding $\mathbf{x_c}$ that is small compared to the distance of $\mathbf{x_c}$ from the critical curve:

$$\mathbf{c}(\varphi) = \mathbf{x_c} + \begin{pmatrix} \rho_1 \cos \varphi \\ \rho_2 \sin \varphi \end{pmatrix} .$$

The ellipse is mapped by the ray-trace equation onto the ellipse

$$\mathbf{d}(\varphi) = \mathbf{y_c} + \begin{pmatrix} (2 - m'/x_c)\rho_1 \cos \varphi \\ \delta \rho_2 \sin \varphi \end{pmatrix}$$

in the source plane, where $\mathbf{y_c}$ is the image of $\mathbf{x_c}$ under the GL mapping. In other words, $\mathbf{c}(\varphi)$ is an image of the source $\mathbf{d}(\varphi)$. Now, let $\mathbf{d}(\varphi)$ be a circle, i.e.,

$$|\rho_2| = \frac{2 - m'/x_c}{\delta} |\rho_1| . \tag{8.19}$$

The image $\mathbf{c}(\varphi)$ of this circular source is a highly elongated ellipse ($|\rho_2| \gg |\rho_1|$) along the x_2-direction, i.e., tangent to the critical curve. Therefore,

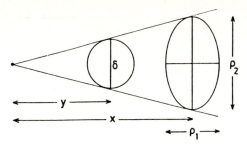

Fig. 8.1. An infinitesimal source of diameter δ and its image, an ellipse with axes ρ_1 and ρ_2, are drawn in the same diagram, for simplicity. Since light rays from a source point are confined to the plane spanned by this point, the center of the symmetric lens, and the observer, the source and its image are seen under the same angle φ from the respective origins (see the text)

critical curves with $m/x^2 = 1$ are called "tangential". Similarly, one can show that, at a radial critical curve, the image of a circular source is highly elongated in the direction perpendicular to the critical curve.

Geometrical meaning of the eigenvalues. In general, the two factors of det A in (8.16) describe the image distortion in the radial and tangential directions. Consider an infinitesimal circular source of diameter δ at $\mathbf{y}$, and an image at $\mathbf{x}$, which is an ellipse with axes ρ_1 in the radial direction and ρ_2 in the tangential direction (see Fig. 8.1). As seen from the origin of the source plane, the source subtends the angle $\varphi = \delta/y$. Since the polar coordinate is unchanged by axially symmetric deflectors, $\varphi = \rho_2/x$, and using y/x from (8.6),

$$\frac{\delta}{\rho_2} = 1 - \frac{m}{x^2} \quad . \tag{8.20}$$

Writing the source diameter as $\delta = \frac{dy}{dx}\rho_1$, we find

$$\frac{\delta}{\rho_1} = 1 + \frac{m}{x^2} - 2\kappa \quad . \tag{8.21}$$

Hence, images are stretched in the radial direction by a factor $\left(\frac{dy}{dx}\right)^{-1} = \left(1 + \frac{m}{x^2} - 2\kappa\right)^{-1}$ and in the tangential direction by $\left(1 - \frac{m}{x^2}\right)^{-1}$. Since a tangential critical curve does not lead to a caustic curve, but the corresponding caustic degenerates to a single point $\mathbf{y} = 0$, the tangential critical curves have no influence on the image multiplicity, except for a source at $\mathbf{y} = 0$. Thus, pairs of images can only be created or destroyed if $\frac{dy}{dx} = 0$.

General properties of axially symmetric lenses. We now describe several properties of symmetric lenses related to the number of images ([DY80.2], [SU86.1]). Let us assume that the lens is transparent and that the surface mass density is piecewise continuous and falls off for large values of x, and hence is also bounded, i.e.,

$$0 \leq \kappa(x) \leq \kappa_{\max} \quad \text{for all } x \;, \tag{8.22a}$$

$$\lim_{x \to \infty} \kappa(x)\, x = 0 \;. \tag{8.22b}$$

It is convenient to define the *mean surface mass density* within x,

$$\bar{\kappa}(x) = \frac{m(x)}{x^2} \;. \tag{8.23}$$

Then:
(a) For a source position $y > 0$, any image with $x > 0$ has $x \geq y$, as follows directly from the lens equation (8.6) and $m(x) \geq 0$.
(b) For sufficiently large values of y, there is exactly one image. Equation (8.22b) requires that there be a constant c and a value a such that, for $|x| > a$, $\kappa(x) < \frac{c}{|x|}$. This bounds the mass, for $|x| > a$: $m(x) < m(a) + 2c(|x| - a)$, so that $\frac{m(x)}{|x|} < d$. For piecewise continuous $\kappa(x)$, the deflection angle $\alpha(x) = \frac{m(x)}{x}$ is everywhere continuous, hence, $\left|\frac{m(x)}{x}\right| < b$ for all x. Thus, for $y \geq b$ the equation $y = x - \frac{m}{x}$ implies $x > 0$, and, because of (a), even $x \geq y$. Moreover, for $|x| > a$, $\bar{\kappa} < \frac{d}{|x|}$ and $\left|\frac{dy}{dx} - 1\right| < \frac{d}{|x|}$. The images of y for $y \geq \max(a,b)$ therefore obey $x \geq y \geq a$, $0 \leq x - y = \frac{m}{x} < b$, i.e., they are contained in a diagonal strip of width b. Since $\frac{dy}{dx} \to 1$ for $|x| \to \infty$, y has exactly one image for large y.
(c) A lens can produce multiple images if, and only if, at least at one point $1 - 2\kappa + \bar{\kappa} < 0$. If $\frac{dy}{dx} = 1 - 2\kappa + \bar{\kappa} \geq 0$ throughout, a lens produces no multiple images, since $y(x)$ increases monotonically. If, on the other hand, there is a point where $\frac{dy}{dx} < 0$, there is at least one local maximum x_1 and one local minimum $x_2 > x_1$ of the curve $y(x)$, since $\frac{dy}{dx} \to 1$ for $|x| \to \pm\infty$. For values of y such that $y(x_2) < y < y(x_1)$, there are at least three images.
(d) A necessary condition for multiple imaging is $\kappa > \frac{1}{2}$ at one point; a sufficient condition is $\kappa > 1$ at one point. The first statement follows from (c) by noting that $\frac{dy}{dx} < 0$ implies $\kappa > \frac{(1+\bar{\kappa})}{2} \geq \frac{1}{2}$. To prove the second statement, let κ have a maximum at x_{m}, $\kappa(x_{\mathrm{m}}) > 1$; then $\bar{\kappa}(x_{\mathrm{m}}) \leq \kappa(x_{\mathrm{m}})$ and $\frac{dy}{dx} < 0$ at x_{m}. The statement then follows from (c).
(e) If the surface mass density does not increase with x, $\kappa'(x) \leq 0$, multiple imaging occurs if, and only if, $\kappa(0) > 1$. Sufficiency was shown already in (d). Now, suppose that $\kappa(0) \leq 1$; then, writing the lens equation in the form $y = x(1 - \bar{\kappa})$, we have for $x \geq 0$: $\frac{dy}{dx} = (1 - \bar{\kappa}) - x\bar{\kappa}'$. Since

$$\bar{\kappa}(x) = 2\int_0^1 du\, u\, \kappa(ux) \;, \quad \text{then} \quad \frac{d\bar{\kappa}}{dx} = 2\int_0^1 du\, u^2\, \kappa'(ux) \leq 0 \;,$$

and $\bar{\kappa}(x) \leq \kappa(0) \leq 1$, we see that $\frac{dy}{dx} \geq 0$, so that no multiple images can occur.

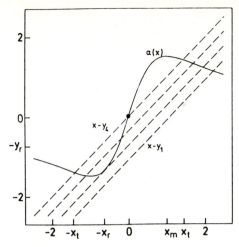

Fig. 8.2. The deflection angle $\alpha(x)$ (solid line), for a typical (centrally condensed) axially symmetric GL, together with a few lines $x - y_i$ (dashed lines), for various source positions y_i. The source at y_1 produces only a single image, whereas a source at y_3 has three images. Two special cases are also shown: the source at $y_2 = y_r$ lies on the radial caustic and the line $x - y_2$ is tangent to $\alpha(x)$ at x_r, the radius of the radial critical curve. The source at $y_4 = 0$ lies on the caustic point and the ring $x = x_t$ is an image of the point $y = 0$; in addition, it has an image at $x = 0$. At the point x_m the deflection angle has its maximum; it is characterized by $\bar{\kappa}(x_m) = 2\kappa(x_m)$.

Multiple image diagram. In Fig. 8.2 we display a typical $\alpha(x)$ curve for a centrally-condensed lens [defined such that $\kappa(x)$ decreases monotonically with x], together with a few $x - y$ lines, for various values of y. Each intersection of $x - y$ with $\alpha(x)$ corresponds to an image. A source at y_1 produces only one image. The point y_2 lies on the caustic corresponding to a radial critical curve, where $1 + \bar{\kappa} - 2\kappa = 0$. A source at y_3 has three images, and $y_4 = 0$ corresponds to the caustic point of a tangential critical curve. The points x_r, x_m and x_t are characterized by $1 + \bar{\kappa}(x_r) - 2\kappa(x_r) = 0$, $\bar{\kappa}(x_m) = 2\kappa(x_m)$ and $\bar{\kappa}(x_t) = 1$. This figure also illustrates the odd-number-of-images-theorem for symmetric lenses, since any straight line intersects an odd function $[\alpha(x) = -\alpha(-x)]$ at an odd number of points, except at caustics. The image with $x > y > 0$ corresponds to a minimum of ϕ and is always magnified. The innermost image corresponds to a maximum and the remaining one to a saddle point. From the figure, it is clear that multiple images can only occur (for centrally-condensed lenses) if $\frac{d\alpha}{dx}(0) = \kappa(0) > 1$.

Point source near $y = 0$. For later applications, we consider now the magnification of a source close to $y = 0$. The tangential critical curve at x_t satisfies $\bar{\kappa}(x_t) = 1$. A source at $y \ll 1$ has two images close to the tangential critical curve at $x = x_t + \Delta x$ and $x = -x_t + \Delta x$, where $\Delta x = y \left(\frac{dy}{dx}\right)^{-1}$. The absolute value of the magnification for each image, $|\mu| = |\det A|^{-1}$ is, according to (8.17),

$$|\mu|^{-1} \simeq |1 - \bar{\kappa}(\pm x_t + \Delta x)| \left|\frac{dy}{dx}\right| \simeq \left|\frac{d\bar{\kappa}}{dx}\right| y \; .$$

Hence, the critical images of a point source at small impact parameter y have a total magnification

$$\mu_p = \frac{g_p}{y} \; , \tag{8.24a}$$

where

$$g_p = 2 \left|\frac{d\bar{\kappa}}{dx}(x_t)\right|^{-1} \; . \tag{8.24b}$$

Extended source near $y = 0$. The magnification of an extended source is given by (6.81). Consider a circular source of radius R at a distance y from the critical point $y = 0$, with a brightness distribution $I(r/R)$, where r is the distance of a source point from the center of the source. If we choose polar coordinates centered on the source, (6.81) becomes

$$\mu_e = \frac{1}{\pi} \left[\int_0^\infty r\,dr\, I(r/R)\right]^{-1} \int_0^\infty r\,dr\, I(r/R) \int_0^\pi d\varphi \frac{g_p}{\sqrt{y^2 + r^2 + 2ry\cos\varphi}}$$

With the substitution $x = r/R$ we can write this equation as

$$\mu_e = \frac{g_p}{R} \zeta\left(\frac{y}{R}\right) \; , \tag{8.25}$$

where

$$\zeta(w) = \frac{1}{\pi} \left[\int_0^\infty x\,dx\, I(x)\right]^{-1} \int_0^\infty x\,dx\, I(x) \int_0^\pi \frac{d\varphi}{\sqrt{w^2 + x^2 + 2wx\cos\varphi}} \; .$$

From equation (2.571.4) of [GR80.3] we find for the inner integral

$$\int_0^\pi \frac{d\varphi}{\sqrt{w^2 + x^2 + 2wx\cos\varphi}} = \begin{cases} 2/x\, K(w/x) & \text{for } w \leq x \\ 2/w\, K(x/w) & \text{for } w \geq x \end{cases} \; ,$$

where $K(x)$ is the complete elliptic integral of the first kind. Thus,

$$\zeta(w) = \frac{2w}{\pi} \frac{\int_0^1 dy\, [y\,I(yw) + y^{-2}\, I(w/y)]\, K(y)}{\int_0^\infty x\,dx\, I(x)} \; . \tag{8.26}$$

The function $\zeta(w)$ can be readily evaluated for any brightness distribution $I(x)$. For the simple case of a source with uniform surface brightness, we find, with equations (5.112.3 & 9) of [GR80.3], that for

$$I(x) = H(1-x): \; \zeta(w) = \begin{cases} \dfrac{4E(w)}{\pi} & \text{for } w \leq 1 \\[1ex] \dfrac{4w}{\pi}\left[E(1/w) - \left(1 - \dfrac{1}{w^2}\right) K(1/w)\right] & \text{for } w \geq 1 \end{cases} \tag{8.27}$$

and $E(x)$ is the complete elliptic integral of the second kind. The function $\zeta(w)$ is plotted

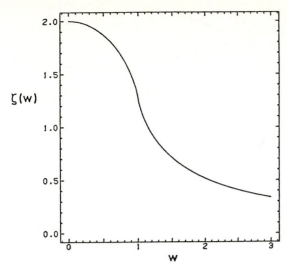

Fig. 8.3. The function $\zeta(w)$ (8.27)

in Fig. 8.3. Its maximum is at $w = 0$ where $\zeta(0) = 2$. For different brightness profiles, see [SC87.6]. Note that the complete elliptic integrals are easily evaluated numerically (see Sect. 17.3 of [AB65.1] and [PR86.1]).

8.1.2 The Schwarzschild lens

Since the Schwarzschild lens ([EI36.1], [RE64.1], [LI64.1]) was already discussed in Chap. 2, we just quote the results, with the notation introduced in Chap. 5. Consider a "point mass" M located at the origin of the lens plane, $\boldsymbol{\xi} = 0$. The surface mass density is

$$\Sigma(\boldsymbol{\xi}) = M\delta^2(\boldsymbol{\xi}) \ .$$

The natural length scale is ξ_0 as given in (2.6b). From (5.4) and (8.3), we obtain

$$m(x) = 1 \ ,$$

and the lens equation is

$$y = x - 1/x \ ,$$

which has two solutions,

$$x_{1,2} = \frac{1}{2}\left(y \pm \sqrt{y^2 + 4}\right) \ , \tag{8.28}$$

i.e., one image on each side of the lens. From (8.16),

$$\mu = \left(1 - \frac{1}{x^4}\right)^{-1} \ ,$$

which can be combined with (8.28) to yield

$$\mu_{1,2} = \pm \frac{1}{4} \left[\frac{y}{\sqrt{y^2+4}} + \frac{\sqrt{y^2+4}}{y} \pm 2 \right] \quad, \tag{8.29a}$$

and the total magnification is the sum of the absolute values of these two magnifications:

$$\mu_p = \mu_1 - \mu_2 = \frac{y^2+2}{y\sqrt{y^2+4}} \quad. \tag{8.29b}$$

From (8.7),

$$\psi(x) = \ln x \quad,$$

and from (5.44) and (8.28), we obtain the time delay for the two images,

$$c\Delta t = \frac{4GM}{c^2} (1+z_d) \tau(y) \quad, \tag{8.30a}$$

where

$$\tau(y) = \frac{1}{2} y \sqrt{y^2+4} + \ln \frac{\sqrt{y^2+4}+y}{\sqrt{y^2+4}-y} \quad. \tag{8.30b}$$

Since the two images are of comparable brightness only if $y \lesssim 1$, and since $\tau(1) \approx 2.08$, we find from (8.30a) that *the time delay is of the same order as the time it takes light to traverse a distance of about one Schwarzschild radius.* The magnification of an extended source by a Schwarzschild lens was discussed in [BO79.1]. For $y \ll 1$, (8.29b) yields $\mu_p \approx 1/y$; hence, for $\mu_p \gg 1$ we obtain (8.25), with $g_p = 1$. More general analytic approximations for $\mu_e(y, R)$ for the Schwarzschild lens are given in [SC87.4]. For a source of uniform surface brightness with radius R, we deduce from (6.81) and (8.29b) that the maximum magnification is

$$\mu_{\max} = \frac{\sqrt{4+R^2}}{R} \quad. \tag{8.31}$$

8.1.3 Disks as lenses

Consider a disk of radius ϱ, surface mass density $\Sigma(\xi)$ and mass $M = 2\pi \int_0^\varrho \Sigma(\xi)\xi d\xi$. We choose as the length scale in the lens plane the Einstein radius (2.6b)

$$\xi_0 = \sqrt{2R_S \frac{D_d D_{ds}}{D_s}}$$

of the disk's total mass; then its dimensionless radius is $x_0 = \varrho/\xi_0$.

Homogeneous disk. First, we treat the *homogeneous* case, $\Sigma = \Sigma_0$. Then, $M = \pi \varrho^2 \Sigma_0$ and, from (5.4), (5.5), $\kappa = \kappa_0 = x_0^{-2}$. The resulting lens mapping is

$$y = \begin{cases} x - (x/x_0^2) & \text{for } |x| \leq x_0 \\ x - 1/x & \text{for } |x| \geq x_0 \end{cases} \qquad (8.32)$$

It can easily be inverted. For $x_0 < 1$, a source at y has three images if $0 \leq y \leq \frac{1-x_0^2}{x_0}$, one inside the disk at $x = \frac{x_0^2}{x_0^2-1} y$ and two outside, given by (8.28); two images if $y = \frac{1-x_0^2}{x_0}$, one at the rim of the disk, the other one outside, at x_0^{-1}; one image if $y > \frac{1-x_0^2}{x_0}$, at $\frac{y}{2} + \sqrt{\frac{y^2}{4} + 1}$. For $x_0 > 1$, there is one image only. It is inside the disk at $x = \frac{x_0^2}{x_0^2-1} y$ if $0 \leq y \leq \frac{x_0^2-1}{x_0}$, outside at $x = \frac{y}{2} + \sqrt{\frac{y^2}{4} + 1}$ otherwise.

Note that in each case the magnification of the image inside the disk is $(1 - \kappa_0)^{-2}$. Its parity is always positive, and it corresponds to a maximum (minimum) of the Fermat potential if $x_0 < 1 (x_0 > 1)$.

In the special case $x_0 = 1$, i.e., $\kappa_0 = 1$, all points of the disk are mapped into the point $y = 0$ on the optical axis. Thus, this "lens" indeed acts as a perfect thin lens the focus of which is at the observer. For a fixed physical mass density Σ_0, the value of κ_0 depends on the distances. The value $\kappa_0 = 1$ corresponds to $\varrho = \xi_0$, i.e., to $2R_S D_d D_{ds} = \varrho^2 D_s$. This equation is (naturally) symmetrical in D_d, D_{ds} and reduces, in the case of Euclidean distances when $D_s = D_d + D_{ds}$, to the elementary formula $\frac{1}{D_d} + \frac{1}{D_{ds}} = \frac{1}{f} =: \frac{2R_S}{\varrho^2}$ for a thin lens of focal length f. If the observer is between the lens and the focus, there can be a single image of the source only, but if the focus is between the lens and the observer, three images are possible. The magnification of the image within the disk is large if the observer is near the focus.

Inhomogeneous disk. In the "interesting" case $x_0 < 1$, the circle of radius $y = \frac{1-x_0^2}{x_0}$ separates the three-image region from the single-image region. The corresponding circle with radius x_0 in the lens plane – the rim of the disk – is the curve where images fuse or are created. However, this circle is not a critical curve in the strict sense since the mapping given by (8.32) is not differentiable there. But suppose we smooth the edge of the disk so that $\kappa(x)$ becomes differentiable. Then there is a point x_c close to x_0 where $\frac{dy}{dx}$ vanishes, and the corresponding circle in the source plane is indeed a fold-type caustic. Let the smoothing be done in such a way that, (i) a central region $|x| \leq x_1$ of the disk remains homogeneous, and (ii) for positive x, the scaled deflection angle $\alpha(x)$ has but one maximum, as in the homogeneous case. *Inhomogeneous disks* of this kind form an open set within the collection of axially symmetric lenses. Their qualitative properties will now be described. We put

$$s = \sqrt{\frac{2R_S D_d}{\varrho^2}} \quad , \quad p = \sqrt{\frac{D_{ds}}{D_s}} \quad ,$$

so that $0 < p < 1$ and $x_0^{-1} = ps < s$. Note that $\frac{2R_S}{\varrho}$ is a measure of the

241

compactness of the lens, s increases with the distance of the lens from the observer, and for fixed s, p increases with the distance of the source from the lens. The following cases are possible:

1) $s \leq 1$. Then $x_0 > 1$ and, as in the homogeneous case, there are no critical points. The observer is so close to the lens that, even for distant sources, the light deflection does not suffice to produce more than one image; the past light cone of the observer is smooth, free of caustics; see Fig. 8.4a,b.

2) $s > 1$. This case is illustrated in Fig. 8.4c. The large caustic C is indicated as a thick line. In three-space, it consists of a conical surface M and an axis A, a half-line meeting M at the vertex V. Sources so close to the lens that $ps < 1$, have one image only; the corresponding source planes do not intersect C. If $ps = 1$, $x_0 = 1$; the source plane intersects C in the vertex V only. The corresponding critical set is the whole homogeneous "core" $|x| \leq x_1$ of the disk; it consists of degenerate critical circles all mapped into V. If $ps > 1$, $x_0 < 1$, and C intersect the source plane in a point (the image of tangential critical circles in the lens plane) and a circle (the image of a radial critical circle). If we consider a sequence of point sources on the optical axis starting closely behind the lens, the observer will see a point until the source has reached V when suddenly the source is seen

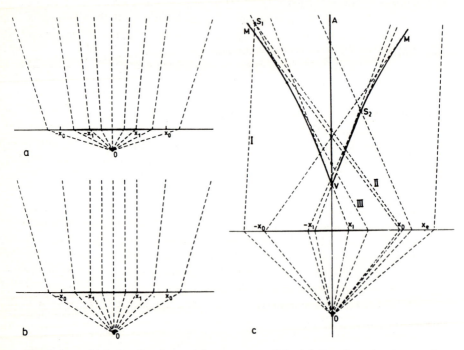

Fig. 8.4a–c. Light bending by an inhomogeneous disk. (**a**) For observers close to the disk, there is no caustic. Followed backwards from O, all rays diverge. (**b**) Rays from a source "at infinity" are focused at the observer by the homogeneous inner part of the disk. (**c**) For observers at larger distances than in case (**b**), the large caustic has the shape of an inverted tent with a pole. Details are explained in the text

as a disk. Sources at larger distances produce a central point-image and an Einstein ring, first inside, then outside the disk.

It is also instructive to locate the points conjugate to O on all light rays reaching O. Let $\mathbf{x}'$ denote the point at which a ray crosses the lens plane. Then if $|\mathbf{x}'| \geq x_e := \left[\frac{2D_s}{D_s - D_d} \int_0^{x_0} x\kappa(x)dx \right]^{1/2}$, the ray does not contain a point conjugate to O; if $x_0 \leq |\mathbf{x}'| < x_e$, it contains exactly one (simple) conjugate point, which is located on the axis A; if $x_1 < |\mathbf{x}'| < x_0$, the ray contains two (simple) conjugate points, one on A and one where the ray touches M; and if $|\mathbf{x}'| \leq x_1$, the ray contains only one (degenerate) conjugate point, the vertex V of the large caustic. Clearly, the latter is the set of points conjugate to O on some ray through O. (Note that the preceding description holds for the rays and the caustic in spacetime although the figure and the discussion refer to its projection into the three-space.)

Figure 8.4c also shows that a source at S_1 has one image of each of the three types, with the appropriate transitions of their respective rays through caustics, as stated in connection with (5.29), while a source on the caustic – at S_2, for example – has two images. One can also verify intuitively the parities of the various images.

The discussion just given remains valid if the homogeneous inner disk shrinks and is removed, $x_1 \to 0$. Then M has a cusp, not only a vertex, at V, and only the central ray passes V. An example is provided by the disk representing a homogeneous, spherical mass [LA71.1]. The reader may also wish to verify that the following two examples, in Sects. 8.1.4 and 8.1.5, are covered, directly or as a degenerate limiting case, by the above discussion.

8.1.4. The singular isothermal sphere

A mass distribution often used in lens modeling is the so-called singular isothermal sphere, with surface mass density

$$\Sigma(\xi) = \frac{\sigma_v^2}{2G}\xi^{-1} \ . \tag{8.33}$$

Although the surface mass density is infinite at $\xi = 0$, its behavior for larger values of ξ seems to approximate the matter distribution of galaxies fairly well, since it yields the observed flat rotation curves of spiral galaxies. Real galaxies cannot follow the density law (8.33), of course, due to the infinite density at the center and the infinite total mass. The latter point is not critical, because a cut-off at a radius ξ_t does not significantly affect the lensing behavior, as long as the cut-off radius is much larger than the radius ξ_0 introduced in (8.34a). A more realistic model for galaxies is given in the next subsection, where the singularity of Σ is removed. The parameter σ_v is the line-of-sight velocity dispersion of stars in the galaxy (cf. [BI87.1]). Choosing

$$\xi_0 = 4\pi \left(\frac{\sigma_v}{c}\right)^2 \frac{D_d D_{ds}}{D_s} \ , \tag{8.34a}$$

we obtain

$$\kappa(x) = \frac{1}{2x} \quad ; \quad \alpha(x) = \frac{x}{|x|} \quad , \tag{8.34b}$$

and

$$y = x - \frac{x}{|x|} \quad . \tag{8.34c}$$

Consider $y > 0$; for $y < 1$, there are two images, at $x = y+1$ and $x = y-1$, i.e., on opposite sides of the lens center. For $y > 1$, only one image occurs, at $x = y+1$. The images at $x > 0$ are of type I, those at $x < 0$ of type II [see (5.29)]. If the singularity in the center were removed, a third image would be produced in the core.

The magnification for an image at x is

$$\mu = \frac{|x|}{|x|-1} \quad ; \tag{8.35a}$$

the circle $|x| = 1$ is a tangential critical curve. From (8.15) we find that $\gamma(x) = \kappa(x) = 1/(2x)$; thus, images are stretched in the tangential direction by a factor $|\mu|$, whereas the distortion factor in the radial direction is unity. The total magnification of a point source is

$$\mu_{\rm p} = \begin{cases} 2/y & \text{for } y \leq 1 \\ (1+y)/y & \text{for } y \geq 1 \end{cases} \quad . \tag{8.35b}$$

Note that for $y \to 1$, the second (inner) image becomes very faint. The deflection potential is $\psi(x) = |x|$, and the time delay for the two images is

$$c\Delta t = \left[4\pi \left(\frac{\sigma_v}{c} \right)^2 \right]^2 \frac{D_{\rm d} D_{\rm ds}}{D_{\rm s}} (1 + z_{\rm d}) \, 2y \quad . \tag{8.36}$$

8.1.5 A family of lens models for galaxies

Ray-trace equation. Most of the lens models considered in the previous subsections do not satisfy the conditions for the application of the general results obtained in Sect. 8.1.1, since they either have a singular or a discontinuous surface mass density distribution. Here, we study a class of models with smooth and non-singular matter distributions, which provide more realistic models for galaxies. Their surface mass density is

$$\Sigma(\xi) = \Sigma_0 \frac{1 + p\,(\xi/\xi_{\rm c})^2}{[1 + (\xi/\xi_{\rm c})^2]^{2-p}} \quad . \tag{8.37}$$

Here, Σ_0 is the central surface mass density, $\xi_{\rm c}$ is the scale over which the distribution falls off (which can be identified with a core radius of the galaxy), and p determines the "softness" of the lens; for $\xi \gg \xi_{\rm c}$, $\Sigma \propto \xi^{2(p-1)}$. We restrict our attention to values $0 \le p \le 1/2$. As for the model in Sect. 8.1.4, the total mass of the lens described by (8.37) is infinite, but a cut-off at sufficiently large radius does not affect the lensing behavior significantly, as long as the tangential critical curve has a radius much smaller than the cut-off radius. For $p = 0$, the distribution is called a Plummer model, whereas for $p = 1/2$, it approximates the isothermal sphere for large ξ [BL87.2].

We choose the length scale as $\xi_0 = \xi_c$, for which the dimensionless surface mass density becomes

$$\kappa(x) = \kappa_0 \frac{1 + p x^2}{(1 + x^2)^{2-p}} \quad ; \tag{8.38}$$

from (8.8), the corresponding deflection potential is

$$\psi(x) = \frac{\kappa_0}{2p} \left[(1 + x^2)^p - 1 \right] \quad , \quad p \ne 0 \quad , \tag{8.39a}$$

which, in the limit $p \to 0$, becomes

$$\psi = \frac{\kappa_0}{2} \ln (1 + x^2) \quad , \quad p = 0 \quad . \tag{8.39b}$$

The corresponding lens equation is

$$y = x - \alpha(x) = x - \kappa_0 \frac{x}{(1 + x^2)^{1-p}} \quad . \tag{8.40}$$

Critical curves and caustics. For $\kappa_0 > 1$, a lens with the surface mass density (8.38) has a tangential critical curve at $x = x_t$, where

$$x_t = \sqrt{\kappa_0^{1/(1-p)} - 1} \quad , \tag{8.41a}$$

and a radial critical curve at $x = x_r$, which is determined from (see (8.17))

$$1 - \kappa_0 \left(1 + x_r^2 \right)^{p-2} \left[1 + (2p - 1) x_r^2 \right] = 0 \quad . \tag{8.41b}$$

The implicit equation (8.41b) for x_r cannot be solved analytically for general p; however, for $p = 0$ and $p = 1/2$ analytic solutions exist:

$$x_r = \sqrt{\sqrt{2\kappa_0 + \frac{\kappa_0^2}{4} - 1} - \frac{\kappa_0}{2}} \quad , \quad \text{for} \quad p = 0 \quad , \tag{8.41c}$$

$$x_r = \sqrt{\kappa_0^{2/3} - 1} \quad , \quad \text{for} \quad p = 1/2 \quad . \tag{8.41d}$$

We plot x_r as a function of κ_0 for several values of p in Fig. 8.5. As expected, x_r increases with κ_0 and with p. The corresponding caustic in the source plane has a radius $|y(x_r)| = y_r$, where

$$y_r = \frac{2(1-p) x_r^3}{1 - (1 - 2p) x_r^2} \quad . \tag{8.42}$$

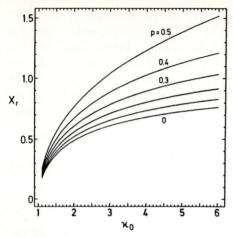

Fig. 8.5. The radius x_r of the radial critical curve as a function of the central surface mass density. The curves are labeled with the value of p

Sources with $|y| < y_r$ have three images, those with $|y| > y_r$ have one image. The three images for $0 < y < y_r$ are at $x > x_t$ (type I image), $-x_t < x < -x_r$ (type II) and $-x_r < x < 0$ (type III). The magnification of an image is

$$\mu = \left[1 - \frac{\kappa_0}{(1+x^2)^{1-p}}\right]^{-1} \left[1 - \frac{\kappa_0}{(1+x^2)^{2-p}}\left[1 + (2p-1)x^2\right]\right]^{-1}. \quad (8.43)$$

Although the lens equation (8.40) cannot be solved analytically, we can easily find the roots numerically. Figure 8.2 was computed for a Plummer model ($p = 0$) and $\kappa_0 = 3$.

The lens equation near critical curves. Since we will use this model in a later chapter, we investigate its behavior at critical curves explicitly. Consider a source at $\Delta y \ll y_r$; it has two critical images near the tangential critical curve at $x = \pm x_t + \Delta x$, where

$$\Delta y = y'(x_t) \Delta x \quad ,$$

the prime denoting differentiation with respect to x. Explicitly,

$$y' = 1 - \kappa_0 \left(1+x^2\right)^{p-2} \left[1 + (2p-1)x^2\right] \quad .$$

To determine the magnifications of these two images, we note that $(\det A)' = (y/x)'y' + (y/x)y''$; since $\det A(\pm x_t) = 0$, neglecting terms of order $(\Delta y)^2$,

$$\mu^{-1}(\pm x_t + \Delta x) = (\det A)(\pm x_t + \Delta x) = (y/x)'(\pm x_t) y'(x_t) \Delta x$$
$$= \pm 2 \kappa_0 (1-p) x_t \left(1 + x_t^2\right)^{p-2} \Delta y \quad . \quad (8.44)$$

A source at $-y_r + \Delta y$ has two images near the radial critical curve at $x_r \pm \Delta x$, where

$$\Delta y = \frac{1}{2} y''(x_r) \Delta x^2 \quad ,$$

and

$$y''(x) = 2\kappa_0 (1-p) x \left(1+x^2\right)^{p-3} \left[3 - (1-2p)x^2\right] ,$$

and the magnifications are

$$\mu^{-1}(x_r \pm \Delta x) = \pm (\det A)'(x_r) \Delta x = \pm \left[1 - \kappa_0 \left(1+x_r^2\right)^{p-1}\right] \sqrt{2y''(x_r) \Delta y} . \quad (8.45)$$

Note that, for $p = 0$ and $p = 1/2$, these equations give a complete analytic description of the critical behavior, since x_r is known (8.41c,d).

8.1.6 A uniform ring

As a final example for circularly-symmetric lenses, we consider a ring of matter with constant surface mass density. Such a GL model may find little astrophysical application, but it has an interesting structure and can be completely analyzed analytically.

The lens equation. Consider a ring with dimensionless surface mass density κ_0; we can always use a length scale in the source and lens planes such that the inner radius of the ring is $x_i = 1$. Therefore, the model is described by κ_0 and the outer radius of the ring, $x_o > 1$. The lens equation then reads:

$$y = \begin{cases} x & \text{for } 0 \le x \le 1 \\ x - \kappa_0 \dfrac{x^2-1}{x} & \text{for } 1 \le x \le x_o \\ x - \dfrac{M}{x} & \text{for } x \ge x_o \end{cases} ,$$

where $M = \kappa_0(x_o^2 - 1)$ is the dimensionless mass of the ring. Rays with $x \le 1$ are not deflected, whereas those with $x \ge x_o$ experience a deflection from the total mass M of the ring, as if it were a point mass. The function $y(x)$ is continuous, but has a discontinuous derivative at $x = 1$ and $x = x_o$, due to the discontinuity in the surface mass distribution. Similarly to the discussion in Sect. 8.1.3, we can always consider the edges of the ring to be "smoothed" in a small transition zone. In that case, the function $y(x)$ becomes a smooth function for all x. In the following, we shall assume that such a smoothing has taken place.

Critical curves and caustics. For $|x| < 1$ and $|x| > x_o$, the function $y(x)$ is monotonic, so no radial critical curves can exist for these values of x. A radial critical curve exists at $x = 1$ if the slope of $y(x)$, which is positive for $x < 1$, changes sign at $x = 1$; this is the case for $\kappa_0 > \frac{1}{2}$. Since $y(x)$ increases monotonically for $x > x_o$, the existence of a radial critical curve at $x = 1$ implies that there must be a second radial critical curve. For $\frac{1}{2} < \kappa_0 < \frac{x_o^2}{(1+x_o^2)}$, this second radial critical curve lies within the ring, namely at $x_r = \sqrt{\kappa_0/(1-\kappa_0)}$, and we have $y_r = y(x_r) = 2\sqrt{\kappa_0(1-\kappa_0)}$; otherwise, it is at x_o, and $y_o = y(x_o) = (1-\kappa_0)x_o + \kappa_0/x_o$ (we will not present the

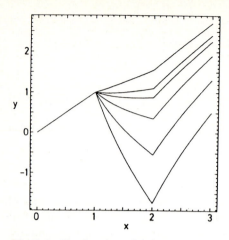

Fig. 8.6. The function $y(x)$ for a ring of constant surface mass density κ_0. The inner radius of the ring is $x_i = 1$, and the outer radius was chosen to be $x_o = 2$. The curves are drawn for six different values of κ_0; they correspond to the six different cases as detailed in Table 8.1

derivation of these and the following results; this is left to the interested reader). Since $y(1) > 0$ and $y(x) > 0$ for large values of x, tangential critical curves (i.e., where $y = 0$) also have to occur in pairs. They exist if $y_o < 0$, one with $1 < x < x_o$, and one with $x > x_o$. The condition $y_o < 0$ is equivalent to $\kappa_0 > \frac{x_o^2}{(x_o^2 - 1)}$. This means that, for sufficiently high surface density in the ring, a source at the origin $y = 0$ produces two rings, and an unperturbed image at $x = 0$.

In Fig. 8.6, we plot the function $y(x)$ for $x_o = 2$ and for various values of κ_0, and in Table 8.1, the six different cases of image configurations are listed. Again, these results can be easily verified by the interested reader.

We have reached the end of our discussion of axially symmetric lens models; we return to them in later chapters, where we discuss the applications of GL theory. Owing to their simplicity, symmetric lens models are frequently used in the study of statistical properties of an ensemble of lenses. In addition, an understanding of symmetric lenses is necessary to study more general lens models, as will be seen in the next section, where lenses with perturbed symmetry are considered. We would like to stress again that one must be cautious in applying symmetric lens models to observed lens cases; their special properties (e.g., the degeneracy of a caustic curve to a point) can be misleading.

Table 8.1. Classification according to the surface mass density κ_0 of a homogeneous ring, for the six different image configurations (column 1). The second and third columns describe the range of image positions for a given source position y. I, R, and O denote, respectively, the range $|x| < 1$ (inner region), $1 < |x| < x_o$ (the ring), and $|x| > x_o$ (outer region). A plus (minus) sign as subscript indicates that the image lies at $x > 0$ ($x < 0$) for positive y

$\kappa_0 < \dfrac{1}{2}$	$0 < y < 1$	I_+
	$1 < y < y_o$	R_+
	$y_o < y$	O_+
$\dfrac{1}{2} < \kappa_0 < \dfrac{x_o}{1+x_o}$	$0 < y < y_r$	I_+
	$y_r < y < 1$	$I_+, 2R_+$
	$1 < y < y_o$	R_+
	$y_o < y$	O_+
$\dfrac{x_o}{1+x_o} < \kappa_0 < \dfrac{x_o^2}{1+x_o^2}$	$0 < y < y_r$	I_+
	$y_r < y < y_o$	$I_+, 2R_+$
	$y_o < y < 1$	I_+, R_+, O_+
	$1 < y$	O_+
$\dfrac{x_o^2}{x_o^2+1} < \kappa_0 < \dfrac{x_o^2}{x_o^2-1}$	$0 < y < y_o$	I_+
	$y_o < y < 1$	I_+, R_+, O_+
	$1 < y$	O_+
$\dfrac{x_o^2}{x_o^2-1} < \kappa_0 < \dfrac{x_o}{x_o-1}$	$0 < y < -y_o$	O_-, R_-, I_+, R_+, O_+
	$-y_o < y < 1$	I_+, R_+, O_+
	$1 < y$	O_+
$\kappa_0 > \dfrac{x_o}{x_o-1}$	$0 < y < 1$	O_-, R_-, I_+, R_+, O_+
	$1 < y < -y_o$	O_-, R_-, O_+
	$-y_o < y$	O_+

8.2 Lenses with perturbed symmetry (Quadrupole lenses)

Motivation and lens equation. Real matter distributions can hardly be described by axi-symmetric potentials, other than in the idealized world of a theorist. No galaxy is expected to be exactly spherically symmetric, and neither is the gravitational field around a black hole, since it is part of a larger system, e.g., a galaxy. The gravitational field of the galaxy perturbs the field of the black hole and disturbs the symmetry. In many cases of astrophysical interest (e.g., a star in a galaxy, a galaxy in a cluster of galaxies) the gravitational field of the perturber changes very little over the relevant length scale of the originally axi-symmetric deflector. To give just one example, consider a star in a galaxy. The natural length scales of the two matter components are given by (2.6b) (for $M = M_{\text{star}}$, $M = M_{\text{galaxy}}$, respectively; that is, they differ by many orders of magnitude. It is thus natural to expand the field of the perturber about the center of the main deflector; the lowest-order, non-trivial term in the expansion is the quadratic term (tidal field). In this section we study the lens action of a symmetric matter distribution which is perturbed by a larger-scale gravitational field, and we assume that the latter is well described by its quadratic Taylor approximation, henceforth called quadrupole terms. Note that the perturbation is *not* assumed to be small; the following discussion is valid for perturbing fields of arbitrary strength. The deflection caused by the perturber is

$$\boldsymbol{\alpha}_p(\mathbf{x}) = \boldsymbol{\alpha}_p(0) + \begin{pmatrix} \Gamma_1 & 0 \\ 0 & \Gamma_2 \end{pmatrix} \mathbf{x} = \boldsymbol{\alpha}_p(0) + \begin{pmatrix} \kappa_p + \gamma_p & 0 \\ 0 & \kappa_p - \gamma_p \end{pmatrix} \mathbf{x} \quad ,$$

where the coordinates are chosen such that the origin coincides with the center of the axi-symmetric matter distribution, and the orientation is chosen to diagonalize the quadrupole matrix (or tidal matrix). Whether the lowest-order Taylor expansion of the perturbing deflection, in the region where the lens equation is considered, is a reasonable approximation has to be decided from case to case. According to (5.23), $(\Gamma_1 + \Gamma_2)/2$ is the local surface mass density of the perturber, and $(\Gamma_1 - \Gamma_2)/2$ its shear. In the following we assume that $\Gamma_1 \neq \Gamma_2$; for $\Gamma_1 = \Gamma_2$ the resulting lens is still symmetric. The ray-trace equation then reads

$$\mathbf{y} = \mathbf{x}[1 - \bar{\kappa}(x)] - \begin{pmatrix} \Gamma_1 & 0 \\ 0 & \Gamma_2 \end{pmatrix} \mathbf{x} \quad , \tag{8.46}$$

where we have translated the origin of the source plane by $\mathbf{y} \to \mathbf{y} + \boldsymbol{\alpha}_p(0)$. Here, $\bar{\kappa}(x) = m(x)/x^2$ is the mean surface mass density of the symmetric lens inside a circle with radius x, and $x = |\mathbf{x}|$.

Geometrical solution. Let us introduce polar coordinates, $\mathbf{y} = y(\cos\vartheta, \sin\vartheta)$, $\mathbf{x} = x(\cos\varphi, \sin\varphi)$, and rewrite (8.46) as

$$y\cos\vartheta = x\cos\varphi[1 - \bar{\kappa}(x) - \Gamma_1] \quad , \tag{8.47a}$$

$$y\sin\vartheta = x\sin\varphi[1 - \bar{\kappa}(x) - \Gamma_2] \quad . \tag{8.47b}$$

We can find the solutions (x, φ) of this system for a given source position (y, ϑ), $y > 0$, by the following pictorial approach [KO87.3]: consider the two curves

$$u_1(\varphi) = \frac{\cos\varphi}{y} \quad ; \quad v_1(\varphi) = \frac{\sin\varphi}{y}$$

and

$$u_2(x) = \frac{\cos\vartheta}{x[1 - \bar{\kappa}(x) - \Gamma_1]} \quad ; \quad v_2(x) = \frac{\sin\vartheta}{x[1 - \bar{\kappa}(x) - \Gamma_2]}$$

in a (u, v)-coordinate plane. The first curve (u_1, v_1) is a circle of radius $1/y$, whereas the second curve (u_2, v_2) is more complicated, and depends on ϑ and the density profile. Points (u, v) where these two curves intersect correspond to solutions of (8.47); in fact, φ is also the polar angle in the (u,v)-plane, and elimination of $\bar{\kappa}(x)$ from (8.47) yields the corresponding value for x,

$$x = \frac{2y\sin(\varphi - \vartheta)}{(\Gamma_2 - \Gamma_1)\sin 2\varphi} \;.$$

For more details of this method, and several applications, see [KO87.3]. We now pursue a more analytic method.

One-dimensional form of the ray-trace equation. As was stated before, the solution of a general lens equation is non-trivial, as one must find all the roots of a two-dimensional equation. The quadrupole lens can be reduced to a one-dimensional equation: rewriting (8.47) as

$$\cos\varphi = \frac{y_1}{x\left[1 - \bar{\kappa}(x) - \Gamma_1\right]} \;, \tag{8.48a}$$

$$\sin\varphi = \frac{y_2}{x\left[1 - \bar{\kappa}(x) - \Gamma_2\right]} \;, \tag{8.48b}$$

we obtain, by adding the squares of equations (8.48),

$$x^2\left[1 - \bar{\kappa}(x) - \Gamma_1\right]^2\left[1 - \bar{\kappa}(x) - \Gamma_2\right]^2 - y_1^2\left[1 - \bar{\kappa}(x) - \Gamma_2\right]^2 \\ - y_2^2\left[1 - \bar{\kappa}(x) - \Gamma_1\right]^2 = 0 \;. \tag{8.49}$$

Solutions such that $x \geq 0$ for (8.49) yield all the image positions with (8.48).

The Jacobian and critical curves. Next, we consider the Jacobian matrix,

$$A = \begin{pmatrix} 1 - \bar{\kappa}(x) - \Gamma_1 - \dfrac{x_1^2}{x}\bar{\kappa}'(x) & -\dfrac{x_1 x_2}{x}\bar{\kappa}'(x) \\[2mm] -\dfrac{x_1 x_2}{x}\bar{\kappa}'(x) & 1 - \bar{\kappa}(x) - \Gamma_2 - \dfrac{x_2^2}{x}\bar{\kappa}'(x) \end{pmatrix} \;, \tag{8.50}$$

where the prime denotes differentiation with respect to x. Its determinant is

$$\det A = (1 - \bar{\kappa} - \Gamma_1)(1 - \bar{\kappa} - \Gamma_2) \\ - x\bar{\kappa}'\left(1 - \bar{\kappa} - \Gamma_2\cos^2\varphi - \Gamma_1\sin^2\varphi\right) \;. \tag{8.51}$$

The critical curves, where $\det A = 0$, satisfy the equation

$$\cos^2\varphi = \frac{1 - \bar{\kappa} - \Gamma_1}{\Gamma_1 - \Gamma_2}\left(\frac{1 - \bar{\kappa} - \Gamma_2}{x\bar{\kappa}'} - 1\right) \;. \tag{8.52}$$

Hence, we can easily obtain the critical curves by computing the right-hand side of (8.52) as a function of x. When its value is between 0 and 1, there are critical points at a distance x from the center. The corresponding value of $\cos^2\varphi$ yields in fact four different critical points, one in each quadrant of the lens plane, due to the symmetry of our lens model with respect to both reflections $(x_1, x_2) \mapsto \pm(x_1, -x_2)$, $(y_1, y_2) \mapsto \pm(y_1, -y_2)$.

Structure of the caustics. Consider the intersection of critical curves with the x_2-axis (i.e., $x_1 = 0, \cos^2 \varphi = 0$). From symmetry, we know that at those points the tangent vector at the critical curves is proportional to (1,0). From (8.50), we see that A is diagonal at the points under consideration. There are two kinds of critical points: those where $1 - \bar\kappa - \Gamma_1 = 0$, henceforth called C points (for "cusp"), since at those points the tangent vector of the critical curve is an eigenvector of A belonging to the eigenvalue 0, which means that the corresponding point on the caustic at $\mathbf{y} = (0, x_2\,(\Gamma_1 - \Gamma_2))$ is a cusp (see Sect. 6.2.2). For $\Gamma_1 \to \Gamma_2$, we have $\mathbf{y} \to 0$, so the corresponding critical curve evolves from the tangential critical curve by distortion of the symmetry. The second kind of critical curves on the x_2-axis are those where $1-\bar\kappa-\Gamma_2-x_2\bar\kappa' = 0$; for these, $A_{22} = 0$, and the tangent vector at the critical curve is not mapped onto 0. Therefore, these points do not give rise to cusps and are called F points (for "fold"). An analogous discussion of critical points on the x_1-axis yields the result that C points satisfy $1 - \bar\kappa - \Gamma_2 = 0$ and F points are described by $1 - \bar\kappa - \Gamma_1 - x_1\bar\kappa' = 0$.

From the preceding discussion, we find that the tangential critical curve of a symmetric lens is distorted by symmetry-breaking in such a way that the degenerate caustic point $\mathbf{y} = 0$ breaks up into a curve with four cusps, at least as long as the perturbation is sufficiently small ($\Gamma_1, \Gamma_2 \ll 1$; see also Sect. 13.2.2). For larger perturbations, the equations $1 - \bar\kappa - \Gamma_i = 0$ ($i = 1, 2$) may have no solutions; in such a case, there are no cusps, as we demonstrate in Sect. 8.2.1.

A special solution of the lens equation. Finally, let us consider the images of a source at $\mathbf{y} = 0$. There are three kinds of solutions of the ray-trace equation: (i) the origin, $x_1 = x_2 = 0$; (ii) the points $(\pm x, 0)$, such that $1 - \bar\kappa(x) - \Gamma_1 = 0$, and (iii) the points $(0, \pm x)$, such that $1 - \bar\kappa(x) - \Gamma_2 = 0$. Note that, for $\Gamma_1 \neq \Gamma_2$, none of these points is on a critical curve, and there is no solution with $x_1, x_2 \neq 0$. Thus, for centrally condensed lenses, with $\kappa(0) > 1$ and sufficiently small perturbations, the source at $\mathbf{y} = 0$ has five images.

In the following two subsections, we consider two special cases of quadrupole lenses, a perturbed Plummer model and the so-called Chang–Refsdal lens, which is a point mass plus a quadrupole term. The treatment in both cases is rather detailed, since these two models are amenable to analytical study, and they show all the essential features of a general lens mapping.

8.2.1 The perturbed Plummer model

We consider a quadrupole lens where the axi-symmetric matter distribution is described by

$$\bar\kappa(x) = \frac{\kappa_0}{1+x^2} \, , \tag{8.53}$$

which corresponds to (8.38) with $p = 0$. Insertion of this function into (8.49) and multiplication by $(1+x^2)^4$ yields a fifth-order equation for x^2; hence, this lens produces at most five images. Let us introduce scaled variables in the source plane, $\tilde{y}_i = \frac{y_i}{(1-\Gamma_i)}$, $i = 1, 2$, and define $\beta_i = \frac{\kappa_0}{(1-\Gamma_i)}$; then, the lens equation becomes

$$\tilde{y}_1 = x_1 \left(1 - \frac{\beta_1}{1 + x^2}\right) \quad ; \tag{8.54a}$$

$$\tilde{y}_2 = x_2 \left(1 - \frac{\beta_2}{1 + x^2}\right) \quad . \tag{8.54b}$$

Critical points on the axes. (a) x_1-axis: C points are described by $1 - \bar{\kappa} - \Gamma_2 = 0$, or $x_1^2 = \beta_2 - 1$. Hence, there exists at most one C point on the positive x_1-axis, which for $\beta_2 > 1$ is at $x_1 = \sqrt{\beta_2 - 1}$; for $\beta_2 < 1$, there is no C point on the x_1-axis. The discussion for F points is somewhat more involved: F points satisfy $1 - \bar{\kappa} - \Gamma_1 - x_1\bar{\kappa}' = 0$, or $(1 + x_1^2)^2 = \beta_1(1 - x_1^2)$. This second-order equation in x_1^2 has the solution

$$x_1^2 = -\frac{2 + \beta_1}{2} \pm \sqrt{\frac{(2 + \beta_1)^2}{4} + \beta_1 - 1} \quad .$$

There are three distinct cases. If $\beta_1 > 1$, the two solutions for x_1^2 are of opposite sign. Only the positive solution corresponds to a real F point. For $0 < \beta_1 < 1$, both solutions are negative. For $-8 < \beta_1 < 0$, the discriminant is negative, so there is no real solution for x_1^2. For $\beta_1 < -8$, the two real solutions are both positive and there are two F points. Hence, we must distinguish the cases $\beta_1 < -8$, $-8 < \beta_1 < 1$, and $1 < \beta_1$.

(b) x_2-axis: the discussion is completely analogous to that for the x_1-axis; one just interchanges the indices. Hence, one C point exists for $\beta_1 > 1$ at $x_2 = \sqrt{\beta_1 - 1}$. There are one, zero, or two F points, depending on whether $\beta_2 > 1$, $-8 < \beta_2 < 1$, or $\beta_2 < -8$, respectively.

Classification of caustic topologies and image multiplicities. Assuming without loss of generality that $\beta_1 > \beta_2$ (interchanging β_1 and β_2 corresponds to an interchange of the x_1 and x_2 axes), we have six qualitatively different topologies for the critical curves of a perturbed Plummer model:

Table 8.2. Image positions as a function of source positions. The symbols Ln and Sn denote the region in the lens and source plane, respectively, according to Fig. 8.7. The roman numerals denote the quadrant of the image for a source in the first quadrant; e.g., L3II is an image in the region L3 in the second quadrant

S1:	L3I,	L3II,	L2II,	L2III,	L1III
S2:	L3I,	L3II,	L2II		
S3:	L3I,	L2III,	L1III		
S4:	L3I				
S5:	2L3I,	L3II,	L2II,	L4I	
S6:	L5I				
S7:	L3I,	L5I,	L6I		
S8:	2L3I,	L4I			

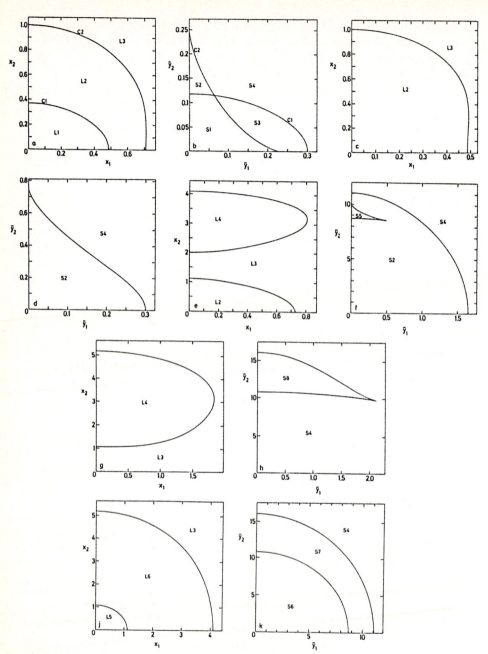

Fig. 8.7a–k. The critical curves and caustics for the perturbed Plummer model. Since the lens is symmetric with respect to reflection at the axes, only one quadrant of each plane is shown. The symbols in the individual regions of both planes are used to identify the image positions, as explained in the text and in Table 8.2. (**a**) $\beta_1 = 2.0$, $\beta_2 = 1.5$, lens plane; (**b**) source plane; (**c**) $\beta_1 = 2.0$, $\beta_2 = 0.5$, lens plane; (**d**) source plane; (**e**) $\beta_1 = 5.0$, $\beta_2 = -20.0$, lens plane; (**f**) source plane; (**g**) $\beta_1 = -2.0$, $\beta_2 = -30.0$, lens plane; (**h**) source plane; (**j**) $\beta_1 = -20.0$, $\beta_2 = -30.0$, lens plane; (**k**) source plane

(1) $\beta_1 > \beta_2 > 1$: there is one C and one F point on either axis. The critical curves are shown in Fig. 8.7a, and the corresponding caustics on the $\tilde{y}$-plane in Fig. 8.7b. Because of the 4-fold symmetry, only the first quadrant is shown in all cases. The critical curves separate different regions in the lens plane denoted by L1, L2, L3; similarly, the caustics lead to four different regions in the source plane. In Table 8.2 we have listed the image regions for sources in one of the areas denoted by S1 to S8. For example, a source (in the first quadrant) in the region S2 will produce three images, one in L3 in the first quadrant (L3I), one in L2 in the second quadrant (L2II), and one in L1 in the second quadrant. If the source moves from S2 to S4 across the critical curve C2, the two images in the second quadrant will fuse and only image L3I is left. The eigenvalues of A on the axes are:

$$x_1\text{-axis:} \quad a_1 = \frac{\kappa_0}{\beta_2}\left(1 - \frac{\beta_2}{1+x^2}\right), \quad a_2 = \frac{\kappa_0}{\beta_1}\left[1 - \beta_1 \frac{1-x^2}{(1+x^2)^2}\right] \quad ; \tag{8.55}$$

$$x_2\text{-axis:} \quad a_1 = \frac{\kappa_0}{\beta_1}\left(1 - \frac{\beta_1}{1+x^2}\right), \quad a_2 = \frac{\kappa_0}{\beta_2}\left[1 - \beta_2 \frac{1-x^2}{(1+x^2)^2}\right] \quad .$$

From these relations, for the case under consideration, L1, L2, L3 correspond to images of type III, II, and I, respectively, in terms of the image classification (5.29).

(2) $\beta_1 > 1 > \beta_2 > -8$: there is one F point on the x_1-axis, and one C point on the x_2-axis. The corresponding caustic has two cusps on the $\tilde{y}_2$-axis (Fig. 8.7c,d). The regions L2 and L3 correspond to images of type II and I (if $\beta_2 > 0$) or type III and II (if $\beta_2 < 0$), respectively. For the image positions, see Table 8.2.

(3) $\beta_1 > 1$, $\beta_2 < -8$: there is one F point on the x_1-axis, and two F and one C points on the x_2-axis (Fig. 8.7e,f). The caustic which encloses S5 has two cusps off the axis. Sources inside S5 have five images (Table 8.2). The regions L2, L3, L4 correspond to, respectively, images of type III, II and I.

(4) $1 > \beta_1 > \beta_2 > -8$: in this case, there is no critical curve. The whole lens plane corresponds to images of type I (for $\beta_1 > \beta_2 > 0$), type II (for $\beta_1 > 0 > \beta_2$), or type III (for $0 > \beta_1 > \beta_2$).

(5) $1 > \beta_1 > -8 > \beta_2$: there are two F points on the x_2 axis (Fig. 8.7g,h); the caustic again has cusps off axis. Regions L3 and L4 correspond to images of type II and I (for $\beta_1 > 0$) or III and II (for $\beta_1 < 0$), respectively.

(6) $-8 > \beta_1 > \beta_2$: there are two F points on either axis (Fig. 8.7j,k). L5, L6 and L3 correspond to images of type III, II and I, respectively.

We must remember that the above analysis is not really complete; e.g., in case (1) the caustic C2 may lie inside C1, so the region S2 would disappear. However, as far as image positions and parities are concerned, the classification is complete.

Note that the perturbed Plummer lens model can produce multiple images, even if $\kappa_0 < 1$; it is clear from Sect. 8.1 that if the perturbation has non-zero surface mass density (i.e., $\Gamma_1 + \Gamma_2 > 0$), the local value of $\kappa = \kappa_0 + (\Gamma_1 + \Gamma_2)/2$ can be increased above 1; but even for a pure shear perturbation ($\Gamma_1 = -\Gamma_2$) multiple images are generated if $\beta_1 = \kappa_0/(1-\Gamma_1) > 1$, i.e., if $\Gamma_1 > 1 - \kappa_0$.

8.2.2 The perturbed Schwarzschild lens ('Chang–Refsdal lens')

We obtain this lens model by perturbation of the gravitational field of a point mass. Such a combination is expected to occur frequently: if a star in a galaxy acts as a lens, the field of the galaxy will perturb that of the star. This is clearly a two length-scale problem, as discussed at the beginning of this section. To distinguish between the different scales, it has become common to denote the lensing by stars in a galaxy as *microlensing* (ML, as before), as opposed to the lensing of the galaxy as a whole, which is sometimes called *macrolensing*. These expressions are motivated by the fact that macrolensing leads to image separations on the order of arcseconds, whereas a microlens produces images separated typically by 10^{-5} arcseconds. For completeness, we should also mention the possibility that globular clusters and molecular clouds in galaxies can lead to lensing. Since the typical masses of these objects are of the order of $10^6 M_\odot$, they could lead to image separations of roughly 10^{-3} arcseconds, an effect that might be called *millilensing*. The lens model considered here is also called Chang–Refsdal lens after its investigators ([CH79.1], [CH84.2]; also [SU85.1]; in [CH84.1], the magnification of extended sources by a Chang–Refsdal lens has been investigated).

The scaled lens equation. For a point mass M, we again choose the length scale ξ_0 according to (2.6b); then, $\bar{\kappa}(x) = 1/x^2$, and the ray-trace equation becomes

$$\mathbf{y} = \mathbf{x} - \frac{\mathbf{x}}{|\mathbf{x}|^2} - \begin{pmatrix} \kappa_c + \gamma & 0 \\ 0 & \kappa_c - \gamma \end{pmatrix} \mathbf{x} \quad , \tag{8.56}$$

where κ_c and γ are the surface mass density and the shear, respectively, of the macrolens at the location of the point mass. Note that the origin in the source plane is chosen such that in the absence of the point mass, there would be an image at $\mathbf{x} = 0$, produced by the macrolens, for a source at $\mathbf{y} = 0$.

As noted before, the quadrupole expansion of the field of the perturber breaks down if the point mass is at a critical curve of the macrolens. Hence, we must assume that $1 - \kappa_c \pm \gamma \neq 0$. Then, we can reduce the lens equation to a one-parameter family of models by the following scaling: define new coordinates $\mathbf{X}$ and $\mathbf{Y}$ in the lens and source plane,

$$\mathbf{X} = \sqrt{|1 - \kappa_c + \gamma|}\, \mathbf{x} \quad ; \quad \mathbf{Y} = \frac{\mathbf{y}}{\sqrt{|1 - \kappa_c + \gamma|}} \quad . \tag{8.57}$$

In terms of these, the lens equation becomes

$$\mathbf{Y} = \epsilon \begin{pmatrix} \Lambda & 0 \\ 0 & 1 \end{pmatrix} \mathbf{X} - \frac{\mathbf{X}}{|\mathbf{X}|^2} \quad , \tag{8.58}$$

where

$$\epsilon = \mathrm{sign}(1 - \kappa_c + \gamma) \tag{8.59a}$$

and

$$\Lambda = \frac{1 - \kappa_c - \gamma}{1 - \kappa_c + \gamma} \quad . \tag{8.59b}$$

Since a change of sign of γ corresponds to an interchange of the axes, one can always choose the sign such that $|\Lambda| \leq 1$. The case $\Lambda = 1$ corresponds to vanishing shear, and will be treated separately at the end of this subsection.

For $|\gamma| \to \infty$ and for $\kappa_c = 1$, $\Lambda \to -1$, and $\Lambda = 0$ is excluded since this would correspond to the case where the point mass is at a critical curve of the macrolens. Thus, we consider the ranges $0 < \Lambda < 1$ and $-1 \leq \Lambda < 0$.

Jacobian and critical curves. Owing to its importance, we discuss the Chang–Refsdal lens in some detail. First, consider the Jacobian matrix

$$\frac{\partial \mathbf{y}}{\partial \mathbf{x}} = A = |1 - \kappa_c + \gamma| \, \tilde{A} = |1 - \kappa_c + \gamma| \frac{\partial \mathbf{Y}}{\partial \mathbf{X}} \quad , \tag{8.60a}$$

where

$$\tilde{A} = \begin{pmatrix} \epsilon \Lambda + \dfrac{X_1^2 - X_2^2}{X^4} & \dfrac{2X_1 X_2}{X^4} \\ \dfrac{2X_1 X_2}{X^4} & \epsilon - \dfrac{X_1^2 - X_2^2}{X^4} \end{pmatrix} . \tag{8.60b}$$

Its determinant is

$$\det \tilde{A} = \frac{1}{X^4} \left[\Lambda \left(X_1^2 + X_2^2 \right)^2 + \epsilon(1 - \Lambda) \left(X_1^2 - X_2^2 \right) - 1 \right] \quad , \tag{8.61}$$

where $X = |\mathbf{X}|$. Hence, the critical curves are Cassini ovals. Before we discuss the critical curves and the corresponding caustics, we consider the critical points on the axis. For $X_2 = 0$, the condition $\det \tilde{A} = 0$ leads to a quadratic equation in X_1^2 with the formal solutions

$$X_1^2 = \frac{1}{2\Lambda} \left[\epsilon(\Lambda - 1) \pm (\Lambda + 1) \right] \quad ; \tag{8.62a}$$

similarly, critical points on the X_2-axis are described by

$$X_2^2 = \frac{1}{2\Lambda} \left[\epsilon(1 - \Lambda) \pm (\Lambda + 1) \right] \quad . \tag{8.62b}$$

The corresponding points of the caustic are

$$Y_1^2 = \frac{\epsilon \pm 1}{2} (\Lambda + 1)^2 - 4\epsilon \Lambda \quad , \tag{8.62c}$$

$$Y_2^2 = \frac{\epsilon \pm 1}{2} \frac{(\Lambda + 1)^2}{\Lambda} - 4\epsilon \quad . \tag{8.62d}$$

Since these equations describe critical points and caustic points only for positive values of the right-hand sides, one finds these points by considering separately the cases $\Lambda > 0$, $\Lambda < 0$, and $\epsilon = \pm 1$; the results are given in Table 8.3.

A critical point on the X_1-axis (X_2-axis) is a C point if $\tilde{A}_{22} = 0$ ($\tilde{A}_{11} = 0$). One can then decide with (8.60b) which of the points described by (8.62) are C points. This information is also provided in Table 8.3. The critical curves and the corresponding caustics are plotted in Fig. 8.8 for the four cases $\Lambda > 0$, $\Lambda < 0$, $\epsilon = \pm 1$ separately. Before discussing them, we turn to the solution of the ray-trace equation.

Table 8.3. Critical points on the axes for the Chang–Refsdal lens

	X_1^2	Y_1^2	X_2^2	Y_2^2
$0 < \Lambda < 1$				
$\epsilon = +1$	1	$(1-\Lambda)^2$ [C]	$1/\Lambda$	$(1-\Lambda)^2/\Lambda$ [C]
$\epsilon = -1$	$1/\Lambda$	4Λ [F]	1	4 [F]
$-1 < \Lambda < 0$				
$\epsilon = +1$	1	$(1-\Lambda)^2$ [C]	—	—
	$-1/\Lambda$	-4Λ [F]	—	—
$\epsilon = -1$	—	—	1	4 [F]
	—	—	$-1/\Lambda$	$-(1-\Lambda)^2/\Lambda$ [C]

Solution of the lens equation. Let us introduce polar coordinates in the lens plane, $X_1 = X \cos\varphi$, $X_2 = X \sin\varphi$; the ray-trace equation becomes

$$Y_1 = X \cos\varphi \, (\epsilon\Lambda - 1/X^2) \quad,$$
$$Y_2 = X \sin\varphi \, (\epsilon - 1/X^2) \quad;$$

solving for $\cos\varphi$, $\sin\varphi$,

$$\cos\varphi = \frac{Y_1}{X(\epsilon\Lambda - 1/X^2)} \quad ; \quad \sin\varphi = \frac{Y_2}{X(\epsilon - 1/X^2)} \quad . \tag{8.63a}$$

Adding the squares of the last two equations yields a fourth-order equation for X^2:

$$\begin{aligned}
\Lambda^2 X^8 &- \left[2\epsilon\Lambda(\Lambda+1) + Y_1^2 + \Lambda^2 Y_2^2\right] X^6 \\
&+ \left[\Lambda^2 + 4\Lambda + 1 + 2\epsilon\left(Y_1^2 + \Lambda Y_2^2\right)\right] X^4 \\
&- \left[2\epsilon(\Lambda+1) + Y_1^2 + Y_2^2\right] X^2 + 1 = 0 \quad,
\end{aligned} \tag{8.63b}$$

which can be solved by standard methods. For any positive solution X^2, we can take the positive value of X and obtain the image position from (8.63a). Since (8.63b) is of fourth order in X^2, there are either zero, two, or four solutions of the lens equation.

Images for a source on one of the axes.
(1) Let $Y_2 = 0$; then we have the following possibilities:
 (a) $X_2 = 0 = \sin\varphi$ implies $Y_1 = \epsilon\Lambda X_1 - 1/X_1$, which has the solutions

$$X_1 = \frac{1}{2\epsilon\Lambda}\left[Y_1 \pm \sqrt{Y_1^2 + 4\epsilon\Lambda}\right], \quad \text{for } Y_1^2 > -4\epsilon\Lambda \quad. \tag{8.64a}$$

 (b) $X_2 \neq 0$ requires $X^2 = \epsilon$ and leads to $Y_1 = \epsilon(\Lambda - 1)X_1$; thus,

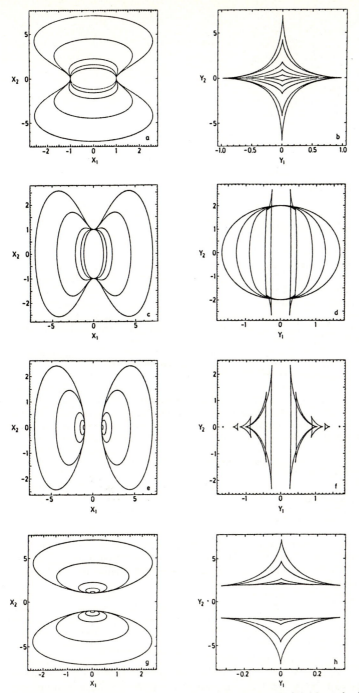

Fig. 8.8. The critical curves and caustics of the Chang–Refsdal lens, for $|\Lambda| = 0.02, 0.05, 0.2, 0.4, 0.7$; (**a**) $\Lambda > 0$, $\epsilon = +1$, lens plane; (**b**) source plane; (**c**) $\Lambda > 0$, $\epsilon = -1$, lens plane; (**d**) source plane; (**e**) $\Lambda < 0$, $\epsilon = +1$, lens plane; (**f**) source plane; (**g**) $\Lambda < 0$, $\epsilon = -1$, lens plane; (**h**) source plane

$$X_1 = \frac{1}{\Lambda - 1} Y_1 \;,$$

$$X_2 = \pm \sqrt{1 - \frac{Y_1^2}{(\Lambda - 1)^2}} \;.$$
for $\epsilon = +1$, $Y_1^2 < (1 - \Lambda)^2$. (8.64b)

(2) Let $Y_1 = 0$; then
 (a) $X_1 = 0 = \cos\varphi$ leads to the solutions

$$X_2 = \frac{\epsilon}{2} \left[Y_2 \pm \sqrt{Y_2^2 + 4\epsilon} \right], \quad \text{for } Y_2^2 > -4\epsilon \;. \tag{8.64c}$$

 (b) $X_1 \neq 0$ implies

$$X_2 = \epsilon \frac{Y_2}{1 - \Lambda} \;,$$

$$X_1 = \pm \sqrt{\frac{\epsilon}{\Lambda} - \frac{Y_2^2}{(1 - \Lambda)^2}} \;.$$
for $\epsilon\Lambda > 0$, $Y_2^2 < \frac{(1 - \Lambda)^2}{\epsilon\Lambda}$. (8.64d)

Discussion of the caustics (Fig. 8.8) and image multiplicities.
(1) $\Lambda > 0$, $\epsilon = +1$: The caustic is a single closed curve with four cusps on the axis, which lie at $Y_{1c} = \pm(1 - \Lambda)$ and $Y_{2c} = \pm(1 - \Lambda)/\sqrt{\Lambda}$. For sources with $Y_2 = 0$, the images (8.64a) always exist, whereas the images (8.64b) are present only for $Y_1^2 < Y_{1c}^2$. Correspondingly, for $Y_1 = 0$ we always have the two images (8.64c), but the images (8.64d) exist only for $Y_2^2 < Y_{2c}^2$. Hence, a source within the diamond-shaped caustic produces four images, whereas sources outside it have two.
(2) $\Lambda > 0$, $\epsilon = -1$: The caustic is a single closed curve which is self-intersecting for sufficiently small Λ; in these cases, the caustic has eight cusps, whereas for larger values of Λ it has none. These self-intersection points of the caustic are swallowtails, as can be easily seen from their shape (see Sect.6.3.2). Sources with $Y_2 = 0$ have images (8.64a) only for $Y_1^2 > Y_{1c}^2 = 4\Lambda$, and sources with $Y_1 = 0$ have images (8.64c) only for $Y_2^2 > Y_{2c}^2 = 4$. The images (8.64b,d) cannot occur. Hence, sources in the inner region of the caustic have no images, those outside have two, and for small Λ, sources in the triangle-shaped regions near the self-intersection points have four images.
(3) $\Lambda < 0$, $\epsilon = +1$: There are two caustics, each with one cusp on the Y_1 axis and two cusps away from the axes. Increasing $|\Lambda|$ leads to a shrinking of the caustics, which also become more separated. For $Y_2 = 0$, the two images (8.64b) exist for $Y_1^2 < (1 - \Lambda)^2$, whereas the images (8.64a) are present for $Y_1^2 > 4|\Lambda|$. For $Y_1 = 0$, the images (8.64c) are present for all Y_2. Thus, sources inside the caustic curves have four, those outside have two images.
(4) $\Lambda < 0$, $\epsilon = -1$: There are two caustics both with one cusp on the Y_2 axis and two cusps off-axis. Increasing $|\Lambda|$ leads to a shrinking of the caustics. For $Y_2 = 0$, there are two images (8.64a); for $Y_1 = 0$, the images (8.64c) are present for $Y_2^2 > 4$, and images (8.64d) are present for $Y_2^2 < (1 - \Lambda)^2/|\Lambda|$. Hence, sources inside the caustic have four, those outside have two images.

Our discussion not only gives a complete and detailed classification for the Chang–Refsdal model, it also illustrates how one can obtain the basic properties of a lens (e.g., positions of critical curves, caustics, image multiplicities). It turned out to be much simpler to derive the critical structure of the lens than to solve the lens equation. This is true for all of the models considered in this chapter.

The case $\gamma = 0$. Finally, we consider the case $\Lambda = 1$, where the lens geometry becomes axially symmetric, i.e., we can use the one-dimensional lens equation

$$Y = \epsilon X - \frac{1}{X} \;. \tag{8.65a}$$

The solutions of this equation are

$$X = \frac{\epsilon Y}{2} \pm \sqrt{\frac{Y^2}{4} + \epsilon} \quad ; \qquad \text{for } Y^2 > -4\epsilon \quad , \tag{8.65b}$$

hence, for $\epsilon = +1$ there are always two solutions, whereas for $\epsilon = -1$, there are two solutions for $|Y| > 2$, and none for $|Y| < 2$. There is a tangential critical curve for $\epsilon = +1$ at $X = 1$ which is mapped onto the degenerate caustic at $Y = 0$, and a radial critical curve at $X = 1$ for $\epsilon = -1$, where the corresponding caustic has radius $Y = 2$.

Vanishing of images. We can easily understand why, for $\epsilon = -1$ and $\Lambda > 0$, there are no images for certain source positions, if we consider the Fermat potential $\phi(\mathbf{X}, \mathbf{Y})$ introduced in Chap. 5. The case $\epsilon = -1$, $\Lambda > 0$ corresponds to a maximum of the macrolens at the location of the point mass. The point mass adds a (positive) logarithmic spike to ϕ. If this spike is too close to that original maximum, it will transform the maximum into a spike without any stationary point left over. If, however, the source is sufficiently misaligned with the lens, the spike will be at a separate position, leaving the maximum and an additional saddle point.

8.3 The two point-mass lens

Motivation. A natural generalization of the Schwarzschild lens is a lens consisting of two point masses. Although such a model cannot be used to describe any observed lens case, it is of more than academic interest, as it helps the understanding of the caustic structure obtained from non-linear ML simulations (see Sect. 11.2.5). Furthermore, this lens model is sufficiently simple that one can derive many properties analytically. In particular, the caustic structure can be determined (nearly) analytically. We briefly describe the shape of the critical curves and caustics for the case that both masses are equal, and use the model to illustrate the imaging of extended sources. Then, we present the results of a study which generalizes this lens model to unequal masses and the inclusion of external shear. Finally, the generalization to N point-mass lenses is considered briefly.

8.3.1 Two equal point masses

Critical structure. For simplicity, consider two point masses of mass $M/2$ each [SC86.4], and use the length scale (2.6b) to normalize the lens equation. Let 2χ denote the (dimensionless) separation of the point masses; we can then classify the possible topologies of the sets of critical curves, according to whether $\chi^2 < 1/8$, $1/8 < \chi^2 < 1$, or $\chi^2 > 1$ (see Fig. 8.9).[5] For $\chi^2 < 1/8$,

---

[5] We are greateful to H. Erdl for preparing this figure for us.

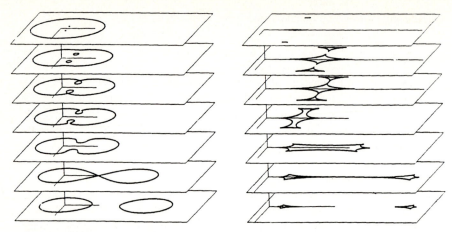

Fig. 8.9. Critical curves (**a**) and caustics (**b**) for the two point-mass lens with equal masses, for seven values of the separation between the two lenses. Each critical curve and caustic is drawn on a plane, for better visualization, and the separation χ of the two lenses increases towards the bottom plane. On the third plane, the separation is $\chi = 1/\sqrt{8}$, and $\chi = 1$ on the sixth plane. The occurrence of metamorphoses at these two parameter values is clearly seen. The number of images is five if the source is inside one of the caustics, and three otherwise

a closed critical curve separates the outer region (of positive parity) from the inner region; for $\chi^2 \to 0$, this critical curve tends toward the critical circle of a Schwarzschild lens of mass M, i.e., $x = 1$. In addition, there are two small critical curves in the inner region, within which the parity is again positive.[6] Thus, there are three closed caustics, one shaped like a diamond, and two almost like triangles. A point source inside one of these caustics has five images, otherwise three. For $1/8 < \chi^2 < 1$, we find a single critical curve enclosing the two masses, and the corresponding caustic has six cusps. The number of images is either three or five, depending on whether the source is outside or inside the caustic. For $\chi^2 > 1$, we obtain two critical curves, each enclosing one mass; in the limit $\chi^2 \to \infty$, they tend to Schwarzschild critical curves with radius $2^{-1/2}$, and the corresponding caustics become points.

The transitions between the three different stable caustic topologies, at $\chi^2 = 1/8$ and $\chi = 1$, occur such that two cusps approach each other, fuse, and disappear. Hence, these transitions are beak-to-beak metamorphoses (see Sect. 6.3.3), the only higher-order catastrophes occurring in this family of lens models.

---

[6] The reason why these two regions of positive parity have to occur can be easily understood: since for point-mass lenses, $\kappa = 0$ everywhere, the only effect on small light bundles is shear. If the shear caused by the two individual stars is equal in magnitude and acting mutually perpendicular on a light bundle, the resulting shear vanishes, so that at such a point, $A = \mathcal{I}$. For the two point-mass lens (with arbitrary mass ratio), two such points exist, and for the present case, they are enclosed by the two critical curves occurring for $\chi^2 < 1/8$.

Imaging of extended sources. In Fig. 8.10, we display the images of an extended source with circular isophotes. Depending on the position of the source relative to the caustic, a variety of image shapes are possible. If the source lies completely inside the caustic (Fig. 8.10a), five separate images are produced, one very close to the line connecting the two masses. The

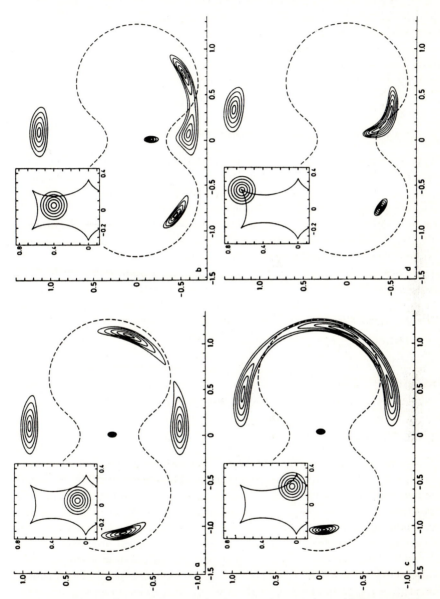

Fig. 8.10. Imaging of extended sources for the two point-mass lens with lens separation $2\chi = 1.0$. The insert shows the isophotes of a circular source, together with part of the caustic. The dashed line is the critical curve. Depending on the source position, the images can have vastly different shapes

two images close to the critical curve are highly elongated and point towards each other. If the source lies on a caustic (Fig. 8.10b), some of the inner isophotes still have five separate images, but those which cross the caustic have fewer. This leads to the formation of images with internal structure. When the source lies close to a cusp, more complicated images are possible, as illustrated in Fig. 8.10c; here, a very elongated curved image with much internal structure is seen. We return to such a case in Chap. 13. In fact, extended source close to, and inside of cusps always lead to highly elongated images; this is the 'canonical' explanation for the formation of luminous arcs – see Sect. 2.5.5. If the source center is on a cusp (Fig. 8.10d), it results in only a single, highly magnified and deformed image close to the critical curve.

8.3.2 Two point masses with arbitrary mass ratio

This lens model has been studied in great detail in [ER90.1]. After considerable algebra, it can be shown that the topology of critical curves and caustics is the same for this case as in the case of equal masses. In particular, only three different caustic topologies occur, and the transition between them is again through a beak-to-beak singularity. Furthermore, the discussion on the image multiplicity is unchanged compared to the case of equal masses: either three or five images can occur.

Consider two point masses with mass $\mu_1 M$ and $\mu_2 M$, such that $\mu_1 + \mu_2 = 1$, and separation $d = 2\chi$, again in units of ξ_0 (2.6b). The generalization of the two bifurcation values $\chi = 1/\sqrt{8}$ and $\chi = 1$ in the case of equal masses ($\mu_1 = \mu_2 = 0.5$) is given by the equations

$$\mu_1 \mu_2 = \frac{(1-d^4)^3}{27\,d^8} \quad , \quad d^2 = \left(\mu_1^{1/3} + \mu_2^{1/3}\right)^3 \quad ,$$

respectively. These two equations are symmetric in μ_1 and μ_2, and yield the correct expressions in the special case $\mu_1 = \mu_2 = 0.5$.

8.3.3 Two point masses with external shear

If an external shear is added [GR89.2], the two point-mass lens becomes much more difficult to handle analytically. The algebraic equation determining the critical curves becomes more complicated, and the corresponding caustics can have points of self-intersection. This latter fact makes the determination of the caustic topologies fairly difficult, since the knowledge of the critical curves does not directly yield the topological properties of the caustics. For instance, it occurs that the critical curve is a single, non-intersecting closed curve, whereas the caustic intersects itself several times, creating a region in the source plane where up to nine images of a source can occur (see Fig. 6, model 11, in [GR89.2]).

Including external shear also leads to the occurrence of higher-order catastrophes. In addition to the beak-to-beak singularity which is also present without shear, swallowtails and butterflies appear for particular values of the parameters of this GL model. Without external shear, the mass ratio of the two lenses has no impact on the topology of caustics (but it does, of course, on their size and their detailed shape); whether this remains true with shear is unclear. A further, semi-analytic study of this lens model can yield considerable insight into the structure and evolution of caustics, as a function of lens parameters.

8.3.4 Generalization to N point masses

Since the two point-mass lens is already fairly complicated to analyze in detail, there is less hope to be able to study the N point-mass lens analytically. However, some properties of this GL model can be derived [WI90.1]; in particular, it is possible to obtain a fairly simple equation for the critical curves, which can readily be applied numerically.

Let M be a reference mass, and consider N point masses with mass $m_i M$ and position $\mathbf{x}_i$, measured in units of ξ_0 (2.6b). The lens equation then becomes

$$\mathbf{y} = \begin{pmatrix} 1 - \kappa_c + \gamma & 0 \\ 0 & 1 - \kappa_c - \gamma \end{pmatrix} \mathbf{x} - \sum_{i=1}^{N} \frac{m_i}{|\mathbf{x} - \mathbf{x}_i|^2} (\mathbf{x} - \mathbf{x}_i) \quad,$$

where the coordinate frame has been oriented such that the shear γ acts along one of the coordinate axes, and κ_c is the smooth surface mass density in units of Σ_{cr} (5.5).

We next write the lens equation in complex form, by defining the complex numbers $X = x_1 + ix_2$, $Y = y_1 + iy_2$,

$$Y = (1 - \kappa_c) X + \gamma X^* - \sum_{i=1}^{N} \frac{m_i}{X^* - X_i^*} \quad,$$

and the asterisk denotes complex conjugation. Formally, we can consider Y as a function of X and X^*, and write:

$$\frac{\partial Y}{\partial x_i} = \frac{\partial X}{\partial x_i} \frac{\partial Y}{\partial X} + \frac{\partial X^*}{\partial x_i} \frac{\partial Y}{\partial X^*} \quad,$$

so that

$$\frac{\partial Y}{\partial x_1} = \frac{\partial Y}{\partial X} + \frac{\partial Y}{\partial X^*} \quad, \qquad \frac{\partial Y}{\partial x_2} = i \left(\frac{\partial Y}{\partial X} - \frac{\partial Y}{\partial X^*} \right) \quad.$$

From the lens equation, we find $\frac{\partial Y}{\partial X} = (1 - \kappa_c)$, so that the preceding equations can be written as

$$\frac{\partial Y}{\partial X^*} = \frac{\partial y_1}{\partial x_1} + i \frac{\partial y_2}{\partial x_1} - (1 - \kappa_c), \qquad \left(\frac{\partial Y}{\partial X^*}\right)^* = (1 - \kappa_c) - \frac{\partial y_2}{\partial x_2} - i \frac{\partial y_1}{\partial x_2} \quad.$$

Multiplying these two equations, and using $\frac{\partial y_1}{\partial x_2} = \frac{\partial y_2}{\partial x_1}$, and $\frac{\partial y_1}{\partial x_1} + \frac{\partial y_2}{\partial x_2} = 2(1 - \kappa_c)$, we obtain

$$\det A = \left(\frac{\partial Y}{\partial X}\right)^2 - \frac{\partial Y}{\partial X^*}\left(\frac{\partial Y}{\partial X^*}\right)^* .$$

The critical curves are determined by setting $\det A = 0$:

$$\left(\frac{\partial Y}{\partial X}\right)^2 = \frac{\partial Y}{\partial X^*}\left(\frac{\partial Y}{\partial X^*}\right)^* ,$$

or

$$\frac{\partial Y}{\partial X^*} = \frac{\partial Y}{\partial X} e^{i\phi} .$$

For each value of the phase ϕ, this equation determines the critical points. Applying this equation to the original lens equation, we obtain (after complex conjugation):

$$\sum_{i=1}^{N} \frac{m_i}{(X - X_i)^2} = (1 - \kappa_c) e^{-i\phi} - \gamma ,$$

which, after clearing fractions, becomes a complex polynomial equation of degree $2N$, for every phase ϕ. Each of the $2N$ solutions determines a critical point, and by varying ϕ through the interval $[0, 2\pi]$, the critical curves are obtained. Since the roots of this polynomial depend smoothly on ϕ, they need to be searched with a root solver only for one value of ϕ; the solutions for other values of ϕ can then be found by a Newton iterative scheme. In [WI90.1], this method was applied to determine the critical curves (and the corresponding caustics) for up to $N = 600$ lenses.

Furthermore, one can show that the maximum number of images for a source lensed by N point masses is $N^2 + 1$, if there is no shear ($\gamma = 0$), and $(N+1)^2$ for cases including shear [WI90.1].

In Sect. 11.2, we consider a star field with an infinite number of point-mass lenses of given density, including external shear and convergence due to a smooth mass density. For such a 'random star field', several useful statistical properties can be derived.

8.4 Lenses with elliptical symmetry

The motivation to study the lens action of elliptical lenses is provided by the fact that many galaxies appear to have elliptical isophotes. Hence, an elliptical lens may be a realistic model for observed GL systems, although for at least some of the lens candidates such a model is still too simple. Deflector models with elliptical surface mass density have been studied in detail ([BO73.1], [BO75.1], [FA83.1], [BR84.1], [SC90.4]); as will be shown in Sect. 8.4.1 these models are fairly difficult to analyze. However, since the light distribution of real galaxies is not purely elliptical (e.g., the ellipticity of the isophotes of "elliptical" galaxies varies, or the orientation of the axes of the isophotes changes, for increasing distance from the center) it is unclear at the moment whether the description of lensing galaxies by elliptical matter distributions is realistic. In addition, there is, of course, no guarantee that the light distribution of an observed galaxy traces the matter distribution. Since we will see in Sect. 8.4.1 that the lens equation for elliptical matter distributions is rather complicated and can only be solved numerically, one might be tempted, in view of the uncertainties just mentioned, to use a simpler description for lensing galaxies, one for which analytical results can be obtained. In fact, such an approach has been followed by using potential wells where the lines of constant $\psi(\mathbf{x})$ are ellipses ([KO87.1], [BL87.2], [KO87.2]). However, as we will see in Sect. 8.4.2, the corresponding isodensity curves have unphysical shapes, which do not resemble the light distribution of observed galaxies. Therefore, it is controversial at the moment whether such lens models, with elliptical isopotential curves, could be used for modeling in gravitational lensing studies. In any case, owing to their relative simplicity, we will treat them in Sect. 8.4.2. In Sect. 8.4.3 we briefly describe a method how one can deal with (nearly) elliptical matter distributions in 'practical work'.

8.4.1 Elliptical isodensity curves

Let us define the coordinates ρ and φ in the lens plane by

$$x_1 = \rho \cos\varphi \quad , \qquad x_2 = \rho \cos\beta \, \sin\varphi \quad ,$$

where $\sin\beta$ defines the ellipticity of the isodensity contours. Then, $\kappa = \kappa(\rho)$. As noted in Sect. 5.1, this symmetry has been treated by using a complex formulation, for which the ray-trace equation is given by (5.14). To compute the complex scattering function $I_c(x_c)$, we let $dx'_c = \rho' \cos\beta \, d\rho' d\varphi'$. If the impact point of a ray is described by (ρ, φ), we obtain from (5.14b)

$$I_c(x_c) = \frac{\cos\beta}{\pi} \int_0^\infty \rho' d\rho' \kappa(\rho') \int_0^{2\pi} \frac{d\varphi'}{x_c - \rho' \cos\varphi' - i\rho' \cos\beta \sin\varphi'} \quad .$$

The φ'-integral can be evaluated with the substitution $\omega = \tan(\varphi'/2)$ and employing complex integration theory ([BO75.1], [BR84.1]). It turns out that, for $\rho' > \rho$, the inner integral vanishes; one finds

$$I_c(x_c) = \text{sign}(x_c) \, 2\cos\beta \int_0^\rho \frac{\rho' d\rho' \kappa(\rho')}{\sqrt{x_c^2 - \rho'^2 \sin^2\beta}} \quad , \tag{8.66a}$$

where

$$\text{sign}(x_c) = \frac{\sqrt{x_c^2}}{x_c} \quad . \tag{8.66b}$$

Although it is possible to evaluate the ρ' integral in (8.66a) for simple forms of κ, the problem is relatively involved and an analysis of the lens equation can only be done numerically [BO75.1].

It turns out that separating the deflection angle into real and imaginary parts is extremely difficult; in fact, the components of the deflection angle (8.66) have not been published as far as we know. Recently, however, an alternative derivation of the deflection angle for elliptical lenses was published [SC90.4], which uses results from potential theory of gravitating triaxial ellipsoids [CH87.2]. The derivation is a beautiful application of geometrical arguments combined with analytic results from the theory of ellipses, but it is rather technical and will not be reproduced here. The interested reader is referred to [SC90.4]; we merely quote the result:

$$\alpha_1(x_1,x_2) = 2\cos\beta \int_0^{\rho(x_1,x_2)} \rho'\,d\rho'\,\kappa(\rho')\,\frac{p^2(\rho')}{a(\rho')\,b(\rho')}\,\frac{x_1}{a^2(\rho')} \quad , \tag{8.67a}$$

$$\alpha_2(x_1,x_2) = 2\cos\beta \int_0^{\rho(x_1,x_2)} \rho'\,d\rho'\,\kappa(\rho')\,\frac{p^2(\rho')}{a(\rho')\,b(\rho')}\,\frac{x_2}{b^2(\rho')} \quad , \tag{8.67b}$$

where

$$\rho(x_1,x_2) = \sqrt{x_1^2 + \frac{x_2^2}{\cos^2\beta}} \quad , \tag{8.67c}$$

$$a(\rho') = \sqrt{\rho'^2 + \lambda(\rho')} \quad ; \quad b(\rho') = \sqrt{\rho'^2\cos^2\beta + \lambda(\rho')} \quad , \tag{8.67d}$$

and $\lambda(\rho')$ and $p(\rho')$ are determined from

$$\frac{x_1^2}{\rho'^2 + \lambda} + \frac{x_2^2}{\rho'^2\cos^2\beta + \lambda} = 1 \quad ; \quad \frac{1}{p^2} = \frac{x_1^2}{a^4} + \frac{x_2^2}{b^4} \quad . \tag{8.67e}$$

These equations can be used directly for numerical work.

Since lenses with elliptical mass distributions are hardly used, except for (numerical) model fitting, we do not analyze them in more detail. Their overall properties, e.g., the shape of critical curves and caustics, do agree qualitatively with those of the models we discuss next.

8.4.2 Elliptical isopotentials

Lens equation and reduction to one-dimensional form. Consider the case where the deflection potential ψ (5.9) depends only on the variable u, where

$$u = (1-\epsilon)x_1^2 + (1+\epsilon)x_2^2 \tag{8.68a}$$

is constant on ellipses, and ϵ is the ellipticity of the isopotentials. Denoting differentiation with respect to u by a prime, we find from (5.8) for the deflection angle

$$\boldsymbol{\alpha} = 2\psi' \begin{pmatrix} (1-\epsilon)x_1 \\ (1+\epsilon)x_2 \end{pmatrix} , \qquad (8.68b)$$

and the lens equation is

$$y_1 = x_1 - 2(1-\epsilon)\,\psi'(u)\,x_1 \ , \qquad (8.69a)$$
$$y_2 = x_2 - 2(1+\epsilon)\,\psi'(u)\,x_2 \ . \qquad (8.69b)$$

Since the standard problem in gravitational lensing is the determination of all image positions **x** for a given source position **y**, it is important to note that this system of equations can be reduced to a single one-dimensional equation. Eliminating ψ' from the system of equations (8.69), we obtain

$$x_2 = \frac{y_2 x_1 (1-\epsilon)}{(1+\epsilon)y_1 - 2\epsilon x_1} \ . \qquad (8.70)$$

Substituting (8.70) in (8.68a) and inserting the result into (8.69a) yields an equation which depends only on x_1. If a root x_1 of this equation is found, we compute the corresponding value of x_2 using (8.70).

Jacobian and critical curves. Next, we consider the Jacobian matrix $A = \partial \mathbf{y}/\partial \mathbf{x}$, whose elements can be derived by differentiation of (8.69):

$$\begin{aligned} A_{11} &= 1 - 2(1-\epsilon)\psi' - 4(1-\epsilon)^2 x_1^2\, \psi'' \ , \\ A_{12} &= A_{21} = -4\left(1-\epsilon^2\right) x_1 x_2\, \psi'' \ , \\ A_{22} &= 1 - 2(1+\epsilon)\psi' - 4(1+\epsilon)^2 x_2^2 \psi'' \ . \end{aligned} \qquad (8.71)$$

We can then compute the determinant:

$$\det A = 1 - 4\psi' + 4\left(1-\epsilon^2\right)\psi'^2 + 8\left(1-\epsilon^2\right)u\psi'\psi'' - 4(u+\epsilon v)\psi'' \ , (8.72)$$

where

$$v = (1+\epsilon)x_2^2 - (1-\epsilon)x_1^2 \ . \qquad (8.73)$$

Thus, $\det A$ has been expressed with the two variables u, v which satisfy $|v| \leq u$. Owing to the symmetry of the lensing geometry, there are four points (x_1, x_2) for any pair (u,v), i.e.,

$$x_1 = \pm\sqrt{\frac{u+v}{2(1+\epsilon)}} \quad ; \quad x_2 = \pm\sqrt{\frac{u-v}{2(1-\epsilon)}} \ .$$

One can thus easily obtain the critical curves of these GL models by solving the equation $\det A = 0$ for v as a function of u. Those solutions which satisfy $|v| \leq u$ correspond to a critical point.

Surface mass density. The surface mass density κ can be obtained from the trace of A (5.26):

$$\kappa = 2\psi' + 2(u+\epsilon v)\psi'' \ . \qquad (8.74)$$

In particular, we see that, for $\epsilon = 0$, det A and κ depend only on u, in which case u is the square of the distance of the point (x_1, x_2) from the origin. Note that κ is *not* constant on ellipses.

Example: A family of models. Let us now consider a family of lens models which generalizes the axi-symmetric models of Sect. 8.1.5. Consider the potential

$$\psi(u) = \frac{\kappa_0}{2p} \left[(1+u)^p - 1 \right] \quad , \tag{8.75}$$

which, for $\epsilon = 0$, reduces to (8.39). The surface mass density for these models is

$$\kappa = \kappa_0 (1+u)^{p-2} \left[1 + pu - (1-p)\epsilon v \right] \quad ; \tag{8.76}$$

hence, $\kappa_0 = \kappa(0)$ is the surface mass density at the origin. The values of κ can become negative for some models, for

$$v > \frac{1+pu}{\epsilon(1-p)} \quad .$$

Since from the definition of u and v, $|v| \leq u$, we can calculate the smallest value of u where the above inequality can be satisfied: $u_{\min} = \left[\epsilon(1-p) - p \right]^{-1}$. Since $u_{\min}$ must be positive, negative values for κ are attained for models such that $\epsilon > \epsilon_{\text{cr}} = p/(1-p)$. The line $\kappa = 0$ in these models intersects the x_2-axis at $x_2 = \sqrt{u_{\min}/(1+\epsilon)}$ and behaves for large distances from the origin as

$$x_2 = \sqrt{\frac{(1-\epsilon)\left[\epsilon(1-p) + p\right]}{(1+\epsilon)\left[\epsilon(1-p) - p\right]}} x_1 \quad .$$

In Fig. 8.11, we plot the lines of constant κ for an extreme model with $p = 0$ and $\epsilon = 0.4$, to illustrate that the use of an elliptical potential can in some cases correspond to unusual matter distributions. Note that, for $\kappa \gtrsim \kappa_0/5$, the isodensity contours are nearly elliptical, whereas for smaller values of κ, they behave very differently. The range within which the isodensities are elliptical increases with decreasing ϵ and increasing p; κ attains negative values above the line $\kappa = 0$. There is a minimum for κ on the x_2-axis. It has been argued that these elliptical isopotentials approximate elliptical matter distributions in their lensing behavior, if one is only interested in the lens mapping for small distances from the lens center. This, however, neglects the fact that the deflection due to matter shells outside the point where the deflection is calculated does *not* average out to zero, as happens to be the case with elliptical (and spherical) matter distributions. Therefore, the inner part of these elliptical isopotentials cannot be considered as an approximation to the corresponding elliptical mass densities. We suspect, however, that the differences between these GL models are not larger than the uncertainties with which the parameters of physical lenses are known

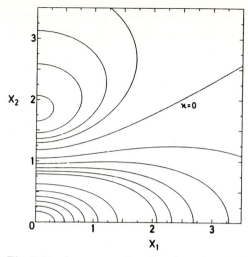

Fig. 8.11. A contour plot of κ for a lens with elliptical isopotentials of the form given by (8.75), with $p = 0$, $\kappa_0 = 5$ and $\epsilon = 0.4$. The contours are for $\kappa = 3.5, 3.0, 2.5, 2.0, 1.5, 1.0, 0.5, 0.4, 0.3, 0.2, 0.1, 0.0, -0.05\ -0.07, -0.09, -0.11$, respectively

(or can be determined). The qualitative lensing behaviors of both types of elliptical lenses are quite similar.

Although the lens models described by (8.75) have been used for statistical considerations ([BL87.2], [KO87.2]) they have not been analyzed in detail as an individual lens. Since (8.75) is a three-parameter model (κ_0, p, ϵ) we can expect an interesting variety of critical structures.

8.4.3 A practical approach to (nearly) elliptical lenses

In the last two subsections, we have seen that, on the one hand, truly elliptical matter distributions lead to fairly complicated lens equations; in particular, the deflection angle at each point must be calculated by means of two one-dimensional integrals over messy expressions. On the other hand, elliptical potentials are fairly easily handled, but describe mass distributions which are not realistic for large distances from the center of the lens. A third way to deal with elliptical lenses will be described here; we only present the general ideas and refer the interested reader to the original paper [SC91.1]. The starting point is that the matter distribution of real galaxies is not known to be really elliptical, as noted in the introduction to this section. On the other hand, we do not expect the (projected) mass density to have dumbbell shaped contours. Therefore, the strategy of the method presented here is the construction of lens models which are approximately elliptical (in fact, they can resemble a truly elliptical mass distribution as closely as desired), and for which the deflection angle can be easily calculated (numerically).

Multipole expansion. Consider an elliptical mass distribution, as was done in Sect. 8.4.1, described by the surface mass density $\kappa(\rho)$. According to (5.13), the deflection potential $\psi(\mathbf{x})$ satisfies a Poisson equation. Let r and φ denote polar coordinates in the lens plane; due to the symmetry of the mass distribution, we can expand

$$2\kappa(r,\varphi) = \sum_{n=0}^{\infty} \kappa_n(r) \cos(2n\varphi) \quad , \tag{8.77a}$$

and, accordingly,

$$\psi(r,\varphi) = \sum_{n=0}^{\infty} \psi_n(r) \cos(2n\varphi) \quad . \tag{8.77b}$$

Equation (5.13) then becomes, for $n \geq 0$,

$$\frac{d^2\psi_n}{dr^2} + \frac{1}{r}\frac{d\psi_n}{dr} - \frac{4n^2}{r^2}\psi_n = \kappa_n(r) \quad .$$

Since the solutions of the homogeneous version of the foregoing equation are easily found [$\psi_n(r) = r^{\pm 2n}$ for $n \geq 1$, and $\psi_0(r) = \text{const.}$ and $\psi_0(r) = \ln(r)$], it is easy to find the general solution, using the method of Greens functions (see, e.g., [AR85.1], Sect. 8.7),

$$\psi_0(r) = \ln r \int_0^r dr'\, r'\, \kappa_0(r') + \int_r^\infty dr'\, r'\, \ln r'\, \kappa_0(r') \quad , \tag{8.78a}$$

and for $n \geq 1$

$$\psi_n(r) = -\frac{1}{4n} r^{-2n} \int_0^r dr'\, r'^{(2n+1)}\, \kappa_n(r') \\ -\frac{1}{4n} r^{2n} \int_r^\infty dr'\, r'^{(1-2n)}\, \kappa_n(r') \quad . \tag{8.78b}$$

The deflection angle is then calculated from (5.8) to be

$$\alpha_1 = \psi_0'(r) \cos\varphi \\ + \sum_{n=1}^{\infty} \left[\psi_n'(r) \cos(2n\varphi) \cos\varphi + \frac{2n\psi_n(r)}{r} \sin(2n\varphi) \sin\varphi \right] \quad ,$$

$$\alpha_2 = \psi_0'(r) \sin\varphi \\ + \sum_{n=1}^{\infty} \left[\psi_n'(r) \cos(2n\varphi) \sin\varphi - \frac{2n\psi_n(r)}{r} \sin(2n\varphi) \cos\varphi \right] \quad , \tag{8.79}$$

These equations are still exact; the approximation one can introduce now is to perform the multipole expansion up to a finite order $n \leq N$; in this case, the deflection angle is exact for a matter distribution which results if the expansion (8.77a) is also carried out up to order N. Hence, by choosing N appropriately, one can approximate the elliptical mass distribution as closely as desired, and for this approximate mass distribution, the deflection angle is computed exactly. The numerical problem reduces to the (in general, numerical) calculation of the functions $\psi_n(r)$, and the deflection angle is computed via (8.79) by interpolating in tables of ψ_n. Note that it is much more convenient to interpolate $N+1$ one-dimensional functions than interpolating two two-dimensional functions.

A family of mass distributions. Consider the mass distribution

$$\kappa = \frac{\kappa_0}{(1+\rho^2/x_c^2)^\nu} \quad , \tag{8.80a}$$

where x_c is the 'core size', κ_0 is the central surface mass density, and the coefficient ν describes the behavior of κ for large distances from the lens center; in particular, for $\cos\beta = 1$ and $\nu = 1/2$, the distribution (8.80a) resembles that of an isothermal sphere with finite core radius. Defining $y = \frac{r^2}{\cos^2\beta\, x_c^2}$, (8.80a) can be written as

$$\kappa = \frac{\kappa_0}{(1+y)^\nu} \left[1 - \frac{y}{1+y}\sin^2\beta\,\cos^2\varphi \right]^{-\nu} \quad .$$

Since the second term in the bracket is smaller than one, the bracket can be expanded, resulting in an (infinite) series. This series can then be used to evaluate the multipole components $\kappa_n(r)$; the result is

$$\kappa(\mathbf{x}) = \frac{\kappa_0}{2}\frac{1}{(1+y)^\nu} \sum_{n=0}^{\infty} \cos(2n\varphi) \sum_{m=n}^{\infty} v(n,m) \left(\frac{y}{1+y}\right)^m \sin^{2m}\beta, \tag{8.80b}$$

where the coefficients $v(n,m)$ are listed in [SC91.1]. An approximation for κ is obtained by truncating both summations in (8.80b), at $n \leq N$ and $m \leq M$, respectively. For the resulting matter distribution, the exact deflection angle can be calculated as a two-dimensional sum. Explicit equations for this procedure are detailed in [SC91.1]. The calculation of the deflection angle can be performed quickly and efficient, and 'reasonable' matter distributions are obtained even for fairly small values of N and M. We therefore consider this method for dealing with (nearly) elliptical lenses as highly efficient in numerical work, and it should suffice for most, if not all applications.

Finally, we want to mention that a combination of matter distributions of the form (8.80a) can be used to construct more complicated lenses. For example, consider the distribution

$$\kappa = \frac{\kappa_0}{(1+\rho^2/x_c^2)^\nu} - \frac{\kappa_0}{(x_t^2/x_c^2+\rho^2/x_c^2)^\nu} \quad ;$$

for values $\rho \ll x_t$ this distribution behaves like that in (8.80a), whereas for 'separations' larger than the truncation values x_t, $\kappa \propto \rho^{-2(\nu+1)}$. Thus, such a combination can be used to construct lenses with several scales and with a finite total mass.

8.5 Marginal lenses

General description. In order to produce multiple images of a source, a GL must be sufficiently 'strong'. For example, centrally-condensed symmetric lenses must have a central surface mass density $\kappa_0 > 1$ to produce multiple images. The strength of a lens depends on its physical matter distribution and its distance to the source and the observer, all of which affect the dimensionless surface mass density κ. It is interesting to study the transition of a lens from "non-critical" to "critical" in the sense that, as the strength of the lens is increased, there will be a value of the strength parameter above which the lens can produce multiple images. To be more specific, consider a lens with deflection potential $(1+\delta)\psi(\mathbf{x})$. For $\delta \gtrsim -1$, this lens is weak, and the Jacobian determinant of the lens mapping is positive everywhere. For increasing δ, there will be – for a generic potential $\psi(\mathbf{x})$ – a value $\delta = \delta_0$ for which det A vanishes at exactly one point $\mathbf{x}_0$, being positive for all $\mathbf{x} \neq \mathbf{x}_0$. For values of $\delta > \delta_0$ there will be a region in the lens plane where det $A < 0$, a necessary and sufficient condition for the generation of multiple images. By choosing the normalization of ψ appropriately, we can set $\delta_0 = 0$. For $\delta \gtrsim 0$, the lens will be just able to split images. Such lenses have been termed marginal [KO87.1].

Translating the origin of the coordinate frames, we can take $\mathbf{x}_0 = 0$. The potential in a neighborhood of $\mathbf{x}_0$ can be expanded as was done in Chap. 6 [KO87.1], where now, an additional small parameter δ enters the model. It is clear that the corresponding singularity around $\mathbf{x}_0$ is a lips catastrophe (Sect. 6.3.3). Here, we illustrate the features of marginal lenses with the special case of a lens with elliptical isopotentials, a case that reveals all the important properties.

Illustrative case: Elliptical potentials. Consider a potential $\psi(u)$, where u is given by (8.68a). Using the expansion $\psi'(u) = \psi'_0 + u\psi''_0$, where ψ'_0, ψ''_0 denote the first and second derivative of ψ at $u = 0$, and scaling ψ with $(1+\delta)$, we find from (8.72), up to first order in u and δ

$$\det A = 1 - 4(1+\delta)\psi'_0 + 4\left(1-\epsilon^2\right)(1+2\delta)\psi'^2_0 \\ + 16\left(1-\epsilon^2\right)u\psi'_0\psi''_0 - 4(2u+\epsilon v)\psi''_0 \quad , \tag{8.81}$$

and v is given by (8.73).

For $\delta = 0$ and $u = 0$, we require det $A = 0$, which leads to $\psi'_0 = [2(1 \pm \epsilon)]^{-1}$. Since we want to investigate the situation where the lens is capable of producing multiple images for the smallest strength, we must choose the smaller value (for $\epsilon > 0$), $\psi_0' = [2(1+\epsilon)]^{-1}$.[7] This corresponds to the smaller eigenvalue of A becoming zero while the other remains positive. It is important to have ϵ strictly positive; otherwise, the two eigenvalues are the same. Inserting this value of ψ_0' into (8.81) yields

$$\det A = -\frac{2\delta\epsilon}{1+\epsilon} - 4\epsilon\,(2u+v)\,\psi_0'' \quad . \tag{8.82}$$

The requirement $\psi_0'' < 0$ guarantees that $\det A$ has a minimum at $u = 0$ for $\delta = 0$ (in fact, for all δ), since $2u + v \geq 0$. For $\delta < 0$, $\det A > 0$ for all values of u and v, whereas for $\delta > 0$, the determinant vanishes on the ellipse

$$(1-\epsilon)x_1^2 + 3(1+\epsilon)x_2^2 = a^2 \equiv \frac{\delta}{-2(1+\epsilon)\psi_0''} > 0 \quad , \tag{8.83}$$

which is the critical curve in the lens plane. The latter intersects the x_1-axis at

$$\mathbf{x} = \left(\pm\frac{a}{\sqrt{1-\epsilon}}, 0\right) \quad , \tag{8.84a}$$

and the x_2-axis at

$$\mathbf{x} = \left(0, \pm\frac{a}{\sqrt{3(1+\epsilon)}}\right) \quad . \tag{8.84b}$$

From (8.71), we find that the Jacobian matrix on the axes is diagonal and has components

$$A_{11} = \frac{2\epsilon}{1+\epsilon} - 2(1-\epsilon)\left[u + 2(1-\epsilon)x_1^2\right]\psi_0'' \quad ,$$
$$A_{22} = -\delta - 2(1+\epsilon)\left[u + 2(1+\epsilon)x_2^2\right]\psi_0'' \quad .$$

From these equations, we see that the tangent vector to the critical curve at its intersection with the x_1-axis is mapped onto the null vector, i.e., the image points of (8.84a) are cusps. The source positions corresponding to the points (8.84a,b) are, respectively,

$$\mathbf{y} = \left(\pm\frac{a}{\sqrt{1-\epsilon}}\frac{2\epsilon}{1+\epsilon}, 0\right) \quad , \tag{8.85a}$$

$$\mathbf{y} = \left(0, \mp\frac{2}{3}\delta\frac{a}{\sqrt{3(1+\epsilon)}}\right) \quad . \tag{8.85b}$$

---

[7] the other value also corresponds to a lips singularity

It should be stressed again that it is essential to have $\epsilon > 0$; the transition $\epsilon \to 0$ does not lead to meaningful results, since the separation of eigenvalues is required in this treatment.

The ray-trace equation for the case at hand reads

$$y_1 = x_1 \left[\frac{2\epsilon}{1+\epsilon} - 2(1-\epsilon) \, u \, \psi_0'' \right] \quad , \tag{8.86a}$$

$$y_2 = x_2 \left[-\delta - 2(1+\epsilon) \, u \, \psi_0'' \right] \quad , \tag{8.86b}$$

as is easily derived from (8.69). We consider the cases where the source is on one of the two axes, and determine the image positions only to leading order.

The case $y_2 = 0$. In this case, there is one image on the axis with

$$x_1 = \frac{1+\epsilon}{2\epsilon} y_1 \quad ; \quad x_2 = 0 \quad .$$

In addition, there can be two additional images with $x_2 \neq 0$, implying $u = a^2$. Since again, $x_1 = (1+\epsilon)y_1/(2\epsilon)$, we obtain the inequality

$$u = a^2 > (1-\epsilon)x_1^2 = (1-\epsilon)\frac{(1+\epsilon)^2 y_1^2}{4\epsilon^2} \quad ;$$

thus, these two images occur only if the source lies between the two points (8.85a), in which case the images are at

$$x_1 = \frac{1+\epsilon}{2\epsilon} y_1 \quad , \quad x_2 = \pm \frac{1}{\sqrt{1+\epsilon}} \sqrt{a^2 - \frac{(1-\epsilon)(1+\epsilon)^2}{4\epsilon^2} y_1^2} \quad .$$

The case $y_1 = 0$. In this case there are only solutions with $x_1 = 0$; then, (8.86b) becomes

$$y_2 = x_2 \left[-\delta - 2(1+\epsilon)^2 \psi_0'' \, x_2^2 \right] \quad .$$

This third order equation has three solutions if the source lies between the points (8.85b), and one solution otherwise. Hence, for three solutions, y_2 is of order $\delta^{3/2}$, so the corresponding image coordinates are of order $\delta^{1/2}$.

We thus obtain the following quantitative picture of a marginal lens: the critical curve is an ellipse with linear extension of order $\delta^{1/2}$. The caustic is shaped as lips, where the two cusps are separated by a distance of order $\delta^{1/2}$ and the two "lips" by a maximum distance of order $\delta^{3/2}$. Multiple images have separations of order $\delta^{1/2}$, and those occurring in pairs are highly magnified, $\mu \sim \delta^{-1}$.

This concludes our survey of simple lens models; we will use the results from this chapter later, when we consider the applications of GL theory. The selection of models in this chapter arose by considering which types of lenses are sufficiently simple to be analyzed in some detail and by the availability of these studies in the literature. Hence, our selection by no means implies that these lenses can be used to model any real GL systems. Model-fitting usually proceeds by trial and error, i.e., one chooses a (generally complicated) lens model and varies its parameters until an acceptable fit to the observations is obtained. We describe several numerical methods appropriate for GL model-fitting in Chap. 10.

8.6 Generic properties of "elliptical lenses"

The 'simple' lens models discussed in this chapter have all been used for modeling of observed GL systems, with some modifications. Depending on the available observed details of the lens systems, such models can successfully describe the observations. For example, if two images of a QSO are observed, together with an approximate position of the lens center, this system can be modeled by a axially-symmetric GL model if the lens center lies on the line connecting the two images. If, on further observations, it turns out that the actual arrangement of images and lens is not truly colinear, a more complicated model must be used: either a symmetric lens with some superposed shear (i.e., quadrupole lenses as described in Sect. 8.2) or one of the various descriptions for lenses with elliptical symmetry, as discussed in Sect. 8.4. In general, all these various models will be successful in describing the observations; in fact, even for GL systems with more observables, a good fit can be achieved with models from each of the aforementioned classes of lenses.

There are two principal reasons for this freedom in choosing a class of lens models: first, the observational constraints are typically not strong enough to distinguish between these classes of models, and second, the general properties of lenses from these classes are fairly similar. In this final section, we therefore want to describe some of the qualitative features of such GL models, which in the following will be termed "elliptical lenses"; they include the quadrupole lenses, lenses with elliptical isodensities or with elliptical isopotentials, and the lens models of Sect. 8.4.3. However, the properties to be described below will not be restricted to these models, which, after all, still have a high degree of symmetry, but will be valid for a much larger class of GL models.

8.6.1 Evolution of the caustic structure

Consider a matter distribution with fixed surface mass density $\Sigma(\boldsymbol{\xi})$, and consider the separation between lens and source as a variable; e.g., let $p = D_{\mathrm{ds}}/D_{\mathrm{s}}$. We assume that the mass distribution is of one of the types mentioned above, and that Σ is finite everywhere and decreases outwards.

For sufficiently small p, the lens is too weak to produce critical curves, and thus multiple images. For $p = p_1$, the Jacobian matrix at the center of the lens will have a zero eigenvalue, whereas the second eigenvalue is positive. This characterizes a lips catastrophe, as described in the previous section. For increasing p, the size of the lips caustic grows, and the Jacobian matrix at the center has one eigenvalue of either sign. At $p = p_2$, the larger of the two eigenvalues becomes zero, leading again to a lips catastrophe. For $p \gtrsim p_2$, there are now two lips-shaped caustics, one inside the other, and oriented mutually perpendicularly. Increasing p further, the size of the inner caustic grows, i.e., its two cusps approach the outer caustic. At $p = p_3$, the cusps 'touch' the outer caustic, and at these two points, hyperbolic umbilics occur (see Sect. 6.3.1). For $p > p_3$, there are two closed caustic curves, one with four and one with no cusps; we will call the first one the "tangential" caustic (since this caustic occurs if the tangential critical circle of a symmetric lens is slightly perturbed) and the latter one "radial" caustic. For $p > p_3$, no more metamorphoses occur in general (although for specific models, swallowtails can possibly be formed; we will neglect this possibility here); nevertheless, the qualitative arrangement of the caustics changes for increasing p: for $p \gtrsim p_3$, the two cusps which were formed during the hyperbolic umbilic transition lie inside the radial caustic, whereas the other two cusps of the tangential caustic lie outside the radial one. At $p = p_4$, these two latter cusps cross the radial caustic, so that for $p > p_4$ the tangential caustic lies completely inside the radial one.

If a cusp lies outside the radial caustic, a source close to it will have one or three highly-magnified images, without any additional images; we term such a cusp "naked". Naked cusps probably are relevant for explaining some of the luminous arcs; as we will demonstrate in Sect. 13.2, the generic model for the occurrence of arcs is that of an extended source just inside a cusp. In order to avoid additional images of the corresponding source, the cusp must be naked.[8] Straight arcs can occur if an extended source lies within a lips caustic.

8.6.2 Imaging properties

It is usually assumed that the central part of galaxies is sufficiently compact, so that for typical lens and source redshifts, $\kappa \gg 1$ at the center of the lens. In fact, the observational evidence for the occurrence of even number of images (in contrast to the expected odd number) supports this view, as it is usually assumed that the 'missing' image is situated close to the center of the lens, where the surface mass density is so large as to cause a strong demagnification of the corresponding image. Assuming this to be the case, the typical caustic of an "elliptical lens" is that corresponding to the case $p > p_4$, i.e., where the tangential caustic with its four cusps lies entirely inside the radial caustic. Such a gravitational lens model can in fact account

---

[8] Note, however, that it is difficult to prove observationally that the source corresponding to an arc has no additional images, due to the lacking decisive features singling out a particular high-redshift galaxy amongst others. Therefore, not all arcs must be produced by naked cusps, and for the arc in cluster Cl2244−02, possible additional images have been suspected [HA89.1].

for nearly all observed cases of multiply imaged QSOs (with the exception of 2016+112, which seems to require a more complicated lens model – see Sect. 2.5.2), as has been pointed out in [GR89.3]. We consider this case now in somewhat more detail.

If a source is situated close to the center of the source plane, it will have five images, one of which will be close to the center of the lens and thus strongly demagnified, whereas the other four will lie more or less symmetrically around the lens center. The more symmetric the image arrangement is, the less are the differences in the magnification of these four images. The two GL systems 2237+0305 and 1413+117 (for references, see Sect. 2.5) are probably typical examples for such a lensing configuration. Moving the source closer to the tangential caustic, but not close to one of its cusps, two of the four images will come closer together and become brighter. The GL system 1115+080, and probably also 0414+053 [HE89.1] can be accounted for with such a lensing geometry.

If the source lies close to, but on the inside of a cusp, three bright images will be very close together; depending on the resolution of the observations, such a triple image is hard to resolve, so that it appears to an observer as a single, very bright image. It is unclear at the moment whether such a system has been found already. Due to the large overall magnification, such GL system should be included in flux-limited samples preferentially (see the discussion on amplification bias in Sect. 12.5), and one might therefore suspect that some of the GL candidate systems with large flux ratios of the images (most noticibly, UM425=1120+019 [ME89.1]) can be accounted for by such a lens arrangement. In all the three cases just described, the fifth image is usually much fainter than the other images, if the lens is sufficiently compact, and can therefore easily escape detection.

Consider next the case that a source lies outside the tangential caustic, but inside the radial caustic. If it is not too close to the latter, two of its three images will have comparable magnification, whereas the third close to the center of the lens will still be strongly demagnified. This is the lensing geometry which is expected to occur most frequently, due to its relatively large probability. The existing sample of GL systems, however, does not contain a large fraction of such systems (one example is 0957+561); we interpret this discrepancy as being due to strong selection biases, to be explained in more detail in Sect. 12.5. Briefly, optical QSO surveys bias against GL systems with two images of comparable brightness, since one of the selection criteria is the stellar appearance of the source [KO91.1]. Convolving a pair of images of comparable brightness with the seeing disc, it appears elliptically, and thus extended, and will be excluded from further examination (by spectroscopic means). Furthermore, the total magnification of such systems is much smaller than for the cases described above, and thus, the amplification bias is less effective for such GL systems than for the 4-image cases.

Moving the source closer to the radial caustic, the image with negative parity moves towards the radial critical curve and becomes weaker.

Depending on the separation from the radial caustic, the third image can become of comparable brightness to the negative parity image, but for this to occur, these two images must be fairly close together, so that they are difficult to resolve observationally [note that a pair of images near a radial critical curve need not necessarily be highly magnified; whereas the scaling $\mu \propto (\Delta\theta)^{-1}$ is of course also valid for radial critical curves, the constant of proportionality can be fairly small for realistic lens models]. If the double nature of the B image in 2345+007 can be confirmed, it can be accounted for by such an arrangement (although the corresponding lens mass must be extremely large in that system, in which a very strict upper limit to the lens brightness has been derived, see Sect. 2.5.1). Also, the two GL candidates 1635+267 and 0023+171 can be of that type.

The foregoing discussion has shown that a single "elliptical lens" can account for nearly all types of observed gravitationally lensed QSOs (and it can also account for the observed radio rings). From that we conclude that the existing GL systems pose no problem in understanding their qualitative features (again, except for 2016+112) with a simple lens model, although the lens parameters required from observations, in particular the lens mass, appear in some cases to be fairly unusual. On the other hand, since most multiple QSO GL systems are so easily reproduced by simple models, this modeling procedure can not infer much information about the matter distribution of the lenses; the only solid model parameter is the mass inside a circle traced by the observed images.

9. Multiple light deflection

If the light emitted by a source is deflected by a single matter concentration in the universe, the theory for a single geometrically-thin mass distribution as developed in Chap. 5 applies. However, one can easily imagine situations where the light is deflected several times on the way from the source to the observer. For example, a distant QSO might be lensed by two galaxies at quite different redshifts; in particular, in the spectra of high redshift QSOs, one often finds absorption lines which correspond to intervening matter at several redshifts. If the absorbing matter forms part of a sufficiently large mass concentration (e.g, a galaxy), the multiple absorption lines could indicate that the light from the QSO is deflected more than once. In addition, microlensing need not be confined to objects at a single redshift. Furthermore, if the clumpy universe described in Chap. 4 does approximate the matter distribution in our universe, clumps at all redshifts will influence light propagation. We describe numerical studies of this process in a later chapter. As it turns out, the multiple lens-plane theory ([BL86.1], [KO87.5]) is very convenient for the study of light rays (and bundles) through a clumpy universe.

Consider two matter distributions that are geometrically thin, so that the single lens-plane deflection law is valid for each individual distribution. If the separation D between the two matter components is sufficiently large, a light ray is deflected by each lens separately, i.e., the deflection of a ray at one lens is not influenced by the other lens. On the other hand, if D is sufficiently small, the two lenses will act as if they were a single geometrically-thin lens, and the single lens-plane theory applies, for a mass distribution equal to the sum of the two components. The actual light ray, when deflected by a single matter distribution, is a curved path, which becomes asymptotically straight at large distances from the deflector. The assumption regarding the validity of the two lens-plane theory is that the light ray, after being deflected by the first matter component, is well in an asymptotic regime before the second component's deflection comes into play. However, one can easily imagine arrangements where neither the two lens-plane theory, nor the single lens-plane theory is applicable, for instance, any case where the deflection of a light ray by one of the two components is affected by the other. So far, no calculations have been published that show whether such situations occur with any significance, and if so, whether the breakdown of the multiple lens-plane theory is at all relevant. For example, one can place two point masses in such a configuration that neither the single, nor the multiple-lens ray-trace equation would yield the correct relation between the direction of a light ray reaching the observer and the position vector for the corresponding source. If the lens theory breaks down only for highly demagnified rays, however, such a failure is irrelevant in practice. There is no straightforward quantitative measure of the degree of breakdown of the GL theory; one would have to integrate the equation of motion for a photon in a specific deflector arrangement, and to compare it with the prediction of the theory. For those light rays where the difference in the results is appreciable, one would have to test their relevance, i.e., whether their associated magnification suffices to be detectable in practice.[1] Owing to the lack of a thorough study, we will simply assume that the multiple lens-plane theory (which contains the single lens-plane theory as a special case) can be used in all astrophysically relevant cases, but one should keep in mind that this hypothesis remains formally untested. Multiple deflection theory in terms of optical scalars (see Sect. 3.4.2) has been investigated in, e.g., [GU67.2], [DY77.1], [TO90.1].

In this chapter, we develop the multiple lens-plane theory. The corresponding ray-trace equation can easily be derived using geometry only, as for the single lens. The main new feature of the lens mapping is that its Jacobian matrix is no longer symmetric; i.e., it is not a gradient mapping from $\mathbb{R}^2$ to $\mathbb{R}^2$. This implies a net rotation of the images with respect to the source. We derive the time-delay function for multiple deflection and reobtain the lens equation from Fermat's principle. Finally, the theory is applied to a situation where all deflections but one can be described by a quadrupole term [KO87.5]. Since in most cases of relevance, a multiply-imaged source will be influenced mainly by one (strong) lens and, in addition, by several weak lenses, this "generalized quadrupole lens" theory can be expected to be of direct relevance to observed lens cases.

9.1 The multiple lens-plane theory

9.1.1 The lens equation

Consider N matter distributions, characterized by their surface mass densities Σ_i, at redshifts z_i, $i = 1, ..., N$, ordered such that for $i < j$, $z_i < z_j$. Let the source have a redshift $z_s > z_N$. In Fig. 9.1, we display the geometry for two lens planes. From the geometry, the ray-trace equation is read off to be

---

[1] A numerical study of the kind described above has been performed for the case of two point masses, with the result that, for all light rays not strongly demagnified, the GL approximation is extremely well satisfied (P. Haines, private communication).

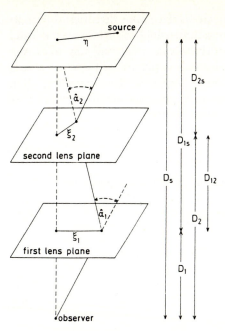

Fig. 9.1. Positions and ray-paths for two lenses at different redshifts

$$\eta = \frac{D_s}{D_1}\xi_1 - \sum_{i=1}^{N} D_{is}\hat{\alpha}_i(\xi_i) \quad , \tag{9.1}$$

where η and ξ_i denote the position vectors in the source plane and the i-th lens plane, respectively, and $\hat{\alpha}_i(\xi_i)$ is the deflection a light ray undergoes if it traverses the i-th lens plane at ξ_i. The impact vectors ξ_i are obtained recursively from

$$\xi_j = \frac{D_j}{D_1}\xi_1 - \sum_{i=1}^{j-1} D_{ij}\hat{\alpha}_i(\xi_i) \quad . \tag{9.2}$$

The distance factors are defined such that $D_{ij} = D(z_i, z_j)$ is the angular diameter distance of the j-th plane from the i-th plane, $D_i = D(0, z_i) \equiv D(z_i)$ is the angular diameter distance of the i-th lens plane from the observer, and the index s denotes the source plane. Equation (9.2) agrees with (9.1) when $j = s = N + 1$, of course.

Conversion to angular variables. Throughout this chapter we use angular variables to convert the lens equation to dimensionless form. Hence, defining $\mathbf{x}_i = \xi_i/D_i$ for $1 \leq i \leq N+1$, the lens equation (9.2) becomes

$$\mathbf{x}_j = \mathbf{x}_1 - \sum_{i=1}^{j-1} \frac{D_{ij}}{D_j}\hat{\alpha}_i(D_i\mathbf{x}_i) \quad . \tag{9.3}$$

The deflection angles are derived from the surface mass densities as in (5.2). As in Chap. 5, it is convenient to define a dimensionless surface mass density. While for the single lens-plane theory there was a 'natural' way to convert Σ to κ, this is no longer true for multiple lens planes, as there are several distance factors which can be used, in principle, in this conversion. Here we use a definition of κ that will lead to a ray-trace equation which is as similar as possible to the single lens-plane theory, viz.

$$\kappa_i(\mathbf{x}_i) = \frac{4\pi G}{c^2} \frac{D_i D_{is}}{D_s} \Sigma_i(D_i \mathbf{x}_i) \quad . \tag{9.4}$$

With these values of κ_i, we define the scaled deflection potentials

$$\psi_i(\mathbf{x}_i) = \frac{1}{\pi} \int_{\mathbf{R}^2} d^2 x' \, \kappa_i(\mathbf{x}') \, \ln|\mathbf{x}_i - \mathbf{x}'| \tag{9.5}$$

and the scaled deflection angles $\boldsymbol{\alpha}_i = \nabla \psi_i$, defined in analogy with (5.7). The relation between $\hat{\boldsymbol{\alpha}}_i$ and $\boldsymbol{\alpha}_i$ is provided by

$$\boldsymbol{\alpha}_i = \frac{D_{is}}{D_s} \hat{\boldsymbol{\alpha}}_i \quad . \tag{9.6}$$

In terms of the $\boldsymbol{\alpha}_i$, the lens equation (9.3) becomes

$$\mathbf{x}_j = \mathbf{x}_1 - \sum_{i=1}^{j-1} \beta_{ij} \boldsymbol{\alpha}_i(\mathbf{x}_i) \quad , \tag{9.7a}$$

where

$$\beta_{ij} = \frac{D_{ij} D_s}{D_j D_{is}} = \frac{\chi_{is}}{\chi_{ij}} \quad , \qquad \chi_{ij} = \left[\chi(z_i) - \chi(z_j)\right]^{-1} \quad , \tag{9.8}$$

where the function $\chi(z)$ was defined in Sect. 4.6. In particular, $\beta_{is} = 1$, so for $j = N+1 = s$, (9.7a) becomes

$$\mathbf{y} \equiv \mathbf{x}_{N+1} = \mathbf{x}_1 - \sum_{i=1}^{N} \boldsymbol{\alpha}_i(\mathbf{x}_i) \quad ; \tag{9.7b}$$

hence, if all but one of the Σ_i's vanish, (9.7b) reduces to the single lens-plane equation.

The equations (9.7a) describe the impact parameters of a light ray in all the lens planes as functions of $\mathbf{x}_1$, which is the independent variable, since it represents the angular position of an image on the observer's sky. The ray-trace equation (9.7b) is a mapping from the image plane onto the source plane similar to that of a single lens-plane. The two theorems of Sect. 5.4 are valid also in the presence of several lens-planes, as the general arguments presented there show, with the provisos stated there (see also [PA88.1]). In fact, the proof of the odd number theorem proceeds in the same way as that for a single lens plane, given in Sect. 5.4.1; the only modification being that the vector field $\mathbf{X}$ used in the proof must now be replaced by $\mathbf{X}(\mathbf{x}_1) = \mathbf{y}(\mathbf{x}_1) - \mathbf{y}_0$, and the fact that at a stationary point corresponding to a non-critical image, such a vector field has index ± 1, namely $+1$ at sinks, sources, and centers, -1 at saddle points. This can be seen by considering the decomposition (9.13) of a general matrix A into a symmetric one and a rotation matrix. The latter rotates all vectors locally by the same amount and thus does not influence the index. In fact, the index is $+1$ (-1) if $\det A > 0$ ($\det A < 0$). We could also generalize the classification of ordinary images from (5.29) to the multiple deflection situation, using only the symmetric factor of A, or, equivalently, the signs of the λ's.

9.1.2 The magnification matrix

We consider next the Jacobian matrix A of the mapping (9.7b). First, we define the Jacobian matrices of the mappings (9.7a),

$$A_i = \frac{\partial \mathbf{x}_i}{\partial \mathbf{x}_1} \quad , \tag{9.9}$$

as well as the derivatives of the deflection angles,

$$U_i = \frac{\partial \boldsymbol{\alpha}_i}{\partial \mathbf{x}_i} \quad . \tag{9.10}$$

Then we find from (9.7b) that

$$A \equiv \frac{\partial \mathbf{y}}{\partial \mathbf{x}_1} = \mathcal{I} - \sum_{i=1}^{N} \frac{\partial \boldsymbol{\alpha}_i}{\partial \mathbf{x}_1} = \mathcal{I} - \sum_{i=1}^{N} \frac{\partial \boldsymbol{\alpha}_i}{\partial \mathbf{x}_i} \frac{\partial \mathbf{x}_i}{\partial \mathbf{x}_1} = \mathcal{I} - \sum_{i=1}^{N} U_i A_i \quad ; \tag{9.11}$$

here, $\mathcal{I}$ is the two-dimensional identity matrix. The A_i can be obtained by recursion, since (9.7a) yields

$$A_j = \mathcal{I} - \sum_{i=1}^{j-1} \beta_{ij} U_i A_i \quad , \tag{9.12}$$

and $A_1 = \mathcal{I}$. Of course, for $j = N + 1$, (9.12) agrees with (9.11).

The matrices A_i, in particular A, need not be symmetric, since the product of two symmetric matrices in general is not symmetric. However, we can decompose A into a rotation matrix and a symmetric matrix:

$$\begin{aligned} A &= R(-\varphi) \begin{pmatrix} \lambda_1 & 0 \\ 0 & \lambda_2 \end{pmatrix} R(\vartheta) \\ &= R(\vartheta - \varphi) R(-\vartheta) \begin{pmatrix} \lambda_1 & 0 \\ 0 & \lambda_2 \end{pmatrix} R(\vartheta) \quad , \end{aligned} \tag{9.13}$$

where the R's are rotation matrices,

$$R(\varphi) = \begin{pmatrix} \cos\varphi & \sin\varphi \\ -\sin\varphi & \cos\varphi \end{pmatrix} \quad .$$

If A is symmetric, $\varphi = \vartheta$; in general, $\varphi - \vartheta$ is the net rotation of the image. The two invariants $\lambda_{1,2}$ of A describe the image distortion produced by lensing. In particular, the image for a small circular source in general will be an ellipse, whose axes are stretched by factors $1/\lambda_{1,2}$ compared to the radius of the source. As in the single lens-plane theory, the magnification is $\mu = 1/\det(A)$.

It is useful to note that $\det A = \lambda_1 \lambda_2$, $\mathrm{tr}\left(AA^T\right) = \lambda_1^2 + \lambda_2^2$, where $\left(A^T\right)_{ij} = A_{ji}$, which determines the λ_i up to numeration and a common sign. Also,

$$A^T A = R(\varphi - \vartheta) A A^T R(\vartheta - \varphi) \quad ; \quad 0 \leq \varphi - \vartheta < \pi \ .$$

This determines $\varphi - \vartheta$ uniquely. From (9.13), we find from the invariance of the trace under rotations,

$$\mathrm{tr} A = \mathrm{tr}\left(\begin{pmatrix} \lambda_1 & 0 \\ 0 & \lambda_2 \end{pmatrix} R(\vartheta - \varphi)\right) = (\lambda_1 + \lambda_2) \cos(\vartheta - \varphi) = a_1 + a_2 \ ,$$

where a_1 and a_2 are the eigenvalues of A, given by

$$a_1 - a_2 = \sqrt{(\lambda_1 + \lambda_2)^2 \cos^2(\vartheta - \varphi) - 4\lambda_1 \lambda_2}$$

and the foregoing equation.

Special case: Two lens planes. To illustrate how cumbersome the equations may become, we write down the magnification for the case of two lens planes. Let the matrices $U_{1,2}$ in the two lens planes be

$$U_i = \kappa_i \mathcal{I} + \gamma_i \begin{pmatrix} \cos \vartheta_i & \sin \vartheta_i \\ \sin \vartheta_i & -\cos \vartheta_i \end{pmatrix} \ ,$$

where κ_i describes the local surface mass density, γ_i is the amount of shear, and ϑ_i describes the direction of the shear. Then,

$$\begin{aligned} \mu^{-1} = &\, (1 - \kappa_1 - \kappa_2)^2 - \gamma_1^2 - \gamma_2^2 - 2\gamma_1 \gamma_2 \cos(\vartheta_1 - \vartheta_2) \\ &+ 2\beta_{12} \big[(1 - \kappa_1 - \kappa_2)\kappa_1 \kappa_2 + \gamma_1 \gamma_2 \cos(\vartheta_1 - \vartheta_2) \\ &+ \kappa_2 \gamma_1^2 + \kappa_1 \gamma_2^2\big] + \beta_{12}^2 \left(\kappa_1^2 - \gamma_1^2\right)\left(\kappa_2^2 - \gamma_2^2\right) \ . \end{aligned} \quad (9.14)$$

This expression is complicated; we shall not analyze it in full generality. However, there are three special cases amenable to a simple interpretation.

Case 1: $\beta_{12} = 0$, equivalent to having a single lens plane with matter density $\kappa = \kappa_1 + \kappa_2$ and shear $\gamma^2 = \left[\gamma_1^2 + \gamma_2^2 + 2\gamma_1 \gamma_2 \cos(\vartheta_1 - \vartheta_2)\right]$. From (9.14) we see that in this case

$$\mu^{-1} = (1 - \kappa)^2 - \gamma^2 \ ,$$

which is the correct result for a single lens.

Case 2: $\gamma_1 = \gamma_2 = 0$, both shear components vanish, and we have a combination of anastigmatic lenses, which also represents an anastigmatic lens (i.e., the shear for the combined lens also vanishes). Then,

$$\mu^{-1} = (1 - \kappa_1 - \kappa_2 + \beta_{12}\kappa_1\kappa_2)^2$$

This case is thus equivalent to a single lens with surface mass density $\kappa = \kappa_1 + \kappa_2 - \beta_{12}\kappa_1\kappa_2$. Note, however, that κ can become negative. This is the case if the first lens produces a focus between the two lens planes (i.e., $\beta_{12} > 1/\kappa_1$) and if, in addition, the second lens is sufficiently strong, $\kappa_2 > \kappa_1/(\beta_{12}\kappa_1 - 1)$.

Case 3: $\kappa_1 = \kappa_2 = 0$: In this case, the light beam is affected by shear only, and

$$\mu^{-1} = \left(1 - \beta_{12}\gamma_1^2\right)\left(1 - \beta_{12}\gamma_2^2\right) - (1 - \beta_{12})\gamma^2$$

(γ as in case 1). One of the new features of the multiple lens-plane geometry is that a combination of two shear terms can lead to a convergence and thus mimic a local surface mass density. For $\gamma_1 = \gamma_2$ and $|\vartheta_1 - \vartheta_2| = \pi$, the Jacobian matrix is $A = (1 - \beta_{12}\gamma_1^2)\mathcal{I}$ and is thus locally equivalent to a single lens with surface mass density $\beta_{12}\gamma_1^2$ and vanishing shear. This example shows that it will be difficult to tell from observations of a multiply imaged source whether more than one deflector is responsible for the splitting.

9.1.3 Particular cases

Unfortunately, the foregoing equations have not been used extensively so far to investigate the properties of a multiple lens-plane geometry in detail. Although for a single lens plane we understand the behavior of some lens models very well (see Chap. 8), that is not true for multiple lens planes (for an exception, see below). One of the underlying reasons is the large number of free parameters. Even in the simplest case, where one treats the lens action of two point masses at different distances, one has three free parameters (in addition to the source position), namely the transverse angular separation of the masses, their mass ratio, and a parameter describing their separation along the line-of-sight. As it turns out, even this simple situation presents many analytical problems, and the volume of parameter space is rather large for any numerical study. On the other hand, we believe that it is necessary to understand some simple multiple-lens configurations to find out what can happen for multiple deflections. As will be shown in a later chapter, numerical simulations suggest that the structure of caustics may be significantly different for multiple lens planes.

Recently, a lens system consisting of two circular matter distributions of the form introduced in Sect. 8.1.5 was studied in detail [KO88.1]. This paper contains various plots of the caustics of this lens system, as well as a few examples for imaging of extended sources. From the typical transverse separation of the two lenses, such that they mutually affect each other's lensing behavior, the authors conclude that multiple deflection should play a role in about 1–10% of all GL systems. Whereas this paper yields much information about the probabilities for image multiplicities and magnifications, it is not aimed towards a detailed investigation of the possible image and caustic configurations, comparable to the classification performed for some simpler lens models (e.g., the quadrupole lenses in Sect. 8.2 or the two point-mass lens in Sect. 8.3).

Generalizing the two point-mass lens, described in Sect. 8.3.2, recent work [ER90.1] considered a GL consisting of two point masses at different redshift. Whereas the addition of another lens parameter (a combination of the distances to the two masses) leads to an enormous complication of the analysis, it was nevertheless possible to perform a thorough study of the caustic structure of this GL model. In particular, a complete classification of the caustics was obtained, in the form of a 17-th order polynomial, which is readily solved numerically. From that it was shown that there are five different topologies of caustics, separated by four two-dimensional surfaces in the three-dimensional parameter space. These are the bifurcation surfaces. As for the two point-mass lens in one lens plane, all these bifurcations are beak-to-beak singularities (see Sect. 6.3.3). In addition, there is a surface in parameter space on which the GL obtains two elliptic umbilics. One might wonder how this can happen, as we have pointed out that at umbilics both eigenvalues of the Jacobian matrix A vanish. In the case of a single lens plane, this can only occur if the local surface mass density $\kappa = 1$, but as we discussed in the previous subsection, for two lens planes with $\kappa = 0$ each, an 'equivalent total κ' can be mimicked. Noticibly, the ratio of the two lens masses and the distance between the two lenses (as long as it is non-zero) seem to be less important parameters than the transverse separation of the lenses, in the sense that, for any fixed mass ratio and any distance, a variation of the separation yields all possible caustic topologies. As far as we are aware, the paper just discussed is the first in which catastrophe theory has been extensively used for a classification of a parametrized lens model.

9.2 Time delay and Fermat's principle

Derivation of an expression for the time delay. For a single lens-plane, we have shown that the lens equation can be derived from a scalar function $\phi(\mathbf{x}, \mathbf{y})$ (Fermat potential) which is essentially the light-travel-time along a ray from the source at $\mathbf{y}$ which traverses the lens plane at $\mathbf{x}$. It is clear that a similar function exists in a multiple deflection situation, but there the time delay depends on all variables $\mathbf{x}_i$, $1 \leq i \leq N+1$, which are necessary to describe a piecewise straight, kinematically possible, deflected light ray. We now compute the analogue of (5.45) for several lens planes.

Consider first a light ray which originates in the j-th plane at $\mathbf{x}_j$, and is deflected only in the i-th plane, where the impact vector is $\mathbf{x}_i$. From (5.45) we see that its time delay relative to the undeflected ray is, up to an unimportant additive constant,

$$T_{ij}(\mathbf{x}_i, \mathbf{x}_j) = \frac{1+z_i}{c} \frac{D_i D_j}{D_{ij}} \phi_{ij}(\mathbf{x}_i, \mathbf{x}_j) \quad , \tag{9.15}$$

where ϕ_{ij} is defined in (5.11) with $\mathbf{x}_i$, $\mathbf{x}_j$ substituted for $\mathbf{x}$, $\mathbf{y}$, and z_i, z_j instead of z_d, z_s.[2] One must be careful when analyzing (9.15), because ϕ_{ij} contains the dimensionless surface mass density, which in turn depends on the conversion factor as discussed after (9.3). It is therefore useful to rewrite (5.12) in terms of the physical surface mass density,

$$\phi_{ij}(\mathbf{x}_i, \mathbf{x}_j) = \frac{1}{2}(\mathbf{x}_i - \mathbf{x}_j)^2 - \frac{4GD_iD_{ij}}{c^2 D_j} \int_{\mathbb{R}^2} d^2 x' \, \Sigma_i(\mathbf{x}' D_i) \ln |\mathbf{x}_i - \mathbf{x}'| \quad .$$

Comparison of this equation with (9.4) and (9.5) then yields

$$T_{ij}(\mathbf{x}_i, \mathbf{x}_j) = \frac{1+z_i}{c} \frac{D_i D_j}{D_{ij}} \left[\frac{1}{2}(\mathbf{x}_i - \mathbf{x}_j)^2 - \beta_{ij} \psi_i(\mathbf{x}_i) \right] \quad , \tag{9.16a}$$

where the coefficient β_{ij} is defined by (9.8), or

$$H_0 T_{ij} = \frac{1}{2} \chi_{ij} (\mathbf{x}_i - \mathbf{x}_j)^2 - \chi_{is} \psi_i(\mathbf{x}_i) \quad . \tag{9.16b}$$

Note that the potential time delay [i.e., the second term in (9.16)] does not depend on z_j.

The time delay for a light ray characterized by $\mathbf{x}_i$, $1 \leq i \leq N+1$, relative to the undeflected ray ($\mathbf{x}_i = \mathbf{x}_{N+1} = \mathbf{y}$ for all $1 \leq i \leq N$), is determined by replacing the deflected ray successively by "straighter" rays, by "cutting out the edges" (see Fig. 9.2). For example, the light-travel-time along the actual ray is larger by $T_{12}(\mathbf{x}_1, \mathbf{x}_2)$ than that of a ray for which $\mathbf{x}_1 = \mathbf{x}_2$ and $\psi_1(\mathbf{x}_1) \equiv 0$. By summation, we find that the time delay for the actual ray is, up to an additive constant which does neither depend on the $\mathbf{x}_i$'s nor on $\mathbf{y}$ (see discussion in Sect. 4.3),

$$T(\mathbf{x}_1, ..., \mathbf{x}_N, \mathbf{y}) = \sum_{i=1}^{N} T_{i,i+1}(\mathbf{x}_i, \mathbf{x}_{i+1}) \quad ; \quad \mathbf{x}_{N+1} = \mathbf{y} \quad . \tag{9.17}$$

For a fixed source position, this function depends on the N impact vectors in the lens planes. Fermat's principle states that the physical light rays are those along which T is stationary, with respect to all arguments. Hence, from (9.16b) and (9.17):

$$\frac{\partial}{\partial \mathbf{x}_j} T(\mathbf{x}_1, ..., \mathbf{x}_N, \mathbf{y})$$
$$= \frac{\partial}{\partial \mathbf{x}_j} \sum_{i=1}^{N} \left[\frac{1}{2} \chi_{i,i+1} (\mathbf{x}_i - \mathbf{x}_{i+1})^2 - \chi_{is} \psi_i(\mathbf{x}_i) \right] = 0 \quad , \tag{9.18}$$

for $1 \leq j \leq N$. In fact, we will show now that (9.18) is equivalent to (9.7).

---

[2] In this context, the indices of ϕ and T do, of course, not indicate partial differentiations.

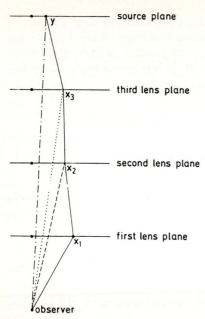

Fig. 9.2. For the determination of the time delay, a multiply-deflected light ray (solid line) is replaced by "straighter" rays (dashed, dotted, and dashed-dotted lines). If we denote the time delay for a ray with source position $\mathbf{y}$ and impact vector in the i-th plane $\mathbf{x}_i$ by $T(\mathbf{x}_1, \mathbf{x}_2, \mathbf{x}_3, \mathbf{y})$, we find that the time delay for the dashed ray is $T(\mathbf{x}_2, \mathbf{x}_2, \mathbf{x}_3, \mathbf{y}) = T(\mathbf{x}_1, \mathbf{x}_2, \mathbf{x}_3, \mathbf{y}) - T_{12}(\mathbf{x}_1, \mathbf{x}_2)$. Correspondingly, we can obtain relations for the time delay of the dotted and dashed-dotted ray. The final result is given by (9.17).

Equivalence of (9.7) and (9.18): Fermat's principle. Written out and rearranged, the equations (9.18) read as follows $[\boldsymbol{\alpha}_i \equiv \boldsymbol{\alpha}_i(\mathbf{x}_i)]$:

$$
\begin{aligned}
j = 1: \quad & \mathbf{x}_2 = \mathbf{x}_1 - \frac{\chi_{1s}}{\chi_{12}} \boldsymbol{\alpha}_1 \;, \\
2 \le j \le N: \quad & \mathbf{x}_{j+1} = \frac{\chi_{j-1,j}}{\chi_{j-1,j+1}} \mathbf{x}_j - \frac{\chi_{j-1,j}}{\chi_{j,j+1}} \mathbf{x}_{j-1} - \frac{\chi_{js}}{\chi_{j,j+1}} \boldsymbol{\alpha}_j \;.
\end{aligned}
\qquad (9.19)
$$

We want to point out that the equations (9.19) provide a relation between three consecutive impact parameters, governed by the deflection angle in the middle plane.

Clearly, these equations determine $\mathbf{x}_2, ..., \mathbf{x}_{N+1}$ uniquely in terms of $\mathbf{x}_1$. In fact, using the definition (9.8) of the χ_{ij}, the last two equations imply that, for $2 \le j \le N+1$,

$$
\mathbf{x}_j = \mathbf{x}_1 - \sum_{i=1}^{j-1} \frac{\chi_{is}}{\chi_{ij}} \boldsymbol{\alpha}_i = \mathbf{x}_1 - \sum_{i=1}^{j-1} \beta_{ij} \boldsymbol{\alpha}_i \;,
$$

which is indeed (9.7). Hence, the multiple lens-plane lens equations can be derived from Fermat's principle. Moreover, the foregoing equations enable one to compute the time delays for different images of a source.

9.3 The generalized quadrupole lens

Many common astrophysical cases of light deflection involve vastly different length scales, e.g., deflection by a star which resides in a galaxy, or deflection by a galaxy which is part of a cluster of galaxies. In these two cases, where the two lens components are at the same redshift, the quadrupole lens approximation as described in Sect. 8.2 is most appropriate.

Light rays are affected by matter concentrations near their path from the source to the lens and/or from the lens to the observer; such perturbations may be described in a quadrupole approximation. The perturbations do not necessarily have to be small; the only requirement is that the deflection they cause vary on length scales which are much larger than the typical length scale in the main lens. These properties then lead to the generalized quadrupole lens [KO87.5].

We consider N lens planes, and denote the main lens by the subscript l. The deflection angle in all lens planes but the l-th is described by the quadrupole term

$$\boldsymbol{\alpha}_i(\mathbf{x}_i) = \boldsymbol{\alpha}_i^0 + U_i \mathbf{x}_i \quad , \tag{9.26}$$

where $\boldsymbol{\alpha}_i^0$ is a constant vector and U_i is a matrix as in (9.10). Inserting (9.26) into the ray-trace equation (9.7), we find that, for $j \leq l$,

$$\mathbf{x}_j = B_j \mathbf{x}_1 - \mathbf{u}_j \quad , \tag{9.27}$$

and the matrices B_j and vectors $\mathbf{u}_j$ are determined by recursion:

$$B_j = \mathcal{I} - \sum_{i=1}^{j-1} \beta_{ij} U_i B_i \quad , \tag{9.28a}$$

$$\mathbf{u}_j = \sum_{i=1}^{j-1} \beta_{ij} \left(\boldsymbol{\alpha}_i^0 - U_i \mathbf{u}_i \right) \quad , \tag{9.28b}$$

with $B_1 = \mathcal{I}$, $\mathbf{u}_1 = 0$. After comparison of (9.27) with (9.9), we see that $B_j = A_j$ for $j \leq l$. In particular, for the impact vector in the main lens plane,

$$\mathbf{X} \equiv \mathbf{x}_l = B_l \mathbf{x}_1 - \mathbf{u}_l \quad . \tag{9.29}$$

For $j > l$,

$$\mathbf{x}_j = B_j \mathbf{x}_1 - \mathbf{u}_j - C_j \boldsymbol{\alpha}_l(\mathbf{X}) \quad , \tag{9.30}$$

which may then be inserted into the lens equation to yield the recursion

$$B_j = \mathcal{I} - \sum_{i \neq l}^{j-1} \beta_{ij} U_i B_i \quad , \tag{9.31a}$$

$$\mathbf{u}_j = \sum_{\substack{i=l \\ i \neq l}}^{j-1} \beta_{ij} \left(\boldsymbol{\alpha}_i^0 - U_i \mathbf{u}_i \right) , \qquad (9.31b)$$

$$C_j = \beta_{lj} \mathcal{I} - \sum_{i=l+1}^{j-1} \beta_{ij} U_i C_i , \qquad (9.31c)$$

with $C_{l+1} = \beta_{l,l+1} \mathcal{I}$. In particular, for the source position $\mathbf{y} \equiv \mathbf{x}_{N+1}$, we find

$$\mathbf{y} = B_s \mathbf{x}_1 - \mathbf{u}_s - C_s \boldsymbol{\alpha}_l(\mathbf{X}) , \qquad (9.32)$$

where, since $\beta_{is} = 1$,

$$B_s = \mathcal{I} - \sum_{\substack{i=l \\ i \neq l}}^{N} U_i B_i , \qquad (9.33a)$$

$$\mathbf{u}_s = \sum_{\substack{i=l \\ i \neq l}}^{N} \left(\boldsymbol{\alpha}_i^0 - U_i \mathbf{u}_i \right) , \qquad (9.33b)$$

$$C_s = \mathcal{I} - \sum_{i=l+1}^{N} U_i C_i . \qquad (9.33c)$$

In the following, we assume that the matrices B_l, B_s and C_s are non-singular. Then, the mapping (9.29) between $\mathbf{x}_1$ and $\mathbf{X}$ is a diffeomorphism, and we can replace the independent variable $\mathbf{x}_1$ by $\mathbf{X}$, $\mathbf{x}_1 = B_l^{-1}(\mathbf{X} + \mathbf{u}_l)$. Let us insert this into (9.32), and apply the following linear transformation in the source plane:

$$\mathbf{Y} = C_s^{-1}(\mathbf{y} + \mathbf{u}_s - B_s B_l^{-1} \mathbf{u}_l) , \qquad (9.34)$$

we find

$$\mathbf{Y} = C\mathbf{X} - \boldsymbol{\alpha}_l(\mathbf{X}) , \qquad (9.35)$$

where

$$C = C_s^{-1} B_s B_l^{-1} . \qquad (9.36)$$

Equation (9.35) is equivalent to equation (6.13) in [KO87.5]. The form of (9.35) parallels that for a single lens-plane for a lens whose deflection is described by $\boldsymbol{\alpha}_l$ plus a quadrupole term. To demonstrate the equivalence, we must prove that C is symmetric, as can be easily shown [KO87.5]. Since the only independent variable is $\mathbf{X}$, there must be a function $\phi(\mathbf{X}, \mathbf{Y})$, proportional to the time delay for a ray from the source at $\mathbf{Y}$, which intersects the l-th lens plane at $\mathbf{X}$, relative to the case $\boldsymbol{\alpha}_l \equiv 0$. From Fermat's principle, we know that $\partial \phi / \partial \mathbf{X} = 0$ is equivalent to the lens equation (9.35). Hence, $\phi(\mathbf{X}, \mathbf{Y})$ must satisfy the pair of differential equations

$$\frac{\partial \phi}{\partial X_i} = (C\mathbf{X})_i - Y_i - \frac{\partial \psi_l}{\partial X_i} \quad . \tag{9.37}$$

Differentiation of (9.37) for $i = 1$ with respect to X_2, and correspondingly for $i = 2$ with respect to X_1, yields the integrability condition $C_{12} = C_{21}$; hence, C is symmetric. Therefore,

$$\phi(\mathbf{X}, \mathbf{Y}) = \frac{1}{2}(C\mathbf{X} - \mathbf{Y})\left[C^{-1}(C\mathbf{X} - \mathbf{Y})\right] - \psi_l(\mathbf{X}) \quad . \tag{9.38}$$

Owing to the symmetry of the matrix C, equation (9.35) is mathematically equivalent to the single lens-plane quadrupole equation. In particular, if the main deflector is circularly symmetric, the formalism developed in Sect. 8.2 applies to the generalized quadrupole lens.

10. Numerical methods

Several times, we have run into problems where analytical methods are no longer sufficient to analyze a GL system. Our "standard" problem is the inversion of the lens equation, i.e., to find all the images **x** for a given source position **y**. Only in the simplest cases can this be done analytically; in general, one must find all the roots of a two-dimensional system of equations. This is not trivial, in particular because the number of solutions is not known a priori. There are other problems where numerical analysis must be applied, e.g., if one wants to determine the total magnification for an extended source, as a function of its position. This problem is computationally simpler than the standard problem.

In this chapter, we discuss some of the methods which have been found useful in 'numerical GL theory'. We assume that the reader is familiar with standard numerical methods, and we do not discuss problems solvable with standard software. For those cases, the reader is referred to the NAG library of numerical routines, or to the delightful 'cookbook' by Press et al. [PR86.1]. The methods considered here will be applicable to many problems one may encounter in numerical GL theory, but, as usual, they should be applied with due caution.

The choice of method depends strongly on the goal of the user. It will be shown below that if one is interested in producing a plot of the images of an extended source, simple routines are readily written, based on contour plot routines. Other methods must be used if one wants to compute all image positions that correspond to a given point source. There are interactive procedures which are very useful for model fitting in observed lens cases. For statistical calculations, such as the determination of cross-sections (see next chapter) still other methods will be used. We do not intend to provide the reader with explicit source codes of routines, since these depend very much on the specific problem at hand; rather, we outline the essential ideas which may be useful to those who want to use numerical methods in GL studies.

In Chap. 8, we saw that many GL problems can be reduced to a single equation whose roots are to be found. Since this problem is much simpler than a general, two-dimensional search for images, we treat this case first in Sect. 10.1. In Sect. 10.2, we present a simple method for finding the images of an extended source. An interactive routine to find the images in a given lens situation is described in Sect. 10.3. For certain purposes, one seeks to obtain the images of a large number of source positions simultaneously; in this case, one may use a grid search for images, as outlined in Sect. 10.4. Usually, such a grid search only yields approximate image positions; in Sect. 10.5, we describe how one obtains the exact positions once approximate values are found. It will be seen that this method is also useful in cases where one knows the images for one source position, and wants to compute the images of a nearby source. To find the total magnification as a function of the source position, one may use the ray-shooting procedure described in Sect. 10.6. Finally, we consider in Sect. 10.7 a method to obtain a lens model for extended sources with corresponding resolved images, as is the case for the radio ring image MG1131+0456 described in Sect. 2.5.6.

10.1. Roots of one-dimensional equations

As we saw in Chap. 8, the lens equation for many lens models can be reduced to a one-dimensional form. This is trivially true for symmetric lenses, it is also true for quadrupole lenses (Sect. 8.2) and for lenses with elliptical potentials (Sect. 8.4.2). Here, we consider the problem of finding all the roots of the equation $f(x) = 0$, and we assume that f is continuous and sufficiently differentiable on the interval of interest. If the function f is a polynomial, it is more convenient to use one of the standard routines to find all the roots.

Let us suppose that, for $a < b$, $f(a)f(b) < 0$; then, from continuity, it is clear that f has at least one root x in the interval $a < x < b$. There are standard routines (e.g., ZBRENT in [PR86.1]) which find at least one root with any desired precision, limited only by one's machine, and provided that the function f can be computed with sufficient precision. Thus, once an isolated root is bracketed, it can be found easily.

A search for all the roots of an equation starts with a choice of a range of x-values, divided into a number of intervals. Two difficulties arise immediately. First, one must insure that there are no roots outside the chosen range. Of course, a very large range does not necessarily include all roots, and computer time increases with the number of intervals checked for a change of sign. Fortunately, there is an easier route that can be followed in the GL context. A lens with a finite mass distribution will not produce images arbitrarily far away from the center of the lens (see Sect. 5.4). For large values of $|\mathbf{y}|$, there will be one image at $\mathbf{x} \approx \mathbf{y}$, and possibly others near the center of the lens. Thus, one can usually find a range of values of x that includes all physically feasible roots. Second, suppose there are two roots in the interval $a < x < b$; then, $f(a)f(b) > 0$, and the sign-change criterion will not detect these roots. Increasing the number of subintervals will lower the likelihood of missing images, but it cannot prevent it. If the range for x were of order unity, one would need 10^5 subintervals to find all the roots, provided they are separated by at least 10^{-5}. Of course, there is no guarantee that this is the case. In particular, there will be a close pair of images if the source is sufficiently near a caustic, which corresponds to two close roots. Additionally, the amount of computing time will be excessive for very fine divisions of the range of interest.

However, since we have assumed that the function under consideration is sufficiently differentiable, one can use the property that, between any two roots of f, there is at least one root of the first derivative f'. Suppose one knew all roots of f'; then, one could easily find all the roots of f by applying the sign criterion to the subdivision of the range given by the roots of f'. The roots of f' can in principle be found by a similar interval search for f'', and so on. But what if the highest derivative considered has closely spaced roots? Then the method we propose does not work!

However, for the lens problems under consideration, not more than three images will be very close together, in general. As discussed in Chap. 6, the generic catastrophes of the lens equation either lead to two (fold) or three (cusp) closely-spaced images. Hence, for most situations we can expect at most three roots of f close together, or two roots of f', or a single root of f''. For such cases, the method works well.

Let us consider a range $x_1 < x < x_2$, divided into N intervals. We compute f and all its derivatives up to order n at all dividing points, and we apply the sign-change criterion to the highest derivative. If any roots are found, we add them to the grid of interval points, and we search for roots of $f^{(n-1)}$ on the extended grid; any other roots that are found are added to the grid, and so on. Finally, we compute the roots of f itself. The method will certainly fail for functions like $\sin(1/x)$, where the derivatives of all orders have equally closely-spaced roots, but such pathological functions are generally irrelevant to GL problems. The derivatives of f can either be calculated by analytical differentiation, or by using a routine for numerical evaluation of derivatives. Note that the NAG library includes such a routine (D04AAF).

There are cases, however, where one should proceed with particular caution. If a source lies exactly on a caustic, it will have a second-order root of the function f, at the point where its critical image is found. More generally, second-order roots occur in cases where the lens is mirror-symmetric with respect to one or two axes, as for the elliptical and quadrupole lenses considered in Chap. 8. To account for this, the roots of all derivatives should be checked as possible roots of the function itself. In particular, if a root of f' is also a root of f, a second-order root for f has been found. Note, however, that such higher-order roots occur only for special source positions, either close to a fold or cusp, or to a line of symmetry.

It is a worthwhile effort to compute the number of images for some source positions which can be plotted, together with the caustics in the source plane. Since the number of images changes by two if, and only if, the source position changes across a caustic, one can check whether the routine finds all the images. If it succeeds for sources near folds and cusps, and for sources on points and axes of symmetry, it may be assumed that it works well in general.

10.2 Images of extended sources

There is a simple method to plot the images of an extended source [SC87.2]. Consider the function

$$\chi(\mathbf{y}) = (\mathbf{y} - \mathbf{y}_0)^2 \quad ;$$

the points where $\chi =$ const. describe a circle of radius $\sqrt{\chi}$ around $\mathbf{y}_0$. One can consider χ as a function of $\mathbf{x}$ by defining

$$\chi(\mathbf{x}) = [\mathbf{y}(\mathbf{x}) - \mathbf{y}_0]^2 \quad ,$$

where $\mathbf{y}(\mathbf{x})$ is given by the lens equation. Then, all the points $\mathbf{x}$ of constant χ are mapped onto points $\mathbf{y}$ which have a distance $\sqrt{\chi}$ from $\mathbf{y}_0$. In other words, the curves of constant χ in the lens plane are the images of a circle of radius $\sqrt{\chi}$ around $\mathbf{y}_0$ in the source plane. If the latter circle can be considered an isophote of a circular source, one has thus found the corresponding isophotes of the images.

Fortunately, there are standard plot routines for contour plots. One proceeds as follows: the function χ is computed on a grid in the lens plane. This is straightforwardly done by applying the lens equation $\mathbf{y}(\mathbf{x})$. The array χ is then fed to a standard contour plot routine. By choosing the contours to be drawn, one can obtain the images of a source with distinct isophotes. Note that the vector field $\mathbf{y}(\mathbf{x})$ need be computed only once, if plots are needed for several source positions.

This method is not restricted to circular sources. For example, one can consider elliptical sources with the function

$$\chi(\mathbf{y}) = a(y_1 - y_{01})^2 + b(y_2 - y_{02})^2 \quad ;$$

curves of constant χ are then ellipses, and finding the images proceeds in the same way as above. Examples for an application of this method are given in Fig. 8.9 and Fig. 13.7.

As an aside, we would like to note that this method can be profitably combined with color graphics. Coding the values of χ as colors, and assigning each grid point in the lens plane a color corresponding to its value of χ, one obtains an "intensity" map of the images of the source. The color graphics method can be generalized to arbitrary source profiles, i.e., if the brightness distribution of the source is described by a function $I(\mathbf{y} - \mathbf{y}_0)$ and the values of I are coded into colors, one can derive the corresponding brightness distribution of the images. Note that these routines are very robust and not very time consuming.

10.3 Interactive methods for model fitting

One of the classical problems in GL theory is the construction of a GL model which reproduces an observed configuration of images. Usually one takes a multiple component model for the lens, and varies the model parameters (including the source position), until an acceptable fit has been obtained. These models should not only reproduce the positions of the images, but also their brightness ratios and, if observable, the relative magnification matrices for image pairs. Since the number of free parameters is usually large, it is quite difficult to explore parameter space with "blind searches"; rather, one should look for a means to vary the parameters interactively, until one approaches the desired image configuration.

Following [BL86.1] we describe a method to achieve this goal. Consider a GL model with a number of free parameters, and suppose that one can compute the corresponding deflection potential $\psi(\mathbf{x})$. It is then a straightforward matter to compute the Fermat potential $\phi(\mathbf{x}, \mathbf{y})$ for any source position on a grid in the lens plane. The contours of constant ϕ can then be plotted (e.g., on a screen) and minima, maxima and saddle points may be readily identified. If the program is written such that one can interactively change the source position and the parameters of the lens model, one can study the evolution of the image positions as a function of these parameters. One can then perform changes in parameter space that move the image positions towards desired positions. One can also read off approximate values for the magnification of the images, since these depend on the curvature of the Fermat potential at the stationary points.

Another method applicable to interactive model fitting is "cross-hairs imaging" ([NA82.1], [KA88.2]): one simultaneously plots the contours $y_1(\mathbf{x}) = y_{01}$ and $y_2(\mathbf{x}) = y_{02}$. The intersections of these two contours are the image positions of a source at $\mathbf{y} = (y_{01}, y_{02})$. Although it may seem that this second method is faster than the first one, we can argue that the former offers more information in one plot (or iteration step) than the latter, since it provides the global shape of the time-delay surface. Such information can be very helpful for modifying the lens parameters in the direction of improving agreement with the observed image configuration.

Finally, we describe a simple method for model fitting which, in the first step, does not proceed with interactive graphics: suppose one had observed N images of a point source at angular positions $\mathbf{x}_i$ with fluxes S_i, $1 \leq i \leq N$. Assume that a GL model with M parameters p_ν, $1 \leq \nu \leq M$ can account for the observed image configuration. Fixing the p_ν determines the lens mapping, $\mathbf{y}(\mathbf{x}, p_\nu) = \mathbf{x} - \boldsymbol{\alpha}(\mathbf{x}, p_\nu)$. Since the images are assumed to originate from the same source, we require $\mathbf{y}(\mathbf{x}_i, p_\nu) = \mathbf{y}(\mathbf{x}_j, p_\nu)$. Assuming that microlensing is unimportant and the differential absorption of the images is insignificant (or can be corrected for by standard reddening-extinction techniques), the magnification ratios from the lens model should reproduce the observed flux ratios. If we define [SC88.1] the function

$$E(p_\nu) = \sum_{i=1}^{N-1} \Big[|\mathbf{y}(\mathbf{x}_i,p_\nu) - \mathbf{y}(\mathbf{x}_{i+1},p_\nu)| + w_i |\mu(\mathbf{x}_i,p_\nu)S_{i+1} - \mu(\mathbf{x}_{i+1},p_\nu)S_i|\Big] \quad,$$

with $w_i \geq 0$, a perfect match between lens model and observations would yield $E(p_\nu) = 0$. The w_i are weight factors; if it can be assumed that microlensing is important for some (or all) images, the corresponding weight factors w_i should be chosen small in order to assign a low weight to a mismatch between observed and predicted flux ratios. Also, the w_i corresponding to faint images for which the measurement error of the flux is larger should be relatively low. After the w_i are chosen, standard software can be used to search for the minimum of $E(p_\nu)$ (e.g., the simplex downhill method in Sect. 10.4 in [PR86.1]). Once this "best model" is found, it has to be checked whether this model predicts additional, unobserved images of the source. If it does, the model has to be rejected as unacceptable, and a different parametrization should be tried.

10.4 Grid search methods

The numerical procedures described in the last two sections can readily be used to produce informative graphics for any GL system. However, sometimes the solution of the lens equation is only part of a larger computational program, which requires numbers instead of plots. For example, the imaging of extended sources described in Sect. 10.2 does not determine the coordinates of the images of a point source. To obtain these numbers one would like to solve the equation $\mathbf{y}(\mathbf{x}) = \mathbf{y}_0$, i.e., to find all the roots of a two-dimensional system of equations. Of course, this is more complicated than in the one-dimensional case, and no robust algorithm is readily available. As in the one-dimensional case, it is always possible to find one root if one knows approximately where it is (see next section). But there are no guaranteed methods to find all the roots in a given part of the lens plane. The best one can try is scanning a grid for solutions, by analogy with the pedestrian way to solve a one-dimensional equation. We now describe such a method, which has been used with some modification in many papers (e.g., [YO80.1], [BL87.2], [FA91.2]). It yields the image positions for a large number of sources simultaneously, and thus allows a complete analysis of a given lens model, whereas the methods described in the preceding section require computation for any individual source position.

Let us cover the region of interest in the source plane with a number of points $\mathbf{y}_k$ for which solutions of the lens equation are needed, and lay down a rectangular grid $\mathbf{x}(m,n)$ in the lens plane. This grid can readily be mapped onto the source plane, onto the points $\mathbf{y}(m,n)$. For each grid point $\mathbf{x}(m,n)$, two triangles are defined, as shown in Fig. 10.1a. The vertices

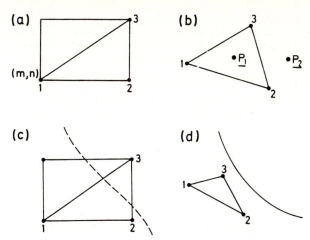

Fig. 10.1a–d. Illustration of the grid search method (see text). (**a**) A rectangular grid cell in the lens plane, labeled by (m, n). The points 1,2,3 define the lower-right triangle, which is mapped onto the points 1,2,3 in the source plane, as seen in (**b**). The point $\mathbf{P}_1$ lies inside this triangle and has thus an image in the corresponding triangle of the lens plane, whereas $\mathbf{P}_2$ lies outside. (**c**) If a critical curve crosses the triangle in the lens plane, the grid search method becomes inaccurate: a point lying between the line connecting points 2 and 3 in the source plane (**d**) and the caustic has in fact two images in the corresponding triangle of the lens plane, but these are not found by the grid search method

$\mathbf{y}(\mathbf{x})$ of the mapped points of such a triangle correspond to a triangle in the source plane. Every source position $\mathbf{y}_k$ can now be checked to see whether it is inside this triangle[1]; if it is, this source position is assigned the label of the corresponding grid point (m, n) in the source plane [and, if desired, the corresponding magnification at $\mathbf{x}(m, n)$, which can also be obtained approximately by the ratio of the areas of the triangles in lens and source plane]. After the whole grid in the lens plane has been mapped, all source points $\mathbf{y}_k$ have one or more labels, which can be then traced back to the corresponding grid points $\mathbf{x}(m, n)$ in the lens plane. Each such label corresponds to an approximate image of the source at $\mathbf{y}_k$.

Our grid labeling may proceed thus: let (m, n) be the grid coordinates for the lower-left vertex of a triangle in the source plane. We define an array $NIMAGE(k, j, NCOUNT)$, such that each point $\mathbf{y}_k$ in the source triangle is assigned the values $NIMAGE(k, 1, NCOUNT) = \pm m$, $NIMAGE(k, 2, NCOUNT) = n$ [where the sign is chosen according to whether the source point $\mathbf{y}_k$ is in the upper or lower triangle corresponding to the grid point (m, n)], and $NCOUNT$ is an index raised for every new source triangle that is found to contain the source under consideration. Since a typical lens produces at most, say, seven images, the amount of storage needed is 14 times the

---

[1] One method that serves this purpose [FA83.1] is the following: consider a point $P = \mathbf{y}_k$ and the vectors $\mathbf{d}_i$ from the vertices of the triangle to P. The cross product of two vectors $\mathbf{X}$, $\mathbf{Y}$ is $\mathbf{X} \times \mathbf{Y} = X_1 Y_2 - X_2 Y_1$. From Fig. 10.1b, it is clear that the products $\mathbf{d}_1 \times \mathbf{d}_2$, $\mathbf{d}_2 \times \mathbf{d}_3$, and $\mathbf{d}_3 \times \mathbf{d}_1$ all have the same sign if the point P lies inside the triangle; this is the case for the point P_1 in Fig. 10.1b, for which all these products are positive. For the point P_2, which lies outside the triangle, the product $\mathbf{d}_2 \times \mathbf{d}_3$ is negative, and the other two are positive.

number of gridpoints in the source plane. The reason for mapping triangles instead of rectangles is that a triangle on the image plane is mapped onto a convex region in the source plane, but a rectangle in general is not; it is thus simpler to test whether a point $\mathbf{y}_k$ lies within a region defined by three points than in one defined by four points.

After the whole grid in the lens plane is mapped, one can find the approximate position of the images for every source point $\mathbf{y}_k$ from the array $NIMAGE$. The quality of the approximation of the images depends on the coarseness of the grid in the lens plane. One must, therefore, reach a compromise between computer time and accuracy. If more accurate image positions are needed, the approximate positions from the grid search can be used as a starting point for the 'transport of images' that we describe in the following section.

One of the difficulties of the grid search, which is straightforward as described, is that one can miss images very near critical curves, as can be easily seen from Fig. 10.1c,d. Of course, for the same reason one will miss closely-spaced roots in the one-dimensional case. Hence, to find all the images of sources near a caustic, we can extend the method beyond pure grid search, as follows: after the grid search has been performed, we know the number of images for all points $\mathbf{y}_k$ on the source grid. We can then identify those points with neighboring points that have different numbers of images; these points are near to a caustic. For every point $\mathbf{y}_k$ with at least one neighboring point $\mathbf{y}_j$ with more images than $\mathbf{y}_k$, one can 'transport' the images of $\mathbf{y}_j$ to the source position $\mathbf{y}_k$, using the method we describe next. With this prescription, one can find all images of all sources.

10.5 Transport of images

Sometimes we can obtain all of the images for a certain point source at $\mathbf{y}$, but we need the image positions for a nearby source position $\mathbf{y}_0$. Or, we have an approximate solution $\mathbf{x}$ of the lens equation for a source at $\mathbf{y}_0$, and we want to calculate a more accurate image position. These two problems are closely related to each other; let us consider the latter first.

If $\mathbf{x}$ is an approximate solution of the lens equation for a source at $\mathbf{y}_0$, and $\mathbf{x}_0$ is the exact solution close to $\mathbf{x}$, then we can apply Newton's approximation scheme to find a better approximation. Let $\mathbf{y}$ be the point in the source plane that corresponds to the approximate solution, $\mathbf{y} = \mathbf{y}(\mathbf{x})$, and let $A(\mathbf{x})$ be the Jacobian matrix for the lens equation at that point. Then, the difference $\Delta\mathbf{x} = \mathbf{x}_0 - \mathbf{x}$ is related to the difference vector $\Delta\mathbf{y} = \mathbf{y}_0 - \mathbf{y}$, up to first order, by $\Delta\mathbf{x} = A^{-1}(\mathbf{x})\Delta\mathbf{y}$, where A^{-1} is the inverse of the Jacobian matrix. The new image position $\mathbf{x}' = \mathbf{x} + \Delta\mathbf{x}$ will, in general, not coincide with the exact solution $\mathbf{x}_0$, but will be a better approximation than $\mathbf{x}$, provided that $|\Delta\mathbf{y}|$ is sufficiently small and the images under consideration are not near a critical curve. This new position $\mathbf{x}'$ can then be used as a starting point for the next iteration, and we can iterate until the $|\Delta\mathbf{y}|$ becomes as small as desired. In general, this iteration method works quite well and quite quickly. It can be improved by considering the ordinary differential equation $\frac{d\mathbf{x}}{d\lambda} = A^{-1}(\mathbf{x}(\lambda))(\mathbf{y} - \mathbf{y}_0)$; the solution at $\lambda = 1$ is the image of the point $\mathbf{y}$ if we take the initial condition $\mathbf{x}(0) = \mathbf{x}_0$. An approximate solution of this differential equation can be obtained by a fourth-order Runge-Kutta step (see [PR86.1], Sect. 15.1). This method works more efficient than the simple Newton scheme described above.

The first of the problems mentioned above has the same structure: if the images are known for a source point $\mathbf{y}$ and are to be computed for a nearby point $\mathbf{y}_0$, the old images will be approximations for the images of the new source position, and the iteration can be applied to all of the images individually. Such transport of images is an effective method if the image positions have to be calculated for many different source positions.

The transport of images of a source which crosses a caustic can be carried out if the direction of crossing is such that two images fuse; one can easily deal with that problem because the two critical images become more and more magnified as the Jacobian determinant tends to zero. It is straightforward to define a criterion for excluding these two images, e.g., when their magnifications exceed a given threshold. Therefore, the combination of a grid method as described in Sect. 10.4 and the transport of images provides a powerful method for solving the lens equation for a grid of source positions.

10.6 Ray shooting

General description. As we will see in Chap. 11, one sometimes is interested in the total magnification of an extended source seen through a GL. The method described in Sect. 10.2 directly provides graphics from which one can read off an approximate value of the total magnification, by estimating the total area of all images and dividing this area by that of the unlensed source. If we are interested in more quantitative information, we must use a different method. If the total magnification is needed only for a small number of sources, it is easiest to solve the lens equation for a sufficiently large number of source points inside the extended source, to add the absolute values of the magnifications of the individual images for each source point (thereby obtaining the total magnification of the corresponding point source), and integrate over the extended source. Since this method is very time-consuming, it should be used only if the number of sources for which the information is required is modest.

If we need to find the total magnifications for a large number of extended sources, we can use the ray-shooting method ([SC86.4], [SC87.3]). In this method, one defines a set of sources in the source plane (e.g., a set of circular sources characterized by their radii and the positions of their centers) and maps a uniform grid of points $\mathbf{x}$ via the lens equation onto the source plane. Each of these mappings can be thought correspond to a light ray. Once the corresponding point $\mathbf{y}$ is computed, it is checked to see whether it is inside any of the sources under study; if it is, the index of an array, which counts the total number of rays in each source, is increased by one. After the whole grid is mapped, the total magnification of each source can be computed as the ratio of the number of rays that hit it, and the number it would contain in the absence of the GL. If the sources have prescribed brightness profiles, each light ray which hits a source must be multiplied with a weight factor proportional to the local value of the surface brightness. Ray shooting is thus very similar to an experiment in optics, where one shines light through an arrangement of lenses; the magnification then depends on the position in the "source" plane and is proportional to the density of rays.

Sometimes it is more convenient to divide the source plane into small pixels and compute, with the foregoing method, the magnification of these pixels, i.e., the magnification of sources with the size and at the position of the pixels. The magnification for more realistic source shapes and brightness profiles can then be obtained from this pixel field by summing over the magnification of those pixels covered by the source, weighted by the corresponding surface brightness, provided the sources to be considered are larger than the pixels.

Before using the ray shooting method, two points must be considered. First, the size of the grid in the lens plane must be chosen so that no rays are "missing", in the sense that points $\mathbf{x}$ outside the grid are mapped onto one of the sources. If that were the case, one would introduce an error due to the missing flux. Hence, one must determine a region in the lens plane such that no point outside the region is mapped via the lens equation onto one of the sources. Second, the density of the grid must be adapted to the source size. The accuracy of the total magnification depends on the number N of rays collected in the source. The relative error in the magnification is smaller than $\sim 1/\sqrt{N}$; since we assumed a uniform grid, the relative error is approximately $N^{-3/4}$ [KA86.2]. Hence, the smaller the sources, the denser the grid must be. In Chap. 11, we describe several applications of the ray-shooting method; here, it should be noted that it can be used in a very straightforward fashion, even for relatively complicated lenses. It is therefore a method which can never fail if used appropriately.

Ray shooting in microlensing calculations. The most common application of ray shooting has been the study of ML, where one is interested in the spatial distribution of magnifications of sources behind a random star field. We describe the motivation in the following chapter; here, we consider the technical aspects of these calculations. Consider an ensemble of N compact objects, characterized by their positions $\mathbf{x}_i$ and their mass $m_i M_0$, where M_0 is a reference mass. The lens equation then reads

$$\mathbf{y}(\mathbf{x}) = \mathbf{x} - \sum_{i=1}^{N} m_i \frac{\mathbf{x} - \mathbf{x}_i}{|\mathbf{x} - \mathbf{x}_i|^2} - \begin{pmatrix} \kappa_c + \gamma & 0 \\ 0 & \kappa_c - \gamma \end{pmatrix} \mathbf{x} \quad , \tag{10.1}$$

where the sum describes the deflection by the stars, and the last term is a quadrupole contribution due to a larger matter distribution, e.g., from the galaxy that contains the stars. The field of sources is usually chosen to be a pixel field, i.e., a certain region in the source plane (usually square shaped) is divided into pixels, where the light rays are collected.

Although it is straightforward to compute the deflection angle for any ray in the lens plane, the ray-shooting method limits the range of possible parameters of ML situations to be considered, as can be seen as follows. The number of pixels in the source plane is typically of the order of 10^6; to determine the magnification for each pixel with sufficient accuracy, one should have at least 100 rays per pixel on average, which means that at least 10^8 rays must be calculated. In practice, this number is considerably larger, because due to the properties of the Schwarzschild lens a large area of the lens plane must be covered with lenses and rays, to insure that only little flux is lost. Therefore, a large region must be included in the calculation. However, most of the light rays will not hit the area of the source plane under consideration. The number of stars needed in a particular situation depends mainly on the surface mass density of stars and their mass spectrum, and is, for moderate optical depth, of the order of 10^3, but for star fields whose density is near critical, many more stars must be included. Multiplying the numbers above, we find that, in a typical simulation, 10^{12} to 10^{15} individual deflections are needed, but cannot be reasonably calculated on any computer.

The history of ML simulations with ray shooting (a different method [PA86.1] is described briefly in Sect. 11.2.5) is thus a story of increasingly cleverer methods to approximate the lens equation (10.1). The straightforward shooting is confined to about 100 stars due to computing time limitations [KA86.2]. A large amount of time can be saved by noting that the deflection angle of any ray can be split into two contributions, those of 'near' stars and those of the rest; the latter is a relatively smoothly varying function of the ray impact vector and can thus be expanded in a Taylor series ([SC87.3], [KA89.1]). In practice, one divides the lens plane into 'shooting squares'; surrounding each such square a larger square is selected. Stars within this outer square are termed 'near' and their deflection is computed explicitly, whereas the deflection for the rest of the stars is expanded in a Taylor series about the center of the shooting square. This method has been used for up to several thousand stars.

Further progress in ray shooting through a random star field occurred through the use of hierarchical tree methods, which were developed for N-body simulations in stellar dynamics and first applied to microlensing in [WA90.2] and particularly adopted to ML in [WA90.4]. The computing time for this method increases only as $\log N$, in contrast to the method described above, for which the time increases linearly with N. The basic idea of the tree code is as follows: find a square where all the stars are included and divide this square into four equally sized smaller squares. For each of these squares (cells), compute the total mass of the stars included in it, the center of mass, and higher order multipole terms, which are then stored. Each of these four cells, if they contain more than one star, can be subdivided further into four smaller cells, and so on. This process defines a tree of hierarchies. For each cell on each level the mass, center of mass, and multipole components are computed and stored just once. The procedure is stopped when all stars are resolved into cells, i.e., at the highest level of the hierarchy, each cell contains one star. Then, the information for the stars is completely contained in the coefficients of the cells. The total number of cells is in general less than $2N$; N cells contain just one star, for which only the position (which agrees with the center of mass) and mass need be stored.

To compute the deflection for a light ray, one needs the deflection caused by nearby stars very accurately, and thus one needs to go up the hierarchy. On the other hand, a group of stars far away from the ray under consideration can be combined, and the multipole components for such a group suffice to compute their deflection. Hence, the question of how far a group of stars is resolved (how far up the hierarchy one must climb) depends on the distance of this group to the ray under consideration. The threshold may be, for instance, the diameter of a cell divided by the distance to the center of the cell. If this ratio is larger than a given value, the cell is resolved into its four subcells, and so on, until the accuracy estimate is met.

A further increase of the efficiency of this hierarchical tree method is provided by combining it with interpolation. As described so far, the method does not use the information that the deflection angle of very closely-separated rays will be nearly the same. Since a large fraction of the computing time is used in climbing up the hierarchy, this information can be used profitably in the code. Consider a square in the lens plane; for the central ray in this square, one can use the method already described, i.e., fix the level structure of the hierarchy, and then compute the deflection angle from the stored multipole components of the cells explicitly used. This hierarchy can then be held fixed for all rays in this square. In addition, the deflection angle from distant stars will be a smooth function of the ray's impact vector, and can thus be expanded or interpolated. As before, the stars (or cells) are divided into 'near' and 'far' cells, and for the latter ones interpolation is used. An efficient method for interpolation is as follows: let the deflection angle from the 'far' stars be computed for the four rays at the corners of a square. Then, consider an expansion of the gravitational potential ψ of the 'far' stars around the center of this square. The constant term is irrelevant, as it does not contribute to the deflection. The first-order terms have two independent coefficients, ψ_1 and ψ_2 (where subindices denote partial derivatives). The second-order terms have three coefficients, but these are not independent of one another. Since the surface mass density of the stars vanishes, from (5.13), $\psi_{11} = -\psi_{22}$, so that only two coefficients are independent. It is easy to see that for each order of expansion, only two coefficients are independent (e.g., $\psi_{111} = -\psi_{122}$, $\psi_{112} = -\psi_{222}$). Thus, an expansion up to fourth order has eight independent coefficients, and this is the number of known deflection components. Hence, these eight coefficients can be determined from the deflection angles at the corners of the square under consideration, and the deflection from the 'far' stars within the square can then be computed by using this expansion.

The method just described has proved to be extremely efficient and much faster than the other methods; in fact, the number of stars which can be considered with this method is more likely to be limited by the memory of the computer than by CPU time. Actually, the code used and described in [WA90.4] is close to 'optimal', as a considerable fraction of the computing time is used for 'sorting rays into pixels', a part of the program that cannot be vectorized. Therefore, even if a still more efficient method is found to calculate deflection angles, this fundamental limitation will prevent that the calculation can be speeded up by more than a factor of about 2.

10.7 Constructing lens and source models from resolved images

When a point source (i.e., an unresolved source) is multiply imaged by a GL, the observations yield just a handful of constraints on a lens model: it should reproduce the image separation, the image positions relative to the lens (if observed), and the relative magnifications (if one believes that ML is not important). For the 0957+561 system, the VLBI structure of the two images has been resolved, and this turned out to yield a strong constraint on the lens model. It is clear that multiple images of a point source only probe the lens locally, at a few positions. The situation changes drastically if an extended source is imaged, since then the lens is probed 'in parallel' at all image points, and therefore, lensed extended sources are potentially much more valuable for model fitting than lensed point sources.

The advantage of extended sources has been beautifully demonstrated for the radio ring image MG1131+0456 (Sect. 2.5.6): consider the observed image of an extended source, described by a brightness distribution $I(\mathbf{x})$. If we choose a lens model, we have a mapping $\mathbf{y}(\mathbf{x})$. Since the surface brightness of a source is unchanged by lensing, a source point at $\mathbf{y}$ must then have a surface brightness $I(\mathbf{x})$, where $\mathbf{x}$ is a solution of the lens equation for a source at $\mathbf{y}$. In general, however, a point source at $\mathbf{y}$ will have more than one image, say, images at $\mathbf{x}_i$, $i = 1, ..., N$, and in general, $I(\mathbf{x}_i) \neq I(\mathbf{x}_j)$. Hence, all the image points assigned to a given source point with the lens equation must have the same surface brightness, if the guessed lens model is correct.

In practice, one never finds a lens model that meets all requirements for all source points; for example, the observations are noisy, the finite resolution of a telescope will smooth the surface brightness of the image, etc. One can, however, define a quantity that measures the "goodness of fit" of the procedure described above. Consider the source plane to be divided into pixels, labeled (i, j). The source point at (i, j) will have N_{ij} images under the lens mapping; denote the surface brightness of the images by I_{ij}^k,

$k = 1, ..., N_{ij}$. The best guess for the surface brightness of the source point (i,j) is then the mean of the surface brightnesses of the images, denoted by $\bar{I}_{ij}$. If the lens model chosen is the correct one, then $I_{ij}^k = \bar{I}_{ij}$ for all k. That is, the quantity

$$S_{ij} = \sum_{k=1}^{N_{ij}} \left(I_{ij}^k - \bar{I}_{ij}\right)^2$$

is a measure of the 'distance' between the observations and the model predictions at the source point (i,j). Hence, a measure for the total discrepancy between observations and predictions is the sum $S = \sum S_{ij}$ of these quantities over all multiply-imaged source points; for further details, the reader is referred to [KO89.1]. The best lens model is then found by varying the model parameters until S is minimized. Note that this lens model directly determines the brightness distribution for the lensed source. For an application of this method, see Sect. 2.5.6.

For a successful application of the method as described here, the angular resolution of the observed surface brightness distribution is of vital importance [KO89.1]. A generalization of this method which accounts for the finite angular resolution of observations would be highly desirable. Such a generalization could then be used to investigate the lensing nature of some multiple sources with resolved radio components (see, e.g., [SU90.2]).

11. Statistical gravitational lensing: General considerations

Introduction and motivation. The study of isolated GL models has provided us with the basic understanding of the properties of gravitational light deflection, such as the possibility of multiple imaging of sources and the magnification and distortion of the shape of images. Detailed knowledge of the properties of simple lens models is essential for the understanding of observed lens cases. In this and the following chapter, we consider the statistical properties of an ensemble of lenses.

Zwicky [ZW37.2] estimated the probability that a distant point source is multiply imaged by "extragalactic nebulae" using the surface density of these objects on the sky. This first and basic problem of statistical gravitational lensing has since been studied in considerable detail ([LI64.1], [PR73.1], [TU84.1], [DY84.1], [HI87.1], [BL87.2], [KO87.2]). The results of such an investigation depend on the assumed distribution of lens masses and their individual matter distributions. A comparison of these results with a sample of observed lens cases can in principle allow one to constrain the lens contents of the universe. We discuss this problem in more detail in Sect. 12.2.

A second typical problem of statistical lensing is the so-called amplification bias. Let us consider a sample which should include all sources of a certain kind in a region of the sky brighter than a given threshold (a flux-limited sample). From the observed fluxes of the sources and their distances (e.g., determined from their redshifts) we can derive the intrinsic luminosities of the sources. If a source is magnified by a GL, its derived luminosity will not be the true one, but will be higher in general. Moreover, there may be sources in the sample which do not belong there because they are intrinsically too faint to be included, but have been magnified above the threshold of the sample. Since flux-limited samples of extragalactic sources are used to derive information about the evolution of the sources and about the structure of the universe, the magnification can "fool" astronomers. Statistical lensing theories are used to estimate the importance of this effect and its consequences.

Outline of this chapter. The examples just discussed are meant to motivate the treatment that follows. In this chapter, we consider the more general aspects of statistical lens theory, and specific applications are deferred to Chap. 12. The concept of cross-sections is introduced in Sect. 11.1. The basic idea is as in nuclear physics, if one wants to determine the behavior of a particle beam sent through a target. There, one needs the cross-section for the interaction of a particle in the beam with a single atom of the target. Similarly, one utilizes the cross-sections of individual lenses to determine the properties of an ensemble of lenses. A simple ensemble of lenses is a random star field, i.e., a distribution of point masses all at a single redshift; it is investigated in some detail in Sect. 11.2.

We then turn to the application of cross-sections to a cosmological scenario, i.e., randomly distributed lenses throughout the universe (Sect. 11.3). Clearly, the magnification probabilities for sources in the universe are strongly related to the light propagation in a clumpy universe. In Sect. 11.4 we take up the discussion from Sect. 4.5, where the Dyer–Roeder differential equation was discussed. We show that statistical lensing is a useful tool for the investigation of light propagation in a clumpy universe.

11.1 Cross-sections

Example: Magnification by a Schwarzschild lens. Consider a point source at distance D_s, and a point mass at distance D_d from Earth. The separation of the two images and their magnifications depend on the relative alignment of source, observer, and lens. There is a one-to-one relationship between the source position y measured in units of η_0, see (2.6c), and the corresponding total magnification μ_p, see (8.29b), where total magnification means the summed magnifications of the individual images. Thus, for any $\mu_p > 1$ there is a value of y such that, if the distance of the source is less than y, the latter is magnified by more than μ_p:

$$y^2 = 2\left(\frac{\mu_p}{\sqrt{\mu_p^2 - 1}} - 1\right) \; . \tag{11.1}$$

Hence, we can define the dimensionless cross-section $\sigma(\mu_p) = y^2\pi$ for magnification larger than μ_p. The corresponding dimensional cross-section $\hat{\sigma}(\mu_p)$ in the source plane is

$$\hat{\sigma}(\mu_p) = \eta_0^2 \sigma(\mu_p) \; . \tag{11.2}$$

Thus, we have derived the magnification cross-section for a point mass, and a point source.

As a second example, we consider the ratio r of the absolute values of the magnifications (brightness ratio) for the two images produced by a Schwarzschild lens, given by (2.25). Again, r is a monotonic function of the distance y and can thus be inverted. We find that

$$y = r^{1/4} - r^{-1/4} \; ; \tag{11.3}$$

hence, we obtain the cross-section $\sigma(r < r_0)$ for a brightness ratio less than r_0,

$$\sigma(r < r_0) = \pi \left(r_0^{1/2} + r_0^{-1/2} - 2 \right) \quad . \tag{11.4}$$

General definition of a cross-section. Consider a source and a lens, both at fixed distances from Earth. The lens may be described by a set of parameters, and the source is characterized, say, by its size and its brightness profile. If one is interested in a certain property Q of this GL system, one can ask where the source must be in its own plane such that the images have the property Q. Two examples for Q were given above, namely, that the total magnification is larger than μ and that the brightness ratio of the images is smaller than r. More complicated examples of Q will be considered below. The question can be answered through an analysis of the GL model, as demonstrated above for the Schwarzschild lens. One usually finds that the source must be in a certain region of the source plane. The area of that region is then the Q–cross-section $\hat\sigma_Q$ for this lens–source system. In Chap. 5, we introduced dimensionless quantities in the source and lens planes by defining a length scale ξ_0 in the lens plane; correspondingly, one can define a dimensionless cross-section σ_Q, related to $\hat\sigma_Q$ through

$$\hat\sigma_Q = \eta_0^2 \sigma_Q \quad , \tag{11.5}$$

with $\eta_0 = (D_s/D_d)\xi_0$.

Problems in determining cross-sections. The Q–cross-section depends on Q, the lens and source parameters, and the distances of lens and source from the Earth. For a point source lensed by a point mass, it is easy to compute, but in general the explicit expressions for $\hat\sigma$ are difficult to obtain, and in most cases we must determine the cross-sections numerically. Since $\hat\sigma$ depends on a considerable number of variables, this is not a trivial task. We return to that problem below, but let us first consider a special case needed to calculate the expected number of observed lens cases in the universe. In order to detect lensing, we require that the lens produce at least two images of the source. If one searches for images with ground-based optical telescopes, the separation of the images should be larger than the resolution limit of the instrument. If the flux ratio is too large, it is likely that the fainter image will be missed in the search. The dynamic range for which two images can be detected increases with increasing image separation. Thus, in this case, one needs to consider the cross-section for the property Q that the lens produce at least two images whose separation is larger than ϑ and whose brightness ratio is smaller than $r(\vartheta)$. Except for the simplest lens models, these cross-sections cannot be computed analytically. Furthermore, this example shows that the cross-sections one needs for theoretical analysis must include the observational biases to allow a comparison with the results of a survey; however, in most cases the observational bias is only partially known (e.g., the resolution of a telescope depends on the atmospheric conditions at the time of observation). The rest of this section will deal with specific cross-sections, which are then used in Chap. 12 when we discuss the applications of statistical lensing.

11.1.1 Multiple image cross-sections

Splitting of point sources by a point-mass lens. Let us return to the example of cross-sections for multiple images with thresholds for the image separations and the brightness ratios. First, consider the point-mass lens (Sect. 8.1.2.). The relation between the impact parameter of the source and the brightness ratio of the two images is given by (11.3). The distance that separates the two images is obtained from (8.28), $\Delta x = \sqrt{y^2 + 4}$. Hence, the cross-section $\sigma(\Delta x, r)$ for images with brightness ratio less than r and separation greater than Δx is:

$$\sigma(\Delta x, r) = \begin{cases} \pi\left(r^{1/2} + r^{-1/2} - 2\right) & \text{for } \Delta x \leq 2 \, , \\ \pi\left(r^{1/2} + r^{-1/2} - \Delta x^2 + 2\right) & \text{for } 2 \leq \Delta x \leq \sqrt{r^{1/2} + r^{-1/2} + 2}, \\ 0 & \text{for } \Delta x \geq \sqrt{r^{1/2} + r^{-1/2} + 2} \, . \end{cases} \quad (11.6)$$

In many cases, it may suffice to consider the image separation fixed, since for $r < 10$, $2 \leq \Delta x \leq 2.34$; that is, for relatively small brightness ratios, the images have a separation close to 2.

To convert the dimensionless cross-section to a dimensional one, we just note that Δx is related to ϑ, the angular separation of the images, by $\vartheta = \Delta x \, \eta_0/D_s$; hence, the cross-section $\hat\sigma(\vartheta, r)$ for an angular separation greater than ϑ is

$$\hat\sigma(\vartheta, r) = \eta_0^2 \, \sigma\left(\frac{\vartheta D_s}{\eta_0}, r\right) \, .$$

Image splitting by a singular isothermal sphere. Now, we consider the singular isothermal sphere treated in Sect. 8.1.4. The properties of this model are even simpler than those of the Schwarzschild lens, because the image separation does not depend on the impact parameter of the source. With the choice of length scale as in Sect. 8.1.4., the image separation is $\Delta x = 2$. Two images occur only for $y < 1$. From (8.35a), we see that the ratio of the absolute values of magnifications of the images is $r = (1+y)/(1-y)$; hence, the cross-section for two images with brightness ratio less than r is

$$\sigma(r) = \pi \left(\frac{r-1}{r+1}\right)^2 \, . \quad (11.7)$$

In addition, it is of interest to consider the total magnification of the two images, since any sample of sources may contain a significant excess of magnified images due to the corresponding bias. From (8.35b), we find for the cross-section for having two images with total magnification larger than μ:

$$\sigma(\mu) = \frac{4\pi}{\mu^2} \quad \text{for } \mu \geq 2 \, . \quad (11.8)$$

Isothermal sphere plus smooth matter sheet. Since most galaxies are not isolated, but are part of a cluster which contains intracluster gas (visible through its radio and X-ray emission) and dark matter (as inferred from the velocity dispersion of the cluster galaxies), we consider a lens model with a galaxy embedded in a homogeneous disk of matter. This approach is valid since the length-scale over which the density of the intracluster material varies is much larger than the scale of an individual galaxy. Let us assume that the galaxy is a singular isothermal sphere; using the same scaling as in Sect. 8.1.4, the lens equation becomes

$$y = x - \frac{x}{|x|} - \kappa_0 x \, , \quad (11.9)$$

where κ_0 is the dimensionless surface mass density of the disk. For simplicity, we assume that $\kappa_0 < 1$. For $y < 1$, this equation has two solutions, one at $x_1 = (y+1)/(1-\kappa_0) > 0$ and one at $x_2 = (y-1)/(1-\kappa_0) < 0$, whereas for $y > 1$ only x_1 is a solution. With the equations of Sect. 8.1.1, we find that the magnifications of the images are

$$\mu_1 = \frac{x}{(1-\kappa_0)[x(1-\kappa_0)-1]} = \frac{1}{(1-\kappa_0)^2}\frac{(y+1)}{y} \quad , \tag{11.10a}$$

$$\mu_2 = \frac{x}{(1-\kappa_0)[x(1-\kappa_0)+1]} = \frac{1}{(1-\kappa_0)^2}\frac{(y-1)}{y} \quad , \tag{11.10b}$$

and the total magnification of the two images for $y < 1$, $\mu = |\mu_1| + |\mu_2|$ is

$$\mu = \frac{1}{(1-\kappa_0)^2}\frac{2}{y} \quad . \tag{11.10c}$$

From these equations, we see that the cross-section $\sigma(r)$ for the brightness ratio to be less than r is given by (11.7), whereas the cross-section for having two images with total magnification larger than μ is

$$\sigma(\mu) = \frac{4\pi}{(1-\kappa_0)^4\mu^2} \quad \text{for} \quad \mu \geq \frac{2}{(1-\kappa_0)^2} \quad . \tag{11.11}$$

The separation of the images is $\Delta x = 2/(1-\kappa_0)$, independent of the impact parameter. Hence, adding a homogeneous disk to the lens model increases the distances in the lens plane by a factor $(1-\kappa_0)^{-1}$ and the magnifications by a factor $(1-\kappa_0)^{-2}$.

The multiple image cross-sections of the lens models treated in Sect. 8.1.5 and 8.4 were considered by [BL87.2]. We defer the discussion of this case to Sect. 12.2, where we discuss it in connection with its application to observations [KO87.2].

11.1.2 Magnification cross-sections

Magnification of extended sources by a Schwarzschild lens. To study the influence and the magnitude of the amplification bias, we need to consider the probability that a source is magnified, which can be determined from the magnification cross-sections. One example was given at the beginning of this section, namely, the case of a point source lensed by a point mass. As will become clear, the amplification bias depends very critically on the source size if this bias is mainly due to ML; therefore, calculations based on the point-source approximation can yield misleading results. Thus, we first consider the magnification μ_e of an extended source lensed by a point mass. It is a function of the position of the center of the source and its dimensionless radius $R = \rho/\eta_0$, where ρ is the radius. The weighted average of μ_p (8.29b) over the source is μ_e. Here, we consider a circular source with uniform surface brightness; then, the magnification, given in general by (6.81), is computed by introducing polar coordinates in the source plane; the integration over the polar angle can be performed easily:

$$\mu_e(y,R) = \frac{2}{R^2\pi}\left[\int_{|y-R|}^{y+R} dr \frac{r^2+2}{\sqrt{r^2+4}} \arccos\frac{y^2+r^2-R^2}{2ry} \right.$$
$$\left. + \mathrm{H}(R-y)\frac{\pi}{2}(R-y)\sqrt{(R-y)^2+4}\right] \quad , \tag{11.12}$$

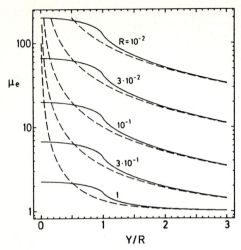

Fig. 11.1. The magnification $\mu_e(y, R)$ of an extended source for a Schwarzschild lens, as a function of the dimensionless position y of the source, for several values of the dimensionless source radius R. The dashed lines show the magnification of a point source

where H(x) is the *Heaviside* step function. The magnification in (11.12) is readily integrated numerically [BO79.1] and is shown in Fig. 11.1. For $R \ll y$, the magnification approaches that of a point source at y, shown as the dashed lines in Fig. 11.1. From (2.28) we find that the maximum magnification of a source with radius R is

$$\mu_{e,\max} = \frac{\sqrt{R^2 + 4}}{R} = \mu_e(0, R) \quad , \tag{11.13}$$

due to the monotonic decrease of μ_e with y. This also allows us to invert (11.12) to solve for $y(\mu_e, R)$; then, the magnification cross-section is

$$\sigma(\mu_e, R) = \pi y^2(\mu_e, R) \quad . \tag{11.14}$$

Approximations and practical calculation of $\mu_e(y, R)$. Approximations to (11.12) given in [SC87.4] provide an accurate representation of $\mu_e(y, R)$ for a large fraction of the (y, R) parameter plane. For small y, the point-source magnification (8.29b) can be approximated by $\mu_p \approx 1/y$; this yields an approximate value for μ_e,

$$\mu_e(y, R) \approx \frac{1}{R} \zeta\left(\frac{y}{R}\right) \quad , \tag{11.15}$$

as shown in Sect. 8.1.1, and the function $\zeta(w)$ is given in (8.27). In Fig. 11.2 we show the contours of constant relative error introduced if one uses (11.15) instead of (11.12). We see that the fractional error is a monotonic function of y/R for fixed R, and increases for fixed y/R approximately as R^2. In particular, for $R \lesssim 0.05$, the fractional error is smaller than about 10^{-2}, for $y \lesssim 5R$. For larger values of y, the point source magnification provides a good approximation to μ_e.

In statistical lens calculations, $\mu_e(y, R)$ needs to be computed frequently. It is thus impractical to solve the integral (11.12) repeatedly. One can proceed as follows: μ_e may be computed on a grid of appropriate size and mesh density in (y, R)-space. Then,

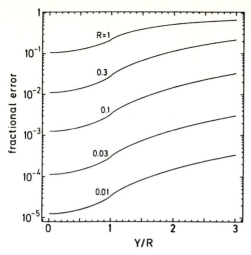

Fig. 11.2. Contours of relative error introduced by the approximation (11.15) for the magnification $\mu_e(y, R)$ of an extended source of radius R at a distance y from the optical axis

the value of $\mu_e(y, R)$ is obtained by two-dimensional interpolation. If a uniform grid is chosen, bilinear interpolation [PR86.1] is very fast because one does not have to search in the grid. The interpolation is not sufficiently accurate for high magnifications, since then μ_e is a rapidly varying function of both of its parameters. However, as we have just discussed, for those cases the approximate result (11.15) can be used. Since the table is computed only once, this is a very efficient way to calculate μ_e.

By numerical inversion of μ_e, one obtains the magnification cross-section $\sigma(\mu_e, R)$. In Fig. 11.3, we show the ratio $q = \sigma(\mu_e, R)/\sigma(\mu_e, 0)$ of the cross-sections of an extended source and a point source. The relation (11.15) suggests that $R\mu_e$ should be used as the independent variable. Note the cut-off in the cross-section at the maximum magnification $\mu_{e,\text{max}}$; for magnifications $\mu_e \lesssim 0.8\mu_{e,\text{max}}$ the cross-section for extended sources is larger than that for point sources.

It is straightforward to extend the foregoing discussion to magnification cross-sections for axially symmetric lenses and extended sources; the symmetry of the lens allows one to invert the magnification function easily, since it depends only on one impact parameter, in contrast to non-axially-symmetric lenses, where μ_e depends on a two-dimensional impact vector. It is also straightforward to account for different brightness profiles of the source [SC87.6].

Magnification cross-sections for a Chang–Refsdal lens with $\kappa_c = 0$. We now turn to the Chang–Refsdal lens model discussed in Sect. 8.2.2. The model is relevant, because there are no isolated lenses: stars are part of a galaxy and their axially-symmetric gravitational field is perturbed by the galaxy as a whole, or the symmetry is broken by matter inhomogeneities at different redshifts.

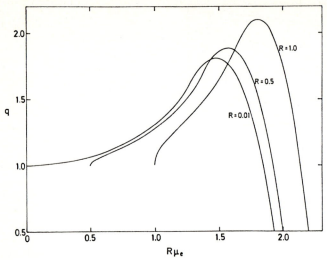

Fig. 11.3. The ratio q of the cross-section $\sigma(\mu_e, R)$ for a source of radius R to be magnified by more than μ_e, and the corresponding cross-section $\sigma(\mu_e, 0)$ for a point source

The magnification cross-section for an extended source and a Chang–Refsdal lens can be calculated with the ray-shooting method described in Sect. 10.6 [SC87.4]. For numerical convenience, we choose a set of square-shaped sources, each of which corresponds to a pixel in the source plane. The explicit solution of the lens equation allows us to determine the region of the source plane where the magnification is larger than a predefined threshold, and we can also find the corresponding region of the lens plane that needs to be mapped. By sorting the rays into pixels, we find the magnifications of the corresponding sources. We can then average over pixels to calculate the magnifications of larger sources. The number of sources with magnification larger than μ_e is proportional to the magnification cross-section $\sigma(\mu_e, b, \gamma)$, where b is the size of the source and γ is the shear parameter [compare with (8.56); for the moment we consider only $\kappa_c = 0$].

In Fig. 11.4, we display the ratio $q = \sigma(\mu_e, b, \gamma)/\sigma(\mu_e, 0, 0)$ of the cross-sections for an extended source in Chang–Refsdal geometry and a point source in Schwarzschild geometry, for $b = 0.1$ and $b = 0.2$ and several values of γ, as a function of μ_e. Again, we see that the cross-section for extended sources cuts off at $\mu_{e,\max}$, but now this cut-off depends on the shear. The reason is clear: the shear determines the structure of the caustics in the source plane. The highest magnifications are obtained for sources very close to a cusp of the caustics. Hence, the function q for $\mu_e \lesssim \mu_{e,\max}$ reflects the properties of the cusp, which in turn depend on γ. For magnifications not close to $\mu_{e,\max}$ the cross-section for an extended source and a Chang–Refsdal lens is larger than for a point source and a Schwarzschild lens. The behavior of the curves at $\mu_e \approx 1$ is due to the fact that the magnification for large impact parameters does not tend to one, but to $(1 - \gamma^2)^{-1}$.

Scaling for the case $\kappa_c \neq 0$. The restriction $\kappa_c = 0$ can be overcome by recalling the scaling procedure introduced in Sect. 8.2.2; there, we saw that the properties of the Chang–Refsdal model depend only on the two param-

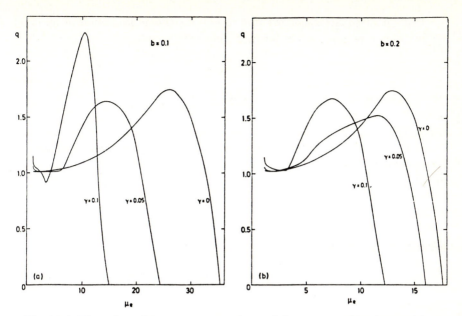

Fig. 11.4. The ratio q of the cross-section $\sigma(\mu_e, b, \gamma)$ for a source of size $b = 0.1$ (**a**) and $b = 0.2$ (**b**) to be magnified by more than μ_e by a Chang–Refsdal lens with shear γ, and the corresponding cross-section $\sigma(\mu_e, 0, 0)$ for a point source and a Schwarzschild lens.

eters Λ and ϵ, the latter one only taking the values ± 1. For a lens with smooth surface mass density $\kappa_c < 1$, and for $\gamma \geq 0$ (changing the sign of γ is equivalent to interchanging the two coordinate axes), we see from (8.59a) that $\epsilon = +1$. We treated the Chang–Refsdal model in terms of scaled coordinates $\mathbf{X}, \mathbf{Y}$ in the source and lens plane, see (8.57).

Let us consider a model with given γ and κ_c. An equivalent model in terms of scaled coordinates is parametrized by Λ (8.59b). A source of unscaled radius R will have a scaled radius of

$$\tilde{R} = \frac{R}{\sqrt{1 - \kappa_c + \gamma}} \quad , \tag{11.16a}$$

as can be seen from the scaling (8.57) of the source-plane coordinates. The transformation (8.60a) of the Jacobian matrices from scaled to unscaled coordinates implies that the scaled and unscaled magnifications are related by

$$\tilde{\mu}_e = (1 - \kappa_c + \gamma)^2 \mu_e \quad . \tag{11.16b}$$

In addition, the scaling in the source plane implies that cross-sections are related by

$$\tilde{\sigma} = \frac{\sigma}{1 - \kappa_c + \gamma} \quad . \tag{11.16c}$$

This set of equations allows one to express the cross-section of an extended source with $\kappa_c \neq 0$ in terms of that with $\kappa_c = 0$:

$$\sigma(\mu_e, \gamma, \kappa_c, R)$$
$$= (1 - \kappa_c + \gamma)\,\tilde{\sigma}\left((1 - \kappa_c + \gamma)^2 \mu_e,\, \Lambda,\, \frac{R}{\sqrt{1 - \kappa_c + \gamma}}\right) \quad, \tag{11.17}$$

where Λ is given in (8.59b). Since the cross-section $\tilde{\sigma}$ does not depend on γ and κ_c individually, but only on their combination Λ, we can calculate it for vanishing κ_c by choosing a value of $\gamma = \gamma_\Lambda$ that yields the same value of Λ:

$$\gamma_\Lambda = \frac{1 - \Lambda}{1 + \Lambda} \quad ; \tag{11.18}$$

hence, (11.16c) yields

$$\tilde{\sigma}(\tilde{\mu}_e, \Lambda, \tilde{R}) = \frac{1}{1 + \gamma_\Lambda}\,\sigma\left(\frac{\tilde{\mu}_e}{(1 + \gamma_\Lambda)^2},\, \gamma_\Lambda,\, 0,\, \sqrt{1 + \gamma_\Lambda}\,\tilde{R}\right) \quad . \tag{11.19}$$

Combining (11.17) and (11.19), and noting that $\gamma_\Lambda = \frac{\gamma}{(1-\kappa_c)}$ and $\frac{(1-\kappa_c+\gamma)}{(1+\gamma_\Lambda)} = 1 - \kappa_c$, we obtain the scaling relation

$$\sigma(\mu_e, \gamma, \kappa_c, R) = (1 - \kappa_c)\,\sigma\left((1 - \kappa_c)^2 \mu_e,\, \frac{\gamma}{(1-\kappa_c)},\, 0,\, \frac{R}{\sqrt{1 - \kappa_c}}\right) \tag{11.20}$$

for the cross-sections for Chang–Refsdal models with zero and non-zero smooth surface mass density. Later, a similar scaling relation will be presented for the properties of a random star field subject to shear and convergence.

The limiting case $\mu \gg 1$ for point sources. To conclude this section, we consider the high-μ tail of the magnification cross-sections for point sources. The basic idea is that high magnifications only occur at caustics. So the high-μ area forms a strip along the caustics in the source plane. Since the properties of the lens mapping near caustics are determined by the universal behavior studied in Chap. 6, we can use it to find a general expression for the high-magnification cross-sections. A source near a caustic will have two images close to, and on opposite sides of, a critical curve in the lens plane; in addition, there are other non-critical images with magnification on the order of one or smaller. Thus, the total magnification μ_p for a point source near a caustic is dominated by the magnification μ of the two critical images, which are equally bright and of opposite parity.

Let us consider a critical curve in the lens plane, $\mathbf{x}_c(\lambda)$. Let us also consider a point $\mathbf{x}$ near that curve, with $\mu > 0$, at a distance z from its nearest point $\mathbf{x}_c$ on the curve. Since the determinant of the Jacobian matrix A vanishes on the critical curve, its gradient is perpendicular to this curve. Thus, the magnification at $\mathbf{x}$ is approximately

$$\mu \cong \frac{1}{|\nabla \det A(\mathbf{x}_c)| \, z} \, .$$

The area on one side of the closed critical curve where the magnification is larger than μ is thus

$$a_x(\mu) \cong \oint d\lambda \frac{dl}{d\lambda} \frac{1}{|\nabla \det A(\mathbf{x}_c(\lambda))| \, \mu} \equiv \frac{C}{\mu} \, , \tag{11.21a}$$

where the integral is computed along all critical curves in the lens plane. Thus, we see that the area on one side of the critical curves where the magnification is between μ and $\mu + d\mu$ is

$$\left| \frac{da_x(\mu)}{d\mu} \right| d\mu = \frac{C}{\mu^2} d\mu \, .$$

This area is mapped onto the area $\left| \frac{d\sigma(\mu_p)}{d\mu_p} \right| d\mu_p$ in the source plane, where $\mu_p = 2\mu$, $d\mu_p = 2d\mu$, since each point source near the caustic has two images with equal magnification. Recall that the magnification is the area distortion of the lens mapping; then,

$$\left| \frac{d\sigma(\mu_p)}{d\mu_p} \right| d\mu_p = \frac{1}{\mu} \left| \frac{da_x(\mu)}{d\mu} \right| d\mu = \frac{4C}{\mu_p^3} d\mu_p \, .$$

The area in the source plane where the point-source magnification is larger than μ_p is

$$\sigma(\mu_p) = \frac{2C}{\mu_p^2} \, . \tag{11.21b}$$

Hence, *the high-magnification tail of the cross-section for point sources behaves as $\propto \mu_p^{-2}$*, independently of the lens model.

In the derivation of (11.21) we have implicitly assumed that the only singularities of the lens mapping are folds. This, however, is not a necessary assumption: as we have stated in Sect. 6.2.2, the magnification of the two images of equal parity of a source positioned inside a cusp equals, up to the sign, the magnification of the third image of opposite parity. A little thought then shows that this property suffices to guarantee the validity of (11.21) even in the presence of cusps.

The next-order term for the magnification cross-section of point sources in the limit $\mu_p \to \infty$ is due to sources outside of cusps. A simple scaling argument applied to the lens mapping near cusps as given in Sect. 6.2.2 shows that this term will be of the form $\propto \mu_p^{-5/2}$. Without derivation, we quote the contribution to that term from a single cusp (see [SC91.4]):

$$C_c^{(i)} = \frac{8\sqrt{2}}{15} \frac{1}{|bc|} \frac{1}{\sqrt{|2ac + b^2|}} = \frac{8\sqrt{6}}{15} \sqrt{\frac{|\mathrm{tr} A|}{|\mathbf{T}| |(\mathbf{T} \cdot \nabla) \cdot (A\mathbf{T})|}} \, , \tag{11.22a}$$

where the notation is as in Sect. 6.2.2. We have expressed $C_c^{(i)}$ in terms of the (coordinate-dependent) coefficients a, b, and c, as well as in coordinate-free invariants of the lens mapping near cusps. In total, the magnification cross-section for a point source in the limit $\mu_p \to \infty$ is given as

$$\sigma(\mu_p) = \frac{2C}{\mu_p^2} + \sum_i C_c^{(i)} \mu_p^{-5/2} \quad , \tag{11.22b}$$

where the sum extends over all cusps of the lens mapping.

For the Chang–Refsdal lens the value of C in (11.21) can be calculated analytically ([SC87.6], Appendix B); for $\kappa_c = 0$,

$$C = \frac{1}{|1-\gamma^2|} \frac{1}{|1-\gamma|} \mathrm{E}\left(\sqrt{\frac{4\gamma}{(1+\gamma)^2}} \right) \quad , \tag{11.23a}$$

where $\mathrm{E}(x)$ is the complete elliptic integral of the second kind, and, for $\gamma < 1$ [SC91.4],

$$C_c = \frac{\sqrt{8}}{15\sqrt{\gamma}} \left[\frac{1}{(1-\gamma)^{5/2}} + \frac{1}{(1+\gamma)^{5/2}} \right] \quad . \tag{11.23b}$$

The corresponding result for a model with surface mass density $\kappa_c \neq 0$ can easily be obtained with the scaling relation (11.20). Equation (11.23a) was checked numerically in [NI84.1] up to first order in γ; we have recently verified (11.23a,b) numerically for arbitrary γ.

11.2 The random star field

Statistical homogeneity. If a QSO is observed through a galaxy, its light is expected to be influenced by the stars in that galaxy. For typical cosmological situations, i.e., where the QSO and the lens have appreciable redshifts (say, $z_d \sim 0.5$, $z_s \sim 2.0$), we have a typical *two length-scale problem* : the relevant length-scale in a galaxy is measured in kiloparsecs, as is the distance between possible multiple images. On the other hand, the radius ξ_0 of the Einstein ring for a typical star is of the order of a few milliparsecs. Therefore, to estimate the effect of stars on a light bundle, we can consider the star field to be *statistically homogeneous*, so that the mean surface density of stars does not vary over the length scale where the influence of the stars (ML) on the light bundle is important.

In this section, we investigate the lensing properties of a random star field, described by the number density of its stars, and by "macroscopic" shear and convergence parameters, the latter arising from smoothly-distributed matter (e.g., interstellar gas, dust, dark matter) and the former from the tidal field of the deflection angle for the galaxy as a whole (note that the shear of a homogeneous disk vanishes). The assumption that the stars are randomly distributed does not take into account the clustering of stars; the probability that two stars are close together is much higher than for a uniformly random distribution because double stars are gravitationally bound, and stars are formed in groups. The effects of clustering are unknown; we expect, however, that there will be no qualitative changes in the results.[1] In any case, the *random star field* is a few-parameter lens model, which can be studied in great detail, numerically as well as analytically. Its value for statistical gravitational lensing is similar to that of the infinite homogeneous unmagnetized plasma for plasma physics.

In Sect. 11.2.1 we analyze the probability distribution for the deflection angle of a light ray traversing the star field, and study in Sects.11.2.2 and 11.2.5 the magnification probability for sources behind such a star field. Since this is the simplest statistical ensemble of lenses we can study, we do so in some detail; in particular, we will see the influence of the non-linearity of statistical lensing: the action of an ensemble of lenses is different from the sum of the action of its components. The random star field is the only situation for which fully non-linear results have been derived; for a summary, see [SC90.1].

Markov's method. Before treating one of the special problems just mentioned, we present a unified method of analysis, following [CH43.1]. Consider the probability distribution for an 2-component quantity $\mathbf{g}$, which is the sum of random quantities $\mathbf{g}_j$. Thus,

$$\mathbf{g}_N = \sum_{j=1}^{N} \mathbf{g}_j \ .$$

If the random variables $\mathbf{g}_j$ are mutually statistically independent, as is the case in our applications, the probability density $p^{(N)}$ for $\mathbf{g}_N$ is given by the convolution integral

$$p^{(N)}(\mathbf{g}) = \int \prod_{j=1}^{N} d^2 g_j \, p_j(\mathbf{g}_j) \, \delta\left(\mathbf{g} - \sum_{i=1}^{N} \mathbf{g}_i\right) \tag{11.24}$$

of the probability densities p_j for the $\mathbf{g}_j$'s. Markov's method to determine $p^{(N)}$ is based on the fact that the inverse Fourier transform

$$Q_N(\mathbf{t}) = \int d^2 g \, p^{(N)}(\mathbf{g}) \, e^{i\mathbf{t}\cdot\mathbf{g}}$$

of $p^{(N)}$, in probability theory called the *characteristic function* of $p^{(N)}$, is equal to the product

---

[1] For instance, if the components of a double star are separated by a distance much smaller than their Einstein radius ξ_0, they would act on most light rays as a single object with the sum of their masses.

$$Q_N(\mathbf{t}) = \prod_{j=1}^{N} q_j(\mathbf{t}) \tag{11.25}$$

of the characteristic functions q_j of the p_j's; this follows immediately from (11.24) and the definition of the characteristic functions. In the particularly simple case of interest to us, the functions $\mathbf{g}_j$ are all equal, $\mathbf{g}_j = \mathbf{g}$; moreover, $\mathbf{g} = \mathbf{g}(\boldsymbol{\xi})$ depends on the position $\boldsymbol{\xi}$ of an individual star. Therefore,

$$Q_N(\mathbf{t}) = (q(\mathbf{t}))^N , \tag{11.26a}$$

where

$$q(\mathbf{t}) = \int d^2g \, p(\mathbf{g}) \, e^{i\mathbf{t}\cdot\mathbf{g}} = \frac{1}{\pi \Xi^2} \int_{|\boldsymbol{\xi}| \le \Xi} d^2\xi \, e^{i\mathbf{t}\cdot\mathbf{g}(\boldsymbol{\xi})} , \tag{11.26b}$$

where Ξ is the radius of the star field, and, by Fourier's inversion theorem,

$$p^{(N)}(\mathbf{g}) = \frac{1}{(2\pi)^2} \int d^2 t \, Q_N(\mathbf{t}) \, e^{-i\mathbf{t}\cdot\mathbf{g}} . \tag{11.27}$$

To obtain the second equation of (11.26b), we used that the probability density for $\boldsymbol{\xi}$ is given by $(\pi \Xi^2)^{-1}$ if $|\boldsymbol{\xi}| \le \Xi$, and zero otherwise, which implies that $p(\mathbf{g}) \, d^2g = \frac{d^2\xi}{\pi \Xi^2}$. The equations (11.26), (11.27) allow the determination of the probability distribution of a sum of mutually independent, equally distributed random variables.

In the case of interest to us, we are concerned with a geometrically-thin field of stars, each with mass M, with mean surface mass density Σ_*; the radius Ξ of the star field, its surface number density n and the total number of stars N are related through

$$\Sigma_* = Mn , \quad N = n\Xi^2 \pi . \tag{11.28}$$

We take $\mathbf{g}$ to be the deflection of a light ray, caused by the star field, or the shear acting on the light bundle. The deflection is the sum of the deflection angles of individual stars, and similarly, the total shear is the sum of the shears of single stars. The statistics for these variables will be treated below.

11.2.1 Probability distribution for the deflection

A light ray traversing our random star field will experience a deflection which depends on the relative position of the stars. If one of the stars is sufficiently near the ray, the deflection is large, whereas it can be expected to be small if the ray is well away from all the stars. Here, we consider the corresponding probability distribution for the deflection. The analysis [KA86.1] will enable us to study the statistical properties of the structure of a macroimage produced by a lens (galaxy) composed of stars; furthermore, the transition between those cases where a single star is mainly responsible for the deflection, and those where the compound effects of an ensemble of stars is relevant, can be seen clearly. The latter result demonstrates the importance of effects which are due to the nonlinearities occurring in (11.24) and (11.26a).

Calculation of the characteristic function. The deflection angle $\hat{\alpha}$ of a light ray through the center of a field of stars at coordinates $\boldsymbol{\xi}_i$ is

$$\hat{\boldsymbol{\alpha}} = \frac{4GM}{c^2} \sum_{i=1}^{N} \frac{\boldsymbol{\xi}_i}{|\boldsymbol{\xi}_i|^2} = \sum_{i=1}^{N} \hat{\boldsymbol{\alpha}}_i(\boldsymbol{\xi}_i) \quad ; \tag{11.29}$$

we have assumed that each star has mass M. This equation shows that the assumptions which lead to (11.27) are satisfied, and we can apply it by setting $\mathbf{g}_N = \hat{\boldsymbol{\alpha}}$ and $\mathbf{g} = 2R_S \boldsymbol{\xi}/|\boldsymbol{\xi}|^2$. In polar coordinates, $\mathbf{t} = (t \cos \beta, t \sin \beta)$, $\boldsymbol{\xi} = (\xi \cos \varphi, \xi \sin \varphi)$, (11.26b) becomes

$$q(\mathbf{t}) = \frac{1}{\pi \Xi^2} \int_0^{\Xi} \xi \, d\xi \int_0^{2\pi} d\varphi \, e^{i(4GM/c^2)(t/\xi) \cos(\varphi - \beta)} \quad . \tag{11.30a}$$

The integral over angles can be performed with equation (3.715.18) of [GR80.3] which yields

$$q(\mathbf{t}) = \frac{2}{\Xi^2} \int_0^{\Xi} \xi \, d\xi \, J_0\left(a \frac{\Xi}{\xi}\right) \quad , \tag{11.30b}$$

where

$$a = \frac{4GM}{c^2 \Xi} t \quad , \tag{11.31}$$

and $J_\nu(z)$ is the ordinary Bessel function of index ν. With the substitution $x = \Xi/\xi$, (11.30b) becomes

$$q(\mathbf{t}) = 2 \int_1^{\infty} \frac{dx}{x^3} J_0(ax) = J_0(a) - \frac{a}{2} J_1(a) - \frac{a^2}{2} \int_1^{\infty} \frac{dx}{x} J_0(ax) \quad , \tag{11.32}$$

where in the last step we have integrated by parts twice, using the relations (9.1.30) of [AB65.1]. The final integral is given by (11.1.20) of [AB65.1]. Thus, we can calculate $q(\mathbf{t})$ exactly; however, the resulting expression cannot be easily exponentiated as required by (11.26a), except for small values of a. Note that the restriction to small a means that the Fourier transform $Q(\mathbf{t})$ of $p(\hat{\boldsymbol{\alpha}})$ will be determined only for sufficiently small values of t. However, as we demonstrate below, this does not impair the analysis of (11.26). Thus, restricting our attention to $a \ll 1$, we find from (11.32), up to second order in a:

$$q(\mathbf{t}) = 1 - \frac{a^2}{2} \ln(B/a) + \mathcal{O}(a^4) = e^{-\frac{a^2}{2} \ln(B/a)} + \mathcal{O}(a^4) \quad , \tag{11.33}$$

where $B = 2e^{1-\gamma} \approx 3.05$, and $\gamma \simeq 0.577$ is Euler's constant. This form of q allows us to calculate Q with (11.26a),

$$Q(\mathbf{t}) = e^{-\frac{Na^2}{2} \ln(B/a)} \quad . \tag{11.34}$$

Since there is no preferred direction in the problem, Q depends only on the magnitude t of $\mathbf{t}$ via (11.31).

Calculation of the deflection probability. Insertion of (11.34) into (11.27) yields, with $\hat{\alpha} = |\hat{\boldsymbol{\alpha}}|$, denoting for simplicity $p^{(N)}$ with p,

$$p(\hat{\boldsymbol{\alpha}}) = \frac{1}{(2\pi)^2} \int_0^\infty t\,dt\,Q(t) \int_0^{2\pi} d\varphi\,e^{-i\hat{\alpha}t\cos\varphi} \qquad (11.35a)$$
$$= \frac{1}{2\pi} \int_0^\infty t\,dt\,Q(t)\,J_0(\hat{\alpha}t) \quad,$$

where (3.715.18) of [GR80.3] was used in the last step. Since we do not know $Q(t)$ for large t, we now write

$$p(\hat{\boldsymbol{\alpha}}) = \frac{1}{2\pi} \int_0^{t_{\max}} t\,dt\,Q(t)\,J_0(\hat{\alpha}t) \quad. \qquad (11.35b)$$

Later we shall see that there is a natural choice for the value of the parameter $t_{\max}$, and that the substitution of (11.35b) for (11.35a) is a reasonable approximation.

Note that the mean distance between stars in the field is $\Xi N^{-1/2}$; hence, there is a natural scale for the deflection angle, namely, the deflection due to a single star for impact parameter equal to the mean separation between stars:

$$\hat{\alpha}_0 = \frac{4GM\sqrt{N}}{c^2 \Xi} = \frac{4GM}{c^2}\sqrt{\pi n} \quad. \qquad (11.36)$$

With the substitution $x = \hat{\alpha}_0 t$, we obtain from (11.34) and (11.35):

$$p(\hat{\boldsymbol{\alpha}}) = \frac{1}{2\pi\hat{\alpha}_0^2} \int_0^{x_{\max}} x\,dx\,J_0\left(x\frac{\hat{\alpha}}{\hat{\alpha}_0}\right) e^{-(x^2/2)\ln(B\sqrt{N}/x)} \quad. \qquad (11.37)$$

The exponential factor in the integrand first decreases, has a minimum at $x_0 = B\sqrt{N}e^{-1/2} \approx 1.85\sqrt{N}$ and increases for larger values of x. At x_0, the value of the exponential factor is $e^{-B^2N/4e} \approx 10^{-0.372N}$, extremely small for reasonable values of N. Therefore, the choice $x_{\max} = x_0$ is a natural one; owing to the small value of the integrand in the neighborhood of x_0, the exact value of $x_{\max}$ is unimportant. Also note that $x < x_0$ corresponds to $a < 1$, as required for the validity of (11.34). The rise of the integrand beyond $x = x_0$ is an artifact due to neglected terms of $\mathcal{O}(a^4)$ in (11.34).

The step from (11.35a) to (11.35b) amounts to the multiplication of the integrand by Heaviside's step function $H(t_{\max} - t)$. Hence, the left-hand side of (11.35b) is actually not $p(\hat{\boldsymbol{\alpha}})$, but the convolution of $p(\hat{\boldsymbol{\alpha}})$ with the inverse Fourier transform $\hat{H}$ of H, the so-called Dirichlet factor whose width is about $1/t_{\max} \sim \hat{\alpha}_0/\sqrt{N}$, much smaller than $\hat{\alpha}_0$ for large N. Thus, as long as $p(\hat{\boldsymbol{\alpha}})$ does not change appreciably on the scale $\hat{\alpha}_0/\sqrt{N}$, (11.35b) holds approximately.

In general, the integral must be computed numerically, but from (11.37), we can extract some information directly. First, $p(\hat{\alpha})$ depends on the stellar mass, as expected from the occurrence of a natural scale $\hat{\alpha}_0$ in the problem which itself depends on M. Second, the probability distribution is not independent of N (for constant $\hat{\alpha}_0$); in particular, the limit of $p(\hat{\alpha})$ for $N \to \infty$ (for fixed n) does not exist. This result, which may be surprising at first, is quite similar to the infrared divergence one has in the scattering of charged particles in a plasma. However, there one has a natural cut-off in the range for which the plasma particles act on the charge, namely the Debye length. For gravitation, there is no such length, since there are no negative deflectors and thus no shielding.

The "infrared" divergence. We can explain this divergence as follows. Suppose we add to our star field a ring of stars with outer radius $b\Xi$ and number density n. The mean squared deflection $\langle \hat{\alpha}^2 \rangle$ caused by these $N_r = nA$ stars (A is the surface area of the ring) is twice the mean squared x-component of the deflection, due to the isotropy of the problem. We have

$$\langle \hat{\alpha}_x^2 \rangle = \left(\frac{4GM}{c^2}\right)^2 \left[\prod_{i=1}^{N_r} \frac{1}{A} \int_\Xi^{b\Xi} \xi_i\, d\xi_i \int_0^{2\pi} d\varphi_i\right] \left[\sum_{k=1}^{N_r} \frac{\cos\varphi_k}{\xi_k}\right]^2 . \qquad (11.38)$$

The integral of the cross terms in the squared sum vanishes, since the positions of the stars are statistically independent. The only terms that survive are the squared terms, $(\cos\varphi_k/\xi_k)^2$. The factors in the integral with $i \neq k$ reduce to unity, whereas the factors with $i = k$ reduce to π after the angular integration is carried out. All N_r terms are equal, so we obtain

$$\langle \hat{\alpha}_x^2 \rangle = \frac{\pi}{A} \left(\frac{4GM}{c^2}\right)^2 N_r \int_\Xi^{b\Xi} \frac{d\xi}{\xi} = \hat{\alpha}_0^2 \ln b \; ; \qquad (11.39)$$

we used (11.28) and (11.36) to calculate (11.39). We see that increasing the size of the ring leads to a diverging mean square deflection angle, which is a clear sign of the (logarithmic) infrared divergence. However, this divergence does not lead to physical problems, since star fields occurring in nature are all finite, and they cannot be approximated by a homogeneous field over too large a range. Below, we discuss the number of stars required in practice.

Results. From (11.37), we see that the probability is isotropic, as expected. We show the distribution for several values of N in Fig. 11.5, on a logarithmic scale. For small values of $\hat{\alpha}/\hat{\alpha}_0$ the distribution is Gaussian, as can be expected from the central limit theorem. Explicitly, we have ([KA86.1], see also [DE88.2])

$$p(\hat{\alpha}) \approx \frac{1}{\pi\Delta} e^{-\hat{\alpha}^2/\Delta} , \qquad (11.40a)$$

where

$$\Delta = 2\hat{\alpha}_0^2 \ln\left(B\sqrt{N}\right) . \qquad (11.40b)$$

For large values of the deflection angle,

$$p(\hat{\alpha}) \approx \frac{\hat{\alpha}_0^2}{\pi\hat{\alpha}^4} . \qquad (11.41)$$

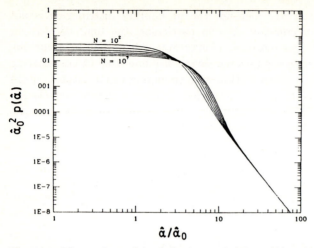

Fig. 11.5. The product of the deflection probability $p(\hat{\alpha})$ and the square of the characteristic angle $\hat{\alpha}_0$ as a function of $\hat{\alpha}/\hat{\alpha}_0$, for various values of the number N of stars in the field under consideration. For small values of $\hat{\alpha}/\hat{\alpha}_0$, the probability distribution is Gaussian, but for large deflection angles, it decreases as $\hat{\alpha}^{-4}$ (from [KA86.1])

This result, which can be obtained from an asymptotic approximation of (11.37) [KA86.1], has a very simple origin, since for large deflection angles, a single star prevails. The impact parameter for a star to cause a deflection $\hat{\alpha}$ is $\xi = 4GM/(c^2\hat{\alpha})$. Hence, the probability that the light ray is deflected by a single star, by more than $\hat{\alpha}$, is

$$P(\hat{\alpha}) = n\pi\xi^2 = n\pi \left(\frac{4GM}{c^2}\right)^2 \frac{1}{\hat{\alpha}^2} = \frac{\hat{\alpha}_0^2}{\hat{\alpha}^2} \quad . \tag{11.42}$$

Since this cumulative probability is related to the probability distribution by

$$P(\hat{\alpha}) = 2\pi \int_{\hat{\alpha}}^{\infty} d\phi \, \phi \, p(\phi) \quad ,$$

we recover the result (11.41) by differentiation of this last equation, using (11.42). Hence, the probability distribution for large deflection angles can be obtained from a linear analysis, whereas for small deflections, the combined effect of many stars must be taken into account. As can be seen from Fig. 11.5, the larger the star field, the larger $\hat{\alpha}$ must be for the linear result to be valid, since then the combined effects of the other stars become more important. In [KA86.1], the deflection probability for a random star field with stars having a mass spectrum is derived.

Morphology of a microlensed macroimage. We now consider a deflection $\hat{\alpha}$ as a random deflection angle caused by a star field. Let us consider a light ray traversing a random star field at an off-center position. Then, there will be two contributions to the deflection angle: first, the random contribution as discussed above, and, in addition, a systematic contribution to the deflection due to the stars inside the circle with radius equal to the impact parameter measured from the center of mass of the field. Clearly, this latter effect causes the mean value of the deflection to be different from zero.

We can now write the lens equation for a random star field [compare (8.46)]:

$$\frac{D_d}{D_s}\eta = \begin{pmatrix} 1-\kappa-\gamma & 0 \\ 0 & 1-\kappa+\gamma \end{pmatrix}\xi - \frac{D_d D_{ds}}{D_s}\hat{\alpha} \, , \tag{11.43}$$

where γ is the macroscopic shear acting on the star field (e.g., due to the global quadrupole field of the galaxy that contains the star field), $\kappa = \kappa_c + \kappa_*$ is the total convergence due to the smooth surface mass density κ_c and the dimensionless density $\kappa_* = \Sigma_*/\Sigma_{cr}$ [see (5.5)] of the stars, and $\hat{\alpha}$ is the random contribution to the deflection. Suppose that a point source is at $\eta = 0$; for this source to be seen at ξ, the random deflection angle at ξ, $\hat{\alpha}(\xi)$, must satisfy the lens equation (11.43). Therefore, the flux $S(\xi)d^2\xi$ from a surface element $d^2\xi$ surrounding ξ is proportional to the probability $p(\hat{\alpha}(\xi))d^2\xi$ that the deflection angle satisfies (11.43) at that point,

$$\frac{S(\xi)}{S_0} = \left(\frac{D_s}{D_d D_{ds}}\right)^2 p(\hat{\alpha}(\xi)) \, , \tag{11.44}$$

where S_0 is the flux from the unlensed source. Since $p(\hat{\alpha})$ is normalized, the integrated flux is

$$\frac{\int S(\xi)\,d^2\xi}{S_0} = \frac{1}{|(1-\kappa)^2 - \gamma^2|} \, , \tag{11.45}$$

i.e., the magnification the image would have in the absence of stars. The law of flux conservation implies that the mean magnification here must be the same. The microlensed image comprises a large number of very faint microimages; if these images were smoothed out, an image with elliptical isophotes would result, with an axial ratio of $\frac{(1-\kappa-\gamma)}{(1-\kappa+\gamma)}$.

Estimate of the necessary number of stars in microlensing simulations. Let us finally consider the question, relevant for later discussion, of how many stars lie within the isophote of the image which contains 99% of the flux. From the behavior of $p(\hat{\alpha})$ for large deflections (11.42), 99% of all rays have a random deflection smaller than $10\hat{\alpha}_0$. The corresponding area in the lens plane is an ellipse with semiaxes

$$\xi_1 = \frac{1}{|1-\kappa-\gamma|}\frac{D_d D_{ds}}{D_s}10\hat{\alpha}_0 \, , \quad \xi_2 = \frac{1}{|1-\kappa+\gamma|}\frac{D_d D_{ds}}{D_s}10\hat{\alpha}_0 \, ; \tag{11.46}$$

the area of the ellipse is $\pi \xi_1 \xi_2$, and the number of stars within that area is $N_{99} = \pi \xi_1 \xi_2 n$. Using (11.28) and (11.36) this becomes

$$N_{99} = 100 \langle \mu \rangle \kappa_*^2 \qquad (11.47)$$

(compare [KA86.1], [SC87.3]), and $\langle \mu \rangle$ is the magnification in (11.45) due to the smoothed-out matter distribution, which equals the mean magnification. Thus, the number of stars that must be taken into account for numerical estimates of the properties of a random star field can be considerable, particularly if the mean magnification is high. We can also obtain a prediction for the range over which a star field should be homogeneous, to satisfy the assumptions we made in this section. If the length scale over which the conditions of the star field vary is much larger than ξ_1, ξ_2 of (11.46), the results derived here apply.

11.2.2 Shear and magnification

A light bundle traversing a star field experiences shear due to the tidal forces produced by the stars. In this subsection, we derive the probability distribution for this shear, and the corresponding distribution for the magnification ([NI84.1], [SC87.5]).

The shear due to a random star field. Let us consider again a circular star field with N stars of mass M and dimensionless surface mass density κ_*. We use the dimensionless lens equation by scaling with ξ_0 defined in (2.6b). Let the positions of the stars be $\mathbf{x}_i$, $i = 1, ..., N$; the lens equation then reads

$$\mathbf{y} = \mathbf{x} - \sum_{i=1}^{N} \frac{\mathbf{x} - \mathbf{x}_i}{|\mathbf{x} - \mathbf{x}_i|^2} \qquad (11.48)$$

At the center of the star field, $\mathbf{x} = 0$, the Jacobian matrix of the mapping (11.48) is

$$A = \frac{\partial \mathbf{y}}{\partial \mathbf{x}} = \begin{pmatrix} 1 + S_1 & S_2 \\ S_2 & 1 - S_1 \end{pmatrix}, \qquad (11.49a)$$

where

$$S_1 = \sum_{i=1}^{N} \frac{\cos 2\varphi_i}{x_i^2} \quad ; \quad S_2 = \sum_{i=1}^{N} \frac{\sin 2\varphi_i}{x_i^2} \qquad (11.49b)$$

are the components of the shear, which are sums of components of the shear due to each star. In (11.49b) we used polar coordinates, $\mathbf{x}_i = (x_i \cos \varphi_i, x_i \sin \varphi_i)$.

Calculation of the characteristic function. From (11.49), we find that the assumptions which lead to (11.27) are satisfied; thus, it can be used to compute the probability $p(S_1, S_2) dS_1 dS_2$ that the components of the shear lie within dS_i of S_i. The dimensional radius Ξ of the star field in (11.26b) is now replaced by the dimensionless radius $X = \Xi/\xi_0$. From (11.28), we find that $X^2 = NM/(\pi \xi_0^2 \Sigma_*)$, and by inserting the definition of ξ_0 and using (5.5), $X = \sqrt{N/\kappa_*}$. Therefore, (11.26b) becomes

$$q(\mathbf{t}) = \frac{1}{\pi X^2} \int_0^X x\, dx \int_0^{2\pi} d\varphi\, e^{i(t/x^2)\cos(2\varphi - \beta)} \quad,$$

where we have introduced $\mathbf{t} = (t\cos\beta, t\sin\beta)$. With (3.715.18) of [GR80.3], this becomes

$$q(\mathbf{t}) = \frac{2}{X^2} \int_0^X x\, dx\, J_0\left(\frac{t}{x^2}\right) = \frac{t}{X^2} \int_{t/X^2}^{\infty} \frac{dz}{z^2} J_0(z) \quad, \tag{11.50}$$

with the substitution $z = t/x^2$ in the last step. If the integral in (11.50) is rewritten as

$$\int_{t/X^2}^{\infty} \frac{dz}{z^2} J_0(z) = \int_{t/X^2}^{\infty} \frac{dz}{z^2} [J_0(z) - 1] + \int_{t/X^2}^{\infty} \frac{dz}{z^2} \quad, \tag{11.51}$$

we find that, in the limit $X \to \infty$, the first term on the right-hand side of (11.51) becomes -1 (see (11.4.18) of [AB65.1]), whereas the second is trivially integrated. Combining (11.50) and (11.51),

$$q(\mathbf{t}) \approx 1 - t/X^2 \approx e^{-t/X^2} \quad, \tag{11.52}$$

so that, according to (11.26a),

$$Q(\mathbf{t}) = e^{-Nt/X^2} = e^{-t\kappa_*} \quad. \tag{11.53}$$

Thus, it is possible to consider the "thermodynamic" limit $N \to \infty$ by keeping κ_* fixed; indeed, $Q(\mathbf{t})$ depends only on κ_* in this limit. Furthermore, $Q(\mathbf{t})$ is independent of the mass of the stars; this suggests that, in contrast to the distribution of deflection angles, the shear probability is independent of the mass distribution of the stars.

The shear distribution. Insertion of (11.53) into (11.27) yields, after use of (3.715.18) of [GR80.3],

$$p(S_1, S_2) = \frac{1}{2\pi} \int_0^{\infty} t\, dt\, e^{-\kappa_* t} J_0(St) \quad,$$

where $S = \sqrt{S_1^2 + S_2^2}$ is the magnitude of the shear. Finally, applying (6.621.4) of [GR80.3],

$$p(S_1, S_2) = \frac{1}{2\pi} \frac{\kappa_*}{(\kappa_*^2 + S^2)^{3/2}} \quad, \tag{11.54}$$

([NI84.1], [SC87.5]) a result *independent of the mass spectrum of the stars*.

The magnification distribution. The distribution for the shear can be used to compute the probability $p(\mu)d\mu$ that the magnification of a light ray traversing a random star field is within $d\mu$ of μ, by noting that $\mu = (1 - S^2)^{-1}$; thus,

329

$$p(\mu) = \int_{-\infty}^{\infty} dS_1 \int_{-\infty}^{\infty} dS_2 \; p(S_1, S_2) \; \delta\left(\mu - \frac{1}{1-S^2}\right) \; .$$

Transforming this into an integral over S^2, the δ-function can be integrated, to yield

$$p(\mu) = \frac{1}{2\mu^2} \frac{\kappa_*}{\left[\kappa_*^2 + 1 - 1/\mu\right]^{3/2}} \; , \quad \text{for} \quad \mu \geq 1 \quad \text{and} \quad \mu < 0 \; ,$$

$$p(\mu) = 0 \; , \quad \text{for} \quad 0 \leq \mu < 1 \; .$$

(11.55)

This last result should have been expected, since a star field can only lead to images of a source which are of Type I [see (5.29)], i.e., with $\mu \geq 1$, or of Type II, $\mu < 0$. Type III images require a local (smooth) surface mass density $\kappa_c > 1$, and in the situation considered here, $\kappa_c = 0$.

11.2.3 Inclusion of external shear and smooth matter density

A scaling relation. Let us consider now a star field, subject to external shear γ and convergence due to a smooth surface mass density κ_c. We can always rotate the coordinate frame to diagonalize the shear matrix; the lens equation becomes

$$\mathbf{y} = \begin{pmatrix} 1 - \kappa_c + \gamma & 0 \\ 0 & 1 - \kappa_c - \gamma \end{pmatrix} \mathbf{x} - \sum_{i=1}^{N} m_i \frac{\mathbf{x} - \mathbf{x}_i}{|\mathbf{x} - \mathbf{x}_i|^2} \; , \qquad (11.56)$$

where $m_i M$ is the mass of the i-th star, and the corresponding Jacobian matrix at $\mathbf{x} = 0$ is

$$A = \begin{pmatrix} 1 - \kappa_c + \gamma + S_1 & S_2 \\ S_2 & 1 - \kappa_c - \gamma - S_1 \end{pmatrix} \; , \qquad (11.57)$$

where S_1 and S_2 are the shear components produced by the stars. As in the case of an isolated Chang–Refsdal lens where there is a scaling for magnification cross-sections for models with different κ_c, (11.20), there is a similar scaling for the random star field considered here ([PA86.1],[KA86.2], [SC87.3]). Let

$$\mathbf{X} = \sqrt{|1 - \kappa_c|} \; \mathbf{x} \; ,$$

$$\mathbf{X}_i = \sqrt{|1 - \kappa_c|} \; \mathbf{x}_i \; ,$$

$$\mathbf{Y} = \frac{1}{\sqrt{|1 - \kappa_c|}} \; \mathbf{y} \; ;$$

in terms of these scaled variables, the lens equation (11.56) becomes

$$\mathbf{Y} = \begin{pmatrix} \epsilon + \Gamma & 0 \\ 0 & \epsilon - \Gamma \end{pmatrix} \mathbf{X} - \sum_{i=1}^{N} m_i \frac{\mathbf{X} - \mathbf{X}_i}{|\mathbf{X} - \mathbf{X}_i|^2} \; , \qquad (11.58a)$$

where $\epsilon = \text{sign}(1 - \kappa_c)$ and

$$\Gamma = \frac{\gamma}{|1 - \kappa_c|} \quad . \tag{11.58b}$$

There is a fundamental reason for this scaling to break down for $\kappa_c = 1$, although the behavior of the lens equation (11.56) is not singular. From (11.57), we see that, for $\kappa_c = 1$, the trace of the Jacobian matrix vanishes identically. This means that at any critical point with $\det A = 0$, both eigenvalues of A vanish. From our discussion in Chap. 6, we infer that such critical points are (elliptic) umbilics. In particular, for $\kappa_c = 1$, there are no fold or cusp catastrophes, which means that this case is qualitatively different from those with $\kappa_c \neq 1$. This can also be seen in the magnification patterns shown in Fig. 11.7.

In the following, we confine our treatment to the case $\kappa_c < 1$, so that $\epsilon = +1$. Then, the two lens equations (11.56) and (11.58a) agree for $\kappa_c = 0$, and their equivalence leads to the following correspondence between a model with $\kappa_c \neq 0$ and one with $\kappa_c = 0$, the latter one denoted by quantities with a tilde. A physical length ξ in the lens plane is related to a dimensionless length x by $\xi = \xi_0 x$; similarly, $\xi = \tilde{\xi}_0 X$; hence

$$\tilde{\xi}_0 = \frac{\xi_0}{\sqrt{1 - \kappa_c}} \quad . \tag{11.59a}$$

Correspondingly, length-scales in the source plane are related through

$$\tilde{\eta}_0 = \sqrt{1 - \kappa_c}\, \eta_0 \quad . \tag{11.59b}$$

The point-source magnifications satisfy

$$\tilde{\mu} = (1 - \kappa_c)^2 \mu \quad , \tag{11.59c}$$

a relation that also holds for the magnification of extended sources, provided that the dimensionless source sizes satisfy

$$\tilde{R} = \frac{R}{\sqrt{1 - \kappa_c}} \quad . \tag{11.59d}$$

The N stars are distributed inside a circle, with radius lowered by a factor $\sqrt{1 - \kappa_c}$; therefore, the densities of stars in both models are related through

$$\tilde{\kappa}_* = \frac{\kappa_*}{1 - \kappa_c} \quad . \tag{11.59e}$$

From these scaling relations, we deduce that the probability for a source of radius R to be magnified by more than μ_e satisfies

$$P(\mu_e, \kappa_*, \kappa_c, \gamma, R)$$
$$= P\left(\mu_e(1 - \kappa_c)^2, \frac{\kappa_*}{(1 - \kappa_c)}, 0, \frac{\gamma}{(1 - \kappa_c)}, \frac{R}{\sqrt{1 - \kappa_c}}\right) \quad , \tag{11.59f}$$

for a random star field characterized by κ_*, κ_c and γ. Finally, relation (11.20) for the cross-sections is also valid for the random star field.

Magnification probabilities: Linear approach. To estimate the amplification bias in source counts, which will be described in Chap. 12, we need the probability $P(\mu_e)$ that a source seen through a star field is magnified by more than μ_e. The calculation of $P(\mu_e)$ can only be treated numerically in general, and we will describe some of the results further below. However, we can perform a linear analysis for $P(\mu_e)$, based on the assumption that every star in the field acts as if it were an isolated lens. In other words, this approximation assumes that the magnification cross-section for a star field equals the sum of the magnification cross-sections for all the stars in the field. This assumption can be expected to be reasonable if the density of stars, κ_*, is sufficiently small; in that case, the mean distance between stars is sufficiently large for disturbances of the gravitational field of one star by that of the others to be neglected. Equivalently, one can consider this approximation to be valid if the cross-sections for the individual stars in the field do not overlap. With this assumption, the analysis proceeds as follows: Consider an area $\mathcal{A}_x$ of the lens plane; then, there are $N = \kappa_* \mathcal{A}_x/\pi$ stars within that area. Each star has a cross-section $\sigma(\mu_e, \gamma, \kappa_c, R)$ for magnifying a source of size R by more than μ_e, if γ, κ_c are the external shear and convergence. The area $\mathcal{A}_x$ is mapped by the lens equation onto the area $\mathcal{A}_y = \mathcal{A}_x/\langle \mu \rangle$, where

$$\langle \mu \rangle = \frac{1}{|(1-\kappa_c-\kappa_*)^2 - \gamma^2|} \tag{11.60}$$

is the mean magnification due to the star field and thus, by definition, the area distortion of the lens mapping. The total cross-section within $\mathcal{A}_y$ is $\sigma_{\text{tot}} = N\sigma$, and therefore the magnification probability is

$$P(\mu_e) = \frac{\sigma_{\text{tot}}}{\mathcal{A}_y} = \frac{N\sigma \langle \mu \rangle}{\mathcal{A}_x} = \frac{\langle \mu \rangle \kappa_*}{\pi} \sigma(\mu_e, \gamma, \kappa_c, R) \quad .$$

Using (11.20), this becomes

$$P(\mu_e) = \frac{\langle \mu \rangle \kappa_*(1-\kappa_c)}{\pi} \sigma\left((1-\kappa_c)^2 \mu_e, \frac{\gamma}{1-\kappa_c}, 0, \frac{R}{\sqrt{1-\kappa_c}}\right) \quad . \tag{11.61}$$

Note that this probability function satisfies the scaling relation (11.59f). In many applications, the effect of the external shear is neglected for simplicity; if we use (11.14), (11.61) becomes with this approximation

$$P(\mu_e) = \langle \mu \rangle \kappa_*(1-\kappa_c) \, y^2 \left((1-\kappa_c)^2 \mu_e, \frac{R}{\sqrt{1-\kappa_c}}\right) \quad , \tag{11.62}$$

and the function $y(\mu_e, R)$ is the inverse of (11.12).

A more sophisticated treatment of the problem would assign a random shear S_1, S_2 to each star in the field, according to the probability density of (11.54), and use the cross-section for a Chang–Refsdal lens with an effective shear parameter $\sqrt{(S_1+\gamma)^2 + S_2^2}$. Such a "quasi-linear" approach has not been attempted. Before it can be applied, one needs a table of Chang–Refsdal cross-sections $\sigma(\mu_e, \gamma, 0, R)$, which has not yet been systematically calculated. Furthermore, since there is a finite probability that the random shear takes values such that the corresponding Chang–Refsdal lens is no longer defined [when $(S_1+\gamma)^2 + S_2^2 = 1$, see Sect. 8.2.2], an approximation must be derived for these parameter values.

Magnification probability for large μ_p: Asymptotic behavior. There is one fully non-linear result that may be derived analytically [SC87.8], namely, the magnification probability distribution $p(\mu_p)$ for point sources in the limit $\mu_p \to \infty$. The general form must be $p(\mu_p) \propto \mu_p^{-3}$, as follows from (11.21b). Now we determine the amplitude of this power law.

The magnification probability *for point images* (i.e., individual microimages) can be obtained from the probability distribution of the shear, (11.54); since $\mu = (\det A)^{-1}$ and (11.57),

$$\frac{1}{\mu} = (1 - \kappa_c)^2 - (\gamma + S_1)^2 - S_2^2 \quad ,$$

so

$$p(\mu) = \int_{-\infty}^{\infty} dS_1 \int_{-\infty}^{\infty} dS_2 \, p(S_1, S_2) \, \delta\left[\mu - \frac{1}{(1 - \kappa_c)^2 - (\gamma + S_1)^2 - S_2^2}\right] \quad .$$

If we define new variables, $x = (\gamma + S_1)$, $y = S_2$, $r^2 = x^2 + y^2$ and $\tan\varphi = y/x$, this becomes

$$p(\mu) = \frac{1}{2} \int_0^{\infty} dr^2 \, \delta\left[\mu - \frac{1}{(1-\kappa_c)^2 - r^2}\right] \int_0^{2\pi} d\varphi \, \frac{1}{2\pi} \frac{\kappa_*}{\left[\kappa_*^2 + r^2 + \gamma^2 - 2r\gamma\cos\varphi\right]^{3/2}} \quad .$$

The inner integral can be calculated by substituting $z = \cos\varphi$ and applying (3.133.3) of [GR80.3]; the remaining integration is then performed as for (11.55). We find

$$p(\mu) = \frac{\kappa_*}{\pi\mu^2} \frac{1}{(u-v)\sqrt{u+v}} \, E\left(\sqrt{\frac{2v}{u+v}}\right) \quad , \tag{11.63a}$$

where $E(x)$ is the complete elliptic integral of the second kind as defined in §8.11 of [GR80.3], and

$$u = \kappa_*^2 + \gamma^2 + (1 - \kappa_c)^2 - \frac{1}{\mu} \quad , \tag{11.63b}$$

$$v = 2\gamma\sqrt{(1 - \kappa_c)^2 - \frac{1}{\mu}} \quad . \tag{11.63c}$$

The result (11.63a) is valid for $\mu > (1 - \kappa_c)^{-2}$ and for $\mu < 0$, for the same reasons as for (11.55). Also, note that the probability (11.63) satisfies the scaling law (11.59f).

The magnification probability for individual images is not easily related to the magnification probability $p(\mu)$ for sources, because of multiple imaging. Only for the special case of highly magnified sources can these two probabilities be simply related, because then, magnifications are dominated by the two highly magnified images next to a critical curve – compare the discussion leading to (11.21). Thus, consider an area $\mathcal{A}_x$ in the lens plane; inside this area, microimages are magnified by a factor within $d\mu$ of μ in an area $d a_x(\mu) = \mathcal{A}_x p(\mu) d\mu$. The area $\mathcal{A}_x$ is mapped onto the area $\mathcal{A}_y = \mathcal{A}_x / \langle\mu\rangle$, where $\langle\mu\rangle$ is given by (11.60). The probability distribution for the magnification μ_p of a point source is obtained following the same steps which lead from (11.21a) to (11.21b).

The result is

$$p(\mu_p) = \frac{q(\kappa_*, \kappa_c, \gamma)}{\mu_p^3} \quad \text{for} \quad \mu_p \to \infty \quad , \tag{11.64a}$$

where

$$q(\kappa_*, \kappa_c, \gamma) = \frac{4\kappa_* \langle \mu \rangle}{\pi} \frac{1}{(u-v)\sqrt{u+v}} E\left(\sqrt{\frac{2v}{u+v}}\right) \quad , \quad (11.64b)$$

and

$$u = (1-\kappa_c)^2 + \kappa_*^2 + \gamma^2 \quad ; \quad v = 2\gamma(1-\kappa_c) \quad . \quad (11.64c)$$

Again, we note that (11.64) is compatible with (11.59f). As is clear from the discussion following (11.21b), the next order term of $p(\mu_p)$ will be $\propto \mu_p^{-7/2}$.

Comparison with the linear result. The foregoing fully nonlinear result can be compared with the corresponding linear one, expressed in (11.61), (11.21b) and (11.23a):

$$p_{\text{linear}} = \frac{q_L(\kappa_*, \kappa_c, \gamma)}{\mu_p^3} \quad ,$$

$$q_L(\kappa_*, \kappa_c, \gamma) = \frac{4\kappa_* \langle \mu \rangle}{\pi} \frac{1}{|1-\kappa_c-\gamma|^2} \frac{1}{|1-\kappa_c+\gamma|} E\left(\frac{2\sqrt{\Gamma}}{1+\Gamma}\right) \quad ,$$

$$\Gamma = \frac{\gamma}{|1-\kappa_c|} \quad .$$

These two probabilities are most easily compared for the special case $\kappa_c = \gamma = 0$:

$$q(\kappa_*, 0, 0) = 2\kappa_* \langle \mu \rangle \left(1 + \kappa_*^2\right)^{-3/2} \quad ,$$

$$q_L(\kappa_*, 0, 0) = 2\kappa_* \langle \mu \rangle \quad .$$

Hence, in this case the linear result *overestimates* the high-magnification tail of the probability distribution for point sources; however, for small values of κ_* these two distributions agree very well. For further discussions, see [SC87.8].

11.2.4 Correlated deflection probability

In subsection 11.2.1 above we saw that the probability distribution for the deflection angle of a light ray in a random star field has no "thermodynamic" limit, i.e., the probability distribution depends on the number N of stars in the field. On the other hand, for the probability distribution of the shear, the "thermodynamic" limit $N \to \infty$ exists. The different behaviors in these two cases are due to the different dependence of the deflection and the shear on the separation of a point mass from the light ray: whereas the deflection decreases only as $(\text{distance})^{-1}$, the shear falls off as $(\text{distance})^{-2}$. Although the physical origin of the "infrared divergence" is clear, it is not very important in practice: the deflection caused by very distant stars, i.e., those responsible for the divergence, varies slowly over a very large distance. Therefore, all the rays near the one under consideration will experience the same deflection due to distant stars. This means that the contribution of the deflection caused by these stars can be considered to be due to the host macrolens for the microlenses.

To extract the microlens contribution of the deflection probability from that of the macrolens, we next consider the difference in the deflection angle of two rays in a star field described by κ_c, κ_*, and γ, and we will assume for simplicity that all stars have the same mass.

Let s be the separation of two light rays, measured in units of the length-scale ξ_0 (2.6b). The difference $\boldsymbol{\Delta\alpha} = \boldsymbol{\alpha}(\mathbf{x}_a) - \boldsymbol{\alpha}(\mathbf{x}_b)$ of the deflection angles, where $s = |\mathbf{x}_a - \mathbf{x}_b|$, is then

$$\boldsymbol{\Delta\alpha} = \begin{pmatrix} \kappa_c + \gamma\cos(2\beta) & \gamma\sin(2\beta) \\ \gamma\sin(2\beta) & \kappa_c - \gamma\cos(2\beta) \end{pmatrix}(\mathbf{x}_a - \mathbf{x}_b)$$
$$+ \sum_{i=1}^{N}\left(\frac{\mathbf{x}_a - \mathbf{x}_i}{|\mathbf{x}_a - \mathbf{x}_i|^2} - \frac{\mathbf{x}_b - \mathbf{x}_i}{|\mathbf{x}_b - \mathbf{x}_i|^2}\right) \;.$$

Here, β is the direction of the shear caused by the macrolens, and the $\mathbf{x}_i$ are the positions of the N stars in the field. The systematic difference in the deflection caused by the the macrolens is described by the first term in the preceding equation; here, we concentrate only on the random part, denoted by $\boldsymbol{\Delta}$:

$$\boldsymbol{\Delta} = \sum_{i=1}^{N}\left(\frac{\mathbf{x}_a - \mathbf{x}_i}{|\mathbf{x}_a - \mathbf{x}_i|^2} - \frac{\mathbf{x}_b - \mathbf{x}_i}{|\mathbf{x}_b - \mathbf{x}_i|^2}\right) \;.$$

A scaling relation. The number N of stars in our field is related to its dimensionless radius X through $N = X^2\kappa_*$. Now, suppose that all distances in the star field are multiplied by a factor U; the parameters of this scaled field will be denoted by a tilde. We find:

$$\tilde{X} = UX \;,$$
$$\tilde{\mathbf{x}}_i = U\mathbf{x}_i \;,$$
$$\tilde{s} = Us \;,$$
$$\tilde{\kappa}_* = \frac{\kappa_*}{U^2} \;,$$

and therefore

$$\tilde{\boldsymbol{\Delta}} = \frac{\boldsymbol{\Delta}}{U} \;.$$

In particular, we can choose U such that $U = \sqrt{\kappa_*}$; then, the foregoing relations immediately imply the following scaling law for the probability $P(\Delta)$ that the random component of the difference in the deflection angle is larger than Δ:

$$P(\Delta, \kappa_*, s) = P\left(\frac{\Delta}{\sqrt{\kappa_*}}, 1, \sqrt{\kappa_*}\, s\right) \;. \tag{11.65}$$

Therefore, in the following we can always assume, without loss of generality, that $\kappa_* = 1$. Although all the assumptions that validate Markov's method are satisfied, this method is very difficult to apply here [DE87.1], as the function q of (11.26b) cannot be calculated analytically. Below, we present some results derived by a Monte-Carlo simulation, but first, we discuss some limiting cases.

The case of small separation, $s \ll 1$. If the two light rays are very close together, the probability that a star is closer to them than, say, twice their separation, is small. In this case, one can expect that the difference in their deflection can be described by an expression linear in the separation. To confirm that, let us expand the random component of the deflection angle, $\boldsymbol{\alpha}_r$, about the origin of the star field,

$$\boldsymbol{\alpha}_r(\mathbf{x}) = \boldsymbol{\alpha}_r(\mathbf{x} = 0) + \begin{pmatrix} S_1 & S_2 \\ S_2 & -S_1 \end{pmatrix} \mathbf{x} + \mathcal{O}\left(|\mathbf{x}|^2\right) \quad ,$$

where the shear components S_1, S_2, caused by the star field are given in (11.49b). From this equation, we find that the difference in the random deflection angle is given by

$$\Delta = |\boldsymbol{\Delta}| = \sqrt{S_1^2 + S_2^2}\, s \quad .$$

This linear description for Δ certainly fails if a star is closer to the two light rays than a distance approximately equal to their mutual separation, but in this case, one can assume that the difference in the deflection angle becomes large. Hence, the linear approximation is valid for small Δ. Since the probability that the random shear is larger than S is $1/\sqrt{1+S^2}$, as can be derived from (11.54) with $\kappa_* = 1$, the preceding linear relation yields

$$P(\Delta) = \frac{1}{\sqrt{1 + (\Delta/s)^2}} \quad . \tag{11.66a}$$

The case of large deflection difference, $\Delta \gg 1$. A light ray undergoes a large deflection angle if it passes very close to a star. The probability for a single light ray to be deflected by more than $\alpha \gg 1$ is $1/\alpha^2$, as follows from (11.42). Since this probability is small, it is highly improbable that a second light ray will also be strongly deflected. Hence, from this argument we can conclude that the probability for a deflection difference of more than Δ is

$$P(\Delta) = \frac{2}{\Delta^2} \tag{11.66b}$$

for $\Delta \gg 1$, independent of the separation of the rays, as long as $s\Delta \gg 1$; the factor 2 in (11.66b) stems from the fact that either ray can undergo a large deflection.

The case of large separation, $s \gg 1$. If the two light rays are widely separated, there should be no correlation between the two deflections, except for the fact that the rays are focused towards each other; this is due to the fact that there is matter (stars) between the two rays, which acts as a focusing lens. This, however, is not a truly random contribution, and so it is useful to consider the deflection difference $\boldsymbol{\Delta}_c = \boldsymbol{\Delta} - \kappa_*(\mathbf{x}_a - \mathbf{x}_b)$, where the systematic component has been taken out. Owing to the missing correlation of the deflection angles of two well-separated rays, the probability distribution for $\boldsymbol{\Delta}_c$ should tend towards a distribution which is independent of s,

$$P(\Delta_c, s) \to P_0(\Delta_c) \quad \text{for} \quad s \to \infty \quad . \tag{11.66c}$$

Results from numerical simulations. The probabilities $P(\Delta)$ were computed in a Monte-Carlo simulation: N stars were randomly distributed in a circle of radius $X = \sqrt{N}$, and the difference of the deflection angles for a pair of rays in the middle of the circle, and separated by s, was computed. By calculating this difference of deflection for a large number of randomly generated star fields, the probabilities can be determined. To insure that the finiteness of the star field has no influence on the results, X must be much larger than the separation s; in all the cases shown here, we have tested the sensitivity of our results to the extent of the field by repeating the simulations with twice the number of stars.

In Fig. 11.6a, we show $P(\Delta)$ for several values of s, as a function of Δ/s. As predicted, the curves agree with the approximate result (11.66a) for small Δ and small s; in fact, the difference between the curve for $s = 0.2$ and (11.66a) is smaller than the thickness of the curves in the figure. Also, note how closely the curve for $s = 0.5$ follows this analytic behavior. Fig. 11.6b shows $P(\Delta_c)$ for various values of s. For large Δ_c, the probability follows the asymptotic power law (11.66b), plotted as the dashed line. Note the small difference between the curves for $s = 4$ and $s = 8$; in fact, the result for still larger s is basically identical to that for $s = 8$, which therefore is the theoretically predicted function $P_0(\Delta_c)$ (11.66c).

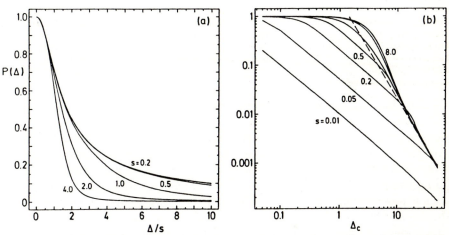

Fig. 11.6. (a) The probability $P(\Delta)$ that the difference of the random deflection angles for two light rays, separated by s, is larger than Δ, as a function of Δ/s, for $s = 0.2, 0.5, 1.0, 2.0, 4.0$. The density of the random star field is $\kappa_* = 1$; for other values of κ_*, the scaling equation (11.65) can be used. The uppermost curve, for which $s = 0.2$, cannot be distinguished from the analytically predicted behavior (11.66a). (b) The corresponding probability $P(\Delta_c)$ for the deflection difference defined in the text, as a function of Δ_c, for $s = 0.01, 0.05, 0.2, 0.5, 1.0, 4.0, 8.0$. The dashed line is the predicted asymptotic behavior (11.66b). The uppermost curve ($s = 8.0$) does not change markedly if the separation of the two light rays is increased; it corresponds to the limiting curve $P_0(\Delta_c)$ (11.66c).

11.2.5 Spatial distribution of magnifications

Motivation. With the formalism of Sect. 11.2.3, we can study the amplification bias in observations of compact extragalactic sources, as will be described in Sect. 12.5. If we consider a source lensed by a galaxy containing stars, its magnification will be a function of time, as the relative position of source and lens changes. To investigate this situation, it is not sufficient to consider only magnification probabilities, but the correlation of magnification factors as a function of relative source position is also required. More specifically, knowing the magnification as a function of the relative position of a source seen through a random star field, we can derive model light curves for a source which changes its position with time.

Point sources seen through a star field. One method to obtain model light curves of point sources seen through a star field consists of solving the lens equation explicitly for an ensemble of point-mass lenses [PA86.1]. The corresponding numerical procedure is quite complicated, since all microimages must be found.

> The total number of such microimages is not known, but is comparable to, and in general slightly larger than, the number of stars considered (due to the fact that a star far away from the line-of-sight to the source will produce a – generally faint – microimage). To avoid missing any microimage, one must make a number of guesses for the image positions. An obvious guess can be made for stars relatively far away from the line-of-sight, which have to produce a nearby image. Other guesses include the midpoint between close pairs of stars; in addition, a fine regular grid in the lens plane must be checked for possible images. Starting from these guessed positions, the iteration method described in Sect. 10.5 can be used. The corresponding solutions of the lens equation can then be compared with all previously obtained solutions and stored, together with their magnifications, if they are different from previous values. The total magnification is obtained by adding up the magnifications for all the images.

The disadvantages of this method are that it is very time consuming, and can thus be applied to star fields of moderate optical depth only, and, since the total number of images is not known a priori, one cannot be sure to have obtained all the images. On the other hand, the method has some advantages over the one we describe next, because it yields the image positions and their individual magnifications, and it is straightforward to include proper relative motion of the stars in the lensing galaxy by assigning to each star, in addition to its random position, a random velocity vector. In addition, the method described here is the only one known which can produce light curves for infinitesimally small sources.

Extended sources seen through a star field. A different approach is provided by the ray-shooting method described in Sect. 10.6. Although the latter method can deal only with extended sources, it is much more powerful than the method described in the preceding paragraph, because it allows a simultaneous calculation of the magnification for a large number of source positions. The drawback of the ray-shooting method is that it is only effective if the star field is static; hence, its results can only be considered realistic if the relative velocity of the source and the lens is much larger than the velocity dispersion of the stars in the field.[2]

---

[2] This assumption should be fairly well satisfied in general. For instance, the velocity of the Earth with respect to the microwave background radiation is about 600 km/s, appreciably larger than the velocity dispersion in elliptical galaxies. If the lensing galaxy is part of a cluster, it will typically have a transverse velocity much higher than the velocity dispersion of its stars.

For extended sources, we utilize the following procedure: a region in the source plane is divided into a large number of pixels, which can be either square or circular. A star field is created using a random number generator, and a large number of light rays is mapped from a uniform, rectangular grid in the lens plane onto the source plane. These rays are then collected in the pixels. The total number of rays in a pixel is directly proportional to the total magnification for a source at the position, and the size and shape of, the pixel. The size of the star field must be chosen such that the total flux of all images outside the mapped grid is vanishingly small; an estimate for the number of stars that should be used is given by (11.47) (see also [SC87.3]). Typically, the magnification for each pixel is evaluated to an accuracy of 1% in recent papers (e.g., [WA90.1]). The number of stars can become very large if large values of $\langle\mu\rangle \kappa_*^2$ are considered. This makes the method time-consuming as well, since a very large number of rays must be mapped, and each deflection angle is a sum of individual deflections due to a large number of stars. To keep the problem tractable, several refinements over a brute force calculation of the deflection angle for each individual light ray have been developed and used, as described in Sect. 10.6.

Magnification pattern. In Fig. 11.7, we display some selected results from ray shooting simulations[3]; a more thorough and systematic study can be found in [WA90.4]. The frames of this figure show the magnification as a function of relative source position. The magnification relative to the mean magnification $\langle\mu\rangle$ is coded in colors, where we define $\Delta m = -2.5 \log \frac{\mu}{\langle\mu\rangle}$. The boundary between green and red corresponds to the mean magnification $\langle\mu\rangle$; the correspondence between magnification and color is described in the caption of Fig. 11.7. Each frame corresponds to different parameters of the star field, which are listed in Table.11.1, together with some additional information.

All but two frames show results for microlenses of the same mass M_0, whereas in frames (k) and (l) a mass spectrum of the form $n(M) \propto M^{-2.35}$ for $m_{\min} \leq (M/M_0) \leq 1$ is considered, with $m_{\min}$ given in Table 11.1. The length of all frames is $20\xi_0$ (see (2.6b)).

Frames (a)–(e) have $\kappa_c = \gamma = 0$, and display the magnification patterns for increasing optical depth κ_*. For $\kappa_* = 0.2$ [frame (a)], most of the area is green, which corresponds to a magnification slightly in excess of 1, or $\Delta m \sim 2.5 \log \langle\mu\rangle$. Individual caustic figures are easily identified; e.g., the structure below the middle of frame (a) stems from a double star (see Sect. 8.3, in particular, Fig.8.9). The caustics of individual stars, perturbed by the shear from the other stars, can also be identified. However, the more complicated caustic structure below, and to the left of the frame's center, cannot be easily traced back to an individual group of stars; this caustic is produced by the collective effects of many stars. It is surprising that even for such a low value of κ_*, fairly large regions filled with caustics can be seen.

This effect is much more pronounced for higher values of κ_*: in frame (b) ($\kappa_* = 0.5$), individual caustics can no longer be traced back to individual stars; rather, the caustics form a connected structure which is larger than the size of the frame. If the resolution of this frame were higher, we could see higher-order catastrophes, such as beak-to-beaks, swallowtails,

---

[3] We are grateful to J. Wambsganss for preparing this figure for us.

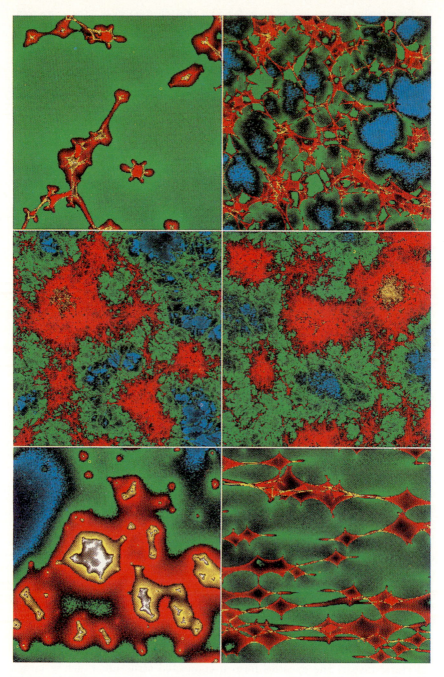

Fig. 11.7a–l. Microlensing magnification patterns for parameter values as given in Table 11.1. Each frame consists of 500^2 pixels, and has a side length of $20\xi_0$. The magnification is coded by colors; with $\Delta m = -2.5\log(\mu/\langle\mu\rangle)$, they are: $\Delta m \geq 3$: black; $2 \leq \Delta m \leq 3$: blue; $1 \leq \Delta m \leq 2$: grayish blue; $0 \leq \Delta m \leq 1$: green; $-1 \leq \Delta m \leq 0$:

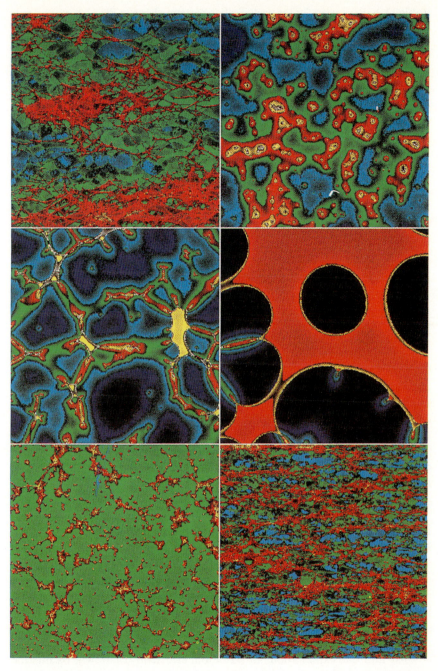

red; $-2 \leq \Delta m \leq -1$: yellow; $\Delta m \leq -2$: white. Thus, the mean magnification $\langle \mu \rangle$ corresponds to the boundary between green and red. Part 1 of the figure contains frames (**a**)–(**f**), part 2 frames (**g**)–(**l**), labeled from upper left to lower right

Table 11.1. Parameters for the magnification patterns shown in Fig. 11.1. κ_*, κ_c, and γ denote, respectively, the dimensionless surface mass density in stars and smooth matter, and the external shear. For frames (a)–(j), all stars are assumed to have the same mass, whereas in frames (k) and (l), a mass spectrum of the form $n(m) \propto m^{-2.35}$ between $m_{\min} \leq m \leq 1$ was assumed. N_* is the number of stars considered for the frame, $\langle\mu\rangle_{\text{num}}$ is the mean magnification as determined from the frame, and $\langle\mu\rangle$ is the theoretical mean magnification

	κ_*	κ_c	γ	$m_{\min}$	N_*	$\langle\mu\rangle_{\text{num}}$	$\langle\mu\rangle$
a	0.2	0.0	0.0	1.0	163	1.41	1.56
b	0.5	0.0	0.0	1.0	987	3.46	4.00
c	0.8	0.0	0.0	1.0	9343	24.68	25.00
d	1.2	0.0	0.0	1.0	19827	28.05	25.00
e	5.0	0.0	0.0	1.0	570	0.11	0.06
f	0.2	0.0	0.5	1.0	568	2.38	2.56
g	0.5	0.0	0.4	1.0	11693	10.57	11.11
h	0.5	1.0	0.0	1.0	987	4.45	4.00
i	0.5	1.5	0.0	1.0	269	0.97	1.00
j	0.5	3.5	0.0	1.0	41	0.11	0.11
k	0.2	0.0	0.0	0.01	5257	1.56	1.56
l	0.36	0.0	0.44	0.02	494005	4.47	4.63

and butterflies (see Chap. 6). Between the caustics there are regions of small magnification, with $\Delta m > 1$. The fact that caustics do strongly cluster means that the region displayed in the figure can not be considered an *average* part of the source plane, but rather a *typical* region. This can also be seen from the fact that the mean magnification over the frames displayed differ from the theoretical mean magnification $\langle\mu\rangle$ (see Table 11.1): this is *not* due to numerical inaccuracy, but to the fact that the frames are too small to cover a fair sample of the source plane.

For optical depth closer to one (frames (c) and (d), with $\kappa_* = 0.8$ and 1.2, respectively), the clustering of caustics becomes much stronger. In fact, individual caustics can hardly be recognized in these two frames, since their density is very high. As a large number of stars was used in the calculation of these two frames, one would expect to obtain a fairly uniform magnification pattern; yet the contrary is true: the occurrence of large regions with high magnification, and correspondingly large regions with low μ clearly shows the high non-linearity of the formation of the magnification pattern.[4] Although the origin of this clustering is not completely understood, we can at least find one explanation: in a random star field, there will be fluctua-

---

[4] A perhaps more familiar example of this effect is the formation of large-scale structure in the universe [BO88.1], where large coherent regions with high density are formed out of a random Gaussian distribution. In fact, some of these 'pancake' simulations are fairly similar to those presented here, with the deflection distribution replaced by a distribution of initial velocities [BU89.1].

tions in the mean star density. Consider a region where the density is higher than the mean. There, more critical curves will exist than on the average. In addition, the overdensity will cause a net focusing in mapping the critical curves onto the source plane, thus leading to a region of high caustic density. Note that the frame with $\kappa_* = 1.2$ is nearly indistinguishable from that with $\kappa_* = 0.8$. In fact, it seems to be true (although it is difficult to quantify this) that frames with $\kappa_* = 1 + \epsilon$ are very similar to those with $\kappa_* = 1 - \epsilon$, for $\epsilon \ll 1$.

Approaching the 'critical' optical depth $\kappa_* = 1$ even further, the coherent regions of higher than average magnification become even larger, and a symmetry between regions of $\mu > \langle\mu\rangle$ and $\mu < \langle\mu\rangle$ seems to develop. At the same time, the differences between highest and lowest magnification become smaller (for a given source size, e.g., the size of one pixel). In fact, this latter result can be derived analytically [DE87.1]: in Fig. 11.8, we show the 'normalized' magnification fluctuation $\langle\mu^2\rangle / \langle\mu\rangle^2$ for a Gaussian source behind a random star field with $\kappa_c = \gamma = 0$, as a function of κ_*. The fluctuations are small for small κ_*, have a maximum at medium values of κ_*, and approach unity for $\kappa_* \to 1$; for higher optical depth, the fluctuations increase again. As expected, the fluctuations are larger for smaller sources. The behavior shown in Fig. 11.8 has been confirmed with numerical ML simulations [WA90.4].

For optical depth much larger than one – as in frame (e) – the magnification pattern becomes more structured again, with fairly large regions with

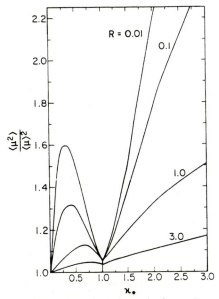

Fig. 11.8. The fluctuation $\langle\mu^2\rangle / \langle\mu\rangle^2$ of the magnification of a Gaussian source with dimensionless radius R seen through a star field with vanishing shear, as a function of the optical depth κ_* (from [DE87.1])

$\Delta m < 0$. Note, however, that for the parameters of frame (e), $\langle \mu \rangle = 0.06$, so that even the white regions in frame (e) do not correspond to a high magnification relative to an unlensed source.

By taking into account a finite external shear γ, the magnification patterns are no longer statistically isotropic. As can be seen in frame (f), there is a preferred direction: the caustics are primarily stretched horizontally. Many of the individual caustics seen in frame (f) can be clearly recognized as those of the Chang–Refsdal lens (see Sect. 8.2.2). The effect of an external shear is also seen in frame (g), but less pronounced.

A finite smooth surface mass density $\kappa_c < 1$ does not lead to different magnification patterns, due to the scaling introduced in Sect. 11.2.3. However, for $\kappa_c \geq 1$, new features appear. In particular, the case $\kappa_c = 1$ is special: as mentioned before, in this case no folds and cusps can occur, since every critical point necessarily leads to an umbilic. For this reason, the pattern of frame (h) differs from all the others shown in Fig. 11.7. For $\kappa_c > 1$, the magnification pattern appears like a honeycomb structure, with large holes of very low magnification, and the 'walls' are high-magnification regions; this is most clearly seen in frame (j), but also visible in frame (i). Note that the magnification distribution in frame (j) is peculiar: most of the area is either 'red' or 'black', meaning that the magnification probability is clearly bimodal.

Finally, frames (k) and (l) show the effects of a mass spectrum of lenses. The optical depth in frame (k) is the same as in frame (a), but the magnification patterns look fairly different. This difference in not only a matter of scale [in the sense that (a) can be considered as a blown-up version of (k)], but due to the mass spectrum, there is not a single preferred length-scale in frame (k). The parameters for frame (l) are those relevant for image A in the 2237+0305 GL system, which will be discussed in detail in Sect. 12.4.2. Model light curves, which are obtained if a source is assumed to move relative to the magnification pattern, will be discussed in Sect. 12.4.1.

We also want to note that all the caustics shown in Fig. 11.7 are convex as viewed from that side where the number of images is larger by two than on the other side. This property was derived at the end of Sect. 6.2.1.

11.3 Probabilities in a clumpy universe

In the preceding section, we considered an ensemble of lenses all at the same distance from the observer. We discuss here the case of cosmologically distributed lenses, and we do not restrict the treatment to a special class of lenses. The main result of this section is an equation for the probability that a source at redshift z_s undergoes a lensing event with preassigned properties. As in Sect. 11.1, we are interested in lensing events with a certain property Q, e.g., where the total magnification is larger than μ, or where the image separation is above the resolution limit of a given telescope.

Description of the cosmological model. In this section, we assume that the Dyer–Roeder description of light propagation in a clumpy universe is at least a valuable working hypothesis for the treatment of gravitational lensing. For a discussion of the assumptions in this model, see Sect. 4.5.

There are then two useful definitions of angular-diameter distances. First, the distance $D(z) = \frac{c}{H_0} r(z)$, the ratio of physical impact parameter ξ at redshift z and the corresponding angular separation θ, and $r(z)$ is the solution of the Dyer–Roeder equation (4.52) with $r(0) = 0$, $r'(0) = 1$. Second, since the large-scale geometry of the universe is assumed to be of the smooth Friedmann–Lemaître type with the same mean matter density, the area of a sphere of constant redshift z is

$$\mathcal{A}(z) = 4\pi\, D_1^2(z) \quad,$$

where $D_1(z) = \frac{c}{H_0} r_1(z)$ is the angular-diameter distance in a smooth ($\tilde{\alpha} = 1$, where $\tilde{\alpha}$ is the smoothness parameter; cf. Sect. 4.5) universe, and $r_1(z)$ is given by (4.57a).

Let us consider a clumpy universe, characterized by the Hubble constant H_0, the density parameter $\Omega = \rho_0/\rho_{\mathrm{cr}}$, i.e., the ratio of the mean density of the universe today and the critical density (4.36), and the smoothness parameter $\tilde{\alpha}$. Here and in the following, the cosmological constant is set to zero.

The lens population. Let us distribute an ensemble of lenses throughout the universe, and let their properties be described by a set of parameters collectively denoted by χ. These parameters can be total mass and core radius of a galaxy, or simply the mass of point lenses. Let $n_{\mathrm{L}}(\chi, z) d\chi$ be the density of lenses at redshift z with parameters within $d\chi$ of χ. In the following, we assume that the population of lenses does not evolve, i.e., their *comoving* density is constant,

$$n_{\mathrm{L}}(\chi, z) = (1+z)^3\, n_{\mathrm{L}}(\chi) \quad,$$

and $n_{\mathrm{L}}(\chi)$ describes the density of lenses today, i.e., at zero redshift. Furthermore, we assume that all clumps in the universe can be considered as lenses, so that their total mass density today is $\rho_0(1 - \tilde{\alpha})$. This corresponds to

$$\int d\chi\, M(\chi)\, n_{\mathrm{L}}(\chi) = (1 - \tilde{\alpha})\Omega \rho_{\mathrm{cr}} \quad, \tag{11.67}$$

where $M(\chi)$ is the mass of a lens described by the parameters χ.

Lensing cross-section and probability. In Sect. 11.1 we defined the Q-cross-section $\hat{\sigma}$ for a lens, which depends in general on the property Q, the source and lens redshifts, z_{s} and z_{d}, respectively, and the parameters χ of the lens, $\hat{\sigma} = \hat{\sigma}(Q, z_{\mathrm{d}}, z_{\mathrm{s}}, \chi)$.

Consider a shell at redshift z with thickness dz; the corresponding physical thickness is $\left(\frac{c}{H_0}\right)\left(\frac{dr_{\mathrm{prop}}}{dz}\right) dz$ [see (4.47b)], and its volume is

$$dV = \mathcal{A}(z)\left(\frac{c}{H_0}\right)\frac{dr_{\text{prop}}}{dz}dz \quad.$$

The number of lenses with parameters within $d\chi$ of χ in this shell is

$$dN = dV(1+z)^3 \, n_L(\chi)\, d\chi \quad,$$

and these lenses have a cross-section $d\hat{\sigma}$ on the sphere at $z = z_s$, where

$$d\hat{\sigma} = \hat{\sigma}(Q, z, z_s, \chi)\, dN \quad.$$

The total cross-section on the source sphere is obtained by integration over the lens redshift and the parameter space,

$$\hat{\sigma}_{\text{tot}}(Q, z_s) = \int_0^{z_s} dz \int d\chi \, \frac{d\hat{\sigma}}{dz\, d\chi} \quad,$$

which is valid as long as the cross-sections of individual lenses do not overlap. In particular, this means that σ_{tot} must be much smaller than the area $\mathcal{A}$ of the source sphere. Using (4.47b) and the equations above,

$$\hat{\sigma}_{\text{tot}}(Q, z_s) = 4\pi \left(\frac{c}{H_0}\right)^3$$
$$\times \int_0^{z_s} dz \int d\chi \, \hat{\sigma}(Q, z, z_s, \chi)\, n_L(\chi)\, r_1^2(z)\, \frac{1+z}{\sqrt{1+\Omega z}} \quad.$$

The probability for a source at redshift z_s to undergo a lensing event with property Q is obtained by dividing the total cross-section by the area $\mathcal{A}(z_s)$ of the source sphere,

$$P(Q, z_s) = \left(\frac{c}{H_0}\right)\frac{1}{r_1^2(z_s)}$$
$$\times \int_0^{z_s} dz\, r_1^2(z)\, \frac{1+z}{\sqrt{1+\Omega z}} \int d\chi\, \hat{\sigma}(Q, z, z_s, \chi)\, n_L(\chi) \quad; \tag{11.68}$$

the condition of non-overlapping cross-sections restricts the validity of (11.68) to $P \ll 1$.

Discussion of (11.68). Most papers in the literature on statistical gravitational lensing do not distinguish between the two different angular-diameter distances relevant to a clumpy universe; in those papers, the $r_1(z)$-factors in (11.68) are replaced by $r(z)$. As was pointed out in [EH86.1] the distinction is necessary for a *self-consistent treatment of probabilities in a clumpy universe*. This paper also explains the absence of such a distinction in the "older" literature, based on a derivation of the probability in which the random variable is the line-of-sight to a source. However, lines-of-sight to the sources are not equivalent, since some happen to be within empty cones (see Sect. 4.5) – those correspond to sources with a small probability of associated lensing – and some are not. The random variable used to derive (11.68) is the position of the source on the sphere $z = z_s$. Since

$$\langle \mu \rangle (z) = \left(\frac{r(z)}{r_1(z)} \right)^2 \tag{11.69}$$

– see (4.82) – is an increasing function of z, the old equation for P *underestimates* the lensing probabilities.

The relative error introduced by neglecting the distinction of r and r_1 is, however, not larger than about a factor of 1.5 in all cases where (11.68) applies, the reason for this being the dependence of $\langle \mu \rangle$ on $\tilde{\alpha}$ and Ω [PE86.1]: if the density of lenses $[\propto (1 - \tilde{\alpha})\Omega]$ is too large, the linear treatment adopted for deriving (11.68) becomes invalid anyway.

Example: Magnification of point sources by point masses. Equation (11.68) is the main result of this section; its application requires the specification of the lens population $n_L(\chi)$ and the knowledge of the cross-section. Specific cases are discussed in the following chapter; here we want to consider the simplest case, namely, the magnification probability for point sources in a universe filled with point masses of mass M (see [CA82.1], [EH86.1], [SC87.1], [IS89.1]). In this case, the cross-section is given by (11.2), which after insertion into (11.68) yields

$$P(\mu_p, z_s) = \frac{4GM}{c^2} \left(\frac{c}{H_0} \right)^2 n_L \, \pi y^2(\mu_p) \left[\frac{r(z_s)}{r_1(z_s)} \right]^2$$

$$\times \int_0^{z_s} dz \frac{1+z}{\sqrt{1+\Omega z}} \left[\frac{r_1(z)}{r(z)} \right]^2 \frac{r(z)\,r(z,z_s)}{r(z_s)} \, ;$$

here, n_L is the present density of the point masses, and $y^2(\mu_p)$ is given by (11.1). Using (11.67) in the form $n_L M = (1 - \tilde{\alpha})\Omega \rho_{cr}$ and (4.36), this becomes

$$P(\mu_p, z_s) = y^2(\mu_p) \, \tau_p(z_s) \quad , \tag{11.70a}$$

where

$$\tau_p(z_s) = \frac{3}{2} (1 - \tilde{\alpha}) \, \Omega \, \langle \mu \rangle (z_s)$$

$$\times \int_0^{z_s} dz \frac{1+z}{\sqrt{1+\Omega z}} \frac{1}{\langle \mu \rangle (z)} \frac{r(z)\,r(z,z_s)}{r(z_s)} \tag{11.70b}$$

is a function of redshift only. It is easy to see that – for point sources – (11.70a) is also valid for a population of point lenses with a mass spectrum, as long as (11.67) is satisfied [RE70.1]. The factorization of $P(\mu_p, z_s)$ makes it very convenient for studies of the amplification bias. For extended sources, the situation becomes more complex (see Sect. 12.5).

In Fig. 11.9 we have plotted the function $\frac{\tau_p(z_s)}{[\Omega(1-\tilde{\alpha})]}$ for various cosmological models. Since the cosmological density of lenses is proportional to $\Omega(1 - \tilde{\alpha})$, these curves are normalized to the same lens density. The differences in the curves thus reflect the different laws of light propagation

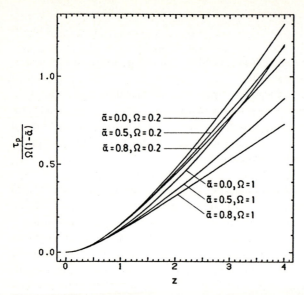

Fig. 11.9. The optical depth τ_p for point-mass lenses, defined in (11.70b), scaled by $\Omega(1 - \tilde{\alpha})$, which is proportional to the density of lenses in the universe, as a function of the source redshift z, for several cosmological models characterized by Ω and $\tilde{\alpha}$

in a clumpy universe. We see that the normalized optical depth is largest for cosmologies with the least matter in the beam ($\tilde{\alpha} = 0$, $\Omega = 0.2$) and smallest for those with the most smoothly-distributed matter.

11.4 Light propagation in inhomogeneous universes

Outline of the problem. In Sect. 4.5, we saw that calculations relevant to the propagation of light rays in a clumpy universe face two essential difficulties. First, we do not know any solution of Einstein's equation capable of describing a clumpy universe. This problem led to the popular assumption that, on large scales, the universe may be described by the homogeneous isotropic Robertson–Walker metric. Owing to the lack of any alternative, we have adopted this point of view as a working hypothesis in this book.

The second difficulty concerns the propagation of light in the space-variant gravitational field caused by clumps in an inhomogeneous universe. For light bundles well away from all matter concentrations, the Dyer–Roeder differential equation (4.52) describes the evolution of the cross-sectional area. To derive this equation, it was assumed that the shear on the light bundles vanishes. However, it is by no means obvious that this is a good approximation for the majority of light rays. Of course, this is a statistical problem.

Transformation into a lensing problem: General idea. GL theory is well suited to study the second of the problems we have just mentioned, since it accounts explicitly for the deflection caused by the clumps. Two different methods of attacking this problem have been followed in the literature: the evolution of the cross-sectional area of light bundles as they propagate was studied in [RE70.1], [SC88.2], [KA88.3], and [JA89.1], and a generalization of the ray-shooting method (see Sects. 10.6, 11.2.5) was considered in [SC88.3]. With the first approach, one obtains information about the statistics of magnifications along light rays, whereas the latter provides the distribution of magnifications of (extended) sources. Here, we describe both methods, the former in more detail, because it yields several interesting analytic results. In particular, we present an alternative form and derivation of the Dyer–Roeder equation.

According to our cosmological assumption, the large-scale properties of the universe are based on the Robertson–Walker metric, independent of the small-scale distribution of matter. Thus, if we concentrate some of the mass into clumps, the mean properties of the universe do not change. In particular, if we consider a 'fat' light bundle which contains many clumps, the ratio of its cross-sectional diameter and its angular diameter will be approximately the angular-diameter distance of a *smooth* Friedmann–Lemaître universe; the Dyer–Roeder angular-diameter distances are relevant only for small bundles which can "slip" between the clumps. Since the basic assumption underlying the derivation of the Dyer–Roeder angular-diameter distance is that shear vanishes along ray bundles, and since gravity is a long-range force (so there is non-zero shear acting on every light bundle), the Dyer–Roeder angular diameter distance can be considered only as a limiting case: no light bundle can be focused less than predicted by the Dyer–Roeder differential equation.

Thus, if we add a hypothetical mass distribution of zero total mass to the matter contents of a smooth Friedmann universe, the average angular-diameter distance remains unchanged. This hypothetical mass distribution can be a distribution of lenses accompanied by a homogeneous negative mass density, such that the mean densities of both components add to zero. The resulting total mass distribution[5] will then be a model for a clumpy universe, where the light propagation can be studied by accounting for the deflection caused by the clumps.

Construction of a lens model for the clumpy universe. Let us divide our hypothetical matter distribution into thin spherical shells centered on the observer. Consider the shell between redshifts Z_{i-1} and $Z_i = Z_{i-1} + \Delta Z$; the surface mass density of clumps in this shell is

---

[5] which must be non-negative everywhere, in the sense that the absolute value of the smooth negative surface mass density must be smaller than the density of the smooth Friedmann–Lemaître universe

$$\Sigma_i = \int_{Z_{i-1}}^{Z_i} dz \, \frac{dr_{\text{prop}}}{dz} \, \rho_{\text{cl}}(z) \; ,$$

where $\rho_{\text{cl}}(z) = (1 - \tilde{\alpha})(1 + z)^3 \Omega \rho_{\text{cr}}$ is the volume mass density in clumps. The clumps are then projected onto the sphere at $z_i = Z_i - \Delta Z/2$. Thus, if we choose a sequence of redshifts $0 = Z_0 < Z_1 < ... < Z_N = z_s$ between the observer and a source at z_s, the foregoing construction leads to an N lens plane problem (Chap. 9). The deflection angle at $\mathbf{x}_i$ in the i-th lens plane is then

$$\hat{\boldsymbol{\alpha}}_i = \hat{\boldsymbol{\alpha}}_i^{\text{cl}} - \frac{4G}{c^2} \pi \Sigma_i \boldsymbol{\xi}_i \; ,$$

where $\hat{\boldsymbol{\alpha}}_i^{\text{cl}}$ is the deflection caused by the clumps and the second term accounts for the negative surface mass density. By construction, the mean deflection angle is zero.

As in Chap. 9, we use angular variables, i.e., $\mathbf{x}_i = \boldsymbol{\xi}_i / D_1(z_i)$, $\mathbf{y} = \boldsymbol{\eta}/D_1(z_s) \equiv \mathbf{x}_{N+1}$; then, equations (9.3) to (9.8) apply. If the Z_i are chosen such that $\Delta Z \ll 1$, we find for the dimensionless surface mass density (9.4)

$$\kappa_i = \frac{3}{2} \Omega (1 - \tilde{\alpha}) \frac{1 + z_i}{\sqrt{1 + \Omega z_i}} \, \frac{r_1(z_i) r_1(z_i, z_s)}{r_1(z_s)} \Delta Z \equiv \kappa_i' \Delta Z \; . \qquad (11.71)$$

11.4.1 Statistics for light rays

The general case. The preceding discussion provides a recipe for numerical simulations of the magnification statistics of infinitesimally small light bundles: after choosing the sequence Z_i, lenses can be distributed in each lens plane according to the physical model under consideration (in most of what follows, we assume that the lenses are distributed randomly). A light bundle traversing this model universe will experience convergence and shear in every lens plane; in addition, it will be deflected. This deflection angle, however, is irrelevant as far as its magnification is concerned, for as long as the lenses in each lens plane are spatially independent, the probability distribution for shear and convergence in a lens plane is independent of the position of the light bundle.

The probability distribution for shear and convergence in each lens plane is obtained either from an analytic formula (as is the case for randomly distributed point-mass lenses – as will be seen in the following paragraph), or must be obtained numerically. In the latter case, one considers a circular area $\mathcal{A}_i$ centered on the light bundle in each lens plane, within which the lenses are distributed with a random number generator. Since the mean mass density of lenses is κ_i, the number of lenses in $\mathcal{A}_i$ can be obtained from a Poisson distribution. The areas $\mathcal{A}_i$ must be circular, since other shapes would introduce an artificial shear (as can easily be seen by considering, e.g., an elliptical distribution). For a light ray in the i-th lens plane, we can then compute the matrix U_i (9.10), and from (9.11) and (9.12) the

Jacobian matrix A of the lens mapping can be determined, and hence, the properties of the magnification and shape distortion of the light bundle can be derived. Such an approach was followed for point lenses [RE70.1] and isothermal galaxies ([KA88.3], [JA89.1]); we describe some results of the latter work at the end of this subsection.

The method for point-mass lenses. For point-mass lenses, there is a direct way to compute a statistical distribution for magnifications of ray bundles [SC88.2]. Since only differential deflection angles enter in A, and since the only effect of a random star field on a light bundle is its shear, whose probability distribution is known analytically (11.54), we do not need to generate random star fields numerically. The matrices U_i are

$$U_i = -\kappa_i \mathcal{I} + \mathcal{S}_i \quad , \tag{11.72a}$$

where the first term accounts for the (negative) convergence caused by the smooth negative surface mass density, and $\mathcal{S}_i$ is the shear matrix,

$$\mathcal{S}_i = \begin{pmatrix} S_{i,1} & S_{i,2} \\ S_{i,2} & -S_{i,1} \end{pmatrix} \tag{11.72b}$$

[compare with (11.49)], whose components are distributed according to (11.54). From that equation, by integration, the probability $P(S_i)$ for the shear to be greater than $S_i = \sqrt{S_{i,1}^2 + S_{i,2}^2}$ is

$$P(S_i) = \frac{\kappa_i}{\sqrt{\kappa_i^2 + S_i^2}} \quad .$$

Hence, in the numerical simulation one generates two random numbers η_1, η_2, uniformly distributed in the interval $[0, 1]$, for each lens plane; then, the shear components

$$S_{i,1} = \kappa_i \frac{\sqrt{1 - \eta_1^2}}{\eta_1} \cos(2\pi \eta_2) \quad ,$$

$$S_{i,2} = \kappa_i \frac{\sqrt{1 - \eta_1^2}}{\eta_1} \sin(2\pi \eta_2)$$

are distributed according to the probability density (11.54) (see, e.g., [HA64.1]) and are assigned to each lens plane. Finally, A is computed according to (9.11) and (9.12).

Different definitions of magnification. In our case, we must distinguish carefully between the magnification $\bar{\mu}$ relative to the smooth Friedmann universe and that relative to the empty cone in a clumpy universe, μ. These two are always related through

$$\mu = \langle \mu \rangle (z_s) \, \bar{\mu} \quad , \tag{11.73a}$$

as should be clear from flux conservation (Sect. 4.5.1), and $\langle\mu\rangle(z)$ is given by (11.69). According to our construction of a clumpy universe, we always work with the angular-diameter distances for a smooth universe, and since the mean convergence of large light bundles relative to that of a Friedmann–Lemaître universe vanishes, the inverse of the Jacobian determinant equals $\bar{\mu}$.

Correspondingly, the magnifications $\bar{\mu}_p$, μ_p for a source, relative to the smooth and clumpy universe, respectively, are related through

$$\mu_p = \langle\mu\rangle(z_s)\,\bar{\mu}_p \quad, \tag{11.73b}$$

and satisfy, due to flux conversation,

$$\langle\bar{\mu}_p\rangle = 1 \quad; \qquad \langle\mu_p\rangle = \langle\mu\rangle(z_s) \quad, \tag{11.73c}$$

averaged over source positions.

Alternative derivation of the Dyer–Roeder differential equation. In our clumpy universe, the divergence of light bundles in empty cones is due to the negative surface mass densities κ_i in the lens planes; therefore, we should be able to derive the Dyer–Roeder angular-diameter distance from our model. This is indeed the case, as we show next.

Empty cones in the clumpy universe model are characterized (or better, defined) by vanishing shear. Thus, consider the multiple lens plane equation (9.12) for the case that all $\mathcal{S}_i = 0$; then, all the A_i (9.9) are diagonal, $A_i = a_i\,\mathcal{I}$, and with (11.72a),

$$a_j = 1 + \sum_{i=1}^{j-1} \beta_{ij}\kappa_i a_i \quad. \tag{11.74a}$$

Here and in the following, it is convenient for analytic considerations to take the continuum limit, $\Delta Z \to 0$. Writing $A(z_i) = A_i = a(z_i)\mathcal{I}$, from (11.71) and (11.74a) we have

$$a(z) = 1 + \int_0^z dy\; \beta(y,z)\,\kappa'(y)\,a(y) \quad, \tag{11.74b}$$

where

$$\beta(y,z) = \frac{r_1(y,z)\,r_1(z_s)}{r_1(z)\,r_1(y,z_s)} \quad;$$

note that the combination $\beta(y,z)\kappa'(y)$ does not contain z_s anymore. Multiplying (11.74b) by $r_1(z)$ and letting $R(z) = r_1(z)\,a(z)$, we obtain:

$$R(z) = r_1(z) + \frac{3}{2}\Omega(1-\tilde{\alpha})\int_0^z dy\; r_1(y,z)\frac{1+y}{\sqrt{1+\Omega y}}\,R(y) \quad. \tag{11.75}$$

Note that $R(z)$ is the angular-diameter distance in units of c/H_0, along the light bundle under consideration. Differentiation with respect to z yields:

$$R'(z) = r_1'(z) + \frac{3}{2}\Omega(1-\tilde{\alpha})\int_0^z dy\;\frac{dr_1(y,z)}{dz}\frac{1+y}{\sqrt{1+\Omega y}}\,R(y) \quad, \tag{11.76a}$$

$$R''(z) = r_1''(z) + \frac{3}{2}\Omega(1-\tilde{\alpha})\left\{\left[\frac{dr_1(y,z)}{dz}\right]_{y=z}\frac{1+z}{\sqrt{1+\Omega z}}R(z)\right.$$
$$\left.+\int_0^z dy\,\frac{d^2r_1(y,z)}{dz^2}\frac{1+y}{\sqrt{1+\Omega y}}R(y)\right\} \quad .$$
(11.76b)

If we multiply the last equation by $(1+z)(1+\Omega z)$, the second derivative of r_1 in the integral can be replaced by the Dyer–Roeder equation (4.48) for $\tilde{\alpha}=1$ (which is equivalent to the redshift-distance relation in a smooth Friedmann–Lemaître universe, as shown in Sect. 4.5), we then use $[dr_1(y,z)/dz]_{y=z} = (1+z)^{-2}(1+\Omega z)^{-1/2}$ [see (4.53)], to obtain, with (11.76a,b),

$$(1+z)(1+\Omega z)\left[R''(z) - r_1''(z)\right] = \frac{3}{2}\Omega(1-\tilde{\alpha})R(z)$$
$$-\left(\frac{7}{2}\Omega z + \frac{\Omega}{2} + 3\right)\left[R'(z) - r_1'(z)\right] - \frac{3}{2}\Omega\left[R(z) - r_1(z)\right] \quad .$$

From the Dyer–Roeder equation for $\tilde{\alpha}=1$ (again, which we know to be valid since it is equivalent to the distance-redshift relation in a smooth Friedmann universe) we see that all terms involving r_1 vanish, and that the remaining equation is the Dyer–Roeder equation for $R(z)$. Since $R(0)=0$ and $R'(0)=r_1'(0)=1$, we find that $R(z)=r(z)$.

Hence, the integral equation (11.75) is equivalent to the Dyer–Roeder equation. Thus, as expected, it is possible to derive the law of propagation of light rays in empty cones solely from the continuum limit of the multiple lens-plane theory, without referring to the focusing equation (3.64).

Dependence of $r(z,\tilde{\alpha})$ on $\tilde{\alpha}$. Equation (11.75) is a Volterra integral equation of the second kind [TR85.1] with the standard iterative solution

$$R(z) \equiv r(z) = r_1(z) + \int_0^z \sum_{i=1}^\infty \left[\frac{3}{2}\Omega(1-\tilde{\alpha})\right]^i K_i(y,z)\,r_1(y)\,dy \quad ,$$

where $K_1(y,z) = (1+y)(1+\Omega y)^{-1/2}\,r_1(y,z)$ for $y \leq z$, zero otherwise, and the kernel satisfies

$$K_{i+1}(y,z) = \int_y^z K_1(y,x)\,K_i(x,z)\,dx \quad .$$

Since for all i, $K_i(y,z) \geq 0$ and $r_1(y) \geq 0$, $r(z)$ decreases with increasing $\tilde{\alpha}$; hence, any two solutions of the Dyer–Roeder equation, $r(z,\tilde{\alpha}_1), r(z,\tilde{\alpha}_2)$, satisfy

$$r(z,\tilde{\alpha}_1) \leq r(z,\tilde{\alpha}_2) \quad \text{for} \quad \tilde{\alpha}_1 \geq \tilde{\alpha}_2 \quad .$$

Analytic results for weakly-sheared images. We now consider light rays that experience only weak shear. For them, the non-linear shear terms, S_iS_j, may be neglected in the matrices A_i (9.9), and from (9.11), (9.12) and (11.72),

$$A_j = a_j \mathcal{I} - \sum_{i=1}^{j-1} w_{ij} \mathcal{S}_i + \mathcal{O}\left(\mathcal{S}^2\right)$$

with coefficients w_{ij} which can be determined from the a_i and β_{ij}. In particular, the Jacobian matrix A takes the form

$$A = a(z_s)\mathcal{I} - \begin{pmatrix} S_1 & S_2 \\ S_2 & -S_1 \end{pmatrix} + \mathcal{O}\left(\mathcal{S}^2\right) \quad . \tag{11.77}$$

As is shown in [SC88.2], one can derive the probability distribution $p(S_1, S_2)$ for the shear coefficients,

$$p(S_1, S_2) = \frac{1}{2\pi} \frac{Y(z_s)}{\left[Y^2(z_s) + S_1^2 + S_2^2\right]^{3/2}} , \tag{11.78a}$$

by considering the continuum limit $\Delta Z \to 0$, where the function $Y(z)$ is determined from the integral equation

$$Y(z) = a(z) - 1 + \int_0^z dy \, \kappa'(y) \, \beta(y, z) \, Y(y) \quad , \tag{11.78b}$$

and $a(z) = r(z)/r_1(z) = \sqrt{\langle \mu \rangle (z)}$. This last function is shown in Fig. 11.10 for various cosmological models.

The shape of $p(S_1, S_2)$ is similar to that of (11.54), with κ_* replaced by $Y(z_s)$, so $Y(z)$ can be interpreted as an effective dimensionless density of clumps in a clumpy universe. Using the same method that leads from (11.54) to (11.55) we find the probability $p_{\bar{\mu}}(\bar{\mu})d\bar{\mu}$ that the magnification is within $d\bar{\mu}$ of $\bar{\mu}$,

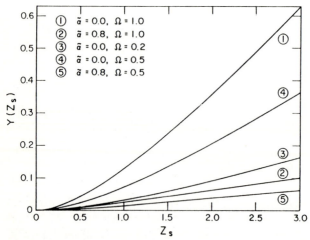

Fig. 11.10. The function $Y(z)$ defined in (11.78b), for several combinations of the cosmological parameters Ω and $\tilde{\alpha}$

$$p_{\bar{\mu}}(\bar{\mu}) = \frac{1}{2\bar{\mu}^2} \frac{Y(z_s)}{\left[Y^2(z_s) + \langle \mu \rangle (z_s) - 1/\bar{\mu}\right]^{3/2}} \quad ; \qquad (11.79)$$

this form is valid for light rays which are only weakly sheared.

Comparison with numerical simulations. We can compare (11.78) and (11.79) with the results of a numerical simulation. However, since the numerically-determined matrix A is not symmetric in general, we must define a shear S, in terms of the components of A, which is as closely related as possible to the components appearing in (11.77); this is most naturally done by defining

$$S^2 = \frac{1}{4}(A_{11} - A_{22})^2 + \frac{1}{4}(A_{12} + A_{21})^2 \quad ,$$

which for a single lens plane yields $S^2 = \gamma^2$, and which has the advantage that terms quadratic in the S_i do not contribute to S^2. In Fig. 11.11a, we compare the results for a ($\tilde{\alpha} = 0$, $\Omega = 1$) cosmological model, with $z_s = 2$ [corresponding to $Y(z_s) \simeq 0.357$]; in this case, 10^5 light rays were traced through 50 lens planes. We see that these two distributions agree very well. However, the agreement vanishes in the probability for the magnifications, shown in Fig. 11.11b; this is due to the fact that terms quadratic in the S_i appear linearly in det A. Hence, if the terms linear in S_i are nearly zero, the quadratic terms become dominant, which leads to a considerable deviation of the numerically determined probability from the analytic estimate for $\mu \gtrsim 1$, i.e., for those magnifications where (11.79) could have been expected to be a reasonable approximation. This again shows the importance of non-linear effects in statistical lensing.

Another indication for the relevance of non-linear effects is shown in Fig. 11.12, where, for one hundred light bundles with $1 \leq \mu \leq 2$, the axial ratio ϵ and magnification is plotted and compared to the unique relation between those quantities for a single lens plane, where, from the Jacobian matrix in its diagonalized form,

$$A = \begin{pmatrix} 1-\gamma & 0 \\ 0 & 1+\gamma \end{pmatrix} \quad ,$$

we find that $\mu = (1-\gamma^2)^{-1}$ and $\epsilon = \frac{(1-\gamma)}{(1+\gamma)}$. The numerically-determined points have a tendency to cluster around the line $\mu = \frac{(1+\epsilon)^2}{(4\epsilon)}$, but there is a large spread in the distribution, again due to non-linear terms in A.

The limit of high magnifications. Owing to multiple imaging of sources, in general it is impossible to transform the statistics for light rays into a distribution of source magnifications. Only for highly magnified *point sources* do we have a clear correspondence, as their total magnification is dominated by two highly magnified images close to a critical curve (Chap. 6). From the discussion in Sect. 11.1.2 we know that the probability $P(\bar{\mu})$ for

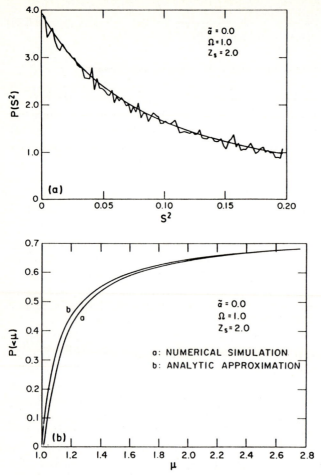

Fig. 11.11. (a) The probability distribution $p(S^2)$ (11.78a) for weakly-sheared images, as a function of S^2, for $z_s = 2$, $\Omega = 1$, and $\tilde{\alpha} = 0$; for these parameters, $Y(z_s) \approx 0.357$. This is compared with the results of a simulation of 50000 infinitesimal light bundles propagated through the universe. The agreement between the simulation and the theoretical curve is excellent. (b) For the same parameters, the probability $P(< \mu)$ that the magnification along an infinitesimal light bundle is between 1 and μ, as a function of μ. Curve (a) is the result of a numerical simulation, whereas curve (b) corresponds to the analytic result (11.79), with $\bar{\mu}$ replaced by μ, using (11.73). In this case, the agreement is much worse, due to the importance of non-linear shear terms in μ

the magnification of a light ray to be larger than $\bar{\mu}$ is $P(\bar{\mu}) = C/\bar{\mu}$ for $\bar{\mu} \gg 1$. Comparison of (11.21a) and (11.21b) then implies that the probability $P_{\mathrm{p}}(\bar{\mu}_{\mathrm{p}})$ that a point source is magnified by more than $\bar{\mu}_{\mathrm{p}}$ is $P_{\mathrm{p}}(\bar{\mu}_{\mathrm{p}}) = 2C/\bar{\mu}_{\mathrm{p}}^2$. Using (11.73b), the probability $P_{\mathrm{p}}(\mu_{\mathrm{p}})$ becomes

$$P_{\mathrm{p}}(\mu_{\mathrm{p}}) = 2C\left[\langle\mu\rangle(z_{\mathrm{s}})\right]^2 \mu_{\mathrm{p}}^{-2} \equiv \frac{\tau}{\mu_{\mathrm{p}}^2} \ . \tag{11.80}$$

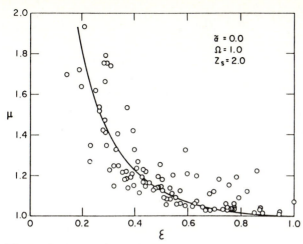

Fig. 11.12. For 100 light bundles with $1 \leq \mu \leq 2$, the magnification μ is shown in the ordinate; the abscissa is the axial ratio ϵ, for the same parameters as in Fig. 11.11. The solid line represents the dependence $\mu = (1 + \epsilon)^2/(4\epsilon)$, which would be expected for a light bundle through a star field

The value of C can be obtained directly from the simulation. In Fig. 11.13 we compare the value of τ with that of the linear treatment, τ_p (11.70b), for an $\tilde{\alpha} = 0$, $\Omega = 1$ cosmological model. Again, we find that the linear analysis *underestimates* the high-magnification probability, and the discrepancy increases with redshift, in agreement with [PE86.1]. The reason for the difference between the linear and non-linear results is that, in the linear treatment, each point mass was considered as a Schwarzschild lens. However, because of the shear due to the other lenses near the light beam, the symmetry is perturbed, so that each lens would be better described as a Chang–Refsdal lens (Sects. 8.2.2, 9.3), with the shear parameter determined by the position of the other lenses (or, statistically, from a probability distribution). From (11.23a) we see that the high-magnification cross-section for a Chang–Refsdal lens is larger than that for a Schwarzschild lens, and this accounts for most of the difference between τ and τ_p.

It is worthwhile at this point to take up the discussion following (11.68). As we remarked there, the distinction between the angular-diameter distances in a smooth and clumpy universe are essential to provide a self-consistent expression for lensing probabilities. The difference between $\tau_p(z_s)$ as given in (11.70b) and the corresponding expression calculated without the factor $\langle \mu \rangle (z_s)/\langle \mu \rangle (z)$ in the integrand amounts to about a factor 1.5 at $z_s = 2$ in a $\Omega = 1$, $\tilde{\alpha} = 0$ cosmological model. Hence, the self-consistent prescription (11.68) yields higher lensing probabilities than the older expression.

The probability (11.68) was derived under the assumption that each lens acts independently of all the other lenses in the universe. Such a linear approach necessarily becomes invalid if the lensing probabilities are not

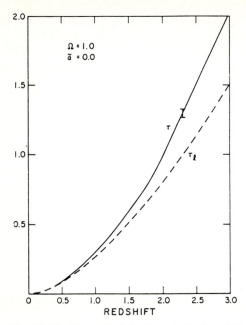

Fig. 11.13. The optical depth τ, defined such that τ/μ^2 is the probability that a point source is magnified by more than μ, as a function of the source redshift for a universe with $\Omega = 1$, $\tilde{\alpha} = 0$. The dashed curve (τ_l) is the linear result (11.70b), and the solid curve (τ) represents the results of numerical simulations. The error bar indicates the typical statistical error. Clearly, the linear result underestimates τ

negligibly small. An expression for the lensing probability of point masses, which accounts approximately for the possibility that there is more than one lens close to the line-of-sight to a source, was published in [CA82.1] (see also [WU90.1], [WU90.2]). As discussed in [IS89.1], this correction to the purely linear equation leads to differences of the same order as those from the self-consistent expression (11.68) and the old one. We interpret the results in [IS89.1] that the combination of the self-consistent probability and the approximate account of multiple deflection should yield a much better approximation to the numerically-determined optical depth τ than either of these two methods alone.

Weakly-magnified bundles. It is instructive to consider the probability distribution for light rays which are only weakly magnified, $\mu \gtrsim 1$. In the Dyer–Roeder picture, they correspond to the empty cones. Assuming that all weakly magnified images correspond to sources which have only this one main image (in addition, there are many very weak images, close to each point mass) we can derive the probability $P(1 \leq \mu_p \leq \mu_0)$ that a source is magnified by less than μ_0. In particular, we can determine the value for μ_m such that $P(1 \leq \mu_p \leq \mu_m) = m/100$. This definition means that $m\%$ of all sources are magnified by less than μ_m. Of course, the preceding assumption leads to an underestimate of μ_m. For two cosmological models, we show μ_m

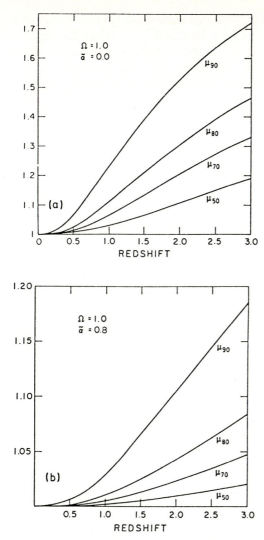

Fig. 11.14. The magnifications μ_m, defined such that $m\%$ of all sources have a magnification between 1 and μ_m, as a function of the source redshift, for a universe with (**a**) $\Omega = 1$, $\tilde{\alpha} = 0$; (**b**) $\Omega = 1$, $\tilde{\alpha} = 0.8$

as a function of the source redshift in Fig. 11.14. The result is surprising: in an $\Omega = 1$, $\tilde{\alpha} = 0$ universe, only 50% of all sources at redshift 2 are magnified by less than 1.1. This shows that the concept of empty cones in a clumpy universe must be considered with great care. When the optical depth τ (or τ_p) is a fair fraction of unity, empty cones become rare.

Isothermal spheres as lenses. Numerical simulations of light propagation in a clumpy universe, where the matter inhomogeneities are modeled as truncated singular isothermal spheres (see Sect. 8.1.4) were performed with the method described at the beginning of this subsection ([KA88.3], see also [JA89.1], [JA91.1]). Calculations were done in a flat, Euclidean background metric; the demagnification of a light beam unaffected by shear and Ricci focusing, relative to the Euclidean case (where light rays are straight), was taken as a distance measure. The authors argue that the statistical distribution of magnifications depends mainly on this attenuation. In a clumpy universe, their distance measure is

$$D_c = 2.5 \log \langle \mu \rangle (z_s) \quad,$$

where $\langle \mu \rangle$ is given in (11.69). In addition to D_c, the probability distribution depends on the combination $\hat{f}$ of lens mass M, radius Ξ and Euclidean distance D to the sources, where

$$\hat{f} = \frac{c^2 \Xi^2}{4GMD} \quad,$$

and D is a unique function of D_c. They find that the probability $P(\bar{\mu}_p)$ that a source is magnified by more than $\bar{\mu}_p$ relative to the smooth Euclidean space is, for $\bar{\mu}_p \gg 1$,

$$P(\bar{\mu}_p) = \frac{\kappa}{\bar{\mu}_p^2} \quad,$$

in agreement with (11.80). The coefficient κ is displayed in Fig. 11.15, as a function of $\hat{f}$, for several values of D_c. It is striking that κ has no maximum for $\hat{f} = 0$ (which corresponds to point-mass lenses), but does for moderate values of $\hat{f}$, e.g., at $\hat{f} \approx 0.25$ for small values of D_c. This means that

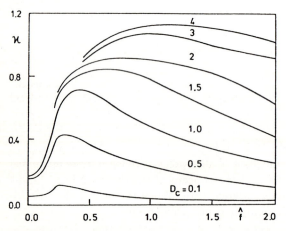

Fig. 11.15. The optical depth κ for high magnification due to isothermal spheres (as defined in the text), as a function of the dimensionless measure $\hat{f}$ for the radius of these spheres, for several values of the distance parameter D_c. The case $\hat{f} = 0$ corresponds to point-mass lenses (from [KA88.3])

isothermal spheres can be more effective in magnifying point sources than point-mass lenses of the same mass; however, for large values of $\hat{f}$, the high magnification probability decreases again. This apparently is in conflict to a claim [VI85.1] that point-mass lenses are more effective than extended ones; however, the values of $\hat{f}$ for which the optical depth κ is larger than that for point-mass lenses are rather small, so for real galaxies, this increase is unlikely to occur.

Linear analysis. To understand the qualitative features of Fig. 11.15, we now perform a linear analysis for the high-magnification probabilities in a universe filled with singular, truncated isothermal spheres. Using (2.6b) as the length scale in the deflector plane, where M is now the total mass of each isothermal sphere, the lens equation becomes

$$y = \begin{cases} x - \dfrac{\text{sign}(x)}{X} & \text{for } x < X \ , \\ x - \dfrac{1}{x} & \text{for } x > X \ , \end{cases} \tag{11.81a}$$

where

$$X = \frac{\Xi}{\xi_0} \tag{11.81b}$$

is the dimensionless radius of the sphere. Using the equations of Sect. 8.1, it is easy to see that the dimensionless cross-section for magnification larger than μ_p (for $\mu_p \gg 1$) is $\sigma(\mu_p) = \pi y^2(\mu_p)$, where

$$y^2(\mu_p) = \begin{cases} \dfrac{4}{X^2 \mu_p^2} & \text{for } X > 1 \ , \\ \dfrac{1}{\mu_p^2} & \text{for } X < 1 \ . \end{cases} \tag{11.82}$$

Thus, for $X < 1$, the cross-section is that for a Schwarzschild lens, since in this case the matter distribution lies within the critical curve at $x = 1$, and the mass distribution inside the Einstein ring is then unimportant. Furthermore, (11.82) shows that the magnification cross-section for an isothermal sphere can be up to four times larger than that of a Schwarzschild lens. We can rewrite (11.81b) as

$$X = \frac{\Xi}{\sqrt{\dfrac{4GM}{c^2}\dfrac{c}{H_0}}} \sqrt{\frac{r(z_s)}{r(z_d)\,r(z_d, z_s)}} \equiv R \sqrt{\frac{r(z_s)}{r(z_d)\,r(z_d, z_s)}} \ , \tag{11.83}$$

where in the last step we defined the redshift-independent normalized lens size R. For sufficiently large R, $X > 1$ for all lens redshifts (e.g. for $R > 1$). For smaller R, there will be two lens redshifts $z_1 < z_2$ where $X = 1$. Thus, from (11.68), (11.67) and (11.82) we have

$$P(\mu_p) = \frac{3}{2}\Omega(1-\tilde{\alpha})\,\langle\mu\rangle\,(z_s)\frac{1}{\mu_p^2}\left[\int_{z_1}^{z_2} dz\,\frac{r(z)r(z,z_s)}{r(z_s)}\frac{1}{\langle\mu\rangle(z)}\frac{1+z}{\sqrt{1+\Omega z}} \right. \\ \left. + \frac{4}{R^2}\left(\int_0^{z_1} + \int_{z_2}^{z_s}\right) dz\,\left(\frac{r(z)\,r(z,z_s)}{r(z_s)}\right)^2 \frac{1+z}{\sqrt{1+\Omega z}}\frac{1}{\langle\mu\rangle(z)} \right] . \tag{11.84}$$

For sufficiently large R, the values of z_1 and z_2 do not exist and a single integral over the second integrand in (11.84) results, from 0 to z_s. On the other hand, for $R \to 0$, $z_1 \to 0$, $z_2 \to z_s$, and (11.84) agrees with (11.70) in this limit, for $\mu_p \gg 1$.

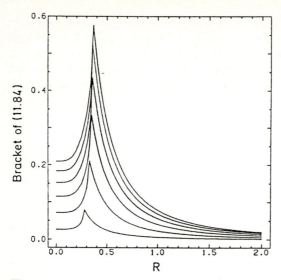

Fig. 11.16. The value in the brackets of (11.84) as a function of R, the measure for the radius of the isothermal spheres, for different values of the source redshift, $z_s = 0.5$ (lowest curve), $1.0, 1.5, 2.0, 2.5, 3.0$ (upper curve). The spike in the curves corresponds to the value of R where z_1 and z_2 [as defined before (11.84)] become equal. For simulations in a clumpy universe, these spikes would be washed out due to the statistical influence of other lenses

We show the value in the bracket of (11.84) as a function of R for several values of z_s in Fig. 11.16. The curves show the same qualitative features as those in Fig. 11.15, which were obtained from the numerical simulations, in that they have a maximum at a finite value of R, near $R \approx 0.35$. There, the magnification probability can be nearly three times as high as for point-mass lenses. Beyond this value of R, the probability decreases as R^{-2}. Hence, in order to have an increased high magnification probability, the value of R must be less than, say, 0.5. Expressing R in useful units,

$$R \approx \left(\frac{\Xi}{30\text{kpc}}\right)\left(\frac{M}{10^{12} M_\odot}\right)^{-1/2},$$

we see that the galaxies must be quite compact to achieve a larger high-magnification probability than point-mass lenses.

Stellar-mass lenses versus galaxies. We should point out that the question of whether point masses or extended lenses have a larger magnification cross-section is not very relevant in practice, due to the nature of the known astrophysical objects which can account for lensing. The only objects known and sufficiently abundant are, on the one hand, objects of stellar mass, which can be described as Schwarzschild lenses and, on the other hand, objects of galactic mass $\approx 10^{11} M_\odot$, for which an extended lens model must be used. The question of what type of lens is relevant for the amplification bias depends on the source size. If the source has an extent comparable to the lensing cross-section for an individual compact object, it will "see" the granularity of the matter distribution in galaxies, and a description of the lens as a smooth isothermal sphere may be inadequate, since it does not account for the small-scale structure of the mass distribution. In this case, it is probably more realistic to treat the individual stars in the galaxy as isolated Schwarzschild lenses. On the other hand, if the source is much larger than the characteristic scale of a Schwarzschild lens, it will average out the effects of the clumpiness in the mass distribution, and the large-scale lens model applies.

Clumpy galaxies as lenses. With a combination of the methods for point-mass lenses and extended lenses (e.g., isothermal spheres), it is possible to consider light propagation in a universe filled with galaxies, which themselves are made of stars [FR90.1]. From the random distribution of galaxy lenses, the shear and local convergence in each lens plane can be obtained by the method described at the beginning of this section. If a fraction ν of the mass in galaxies is in the form of compact objects, the value for the convergence can be translated into a value for the local density of the star field. Then, by using the probability distribution (11.54) for the random shear, one can take the action of the stars on the light bundle into account: the total shear is then obtained as the (vector) sum of the shear from the other galaxies in the field, the shear caused by the galaxies through which the ray under consideration propagates, and the random shear caused by the local star field. The only result we want to quote here from [FR90.1] is that the probability for high magnifications increases if the compact objects are in galaxies, rather than distributed randomly (as was assumed in Fig. 11.13). The ratio of high-magnification probabilities is one for very small galaxies (in that case, the galaxies are so compact that each one can be considered as a large point mass, i.e., their inner structure is unimportant) and very large galaxies (here, the galaxies are so large that a light ray typically traverses many galaxies; the clustering of stars into galaxies is unimportant in this limit), and takes a maximum for galaxies with intermediate radius. The main conclusion is that a distribution of lenses with spatial correlations will yield lensing probabilities different from those obtained for a random distribution of lenses.

Further studies of light propagation in inhomogeneous universes. As we described at the beginning of this section, the propagation of light in a clumpy universe can be considered as a multiple gravitational lens problem. Alternatively, starting from an approximate metric of an inhomogeneous universe (or an exact metric, as in the case of the Swiss–Cheese model – see Sect. 4.5.3), one can integrate the null-geodesics of this metric. Unfortunately, we are not aware of any detailed comparison between the results obtained by these two different approaches, but are convinced that in all cases of interest, the lens approach yields excellent approximations. Since the numerical integration of the geodesic equation is much more time-consuming than the lens approach, the results obtained by following the geodesics have fairly limited statistics (see, e.g., [WA90.3], [FU89.2]).

Lensing by large-scale matter distributions. So far, we have concentrated on the lensing properties of compact objects in an inhomogeneous universe; such compact objects can yield multiple images and form caustics. In our current understanding of the universe, matter is distributed inhomogeneously on very large scales, as is both observed and predicted from theory. One can thus ask which effects these larger-scale inhomogeneities have on the propagation of light bundles.

This problem was treated in [GU67.2], using an approximation to the optical scalar equations (see Sect. 3.4.2) valid for small deviations of light bundles from that in smooth Friedmann–Lemaître universes. This method was further elaborated in [BA91.1], where simulations of the large-scale matter distribution in the universe according to the cold dark matter (CDM) model are performed. For the same cosmological model of the large-scale matter distribution, a multiple lens plane approach as described here was applied in [JA90.1], and in [BA91.2] for a hot dark matter (HDM) model (for a summary of the cosmological models, see [BO88.1]). Although the large-scale matter distribution in HDM and CDM models are fairly different, their lensing behavior are found to be very similar, despite the fact that in [BA91.2], the correlation between shear and convergence has been explicitly accounted for. This is due to the fact that for smooth matter distributions, the effect of the shear on the magnification is much smaller than the convergence due to matter in the beam ([LE90.2], [LE90.3]).

The dispersion in magnification of light bundles in the models mentioned above is fairly small, because in the HDM model used in [BA91.2], no compact structures are present, whereas in the CDM model used in [JA90.1], structures were smoothed over a length scale of about 1 Mpc, due to the finite resolution of the simulations. In particular, no light bundle has been found to be overfocused, even if followed to very large redshifts. If structure on smaller scales is included, however, larger magnification effects can occur [BA91.1]. If we assume that the matter models used in the studies mentioned above yield a realistic description of the large-scale distribution of (dark) matter in the universe, we can conclude that clusters of galaxies are the largest objects which can lead to appreciable lensing effects – larger mass concentrations are not compact enough.

In [JA91.1], the light propagation in a universe with realistic large-scale matter distribution, distributed isothermal galaxies, and an approximate consideration of microlensing was treated. One particular result of this study concerns the magnification probabilities in a universe with randomly distributed galaxies compared to correlated galaxy positions: in the latter case, high magnifications are more probable. This result is in accord with that referred to earlier [FR90.1].

11.4.2 Statistics over sources

To compute statistics for source magnifications, the ray-shooting method as described in Sects. 10.6 and 11.2.5 was applied [SC88.3] with some modifications, appropriate for the multiple lens plane situation. The basic result from these simulations, as in the single lens plane case, is a two-dimensional magnification pattern, which can be found in [SC88.3].

Comparison with single lens plane calculations. Three features of the magnification distribution are most striking. First, critical structure is apparent on various scales. This is because the natural length-scale in the source plane of a Schwarzschild lens is a function of the lens redshift. Thus, large structures are due to lenses with low redshift, whereas smaller structures are caused by lenses closer to the source plane. Second, it seems that the structures are less clustered than those for a single random star field. A third feature different from those of the single lens plane simulations is the occurrence of differently shaped caustics. As we saw in Sect. 6.2.1, point-mass lenses on a single lens plane always lead to caustics where the infinite point-source magnification lies on the convex side of the caustics. This is no longer true for point lenses on several lens planes, as was shown in [ER90.1]. Therefore, the morphology of the magnification pattern in this case is quite different from those shown in Fig. 11.7.

Magnification probabilities. We performed 10 calculations with the same physical parameters but different realizations of the random distribution of lenses. In each of the simulations, 1024^2 pixels were used in the source

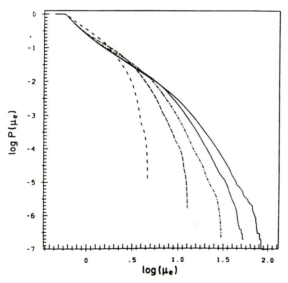

Fig. 11.17. The probability $P(\mu_e)$ for magnification larger than μ_e in a clumpy universe, for five different source sizes: $b = 0.0125$ (solid curve), $b = 0.0375$ (dotted), $b = 0.125$ (dashed-dotted), $b = 0.375$ (long-dashed), and $b = 1.25$ (short-dashed); here, the source size is measured in units of $\eta_0 = \sqrt{4GMD_s/c^2}$. For this figure, ten different simulations of a clumpy universe at redshift $z_s = 2$ were performed, and the results were added together

plane, so the total number of source positions is $\approx 10^7$. From such data, we obtained the probability that a source is magnified by more than $\bar{\mu}$ relative to the smooth universe. This probability is shown in Fig. 11.17, for different values of the source size. As expected, the probabilities cut off at lower values of $\bar{\mu}$ for larger sources. We shall see later that the qualitative features of these curves can also be obtained from a linear analysis, which will show that $P(\bar{\mu})$ behaves as $\bar{\mu}^{-2}$ for moderate values of $\bar{\mu}$ and falls off like $\bar{\mu}^{-6}$ for large $\bar{\mu}$.

11.5 Maximum probabilities

Motivation. The magnification due to gravitational light deflection can lead to the amplification bias, which will be discussed in detail in the next chapter. Briefly, intrinsically faint sources appear in flux-limited samples because the magnification due to lensing has pushed these sources above the flux threshold. Mathematically, the observed source counts are a convolution of the intrinsic luminosity function of the sources and the magnification probability. The latter quantity depends on the nature of the lenses: the magnification probability for a compact source seen through a galaxy depends on whether this galaxy contains compact objects or mainly smooth matter. Therefore, theoretical considerations of the amplification bias are burdened with the uncertainty of the nature of the lens.

Nevertheless, it is possible to find the maximum effect of the amplification bias [KO89.2], independent of the details of the lensing matter, as will be shown in the following chapter. In preparation, we now prove a theorem on maximum probabilities.

Mathematical formulation of the problem. Consider the functional

$$J(p) \equiv \int_A^\infty d\mu\, p(\mu)\, f(\mu) \quad , \tag{11.85}$$

where p is a probability distribution that vanishes for $\mu < A$, and its mean $\langle \mu \rangle > A$ is assumed to be known; we consider the set of probability distributions $p(\mu)$ with the following properties:

$$p(\mu) \geq 0 \quad ; \tag{11.86a}$$

$$\int_A^\infty d\mu\, p(\mu) = 1 \quad ; \tag{11.86b}$$

$$\int_A^\infty d\mu\, p(\mu)\, \mu = \langle \mu \rangle \quad . \tag{11.86c}$$

The function $f(\mu)$ in (11.85) is a function, assumed to be continuous, positive semidefinite, monotonically increasing, and such that $f(\mu)/\mu \to 0$ for $\mu \to \infty$. In fact, much weaker conditions on f are sufficient, but they are best formulated further below. The problem is now whether it is possible from the constraints (11.86) to obtain the maximum of the functional J for a given function f.[6]

It is easy to obtain an upper bound for J: let $\tilde{f}(\mu) = a + b\mu$ be a linear function which satisfies $\tilde{f}(\mu) \geq f(\mu)$ for all $\mu \geq A$; then

$$J(p) \leq \tilde{J} = \int_A^\infty d\mu\, p(\mu)\, \tilde{f}(\mu) = a + b\langle\mu\rangle \quad,$$

where we made use of (11.86b,c). Therefore, for each element of the two-dimensional set $B \subset \mathbb{R}^2$:

$$B = \left\{(a,b) \mid \tilde{f}(\mu) = a + b\mu \geq f(\mu) \quad \text{for} \quad \mu \geq A\right\} \quad,$$

an upper limit for J can be obtained. Not all these upper limits are equally 'good'; in particular, if two elements (a_1, b), (a_2, b) of B have $a_1 < a_2$, the bound from (a_1, b) is more stringent. The lowest-lying lines which are elements of B are those which 'touch' $f(\mu)$; we therefore define

$$B_1 = \{(a,b) \in B \mid \exists \mu_1 > A : a + b\mu_1 = f(\mu_1)\} \quad.$$

For example, the line L_2 in Fig. 11.18 is an element of B_1. If $f(\mu)$ has a continuous first derivative, each element of B_1 is uniquely defined by μ_1, the line being the tangent of $f(\mu)$ at μ_1. Furthermore, lines are singled out which touch the curve at least at two points, such as L_3 and L_4 in Fig. 11.18. These form a set B_2, defined by

$$B_2 = \{(a,b) \in B_1 \mid \exists \tilde{\mu} > \mu' \geq A : a + b\mu' = f(\mu'),\ a + b\tilde{\mu} = f(\tilde{\mu})\} \quad.$$

For the function $f(\mu)$ displayed in Fig. 11.18, L_3 and L_4 are the only elements of B_2.

The function f thus has the following property:

(P): For every $\mu_m > A$, one can either find an element of B_2 such that $\mu' \leq \mu_m \leq \tilde{\mu}$, or an element of B_1 with $\mu_1 = \mu_m$. In the first case, the element of B_2 is unique, in the second case, the element is unique if f has a continuous derivative at μ_m.

The validity of this property is geometrically obvious; one can construct an element of B singled out by (P) in the following way: consider a one-dimensional "mountain" bounded from above by $f(\mu)$. Take a rod such as

---

[6] We acknowledge helpful discussions with M. Lottermoser regarding the derivation of the following results.

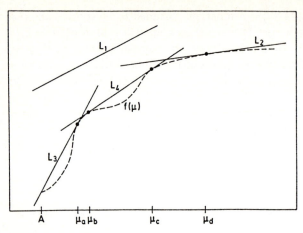

Fig. 11.18. A function $f(\mu)$ (dashed line), together with four linear functions L_i, which belong to the set B. The line L_1 corresponds to a 'weak' upper bound of f, and thus to a 'weak' upper limit on the functional J (11.85). L_2 is an element of B_1, but not of B_2, and is tangent to f at $\mu = \mu_d$. L_3 and L_4 belong to the set B_2, with $(\mu', \tilde{\mu})$ given by (A, μ_a) for L_3 and (μ_b, μ_c) for L_4. The linear function singled out by the property (P) in the text depends on μ_m: for $A \leq \mu_m \leq \mu_a$, it is L_3, for $\mu_b \leq \mu_m \leq \mu_c$, it is L_4, and for all other $\mu_m \geq A$, it is the tangent at $f(\mu_m)$

L_1 in Fig. 11.18 and tie a rope at the point $a + b\mu_m$. Then pull the rope downwards; where it comes to a stop, the rod represents the line singled out in (P). We characterize this line by (a_0, b_0); if $(a_0, b_0) \in B_2$, it is uniquely determined, whereas if $(a_0, b_0) \notin B_2$, $a_0 + b_0 \mu_m = f(\mu_m)$ is unique.

In the following, we relax the conditions on f, requiring only that f be positive semidefinite, that it has a finite number of discontinuities, and that it satisfies property (P). The function $\cos^2 \mu$ would thus be included [the line in (P) being $\tilde{f}(\mu) = 1$], whereas the function $e^{-\mu}$ would not – the tangent at a point μ_m does not belong to B, and the set B_2 is empty.

We now prove the following theorem:

Theorem on maximum probabilities. For a probability distribution characterized by (11.86) and a function $f(\mu)$ with property (P),

$$J_m = a_0 + b_0 \langle \mu \rangle \quad ,$$

with a_0, b_0 as described above for $\mu_m = \langle \mu \rangle$, is not only an upper bound for $J(p)$, but also the maximum of $J(p)$, i.e., there exists a probability distribution $p_0(\mu)$ such that $J(p_0) = J_m$.

The proof is arrived at by constructing $p_0(\mu)$ explicitly. Let the line singled out by (P) for $\mu_m = \langle \mu \rangle$ be an element of B_2 with μ', $\tilde{\mu}$; then,

$$a_0 = \frac{\tilde{\mu} f(\mu') - \mu' f(\tilde{\mu})}{\tilde{\mu} - \mu'} \quad ; \quad b_0 = \frac{f(\tilde{\mu}) - f(\mu')}{\tilde{\mu} - \mu'} \quad ;$$

and

$$J_{\mathrm{m}} = \frac{(\langle\mu\rangle - \mu')f(\tilde{\mu}) + (\tilde{\mu} - \langle\mu\rangle)f(\mu')}{\tilde{\mu} - \mu'} .$$

In this case, the distribution

$$p_0(\mu) = \frac{\tilde{\mu} - \langle\mu\rangle}{\tilde{\mu} - \mu'} \delta(\mu - \mu') + \frac{\langle\mu\rangle - \mu'}{\tilde{\mu} - \mu'} \delta(\mu - \tilde{\mu})$$

is positive semidefinite, and $J(p_0) = J_{\mathrm{m}}$.

In the second case, where $(a_0, b_0) \notin B_2$, we have $a_0 + b_0 \langle\mu\rangle = f(\langle\mu\rangle)$, i.e.,

$$J_{\mathrm{m}} = f(\langle\mu\rangle) ,$$

and the distribution

$$p_0(\mu) = \delta(\mu - \langle\mu\rangle)$$

obviously has $J(p_0) = J_{\mathrm{m}}$. This completes the proof; for an alternative proof, see [KO89.2].

Thus, if no further information about $p(\mu)$ is available, J_{m} is the best upper bound for $J(p)$, as it is the maximum. The corresponding maximum probability distribution $p_0(\mu)$ is either a single delta-function at $\langle\mu\rangle$, or the sum of two delta-functions at μ' and $\tilde{\mu}$, such that $\mu' \leq \langle\mu\rangle \leq \tilde{\mu}$.

12. Statistical gravitational lensing: Applications

The discussion in the preceding chapter focused on the technical aspects of statistical lensing. We now turn to the applications of statistical lensing, making extensive use of the results obtained in Chap. 11.

We start in Sect. 12.1 with a preliminary discussion of the amplification bias, which forms the basis for the following sections. Although the amplification bias is not restricted to QSOs alone, we concentrate on these sources, since they have been most extensively treated in the literature. We also expect the amplification bias to be most important for QSOs. Therefore, a general discussion of the source counts and luminosity functions of QSOs concludes this section.

We then turn to the statistics of multiply-imaged QSOs, i.e., what is the probability of finding multiply-imaged QSOs in a flux-limited sample? This question was partially answered as early as 1937 [ZW37.2]. We will see that more realistic treatments of this problem do not yield very different results; however, the uncertainties are still considerable, owing to our lack of knowledge about the lensing population in the universe.

In Sect. 12.3, we describe two sets of observations which suggest that the angular positions of high-redshift QSOs are associated with those of lower-redshift galaxies. Observations in the first set, which indicate an overdensity of QSOs around low-redshift galaxies ($z \lesssim 0.02$) have in the past raised some doubts as to whether the redshifts of QSOs are indeed a measure of their distance from the Earth. The second, more recent set of observations has revealed an overdensity of galaxies around high-redshift QSOs. We investigate the question of whether these associations can be explained by the magnification of the background QSO by the foreground galaxies; the results show that the first set of observations cannot be explained by lensing, whereas the overdensity of galaxies around high-redshift QSOs can probably be explained by lensing, if the amplification bias for the population of QSOs as a whole is taken into account.

Owing to the small size of the emitting region of the optical continuum radiation in QSOs, individual stars (in galaxies) can significantly magnify the flux from this component. We investigate this microlensing (ML) effect in Sect. 12.4, through a description of possible variability of lensed QSOs due to time-varying magnification. The special case of ML in the GL system 2237+0305 is studied in detail. This system also forms an ideal target to observe the influence of ML on the profile of the broad emission lines in

QSOs; this effect, which we describe next, can yield information about the structure of the broad-line region in QSOs, and is a potential diagnostic for ML. We conclude Sect. 12.4 by discussing whether ML can change the classification of AGNs; in particular, we consider the possibility that at least some BL Lac objects can be interpreted as microlensed QSOs.

We analyze in detail the amplification bias in source counts in Sect. 12.5. As can be seen from (11.70), the investigation of the amplification bias is simplest if one considers point sources. It turns out that this analysis can yield misleading results and that a non-zero source size must be explicitly accounted for. If the (intrinsic) luminosity function of sources is sufficiently steep, source counts can be heavily influenced, or even dominated, by lensing. We discuss whether the observations provide evidence that the amplification bias affects the source counts of QSOs.

Since the shape of light bundles is affected by the clumpiness of the matter distribution in the universe, intrinsically circular sources appear distorted in shape; for small distortions, the images of such sources appear elliptical. Other signatures of deformation include the bending of an intrinsically linear source, such as a radio jet. If we had enough information about the intrinsic shape of sources, we could in principle constrain the degree of inhomogeneity of matter in the universe. This possibility will be analyzed in Sect. 12.6.

All the topics mentioned above suffer from the same intrinsic difficulty because their application depends on the knowledge of the intrinsic properties of sources. For example, the importance of the amplification bias depends critically on the steepness of the intrinsic luminosity function of the sources, which is not directly observable if the amplification bias is significant, but in this case could only be inferred by interpreting the source counts. If the latter are affected by lensing, there is a delicate interplay between the lensing properties and the luminosity function which cannot easily be unfolded. A much more preferable application would be to sources for which sufficient information can be obtained so that a lensing event could be readily identified. In other words, if there were a type of cosmic source with a universal property (such as fixed intrinsic luminosity; such a source is called "standard candle"), we could infer from observations of it whether it is lensed or not. Perhaps the best standard candles known, which are bright enough to be observed at appreciable redshifts, are supernovae. We present a statistical analysis of lensing of supernovae in Sect. 12.7; a sample of lensed supernovae would yield information about the density of compact objects in the universe. Further applications of the theory of statistical lensing will be summarized in Sect. 12.8, including the hypothesis of an extragalactic origin for the γ-ray burst sources and a method for detecting compact dark objects in the halo of our Galaxy.

12.1 Amplification bias and the luminosity function of QSOs

In this section, we consider the amplification bias in source counts in a fairly simple way; a more detailed discussion is given in Sect. 12.5. In particular, we focus our attention on point sources, since they have the highest probability of being magnified by a large factor. Since the probability for lensing increases with the distance (redshift) of the sources, QSOs are the natural candidates for sources in GL systems, as they are the most distant objects known, and, at the same time, their emitting regions are thought to be very compact. In fact, both theoretical arguments and the variability of QSOs indicate that at least the optical continuum radiation comes from a region not much larger than a light-day across; the same is true for their X-ray emission.

As we will see, the amplification bias depends strongly on the luminosity function of the sources; therefore, a brief discussion of QSO source counts is given in Sect. 12.1.2.

12.1.1 Amplification bias: Preliminary discussion

Effects of lensing on source counts. Consider a population of sources in the universe; we will call them QSOs, for simplicity, but our considerations also apply to other types of objects. Since the amplification bias is expected to be mostly relevant for high-redshift and high-luminosity objects, the terminology is justified. We define the dimensionless luminosity function Φ_Q of QSOs such that the *comoving* number density of QSOs at redshift z with luminosity greater than L is

$$\left(\frac{H_0}{c}\right)^3 \Phi_Q(l, z) \quad ,$$

where the definition of l,

$$l = \frac{L}{(c/H_0)^2} \quad , \tag{12.1}$$

converts the luminosity into a quantity with the dimension of flux.[1] The number of sources per unit solid angle in the redshift interval dz about z with (scaled) luminosity larger than l is then

---

[1] The quantity L denotes the luminosity of the sources in a given frequency interval; e.g., for an optical survey, L is the luminosity in the frequency range of the detection system, shifted to the redshift of the source. Hence, we do not apply any redshift corrections, but assume that these are already included in the definition of L. Although this procedure is not very useful if one is interested in the physical properties of the sources, it is convenient for our purposes, since it is that luminosity that is inferred directly from observations, and does not require assumptions on the spectral properties of the source.

$$dn(l, z) = V(z)\, \Phi_{\mathrm{Q}}(l, z)\, dz \quad , \tag{12.2a}$$

with the 'volume factor'

$$V(z) = r_1^2(z)\, \frac{1+z}{\sqrt{1+\Omega z}} \quad , \tag{12.2b}$$

where $r_1(z)$ is the angular-diameter distance in a smooth Friedmann–Lemaître universe, as given by (4.57a). Let us denote by $P(\mu, z)$ the probability that a source at redshift z is magnified by more than μ, and let $p(\mu, z) = -dP(\mu, z)/d\mu$ be the corresponding differential probability. A source with luminosity L is observed to have flux S, where

$$S = \mu \frac{L}{4\pi D_{\mathrm{L}}^2(z)} = \mu \frac{l}{4\pi r^2(z)(1+z)^4} \equiv \mu \frac{l}{\zeta(z)} \quad , \tag{12.3}$$

where $D_{\mathrm{L}}(z)$ is the luminosity distance for redshift z, related to the angular-diameter distance $r(z)$ by (3.80), and in the last step the function $\zeta(z)$ is defined to relate the scaled luminosity l with the flux S. The observed number of sources per solid angle with flux greater than S in the redshift interval dz about z, $dn_{\mathrm{obs}}(S, z) = \frac{dn_{\mathrm{obs}}(S,z)}{dz}\, dz$, is related to the number density $dn(l, z)$ through

$$\frac{dn_{\mathrm{obs}}(S, z)}{dz} = \int_1^\infty d\mu\, p(\mu, z)\, \frac{dn}{dz}\left(\frac{\zeta(z) S}{\mu}, z\right) \quad , \tag{12.4a}$$

or

$$\frac{dn_{\mathrm{obs}}(S, z)}{dz} = V(z) \int_1^\infty d\mu\, p(\mu, z)\, \Phi_{\mathrm{Q}}\left(\frac{\zeta(z) S}{\mu}, z\right) \quad . \tag{12.4b}$$

Properties of the magnification probability. The probability density $p(\mu, z) \geq 0$ must satisfy two integral constraints. First, it must be normalized,

$$\int_1^\infty p(\mu, z)\, d\mu = 1 \quad . \tag{12.5a}$$

The second constraint is due to flux conservation in a clumpy universe, as discussed in Sect. 4.5.1, i.e.,

$$\int_1^\infty p(\mu, z)\, \mu\, d\mu = \langle \mu \rangle (z) \quad , \tag{12.5b}$$

where $\langle \mu \rangle (z)$ is given by (11.69), the magnification of a light beam in a smooth universe relative to that for an empty cone in a clumpy universe. Note that the lower limit on the μ-integrations is due to the theorem proved in Sect. 5.4.

The magnification probabilities can be determined in the linear approximation, where each lens acts as if it were isolated from all others, using the method described in Sect. 11.3; see (11.68). However, this equation is valid only if $P(Q)$ is much less than one; otherwise, the cross-sections of the individual lenses start to overlap, and the linear approach can no longer be valid. For the problem considered here, this means that we determine $P(\mu)$ only for sufficiently large values of μ from a linear analysis, for $\mu \geq \mu_0$, say. The probability for smaller values of μ can only be determined from a nonlinear analysis, such as the numerical simulations described in Sect. 11.4.2. However, for those cases where the amplification bias is important, this is due to (moderately) large magnifications; thus, the exact form of $P(\mu)$ for $\mu \leq \mu_0$ is not important. The approach we follow is the simplest which is consistent with the integral constraints (12.5); i.e., we use a delta-function for $p(\mu)$ for $\mu \leq \mu_0$. Hence,

$$p(\mu, z) = \begin{cases} p_\mathrm{L} \delta(\mu - \mu_\mathrm{L}) & \text{for } 1 \leq \mu \leq \mu_0 \\ -\dfrac{dP(\mu, z)}{d\mu} & \text{for } \mu \geq \mu_0 \end{cases} \quad , \tag{12.6a}$$

where $P(\mu, z)$ is obtained from the linear analysis. The values of $\mu_\mathrm{L}(<\mu_0)$ and p_L are determined from the integral constraints (12.5):

$$p_\mathrm{L} = 1 - P(\mu_0, z) \quad ; \quad \langle \mu \rangle (z) = p_\mathrm{L} \mu_\mathrm{L} + \int_{\mu_0}^{\infty} p(\mu, z)\, \mu\, d\mu \quad . \tag{12.6b}$$

Illustrative example. We now consider a simple model for a luminosity function and for the magnification probability. Our example will demonstrate the importance of the steepness of the luminosity function, and it reveals the value of the slope of Φ_Q for which the amplification bias becomes potentially important. We consider sources at fixed redshift z_s, so that $\zeta \equiv \zeta(z_\mathrm{s})$, and drop the dependence on z_s in the following. Let Φ_Q be a power law with a break at l_0,

$$\Phi_\mathrm{Q}(l) = \begin{cases} (l/l_0)^{-\beta} & \text{for } l \leq l_0 \\ (l/l_0)^{-\alpha} & \text{for } l \geq l_0 \end{cases} \quad , \tag{12.7a}$$

with $\beta < 2$ and $\alpha \geq \beta$; the normalization of Φ_Q is unimportant for the following consideration. The first restriction is needed to obtain finite results for dn_obs, but provides no serious constraint, since the luminosity function must be flat for faint sources; the second restriction simply means that the luminosity function is steeper for high luminosities. Note that a function similar to that in (12.7a) is frequently used to describe the QSO distribution. The following calculations assume $\alpha \neq 2$, but it is easy to consider this special case separately. We consider the simplest case for the magnification probability, i.e., for point sources, for which

$$p(\mu) = \begin{cases} p_\mathrm{L} \delta(\mu - \mu_\mathrm{L}) & \text{for } 1 \leq \mu \leq \mu_0 \\ 2\tau_\mathrm{p} \mu^{-3} & \text{for } \mu \geq \mu_0 \end{cases} \quad , \tag{12.7b}$$

which is a reasonable approximation to (11.70) for magnifications larger than about two. We therefore arbitrarily set the threshold $\mu_0 = 2$. Then, from (12.6b),

$$p_\mathrm{L} = 1 - \tau_\mathrm{p}/4 \quad ; \quad \mu_\mathrm{L} = \frac{\langle \mu \rangle - \tau_\mathrm{p}}{1 - \tau_\mathrm{p}/4} \quad . \tag{12.7c}$$

As already mentioned, we consider sources at fixed redshift, so that $\langle \mu \rangle$, τ_p, and ζ are constant. Using (12.7), we can calculate dn_obs from (12.4) by simple integration. For $\zeta S < 2l_0$,

$$\frac{dn_{\text{obs}}(S)}{dz} = V \left[p_L \mu_L^\beta + \frac{2^{\beta-1} \tau_P}{2-\beta} \right] \left(\frac{\zeta S}{l_0} \right)^{-\beta} , \qquad (12.8a)$$

where $V = V(z_s)$. For $\zeta S > 2 l_0$,

$$\frac{dn_{\text{obs}}(S)}{dz} = V \left[\left(p_L \mu_L^\alpha - \frac{2^{\alpha-1} \tau_P}{\alpha-2} \right) \left(\frac{\zeta S}{l_0} \right)^{-\alpha} \right.$$
$$\left. + 2\tau_P \left(\frac{1}{\alpha-2} + \frac{1}{2-\beta} \right) \left(\frac{\zeta S}{l_0} \right)^{-2} \right] . \qquad (12.8b)$$

In a smooth Friedmann–Lemaître universe, every source is magnified by $\langle\mu\rangle$ compared to the case of an empty cone in a clumpy universe; hence, the number counts in a smooth universe can be obtained from (12.4) by setting $p(\mu, z) = \delta(\mu - \langle\mu\rangle(z))$. Thus, in general,

$$\frac{dn_0(S, z)}{dz} = V(z) \, \Phi_Q \left(\frac{\zeta(z) S}{\langle\mu\rangle(z)}, z \right) . \qquad (12.9)$$

For the ratio of the observed number of sources in a clumpy universe to that in a smooth universe, we obtain for $\zeta S < 2 l_0$,

$$\frac{dn_{\text{obs}}}{dn_0} = \left(1 - \frac{\tau_P}{4}\right)^{1-\beta} \left(1 - \frac{\tau_P}{\langle\mu\rangle}\right)^\beta + \frac{2^{\beta-1}}{2-\beta} \frac{\tau_P}{\langle\mu\rangle^\beta} , \qquad (12.10a)$$

and for $\zeta S / l_0 > \max(2, \langle\mu\rangle)$,

$$\frac{dn_{\text{obs}}}{dn_0} = \frac{1}{\langle\mu\rangle^\alpha} \left(p_L \mu_L^\alpha - \frac{2^{\alpha-1} \tau_P}{\alpha-2} \right) + \frac{2\tau_P}{\langle\mu\rangle^\alpha} \left(\frac{1}{\alpha-2} + \frac{1}{2-\beta} \right) \left(\frac{\zeta S}{l_0} \right)^{\alpha-2} . \qquad (12.10b)$$

Interpretation of the example. Several conclusions can be drawn directly from our example. For low values of the flux, $\zeta S < l_0$, all observed sources are from the "flat" part of the luminosity function. The observed source counts form a power law with the same slope as that of the intrinsic luminosity function, and the ratio of the source counts in a clumpy universe to those in a smooth universe (12.10a) does not depend on S. For $\beta < 1$, the source counts in a clumpy universe are reduced, whereas for $\beta > 1$, they are increased compared to those in a smooth universe. This behavior is not particular to our assumption (12.7b), but it is easy to see from (12.4), (12.5b), and (12.9) that *for a power law of Φ_Q with index -1, source counts in a clumpy and in a smooth universe coincide*. Hence, the amplification bias for low-luminosity sources can lead either to an increase, or to a decrease of observed sources, depending on β. Only for $\beta \lesssim 2$ is the amplification bias large, as can be seen from (12.10a).

For $\zeta S > 2 l_0$, the source counts (12.8b) are a sum of two power laws, one with the intrinsic slope $-\alpha$, and one with slope -2. Depending on whether α is smaller or greater than 2, one of these power laws will be dominant for very large fluxes S. For $\alpha < 2$, the shape of the source counts will be dominated by the intrinsic slope, and the amplitude depends on how close α is to 2. This means that, for large S, most of the observed sources have luminosity $l > l_0$. On the other hand, if $\alpha > 2$, the source counts are dominated by the S^{-2}-law in the limit $S \to \infty$, and the amplification bias depends sensitively on both α and β. If β is close to 2, there are many weak sources which provide a large reservoir from which to magnify sources above the flux threshold. In this case, most of the sources with flux $S \gg l_0/\zeta$ have luminosity $l < l_0$. The noticeable fact to be stressed here is that, for $\alpha > 2$, the source counts for large S have a behavior $\propto S^{-2}$, independent of the intrinsic slope α. Our analysis has also shown that the source counts can be dominated by lensing only if $\alpha > 2$, except for the case $\beta \lesssim \alpha \lesssim 2$, which requires a fine-tuning of those parameters for lensing to be dominant. Thus, in the generic case, if the amplification bias dominates the source counts, these must have a slope of -2. This conclusion is independent of the special assumptions in our model, but is valid in general, *if one considers point sources* [VI85.1]. Note that the origin of this conclusion is the fact that the magnification probability $p(\mu) \propto \mu^{-3}$ for large μ, and this behavior is generic for (nearly) all lens models [see Chap. 6, also (11.22)]. The value above which the S^{-2}-behavior is dominant (for $\alpha > 2$) depends, among other factors, on the amplitude τ_p of the μ^{-3}-tail of the magnification probability distribution.

We show later on that the foregoing result can easily yield misleading conclusions: an immediate consequence would be that source counts steeper than S^{-2} cannot be dominated by lensing, but this is incorrect in general. It is important to consider extended sources, as we can easily see by comparing the length-scale for a point-mass lens in the source plane with the size of QSOs determined from their variability. The dimensionless source size is of the order of about one-tenth, depending on the redshift and on the lens mass, and this is by no means negligible. In particular, we show in Sect. 12.5 that it is possible to obtain a large range of slopes for lens-dominated source counts if we take into account the non-zero size of the sources and a spectrum of masses for the lenses. Thus, the conclusions drawn from the consideration of point sources must be critically judged.

These remarks apply if the amplification bias is (mainly) due to ML. Whether or not this is the case is unclear. It appears that the number density of galaxies in the universe is too small to cause significant amplification bias [OS86.1], but a combination of galaxies and ML may lead to appreciable amplification bias, as we demonstrate in Sect. 12.5. Also, the $p(\mu) \propto \mu^{-3}$-law is valid only asymptotically. If the amplification bias is dominated by magnifications of order $\mu \sim 10$, the foregoing conclusions do not strictly apply, since for those values of μ, the μ^{-3}-dependence is not necessarily expected.

Another conclusion that can be drawn from our example is the redshift dependence of the amplification bias. If the source counts are dominated by lensing, the magnitude of the bias is roughly proportional to τ_p (12.10). Since τ_p increases with redshift, the amplification bias is more important for high-redshift sources, as was to be expected.

12.1.2 QSO source counts and their luminosity function

General remarks. Our preceding discussion of the amplification bias shows that its magnitude depends on the steepness of the intrinsic luminosity function of the sources. Neglecting the possibility of lensing, the luminosity function may be derived directly from source counts, for a chosen cosmological model. Owing to the possible amplification bias, the determination of the luminosity function from source counts is an inverse problem. However, if the probability of significant lensing is small, the observed source counts should closely reflect the intrinsic luminosity function. Whether or not the optical depth for magnification due to gravitational light deflection is appreciable in our universe is not clear, but there is no direct evidence for a large effect. Therefore, it is usually assumed that the effects of the amplification bias are small, at least for faint or moderately luminous sources. There are strong indications, however, that for the high-brightness tail the amplification bias substantially influences the source counts, as we discuss in more detail below (Sect. 12.5).

Studying the luminosity function of QSOs is important for several reasons. First of all, they are the objects with the highest redshifts known, and thus, potentially, they probe the large-scale structure of the early universe. The luminosity function would constrain models for the evolution of QSOs, which may be closely related to the evolution of galaxies, and would yield insight into the physical mechanism responsible for the activity in these objects. Since QSOs are X-ray emitters, they contribute to the extragalactic X-ray background. At present, it is not clear whether they can account for all of the background radiation, but it seems that they contribute at least an appreciable fraction [TU86.2].

For the purposes of gravitational lensing, we are more interested in the source counts which should be matched by theoretical models. Hence, statistical lensing should start from a trial luminosity function and investigate the corresponding observed number of sources as a function of flux and redshift. The example in the last subsection demonstrated that highly lens-dominated source counts depend only weakly on the assumed intrinsic luminosity function. Thus, even from this simple consideration, one sees that lensing can (in principle) hide the true distribution of QSOs, depending on their intrinsic luminosity function.

Observational problems. The difficulty of obtaining QSO counts is that only a very small fraction of all objects in the sky, at a fixed apparent magnitude, are QSOs. QSOs cannot be distinguished from stars by morphology. Typically, searches proceed in two steps: first, candidates are selected, either by their color (QSOs with redshift less than ≈ 2.3 are blue) or by the presence of strong emission lines. In the second stage, high-resolution spectra are taken to confirm the nature of the candidates. The first step is crucial: with the color criterion, one misses relatively red objects, but for $z \leq 2.3$, those seem to be fairly rare. The second method is likely to miss QSOs with weak emission lines, i.e., objects dominated by their continuum emission.

The goal of a QSO search is the construction of a complete sample, i.e., to find all objects up to a predetermined flux threshold in a certain area in the sky. It is well known, however, that there are many observational biases which tend to produce incomplete samples. We will not discuss these difficulties here, since there are a number of excellent reviews (e.g., [VE83.1], [SE84.1], [GR86.2], [SC87.9]; see also [WA85.1]).

Two points should be clear: first, we are more likely to miss faint objects than bright objects in the search area. To compensate, the flux limit of the sample is higher than the limit of the photographic plate. Second, since all QSOs are variable to some degree, their inclusion in the sample depends on their brightness at the time of the first selection. Since their brightness is estimated in the second stage of the observations, sources which have become fainter in the meantime are left out of the sample, but those which have brightened over the threshold are not included in the final sample, since they were dismissed earlier. There is a single cure in this case, which is to increase the threshold of the final sample over the preliminary flux limit. It is clear that both of the effects mentioned, if not accounted for properly, lead to a decrease of faint objects in the sample, and thus they tend to flatten the observed source counts.

Selected observational results. For our purposes, it suffices to note the following: source counts of QSOs, irrespective of their redshift, show a steep increase towards fainter magnitudes. Up to an apparent magnitude of $B \approx 19.8$, the source counts can be described by a power law in S of the form $n(S) \propto S^{-2.6}$ [MA84.1], with ≈ 20 QSOs per square degree brighter than $B = 19.8$. For fainter B magnitudes, the number counts flatten considerably; in fact, from the X-ray background, one can constrain the number of faint QSOs. Number counts at fixed redshift are more difficult to obtain, since the total number in each redshift bin is small. It was shown recently [MA85.1] that the source counts at fixed redshift also show a behavior $dn_{\mathrm{obs}}(S,z)/dz \propto S^{-2.6}$ for the brighter sources and a turnover (flattening) for fainter ones. If the observed number counts are interpreted by neglecting lensing, these results imply a functional form for Φ_Q of $\Phi_Q(l,z) \propto Z(z) l^{-2.6}$ for the brighter sources, and $Z(z)$ describes the evolution of the QSOs with redshift. It is a strongly increasing function, at least up to redshifts $z \approx 2$, i.e., the QSO population appears to evolve strongly with redshift [SC83.1]: at high redshift, QSOs are much more abundant than locally. More recent QSO counts (e.g., [BO88.3]) confirm these results.

Counting sources in the radio and X-ray regions of the spectrum is less burdensome, because a large fraction of detected radio and X-ray sources away from the galactic plane are AGNs. For a determination of the nature of these sources, however, optical identifications and spectra are needed. Even the best known radio catalogue (the Third Cambridge Radio Sample, or 3CR) is not yet fully identified optically. We will not discuss radio surveys in this book, since it seems that the radio-emitting region of AGNs is larger than those in the optical or in X-rays, although recently it was demonstrated

[QU89.1] that flat-spectrum radio-loud QSOs can vary on time-scales of a day. For X-ray selected AGNs, analysis of the Medium Sensitivity Survey ([MA82.3], [ST83.1]) yields the following results: the X-ray luminosity function for high-luminosity objects is a power law with the same slope as in the optical, i.e., it varies as $S^{-2.6}$, but flattens considerably for fainter sources. There is a clear cosmological evolution of the comoving density of X-ray selected AGNs, but it is not as strong as for optically selected ones [MA84.1].

12.2 Statistics of multiply imaged sources

In this section we discuss the question of how many distant sources should be expected to have multiple images. We would like to know the expected number of sources in a given sample with multiple images that can be resolved. As we will see, this question cannot be answered easily (e.g. [PR73.1], [TY83.1], [TU84.1], [DY84.1], [BL87.2], [KO87.2]). The two main difficulties in these studies are our poor knowledge of the mass distribution in the universe and the potential influence of the amplification bias, which in turn depends on the intrinsic luminosity function of sources. Therefore, we will concentrate on a few idealized situations which can be studied in detail and which show the qualitative influence of the effects mentioned above.

In addition to the difficulties we have mentioned, at least two observational biases are present. First, for multiple images to be detected, their angular separation must be larger than the resolution limit of the telescope. For radio-interferometric observations, the resolution is well known in general; however, radio sources can be extended, so multiple images are not easily identified. For Earth-bound optical observations, the resolution depends strongly on the atmospheric conditions (seeing) at the telescope site. Since this changes with time, the optical resolution bias is not well determined. Second, the maximum flux ratio for which multiple images can be detected depends on the dynamic range of the observations, which in turn depends on the apparent magnitude of the images (the dynamic range is larger for brighter sources) and also on the angular separation. Again, this bias can be controlled better for radio than for optical observations. A further bias has been pointed out recently [KO91.1] in that existing QSO samples are likely to be biased *against* multiple images, since one of the defining criteria of a QSO is its stellar appearance on optical plates. It is clear that a model incorporating all these biases poses a complex problem.

The first estimate of the probability for multiple imaging was given by F. Zwicky in 1937 [ZW37.2]: from his estimate of the fraction of the sky covered by "extragalactic nebulae" (galaxies) he concluded that about one out of every four hundred distant sources should be multiply imaged, a value quite consistent with that derived through much more elaborate work (e.g., [TU84.1]).

In Sect. 12.2.1 we consider the probability that a source is multiply imaged in a universe filled with point masses [PR73.1]; extended lenses, represented as isothermal spheres, are considered in Sect. 12.2.2. The final subsections discuss the influence of more complicated lens models on multiple-image statistics, and present a summary of current lens surveys.

12.2.1 Statistics for point-mass lenses

We now calculate the probability $P(\vartheta, q, z_s)$ that a (point) source at redshift z_s has two images with angular separation larger than ϑ, and brightness ratio less than q, in a universe filled with point-mass lenses [PR73.1]. This probability is not directly related to the question of how many multiply-imaged sources are in a certain source sample, since the latter problem involves the amplification and observational biases previously discussed, which will be neglected here. Since the lens population assumed here is not realistic as a model for the matter contents of the universe, the considerations given below should serve only as an illustrative example, whose qualitative features can help as a guideline for further investigation.

Derivation of the lensing probability. Let $n_L(M)dM$ be the comoving number density of point-mass lenses with mass within dM of M; we assume that the lens population is nonevolving, i.e., the comoving density is independent of redshift. In this case, (11.68) applies directly. The dimensionless cross-section $\sigma(\Delta x, q)$ for an isolated Schwarzschild lens to produce two images with separation larger than Δx, and flux ratio smaller than q, is given by (11.6). The angular separation ϑ is related to Δx by $\vartheta = \xi_0 \Delta x / D_d$, where ξ_0 (2.6b) depends on the lens mass and on the distances from the Earth to the lens and to the source. From (11.5), we have for the dimensional cross-section

$$\hat{\sigma}(\vartheta, q, z, z_s, M) = \left[\frac{r(z_s)}{r(z)}\right]^2 \xi_0^2(z, z_s, M) \, \sigma\left(\frac{D(z)\vartheta}{\xi_0}, q\right) \ .$$

Insertion of this expression into (11.68) for the lensing probabilities in an inhomogeneous universe yields

$$\begin{aligned}P(\vartheta, q, z_s) = \left(\frac{c}{H_0}\right)^2 \langle\mu\rangle(z_s) \int_0^{z_s} \frac{dz}{\langle\mu\rangle(z)} \frac{1+z}{\sqrt{1+\Omega z}} \frac{r(z)r(z,z_s)}{r(z_s)} \\ \times \int dM \, n_L(M) \, \frac{4GM}{c^2} \, \sigma\left(\frac{D(z)\vartheta}{\xi_0}, q\right) \ .\end{aligned} \quad (12.11)$$

We write this equation in a more transparent form, using a reference mass M_0:

$$m = M/M_0 \ . \quad (12.12a)$$

Furthermore, we define a density n_0 and a normalized mass distribution $N(m)$ by

$$\int dM\, n_{\rm L}(M)M = n_0 M_0 \quad ; \quad N(m) = M_0 \frac{n_{\rm L}(M_0 m)}{n_0} \quad . \tag{12.12b}$$

Combining these equations, we find that $\int N(m)\, m\, dm = 1$. The reference mass M_0 introduces a "natural" angular scale Θ_0, defined as

$$\Theta_0 = \sqrt{\frac{4GM_0}{c^2}\frac{H_0}{c}} \quad , \tag{12.13a}$$

or, in useful units,

$$\Theta_0 \approx 5.8 \times 10^{-12} \sqrt{\frac{M_0}{M_\odot}} h_{50}\ {\rm rad} \approx 1.2 \times 10^{-6} \sqrt{\frac{M_0}{M_\odot}} h_{50}\ {\rm arcsec}. \tag{12.13b}$$

In terms of Θ_0, the first argument of σ in (12.11) becomes

$$\frac{D(z)\vartheta}{\xi_0} = \sqrt{\frac{r(z)\, r(z_{\rm s})}{m\, r(z, z_{\rm s})}}\frac{\vartheta}{\Theta_0} \quad .$$

Inserting the preceding equations into (12.11), and assuming that the lenses under consideration are the clumps of the inhomogeneous universe, i.e., (11.67) holds: $M_0 n_0 = (1 - \tilde\alpha)\Omega\rho_{\rm cr}$, we find

$$P(\vartheta, q, z_{\rm s}) = \frac{3}{2\pi}(1-\tilde\alpha)\,\Omega\,\langle\mu\rangle\,(z_{\rm s})\int_0^{z_{\rm s}}\frac{dz}{\langle\mu\rangle(z)}\frac{1+z}{\sqrt{1+\Omega z}}\frac{r(z)\,r(z,z_{\rm s})}{r(z_{\rm s})}$$

$$\times \int dm\, N(m)\, m\, \sigma\left(\sqrt{\frac{r(z)\,r(z_{\rm s})}{m\,r(z,z_{\rm s})}}\frac{\vartheta}{\Theta_0}, q\right) \quad . \tag{12.14}$$

Simple approximation. Before we discuss results from the integration of (12.14), some simple analytic considerations will permit a better understanding of this equation: the dimensionless image separation Δx for a Schwarzschild lens is always near 2 for small q; e.g., for $q < 10$, $\Delta x \lesssim 2.34$. Considering only small q, the cross-section σ can be approximated by

$$\sigma(\Delta x, q) \approx \pi y^2(q)\, {\rm H}(2 - \Delta x) \quad ,$$

where $y(q)$ is given by (11.3), and H(x) is the Heaviside step function. Using this approximation, (12.14) can be rewritten as

$$P(\vartheta, q, z_{\rm s}) = \frac{3}{2}(1-\tilde\alpha)\,\Omega\, y^2(q)\,\langle\mu\rangle\,(z_{\rm s})\int_0^{z_{\rm s}}\frac{dz}{\langle\mu\rangle(z)}\frac{1+z}{\sqrt{1+\Omega z}}\frac{r(z)\,r(z,z_{\rm s})}{r(z_{\rm s})}\nu[m_{\rm min}(z)]\, ,$$

where $\nu(m_{\rm min})$ is the fraction of mass in lenses with $m > m_{\rm min}$, and

$$m_{\rm min} = \frac{r(z)\, r(z_{\rm s})}{r(z, z_{\rm s})}\left(\frac{\vartheta}{2\Theta_0}\right)^2 \quad ,$$

or, in useful units,

$$M_{\rm min} = M_0 m_{\rm min} = 1.8 \times 10^{11}\frac{r(z)\, r(z_{\rm s})}{r(z, z_{\rm s})}\frac{1}{h_{50}}\left(\frac{\vartheta}{1\ {\rm arcsec}}\right)^2 M_\odot \quad .$$

382

Hence, only lenses with mass larger than M_{min} at redshift z are able to split the images of a source at z_s with angular separation $> \vartheta$. M_{min} is a monotonically increasing function of z, i.e., for fixed mass, nearby lenses cause larger image separations than distant ones.

Numerical results. Some results of the analysis of (12.14) are displayed in Fig. 12.1. In Fig. 12.1a, the probability $P(\vartheta, q, z_s)$ is shown for three values of the source redshift and two values of the limiting flux ratio q, as a function

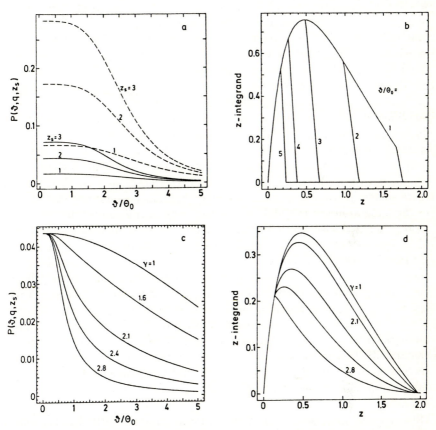

Fig. 12.1. (a) The probability $P(\vartheta, q, z_s)$ that a source at redshift z_s has multiple images with angular separation larger than ϑ and flux ratio less than q, as a function of ϑ/Θ_0. The solid curves are for $q = 5$, the dashed ones for $q = 20$. In both cases, source redshifts $z_s = 1, 2, 3$ are considered. The cosmological model parameters are $\Omega = 1$, $\tilde{\alpha} = 0.8$, i.e., 20% of the critical mass in the universe is assumed to be concentrated in compact objects. We assumed that all lenses have the same mass M_0, from which the angular scale Θ_0 was obtained with (12.13a). (b) The z-integrand in (12.14), for $z_s = 2$, $q = 10$, and $\vartheta/\Theta_0 = 1, 2, 3, 4, 5$. The cosmological model $\Omega = 1$, $\tilde{\alpha} = 0$ was chosen, and all lenses have mass M_0. Note that the integrand – and therefore the redshift distribution of the lenses – is strongly peaked for larger values of ϑ/Θ_0. (c) The probability $P(\vartheta, q, z_s)$ as a function of ϑ/Θ_0 for a power-law distribution of the lens masses, $N(m) \propto m^{-\gamma}$. The different curves are (from top to bottom) for $\gamma = 1, 1.6, 2.1, 2.4, 2.8$. The values $q = 5$, $\Omega = 1$, and $\tilde{\alpha} = 0.8$ were used, as well as $m_{min} = 3 \times 10^{-2}$, $m_{max} = 10$ for the mass range. (d) The z-integrand in (12.14) for the model in (c), with $\vartheta/\Theta_0 = 1$, for the five values of γ as in (c). Owing to the large range in lens masses, the distribution of lens redshifts is much broader than in (b)

of ϑ/Θ_0. We assumed in this case that the lenses all have the same mass M_0, i.e., $N(m) = \delta(m-1)$. For small ϑ/Θ_0, the curves are very flat, indicating that all lenses which produce images with flux ratio below the threshold q are also able to cause sufficient image splitting. For increasing ϑ, lenses close to the source do not contribute to the probability, since their images have too small a separation. That is, increasing ϑ implies the exclusion of lenses at increasingly lower redshifts. This is clearly seen in Fig. 12.1b, where the z-integrand of (12.14) is plotted for several values of ϑ/Θ_0. The breaks in these curves are due to the lack of smoothness of $\sigma(\Delta x, q)$ as a function of Δx. As was to be expected, P increases with increasing z_s and increasing q. The typical width of these curves is roughly $2\Theta_0$. As we can see from Fig. 12.1b, the expected lens redshift decreases with increasing ϑ.

To see the influence of a mass spectrum of lenses, we display $P(\vartheta, q, z_s)$ in Fig. 12.1c, for a distribution $N(m) \propto m^{-\gamma}$ for $m_{\min} \leq m \leq m_{\max}$, and for different values of γ, as a function of ϑ/Θ_0. Here, $m_{\min} = 3 \times 10^{-2}$ and $m_{\max} = 10$, and $q = 5$, $z_s = 2$. At low values of ϑ/Θ_0, the probability is nearly independent of γ, since all lenses with flux ratio smaller than q lead to sufficient image splitting. For increasing ϑ, lower mass clumps at $z \lesssim z_s$ fail to produce the splitting, i.e., if the lenses were displayed in a (z, m)-diagram, the corner at $z \lesssim z_s$ and $m \gtrsim m_{\min}$ would not contribute to the integral in (12.14), and its area increases with ϑ. Since this corner is more populated with lenses for larger values of γ, P is a decreasing function of γ. The z-integrand of (12.14) displayed in Fig. 12.1d illustrates this fact: for larger values of γ, the contribution from higher-redshift lenses is more suppressed. The tail in the z-distribution is due to high-mass lenses, which are missing in the case shown in Fig. 12.1b. Note that all curves in Fig. 12.1d agree for $z < z_*$, meaning that lenses at z_* with $m = m_{\min}$ start to fail to produce sufficient angular separation. The figure also shows that, as long as all lenses produce image splitting above the threshold, the contribution to the probability is independent of the mass spectrum, since the lensing cross-section is proportional to the mass, and thus independence is guaranteed by the normalization condition (12.12b).

Observational prospects. In their paper, Press and Gunn [PR73.1] discussed the possibility of detecting compact objects in the universe by means of observable image splittings produced by such objects. The highest available resolution for observations with VLBI techniques could detect double images separated by a few times 10^{-4} arcsec, i.e., they can potentially detect compact lenses of mass $\gtrsim 10^4 M_\odot$; however, it will be difficult in general to distinguish between lens-induced and intrinsic source structure. VLA and optical observations can probe for compact lenses in excess of $\sim 10^{11} M_\odot$; in fact, surveys are underway, and preliminary results are described further below. Detection of low mass ($\sim 1 M_\odot$) compact objects can be indirectly provided by the corresponding variability of compact sources due to the change of relative position of source and lens in time; we discuss such ML possibilities in Sect. 12.4.

Recently, two samples of compact extragalactic radio sources observed with VLBI have been used to constrain the density of compact objects in the universe [KA91.1], using the methods described above. The available VLBI observations used in this study have a typical angular resolution of a few milliarcseconds and are thus sensitive to detect image splitting by compact objects with mass greater than $\sim 10^7 M_\odot$. Since VLBI observations have a restricted angular field of view (of about one arcsecond), these observations are not sensitive to image splitting by lenses with mass $\gtrsim 10^{10} M_\odot$.

In the sample of 48 objects used in [KA91.1], six are classified as compact double sources. They are rejected by the authors as possible double images, since these particular sources are morphologically different from all other types of compact radio sources in the sample (low polarization, little variability, no prominent large-scale radio structure). From the lack of any double structure in the remaining sample, a conservative upper limit on the density of compact objects in the mass range $10^7 M_\odot \lesssim M \lesssim 10^9 M_\odot$ of $\Omega \lesssim 0.4$ is derived. This limit can be greatly improved, however, if the Very Long Baseline Array is completed (the VLBA is an array of 10 radio antennas to be used exclusively – and thus routinely – for VLBI observations). In fact, once about 1000 high-redshift sources are observed, the upper limit could in priciple be lowered to 10^{-3} – or compact objects in the above quoted mass range can actually be detected if they exist!

12.2.2 Statistics for isothermal spheres

Preliminaries. Let us now consider the statistics for multiply-imaged sources in a universe where matter concentrations (galaxies) can be represented as singular isothermal spheres (see Sect. 8.1.4). This problem was studied in considerable detail in [TU84.1], including the effects of amplification bias, selection effects, and the influence of intracluster gas on the lensing properties of galaxies. Some of the results from [TU84.1] are discussed at the end of this subsection. Here, as in most published work considering lensing by galaxies, the clustering of galaxies is neglected. (Gravitational lens statistics with correlated deflectors have been treated approximately in [AN86.1], using Monte-Carlo methods and a highly simplified procedure for determining multiple images, concluding that the correlation of lensing galaxies does hardly affect the multiple image statistics at all. This has to be compared with the treatment in [KO88.1] where it was concluded that a few percent of all multiple QSOs should be affected by more than one deflector. The effect of embedding lens galaxies in dark matter concentrations will be discussed explicitly below.)

Our approach is slightly different from that in [TU84.1], and is not as strongly adapted to the observational situation, but it includes all important features affecting the lensing statistics and investigates the influence of the amplification bias on multiply imaged QSOs in more detail than in [TU84.1]. As we demonstrate below, the amplification bias may be an extremely important effect that dominates the statistics. It can be taken into proper account if it is caused by galaxy lenses only. However, due to the small size of the emitting region for the optical QSO continuum, single stars may dominate the amplification bias; i.e., ML can be more important than the magnification caused by the galaxy itself, at least if the amplification bias is large. Therefore, the combination of lensing by the galaxy (for image splitting) and by the compact objects in the galaxy (for magnification effects) should be considered. Several results of such a study are described at the end of this subsection. We focus our interest mainly on the qualitative aspects of the statistics, rather than on detailed comparisons with observations, which is not justified yet due to the lack of a well-defined sample of GL systems.

As discussed in [TU84.1], representing galaxies as singular isothermal spheres may be a sufficient approximation for our purposes, for several reasons. First, the matter distribution outside the radius where multiple images can occur is unimportant; thus, the mass distribution in the extended halos of galaxies does not affect the lensing properties under consideration. Second, measurements of the rotation curves show that the density law $\rho \propto r^{-2}$ extends over a wide range in spiral galaxies and is also suggested for ellipticals (for references, see [TU84.1]). Third, if the core radius is much smaller than the typical impact parameter ξ_0 (8.34a), it does not influence the lensing properties markedly. In Sect. 12.2.3, we discuss the influence of a non-zero core, as well as other modifications of the lens model.

Cross-sections and lensing probabilities. The lens model we consider here has two parameters, the line-of-sight velocity dispersion σ_v, and the surface mass density Σ of the intracluster gas at the position of the galaxy [compare equations (11.7–11)]; the latter quantity corresponds to a dimensionless surface mass density $\kappa_0 = \Sigma/\Sigma_{cr}$ (5.5). Scaling the length in the lens plane by ξ_0 (8.34a), the magnifications of the two images, which occur for an impact parameter $y < 1$, are given by (11.10a,b), and the total magnification is given by (10.10c). The image separation Δx and the relation between the flux ratio q of the images and the source position y are

$$\Delta x = \frac{2}{1-\kappa_0} \quad ; \quad y = \frac{q-1}{q+1} \quad .$$

From these relations, we obtain the cross-section for a lens to form images with separation larger than Δx, flux ratio smaller than q and magnification larger than μ:

$$\sigma(\Delta x, q, \mu) = \mathrm{H}\left(\frac{2}{1-\kappa_0} - \Delta x\right)\pi y^2(q,\mu) \quad , \tag{12.15a}$$

where

$$y = \min\left(\frac{q-1}{q+1}, \frac{1}{(1-\kappa_0)^2}\frac{2}{\mu}\right) \quad . \tag{12.15b}$$

On the other hand, the cross-section for a lens to produce a total magnification larger than μ, irrespective of Δx and q, is

$$\sigma(\mu) = \pi y^2(\mu) = \pi \begin{cases} \left(\frac{2}{(1-\kappa_0)^2\mu}\right)^2 & \text{for } \mu \geq \frac{2}{(1-\kappa_0)^2} \\ \left(\frac{1}{(1-\kappa_0)^2\mu-1}\right)^2 & \text{for } \mu \leq \frac{2}{(1-\kappa_0)^2} \end{cases} \quad , \tag{12.15c}$$

as follows from (11.10). The dimensional cross-sections $\hat{\sigma}$ are related to the dimensionless cross-sections above via (11.5); using (8.34a), we have

$$\hat{\sigma} = \left(\frac{c}{H_0}\right)^2 r^2(z, z_s) \Theta^2 \sigma \quad , \tag{12.16a}$$

where z, z_s are the redshifts of the lens and the source, respectively. Also, the deflection angle for an isothermal sphere Θ is

$$\Theta = 4\pi \left(\frac{\sigma_v}{c}\right)^2 \quad , \tag{12.16b}$$

and the angular separation ϑ of the two images is

$$\vartheta = \frac{\Delta x \, \xi_0}{D_d} = \frac{r(z, z_s)}{r(z_s)} \Theta \, \Delta x \quad . \tag{12.16c}$$

Inserting (12.16) into (11.68), we find for the probability $P(\vartheta, q, \mu, z_s)$ that a source at redshift z_s is multiply imaged, with image separation greater than ϑ, flux ratio less than q, and magnification greater than μ:

$$\begin{aligned} P(\vartheta, q, \mu, z_s) = &\left(\frac{c}{H_0}\right)^3 \frac{1}{r_1^2(z_s)} \int_0^{z_s} dz \, h(z, z_s) \\ &\times \int d\chi \, n_L(\chi) \, \Theta^2(\chi) \, \sigma\left(\frac{r(z_s)}{r(z, z_s)} \frac{\vartheta}{\Theta}, q, \mu\right) \quad , \end{aligned} \tag{12.17a}$$

where we defined

$$h(z, z_s) = \frac{1+z}{\sqrt{1+\Omega z}} \, r_1^2(z) \, r^2(z, z_s) \quad . \tag{12.17b}$$

Galaxy distribution. To describe the galaxy population, we adopt the Schechter luminosity function [SC80.1]

$$\Phi_G(L) dL = \Phi_* \left(\frac{L}{L_*}\right)^\nu e^{-L/L_*} dL \quad , \tag{12.18a}$$

where L_* corresponds to an absolute B magnitude of -20.8 mag, and $\nu = -1.25$ [KI83.1]; the normalization Φ_* is fixed such that the mean luminosity-density of all galaxies is $\mathcal{L} = 2 \times 10^8 \, L_\odot/\text{Mpc}^3$ in the same band. Integration yields

$$\mathcal{L} = \Phi_* \, L_*^2 \, \Gamma(\nu + 2) \quad ,$$

where $\Gamma(x)$ is the gamma function. Using an absolute blue magnitude of 5.4 mag for the Sun, we obtain[2]

$$N \equiv \left(\frac{c}{H_0}\right)^3 \Phi_* L_* \approx 1.2 \times 10^9 \quad . \tag{12.19}$$

The relation between luminosity and velocity dispersion is given by the Faber–Jackson relation [FA76.1]

---

[2] Note that N is the equivalent number of galaxies per Hubble volume $(c/H_0)^3$, if all galaxies have luminosity L_*.

$$\frac{L}{L_*} = \left(\frac{\sigma_v}{\sigma_{v*}}\right)^4 = \left(\frac{\Theta}{\Theta_*}\right)^2 \quad , \tag{12.18b}$$

where (12.16b) was used in the second step, and $\Theta_* \equiv \Theta(\sigma_{v*})$.

Next, we account for the intracluster gas or dark matter in clusters of galaxies, where the lensing galaxies may reside. Since the distribution of surface mass densities is known very poorly, we will consider a very simple distribution that reveals the qualitative influence of this lens component. This model distribution is assumed to be independent of galaxy luminosity L. We choose as the probability for a galaxy to have an underlying smooth surface mass density within $d\Sigma$ of Σ,

$$p(\Sigma) = \frac{3}{\Sigma_m}\left(1 - \frac{\Sigma}{\Sigma_m}\right)^2 H(\Sigma_m - \Sigma) \quad , \tag{12.20a}$$

where Σ_m is the maximum surface mass density. The dimensionless surface mass density $\kappa_0 = \Sigma/\Sigma_{cr}$ is then found from (5.5) to be

$$\kappa_0 = \frac{r(z)\, r(z, z_s)}{r(z_s)} x\, \kappa_m \quad \text{with} \quad x = \frac{\Sigma}{\Sigma_m}; \quad \kappa_m = \frac{4\pi G}{c^2} \frac{c}{H_0} \Sigma_m \ . \tag{12.20b}$$

We consider only values of κ_m such that $\kappa_0 < 1$ for all redshifts. Combining (12.18–20) we obtain for the lens population

$$n_L(\chi)d\chi = L_*\Phi_*\, \ell^\nu e^{-\ell} d\ell\, 3(1-x)^2 dx \quad , \tag{12.21}$$

with $\ell = L/L_* = (\Theta/\Theta_*)^2$, i.e., $\Theta^2 = \ell\Theta_*^2$. Inserting this expression into (12.17a) yields

$$P(\vartheta, q, \mu, z_s) = \Theta_*^2 N \frac{1}{r_1^2(z_s)} \int_0^{z_s} dz\, h(z, z_s)\, 3\int_0^1 (1-x)^2 dx$$
$$\times \int_0^\infty d\ell\, \ell^{\nu+1} e^{-\ell}\, \sigma\left(\frac{r(z_s)}{r(z,z_s)}\frac{\vartheta}{\Theta}, q, \mu\right) \ . \tag{12.22}$$

Fraction of multiply imaged sources, without amplification bias. If the amplification bias is neglected, the probability (12.22), irrespective of magnification, is the fraction $f(\vartheta, q, z_s)$ of sources at redshift z_s with multiple images separated by more than ϑ and with flux ratio smaller than q. The Heaviside step function in (12.15a) is accounted for by taking the lower limit of the ℓ integration to be

$$\ell_{\min} = \left[\frac{r(z_s)}{r(z,z_s)}\frac{(1-\kappa_0)}{2}\right]^2 \frac{\vartheta^2}{\Theta_*^2} \quad ;$$

thus, by performing the ℓ-integration, we obtain

$$f(\vartheta, q, z_s) = \Theta_*^2 N\, \pi \left(\frac{q-1}{q+1}\right)^2 \frac{1}{r_1^2(z_s)} \int_0^{z_s} dz\, h(z, z_s)$$
$$\times 3\int_0^1 dx(1-x)^2 \Gamma(\nu+2, \ell_{\min}) \quad , \tag{12.23}$$

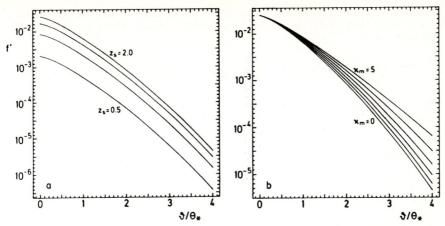

Fig. 12.2a, b. The factor f' as a function of ϑ/Θ_*, where $f(\vartheta, q, z_s) = f'[\Theta_*^2 N\pi(q-1)^2/(q+1)^2]$ is the fraction of sources at redshift z_s which have multiple images with angular separation larger than ϑ and flux ratio less than q. The cosmological model is $\Omega = 1$, $\tilde{\alpha} = 0.9$. (a) Vanishing surface mass density in smooth matter, $\kappa_m = 0$; f' is plotted for different source redshift, $z_s = 0.5, 1.0, 1.5, 2.0$. (b) For $z_s = 2$, f' is plotted for various values of the smooth matter density parameter, $\kappa_m = 0, 1, 2, 3, 4, 5$

where $\Gamma(\alpha, x)$ is the incomplete gamma function [AB65.1]. Thus, the quantity

$$f' = \frac{f}{\left[\Theta_*^2 N\pi \left(\frac{q-1}{q+1}\right)^2\right]}$$

is a function of ϑ/Θ_* only. Note that $\Theta_*^2 \propto \sigma_{v*}^4$ is a sensitive function of σ_{v*}; for $\sigma_{v*} = 220\,\mathrm{km/s}$ [BI87.1], we have $\Theta_*^2 N\pi \approx 0.32$, and $\Theta_* \approx 1.9\,\mathrm{arcsec}$.

In Fig. 12.2a, we display f' as a function of ϑ/Θ_* for various values of the source redshift. The dependence on source redshift is easily understood, since the probability for lensing increases with the number of potential lenses between the source and the observer. In Fig. 12.2b, we show the same function for various values of the maximum smooth surface mass density κ_m. Although f' is independent of κ_m for small angular separations, $\vartheta \ll \Theta_*$, the distribution for large angular separations depends very sensitively on the surface mass distribution. This should be kept in mind in the discussion at the end of this subsection; our poor knowledge of the distribution of mass in which galaxies reside results in one of the largest uncertainties in the theory of the expected frequency of lensing [TU84.1].

Influence of the amplification bias. Now, we consider the influence of the amplification bias on multiple-image statistics. As mentioned earlier, since a realistic estimate of the amplification bias must take ML into account, except for lens searches at radio frequencies (since extragalactic radio sources are expected to have emission regions larger than the typical lensing scale for stellar-mass objects), the following considerations are mainly qualitative but show the importance of properly taking the amplification bias into account.

From (12.4b), the number of sources per unit redshift interval with multiple images with combined flux greater than S and with angular separation of the images $\geq \vartheta$ and flux ratio $\leq q$, is:

$$\frac{dn_{\text{obs}}(S, z_s, \vartheta, q)}{dz_s} = V(z_s) \int_1^\infty d\mu \; p(\vartheta, q, \mu, z_s) \, \Phi_Q\left(\frac{\zeta(z_s)S}{\mu}, z_s\right) \; , \qquad (12.24a)$$

where $V(z)$ is defined in (12.2b), and $p(\vartheta, q, \mu, z_s) d\mu$ is the probability that a source at z_s is 'split' into two images with angular separation larger than ϑ, flux ratio smaller than q, and total magnification within $d\mu$ of μ. This probability is the negative of the partial derivative of (12.22) with respect to μ; $\Phi_Q(l, z_s)$ is the luminosity function of sources, and $\zeta(z)$ was defined in (12.3). Inserting (12.22) into (12.24a) yields

$$\begin{aligned}\frac{dn_{\text{obs}}(S, z_s, \vartheta, q)}{dz_s} &= \frac{1+z_s}{\sqrt{1+\Omega z_s}} 2\pi \Theta_*^2 N \int_0^{z_s} dz \; h(z, z_s) \; 3 \int_0^1 (1-x)^2 dx \\ &\times \int_{\ell_{\min}}^\infty d\ell \; \ell^{\nu+1} \, e^{-\ell} \int_0^{y(q)} y \, dy \; \Phi_Q\left(\frac{\zeta(z_s)S}{\mu(y)}, z_s\right) \; ,\end{aligned} \qquad (12.24b)$$

where we have transformed the μ-integration into an integral over the dimensionless impact parameter y, which can always be done if $\sigma(\mu)$ can be written as $\sigma(\mu) = \pi \, y^2(\mu)$, and $y(\mu)$ is a monotonic function; then, $(d\sigma/d\mu)d\mu = 2\pi \, y(\mu) \, dy$. The Heaviside factor in σ (12.15a) was accounted for by the lower limit of the ℓ-integral, and the limit on q led to the upper limit of the y-integration, with $y(q)$ given just before (12.15a). Correspondingly, $\mu(y) = 2(1-\kappa_0)^{-2}/y$.

For illustrative purposes, we use the luminosity function (12.7a); then, the result of the y-integration for large S consists of two terms, one proportional to $S^{-\alpha}$ and one to S^{-2} – compare the analogous result (12.8b). If we only consider the case for S much larger than the value where the unlensed source counts have a break $[S \gg l_0/\zeta(z_s)]$, and choose $\alpha > 2$, the latter term dominates; thus, we have

$$\int_0^{y(q)} y \, dy \; \Phi_Q\left(\frac{\zeta(z_s)S}{\mu(y)}, z_s\right) \approx \left(\frac{1}{2-\beta} + \frac{1}{\alpha-2}\right) \left[\frac{\zeta(z_s)(1-\kappa_0)^2 S}{2l_0}\right]^{-2} ,$$

which is independent of q, since the integral is dominated, for $S \gg l_0/\zeta(z_s)$, by large magnifications, i.e., small values of y. Then, (12.24) leads to

$$\begin{aligned}\frac{dn_{\text{obs}}(S, z_s, \vartheta, q)}{dz_s} &\approx 2\pi \Theta_*^2 N \frac{1+z_s}{\sqrt{1+\Omega z_s}} \left[\frac{1}{2-\beta} + \frac{1}{\alpha-2}\right] \left(\frac{\zeta(z_s)S}{l_0}\right)^{-2} \int_0^{z_s} dz \; h(z, z_s) \\ &\times 3 \int_0^1 dx \, (1-x)^2 \left[\frac{2}{(1-\kappa_0)^2}\right]^2 \Gamma(\nu+2, \ell_{\min}) \; .\end{aligned} \qquad (12.25)$$

Case (a): Source sample free of amplification bias. To estimate the fraction of multiply-imaged sources in a flux-limited sample, we consider two extreme situations: either the singly-imaged sources are not influenced by lensing at all, or the amplification bias dominates. In the first case, we obtain from (12.9) and for $S > l_0/\zeta(z_s)$

$$\frac{dn_0(S, z_s)}{dz_s} = V(z_s) \left[\frac{\zeta(z_s)S}{\langle\mu\rangle(z_s)l_0}\right]^{-\alpha} .$$

This yields for the fraction $f_a(S, \vartheta, q, z_s)$

$$f_{\rm a}(S,\vartheta,q,z_{\rm s}) = \frac{dn_{\rm obs}(S,z_{\rm s},\vartheta,q)}{dn_0(S,z_{\rm s})} = 2\pi\Theta_*^2 N\left[\langle\mu\rangle(z_{\rm s})\right]^{-\alpha}\left[\frac{1}{2-\beta}+\frac{1}{\alpha-2}\right]$$

$$\times \left(\frac{\zeta(z_{\rm s})S}{l_0}\right)^{\alpha-2}\frac{1}{r_1^2(z_{\rm s})}\int_0^{z_{\rm s}}dz\,h(z,z_{\rm s}) \qquad (12.26a)$$

$$\times\, 3\int_0^1 dx(1-x)^2\left[\frac{2}{(1-\kappa_0)^2}\right]^2\Gamma(\nu+2,\ell_{\min})\ .$$

As expected, this fraction depends on the flux threshold S of the sample. Since we have assumed $\alpha > 2$, the contribution of multiply-imaged sources is larger for high fluxes.

Case (b): Source sample properties dominated by lensing. In our second case, where the flux distribution of the sample is dominated by the amplification bias, we first have to calculate the magnification probability $P(\mu,z_{\rm s})$ from (11.68). Using (12.15c), (12.16a) and (12.21), for large μ, we have

$$P(\mu,z_{\rm s}) = \frac{\tau_{\rm g}}{\mu^2}\ , \qquad (12.27a)$$

where

$$\tau_{\rm g} = \Gamma(\nu+2)\,\pi N\Theta_*^2\,\frac{1}{r_1^2(z_{\rm s})}\int_0^{z_{\rm s}}dz\,h(z,z_{\rm s})\,3\int_0^1 dx\,(1-x)^2\left[\frac{2}{(1-\kappa_0)^2}\right]^2.(12.27b)$$

The x-integral can be solved by using (12.20b):

$$\tau_{\rm g} = \Gamma(\nu+2)\,4\pi N\Theta_*^2\,\frac{1}{r_1^2(z_{\rm s})}\int_0^{z_{\rm s}}dz\,h(z,z_{\rm s})\left[1-\frac{r(z)\,r(z,z_{\rm s})}{r(z_{\rm s})}\kappa_{\rm m}\right]^{-1}\ . \qquad (12.27c)$$

We will assume that (12.27) is valid for all $\mu \geq \mu_0$, where μ_0 is the magnification factor for which the linear analysis for the probability breaks down due to overlapping cross-sections – compare the discussion in Sect. 12.1.1. Following the same steps that led to (12.8), we find for $S \gg l_0/\zeta(z_{\rm s})$:

$$\frac{dn_{\rm obs}(S,z_{\rm s})}{dz_{\rm s}} = V(z_{\rm s})\,2\tau_{\rm g}\left[\frac{1}{\alpha-2}+\frac{1}{2-\beta}\right]\left[\frac{\zeta(z_{\rm s})S}{l_0}\right]^{-2}\ ,$$

and for the fraction $f(\vartheta,q,z_{\rm s})$,

$$f_{\rm b}(\vartheta,z_{\rm s}) = \frac{\int_0^{z_{\rm s}}dz\,h(z,z_{\rm s})\,3\int_0^1 dx\,(1-x)^2\,(1-\kappa_0)^{-4}\,\Gamma(\nu+2,\ell_{\min})}{\Gamma(\nu+2)\,\int_0^{z_{\rm s}}dz\,h(z,z_{\rm s})\left[1-\frac{r(z)\,r(z,z_{\rm s})}{r(z_{\rm s})}\kappa_{\rm m}\right]^{-1}}, \qquad (12.26b)$$

independent of both S and, as remarked before, of q, due to the dominance of high-magnification events.

Numerical results. In Fig. 12.3a,b, we display $f'_{\rm a}/K$, where

$$f'_{\rm a} = \frac{f_{\rm a}}{(\pi N\Theta_*^2)}\quad\text{and}\quad K = \frac{1}{\left[\langle\mu\rangle(z_{\rm s})\right]^\alpha}\left(\frac{1}{2-\beta}+\frac{1}{\alpha-2}\right)\left(\frac{\zeta(z)S}{l_0}\right)^{\alpha-2}\ ,$$

and $f_{\rm b}$, both as a function of ϑ/Θ_*, for a fixed source redshift $z_{\rm s}$ and different values of $\kappa_{\rm m}$. We see in both cases that the influence of a non-zero surface mass density is enhanced relative to the case where no amplification bias is taken into account; this is due to the fact that κ_0 positive not only leads to increased image splitting, but also to additional magnification. For certain values of $\alpha,\beta,\Theta_*^2 N$ and $[\zeta(z_{\rm s})S/l_0]$, the fraction $f_{\rm a}$ may become larger than unity; this then indicates that the assumption under which

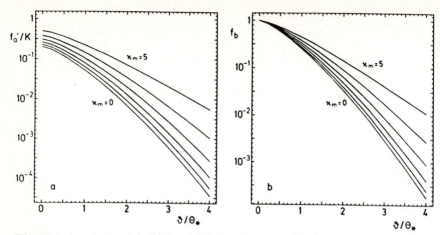

Fig. 12.3. The factors (a) f'_a/K and (b) f_b, where $f_a = f'_a \pi \Theta_*^2 N$ and f_b is the fraction of sources at redshift $z_s = 2$ which have multiple images with angular separation larger than ϑ, as a function of ϑ/Θ_*. The smooth mass density parameter takes the values $\kappa_m = 0, 1, 2, 3, 4, 5$. Note the strong dependence on κ_m

(12.26a) was derived breaks down, i.e., the overall sample of sources can no longer be unaffected by the amplification bias. In the case where the overall sample is lens-dominated, i.e., (12.26b) applies, we see that $f_b(\vartheta = 0) = 1$, which just shows that all sources in the sample are lensed, which is the underlying assumption of (12.26b).

To better understand the influence of the different parameters, we show the fractions (12.23) and (12.26) for $z_s = 2$, $\vartheta = 2\,\mathrm{arcsec}$, as a function of σ_{v*} in Fig. 12.4. Comparison of the curves f and f_b in Fig. 12.4a shows that inclusion of the amplification bias leads to an enormous increase of the fraction of sources lensed. This bias is even more important than the contribution from a positive surface mass density. From Fig. 12.4b, we also see that, when the overall sample is not dominated by lensing, the slope of the luminosity function is important. Furthermore, as the figures show, an increase of σ_{v*} by a factor of two changes the lensed fraction by factors of up to 10^3 for the cases shown.

Further results. The preceding considerations have demonstrated the sensitivity of the expected number of multiply-imaged sources to the various physical parameters such as the velocity dispersion of the galaxies, the influence of the amplification bias, and the distribution of surface mass density in the neighborhood of the lensing galaxies. We now discuss a study where plausible values of the parameters have been assumed, and where additional biases have been taken into account [TU84.1]. Turner and coworkers assumed that 70% of all galaxies are spirals with $\sigma_{v*} = 177\,\mathrm{km/s}$, and that the remaining galaxies are ellipticals with $\sigma_{v*} = 306\,\mathrm{km/s}$. Three percent of all ellipticals, and no spirals, were assumed to be near the core of dense clusters, with a surface mass density which corresponds to $\kappa_m = 4.24$. Most of the quantitative results in [TU84.1] concerning angular separation statis-

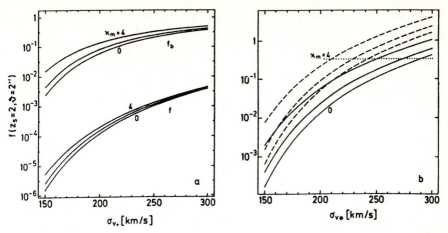

Fig. 12.4a, b. The fraction of sources at redshift $z_s = 2$ with multiple images of angular separation larger than $\vartheta = 2''$ and flux ratio less than $q = 10$, as a function of the velocity dispersion σ_{v*} of an L_* galaxy. (**a**) The fractions f and f_b. (**b**) The fraction f_a, for $\beta = 1$, $\alpha = 2.6$ (solid lines) and $\alpha = 3.4$ (dashed lines), $\zeta S/l_0 = 10$. In all cases, curves for $\kappa_m = 0, 2, 4$ are shown. The curves above the dotted line in (**b**) are of no relevance, since the assumption that the sample as a whole is not influenced by the amplification bias breaks down there

tics depend critically on this – fairly arbitrary – assumption; as we have seen above, multiple image statistics depends sensitively on the assumptions about a smooth surface mass density assisting galaxies in their lensing power. The amplification bias was taken into account with an approximate method, as was the finite resolution bias. Such a bias is extremely important, if one expects to account for the observed distribution of angular separation of multiply-imaged QSOs. In addition, the amplification bias favors the inclusion of highly magnified sources in a sample, and since magnifications as well as image splittings are increased by the finite surface mass density, incorporation of this effect is essential. The QSO population was chosen in accordance with observed samples. The observed distribution of angular separations can reasonably be explained with the model in [TU84.1]: a survey with limiting angular resolution of one arcsecond will detect multiple images in 0.1% to 1% of all high-redshift objects. This fraction is only a lower limit, since the amplification bias was accounted for only approximately. As we shall see in Sect. 12.2.5 below, for the apparently most luminous QSOs, the fraction of observed GL systems is significantly larger than these estimates. The large image separation in observed lens systems should almost always be due to galaxies at the center of rich clusters, as can easily be understood by inspection of Fig. 12.2. Since the z-integrands in (12.23) and (12.26) peak strongly at intermediate redshifts, the lenses should cluster around a redshift of ≈ 0.5. The curves for different redshifts in Fig. 12.2 are almost parallel, so that we expect approximately the same distribution of angular separations for different source redshifts, due to the properties of isothermal spherical lenses.

Inclusion of microlensing. As we have seen, the statistics of multiply-imaged QSOs can be affected by the amplification bias. Since the continuum-emitting regions of QSOs are thought to be smaller than the length-scale η_0 induced by stellar-mass objects, the light bundles "feel" the graininess of the mass distribution in the lensing galaxies. Therefore, to account for the lensing statistics of compact sources, ML by stars in the galaxies should be included.

Since ML does not affect the observed image positions, its main effect is to induce a magnification probability distribution for each light ray traversing a lensing galaxy. In other words, the magnification of an image of a background QSO is no longer a function of the image position alone, but follows from a probability distribution given by the local condition in the star field, as described in detail in Sect. 11.2. A simple model for this combined effect of image splitting by galaxy-mass lenses and the magnification effect of ML was recently published [BA90.1]; in this study, the mass distribution of each galaxy is represented by a singular isothermal sphere, and a fraction ϵ of the matter is concentrated in compact objects. For simplicity, the sources are treated in the point-source approximation. The images of a source are obtained from the lens equation, as described in Sect. 8.1.4, and the parameters of the (random) star field at each image (i.e., κ_*, κ_c, and γ, cf. Sect. 11.2) are then determined. Hence, in the point-source approximation, the high-μ tail of the magnification probability distribution is known [cf. (11.64)]. From the numerical ML simulations, as described in Sect. 11.2.5, one can in principle obtain the full magnification probability distribution; since this would be a considerable numerical effort, a slightly different approach was taken in [BA90.1]: the asymptotic behavior for large μ, together with the mean of the distribution and the lowest possible magnification (from Theorem 2 of Sect. 5.4) sufficiently constrain the distribution, so as to allow a simple analytic approximation, which has the qualitative properties obtained from numerical simulations; in addition, it was checked that variations of the distribution, within the constraints, do not significantly affect the resulting lensing statistics.

The main differences due to ML can be described as follows: for a smooth singular isothermal sphere, large magnifications occur only if the source is near the caustic point, $\mathbf{y} = 0$, with both images close to the critical curve of the lens and with comparable magnification. If ML is considered, single highly-magnified images can occur, i.e., the tight correlation between total magnification and flux ratio of the images is weakened by ML. Hence, if the amplification bias is important for lensing statistics (for which observational evidence will be presented in Sect. 12.5.2), it can be dominated by systems in which a single image is highly magnified. Thus, with the model that includes ML, one obtains a high fraction of multiply-imaged QSOs with large flux ratios.

Since not all compact objects are found within the regions of the lensing galaxies responsible for the image splitting (that is, inside the circle $|\mathbf{x}| = 2$), large magnifications can occur without multiple images. This effect can then lead to an amplification bias without image splitting. Therefore, the more compact objects are outside the image splitting regions of galaxies, the larger is the "contamination" of source counts by singly-imaged QSOs due to the amplification bias. If the bright QSO source counts are affected by the amplification bias, the fraction of multiply-imaged sources can yield an upper limit to the density of compact objects outside (or at large distance from the center of) galaxies. For further results and details, see [BA90.1].

Lensing by dark matter halos. The foregoing analysis assumes that the objects responsible for splitting QSO images are galaxies. However, simulations of the density distribution of the universe in a cold dark matter model suggest that dark matter halos of sufficient compactness can be formed in such a model (see [NA88.1] and references therein). In [NA88.1], a semi-analytic distribution function of such dark halos is derived. The main uncertainty is the core radius of such halos, which cannot be determined by current numerical simulations due to the limited numerical resolution. If, however, the core radii are sufficiently small, then dark matter halos can act as efficient lenses. In particular, the model presented in [NA88.1] can yield a distribution of angular separations of multiply-imaged QSOs which can reproduce the observed distribution, if an amplification bias is taken into account.

12.2.3 Modifications of the lens model: Symmetric lenses

The singular isothermal sphere GL models considered in the last subsection may be too simple to obtain realistic lensing probabilities. Two main modifications must be considered to estimate the reliability of this model. First, one needs to investigate the influence of a non-zero core size for the lenses ([DY84.1], [HI87.1], [BL87.2], [KO87.2]); second, the importance of the ellipticity (or, more general, asymmetry) of the matter distribution of the lenses needs to be studied ([BO76.1], [BL87.2], [KO87.2]); the latter aspect will be considered in the next subsection. It turns out that the influence of a non-zero core size is considerable, as has been found for isothermal galaxies as well as for other types of matter distributions.

Isothermal sphere with non-zero core. Consider the lens action of an isothermal sphere with core radius ξ_c. The matter distribution is given by (8.37) for $p = 1/2$. Whereas the scaling of lengths used in Sect. 8.1.5 is useful for mathematical considerations, we will now use a different scaling which is more adapted to the physical situation to be described. Namely, we use ξ_0 as given by (8.34a) as the length scale in the lens plane. The equations of Sect. 8.1.5 can then be used by replacing x and y by xx_c and yx_c, respectively, where $x_c = \xi_c/\xi_0$ is the dimensionless core radius. Applying this transformation to the lens equation (8.40) yields

$$y = x - x_c \kappa_0 \frac{x}{\sqrt{x^2 + x_c^2}} \quad ,$$

where κ_0 is the dimensionless surface mass density at the center of the lens. If we keep the surface mass density at large separations fixed, it is clear that the value for κ_0 increases with decreasing x_c, that is, the smaller the radius beyond which the $\kappa \propto x^{-1}$ behavior applies, the larger must be the peak value for κ. In order for the ray-trace equation to attain the form corresponding to a singular isothermal sphere for large values of x, we must require

$$x_c \kappa_0 = 1 \quad .$$

The lens has a radial critical curve as given in (8.41d), with the radius of the corresponding caustic circle given by (8.42). In our length scaling this radius is $y_r = \left(1 - x_c^{2/3}\right)^{3/2}$, i.e., sources with impact parameters smaller than y_r will have three images, otherwise only one image occurs. Note that, for $x_c \geq 1$, y_r is no longer defined. Owing to the above relation between core radius and central surface mass density, $x_c \geq 1$ corresponds to $\kappa_0 \leq 1$, and as was shown in Sect. 8.1.1, centrally condensed lenses with $\kappa_0 \leq 1$ are not capable of producing multiple images.

A source at $y = 0$ will have three images (for $x_c < 1$), one at $x = 0$ and the other two at $x = \pm\sqrt{1 - x_c^2}$. Hence, the widest angular separation for the images is $\Delta x = 2\sqrt{1 - x_c^2}$. It turns out that the widest angular separation for the images does not depend strongly on the source position [HI87.1]; we will thus assume here that, whenever multiple images are produced, they have a separation Δx as above.

These simple considerations already show that a non-zero core size reduces the probability of multiple imaging. Interestingly, the maximum source impact parameter for multiple lensing, y_r, is much more dependent on the core radius than the maximum image separation. This will also influence the distribution of expected angular separations for multiply-imaged sources; in fact, this was the main motivation for the consideration of non-zero cores, since the angular distribution obtained from the singular isothermal sphere models have a tendency to be peaked at comparatively low angles. Only by introducing the observational bias can this model be made compatible with the observed distribution of separation angles.

Cross-sections and lensing probabilities. From the above relations, we can easily find the cross-section $\sigma(\Delta x)$ for the lens under consideration to produce multiple images with separation larger than Δx,

$$\sigma(\Delta x) = \pi y_r^2 \, \mathrm{H}\left(2\sqrt{1 - x_c^2} - \Delta x\right) = \pi \left(1 - x_c^{2/3}\right)^3 \mathrm{H}\left(2\sqrt{1 - x_c^2} - \Delta x\right) \quad .$$

This cross-section can now be inserted into (12.22) to obtain the probability $P(\vartheta, z_s)$ that a source at redshift z_s has multiple images with angular separation larger than ϑ. We will do so in the following; for simplicity, we neglect here the possible influence of a smooth background density, i.e., in the notation of the previous subsection, we take $\kappa_m = 0$. Then,

$$x_{\rm c} = \frac{\xi_{\rm c}}{\xi_0} = R_{\rm c} \frac{r(z_{\rm s})}{r(z) r(z, z_{\rm s})} \frac{1}{\sqrt{\ell}} \quad , \tag{12.28a}$$

where

$$R_{\rm c} = \frac{\xi_{\rm c}}{\Theta_*} \frac{H_0}{c} = 1.3 \times 10^{-2} \frac{\xi_{\rm c}}{1\,{\rm kpc}} \left(\frac{\sigma_{v*}}{300\,{\rm km\,s^{-1}}}\right)^{-2} h_{50} \tag{12.28b}$$

is a measure of the core size which no longer depends on the redshift z of the lens; the notation Θ_* is the same as in the previous subsection, compare (12.18b), and $\ell = (\Theta/\Theta_*)^2$, where Θ is a measure of the velocity dispersion of the galaxy (12.16b). The relation between the dimensionless image separation Δx and the corresponding angular separation ϑ is given by (12.16c). The Heaviside factor in the cross-section leads to a requirement on ℓ, i.e., $\Delta x^2 \leq 4\left(1 - x_{\rm c}^2\right)$ is satisfied for $\ell \geq \ell_{\min}$, where

$$\ell_{\min} = \left[\frac{r(z_{\rm s})}{r(z,z_{\rm s})}\right]^2 \left[\frac{1}{4}\left(\frac{\vartheta}{\Theta_*}\right)^2 + \left(\frac{R_{\rm c}}{r(z)}\right)^2\right] \quad . \tag{12.28c}$$

Note that this expression agrees with that given in the last subsection for $R_{\rm c} = 0$ and $\kappa_0 = 0$. Then, from (12.22) we obtain

$$P(\vartheta, z_{\rm s}) = \pi \Theta_*^2 N \frac{1}{r_1^2(z_{\rm s})} \int_0^{z_{\rm s}} dz\, h(z, z_{\rm s}) \int_{\ell_{\min}}^\infty d\ell\, \ell^{\nu+1} e^{-\ell} \left[1 - x_{\rm c}^{2/3}(\ell)\right]^3 . \tag{12.29a}$$

Differentiation of this expression with respect to ϑ/Θ_* then yields the differential probability $p(\vartheta, z_{\rm s})$ for a source at redshift $z_{\rm s}$ to have multiple images with separation ϑ,

$$\begin{aligned} p(\vartheta, z_{\rm s}) &= -\frac{dP(\vartheta, z_{\rm s})}{d(\vartheta/\Theta_*)} \\ &= \frac{1}{2} \frac{\vartheta}{\Theta_*} \frac{\pi \Theta_*^2 N}{r_1^2(z_{\rm s})} \int_0^{z_{\rm s}} dz\, h(z, z_{\rm s}) \left[\frac{r(z_{\rm s})}{r(z, z_{\rm s})}\right]^2 \ell_{\min}^{\nu+1} e^{-\ell_{\min}} \left[1 - x_{\rm c}^{2/3}(\ell_{\min})\right]^3 , \end{aligned} \tag{12.29b}$$

where

$$x_{\rm c}(\ell_{\min}) = \left[1 + \frac{r^2(z)}{4 R_{\rm c}^2}\left(\frac{\vartheta}{\Theta_*}\right)^2\right]^{-1/2} \quad .$$

Influence on lens-redshift distribution. From the expression for $\ell_{\min}$, we notice that the integral in (12.29b) has a very small contribution from low redshifts and from redshifts close to that of the source, $z \approx z_{\rm s}$. The latter suppression also occurs for the singular isothermal sphere model and is due to the fact that, for $z \approx z_{\rm s}$, the first factor in (12.28c) becomes large. The suppression for small redshifts is due solely to the non-zero core size. This can be easily understood by noting that the critical surface mass density $\Sigma_{\rm cr}$ (5.5) becomes large for small redshifts; hence, for a given lens model, implying a given central surface mass density Σ_0, the dimensionless central surface mass density $\kappa_0 = \Sigma_0/\Sigma_{\rm cr}$ becomes small for small redshifts, and multiple imaging is possible only for $\kappa_0 > 1$.

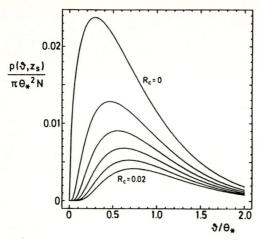

Fig. 12.5. The scaled probability $p(\vartheta, z_s)/(\pi\Theta_*^2 N)$, (12.29b), as a function of ϑ/Θ_*, for $z_s = 2$ and cosmological parameters $\tilde{\alpha} = 0.9$, $\Omega = 1$. A non-zero core radius of the isothermal sphere galaxies was considered; the dimensionless core radius R_c as defined in (12.28b) takes the values $(0,4,8,12,16,20) \times 10^{-3}$. Note that even a small core radius reduces the lensing probability considerably

Distribution of angular separations. In Fig. 12.5, we show $p(\vartheta, z_s)/(\pi\Theta_*^2 N)$ as a function of ϑ/Θ_* for various values of R_c. We see clearly that a non-zero core radius decreases the probability for multiple imaging, and that it also shifts the maximum of the probability distribution to larger values of ϑ. Even for the smallest value of R_c considered, the lensing probability is suppressed by a substantial factor. This result is in qualitative agreement with [HI87.1]; however, in that paper much more drastic reductions of the lensing probability are obtained. This is because in [HI87.1], it was assumed that the luminosity of a galaxy is proportional to the line-of-sight velocity dispersion, whereas we assumed, as in the preceding subsection, the validity of the Faber–Jackson relation, $L \propto \sigma_v^4$ [FA76.1].

Other symmetric lens models. Dyer [DY84.1] considered two different mass distributions for the lensing galaxies, namely King [KI66.1] and de Vaucouleurs [VA48.1] mass distributions. He found that the results obtained are fairly similar. In addition to the parameters considered so far (the value for σ_{v*}, expressed by Dyer as a mass-to-light ratio, with a non-zero core size) he varied a few other parameters; e.g., if one assumes that the distribution of galaxies acting as lenses cuts off at a finite redshift, the mean angular separation to be expected increases, since as a general rule, nearby lenses tend to produce the larger separations, whereas high-redshift lenses create image pairs with small ϑ. The mean angular separation is also increased if there is a low-luminosity cutoff of the lens luminosity function, as expected.

12.2.4 Modification of the lens model: Asymmetric lenses

A further complication in the models was introduced by studies that incorporate an ellipticity of lenses ([BO76.1], [BL87.2], [KO87.2]). The first of these papers uses lenses with elliptical mass distributions with $\Sigma \propto \xi^{-1}$. Multiple image cross-sections were computed numerically, and then convolved with a source distribution. The same idea was developed in the other two papers, where the lens models considered in Sect. 8.4.2 were used, i.e., lenses with elliptical isopotentials. In addition to the cross-sections for multiple imaging, these papers also present detailed discussions of the expected image configurations. We summarize some of their results below. In contrast to our treatment in Sects. 8.1.5 and 8.4, they used the radius of the tangential critical curve $\xi_c x_t$ (8.41a) as the length scale in the lens plane. Hence, in their scaling, the core radius ξ_c has the dimensionless value $\beta = 1/\sqrt{\kappa_0^{1/(1-p)} - 1}$. Thus, β is the ratio between the core radius and the radius of the tangential critical curve. Expressed in terms of β, the central surface mass density becomes $\kappa_0 = \left(1 + \beta^{-2}\right)^{(1-p)}$. Therefore, for all values of β, $\kappa_0 > 1$.

Strong and weak lenses, opposite and allied images. Lenses with large values of β have a central surface mass density just above unity and are thus just able to produce multiple images, at least for a vanishing ellipticity. They are weak (or marginal) lenses, in contrast to strong ($\beta \ll 1$) lenses. For circular potentials, it is useful to consider the two brightest images: if they are on opposite sides of the center of the potential, the image configuration is termed *opposite*, in contrast to the *allied* image case, where the two brightest images are on the same side of the mass distribution. A similar distinction can be made for elliptical potentials. Most of what follows can be directly inferred from the discussion in Chap. 8.

Qualitative discussion of lensing cross-sections. We start by describing the properties of strong ($\beta \ll 1$) circularly symmetric potentials ($\epsilon = 0$). Lenses with singular potentials ($\beta = 0$) always form two images (for $p < 1/2$) which lie on opposite sides of the potential. However, the larger the impact parameter of the source, the larger the flux ratio of the two images. A bright pair of images occurs if the source is near the optical axis; in this case, the images will have a dimensionless separation of 2. For a non-zero core radius, either one or three images are produced, depending on the position of the source. The two brightest images can be either allied or opposite; the latter is the case if the source is close to the optical axis, whereas allied images occur if the source is close to, but inside, the caustic with radius y_r (8.42). Let μ_{12} be the mean of the absolute values of the magnifications of the two brightest images; from (8.44) we can then derive the cross-section $\sigma(\mu_{12})$ for opposite images to be formed with mean magnification larger than μ_{12} (for $\mu_{12} \gg 1$, irrespective of the magnification ratio),

$$\sigma(\mu_{12}) = \frac{\pi\left(1+\beta^2\right)^2}{4(1-p)^2}\frac{1}{\mu_{12}^2} \quad,$$

where the unit of this cross-section is now the square of the radius of the tangential critical curve. Note that the separation of these opposite images is still 2 in units of this radius. Hence, the cross-section is larger for soft potentials ($p \lesssim 1/2$) than for hard ones ($p \gtrsim 0$), and increases with increasing core size. This is not only an intrinsic feature of the lens, but is also due to the scale length selected: for a fixed mass distribution at large x, introduction of a non-zero core reduces the average mean mass density $\bar{\kappa}$, introduced in Sect. 8.1.1, at every point x. Hence, the tangential critical curve moves inward as β is increased. Therefore, for applications the scaling introduced at the beginning of this subsection for the special case of isothermal spheres ($p = 1/2$) is more convenient. For simplicity, we will still follow [BL87.2] here.

An extended core allows the formation of allied images; they appear near the radial critical curve at x_r (8.41c) and close together; it is easy to show from (8.45) that the cross-section for allied images is much smaller than that for opposite images, and that this difference is larger for hard potentials, which is due to the fact that $y''(x_r)$, which enters in (8.45), is larger for smaller values of p. Furthermore, the cross-section for allied images decreases with decreasing β. The image separation of allied images decreases with increasing μ_{12}, as is clear from the behavior of the lens equation near folds, which was discussed in Sect. 6.2.1. For opposed images, the third (faintest) image is close to the center of the potential, and its magnification is roughly $\mu_3 \approx (1-\kappa_0)^{-2}$, whereas for allied images, the third image has a magnification > 1 which is nearly independent of μ_{12}, given by the magnification of the third, non-critical image of a source at y_r.

The situation changes if one considers marginal lenses ($\beta \gtrsim 1$); in this case the properties of the lens do not depend strongly on p, as they are basically determined by the quadratic expansion of the potential at the center. For $\beta \lesssim 1$, the cross-sections for allied and opposed images are equal, as is the magnification μ_3 for the third image in both cases. Note that μ_3 is large for marginal lenses.

Introduction of an ellipticity, $\epsilon \neq 0$, unfolds the critical point at $y=0$ into a caustic with cusps. For small values of ϵ, this caustic is inside the one corresponding to the radial critical curve, but for larger values of ϵ, it can cross the radial caustic (for an example of this, see Fig. 8.7b). A source inside both caustics has five images, one within only one caustic has three, and a source outside the two caustics has one image. For strong lenses, the caustic related to the tangential critical curve has four cusps, whereas for marginal lenses, it has two if $\epsilon \gtrsim 1 - \left(1+\beta^{-2}\right)^{p-1}$ (lips caustic – see Sect. 6.3.3); in this case, the five-image cross-section for marginal lenses is suppressed, since in this case the lens is able to focus the source only in one direction – see Fig. 5.9 for an example.

For strong lenses, the cross-section for five-image geometries is larger than that for three-image geometries with opposed images, for $\mu_{12} \gtrsim (1-p)^{-1}\epsilon^{-1}$. However, the separation of the two brightest images is much smaller in the five-image case than in the opposed image case. In addition, here, the three-image allied case has larger cross-sections for large μ_{12} than the opposed geometry.

In [KO87.2] the results outlined above were used to calculate the expected number of multiply-imaged QSOs on the sky, the angular separation of such images, and their apparent brightness. Qualitatively, their results agree with those obtained in Sect. 12.2.2 above; in particular, they point out that a non-zero core size reduces the multiple image probabilities. Bright multiply-imaged sources should frequently be of the five-image type; however, since the angular separation is smaller in this case than in the three-image opposed case, they are much more difficult to detect. For marginal lenses, the allied three-image geometry can become dominant, but again, the small image separation in this case makes them difficult to detect.

Thus, we see that the lensing statistics depend critically on many parameters which are only poorly determined. Therefore, it will be difficult to compare the results of lens surveys with the theoretical considerations outlined above. In particular, as should be noted again, the amplification bias has only been incorporated approximately, since at least for QSO surveys based on color selection (i.e., using the properties of the optical continuum light), the magnification can be dominated by compact objects in the lensing galaxies.

12.2.5 Lens surveys

Early GL candidates were found by chance. Imaging of QSOs such as 0957+561 revealed multiple images separated by a few arcseconds. The small separations of image pairs aroused further curiosity, and subsequent studies revealed lenses in at least some of the cases (Sect. 2.5). The discovery rate for new systems accelerated as new candidates appeared, from a rate of one new system per year, to several per year, especially after the discovery of arcs (Sect. 2.5.5). Only a few of the more recent examples were found by systematic searches that specifically targeted GLs. Therefore, the current list of GL candidates by no means represents a statistically well-defined sample, and no meaningful statistical conclusions – such as population densities – can be inferred from such data. However, several systematic searches were launched within the past few years, with the aim of building statistically well-defined samples. We describe three such searches in the three following subsections.

The MG survey. A detailed description of the MIT – Green Bank (MG) radio survey can be found in [HE86.2]. The targets for the lens search were extracted from a catalog of nearly 10000 sources found at 5 GHz with the NRAO 300-foot telescope in Green Bank, West Virginia. The brightest

MG sources outside the galactic plane were selected for the lens search observations with the VLA (Sect. 2.5.1), again at 5 GHz. The total number of sources observed with the VLA was about 4000, thus yielding a sample complete to 115 mJy, with resolution between $0\rlap{.}''4$ and $1\rlap{.}''2$.

The VLA observations improved the single-dish, low-resolution radio maps, and permitted the ranking of sources. Furthermore, because of the high sensitivity of the VLA, short "snapshots" sufficed to discern the structure of the sources, so the search became very efficient. The final product was a collection of potential GL candidates with well-defined properties. In particular, the detectable fraction of multiple component images with a given angular separation and given flux ratio, i.e., the selection function, could be estimated reliably. Precise knowledge of the flux ratio limits of images is crucial to such estimates. Since these limits correspond to the well-known limits of dynamic range and resolution of the VLA, a highly reliable selection function can be expected. At that stage, sources with two unresolved components in the field were classified as likely, "class A" candidates; "class B" candidates included all sources with more than one component in the field. A further subdivision depended on whether the separation of the components was larger than $10''$ (these sources were assigned low priority because such separations cannot be caused by galaxies, although, according to [NA84.3], they can be caused by clusters of galaxies), or whether the structure of the source is not colinear; the latter sources were given relatively high priority, since radio sources often show multiple components, and linear structure. Therefore, non-colinear structure was a potential sign of lensing.

Once the VLA list of promising candidates was compiled, the sources were imaged with optical telescopes to ascertain whether radio structures had comparable optical counterparts. If the optical observations revealed such structure, the spectra of the individual components of the source were measured, in order to estimate the redshift of the source. It was then possible to estimate the probability that each source was multiply imaged by a GL. As we pointed out in Sect. 2.4, the verification that a GL is at work is quite difficult, in particular if the lensing object is not seen. The most reliable sign of lensing – besides a pair of sources – is an accurate determination of the redshift of the images and of the lens.[3]

The last step just described is unfortunately the least efficient. Although the VLA observations take little time, optical imaging and spectroscopy are very time-consuming, because most of the sources are very faint. That means that only large telescopes can be used, but these instruments are heavily oversubscribed. Therefore, a long time will probably pass before the survey is completed. Nevertheless, interesting results are already at hand. One multiply imaged QSO (2016+112) and two radio rings (see Sect. 2.5.6) were found through the survey, and additional possible candidates have

---

[3] The probability for chance alignments of radio sources which could be falsly interpreted as GL candidates is fairly small [LA90.2].

emerged (Sect. 2.5). Furthermore, by making extremely conservative assumptions (e.g., that all multiple-component sources are indeed lensed) it is possible to derive upper limits on the density of lenses in a certain mass range (compare Sect. 12.2.1). Whereas these limits are not very stringent (e.g., [HE86.2] concludes that less than $\approx 70\%$ of the critical mass can be in compact objects in the mass range of 10^{12} to $10^{13} M_\odot$), they are likely to improve considerably as the survey approaches completion. In particular, these estimates are based on the assumption that the amplification bias is unimportant, a conservative assumption, in the sense that it leads to the lowest lensing probability for a given matter distribution. In addition, radio-selected lenses are much less likely to be amplification-biased by compact, stellar-mass objects, since their size is thought to be significantly larger than the relevant length scale for such ML. In any case, the number of promising lens candidates that was found is fairly small, which implies roughly that the probability of multiple imaging is about one in a few hundred, as derived in Sect. 12.2.2 and displayed in Fig. 12.4.

APM survey. An automated survey for GLs based exclusively on optical observations is being conducted using the Automated Plate Measuring Facility (APM) in Cambridge, UK [WE88.1]. This machine scans direct Schmidt photographic plates, digitizes them, and identifies sources with significant ellipticities. The survey also included objective-prism plates, which yield low-dispersion spectra of sources on the direct plates. The results reported in [WE88.1] are based on 237000 images of sources with blue magnitudes in the range $15 < m_B < 20$ over 130 square degrees. Of these, 2500 were QSO candidates, which were examined to identify multiple components with separations within the range of $1''$ to $2'$. The selection criteria were the following: components separated by less than 5 arcsec appeared as single images on the plates, but depending on the component separation and brightness ratio, such single images appeared elliptical. Using simulations, the researchers were able to obtain the selection function, and thus determine the angular-separation bias. Objects with separations larger than $5''$ were detected as separate components on the direct plates. In both these cases, the acceptance criteria were that the spectra be QSO-like, and in the second case, in addition, that the spectra of the two separate components be similar. The search yielded a total of 73 GL candidates. The fraction of QSOs among these 73 candidates was estimated to be 20%. The fraction of those which are indeed good lens candidates is yet unknown. Nevertheless, since this optical survey, just as the radio survey described above, has well-controlled selection criteria, it will be very useful for statistical studies once it reaches a more mature stage. One candidate for a multiply imaged QSO [HE89.2] was found in the survey.

HLQ survey. Another survey targets high-luminosity QSOs (HLQs), and is based on a strategy different from that of the other surveys described above ([SU88.1], see also [SU90.1]). Since the amplification bias is likely to be most important for HLQs, we might expect that the fraction of lensed sources is particularly high in a sample of such objects. The HLQ sample consists of 111 QSOs with absolute luminosity $M \lesssim -29.0$ and apparent luminosity $m_V \lesssim 18.5$. Their redshifts are larger than 1.0, and their mean redshift is larger than 2.0. Of these 111 HLQs, 25 were classified as "interesting objects", because they appeared to have multiple images, elongated images, jet-like features, a nearby galaxy, or "fuzz" surrounding the image. It is clear that these selection criteria are not as well defined as in the other two surveys, because they depend critically on the flux of the QSO. Of the 25 "interesting objects", 5 were very good candidates for gravitational lensing. In fact, for three of these objects, 1635+267 [DJ84.1], UM673 [SU87.3], and H1413+117 [MA88.1], the case for a GL is strong (Sect. 2.5). From the histograms in [SU88.2] we see that there is no correlation between the fraction of HLQs classified as "interesting" and redshift, absolute magnitude or apparent magnitude. However, the selection of objects as "interesting" clearly depends strongly on the observing conditions, e.g., the seeing at the time of imaging. In fact, 5 out of 11 sources which were observed under exceptionally good seeing (less than $1''$), are classified as "interesting", and three of those are classified as very promising lens candidates.

It appears that at least one conclusion from these results cannot be avoided: if the fraction of HLQs which are probably lensed is as high as the survey seems to imply, the amplification bias must play a very important role. Since not all magnified sources need be multiply imaged, the results from this survey will provide a lower limit on the importance of the amplification bias. We discuss this point further and in greater detail in Sect. 12.5.

12.3 QSO–galaxy associations

12.3.1 Observational challenges

For more than two decades, the cosmological interpretation of the redshifts of QSOs has been disputed by a few astrophysicists, most noticeably by Halton Arp. Their scepticism about the conventional interpretation of the redshift is based on a number of observations which seem to indicate that some QSOs are associated with foreground galaxies at much smaller redshift than the QSOs (for a description of these observations, and a fascinating, though biased, description of the history of this controversy, see [AR87.1]; for an early history of the redshift controversy, see [FI73.1] and [BU79.1]). There are four examples of nearby galaxies (with redshift $z \sim 0.01$) listed in [AR87.1] which have two or three QSOs within 2 arcmin of their center; these QSOs have a redshift of about unity and are brighter than $m_V \sim 20$. Since the QSO density is about 20 per square degree for QSOs brighter than $m_V \sim 20$, it is highly unlikely to find two or three of them within 2 arcmin around any point. The probability of finding one of these QSOs within 2 arcmin around any point is 0.07; if one assumes that the galaxies have been selected because there was at least one QSO within less than $2'$, the probability of having two around these galaxies is then ~ 0.07, and the probability of finding three is ~ 0.005. In order to judge whether these small probabilities are statistically significant, one would have to know how many galaxies have been searched for associated QSOs. This question has no straightforward answer. It should be mentioned, however, that the one galaxy with three associated QSOs (NGC1073) is one out of only 150 galaxies in the *Hubble Atlas of Galaxies*. There is a vast number of arguments in the literature about whether these associations are statistically significant (for references, see [AR87.1]). We hesitate to add to this discussion here; instead, we will point out additional, independent observations which may suggest that the QSO population near at least some foreground galaxies is different from the average QSO population.

The cross correlation between the *Second Reference Catalogue of Bright Galaxies*, which contains 4364 objects, and the QSO catalogue of Hewitt and Burbidge [HE80.2] which contains 1356 objects, yields strong statistical evidence for a correlation of QSOs at all redshifts and bright galaxies with low (≤ 0.05) redshifts [CH84.3]. One might wonder how a result based on such a large sample of objects can be disputed, but there are problems. For example, the QSO catalogue is by no means an unbiased sample, since QSO searches have been conducted only in certain regions in the sky. Furthermore, radio-loud QSOs are over-represented in the catalogue, since they are more easily found. And, perhaps the most serious problem, a study based on different catalogues of QSOs and galaxies [NI82.1] did not find any statistically significant evidence for associations.

The statistical results from one of the earliest papers on QSO–galaxy associations [BU71.1] have been confirmed by other authors, e.g., [KI74.1]. J.-L. Nieto did not find a significant overdensity of the QSO population as a whole near foreground galaxies ([NI77.1], [NI78.1]), but did find an overdensity for optically variable QSOs [NI79.1] (see also [BA83.1]).

Another strategy to observe QSO–galaxy associations is to look for galaxies around QSOs and see whether the galaxy density is increased relative to the average density of galaxies. For low-redshift QSOs, such an overdensity has been known for a long time [RO72.1], and it is now established that low-redshift QSOs are located in groups and small clusters of galaxies at the same redshift [YE87.1]. J.A. Tyson [TY86.3] found an overdensity of galaxies around both low-redshift and high-redshift QSOs; whereas for the low-redshift QSOs the overdensity can be understood as a physical association of the galaxies with the QSO, with both at the same redshift, the origin of the overdensity of galaxies around high-redshift QSOs is more controversial. If it were interpreted as being due to galaxies in the same cluster as the QSO, it would imply a strong luminosity evolution of the galaxies, i.e., galaxies at high redshift must be much more luminous than they are today. A second interpretation is that the galaxies have lower redshift than the associated QSO; in this case, the overdensity can be interpreted as being due to the amplification bias: the QSOs associated with a galaxy are brightened due to magnification by gravitational light deflection. An independent study conducted by W. Fugmann ([FU88.1], [FU89.1]) yields a statistically significant overdensity of galaxies around high-redshift radio-loud QSOs. He pointed out that this effect is much more pronounced for QSOs with a flat radio spectrum than for steep-spectrum ones. By dividing the sample used by Tyson into these two categories, he obtained a highly significant overdensity of galaxies in the vicinity of flat-spectrum radio QSOs which is unlikely to be due to galaxy evolution.

Recently, R. Webster [WE88.2] reported an observed overdensity of galaxies within 6 arcsec of high-redshift ($z > 0.5$), optically-selected bright QSOs. If one only counts those associations for which a galaxy has been definitely identified, there are 16, compared to 5.5, as expected from the overall galaxy density (which in this sample was obtained from the same plates on which the QSOs were identified). In addition, there are several more possible associations (see [WE90.1] for more details). Several galaxy redshifts are available, showing that they are in the range $0.1 < z < 0.35$, i.e., they are indeed foreground galaxies.

Additional evidence has been derived from an X-ray survey of AGNs. Using the extended *Medium Sensitivity Survey*, J. Stocke and coworkers [ST87.1] found that sources near foreground galaxies have a significantly different redshift distribution than the sample as a whole. In particular, the mean redshift of X-ray selected AGNs near foreground galaxies is twice the mean in the complete sample. From the same sample, it was concluded [RI88.2] that sources near foreground galaxies have a significantly higher intrinsic luminosity than the average X-ray selected AGN. However, keeping in mind that the number of associations (nine) in this sample is small, this result must be considered preliminary; in fact, if the two highest-redshift objects are omitted, the statistical significance of the difference between the redshift distributions vanishes.[4]

Lensing interpretation. The observational results just described, if they turn out to be statistically significant, lead to difficulties with the conventional view of QSOs. There is no plausible way of physically associating an object with redshift 2 with one at redshift 0.05 without drastically changing our view of the origin of the QSO or galaxy redshifts, or both (or of the standard cosmological model). The only possible explanation of these observations not in conflict with the conventional wisdom on QSOs and cosmology is the influence of the gravitational field of foreground galaxies on the light of distant QSOs, i.e., the GL effect. Such an interpretation implies a magnification of the background objects, and therefore intrinsically faint QSOs may be magnified above the flux threshold of a flux-limited sample. In this section, we investigate this idea, but before leaving the topic of difficult-to-explain observations, it should be mentioned that QSO overdensities near

---

[4] For a detailed account of the closest AGN-galaxy pair in that sample, see [GI86.1].

foreground galaxies are not the only kind of unusual occurrences; further "strange" observational facts can be found in [AR87.1].

12.3.2 Mathematical formulation of the lensing problem

Consider a galaxy at redshift z_d and a population of QSOs at redshift $z_s > z_d$. Let $\mathbf{x}$ and $\mathbf{y}$ denote angular variables in the lens and source planes, respectively, and let the mass distribution in the lens plane be described by the surface mass density $\Sigma(\mathbf{x})$; the latter can be transformed into the dimensionless surface mass density $\kappa(\mathbf{x})$, using the equations of Chap. 5. Let us also assume that a fraction ϵ of this matter is concentrated in compact objects, to account for ML.

Depending on $\kappa(\mathbf{x})$, the galaxy may or may not be able to form multiple images of sources at z_s (see Sect. 5.4). Multiple images can occur only within a limited area around the galaxy, which we denote by ω_m. In the following discussion, we exclude this area from our considerations, both to simplify the treatment, and because ω_m is small for realistic galaxy models, in the sense that observed associations of QSOs with foreground galaxies have an angular separation larger than the linear dimension of ω_m.[5]

If we restrict the discussion to the lens plane with ω_m excluded, the ray-trace equation becomes a one-to-one mapping. In particular, the solid angle element d^2x about $\mathbf{x}$ is mapped onto the solid angle $d^2y = d^2x/\langle\mu\rangle(\mathbf{x})$ about $\mathbf{y}$, and $\mathbf{y}$ has no image other than that at $\mathbf{x}$. We can then obtain the (local) magnification probability distribution $p(\mathbf{x}, \mu)$ for a source at $\mathbf{y} = \mathbf{y}(\mathbf{x})$. If the galaxy contains no microlenses ($\epsilon = 0$), this probability is $p(\mathbf{x}, \mu) = \delta(\mu - \langle\mu\rangle(\mathbf{x}))$, where $\langle\mu\rangle(\mathbf{x})$ is the magnification calculated from the galaxy lens model, $\langle\mu\rangle(\mathbf{x}) = |\det(\partial \mathbf{y}/\partial \mathbf{x})|^{-1}$. For $\epsilon \neq 0$, the probability distribution has finite width, with mean $\langle\mu\rangle(\mathbf{x})$, due to local flux conservation.

In the solid angle d^2y about $\mathbf{y}$, there are

$$\frac{dN(l)}{dz_s} = V(z_s)\,\Phi_Q(l, z_s)\,d^2y$$

sources per unit redshift interval brighter than l in scaled units [with the 'volume factor' $V(z)$ defined in (12.2b)]. To each of these sources corresponds one image within the solid angle d^2x about $\mathbf{x}$, and their flux distribution is, according to (12.4),

$$\frac{dN_{\mathrm{obs}}(S, z_s; \mathbf{x})}{dz_s} = \int_1^\infty d\mu\, p(\mathbf{x}, \mu)\, \frac{dN}{dz_s}\left(\frac{\zeta(z_s)S}{\mu}\right) \quad;$$

accordingly, the density of these sources is $dn_{\mathrm{obs}} = dN_{\mathrm{obs}}/d^2x$, or

---

[5] This is not true for the observations of M. Stickel and coworkers [ST88.1], [ST88.2], [ST89.1] who found foreground galaxies nearly centered on high-redshift BL Lac objects; however, since their sample is not statistically well-defined, it is not possible to obtain a value for the corresponding overdensity.

$$\frac{dn_{\text{obs}}(S, z_s; \mathbf{x})}{dz_s} = \frac{V(z_s)}{\langle \mu \rangle (\mathbf{x})} \int_1^\infty d\mu \, p(\mathbf{x}, \mu) \, \Phi_Q\left(\frac{\zeta(z_s)S}{\mu}, z_s\right) \; .$$

Let us consider the solid angle ω that excludes the solid angle ω_m where multiple images can occur. The mean density of sources within ω is obtained by averaging the last expression over solid angle:

$$\frac{dn_{\text{obs}}(S, z_s; \omega)}{dz_s} = \frac{V(z_s)}{\langle \mu \rangle_\omega} \int_1^\infty d\mu \, \Phi_Q\left(\frac{\zeta(z_s)S}{\mu}, z_s\right) p_\omega(\mu) \; , \qquad (12.30a)$$

where

$$p_\omega(\mu) = \frac{\langle \mu \rangle_\omega}{\omega} \int d^2x \, \frac{p(\mathbf{x}, \mu)}{\langle \mu \rangle (\mathbf{x})} \quad ; \quad \frac{1}{\langle \mu \rangle_\omega} = \frac{1}{\omega} \int d^2x \, \frac{1}{\langle \mu \rangle (\mathbf{x})} \; . \qquad (12.30b)$$

The integrals in (12.30b) extend over the solid angle ω. The probability density $p_\omega(\mu)$ is the magnification probability averaged over the solid angle ω, and $\langle \mu \rangle_\omega$ is the mean solid angle distortion due to lensing within the area spanned by ω. The factor $\langle \mu \rangle_\omega$ occurs since the solid angle from which the sources are taken is smaller than ω, due to the average area distortion.

In order to estimate an overdensity of QSOs in the vicinity of foreground galaxies, we must compare the source density in (12.30) with the mean QSO density. Since (12.30a) contains the intrinsic luminosity function of the sources, and the mean QSO density is obtained from source counts, we need to assume a relation between those two functions. This will be discussed in greater detail in Sect. 12.5; here, we make the *assumption* that *the average QSO density is not affected by the amplification bias*, i.e., that source counts (at fixed redshift) directly reflect the intrinsic luminosity function. The validity of this assumption is by no means obvious; in fact, further on we present some indications that QSO source counts *are* affected by the amplification bias. Our assumption is, however, the most conservative one, and we will see when one is forced to discard it.

Given our assumption, the ratio $r(S, z_s; \omega)$ (henceforth called *enhancement factor*) between the source density within the solid angle ω that includes the galaxy and the average source density is

$$r(S, z_s; \omega) = \frac{1}{\Phi_Q\left(\zeta(z_s)S, z_s\right) \langle \mu \rangle_\omega} \int_1^\infty d\mu \, p_\omega(\mu) \, \Phi_Q\left(\frac{\zeta(z_s)S, z_s}{\mu}\right) . (12.31)$$

Note that in all earlier expressions, we have not explicitly mentioned the dependence of $p_\omega(\mu)$ on the lens redshift z_d.

12.3.3 Maximal overdensity

If we now define

$$f(\mu) = \frac{1}{\langle \mu \rangle_\omega} \frac{1}{\Phi_Q\left(\zeta(z_s)S\right)} \Phi_Q\left(\frac{\zeta(z_s)S}{\mu}\right) \; ,$$

equation (12.31) for r is equivalent to (11.85), with $J(p)$ replaced by $r(p_\omega)$, and $\langle\mu\rangle$ of (11.86c) replaced by $\langle\mu\rangle_\omega$. Therefore, we can apply the theorem in Sect. 11.5 on maximal probabilities (or overdensities), provided that the luminosity function Φ_Q is chosen such that $f(\mu)$ satisfies the property (P) of Sect. 11.5.

We choose a luminosity function of the form

$$\Phi_Q\left(\zeta(z_s)S\right) = C \begin{cases} 1 - \beta/\alpha + \beta/\alpha\, s^{-\alpha} & \text{for } s < 1 \\ s^{-\beta} & \text{for } s > 1 \end{cases}, \qquad (12.32)$$

where $s = S/S_b$, and S_b corresponds to the break of the source counts, which occurs at $m_B \approx 19.2$.[6] For the high-flux regime, we choose the power-law slope of $\beta \approx 2.6$, and for the low-flux power-law index, we choose $\alpha \approx 0.8$; note that, for $\alpha < 1$, the function $f(\mu)$ satisfies $\lim_{\mu \to \infty}\left(\frac{f(\mu)}{\mu}\right) = 0$, and thus has the property (P) of Sect. 11.5. The constant of proportionality C cancels out, and is thus unimportant for the following considerations. The form of the luminosity function (12.32) was chosen so as to have a continuous first derivative at the break $s = 1$.

With (12.32), the function $f(\mu)$ becomes

$$f(\mu) = \begin{cases} \dfrac{1}{\langle\mu\rangle_\omega} \dfrac{1 - (\beta/\alpha) + (\beta/\alpha)\, s^{-\alpha}\mu^\alpha}{1 - (\beta/\alpha) + (\beta/\alpha)\, s^{-\alpha}} & \text{for } s \leq 1 \\[1em] \dfrac{\mu^\beta}{\langle\mu\rangle_\omega} & \text{for } 1 \leq \mu \leq s \\[1em] \dfrac{1}{\langle\mu\rangle_\omega} \dfrac{1 - (\beta/\alpha) + (\beta/\alpha)\, s^{-\alpha}\mu^\alpha}{s^{-\beta}} & \text{for } 1 \leq s \leq \mu \end{cases}$$

For $s \leq 1$, the function $f(\mu)$ is concave, i.e., $f'' \leq 0$ for all μ. Hence, the set B_2 as defined in Sect. 11.5 is empty, and the maximal probability distribution is $p_0(\mu) = \delta\left(\langle\mu\rangle_\omega - \mu\right)$. For $s > 1$, the set B_2 contains one element, with $\mu' = 1$ (since the smallest magnification that can occur is 1) and $\tilde{\mu} > 1$ such that

$$f'(\tilde{\mu}) = \frac{f(\tilde{\mu}) - f(1)}{\tilde{\mu} - 1}$$

(compare with the line L_3 in Fig. 11.17). This last equation is transcendental and can only be solved numerically for $\tilde{\mu}$. Then, for $s > 1$, the maximal probability distribution is a single delta-function $p_0(\mu) = \delta\left(\langle\mu\rangle_\omega - \mu\right)$ for $\tilde{\mu} \leq \langle\mu\rangle_\omega$, whereas for $\langle\mu\rangle_\omega < \tilde{\mu}$, it is

$$p_0(\mu) = \frac{\tilde{\mu} - \langle\mu\rangle_\omega}{\tilde{\mu} - 1}\delta(\mu - 1) + \frac{\langle\mu\rangle_\omega - 1}{\tilde{\mu} - 1}\delta(\mu - \tilde{\mu}) \quad .$$

Evaluating the density enhancement factor r (12.31) with the maximal probability p_0, we obtain for the maximal density enhancement r_m:

---

[6] This magnitude is much better determined for the redshift-integrated source counts than for the source counts at a fixed redshift; however, recent QSO samples [BO88.3] indicate that (12.32) is also a fair representation for QSO counts at fixed redshift.

$$r_{\mathrm{m}} = \begin{cases} \dfrac{1}{\langle\mu\rangle_\omega} \dfrac{1-(\beta/\alpha)+(\beta/\alpha)\,s^{-\alpha}\langle\mu\rangle_\omega^\alpha}{1-(\beta/\alpha)+(\beta/\alpha)\,s^{-\alpha}} & \text{for } s \le 1 \ ; \\[1ex] \dfrac{1}{\langle\mu\rangle_\omega} \left[\left(\langle\mu\rangle_\omega - 1\right) \beta\, s^{\beta-\alpha} \tilde\mu^{\alpha-1} + 1 \right] & \text{for } s \ge 1;\ \langle\mu\rangle_\omega \le \tilde\mu \ ; \\[1ex] \dfrac{s^\beta}{\langle\mu\rangle_\omega} \left(1 - \dfrac{\beta}{\alpha} + \dfrac{\beta}{\alpha} s^{-\alpha} \langle\mu\rangle_\omega^\alpha \right) & \text{for } s \ge 1;\ \langle\mu\rangle_\omega \ge \tilde\mu \ . \end{cases} \qquad (12.33)$$

Hence, under the assumption that the overall QSO counts are not affected by the amplification bias, r_{m} (12.33) is the largest density enhancement that can be caused by gravitational lensing. The only lens parameter that enters r_{m} is the mean magnification $\langle\mu\rangle_\omega$ in the solid angle of the lens plane where an overdensity is observed.

The determination of r_{m} in this particular example can be carried out fairly explicitly due to the analytic properties of the function $f(\mu)$. In general, r_{m} must be determined numerically. One method for this is suggested by the fact that the maximizing probability distribution is at most a sum of two delta-functions. Defining the function

$$\mathcal{R}\left(\mu', \tilde\mu\right) := \frac{\tilde\mu - \langle\mu\rangle}{\tilde\mu - \mu'} f(\mu') + \frac{\langle\mu\rangle - \mu'}{\tilde\mu - \mu'} f(\tilde\mu) \ ,$$

one finds r_{m} by maximizing $\mathcal{R}$,

$$r_{\mathrm{m}} = \max_{1 \le \mu' \le \langle\mu\rangle \le \tilde\mu} \mathcal{R}\left(\mu', \tilde\mu\right) \ .$$

In Fig. 12.6, we display r_{m} (solid curves) as a function of $\langle\mu\rangle_\omega$, for various values of $s = (S/S_{\mathrm{b}})$, parametrized such that $(S/S_{\mathrm{b}}) = 10^{-0.4\Delta m}$; hence, Δm is the magnitude difference between the flux threshold B_0 of the QSO sample under consideration and the break magnitude B_{b}. With $B_{\mathrm{b}} = 19.2$ and $B_0 = 18.7$ for the Webster et al. [WE88.2] sample, we find $\Delta m = -0.5$. For sufficiently large $\langle\mu\rangle_\omega$, large overdensities are possible, even for $\Delta m = -0.5$; however, for moderate $\langle\mu\rangle_\omega \lesssim 1.5$, the maximal density enhancement r_{m} is less than 2. Note that r_{m} is a very sensitive function of Δm.

Differentiation of (12.33) yields the result that the function $r_{\mathrm{m}}(\langle\mu\rangle_\omega)$ has a maximum at

$$\langle\mu\rangle_\omega = s \left[\frac{\beta - \alpha}{\beta(1-\alpha)}\right]^{1/\alpha}$$

which corresponds to the value

$$r_{\max} = s^{\beta-1} \beta^{1/\alpha} \left(\frac{1-\alpha}{\beta-\alpha}\right)^{(1/\alpha - 1)} \ .$$

Therefore, $r_{\max}$ is an upper limit to the density enhancement which *is completely independent of the lens properties*. In other words, if one finds a region in the sky where the density enhancement is significantly larger than $r_{\max}$, then either (i) the flux threshold of the sample (i.e., S/S_{b}) is incorrectly determined; (ii) the uncertainties in the overall source counts which lead to the QSO luminosity function are larger than expected; or (iii) our assumption that the overall source counts are unaffected by the amplification bias is wrong. Thus, the existence of an upper limit $r_{\max}$ for the density

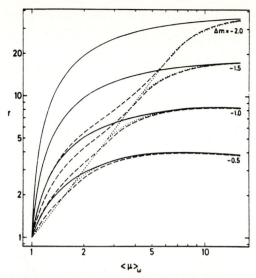

Fig. 12.6. The density enhancement factor r of galaxies around QSOs, as a function of the mean magnification $\langle \mu \rangle_\omega$ within the solid angle which has been investigated for excess galaxies. Four families of curves are drawn, for different values of $\Delta m = B_0 - B_b$, the difference between the flux threshold B_0 of the sample and the break B_b of the source counts (12.32), with $\alpha = 0.8$ and $\beta = 2.6$. The solid curves show the upper limit of the density enhancement r_m (12.33) [KO89.2]. The dotted curves result from assuming the matter in the galaxies to be homogeneously distributed [NA89.1], whereas the dashed curves represent a simplified version of the ML model in [SC89.2]. For small $|\Delta m|$, all three curves agree very well, i.e., the constraint of flux conservation near the break basically determines r. For larger $|\Delta m|$, the upper limit is considerably larger than the two specific model curves, particularly for reasonable values of $\langle \mu \rangle_\omega$. For large values of $\langle \mu \rangle_\omega$, all three curves agree. Note that for $\Delta m = -0.5$, $r \lesssim 4$, independent of $\langle \mu \rangle_\omega$

enhancement which is independent of any lens model may prove that the QSO counts are affected by the amplification bias. Note that the density enhancement of ~ 4 reported in [WE88.2] is already close to $r_{\max}$ for this sample, although the total number of objects in the sample is too small to consider their number to be statistically significant. Before we discuss further the observed overdensity of galaxies around high-redshift QSOs, we consider specific lens models, to estimate more realistic values of the overdensity.

12.3.4 Lens models

General models. We now derive a set of equations to determine the density enhancement factor r of QSOs in the solid angle ω for a general matter distribution $\kappa(\mathbf{x})$ of a galaxy, with a fraction ϵ contained in compact objects. We assume that QSOs can be considered as point sources. From the matter distribution $\kappa(\mathbf{x})$, we can determine the corresponding distribution $\gamma(\mathbf{x})$ of the shear, and the mean magnification $\langle \mu \rangle (\mathbf{x}) = \left[(1-\kappa)^2 + \gamma^2\right]^{-1}$. The asymptotic behavior of the local magnification probability $p(\mathbf{x}, \mu)$ is then determined from (11.64). We assume that a sufficiently good approximation for $p(\mathbf{x}, \mu)$ is given by

$$p(\mathbf{x}, \mu) = \begin{cases} \dfrac{q(\mathbf{x})}{\mu^3} & \text{for } \mu > \langle \mu \rangle (\mathbf{x}) \\ p_\mathrm{L}(\mathbf{x}) \delta \left(\mu - \mu_\mathrm{L}(\mathbf{x}) \right) & \text{for } \mu \leq \langle \mu \rangle (\mathbf{x}) \end{cases}, \qquad (12.34a)$$

where $q(\mathbf{x})$ is given by (11.64b), with $\kappa_* = \epsilon \kappa(\mathbf{x})$, $\kappa_\mathrm{c} = (1 - \epsilon) \kappa(\mathbf{x})$, and $\gamma = \gamma(\mathbf{x})$, and $p_\mathrm{L}(\mathbf{x})$ and $\mu_\mathrm{L}(\mathbf{x})$ are determined from normalization $[\int p(\mathbf{x}, \mu) \, d\mu = 1]$ and flux conservation $[\int p(\mathbf{x}, \mu) \mu \, d\mu = \langle \mu \rangle (\mathbf{x})]$,

$$p_\mathrm{L} = 1 - \frac{q}{2 \langle \mu \rangle^2} \quad ; \quad \mu_\mathrm{L} = \frac{2 \langle \mu \rangle \left(\langle \mu \rangle^2 - q \right)}{2 \langle \mu \rangle^2 - q} \quad . \qquad (12.34b)$$

Note that, for $\epsilon = 0$, we have $\kappa_* = 0$, $q = 0$, and $p(\mathbf{x}, \mu) = \delta \left(\mu - \langle \mu \rangle (\mathbf{x}) \right)$.

Uniform disc. Consider the special case where the region of the lens under consideration is represented as a uniform disc, with surface mass density $\kappa(\mathbf{x}) = \kappa$, and with a fraction ϵ contained in compact objects. Then, $\kappa_*(\mathbf{x}) = \epsilon \kappa$, $\kappa_\mathrm{c}(\mathbf{x}) = (1 - \epsilon) \kappa$, and the shear vanishes, $\gamma(\mathbf{x}) = 0$. The probability distribution $p(\mathbf{x}, \mu)$ does not depend on $\mathbf{x}$, so $p_\omega(\mu)$ is given by (12.34), with

$$q = \frac{2 \epsilon \kappa}{(1 - \kappa)^2} \left\{ [1 - (1 - \epsilon) \kappa]^2 + (\epsilon \kappa)^2 \right\}^{-3/2} \quad .$$

Using this probability distribution and (12.32), we can calculate the density enhancement r as a function of S. Although the resulting expression can be written in explicit form, it is not very informative. We have calculated r for $\epsilon = 0$ and 1, as shown in Fig. 12.6 (dotted and dashed curves, respectively), as a function of $\langle \mu \rangle_\omega = (1 - \kappa)^{-2}$, for several values of Δm. For $\langle \mu \rangle_\omega \gg (S/S_\mathrm{b}) \equiv s$ both values of r agree very well with r_m, whereas for small values of $\langle \mu \rangle_\omega$ both the smooth disc ($\epsilon = 0$) and the ML model $\epsilon = 1$ yield values of r smaller than r_m. In particular, for large s and moderate values of $\langle \mu \rangle_\omega$ the uniform disc model yields *considerably* lower values of r than the maximal value r_m.

Isothermal sphere with $\epsilon = 0$. If the solid angle probed for an overdensity of QSOs is an annular region of an isothermal sphere, between angular separations x_1 and x_2 from the center of the galaxy, we have (compare Sect. 8.1.4): $\kappa(x) = \alpha_0/(2x)$, $\gamma(x) = \alpha_0/(2x)$, with $x = |\mathbf{x}|$, and

$$\alpha_0 = 4\pi \left(\frac{\sigma_v}{c} \right)^2 \left(\frac{D_\mathrm{ds}}{D_\mathrm{s}} \right) \quad .$$

To avoid the region where multiple images can occur, we choose $x_1 > 2\alpha_0$. In fact, we will concentrate on the case $x_1 \gg \alpha_0$; whether the isothermal sphere has a singular mass density at its center or a finite core size is unimportant in this case. The magnification probability is $p(x, \mu) = \delta \left(\mu - \langle \mu \rangle (x) \right)$, with $\langle \mu \rangle (x) = x/(x - \alpha_0)$. Then, from (12.30b),

$$\langle\mu\rangle_\omega = \frac{x_1 + x_2}{x_1 + x_2 - 2\alpha_0} \quad , \qquad (12.35a)$$

and

$$p_\omega(\mu) = \frac{\langle\mu\rangle_\omega}{(x_2^2 - x_1^2)\pi} 2\pi \int_{x_1}^{x_2} dx\, x\, \delta\left(\mu - \frac{x}{x - \alpha_0}\right)\left(\frac{x - \alpha_0}{x}\right) \quad ,$$

which can be easily integrated:

$$p_\omega(\mu) = \frac{2\alpha_0^2\, \langle\mu\rangle_\omega}{(x_2^2 - x_1^2)} \frac{1}{(\mu - 1)^3} \quad \text{for} \quad \mu_2 \leq \mu \leq \mu_1 \quad , \qquad (12.35b)$$

with $\mu_i = x_i/(x_i - \alpha_0)$. From (12.31) and (12.32), the density enhancement r can now be calculated. For the general case, however, the resulting expressions are complicated; therefore, we concentrate on the special case $x_1 \gg \alpha_0$; then, the magnifications $\mu \leq \mu_1$ are close to 1. It can then be expected that r will also be very close to 1. In fact, for $s \geq \mu_1$ we have

$$r = \frac{2\alpha_0^2}{(x_2^2 - x_1^2)} \int_{\mu_2}^{\mu_1} d\mu\, \frac{\mu^\beta}{(\mu - 1)^3} \quad .$$

Given our preceding assumptions, $t := \mu - 1 \ll 1$, and we can therefore approximate $\mu^\beta \approx 1 + \beta t$. The resulting integration becomes trivial and yields:

$$r = 1 + (\beta - 1)\frac{2\alpha_0}{(x_1 + x_2)} \quad . \qquad (12.36)$$

We first note that r does not depend on s: since we have assumed that $s \geq \mu_1$, the magnification is nowhere sufficiently large to probe the luminosity function beyond the turn-over at $s = 1$. Thus, for our case, the magnification integral in (12.31) extends only over the power-law part of Φ_Q. Since a power-law has no inherent scale, it is clear that r must be independent of s.

The fact that $r - 1$ is proportional to $\beta - 1$ is due to the approximation which has been made; however, for a pure power-law luminosity function, the sign of $r - 1$ is always the same as that of $\beta - 1$. In particular, for a power-law with $\beta = 1$, the density enhancement is $r = 1$, independent of the lens model (as follows immediately from flux conservation). If $x_1 \ll x_2$, r is nearly independent of the inner radius x_1 of the annulus and varies as $1/x_2$.

Isothermal sphere with $\epsilon \neq 0$. If we now include ML, we can proceed by evaluating p_ω from (12.30) and (12.34). In general, this leads to a very complicated expression for r; therefore, we concentrate again on $x_1 \gg \alpha_0$, which implies $\kappa(x) \ll 1$ in the annulus. Thus, the values of q in (12.34a) are small, and we can utilize the linear approximation, so that $p(x, \mu)$ is given, for $\mu > \mu_0$, by

$$p(x,\mu) = \frac{\epsilon\,\alpha_0}{x - \alpha_0}\frac{1}{\mu^3} \;,$$

and μ_0 is a threshold above which the power-law behavior $p \propto \mu^{-3}$ is assumed to be valid. The value of μ_0 will be discussed later. We then set

$$p_\omega(\mu) = \begin{cases} C\mu^{-3} & \text{for } \mu > \mu_0 \\ p_L \delta(\mu - \mu_L) & \text{for } \mu \leq \mu_0 \end{cases} \;,$$

where C is obtained from (12.30b):

$$C = \epsilon\,\frac{2\,\alpha_0\,\langle\mu\rangle_\omega}{(x_1 + x_2)} \equiv \epsilon\,t \;,$$

and p_L and μ_L are obtained from normalization and flux conservation,

$$p_L = 1 - \frac{C}{2\mu_0^2} \;;\quad \mu_L = \frac{1}{p_L}\left(\langle\mu\rangle_\omega - \frac{C}{\mu_0}\right) \;.$$

Inserting the probability distribution $p_\omega(\mu)$ into (12.31) and using the luminosity function (12.32), we obtain, for $s \geq \mu_0$,

$$r = \frac{1}{\langle\mu\rangle_\omega}\left(p_L\mu_L^\beta - \frac{C}{\beta-2}\mu_0^{\beta-2}\right) + \frac{C}{\langle\mu\rangle_\omega}s^{\beta-2}D(\alpha,\beta) \;,$$

where

$$D(\alpha,\beta) = \frac{1}{\beta-2} + \frac{1}{2}\left(1 - \frac{\beta}{\alpha}\right) + \frac{\beta}{\alpha}\frac{1}{2-\alpha} \;.$$

Since $D(\alpha,\alpha) = 0$, r is independent of the flux s for a pure power-law luminosity function, as must be the case from the argument given above.

To obtain a more meaningful form for r, we expand r up to first order in the small variable t. Note that $\langle\mu\rangle_\omega = 1 + t$, as can be seen from (12.35a), since the mean magnification $\langle\mu\rangle_\omega$ in the annulus cannot depend on ϵ. Letting $\mu_0 = 1 + t_0$, and considering t_0 to be a small parameter, comparable to t, we have:

$$r - 1 = t\left[(\beta - 1) + \epsilon\left(D(\alpha,\beta)\,s^{\beta-2} - \frac{1+\beta}{2} - \frac{1}{\beta-2}\right)\right] \;. \tag{12.37}$$

The first useful result to note is that, in this approximation, r is independent of t_0, showing that r does not depend (to first order in t) on the arbitrary value of μ_0. Furthermore, for $\alpha = \beta = 1$, the density enhancement factor is $r = 1$, as also follows immediately from flux conservation. Third, for $\epsilon = 0$, we reobtain the result (12.36). Finally, $r - 1$ is proportional to the inverse of the mean radius of the annulus, measured in units of α_0. Note also that $t = 2\bar{\kappa}$, where $\bar{\kappa}$ is the mean surface mass density in the annulus. In fact, one can easily show (with the same method as used in this subsection) that the result (12.37) is valid for any mass distribution, provided that $\kappa(\mathbf{x}) \ll 1$ in the region under consideration. If we specify, according to observations, $\alpha = 0.8$ and $\beta = 2.6$, (12.37) becomes

$$r - 1 = t \left[1.6 + \epsilon \left(3.25\, s^{0.6} - \frac{2}{15} \right) \right] \; .$$

This equation shows that the influence of ML on the overdensity is small for flux-limited samples with a flux threshold near the break in the luminosity function. However, for higher flux thresholds, ML becomes the dominant effect. The effect is larger for higher values of α and β, i.e., for steeper luminosity functions.

12.3.5 Relation to observations

Overdensity of QSOs around nearby galaxies: Statistical significance. We first discuss the observations of an apparent overdensity of QSOs in the vicinity of low-redshift galaxies, mentioned in Sect. 12.3.1, from the perspective of gravitational lensing, following [VI83.1]. Since $z_d \ll 1$, we can set $D_{ds}/D_s \approx 1$; in addition, since the lensing effects become nearly independent of the source redshift, we can replace $\Phi_Q(\zeta(z_s)S)$ by $n_0(S)$ in the preceding subsections, where $n_0(S)$ is the number density of QSOs per unit solid angle with flux greater than S, i.e., the redshift-integrated source counts. The typical angular separation between galaxy and QSOs in the associations reported in [AR87.1] is about 2 arcmin (see, e.g., [AR84.1]); if we set $x_1 \ll x_2 \equiv \theta$, the mean surface mass density in the annulus becomes:

$$\bar{\kappa} = 4\pi \left(\frac{\sigma_v}{c} \right)^2 \frac{1}{\theta} = 4.33 \times 10^{-2} \left(\frac{\theta}{1\,\mathrm{arcmin}} \right)^{-1} \left(\frac{\sigma_v}{300\,\mathrm{km/s}} \right)^2 \; .$$

Suppose we searched for QSOs within an angle θ of N_g galaxies; this corresponds to a total solid angle of $\pi \theta^2 N_g$ on the sky, where, away from all galaxies,

$$N_0(S) = \pi \theta^2 N_g n_0(S)$$

sources brighter than S will be counted. In the area surrounding the galaxies, the corresponding number is $r N_0(S)$, i.e., the number excess of QSOs is $\Delta N = (r-1) N_0(S)$. For such an excess to be observable, ΔN must be larger than the statistical fluctuations $\sqrt{N_0}$, $\Delta N > \sqrt{N_0}$, or $(r-1)\sqrt{N_0} > 1$. With the normalization $n_0(S = S_0) \approx 6\,\mathrm{deg}^{-2}$, we obtain from (12.37):

$$N_g > 2.5 \times 10^4\, \sigma_{300}^{-4}\, \frac{s^{2.6}}{\left[1.6 + \epsilon \left(3.25\, s^{0.6} - 2/15 \right) \right]^2} \; , \qquad (12.38)$$

where $\sigma_v = 300\,\mathrm{km/s} \times \sigma_{300}$. Note that this equation is valid only for $s \gtrsim 1$. Further, note that the limit on N_g is independent of θ. If $\epsilon = 0$ (no ML), (12.38) yields $N_g > (0.98, 388)\sigma_{300} \times 10^4$ for $s = (1, 10)$, and for $\epsilon =$

1, $N_{\rm g} > (0.11, 4.7)\sigma_{300} \times 10^4$.[7] Since a statistically-significant observation means $\Delta N \gtrsim 3\sqrt{N_0}$, one needs to search the vicinities of about 10^4 nearby galaxies for QSOs, up to about 19.5th magnitude, to detect any effect, for the most favorable case, $\epsilon = 1$. Of course, no such program has yet been carried out (and it would be a considerable effort to do so in the future). We thus conclude that the observed associations of QSOs with low-redshift galaxies [AR87.1] cannot be explained by ML in the halos of foreground galaxies. This conclusion was reached in all those papers ([CA81.1], [VI83.1], [ZU85.1], [PE86.1], [SC86.3], [SC87.5]) that assumed that the overall source counts are not affected by lensing. The basic reason is that, although the fractional overdensity is large for bright sources, there are too few of them to lead to statistically significant results. Although the statistics are better for fainter sources, the fractional overdensity is small for $S \gtrsim S_0$.

Modifications of the model. The preceding conclusion does not depend on details of the lens models – the discrepancy is too large between the overdensity claimed [AR87.1] and that obtained from a lensing model. A few modifications have been tried, such as extended sources [ZU85.1], [SC87.5], but the changes in the results are very modest and, in any case, reduce the overdensity relative to the point-source model.

A further modification was considered in [LI88.2], where the assumption that the overall source counts are not significantly affected by the amplification bias was abandoned. We discuss this idea briefly in Sect. 12.5.3.

Overdensity of galaxies near high-redshift QSOs. If we recall the basic reason for the failure of explaining the overdensity of QSOs around nearby galaxies by lensing, we might ask whether it is possible that lensing can explain the related observations mentioned before, i.e., to cause an overdensity of galaxies near QSOs. Note that this is a different situation, at least with respect to observations, for essentially two reasons: galaxies are numerous, so the statistical limitations are much smaller than for QSOs, and second, galaxies are much more readily identified than QSOs, by their morphology (at least if their redshift is sufficiently small), whereas QSO identifications must be verified by time-consuming spectroscopy. As described in Sect. 12.3.1, an overdensity of galaxies near high-redshift QSOs has been observed. We consider the results of Webster et al. [WE88.2] in somewhat more detail here, as this sample is statistically well-defined, and some of the associated galaxies have been observed spectroscopically to verify that they are truly in the foreground (R. Webster, personal communication, and [WE90.1]).

For a flux-limited and statistically complete sample of QSOs and galaxies, the overdensity of galaxies around background QSOs is the same as the overdensity of QSOs around foreground galaxies. In [WE88.2] the density enhancement of galaxies around high-redshift QSOs was claimed to be $r \approx 4$,

---

[7] Note that, although the enhancement factor r increases with s, the bound on $N_{\rm g}$ also increases with s, since bright QSOs are so rare.

although an extension of the sample [WE90.1] yields a somewhat lower value of ≈ 3. In the notation of Sect. 12.3.4, the associations extend from $x_1 = 3''$ to $x_2 = 6''$. If we set $D_{\mathrm{ds}}/D_{\mathrm{s}} \approx 1$, we find that $\langle \mu \rangle_\omega = (1.04, 1.23)$ for $\sigma_v = (200, 300)\,\mathrm{km/s}$. It can then be seen from Fig. 12.6 that even the maximum density enhancement r_m is $\lesssim 1.5$ for $\Delta m \approx -0.5$, whereas the prediction from specific models is even smaller: both a smooth uniform disc model [NA89.1] and a model with ML yield smaller values for r. Hence, although the discrepancy between the observed associations of galaxies and QSOs and GL models appears to be smaller for the Webster et al. sample than for those where QSOs are seen close to bright foreground galaxies [CA88.1], lensing cannot easily explain the associations found in [WE88.2], as pointed out in [HO89.1].

A more detailed analysis of the sample in [WE88.2] proceeds as follows [SC89.2]: consider a galaxy, and QSOs at redshifts much larger than the galaxy's. It is possible to calculate the number of QSOs, at given redshift, seen around this galaxy within a given range of angular separation from its center. Since the observations in [WE88.2] indicate an overdensity for very small angular separations, the mean optical depth can no longer be assumed to be small. Therefore, the amplitude of the μ^{-3}-tail of the local magnification probability distribution for point sources should not be approximated by the linear expression 2κ, but by the fully non-linear result (11.65).

By integrating over the galaxy distribution, described by a Schechter law (12.18a) and the Faber–Jackson relation (12.18b), we obtain the total number of QSOs at fixed redshift, brighter than a limiting magnitude m_Q, which are seen within a certain angular separation from a galaxy brighter than m_G. We obtain the fraction of QSOs close to a nearby galaxy, by dividing this number by the total number of QSOs brighter than m_Q. Let this fraction be denoted by $q(\theta)$, where θ is the angular separation of the galaxy from the QSO. On the other hand, given the distribution of galaxies, we can compute the total number of galaxies brighter than m_G; multiplication of this number by $\pi\theta^2$ yields the fraction q_0 of the sky which lies within θ of a galaxy brighter than m_G. If $q > q_0$, galaxies around QSOs are overdense. One can estimate the number of QSOs that must be observed in order to find a statistically significant excess of galaxies. Suppose we observe N_Q QSOs and finds $N_\mathrm{Q} q(\theta)$ galaxies near them; without lensing, one would find $N_\mathrm{Q} q_0$. The excess $\Delta N = (q - q_0) N_\mathrm{Q}$ must be larger than the statistical fluctuations $\sqrt{q_0 N_\mathrm{Q}}$, which yields

$$N_\mathrm{Q} > \frac{q_0}{(q - q_0)^2} \ .$$

Using the parameters of the sample in [WE88.2], this limit becomes $N_\mathrm{Q} \gtrsim 40$, whereas ≈ 200 QSOs are observed in [WE88.2]. For details, the reader is referred to [SC89.2].

The fractional overdensities calculated in [SC89.2] are smaller than those observed in [WE88.2], in accordance with the preceding considerations. Several factors influence the result, e.g., the assumed value of the velocity dispersion σ_{v*} of an L_* galaxy and the assumed slope α of the faint end of the QSO source counts. Another effect which was neglected is the clustering of galaxies: owing to clustering, finding a second galaxy near a QSO is not independent from the fact that one is already there. This difficulty can be avoided if associations are only counted once, independent of how many galaxies are seen within θ of a QSO. Note that the angle θ within which galaxies are searched for is relatively large in [TY86.3] and [FU88.1], so that in these two samples there is sometimes more than one galaxy around a high-redshift QSO, whereas in [WE88.2], $q(\theta) \ll 1$. However, the following effect of galaxy clustering is more severe: the method outlined above starts with the calculation of the QSO number in the background of galaxies brighter than m_G. If a QSO is magnified by compact objects in a galaxy fainter than m_G, the galaxy itself is not detected. However, owing to clustering, the density of sufficiently bright galaxies around this faint one is larger than the average density. In other words, if there is a group of galaxies with different luminosities, only those QSOs are included in the calculation which are magnified by the brighter galaxies, whereas it may well be that a faint galaxy is the main lens, but because of clustering, different, brighter galaxies are seen near the QSO. It is clear that this effect leads to an underestimate of the excess of galaxies near QSOs, which may partly be the reason for the discrepancy between the simple model and the observations. Furthermore, an additional magnification effect can be due to intracluster matter.

However, as we have seen, there is a maximum value of r that can arise from lensing. Since the observed value is greater than r_m for reasonable values of $\langle\mu\rangle_\omega$, we conclude that *the likely reason for the discrepancy between the observed overdensity and the lensing models is the basic assumption in these models, i.e., that the overall source counts of QSOs are unaffected by amplification bias*. In fact, if further extensions of the observations confirm the claims for the overdensity, we consider this as a *strong indication that the QSO source counts are affected by the amplification bias*.

Overdensity of galaxies around high-redshift flat-spectrum radio QSOs. Correlating the Lick catalogue of galaxies (which provides galaxy counts in $10' \times 10'$ cells, brighter than $m_{pg} = 18.9$) with the position of strong $[S(5\,\text{GHz}) \geq 1\,\text{Jansky}]$, high-redshift ($z \geq 1$) radio QSOs, a highly significant association is obtained (see [FU90.1]; we also refer the reader to this paper for a discussion of earlier studies on this subject and for an account on the associated difficulties). Both, the QSO and the galaxy sample, are unbiased (with the exception that only those radio QSOs from the 1 Jy catalogue have been used whose redshift is known, which excludes preferentially the optically faintest sources), which makes this study much more useful than corresponding ones which use all QSOs in available catalogues, which they have entered for a variety of reasons. Since the galaxies in the Lick catalogue are relatively bright, it is highly unlikely that they have redshifts larger than, say, $z = 0.2$, and the majority of them should be much closer.

This investigation is of large interest for several reasons: it provides independent evidence for the association of high-redshift objects with low-redshift galaxies. Second, no corresponding association between optical QSOs and the Lick galaxies are found in [FU90.1] (as expected from the results in [WE88.2] and the low angular resolution of the Lick catalogue), which means that the association is stronger for radio-loud QSOs than for radio-quiet QSOs – which has been suspected before (see the discussion in Sect. 12.3.1). Third, since an excess galaxy is found in the cells of the high-redshift radio-loud QSOs in about 40% of all cases, the source sample as a whole cannot be unaffected by the amplification bias, if one takes into account that the galaxies in the catalogue are all relatively bright and thus at low redshift.

The fact that strong radio QSOs are apparently associated with foreground galaxies at a higher degree than optically-selected QSOs does not have a simple explanation yet. One possibility would be that the (radio) luminosity function of these sources is steeper at the bright end than that of the optically selected QSOs. A second possibility is "double biasing": if the sample is selected by both its radio flux and its optical flux (the latter being at least partly the case, since the QSO must be sufficiently bright optically to allow a determination of its redshift), the resulting amplification bias is stronger than for samples which are only selected in one spectral band (S. Refsdal, private communication), unless optical and radio luminosity are strongly correlated (which does not seem to be the case). Finally, it is not at all clear that the theoretical treatment in Sect. 12.3.2 applies here; due to the low "angular resolution" of the Lick catalogue, the observed correlation is not necessarily between the QSO and one individual galaxy, but can also be between QSO and groups of galaxies. Therefore, the amplification bias can be caused by larger-scale matter inhomogeneities, and not by individual galaxies themselves, which may only be luminous tracers of the larger-scale matter aggregates.

12.4 Microlensing: Astrophysical discussion

The term *microlensing* (ML) arose from the fact that a single star of about solar mass can lead to images split into 'microimages' separated by microarcseconds. Multiple images due to stellar-mass objects can therefore not be resolved; the only observable effect ML has is the alteration of the apparent luminosity of background sources. The relevant length scale for ML is given by η_0 (2.6c); only sources smaller than η_0 can be significantly affected by ML.

The importance of ML was first pointed out in [CH79.1] and [GO81.1]. The images of a multiply imaged QSO are seen through a galaxy; since galaxies contain stars, stellar-mass objects can affect the brightness of these images. The simplest ML model is a single point-mass lens. In [CH79.1] the point-mass lens with external shear (see Sect. 8.2.2) was developed and later investigated in more detail [CH84.2]; the shear is due to the galaxy, save for the star. The next more complicated model is the two point-mass lens, described briefly in Sect. 8.3. Although more complicated models could be considered in principle, as we have already stated, the number of parameters becomes very large, and the prediction based on a ML model becomes a statistical problem.

When we consider lensing by stars in a galaxy, we are confronted with a typical two-length-scale problem: the length-scale associated with the microlenses is ξ_0 (2.6b), and thus several orders-of-magnitude smaller than the length-scale on which the density of a galaxy varies significantly. Therefore, one can consider the distribution of stars to be locally homogeneous, and the corresponding lens model is that of a random star field. The assumption of randomness in the distribution of stars is not necessarily valid, as stars are not created in isolation. However, the distribution relevant to lensing is that projected on a plane perpendicular to the line-of-sight to a galaxy, and this two-dimensional distribution is more nearly random. Note that double stars with separation smaller than about 100 AU can be considered as a single mass for most lens purposes. We described the lensing properties of random star fields in detail in Sect. 11.2. Here, we want to concentrate on the observable effects of ML.

How to detect microlensing. Since ML is observable only through its effects on the apparent brightness of sources, the methods to detect ML are based on a comparison of flux measurements of at least one GL image. If these measurements are separated in time, so that the light curve of a source is well determined, one can detect variations of the flux caused by ML. However, it is very difficult to separate the intrinsic variability of a source from lensing effects; the only hope is to find features in the light curve characteristic of ML events. If the flux measurements are obtained from two different locations, the relative positions of source, lens, and observer are slightly different for the two measurements. ML could then be identified if the fluxes were different (parallax method, see [GR86.1]). The separation of the two telescopes must be of the order ξ_0 ($\gtrsim 100$ AU), except for 2237+0305, where the lens is nearby, so a baseline of 10 AU would suffice. If the flux measurements are separated in angle, i.e., if a galaxy-mass GL produces multiple images of a source, and the light curves of the images can be obtained, intrinsic variability should appear in all images, separated by the relative time delays. Thus, if variations are observed in one image, but are not matched by variations in other images within a time corresponding to the time delays, one has a clear signature of ML. As will be described below, this method has revealed likely ML in 2237+0305.

A further possibility for detecting ML is provided by the fact that stellar mass objects can magnify only sufficiently compact sources. Thus, if the flux ratios in two spectral bands in two or more images of a multiply-imaged QSO are different, this could be a sign of ML. For example, the broad-line region of QSOs is assumed to be too large to be subject to appreciable ML; so if the line-flux ratio of two images differs from the ratio of optical continuum flux, the latter is likely to be affected by ML in at least one of the images. Another source component useful in this respect is a compact radio core of the QSO. In fact, in the GL system 0957+561, one suspects ML effects from the different flux ratios in optical continuum light, broad-line flux, and in the compact radio component.

12.4.1 Lens-induced variability

One immediate consequence of ML is the expected variation of the flux of lensed sources due to the change of the relative positions of the source, the lens, and the observer with time. For example, a star in a galaxy will have a transverse velocity which is the sum of its peculiar velocity in the galaxy, and the motion of the galaxy in its group or cluster. Furthermore, the source (or its components) can also move. If a large fraction of QSOs are indeed substantially magnified by ML, these variations should be observable. If $\mathbf{v}_o$, $\mathbf{v}_d$, and $\mathbf{v}_s$ are the velocity vectors perpendicular to the optical axis of observer, lens, and source, respectively,[8] the effective transverse velocity vector is [KA86.2]:

$$\mathbf{v} = \frac{1}{1+z_s}\mathbf{v}_s - \frac{1}{1+z_d}\frac{D_s}{D_d}\mathbf{v}_d + \frac{1}{1+z_d}\frac{D_{ds}}{D_d}\mathbf{v}_o \quad , \qquad (12.39a)$$

where this effective velocity is to be understood as follows: for a stationary observer and lens, the position of the source will change according to $\Delta\boldsymbol{\eta} = \mathbf{v}_s \Delta t'$, where $\Delta t'$ is the time measured at the redshift of the source. Due to the cosmic expansion, this time interval is related to that measured by the observer, Δt, through $\Delta t = \Delta t'(1+z_s)$. Thus, the change of the source position in time is $\Delta\boldsymbol{\eta} = \mathbf{v}\,\Delta t$. This equation is also the defining equation for $\mathbf{v}$, in the general case where the observer and the lens are moving.

Variability due to a single star. To find out what kind of variability we may expect, we consider two cases: lensing by an isolated compact object, and lensing by a star field, which is appropriate if a light bundle crosses a galaxy at a position where the density κ_* of microlenses is larger than, say, 0.1. In the first case, the shape of the light curve is determined by a small number of parameters, namely the mass M of the compact object, the radius ρ of the source, the minimum impact parameter η_m of the source's path with respect to the optical axis, and the effective velocity v (we can assume that the source's path is locally a straight line). If $t=0$ denotes the observer's time when the source is closest to the optical axis, its impact parameter as a function of time is $\eta(t) = \sqrt{\eta_m^2 + (vt)^2}$. Conversion to dimensionless

---

[8] These three two-dimensional velocity vectors are defined as the corresponding projections of the three-dimensional velocities onto the plane perpendicular to the optical axis, the line connecting lens and observer.

quantities, using the scaling appropriate for the Schwarzschild lens, yields $y(t) = \sqrt{y_m^2 + \lambda^2}$, where $\lambda = vt/\eta_0$ is a measure of time. In useful units,

$$\lambda = 10^{-2} \left(\frac{v}{300 \,\text{km/s}}\right) \left(\frac{M}{M_\odot}\right)^{-1/2} \left(\frac{r(z_d)}{r(z_s)\,r(z_d, z_s)}\right)^{1/2} \left(\frac{t}{\text{years}}\right). \tag{12.39b}$$

Thus, the magnification as a function of time is $\mu_e = \mu_e(y(t), R)$, (11.12), where R is the source size in units of η_0 (2.6c). It is now easy to determine model light curves from the foregoing equation; we consider here the special case of high magnifications, where the function $\mu_e(y, R)$ can be approximated by (11.15). This yields[9]

$$R\,\mu_e = \zeta\left(\sqrt{\left(\frac{y_m}{R}\right)^2 + \left(\frac{\lambda}{R}\right)^2}\right)$$

In Fig. 12.7, we display $R\mu_e$ as a function of λ/R for several values of y_m/R. Note that $\lambda/R = vt/\rho$; hence, the time is measured in units of the time it takes the source to travel a distance equal to its own radius. From the figure, we see that, for $y_m/R \lesssim 1$, the width of the "light curves" is approximately $2\lambda/R$; hence, the shortest time scale for a source to vary is the time it moves by its own radius. For $y_m/R \gtrsim 1$, the curves are much broader. Using the approximation $\zeta(d) \approx 2/d$ for large arguments,

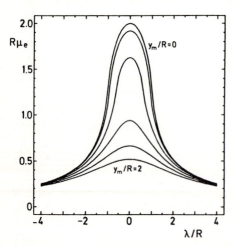

Fig. 12.7. The product $R\mu_e$ of the dimensionless source size and the magnification as a function of the dimensionless time λ scaled by the dimensionless source radius, for several values of y_m/R, the ratio of the minimum impact parameter and the source size. The curves correspond to $y_m/R = 0$ (the upper curve), 0.4, 0.8, 1.2, 1.6, 2.0 (the lower curve). The results are valid for $\mu_e \gg 1$

---

[9] We assume a source with uniform surface brightness.

we find that the width of the curves is roughly $\Delta\lambda \approx y_m$. In the same approximation, the maximum magnification of the source on its path is $\mu_{max} \approx 2/y_m$, so the time scale for variation is approximately $\Delta\lambda \approx 2/\mu_{max}$, valid for $\mu_{max} \lesssim 1/R$. Roughly speaking, the variability time scales we may expect range from the time for a source to travel a distance equal to its own radius to about $\Delta\lambda \approx 1/\mu_e$. If the effective velocity is in the range of 1000 km/s, and the source size is of the order of one light-day ($\approx 3 \times 10^{14}$ cm), the minimum variability time scale is of the order of one year.

Variability due to a lens ensemble. The expected variability of sources lensed by a star field can be obtained directly from the simulations shown in Fig. 11.7; the magnification along a track through these fields yields a model light curve. Smoothing over several pixels, the light curves of more extended sources can be obtained. In Fig. 12.8 we show several sample light curves. The units on the abscissa are again in terms of λ, (12.39b).

Consider first the case without external shear γ, and without smooth surface mass density κ_c. For relatively small optical depth κ_*, the light curves are flat most of the time (i.e., there is only small magnification), interrupted by short phases of rapid variability. For higher optical depth, the phases of variability are much longer; in fact, quiet phases and active phases are rather well separated. This is due to the clustering of caustics, as discussed in Sect. 11.2.5. For near-critical optical depths, $\kappa_* \approx 1$, there are hardly any large changes in magnification, and the light curves are better described as flickering.

For low optical depth, single, closed caustic lines are identifiable (see Fig. 11.7a). If a sufficiently small source crosses such a caustic, its light curve will have a characteristic U-shape, as can be seen in Fig. 12.8: outside the caustic the magnification is small, but it is high at the two points where the source crosses the caustic line, and has a relative minimum between these two points. In order for this U-shape to occur, the source must be much smaller than the typical linear size across a closed caustic. For higher optical depth, caustics start to intersect, and the network of caustics becomes complex. Therefore, no 'standard', simple feature can be identified in the corresponding light curves.

As mentioned before, for optical depths close to unity (cases C and D) the variations are small in amplitude and, for the small sources, fairly rapid. The largest amplitudes occur for the overfocused cases (E, I, J), as can also be easily seen from the corresponding magnification patterns.

If an external shear γ is included (e.g., cases F and G), the magnification pattern is no longer statistically isotropic; light curves for sources moving in the direction of the external shear are different from those for sources moving perpendicular to it. We will not discuss the various possibilities with $\gamma \neq 0$ here; a detailed study of additional magnification patterns, including shear, and their corresponding light curves is given in [WA90.4]. In the next subsection, we consider a special case with finite shear, corresponding to the A image of the GL system 2237+0305, where ML effects have been detected [IR89.1].

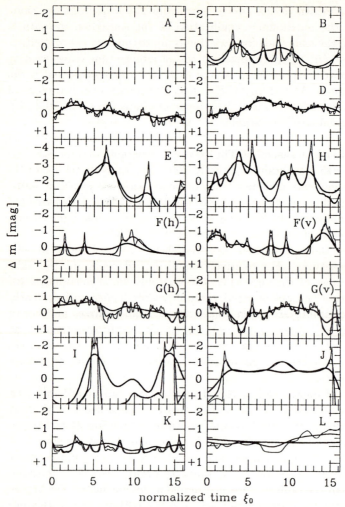

Fig. 12.8. Light curves for sources seen through a random star field. The parameters of the star fields are those given in Table 11.1, for which the magnification patterns have been displayed in Fig. 11.7. For each ML field, labeled A-L, three light curves are plotted, corresponding to sources with Gaussian brightness profiles and widths of 1, 4 and 16 pixels, where one pixel in Fig. 11.7 corresponds to $0.04\xi_0$. For the cases F and G, two sets of light curves are plotted, for source tracks moving horizontally (h) and vertically (v) through the magnification pattern, i.e., parallel and antiparallel to the external shear. The relation between "normalized time" and physical time scale is provided by (12.39b)

Comparison of model light curves with observations is difficult. Only for very few (namely the most spectacular) objects do we have data with appreciable time coverage. Furthermore, the light curves shown in Fig. 12.8 are only *random* light curves, but of course not light curves sampled in any statistically fair fashion. Hence, even if sufficient data were available, it would be difficult to compare them with calculations. For any individual source, it seems impossible to find out whether it is microlensed, at least if no other data (such as multiple imaging) are available. It is not difficult to find individual sources which show variability similar to that of the model light curves; e.g., the highly variable QSO 3C345 varies in a way ([KI89.1], [KI90.1]) one would expect from ML [GR88.1]. Other objects for which ML was invoked to account for the variability, and which were discussed in the literature, include AO0235+164 ([OS85.1], [ST88.1], [KA88.4]), Q1156+295 [SC87.3], and 0846+51W1 ([NO86.1], [ST89.1]). One could exclude ML as the origin of variability for those sources which have been monitored for sufficiently long time so that many "outbursts" are observed, if the variability is intrinsically asymmetric, e.g., if the rise is always faster than the decline, since *ML should yield light curves which are* statistically *symmetric*. For further discussion on the application of microlens-induced variability of QSOs, the reader is referred to [KA86.2].

12.4.2 Microlensing in 2237+0305

The most promising available target for the detection of ML is the 2237 +0305 GL system, for several reasons. The lens has a very small redshift ($z_d \approx 0.039$), which means that $D_d \ll D_s \approx D_{ds}$, so the geometrical factor determining the length-scale η_0 (2.6c) is large, as are the factors entering the equation (12.39a) for the effective velocity. Therefore, any velocity of the lens or observer translates to a large velocity in the source plane. Even more important, however, is the fact that the time delays for the four images are of the order of one day or less. If the source varies intrinsically, flux changes should thus appear in all four images almost simultaneously. If the flux of one image varied, without corresponding changes in the other images, that would be a clear signature of ML. The uniqueness of 2237+0305 as a possible detector of ML was pointed out in several references (e.g., [KA86.2], [KA89.2], [WA90.2]).

The small redshift of the lensing galaxy provides the opportunity to map its light distribution in great detail. Assuming a constant mass-to-light ratio across the galaxy, this brightness distribution yields the surface mass distribution, up to an overall factor (the unknown mass-to-light ratio). One can then try to model the position of the images with this surface mass distribution. This was achieved in [SC88.1]; the resulting lens model then yields the value of the local surface mass density κ and the shear γ for all four images. A different approach was followed in [KE88.1], where an ellipsoidal mass distribution was used to model the improved image positions obtained in [YE88.1]. Table 12.1 provides the values of κ and γ for the four images, together with other quantities to be discussed later. Assuming that all the mass is contained in stars (which appears to be a reasonable approximation, since the four images are seen close to the center of a spiral galaxy), the parameters for the ML models are uniquely determined, up to the mass spectrum of the lenses.

Table 12.1. For the four images of QSO 2237+0305, values of surface mass density κ and shear γ are given, according to the lens model in [SC88.1], together with the magnification and the coefficient C, as used in (12.40)

Image	κ	γ	μ	C
A	0.36	0.44	3.9	1.15
B	0.44	0.28	4.3	1.16
C	0.86	0.47	−5.0	0.80
D	0.60	0.67	−3.5	0.78

A detailed treatment of ML for the four images of 2237+0305 was published in [WA90.2], and particular attention to the image A, for which ML appears to have been observed, was presented in [WA90.1]. Let us discuss these two references. The mass spectrum of the microlenses was assumed to follow a Salpeter mass-function, i.e., $n(M) \propto M^{-2.35}$, with lower and upper cut-off of $0.1 M_\odot$ and $1 M_\odot$, respectively.[10] The Hubble constant was assigned the value $H_0 = 75\,\mathrm{km\,s^{-1}\,Mpc^{-1}}$; for other values of the cut-offs in the mass spectrum and the Hubble constant, simple scaling relations apply to the results quoted below, but, to simplify the discussion, we fix these values at the beginning. One then finds that $\eta_0 \simeq 1.25 \times 10^{17}\,\mathrm{cm}$, for a cosmological model with $\Omega = 1$, $\tilde{\alpha} = 1$. Time-scales will be quoted for the illustrative case that only the observer moves relative to the line-of-sight to the QSO, with a transverse velocity of $600\,\mathrm{km/s}$; a generalization to moving source and lens can easily be obtained from (12.39a).

Magnification patterns corresponding to images B and D are displayed in [WA89.3] and [WA90.2], respectively, and the magnification pattern corresponding to image A is shown in Fig. 11.71. Owing to the external shear, the magnification patterns are anisotropic; in particular, chains of caustics are formed (see Fig. 11.7) in the direction of the external shear. We discussed the origin of these chains in Sect. 11.2.5. The anisotropy of the magnification patterns implies that ML light curves depend on the direction of motion for a source relative to the pattern. In particular, a source moving perpendicular to the chains shows variability on shorter time-scales than one which moves parallel to the external shear.

To characterize the variability time-scale, we may use the autocorrelation function

$$f(\tau) \equiv \frac{\langle \mu(t)\,\mu(t+\tau)\rangle - \langle \mu \rangle^2}{\langle \mu^2 \rangle - \langle \mu \rangle^2} ,$$

where the brackets denote an average over time t along a track of the source through the pattern. Clearly, $f(0) = 1$, and $f(\tau)$ decreases for small values

---

[10] all the results presented in this subsection are based on this mass distribution of the microlenses

of τ. Examples for the function $f(\tau)$ are plotted in [WA89.3] and [WA90.2]; here, we only note that the time τ_h when f has decreased to 0.5 is a measure of the time-scale on which the flux of the corresponding images can be expected to vary. Empirically, for the magnification patterns in [WA90.2],

$$\tau_h \simeq 25 \text{ years } C_i \left(\frac{\rho}{0.01 \text{ pc}}\right)^{1/2} \left(\frac{v_0}{600 \text{ km/s}}\right)^{-1}$$
$$\times \frac{a_0 \, a_{90}}{\sqrt{(a_0 \sin\theta)^2 + (a_{90} \cos\theta)^2}},$$
(12.40)

where ρ is the radius of the source, with Gaussian brightness profile $\propto \exp\left[-r^2/(2\rho^2)\right]$, $a_0 = |1 - \kappa + \gamma|$, $a_{90} = |1 - \kappa - \gamma|$ are the two eigenvalues of the Jacobian matrix corresponding to the macrolens model at the various image positions, θ is the angle between the direction of the transverse velocity of the source and the external shear, and the coefficients C_i are given in Table 12.1 for the four images. This equation quantifies the facts mentioned above, namely, that the variations are faster for a source moving perpendicularly to the external shear; it also shows that the variations are faster for smaller sources. These results are further verified by the model light curves published in [WA89.3] and [WA90.2]; in particular, the model light curves for sources moving perpendicularly to the external shear display quasi-periodic behavior, corresponding to crossings of the chains of caustics.

Since a brightness fluctuation was observed in image A [IR89.1], we want to concentrate now on the ML model for this image. In Fig. 12.10, we display the magnification patterns for this case, with side length $100\eta_0$, $20\eta_0$, $4\eta_0$, and $0.8\eta_0$, each frame having the same source point as its origin. The corresponding light curves are shown in Fig. 12.9, for several source sizes. On the longest time-scale, we note the quasi-periodic behavior of the light curve for the source moving perpendicularly to the external shear. As indicated before, the magnification patterns (at least those with small side length) are typical, rather than average; this of course is also true for the corresponding light curves. The variety of possible flux variations is clearly seen in Fig. 12.9. For small sources, individual events can be distinguished, separated by relatively long intervals of almost constant flux; these events correspond to a source crossing individual caustics. For larger sources, no such isolated events can be found, since their size is larger than the typical spacing between caustics; therefore, their light curves are fairly smooth.

From a sample of 20 light curves, obtained from the frame shown in Fig. 12.10, 'events' were identified by their rate of change of magnification: whenever the magnification changed by more than 0.3 mag within 0.23 years, this change was called an event. Of course, this defining criterion is somewhat arbitrary, but it agrees reasonably well with the amplitude and time-scale of the observed ML event in 2237+0305A. For each such event, the amplitude Δm was defined as the difference between the peak flux and that of the highest of the two minima to the right and left of the maximum,

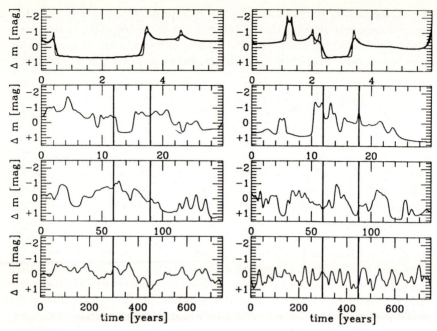

Fig. 12.9. Light curves for different source sizes, corresponding to the A image in 2237+0305, with the parameters described in the text; in particular, a transverse velocity of the observer of 600 km/s was assumed. The light curves on the left side correspond to a source moving parallel to the shear, those on the right to a source moving perpendicularly to the shear. The two light curves shown in the uppermost row correspond to sources with radii of $\rho = 2 \times 10^{14}$ cm and $\rho = 8 \times 10^{14}$ cm, with a thicker line for the larger source. The light curves in the three lower rows correspond to sources with radii $\rho = 2 \times 10^{15}$ cm, 10^{16} cm, and 5×10^{16} cm, respectively. A pair of vertical lines near the middle of these three lower rows indicate the full width of the row just above it. The light curves were obtained from the magnification patterns shown in Fig. 12.10 (from [WA90.1])

and the time Δt was measured as the full width at half maximum. The statistics of events are shown in Fig. 12.11, for two different source sizes and two directions, parallel and perpendicular to the external shear. As can be seen from this figure, events with $|\Delta m| \geq 0.5$ occur on time-scales between less than a month and several months. Therefore, the observed event in 2237+0305A can be easily, but not uniquely, explained by the ML model described here.

We also note that it was not possible to explain the observed ML event in 2237+0305A with a ML model where only $1 M_\odot$ objects act as lenses; furthermore, the source size is constrained to be less than about 2×10^{15} cm. This allows stringent constraints on models for the emitting region of the optical continuum radiation in this source, e.g., it seems to contradict standard thin accretion disks models [RA91.1].

ML in 2237+0305 was expected, and further ML events are likely to occur in the near future in one (or more) of the images. As pointed out in [KA89.2], it is important that such events be recorded with very good

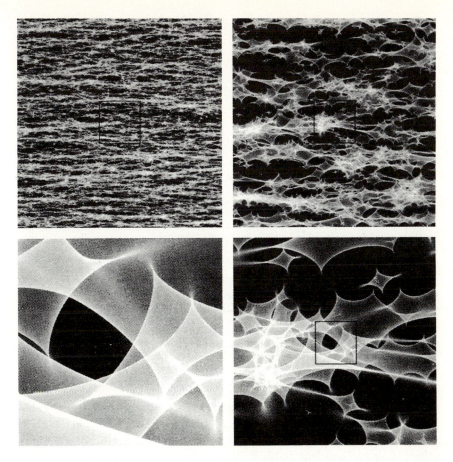

Fig. 12.10. Magnification patterns for the image A of the QSO 2237+0305, on four different frames with side length $100\eta_0$ (upper left), $20\eta_0$ (upper right), $4\eta_0$ (lower right), and $0.8\eta_0$ (lower left), corresponding to 1.5×10^{19}, 3.1×10^{18}, 6.1×10^{17}, and 1.2×10^{17} cm $h_{50}^{-1/2}$. The black squares in the three larger frames show the region which is shown in the next smaller frame. The light curves in Fig. 12.9 were calculated along straight horizontal and vertical tracks through the center of the frames (from [WA90.1])

time-coverage. Only from well-observed light curves can we hope to learn more about the continuum source structure (see Sect. 13.3.2) and the local condition of the ML field. Nevertheless, the likely detection of ML in this GL system demonstrates that this effect, predicted some 10 years ago [CH79.1], is actually important for observations of QSOs.

12.4.3 Microlensing and broad emission lines of QSOs

The broad-line regions of QSOs are substantially larger than the regions where the optical continua are emitted; this follows from variability arguments, which constrain the size of the continuum-emitting region to be of

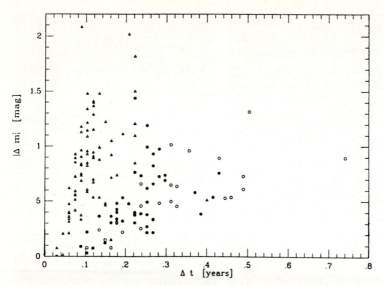

Fig. 12.11. The amplitude $|\Delta m|$ and duration Δt (as defined in the text), for 20 light curves obtained from the magnification pattern in Fig. 12.10. Triangles correspond to a source size of $\rho = 2 \times 10^{14}$ cm, and hexagons correspond to $\rho = 8 \times 10^{14}$ cm. Filled symbols correspond to source motion perpendicular to the external shear, open symbols to parallel motion. As discussed in the text, smaller sources have events of larger amplitude and shorter time-scale than larger sources, and the time-scale is shorter for motion perpendicular to the shear than for parallel motion (from [WA90.1])

the order of a light-day, at least for sources that vary on short time-scales. Models of the broad-line region predict sizes of about 10^{18} cm. This is slightly larger than the scale induced by a stellar mass GL, for canonical values of source and lens redshift. Thus, the broad-line region *as a whole* cannot be strongly magnified by ML. However, this region is believed to have intrinsic structure. It is usually supposed that the broad lines are emitted by a large number of clouds. The dynamics of these clouds is yet unclear; inflow, outflow, irregular, and rotational motion have been considered. ML can lead to a differential magnification of the broad-line region, i.e., one part can be magnified more strongly than another [KA86.2]. For example, if the broad-line region has large-scale rotation, differential magnification can lead to a stronger flux from the receding side of the region than from the approaching clouds. This would amount to an asymmetry of the broad lines and, in addition, to a shift of the central wavelength. In effect, the mean redshifts of these lines can be changed. This could therefore lead to a difference of the redshifts, as measured from broad lines, compared with measurements from narrow lines, which are emitted at much larger distances from the central engine. However, such a redshift difference can also be intrinsic to the QSO, due to radiative transfer effects in the broad-line region, and is thus not necessarily a sign of ML.

However, if a QSO has two (or more) macroimages, ML can lead to different magnifications of the broad-line region in the two images. Consider again the example above, where differential magnification of the broad-line region with large-scale rotation changes the redshift of the lines. If the profiles of the broad lines in the two images agree within stringent observational limits, one can perhaps constrain the amount of large-scale rotation in the broad line region. Similar constraints may apply for inflow or outflow models.

These ideas were first investigated in [NE88.1], where the effects of a single star on the profiles of the broad emission lines were investigated, for several models of the broad-line region. Since the best way to see the effects of ML on the line profiles is a comparison of the profiles in different images of multiply-imaged QSOs, and since the optical depth of the lensing galaxy at the position where multiple images can occur is not small, the single-star model is not a reasonable assumption in those cases. Thus, the effects of a random star field must be considered. These were treated in [SC90.2], where the effects of a random star field with parameters applying to image A of 2237+0305 on two simple models for the geometry and kinematics of the broad-line region were examined in detail. As was the case for the variability of the continuum radiation caused by ML, the 2237+0305 GL system is the best available candidate to see this effect, due to the small redshift of the lens and, correspondingly, the large value of the length-scale η_0 (2.6c) in the source plane. In fact, there is strong indication that the profiles of the MgII line in two images of this QSO are different [FI89.1].

It is a straightforward matter to apply the ML simulations described in Sect. 11.2.5 to the problem at hand, because a model for the broad-line region yields the spectral intensity $I_\nu(\mathbf{y})$ as a function of the projected position $\mathbf{y}$. Therefore, the flux density obtained from this region is

$$\mathcal{F}_\nu = \int d^2y\, \mu(\mathbf{y})\, I_\nu(\mathbf{y}) \quad ,$$

and the ML simulation yields the magnification distribution $\mu(\mathbf{y})$. The unlensed profile is obtained by setting $\mu \equiv 1$.

Two models for the broad line region are considered in [SC90.2]: Keplerian rotation of the line-emitting gas surrounding a black hole of mass $M_\mathrm{h} = M_8 10^8 M_\odot$, and infall with a fraction f of the free-fall velocity. The radial dependence of the emissivity is assumed to be a combination of two power-laws. Five different ML models are considered, with parameters applicable to the A image of 2237+0305 ($\kappa_* = 0.36$, $\gamma = 0.44$ [SC88.1]), which differ in the assumed mass spectrum of the lenses. As a description of the wavelength of the line, the Doppler velocity $V = c(\lambda - \lambda_\mathrm{e})/\lambda$ is used, parametrized as

$$V = \xi\, \sin\vartheta\, \sqrt{M_8}\, h_{50}^{1/4}\, 3400\,\mathrm{km/s}$$

in the case of the rotation model, and

$$V = \xi\, f\, \sqrt{M_8}\, h_{50}^{1/4}\, 4700\,\mathrm{km/s}$$

for the infall model. Here, ϑ is the angle between the rotation axis and the line-of-sight, and λ_e is the rest-wavelength of the emission line under consideration.

In Fig. 12.12, we show several sample line profiles, both for the rotation model and the infall model. The lensed profiles differ from the unlensed profiles considerably. In [SC90.2] it is shown that for the A image of 2237+0305, such differences in the profiles are common: for all rotation models, the relative difference of the lensed from the unlensed profile is larger than 5%, with a typical value of about 10%. Hence, if the broad-line region is indeed in rotation, one should be able to see differences in the profiles of the same line in different images of 2237+0305, as may have been observed [FI89.1].

We have seen that ML may be used to investigate the kinematics of the broad-line region in QSOs. One should bear in mind, however, that the models used in [SC90.2] are probably too simple to allow more quantitative conclusions.

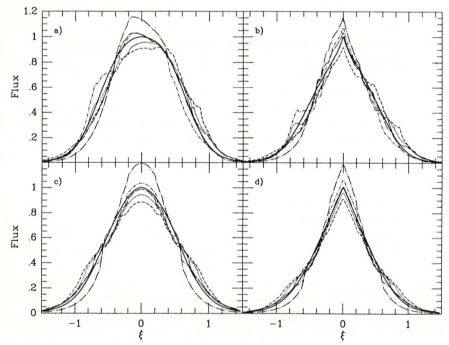

Fig. 12.12a–d. Lensed and unlensed line profiles obtained for various models of the geometry and kinematics of the broad-line region, and five different ML models, corresponding to parameters which apply to the A image in 2237+0305. The heavy solid line in each panel corresponds to the unlensed line profile, and the other five lines show the lensed line profiles obtained for five different ML models, which differ in the assumed mass spectrum of the lenses. The profiles are normalized such that the total line flux is the same as that for the unlensed line. The wavelength is parametrized in terms of the Doppler velocity as measured in the rest frame of the QSO, given here by the dimensionless parameter ξ defined in the text. (**a**) and (**b**) correspond to models with Keplerian rotation, and (**c**) and (**d**) to gravitational infall models (from [SC90.2])

Apart from yielding information about the broad-line regions, observations of profile differences in multiply-imaged QSOs are a convenient means to detect ML (see also [SU87.2]). The method described in the last subsection depends on observing variability in one image, which is not matched in the other images (even by taking a possible time delay into account), and thus has to rely on monitoring the source. A line profile difference, on the other hand, could prove ML in a single observation, at least in those multiply-imaged QSOs where the time delay is much shorter than the timescale on which the profiles can change intrinsically.

12.4.4 Microlensing and the classification of AGNs

The ML effect can change the classification of an AGN. For example, if the active nucleus of a Seyfert galaxy as a whole is magnified, its brightness will outshine the galaxy, and will thus appear stellar-like on an optical plate; hence, such a source would be classified as a QSO [BA68.1]. Another example is that of a QSO, for which only the optical continuum radiation is magnified, but not the broad-line region which is thought to have a larger extent; the spectrum would then be dominated by the continuum, and, depending on the magnification, the source could be classified as a BL Lac object [OS85.1].

In general, if the magnification of the optical continuum source is different from that of the broad-line region, the equivalent width of the broad emission lines is affected. In particular, if the magnification of the continuum source is larger than that of the broad-line region, the QSO will have weaker (in terms of their equivalent width) emission lines. This tendency for bright QSOs to have relatively weaker emission lines has been known for some time [BA77.1] and is known as the Baldwin effect (see also [WA84.1]). We do not mean to imply that the Baldwin effect can be due exclusively to ML, but the effect is similar to that expected from ML.

The idea that some BL Lac objects may be microlensed QSOs [OS85.1] arose from several peculiarities of BL Lac objects. First, some BL Lac objects, when they are in a state of low luminosity, have spectra very similar to those of QSOs but lack broad emission lines when their fluxes are high. Second, the redshift distribution of BL Lac objects is completely different from that of QSOs, since many more of them are found at low redshift, whereas QSOs typically have high redshifts. The redshifts of most BL Lac objects are obtained only indirectly: they appear to reside in a nearby elliptical galaxy, and the redshift determination is made from the galaxy's spectrum. More arguments in favor of the preceding hypothesis can be found in [OS85.1].

The lens hypothesis for BL Lac objects is fairly attractive: the galaxies where they apparently reside are then to be viewed as lenses. ML by compact objects in such lens galaxies can then cause the rapid variability of BL Lac objects, or at least part of it, thus reducing the stringent constraints on source compactness. In addition, the redshift distribution of BL Lac objects can then be similar to that for QSOs, i.e., most of them would be at high redshift, but, due to the lack of observable emission lines, this redshift could not be determined.

An additional argument in favor of the ML origin of some BL Lac objects has been given in [OS90.1]. The redshift distribution of BL Lac objects contained in existing catalogues is highly peaked towards small redshifts, which cannot be solely due to a selection bias. If, on the other hand, the redshift assigned to a BL Lac object is that of a lensing galaxy, the low redshifts can be understood, since determining the redshift of a nearby galaxy is easiest. For a more detailed discussion of the space distribution of BL Lacs, see [OS90.1]. (However, we do not follow the argument given in [OS90.1] about the selection of specific lens redshifts for these type of systems. They argue that lens galaxies with central surface mass density close to $\Sigma_{\rm cr}$ are most efficient. Whereas it is true that the highest magnifications occur for such lenses, it is not true that in such situation, ML is particularly effective in preferentially magnifying the small optical continuum source; as we have seen from the magnification patterns in Fig. 11.7, for optical depth close to the critical one, the relative difference between the magnification of small and large sources becomes smaller than for small and high optical depth.)

In fact, recent observational results support this hypothesis. Of a handful of BL Lac objects with redshifts measured from broad emission lines, three have been found to be apparently associated with foreground galaxies located right in the line-of-sight ([ST88.1], [ST88.2], [ST89.1]). Given the brightness of these galaxies, and their estimated redshifts, lensing is unavoidable, since the surface mass density of the galaxies at the position of the BL Lac image must be a considerable fraction of the critical surface mass density. In fact, for the system AO0235+164 [ST88.1], the galaxy appears to be sufficiently massive to cause multiple images, which are not observed; their absence yields fairly stringent constraints on the core radius of the foreground galaxy [NA90.1].[11] Another aspect worth mentioning is the deconvolution of the BL Lac spectrum into that of the foreground galaxy and the BL Lac itself. Although the (observed) sum of these two spectra yields a steep optical continuum (one of the characteristics of BL Lac objects), deconvolution shows that this is mainly due to the underlying galaxy spectrum, and that the spectrum of the BL Lac itself is much flatter, comparable with the continuum of QSOs [ST89.1].

On the other hand, it should be pointed out that BL Lacs differ from QSOs by more than just the absence of broad emission lines (see [BL90.1] for a discussion). Other differences, such as radio structure and polarization properties, are not easily explained by ML. Since the discussion on whether BL Lac objects are microlensed QSOs (or a subset of them, the optically violently variable QSOs, OVVs) has not reached a conclusion, we leave the subject here and hope that observational and theoretical progress can be made soon.

---

[11] Note, however, that another observation [YA89.1] has failed to detect a foreground galaxy on 0235+164.

12.5 The amplification bias: Detailed discussion

Introduction. Soon after the discovery of QSOs, it was claimed [BA68.1] that these objects are not as luminous as their redshifts and luminosities would suggest, on the basis of the standard Friedmann–Lemaître cosmological model, but that they are in fact gravitationally-lensed nuclei of Seyfert galaxies. Since QSOs appear to be high-redshift, high-luminosity "cousins" of Seyfert nuclei, this suggestion is not unrealistic *a priori*, but it was considered to be very unlikely, due to the low lensing probability for the known population of galaxies (e.g., [SI74.1], [TY81.1], [SE83.1]). Less radical was the claim [TU80.1] that the observed evolution of QSOs with redshift could be due to magnification by GLs. As noted in subsequent papers (e.g., [AV81.1]) this conclusion was quantitatively incorrect, since it was based on an analysis that does not take into account flux conservation (compare the discussion in Sect. 4.5.1). However, the possible importance of the amplification bias has since been taken seriously: as observed samples are flux limited, it is likely that magnified objects are overrepresented.

Detailed analyses of the amplification bias were performed in [PE82.1] and [CA82.1]. In the former paper, extended lenses (galaxies) were considered, whereas the latter investigated the effect of compact objects with stellar masses. For optically-selected QSOs, models with compact, stellar-mass objects are appropriate, since the size of the continuum-emitting region is thought to be smaller than, or comparable to, the relevant length scale of a Schwarzschild lens with stellar mass. Thus, these QSOs are affected by the granularity of the mass distribution in the universe and, in particular, that of the mass distribution of galaxies. An analysis with galaxy-sized objects is relevant for sufficiently extended sources, so that their light is not sensitive to the clumpiness of matter in galaxies, e.g., for extragalactic radio sources. Therefore, the size of a source determines which lens mass-scale must be used for statistical analysis.

12.5.1 Theoretical analysis

The need to consider extended sources. Our preliminary discussion in Sect. 12.1.1 touched on the basic features of the amplification bias. For instance, we saw that this bias can become very important if the luminosity function of the sources is very steep. The fraction of magnified sources is higher in flux-limited samples with a high flux threshold; in particular, the shape of the source counts need not reflect the intrinsic luminosity distribution of the sources. The discussion of Sect. 12.1.1 was confined to point sources, for which the mass spectrum of the lenses is unimportant. However, as follows from (12.8b), the calculated source counts depend very sensitively on the faint end of the luminosity function, since the magnification of point sources can be arbitrarily large. As was shown in Sect. 8.1, the magnification of extended sources is bounded from above. Thus, it is clear that the effects of non-negligible source sizes must be taken into account properly to obtain realistic results. This follows from the fact that the size of the continuum-emitting region in QSOs, as derived from the observed variability, is comparable to the Einstein radius of a Solar-mass lens.

Although most compact objects are likely to be associated with galaxies (e.g., stars, or compact halo objects), they are often treated as if they were isolated. At first sight, this assumption may appear unrealistic; however, most compact objects are well away from the center of their host galaxy, where the mass density in compact objects and the overall gravitational field of the galaxy are small; thus, compact objects act as if they were isolated Schwarzschild lenses. We can also estimate the maximum shear (i.e., the Chang–Refsdal parameter γ, see Sect. 8.2.2) caused by the galaxy such that the compact lens can be treated realistically as a Schwarzschild lens: for $\gamma \ll 1$, the linear size of the caustic is roughly 4γ; if this size is smaller than the dimensionless source size, the shear is unimportant. In any case, in what follows, we neglect the influence of shear on the individual compact lenses.

Magnification probability. We consider the lens effect of a population of compact objects in the universe, described by a mass spectrum $n_\mathrm{L}(M, z) = n_\mathrm{L}(M)(1+z)^3$, where $n_\mathrm{L}(M)$ is the present number density of compact objects (lenses) of mass M; the total mass density of the lenses is $\Omega_\mathrm{L}\rho_\mathrm{cr}$, where ρ_cr is the critical mass density (4.36). Thus,

$$\int dM \, n_\mathrm{L}(M) \, M = \Omega_\mathrm{L}\rho_\mathrm{cr} \quad .$$

Consider sources at redshift z_s, with radius $\mathcal{R}$. Let us define a reference mass M_0 and a dimensionless mass-variable $m = M/M_0$; we also define a normalized number density $N(m)$ of lenses, as described by (12.12). Then, from (11.68) the probability that a source is magnified by more than μ_e is,

$$P(\mu_\mathrm{e}, z_\mathrm{s}) = \frac{3}{2}\Omega_\mathrm{L}\langle\mu\rangle(z_\mathrm{s}) \int_0^{z_\mathrm{s}} \frac{dz}{\langle\mu\rangle(z)} \frac{1+z}{\sqrt{1+\Omega z}} \frac{r(z)r(z,z_\mathrm{s})}{r(z_\mathrm{s})}$$
$$\times \int dm \, m \, N(m) \, y^2(\mu_\mathrm{e}, R(z,m)) \quad , \qquad (12.41a)$$

where $\langle\mu\rangle$ is given by (11.69), and $y(\mu_\mathrm{e}, R)$ is the inverse function of (11.12). The dimensionless source size R, in terms of the physical size $\mathcal{R}$, is

$$R(z, m) = \frac{\mathcal{R}}{\eta_0} = R_0\sqrt{\frac{r(z)}{m\, r(z_\mathrm{s})\, r(z, z_\mathrm{s})}} \quad , \qquad (12.41b)$$

where

$$R_0 = \frac{\mathcal{R}}{\sqrt{(4GM_0/c^2)(c/H_0)}} \qquad (12.41c)$$

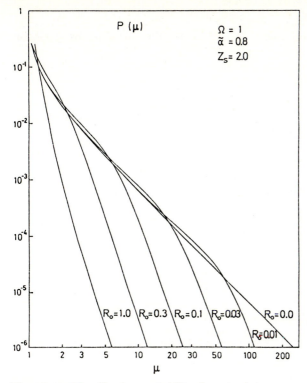

Fig. 12.13. Magnification probability for extended sources at a redshift $z_s = 2.0$ in a universe where all compact objects have the same mass; we assume that the universe is flat ($\Omega = 1$), and that the lenses contain 20% of the mass ($\tilde{\alpha} = 0.8$). The curves are labeled by the value of the dimensionless source radius R_0 (12.41c)

is a redshift-independent measure of the source size. One can see from (12.41b) that lenses with small redshift have a small associated source size; hence, they are mostly important for the high magnifications, whereas lenses with $z \approx z_s$ are able to magnify the source only by a small factor.

Lenses with a single mass. Equation (12.41) is readily evaluated numerically. Let us first consider the case where all lenses have the same mass M_0, i.e., $N(m) = \delta(m-1)$. We show in Fig. 12.13 the probability for this case, where the source redshift is set to $z_s = 2$, and we chose a cosmological model with $\Omega = 1$, $\tilde{\alpha} = 0.8$, $\Omega_L = 0.2$. The result for a point source is the curve $R_0 = 0$, which, for large magnifications, has the form τ_p/μ_p^2, as predicted by (11.70). The probabilities for extended sources follow that of the point source up to a cut-off magnification μ_c, above which they turn over to a μ_e^{-6} law. This cut-off magnification is lower for larger values of R_0.

Derivation of the μ_e^{-6} tail. We can obtain the μ_e^{-6} behavior analytically, using the fact that, for large magnifications, relation (11.15) is valid. Denoting the inverse of the function ζ (8.27) by ζ^{-1}, for large μ_e,

$$y(\mu_e, R) \approx \zeta^{-1}(R\mu_e)\, R \quad .$$

Since the function ζ has a maximum value of 2 (see Fig. 8.3), we require $R\mu_e \leq 2$. Using (12.41b), we find

$$r(z) \leq \frac{4\, r(z_s)\, r(z, z_s)}{R_0^2\, \mu_e^2} \quad .$$

Now, consider the limiting case $\mu_e \to \infty$; then, only lenses at very small redshift can contribute to the integral in (12.41a). Thus, we can replace $r(z)$ by z and $r(z, z_s)$ by $r(z_s)$. Then, the integration must be extended up to

$$z_m = \frac{4\, r^2(z_s)}{R_0^2\, \mu_e^2} \quad .$$

Inserting these approximations into (12.41a), we have, for large μ_e,

$$P(\mu_e, z_s) \approx \frac{3}{2} \Omega_L \langle\mu\rangle (z_s) \left[\frac{R_0}{r(z_s)}\right]^2 \int_0^{z_m} dz\, z^2 \left[\zeta^{-1}(R\mu_e)\right]^2 \quad . \tag{12.42a}$$

The z^2 factor in the integrand suggests that the main contribution to the integral will come from $z \approx z_m$, where the argument of ζ^{-1} is close to 2. We therefore consider the expansion of $\zeta(w)$ about $w = 0$, $\zeta(w) = 2\left(1 - w^2/4\right) + \mathcal{O}\left(w^4\right)$. Thus, we use the approximation

$$\left[\zeta^{-1}(R\mu_e)\right]^2 \approx 4 - 2\, R\, \mu_e \quad .$$

Note that this will lead to an overestimate of the integral in (12.42a). Finally, collecting all terms,

$$P(\mu_e, z_s) \approx \frac{128}{7} \Omega_L \langle\mu\rangle (z_s) \left[\frac{r(z_s)}{R_0}\right]^4 \frac{1}{\mu_e^6} \quad . \tag{12.42b}$$

As noted before, this expression overestimates P due to the approximation we made. It was found numerically that the value in (12.42b) is about 20% too large; in fact, replacing the numerical factor in (12.42b) by the factor 14.25 reproduces the numerical results within $\approx 5\%$ for all values of R_0 and z_s [SC87.7].

Qualitative discussion of the resulting source counts. From the magnification probability, we can calculate a prediction for source counts, using (12.4). We will not do so at this stage, since the assumption that all lenses have the same mass is not realistic. However, the approximate results are: for a luminosity function of the form (12.7a) with $\beta < 2$ and $\alpha > 2$, we expect for $S\zeta(z_s)/l_0 < 1$ no change in the shape of the source counts, compared with the unlensed source counts (12.9), but only a change in amplitude, increasing for $\beta > 1$ and decreasing for $\beta < 1$. For fluxes $S > l_0/\zeta(z_s)$, the observed source counts suffer a transition from $\propto S^{-\alpha}$ to a S^{-2} law. The latter range arises because, for magnifications below the cut-off μ_c, the magnification probability for extended sources is not very different from that for point sources (for which the S^{-2}-law was derived in Sect. 12.1.1). For $S > l_0\, \mu_c/\zeta(z_s)$, the cut-off in the probability is important, and the source counts follow either the intrinsic slope ($S^{-\alpha}$) or the S^{-6} law, whichever is the flatter.

Analytic approximation to the magnification probability. The form of the curves in Fig. 12.13 suggests that $P(\mu_e, z_s)$ can be approximated by

$$P(\mu_e, z_s) \approx \tau_p \, y^2(\mu_e, R = 0) \left[\frac{1}{1 - \nu_1 \mu_e^2 + \nu_2 \mu_e^4} \right] \, , \qquad (12.43a)$$

where the factor within brackets accounts for the deviation of P from the point-source result for large magnifications, and $y(\mu, R = 0)$ is given by (11.1). In particular, by choosing an appropriate value for ν_2, the approximate result for $\mu \to \infty$ can be reproduced:

$$\nu_2 = \frac{\tau_p}{14.25 \, \Omega_L \, \langle \mu \rangle \, (z_s)} \left[\frac{R_0}{r(z_s)} \right]^4 \, . \qquad (12.43b)$$

The coefficient ν_1 can be determined from the requirement that the mean magnification for high-magnification events be independent of source size. Specifically, since a non-zero source size does not influence $P(\mu_e, z_s)$ for $\mu_e \le \mu_0$ (where μ_0 is the usual threshold required to avoid the overlapping of cross-sections; compare the discussion in Sect. 12.1.1), we determine ν_1 by requiring that

$$\langle \mu \rangle_H = \frac{1}{P(\mu_0, z_s)} \int_{\mu_0}^{\infty} d\mu \, p(\mu, z_s) \, \mu$$

be independent of the source size (which enters in the expression for ν_2). Although in general ν_1 must be computed numerically, one can show that, for $\nu_2 \ll 1$, $\nu_1 \simeq \sqrt{\nu_2}$ [SC87.1].

Magnification probability for a lens mass-spectrum. We next consider the magnification probabilities for an ensemble of lenses with a mass spectrum $N(m)$. The preceding results allow a fairly simple analysis, by noting the dependence of (12.43a) on the source size R_0. Since only $R(z, m)$ depends on the lens mass [see (12.41b)], where m and R_0 enter in the combination R_0^2/m, (12.43a) is valid for any m if R_0 is replaced by $R_0/\sqrt{m}$. Thus, if the integrals in (12.41a) are interchanged, the magnification probability becomes

$$P(\mu_e, z_s) = \tau_p \, y^2(\mu_e, 0) \int_{m_{\min}}^{m_{\max}} dm \, m \, N(m) \left[1 - \nu_1(m) \mu_e^2 + \nu_2(m) \mu_e^4 \right]^{-1} \, ,$$

where now the coefficients depend on m; for $\nu_2(m)$, from our discussion we deduce $\nu_2(m) = \nu_2/m^2$, and, using the approximation for $\nu_2 \ll 1$, we have $\nu_1(m) = \sqrt{\nu_2}/m$, where ν_2 is given by (12.43b). Inserting these expressions into the integral, with the substitution $x = \sqrt{\nu_2} \mu_e^2/m$,

$$P(\mu_e, z_s) = \tau_p \, y^2(\mu_e, 0) \, \nu_2 \, \mu_e^4 \int_{x_{\min}}^{x_{\max}} \frac{dx}{x^3} \frac{N\left(\sqrt{\nu_2} \mu_e^2/x\right)}{1 - x + x^2} \, ,$$

where

$$x_{\min} = \sqrt{\nu_2} \frac{\mu_e^2}{m_{\max}} \, , \qquad x_{\max} = \sqrt{\nu_2} \frac{\mu_e^2}{m_{\min}} \, .$$

Let us use the power-law mass spectrum

$$N(m) = \frac{2 - \beta}{m_{\max}^{2-\beta} - m_{\min}^{2-\beta}} \, m^{-\beta} = f_{\text{norm}} \, m^{-\beta} \quad \text{for} \quad m_{\min} \le m \le m_{\max} \, ,$$

where the normalization constant f_{norm} was determined from (12.12b); then,

$$P(\mu_e, z_s) = \tau_p \, y^2(\mu_e, 0) \, f_{norm} \left[\sqrt{\nu_2} \mu_e^2\right]^{2-\beta} \int_{x_{min}}^{x_{max}} dx \, \frac{x^{\beta-3}}{1-x+x^2} \, .$$

The last integral is computed numerically for general integration limits. However, for three special cases, we can obtain good analytic approximations. If $x_{max} \ll 1$, the integrand can be replaced simply by $x^{\beta-3}$. On the other hand, if $1 \ll x_{min}$, the integrand can be approximated by $x^{\beta-5}$. For $x_{min} \ll 1 \ll x_{max}$, the approximation to be used depends on β. For $\beta < 2$, the main contribution comes from small x values, and the integral can be replaced by an integral over $x^{\beta-3}$ from x_{min} to, say, 1. For $\beta > 4$, high x values will dominate the integral, which can then be replaced by an integral over $x^{\beta-5}$ from 1 to x_{max}. For $2 < \beta < 4$, one can replace the integral limits by 0 and ∞, respectively; this definite integral is then a constant C, which is obtained from (3.252.12) of [GR80.3], $C = -2\pi \sin(2\pi\beta/3)/\left[\sqrt{3}\sin(\pi\beta)\right]$. We see that, for $2 < \beta < 4$, C is always positive.

Let the characteristic magnifications be defined as $\mu_1^2 = m_{min}/\sqrt{\nu_2}$, $\mu_2^2 = m_{max}/\sqrt{\nu_2}$; with the approximations discussed above, we have

$$P(\mu_e, z_s) = \begin{cases} \tau_p \, y^2(\mu_e, 0) & \text{for } \mu_e \ll \mu_1 \text{ or } \mu_1 \ll \mu_e \ll \mu_2, \, \beta < 2 \\ \tau_p \, C \, f_{norm} \, \nu_2^{(2-\beta)/2} \, \mu_e^{2-2\beta} & \text{for } \mu_1 \ll \mu_e \ll \mu_2, \, 2 < \beta < 4 \\ \tau_p \, \frac{\beta-2}{\beta-4} \, m_{min}^2 \, \nu_2^{-1} \, \mu_e^{-6} & \text{for } \mu_1 \ll \mu_e \ll \mu_2, \, \beta > 4 \\ \tau_p \, f_{norm} \, \frac{m_{min}^{4-\beta} - m_{max}^{4-\beta}}{\beta-4} \, \frac{1}{\nu_2} \, \mu_e^{-6} & \text{for } \mu_e \gg \mu_2 \end{cases} \quad (12.44)$$

Numerical calculations show that these approximations are good in their respective ranges of application. We show $P(\mu_e, z_s)$ for a special case in Fig. 12.14a, for a mass spectrum characterized by $\beta = 2.3$; this slope corresponds approximately to that of the Salpeter initial mass function [SA55.1]. For the parameters we used in the plot, $\mu_1 \approx 3.3$, $\mu_2 \approx 1.1 \times 10^3$. Below μ_1, the probability is near that for point sources; hence, for low magnifications, the dimensionless source size R is sufficiently small for all lens masses to be assumed to be zero. For $\mu_1 \ll \mu_e \ll \mu_2$, the magnification probability follows a $\mu_e^{2-2\beta}$ law, if $2 < \beta < 4$. If $\beta < 2$, the high mass lenses dominate the mass spectrum, and the source size for these high-mass lenses is sufficiently small for the point-source approximation to apply. On the other hand, if $\beta > 4$, the mass spectrum is dominated by low-mass lenses; in this case, $P(\mu_e, z_s)$ falls off as μ^{-6}, as is the case for $\mu_e \gg \mu_2$. For these two cases, the source size is large, and only lenses with very small redshift can yield large magnifications.

Amplification bias for a mass spectrum. If we select a luminosity function of the form

$$\Phi_Q(l, z_s) = \frac{1}{(l/l_0)^{\alpha_1} + (l/l_0)^{\alpha_2}} \, ,$$

with $\alpha_1 < 2 < \alpha_2$, we can estimate the source counts at fixed redshift from (12.4b); in Fig. 12.14b such a curve is displayed, together with the corresponding source counts for a smooth universe. Note that, for fluxes $S \lesssim l_0/\zeta(z_s)$, these two curves agree, since for the flat part of the luminosity function, lensing is unimportant. For $2 < \beta < 4$, in the range $l_0\mu_1/\zeta(z_s) \lesssim S \lesssim l_0\mu_2/\zeta(z_s)$ the counts follow a $S^{-2(\beta-1)}$ law, if $\alpha_2 > 2(\beta-1)$. For fluxes $S \gg l_0\mu_2/\zeta(z_s)$, the source counts follow either the intrinsic high-luminosity slope, or the S^{-6} law, whichever is flatter.

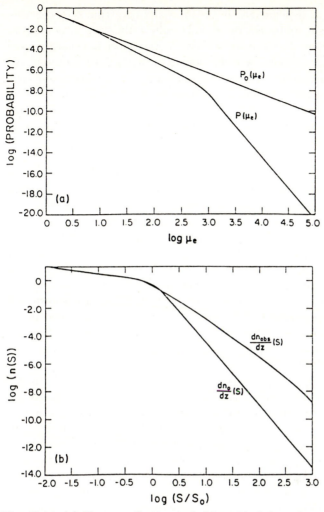

Fig. 12.14. (a) The magnification probability $P(\mu_e)$ for a flat universe where all the matter is concentrated in clumps ($\Omega = 1$, $\tilde{\alpha} = 0$), with a power-law mass spectrum with slope $\beta = 2.3$, and lower and upper cut-offs of $10^{-3} M_\odot$ and $10^2 M_\odot$, respectively. The source size is 10^{15}cm, and the sources are assumed to have redshift $z_s = 2$. For comparison, the probability distribution for point sources, $P_0(\mu)$, is also drawn. Note the two breaks in the distribution, one occurring at $\mu_e \approx 3.3$, and one at $\mu_e \approx 1.1 \times 10^3$. (b) The observed source counts at fixed redshift $z_s = 2$ from the magnification probability in (a), for an assumed luminosity function of the form given in the text, with $\alpha_1 = 0.5$, $\alpha_2 = 4.5$. The corresponding counts one would obtain in a smooth universe, $dn_0/dz_s(S)$, are shown for comparison

We have found that the source counts can be dominated by lensing, without having a S^{-2}-shape, as we concluded from the point-source analysis of Sect. 12.1.1. Hence, the argument that, since the observed source counts have a slope steeper than 2, they cannot be dominated by lensing [VI85.1], is not necessarily valid. In fact, if the source counts are (micro)lens-dominated, one expects them to depend mainly on the mass spectrum of the lenses, provided the intrinsic luminosity function is sufficiently steep. Our analysis has shown, in addition, that if the lenses have a mass spectrum characterized by $\beta = 2.3$, the source counts have the slope 2.6, as observed. Our analysis has neglected, for example, the possible influence of a spectrum of source sizes; in fact, it is expected that QSOs have a significant range of source sizes that may be correlated with their luminosity. We do not consider these questions further, since the amount of unknown parameters would increase drastically.

Deconvolution of the amplification bias. We have used (12.4b) here by "guessing" a luminosity function and investigating the impact of a model magnification probability distribution on the observed source counts. Another procedure would be more appropriate, i.e., starting from the observed counts, to derive the luminosity function Φ_Q, by deconvolution of (12.4b). Ignoring the uncertainty with which the source counts of QSOs at fixed redshift are known, such a procedure could yield, in principle, the intrinsic luminosity function, if we knew the magnification probability distribution.

Such an approach has been followed in [SC91.2]; there, it is shown that the integral equation (12.4b) of the first kind can be transformed into a Volterra integral equation of the second kind, provided $p(1) \neq 0$. For such an integral equation, standard numerical procedures are available, and the deconvolution of (12.4b) can be performed.

As an example of such a procedure, we present in Fig. 12.15a a model for the magnification probability distribution, based on the assumption that the lens population in the universe is described by an optical depth $\tau_g = 0.02$ in galaxies [so they yield, for large μ, $p(\mu) = 2\tau_g/\mu^3$, since the finite extent of the sources is not important for galaxy mass lenses], and by an additional population of compact objects with $\tau = 0.1$, with breaks in the magnification probability distribution at $\mu_1 = 4.5$ and $\mu_2 = 20$ [defined before (12.44)] to account for a mass spectrum of these compact lenses and a finite source size (we consider QSOs at a redshift of, say, $z = 2$). In addition, the larger-scale mass distribution in the universe (e.g., cluster of galaxies which are not sufficiently compact to form caustics, or larger mass concentrations) are accounted for by assuming a Gaussian component of the probability distribution, whose amplitude and width are chosen as to satisfy the integral constraints (12.5).

The observed source counts have been descibed by a double power law, with a (negative) slope of 2.6 at the bright end, and a slope of 0.8 at the faint end; furthermore, it has been assumed that the source counts show an effective cut-off below $S \lesssim 0.1$. This latter assumption appears reasonable physically to account for the finite number of QSOs, but the value of the cut-off is completely arbitrary and has been chosen for convenience. These source counts are displayed by the dashed curve in Fig. 12.15b. The scale of S is arbitrary, but can be related to observed counts by noting that the latter show a break at $B \approx 19.2$, which corresponds to about $S \approx 4$.

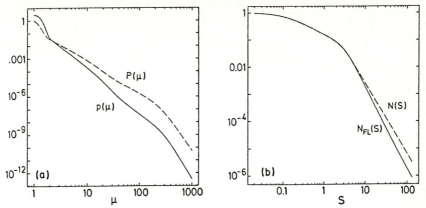

Fig. 12.15. (a) The magnification probability distribution $p(\mu)$ (solid line), for the model described in the text, and the cumulative probability $P(\mu) = \int_\mu^\infty d\mu'\, p(\mu')$ for a magnification larger than μ (dashed line). Beyond $\mu \approx 30$, the contribution from galaxies (and galaxies assisted by their host clusters) dominates. The cut-off at $\mu \gtrsim 300$ has been used just for numerical convenience and does not affect the results to be described below. (b) The dashed line shows the assumed source counts $\frac{dn_{\mathrm{obs}}(S,z)}{dz} \equiv N(S)$, whereas the solid line describes $\frac{dn_0(S,z)}{dz} \equiv N_{\mathrm{FL}}(S)$, which according to (12.9) is directly proportional to the intrinsic luminosity function. These two functions are related via (12.4b) and (12.9), with the magnification probability displayed in (a). The scale of S is arbitrary, but can be scaled by relating $N(S)$ to observed source counts, which show a break at about $B = 19.2$. According to the figure, this break corresponds to about $S = 4$

The result of the deconvolution described above are the unlensed source counts $N_{\mathrm{FL}}(S) \equiv \frac{dn_0(S,z)}{dz}$, related to the luminosity function by (12.9); they are shown by the solid line in Fig. 12.15b. In the flat part, below $S \sim 4$, $N_{\mathrm{FL}}(S)$ and $N(S) \equiv \frac{dn_{\mathrm{obs}}(S,z)}{dz}$ nearly coincide, showing that the amplification bias is very weak for such faint sources (in fact, as can be seen from the preceding discussion, the amplification bias is even negative for these faint sources). For the brighter sources, however, the amplification bias plays a significant role for $N(S)$, as can be seen from the ratio $\frac{N(S)}{N_{\mathrm{FL}}(S)}$.

To show the impact of the amplification bias more clearly, we have displayed in Fig. 12.16 the fraction

$$f(S,\mu) := \frac{1}{N(S)} \int_\mu^\infty d\mu'\, p(\mu')\, N_{\mathrm{FL}}\left(\frac{\langle\mu\rangle\, S}{\mu'}\right)$$

of sources with flux larger than S which are magnified by more than μ, for various values of S, corresponding to blue magnitudes of between $B = 20$ and $B = 16.5$. As expected, f increases with S, i.e., the amplification bias becomes more important for the bright sources. For the highest values of the flux threshold S, the fraction of highly magnified sources is considerable: sources with $B \leq 16.5$ have a probability of $(0.38, 0.16, 0.04)$ to be magnified by more than $\mu = (10, 20, 50)$. As we have remarked before, such high magnifications do not appear unrealistic, since for some of the observed

443

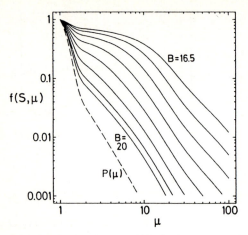

Fig. 12.16. The fraction $f(S,\mu)$ of sources brighter that S which are magnified by more than a factor μ, for the example shown in Fig. 12.15. If we identify the break in the number counts at $B \approx 19.2$ with the flux $S = 4$, the solid curves correspond to (from top to bottom) $B = 16.5, 17, \ldots 20$, and the dashed line is $P(\mu)$, i.e., the probability that any source is magnified by more than μ.

GL systems, the flux ratio of the images is high, pointing towards a large magnification of the total QSO flux.

In [SC91.2], an example is presented where the magnification probability distribution has a larger amplitude; in that case, the observed counts can be obtained from a luminosity function with an efficient cut-off. If this example accounts for the magnification probability in the real universe, all the bright QSOs at high redshifts are strongly magnified. We do, however, not want to conclude this from the currently available theoretical analysis and observational results, but just point out this possibility.

12.5.2 Observational hints of amplification bias

The question of whether source counts are dominated by lensing cannot be answered from theoretical considerations alone. Let us suppose they are not. Then, if our choice of the luminosity function (at a fixed redshift) is reasonable, and if the amplification bias is due to microlensing, we must choose $\alpha_2 = 2.6$ to match the observations. To avoid the dominance of lensing, we either conclude that the mass spectrum of the lenses is too steep, $\beta > 2.3$, or that the source sizes are too large. The latter would mean that one needs, for a mass range up to $10^2 M_\odot$, source sizes larger than 10^{18}cm, which are not likely, since, at least in some QSOs, variability on time scales shorter than a few days has been detected. Of course, there is no indication that the size of the emission region is similar for all types of AGNs. These constraints depend on the assumed density Ω_L of lenses; for small Ω_L, they are weaker than quoted above. The value of Ω_L itself is only weakly constrained. The observed starlight in the universe corresponds to $\Omega_L \approx$

0.01 in main sequence stars. However, compact objects are a promising dark matter candidate [RE82.1], whose dynamical influence is observed in the rotation curves of galaxies and in the dynamics of clusters of galaxies. On the other hand, nucleosynthesis provides additional constraints. To account for the observed amount of He^4, Li^7 and deuterium, the cosmic density in baryons (and black holes which formed after primordial nucleosynthesis) should be less than about 0.2 times the critical density.

The foregoing argument is based on the idea that the high-luminosity end of the luminosity function can indeed be represented as a power law. This need not be the case; there may well be a cut-off in the distribution, which would be difficult to verify observationally, since the statistics become poor for the apparently most luminous QSOs. If there is a (more or less) sharp cut-off in the luminosity function, all sources with apparent luminosity above this cut-off will be magnified, i.e., in this case, the source counts at the upper luminosity end will be completely dominated by the amplification bias.

How do we detect an amplification bias? In general, it is difficult to decide whether source counts are amplification-biased, if no independent information about the intrinsic luminosity function is available. Here, we discuss some possible signatures of the amplification bias, first assuming that it is caused by galaxy-mass lenses.[12]

If the amplification bias is due to galaxy-mass lenses, we expect that some of the magnified sources will be multiply imaged (the so-called magnification-multiplicity conjecture, see, e.g., [KO90.6], [KO91.1]). Thus, searching for multiple images of the apparently most luminous sources of a given kind is a promising search strategy for the amplification bias. As described in Sect. 12.2.5, such a search has been conducted, with a surprisingly high success rate in finding multiply-imaged QSOs [SU88.2] (see also [CR89.1]), a result which is very difficult to understand without assuming that the QSO source counts at high redshift and high luminosity are amplification-biased [BL90.1]. The high flux ratios of the images in UM425 [ME89.1] and 1115+080 [WE80.1], and the lens models for 2237+0305 and 1413+117, also suggest that the total flux of these sources, by which they were selected, is highly magnified, thus providing additional evidence for the bias.

High magnification can also occur without formation of multiple images (or with multiple images that cannot be resolved). Therefore, a valuable search program might call for attempting detections of foreground galaxies, in the vicinity of overly bright, single objects. The statistical overdensity of galaxies near high-redshift QSOs mentioned in Sect. 12.3.5 can be considered evidence in favor of the amplification bias. Also, the detection of galaxies on the line-of-sight towards some high-redshift BL Lac objects (see Sect. 12.4.4) supports the view that they are gravitationally magnified. The difference

---

[12] For an account of the astrophysical consequences of an amplification bias, see [OS86.1].

in the redshift distributions of X-ray selected AGNs with and without a nearby foreground galaxy [ST87.1] may be considered additional evidence, although the statistical significance of this result is not clear.

Quite surprisingly, there is evidence that high-redshift radio galaxies are affected by the amplification bias [HA90.1]: three-quarters of the 3CR radio galaxies having $z \geq 1$ comprise multiple optical components, with sizes and luminosities comparable to those of bright galaxies. The fractional overdensity of (presumably foreground) galaxies surrounding these radio galaxies is about nine, and an overdensity of foreground clusters is also detected, but at lower statistical significance. An analysis of the associations case by case shows that the radio galaxies are in fact magnified (by a factor between ~ 1.6 and ~ 6), and none of the radio galaxies near a foreground galaxy would be in the flux-limited 3CR sample had they not been magnified. The fact that a radio-selected sample can be affected by the amplification bias was not expected [PE82.1].

If extended sources such as the 3CR radio galaxies were subject to magnification, more compact sources, especially QSOs, should be even more strongly affected by the amplification bias. Owing to the small size of their optical continuum-emitting region, they should be affected by ML. Detecting amplification bias due to ML poses a greater challenge than detecting that from galaxy-mass lenses, unless the microlenses are spatially associated with galaxies, because then, one can try to detect the host galaxy. Further signs of ML, such as lens-induced variability, were discussed in Sect. 12.4, together with the difficulties associated with attempting to distinguish variability due to ML from intrinsic variability. The clearest prediction of ML is that the light curve of a source that is microlensed should be statistically symmetric with regard to the time direction.[13] Thus, well-observed light curves of QSOs and BL Lac objects should be checked for such symmetry, which would not occur naturally from intrinsic variations.

Statistical studies of source samples can potentially be used to test whether source counts are affected by ML. For instance, if a sample is selected by color, it includes mainly sources whose continuum is magnified, more than their broad-line regions. Therefore, since the magnification probability increases with redshift, one might expect that higher-redshift sources have a smaller line-to-continuum ratio than lower-redshift ones [CA82.1]. However, even for relatively low QSO redshifts (~ 0.5), the amplification bias can be considerable. Furthermore, if one assumes that there is no significant amplification bias, the QSOs must have undergone strong evolution with redshift. In addition, there is no reason why their intrinsic line-to-continuum ratio should not also evolve, and the same is true for the ratio of optical-to-X-ray emission.

One argument against the dominance of ML is the fact that variable QSOs are, as a rule, radio-loud. Since only $\approx 10\%$ of all QSOs have appre-

---

[13] Even this 'clear prediction' need not necessarily be valid, for if the source is intrinsically asymmetric, systematic statistical asymmetries of ML light curves can be produced.

ciable radio emission, this indicates a strong correlation. The radio emission is usually believed to arise in a region that is too extended to be magnified by ML. On the other hand, optically-variable QSOs usually have a compact radio core. These cores can be quite small as has been determined with VLBI observations. In fact, the recent discovery of variability of compact radio sources with time scales shorter than a day [QU89.1] indicates that radio cores can be much more compact than previously thought. Another possibility is that only sources with a compact radio core are sufficiently small at optical wavelengths to be influenced by ML.

It may well be that we will be able to recognize variability due to ML only after studying ML in multiply-imaged QSOs, where it can be distinguished relatively easily from intrinsic variations. Although the ML simulations described in Sect. 11.2.5 are fairly sophisticated, the calculated light curves (Sect. 12.4) are obtained by assuming a simple geometrical structure for the source. It would not be too surprising if nature were much more complex than assumed in these theoretical studies.

Amplification bias has also be accounted for the understanding of groups of galaxies with discrepant redshifts. The group VV 172 of five galaxies, one of which has more than twice the redshift of the other four, has been investigated by using the magnification effect of the four foreground galaxies on the background object [HA86.2]. This magnification effect increases the probability that a background object is seen projected in this group.

12.5.3 QSO–galaxy associations revisited

As we have shown in this section, it is not necessarily clear that the observed source counts are directly related to the intrinsic luminosity distribution of QSOs. Note that this was one of the assumptions made in Sect. 12.3 when QSO–galaxy associations enhanced by lensing were considered. A natural question to ask is then: is it possible that, starting from a luminosity function which is intrinsically steep, lensing can account for the overall source counts (i.e., with a slope of 2.6 for high fluxes), and *at the same time* produce an overdensity of QSOs in the vicinity of low-redshift galaxies? To cut a long story short, the answer is *no*. The reason is that [LI88.2], if the overall source counts are dominated by the amplification bias at the bright end, then they are proportional to the optical depth τ_p in lenses. On the other hand, the density of QSOs near galaxies is proportional to the sum of τ_p and the contribution τ_g to the optical depth from the compact objects in that galaxy. Thus, the fractional overdensity is proportional to τ_g/τ_p. The optical depth for compact objects in galaxies depends mainly on the velocity dispersion, if one assumes that all the mass is concentrated in compact objects (the most favorable assumption for the effect under investigation); thus, it is fixed by observations. A large overdensity is thus possible only if τ_p is very small. This, however, leads to a contradiction with the observations: the source counts are hardly influenced by lensing at the faint end. At the bright end, the lens-dominated source counts, as shown in this section, have a power law whose slope is determined by the mass spectrum of the lenses, and whose amplitude is proportional to the optical depth in lenses, τ_p. As shown in Fig. 12.14, for a relatively large optical depth, the bright and faint ends of the lens-induced source counts fit together well, but if τ_p is small, the amplitude at the bright end decreases, hardly influencing the faint end. Since these two regions have to be fit together, there would be a "hump" in the source counts near the turnover flux, with amplitude proportional to $1/\tau_p$. To achieve significant overdensities, optical depths as low as 0.05 are required; thus, the source counts would exhibit a rather well-defined feature where they are a factor $\gtrsim 10$ larger at the break than the extrapolation from larger fluxes would predict (J.E. Gunn, private communication). Such a feature would have been observed, but it has not. Thus, even by dropping the assumption that the overall source counts are not influenced by the amplification bias, it does not seem feasible to attribute the "Arp effect" to the GL action by compact objects in foreground galaxies. However, these remarks do not apply to the overdensity of galaxies near high-redshift QSOs, since in this situation, the required amplification bias for the overall source counts is much smaller than for the "Arp associations".

12.6 Distortion of images

So far, we have discussed the statistical influence of the magnification caused by gravitational light deflection on observations of very distant compact sources. Here, we describe a second effect which, although related to the first one, can be observed with different techniques. The image of an extended source can be distorted, as we have seen on different occasions in this book; for instance, a sufficiently small circular source will have elliptical images (see Sect. 5.2).

Our discussion omits technical difficulties and focuses on a physical description of the effects, with examples from the literature. As we show below, the difficulty of detecting the distortion due to lensing is mainly due to the lack of 'simple' sources, e.g., a population of circular sources. The same problem affects the magnification: since no standard candle is known, it is not possible to identify magnification solely from a flux measurement. However, as we will see in Sect. 13.2, there are some images of distant sources ('arcs') whose shape is so extreme that they can readily be identified as candidates for GL distortions.

Consider first the simplest case of a circular source at moderately high redshift ($z \gtrsim 0.3$, say). Since a light bundle emitted by such a source is affected by matter inhomogeneities in the universe, its image will appear elliptical, and the ellipticity will depend on the matter near the line-of-sight to that source. If we had a population of circular sources, such an effect would be easily observed; however, the sky projections of extended sources such as galaxies are not circular in general, but elliptical or irregular. It is thus impossible to distinguish intrinsic from lens-induced ellipticity for any individual object. Thus, the detection of a lens-induced shape distortion becomes a statistical problem.

Galaxy ellipticity due to clusters. Given a model for the matter distribution in the universe, one can in principle calculate the expected distribution of image ellipticities for a population of sources with prescribed intrinsic distribution of ellipticity. Such a calculation was performed in [WE85.1], where a distribution of clusters with power-law functions of radius for the surface mass density represented the lens population. For the calculation of image ellipticities, the two clusters nearest the line-of-sight were taken into account; inclusion of additional clusters hardly influenced the results. The source population was chosen to agree with the distributions in galaxy catalogues, and cosmological evolution for the galaxies was disregarded. The resulting distribution of ellipticities of galaxy images in certain magnitude ranges is shown in Fig. 12.17, for two values of the density Ω in clusters and two values of the power-law index k [$\kappa(x) \propto x^{-k}$, $1 \le k \le 2$]. The distribution for the brighter galaxies (solid line) is hardly influenced by lensing, since these galaxies are predominantly at low redshift, where the optical depth for lensing is small. The most significant change at fainter magnitudes is the depletion of images with small ellipticities. However, for this effect to be observable, the density Ω in lenses must be substantial, since the distribution of intrinsic ellipticities is very broad. In addition, it is difficult to measure small ellipticities for faint galaxies. Therefore, it appears that the ellipticity caused by lensing will be very difficult to distinguish from the broad intrinsic distribution, even in a statistical sense. Note also that a larger value of k increases the lensing effect. This property can easily be understood analytically, as we show next.

Consider an axially symmetric matter distribution and a small circular source with angular diameter $d\beta$ and angular impact parameter β. If θ denotes the angular variable in the lens plane, the ray-trace equation is

$$\beta = \theta - \frac{m(\theta)}{\theta} \quad ,$$

where $m(\theta)$ is the dimensionless mass within a circle with angular radius θ [see (8.3)]. The coordinate θ of an image of the source satisfies the above lens equation. The extent $(d\theta)_r$ of the image in the radial direction is given by

$$d\beta = \left(1 - \frac{m'}{\theta} + \frac{m}{\theta^2}\right)(d\theta)_r \quad ,$$

where the prime denotes differentiation with respect to θ. The angle φ subtended by the source as seen from the origin of the source plane is $\varphi = d\beta/\beta$. Owing to the symmetry of the lens, the image subtends the same angle. Hence, the extent of the image in the tangential direction is

$$(d\theta)_\varphi = \frac{\theta}{\beta} d\beta \quad .$$

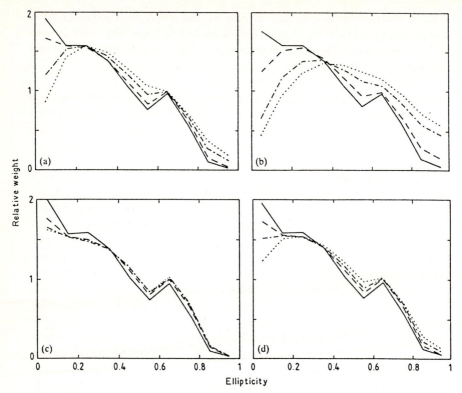

Fig. 12.17a–d. The distribution of ellipticity of galaxy images. In all panels, solid (dashed, dot-dash, dotted) lines correspond to apparent magnitude bins of 15-18 (18-21, 21-24, 24-27) mag. A cosmological density of clusters of $\Omega = 1$ has been assumed for (**a**) and (**b**), and $\Omega = 0.1$ for (**c**) and (**d**). The power-law index k of the radial mass distribution is 1.4 in (**a**) and (**c**), and 1.8 in (**b**) and (**d**). The intrinsic ellipticity distribution of galaxies follows closely the solid line in (**c**) and (**d**). As seen from the figure, gravitational light deflection leads to a lack of galaxy images with small ellipticities. Since faint galaxies are on average at larger distances, their distribution is most strongly affected (from [WE85.1])

The ellipticity of the image, defined as $e = 1 - b/a$, where a and b are the major and minor axis, respectively, of the resulting ellipse, is

$$e = 1 - \frac{(d\theta)_r}{(d\theta)_\varphi} = 1 - \frac{1 - m/\theta^2}{1 + m/\theta^2 - m'/\theta} \ .$$

Consider the foregoing expression for a deflection that occurs well away from the tangential critical curve of the lens, i.e., $m(\theta) \ll \theta^2$; then,

$$e \approx \frac{2m}{\theta^2} - \frac{m'}{\theta} \ .$$

The last equation shows that, for a given mass $m(\theta)$, the ellipticity is largest for a compact lens ($m' = 0$) and smaller for lenses with more extended matter distributions. For a power-law dependence of the surface mass density, $m(\theta) \propto \theta^{2-k}$, and thus $m'/\theta = (2-k)m/\theta^2$; this yields $e \approx k m/\theta^2$. This shows that larger values of k imply larger ellipticities, as seen in Fig. 12.17. The reason is that a larger value of k results in a stronger 'compression' of the image in the radial direction.

As discussed in [WE85.1], the observations of lens-induced ellipticities would be less demanding if one had a population of sources with a narrow distribution of intrinsic ellipticities. It seems that radio galaxies have significantly smaller ellipticity than galaxies in general; very few are found with ellipticity above $e = 0.35$. Hence, a sample of these sources at redshift $z \approx 1$, if observed to show a tail of larger ellipticities, could be used to probe the large-scale matter distribution. However, such conclusions are based on the assumption that any cosmological evolution of source shapes is negligible.

Inclusion of orientation. The foregoing discussion concentrated just on the value of ellipticity, and not on the orientation of GL images. If the image and the lens in a GL system are observed, the orientation of lens-induced distortion is such that the source is elongated in the tangent direction, as shown above (see also Fig. 8.1). An analysis of such properties was undertaken in [TY84.1], where a sample of galaxies was searched for foreground-background pairs, and the orientation and ellipticity of each background source was analyzed. Foreground and background objects were discriminated according to their apparent luminosity, under the assumption that fainter galaxies are further away than brighter ones. The sample comprised $\approx 10^4$ foreground galaxies and ≈ 4 times as many background galaxies. The sample yielded 27802 foreground-background pairs with separations below $63''$. The resulting statistics of ellipticity were then compared with Monte-Carlo simulations with a galaxy distribution drawn from the survey. Such a comparison yields upper limits for the mean mass-compactness of the foreground galaxies, i.e., the mass within a limiting radius. For isothermal spheres, this mass is proportional to the mean line-of-sight velocity dispersion. The upper limit for this dispersion quoted in [TY84.1] is about $170\,\mathrm{km/s}$, less than the values obtained from observations of rotation curves of galaxies. However, in this analysis it was assumed that the background objects are at infinity. This assumption leads to a smaller value of the mean velocity dispersion, since the quantity probed by the observations is $\langle \sigma_v^2 D_{\mathrm{ds}}/D_{\mathrm{s}} \rangle$. As was shown in [KO87.6], correcting for this assumption raises the upper limit on the mean velocity dispersion by up to a factor of 1.7. In particular, the velocity dispersion for an L_* galaxy, the luminosity scale in the Schechter function, can be increased above $300\,\mathrm{km/s}$, now in accordance with other observational limits. Thus, the sample in [TY84.1] does not yield significant constraints on the matter distribution in the universe. However, the ellipticity due to lensing was detected in their sample, as can be seen in their Fig. 1, for small angular separations. Although it has been suspected that these close pairs are contaminated by large numbers of foreground dwarf galaxies [PH85.1], such a possibility can be excluded [TY85.1], based on direct measurements of the angular galaxy-galaxy correlation function.

A further application of the shape distortion due to GLs is the search for very-large-scale matter inhomogeneities. Suppose there is a very large concentration of matter, say a supercluster of galaxies in the sky, which would distort all images of background sources seen through it. If an image is not very near the cluster center, its induced ellipticity is much smaller than the typical intrinsic ellipticity for galaxies; however, images would be preferentially aligned in the direction perpendicular to the radius vector to the center of the matter distribution. Thus, one expects coherence in the ellipticity of galaxy images in a certain region of the sky. A study based on this idea [VA83.1] yields strict upper limits on the coherent, large-scale ellipticity, which in turn leads to a limit for the density contrast in large-scale matter inhomogeneities. We will come back to this issue in Sect. 13.2.

Distortion of radio sources. A related but different problem is the distortion of large-scale radio sources [BL81.1]. 'Classical' radio sources consist of three components, two outer radio lobes and a core. The components in some of these sources are aligned to a remarkable degree. Also, many radio sources contain long and very straight radio jets. A gravitating mass near the line-of-sight towards such a colinear source would bend the structure, thus destroying the alignment of a triple source or leading to a kink in the jet. Using an elegant, though rather involved mathematical technique, Blandford and Jaroszyński [BL81.1] calculated the rms bending angle of an intrinsically colinear, high-redshift triple source to be $\approx 50° \bar{\Omega} \theta^{-0.4}$, where θ is the angular size of the source in arcseconds, and $\bar{\Omega}$ is the cosmological density clustered in galaxies. This effect would be observable if we had sufficient knowledge regarding the intrinsic structure of high-redshift radio sources, but the problem is the same as encountered before: it is very difficult to distinguish misalignment caused by GLs from intrinsic effects. In this context, it is interesting to note that indeed a larger misalignment of source structure is observed for higher redshifts [BA88.2], one of the possible explanations being the effect of the intergalactic medium which disrupts and deflects matter outstreaming from the radio galaxies. Since the density of this medium is expected to be larger at earlier epochs, the observed redshift dependence is to be expected. Perhaps more promising is the occurrence of kinks in jets, which may be caused by the GL effect of a galaxy near the line-of-sight to the jet. Although in this case the kink may also be intrinsic to the source, a measurement of correlation between kinks in jets and the positions of foreground galaxies may be able to isolate lens-induced kinks. For each system, the observed distortion would then yield a mass estimate for the foreground galaxy. Note that such a determination of galaxy masses is quite different from other, kinematical mass determinations, and would thus not only provide an independent check on those methods, but might be applicable to galaxies at large redshifts.

12.7 Lensing of supernovae

Why supernovae? As we mentioned several times before, the main difficulty for detecting the influence of gravitational light deflection on observations is the large spread in source properties. Since QSOs are not standard candles, it is impossible to determine directly whether any given source is magnified; only more indirect signs of magnification, such as variability, can be used to *argue* in favor of magnification. The same is true for the problem considered in the last section: since galaxies are not circular, any ellipticity in the image of a galaxy can be due to intrinsic or to lens-induced ellipticity. In order to detect lensing without spectacular multiple or highly distorted images, one would like to know a population of sources with well-known intrinsic properties. The best standard candles known are supernovae (SN). It seems that at least SN of Type I (SNI) have a very narrow range of peak luminosities (for a review, see [WA86.1]). Supernovae of Type II (SNII) have a larger spread in peak luminosity, as was clearly demonstrated by SN1987a, whose peak luminosity was a factor of about 10 below expectations. However, even for SNII, it may be possible to isolate a subclass which consists of fairly good standard candles, for example, if they are classified according to the velocity of expansion. In what follows, we assume that SN have a well-defined peak luminosity.

Let us consider a supernova that is observed in a distant galaxy. Its peak flux and redshift can be measured, in principle, and thus an intrinsic luminosity can be computed. If the value obtained were a factor of 10 above the value assumed for the standard candle, that would be a good signature of magnification. This idea has been advanced in a series of publications ([WA86.1], [SC87.6], [LI88.1]); we now discuss some of the results.

SN in their brightest phase are very compact sources, with radius $\leq 10^{15}$ cm at peak luminosity. Thus, they can be substantially magnified by stellar-mass objects; more specifically, their magnification statistics can probe the distribution of compact objects in the universe, provided they have a mass $\geq 10^{-3} M_\odot$.[14] Such matter could constitute part of the expected dark matter, as we discussed before. Given the density of compact objects in the universe, one can calculate the probability that a supernova at a certain redshift is magnified by more than a factor μ (see Sect. 12.5). Assuming a distribution of supernova explosions as a function of redshift, and using the fact that the peak luminosity of the class of SN under consideration is the same for all events, it is then straightforward to calculate the fraction of observed SN up to a limiting magnitude (at peak luminosity) that is magnified by more than a factor μ; note again that the assumption of the same intrinsic luminosity for all sources allows to determine the magnification for every individual source, just from the measurement of the peak flux and the redshift. This fraction will be proportional to the density of compact lenses in the universe. Hence, a sample of SN at relatively faint

---

[14] for observational prospects, see below

magnitudes may be used to detect a population of compact objects. In the evaluation of the fraction of magnified sources in a flux-limited survey, it is essential to take into account the redshifting of the spectrum of the source, which shifts the flux density in different spectral bands. This effect is the well-known k-correction; it weakens the flux of SNI relative to the D_L^{-2} distance effect, whereas the contrary is true for SNII (at $z \lesssim 1.5$). The difference in behavior of the k-correction for SNI and SNII causes the fraction of lensed SNI to be about ten times smaller than the fraction of magnified SNII in a sample with limiting magnitude 22.

Selected results. In Fig. 12.18, we display the expected fraction $F(>\mu)$ of SNII in a sample with limiting R magnitude $m_{\rm lim} = 22$, that are magnified by more than a factor μ, where we assumed that the comoving supernova rate evolves with redshift as $(1+z)^\beta$ (i.e., $\beta = 0$ corresponds to no evolution). To obtain these results, we assumed that the compact objects have a mass of $10^2 M_\odot$; however, the precise value of this mass is not important as long as a large mass fraction of the compact objects is contained in those with $M \gtrsim 0.1 M_\odot$. A value of $\Omega_0 = 0.2$ was assumed for the cosmological density of compact objects; the fraction $F(>\mu)$ is proportional to Ω_0. If the supernova rate is not evolving cosmologically ($\beta = 0$), about 1% of all SNII brighter than $m_R < 22$ are magnified by a factor larger than 3; allowing for evolution increases this fraction considerably. This is due to the fact that the magnification facilitates the detection of SN at higher redshift; hence,

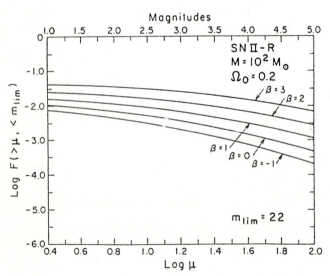

Fig. 12.18. The fraction $F(\mu)$ of SN of type II brighter than R magnitude $m_{\rm lim} = 22$, magnified by more than a factor μ, in a universe with 20% of the critical mass in compact objects. The different curves show the influence of the evolution of the comoving SN-rate [proportional to $(1+z)^\beta$] on the magnified fraction. For this plot, it is assumed that all compact objects have a mass of $10^2 M_\odot$; however, for $\mu \lesssim 50$ and $M \gtrsim 10^{-1} M_\odot$, the result is largely independent of the lens mass

if there are more SN events at larger redshift due to evolution, the behavior shown in Fig. 12.18 is expected.

One might expect that the fraction of magnified SN in a flux-limited sample may be increased by lowering the flux threshold. This is true up to about $m_{\text{lim}} = 23$; for fainter flux thresholds, the gain in redshift at which magnified SN can be observed leads to only a small gain in physical volume. In other words, the gain in volume is balanced by the increase of the luminosity distance. Therefore, a search to fainter than $m_{\text{lim}} = 23$ will yield more faint, unmagnified objects, rather than increase the number of magnified ones; this defines the optimal survey sensitivity. If the supernova rate is evolving, $\beta > 0$, this optimal flux threshold decreases, but not significantly.

Observational prospects. A supernova search is planned by J. McGraw and his colleagues at the University of Arizona; their method is to observe the same strip of the sky every night. Comparison of the data from successive nights will then detect variable objects. Since a supernova increases its brightness by a huge factor ($\sim 10^9$) within one day, these objects can be isolated from the data. This survey will search an area of about $17\,\text{deg}^2$. From the supernova rate observed locally, one can estimate that about 4 SNI, observed in the red, will be found per week, whereas in the same spectral band, roughly 0.6 SNII per week will be detected. Then, from Fig. 12.18, we find the result that, for a non-evolving supernova rate, about $\Omega_0/30$ SNII will be found per week, magnified by more than a factor of 10, or, for $\Omega_0 = 0.2$, about one every three years. From these numbers, it is clear that it will be very difficult to establish a sample of lensed SN to determine the density of compact objects in the universe. In particular, we must keep in mind that, for all SN found, one needs to take a spectrum to determine the class of SN and its redshift, which increases the observational requirements of such a search. On the other hand, this method seems to be the only practical way to detect a cosmologically significant density of dark compact objects of stellar mass. Another search for faint SN is currently carried out by C. Pennypacker at Berkeley. Note that recently, two high-redshift SN have been found, at $z = 0.28$ [HA89.3] and $z = 0.31$ [NO89.1].

Distortion of the light curve, and polarization. Apart from the magnification of the peak flux, a supernova can show other signs of lensing. If the source size at about peak luminosity ($\approx 10^{15}\,\text{cm}$) is not much smaller than the length scale η_0 introduced by a point-mass lens, the magnification will vary with time as the photosphere expands. This variation will change the shape of the light curve of the supernova. Several examples of distorted light curves are given in [SC87.6], both for point-mass lenses, and for more general situations, where the source lies near a caustic of an asymmetric lens. Whether these distortions can be considered as signatures for lensing of SN depends again on whether there is a standard light curve for all events of a certain type (which can be distinguished from the spectrum). This seems indeed to be the case, as discussed in [SC87.6].

Another possible impact of lensing on observations of SN is the occurrence of polarization. Assume the photosphere of a supernova to be spherically symmetric. The light from any point askew the line-of-sight to the center of the source which reaches us is polarized, as the radiation transport is via electron scattering, which results in polarization. However, the assumption of axial symmetry shows that the light integrated over the photosphere is unpolarized, as the contributions from different points cancel out. On the other hand, if the supernova is magnified, and if the magnification varies over the source, there is imperfect cancellation, and a net polarization can occur. Unfortunately, the resulting degree of polarization is small, typically less than 0.5%, which will be very difficult to measure. In addition, if the progenitor of the supernova had angular momentum, there is no reason why the photosphere should be symmetric; thus, any observed polarization may also be intrinsic.

If lensed SN are to be used as cosmological probes, the best method is a search for apparently overluminous events. As discussed before, such a search will be quite time consuming, but on long time-scales it may lead to constraints on the cosmological density of compact objects.

12.8 Further applications of statistical lensing

In this final section, we consider a number of applications of statistical lensing which have been proposed in the literature. In Sect. 12.8.1, the possibility is examined that the missing mass in the halo of our Galaxy, if composed of dark compact objects, can be detected by using their lensing capability. We describe in Sect. 12.8.2 the possible application of gravitational lensing to γ-ray bursts, in particular, for the recurrence of some of the bursts. Sect. 12.8.3 discusses the problem that, for some of the observed lensing candidates, no lens is seen in the vicinity of the images. In such cases, any simple lens model would require a very large mass-to-light ratio in the lens, as discussed in Sect. 2.5.

12.8.1 Gravitational microlensing by the galactic halo

Dark matter in our Galaxy. As discussed in Sect. 4.4, observations indicate that there is dark matter on different scales in the universe. This matter reveals itself by its gravitational action only, and it appears to exist in our immediate neighborhood, as inferred from observations of the dynamics of stars [BA84.1]. Such observations imply that about 50% of the mass in the Solar neighborhood is unobserved. In addition, the flat rotation curve of our Galaxy (and other spiral galaxies) can best be accounted for if it is assumed that the spiral structure is embedded in a spherical halo, whose mass determines the rotation velocity. It thus appears that there are two components of dark matter in our Galaxy: a disk component, with a scale height comparable to the thickness of the galactic disk as observed in stars, and a dark halo, which contains the bulk of the mass and is nearly spherical, but contributes only negligibly to the mass density in the Solar neighborhood.

The nature of dark matter in both components is unknown, but several observational constraints can be obtained. Dark matter in the disk cannot be ordinary baryonic gas, as this would just be another component of the interstellar gas, which can be observed directly (e.g., through its absorption). Also, dark matter in the halo cannot be baryonic gas; in order not to collapse (and form stars) within the lifetime of the Galaxy, it must be in hydrostatic equilibrium, requiring it to have a temperature such that its X-ray emission would conflict with the limits of the X-ray background [HE86.3]. If dark matter in the disk is in compact objects (from "snowballs" to black holes), several stringent limits can be obtained on their masses. The existence of wide binaries in the Galaxy excludes compact objects with mass $\gtrsim 2M_\odot$, as those would disrupt these weakly-bound binaries [BA85.1]. Observations of meteors and craters on the Moon and on planets rule out that these compact objects have a mass in the range $10^{-3}\text{g} \leq M \leq 2 \times 10^{16}\text{g}$ [HI86.1]. Also, since comets have been observed for a long time, a population of cometary-mass objects with the required mass density is ruled out. Objects in the mass range $10^{22}\text{g} \leq M \leq 10^{31}\text{g}$ would be difficult to observe, except at the upper range of this interval ('Jupiters'), where the objects might be detectable as infrared sources. Similar constraints can be obtained for dark matter in the halo [HE86.3], e.g., the halo cannot be composed of remnants of stars, as their evolution would have produced too much mass in heavy elements. This argument is not true if the remnants are black holes which can efficiently swallow the ejecta of their progenitors.

We will discuss below a method to detect compact objects in the halo of our Galaxy with mass in the range of $10^{-7}M_\odot \lesssim M \lesssim 10^2 M_\odot$, due to microlensing induced variability of stars in a nearby galaxy (the lower mass limit above is partly due to the requirement that the source, i.e., the stars in a nearby galaxy, are sufficiently compact for appreciable magnification, and partly from the time resolution of possible observations; the upper limit is obtained from the duration of magnification events – if the lenses are too massive, the duration will exceed several years). A method to detect compact halo objects with mass in excess of $\gtrsim 10^6 M_\odot$ has recently been suggested in [TU90.1].

Lensing by compact halo objects. If the dark matter in the halo of our Galaxy is in compact objects, it may cause gravitational magnification of compact extragalactic sources. As we already estimated in Sect. 2.3, the optical depth for such objects is very small. If their lensing effect is to be observed, one needs a huge number of sources to compensate for the small optical depth. One possibility is the observation of a nearby galaxy, where the individual stars are easily resolved [PA86.4]. Let us briefly estimate the optical depth for lensing of a halo composed of compact objects, which is responsible for the rotation curve of our Galaxy.

The rotational velocity $V_{\rm rot}$ in spiral galaxies is observed to be fairly constant, except in their innermost regions. For a spherical halo, this means that the mass within a radius r from the galactic center, $M(r)$ is given by $M(r) = V_{\rm rot}^2 \, r/G$. Thus, the mass density $\varrho(r)$ of the halo is described by

$$\varrho(r) = \frac{V_{\rm rot}^2}{4\pi G r^2} \;.$$

Consider a point source; if a compact object lies within one Einstein radius of the line-of-sight to this source, it is magnified by more than $\mu \geq 1.34$ (see Sect. 2.3). The Einstein radius ξ_0 (2.6b) for a lens of mass M at distance l from Earth is given by $\xi_0^2 \approx 4GMl/c^2$, where we have assumed that the source is much further away than the lens. The optical depth τ is then the expected number of lenses within an Einstein radius of the line-of-sight, i.e.,

$$\tau = \int_0^{l_{\rm h}} dl \, \pi \, \xi_0^2(l) \, n(l) \;,$$

where $n(l)$ is the density of compact objects at distance l. Note that this integral does not depend on the mass spectrum of the lenses, as $n = \varrho/M$ and $\xi_0^2 \propto M$; thus

$$\tau = \frac{4\pi G}{c^2} \int_0^{l_{\rm h}} dl \, l \, \varrho(l) \;.$$

The upper integration limit $l_{\rm h}$ is the length of the line-of-sight through the halo.

Let us consider the simplest case, where the line-of-sight to the source under consideration is perpendicular to the galactic disk. Let r_0 be our distance to the galactic center; then,

$$\varrho(l) = \left(\frac{V_{\rm rot}^2}{4\pi G}\right) \left(\frac{1}{r_0^2 + l^2}\right) \;.$$

In this case, $l_{\rm h}^2 = r_{\rm h}^2 - r_0^2$, where $r_{\rm h}$ is the radius of the halo. Inserting this value into the integral for τ,

$$\tau = \frac{V_{\rm rot}^2}{c^2} \ln\left(\frac{r_{\rm h}^2}{r_0^2}\right) \;.$$

Since $V_{\rm rot} \approx 200\,\mathrm{km/s}$, we see that the optical depth is of the order of 10^{-6}; thus, one out of a million sources will be magnified, at any given moment, by more than 1.34. For a star to be significantly magnified by a compact object of mass M, its radius R must be smaller than the length scale η_0 (2.6c) induced by the lens. This leads to the constraint:

$$\left(\frac{M}{M_\odot}\right) \gtrsim 3 \times 10^{-9} \left(\frac{R}{R_\odot}\right)^2 \left(\frac{D_{\rm d}}{10\,\mathrm{kpc}}\right) \left(\frac{D_{\rm s}}{100\,\mathrm{kpc}}\right)^{-2} \;.$$

We see that even very small masses can affect stars in a nearby galaxy, although the corresponding optical depth is of order 10^{-6}. The time-scale for the magnification variability is about the time it takes the lens to cross its own Einstein radius, since in the case under consideration, the velocity of the lens is likely to be the main contribution to the effective transverse velocity. Using $V_{\rm rot}$ as a typical lens velocity, the duration time becomes

$$t_{\rm d} = 7\times 10^6\,{\rm s}\left(\frac{M}{M_\odot}\right)^{1/2}\left(\frac{D_{\rm d}}{10\,{\rm kpc}}\right)^{1/2}\left(\frac{V_{\rm rot}}{200\,{\rm km/s}}\right)^{-1}.$$

Hence, for the smallest lens masses that can magnify a star, the lensing duration is of the order of minutes, whereas a Solar-mass object leads to lensing durations of a few months. The rate of events Γ per star monitored is simply $\Gamma = \tau/t_{\rm d}$.

A more detailed investigation of this effect has been given in [GR91.1], where a finite core size of the distribution of compact halo objects is considered (this effect is small for observations of stars in the LMC). In addition, a more realistic velocity distribution of the objects and a finite velocity of the Sun around the Galaxy has been taken into account. The optical depth, mean duration time of events, and rate of events as estimated above are in reasonable agreement with the more sophisticated study in [GR91.1], which also contains probability distributions of event durations, event rates, and the detection efficiency (i.e., the fraction of events which can be detected in a survey of given observation rate and total duration of observation).

Observational prospects. Owing to the small optical depth for halo objects (if they do exist), several million stars in a nearby galaxy must be monitored on various time scales, varying from hours to years, to be sensitive to a large range in lens masses. In fact, such a project seems feasible in the not too distant future, as automatic scanning of photographic plates is developed by various research groups. Then, intrinsic variability of stars must be distinguished from lens-induced variability. Most stars are not variable, however, and most variable stars are periodic. Nevertheless, those stars showing non-periodic (or even non-recurrent) variability have to be distinguished from lensing events, which provides one of the largest difficulties of such an observational project. Note that the light curve induced by a single star has a characteristic shape (Fig. 12.7) and can thus be used to identify a magnified star; in particular, it is symmetric.[15] It will be difficult to obtain statistically relevant observations, but there are hardly any alternatives for detecting dark matter, if it is concentrated in MACHOs (massive compact halo objects) of stellar mass or less. Thus, the observational effort may be rewarded. Also, an observational program will be useful in other respects, such as increasing our knowledge about the distribution of variable stars in nearby galaxies. In fact, several groups are currently planning an observational program of the kind just described.

---

[15] The symmetry is guaranteed only if the compact objects are isolated. If they form double or higher-multiplicity systems, more complicated light curves are expected. However, the shear for an isolated lens due to the other lenses can be completely neglected, as follows from the probability distribution (11.54) and the low optical depth κ_*. An asymmetry of the events can be due to the fact that most stars are in double systems; the light curve obtained by magnifying a double star (or multiple systems) will in general be not symmetric in time. Also, if the two stars have different spectral type, the event may also be chromatic. For a discussion of the observational difficulties, see [GR91.1].

12.8.2 Recurrence of γ-ray bursters

Gamma-ray bursts. Satellites launched to monitor γ-radiation from nuclear explosions detected sudden bursts of radiation whose origin was not on Earth. The detectors aboard these spacecraft had no angular resolution, but the direction could be inferred from measurements by several satellites, due to the difference in arrival times of photons at each satellite. During the short time of a burst (typically a few seconds), a γ-ray burster (GRB) is brighter than any other γ-ray source in the sky, including the Sun, by several orders of magnitude. The power of a GRB usually has a broad peak near 1 MeV, with a small ratio of X-ray to γ-ray luminosity.

The nature of the burst sources is largely unknown. Since the distribution of GRBs is observed to be isotropic, the sources must be either rather local, at distances below the scale-height of the galactic plane (as all known galactic sources are concentrated towards the galactic plane or the galactic center), or at distances much larger than that of the local group. The small rise-time of the bursts (sometimes as short as 0.01s) argues for compact objects. In addition, there have been several claims of detection of two line features, an absorption line at about 70keV, and an emission feature at about 400keV. If confirmed, these two lines could be interpreted as cyclotron absorption in a strong magnetic field typical for neutron stars, and a (gravitationally) redshifted pair-annihilation line, which also could occur in the gravitational field of a neutron star. These arguments in favor of a neutron star are not sufficient to achieve a consensus among the workers in the field; even if the GRBs are from neutron stars, neither their energy source nor their radiation mechanism is well understood. Also, it is quite possible that the GRBs form an inhomogeneous class of objects; e.g., the famous burst from March 5, 1979 is different from the majority of other bursts in nearly all respects. If GRBs are at cosmological distances, the origin of their energy is even more uncertain; several possibilities have been suggested in the literature (see, e.g., [BA87.1], [GO86.3], [PA86.2]).

Burst recurrence and lensing. If GRBs are at cosmological distances, they may be gravitationally lensed [PA86.2]. Since the light-travel-time along different light rays is different, one would expect for the burst to be seen several times, each time at a slightly different position (which cannot be measured with current methods; for sufficient angular resolution of the arrival time measurements, one needs satellites separated by distances of about one astronomical unit). Indeed, apart from the March 5 event, two more recurrent bursters have been observed, i.e., bursts where the directions coincided within the measurement error. In one case, three bursts with a separation of about two days occurred. This is about the time delay expected from a normal galaxy, if the source and the lens are at cosmological distances. In a second case, roughly 50 bursts were discovered within about a year. Interestingly enough, the highest-energy peaks occurred in pairs with small time delay, a feature that would be expected from lensing, where highly magnified images occur in pairs close to a critical curve. These observations motivated a more detailed study [PA87.2], where the image configuration and the mutual time delays for images of a distant source produced by an ensemble of galaxies were investigated. The main results can be described as follows: for small optical depth in these galaxies, only a few images appear, and they are well ordered by flux, i.e., the main image (with the largest magnification) has the smallest light-travel-time, followed by successively weaker images. For large optical depth, the time history of the images is rather chaotic, and many images of comparable magnification occur. In particular, if the lenses are not too compact (in the sense that their masses lie entirely within their Einstein radii), several close pairs of images occur. If the galaxies are clustered, this yields a clustering of images in time. From a comparison of the time-scale in the numerical simulations and the observed burst history, the mass of the galaxies is estimated, yielding about $2 \times 10^{10} M_\odot$, quite reasonable for galaxies.

The hypothesis of a GL origin for the recurrence of some GRBs leads to two predictions which can in principle be checked with observations. First, the shapes of the light curves and the spectra of the individual bursts should be identical (except for an overall magnification factor), at least if the emission of the burster is approximately isotropic and the source size is negligible. The light curves and spectra of the individual bursts of the two recurrent bursters mentioned above seem to agree. Second, the angular position of the bursts should be slightly different. As already noted, this requires a spacecraft in orbit around the Sun.

Of course, the recurrence may have a completely different origin, perhaps one intrinsic to the source. The lensing hypothesis is interesting, but will be difficult to verify in the near future; by then, our knowledge about the nature of GRBs may have increased.

12.8.3 Multiple imaging from an ensemble of galaxies, and the 'missing lens' problem

Motivation. As described in Sect. 2.5, there are multiple QSOs which are suspected to be GL systems, but for which no deflector is observed near the images. Observational constraints on the brightness of the (suspected) deflector are in some cases very stringent, since any simple lens (such as a centrally-condensed symmetric – or nearly so – galactic deflector) must have an extremely large mass-to-light ratio (of the order of several hundred in Solar units). Of course, this is an interesting situation, because these GL candidates may be caused by a concentration of dark matter. However, before such a conclusion can be drawn, one should investigate whether there are lensing situations where a fairly large image splitting can occur without a deflector near the images, or with a deflector of a considerably smaller mass than suggested by simple lens models.

One possibility that has been suggested [KO87.1] is 'shear-induced' lensing, i.e., multiple images that are created not by a centrally condensed mass, where the images of the source are seen on both sides of this deflector, but by the combined action of several deflectors, which can be relatively far from the images, but which generate a strong shear that can split images. Let us consider a specific example. The best way to visualize such a process is the Fermat potential $\phi(\mathbf{x}, \mathbf{y})$ introduced in Sect. 5.1. For a fixed source position $\mathbf{y}$, this potential can be considered as a surface in three-dimensional space, and the images of the source are found at the stationary points of this surface [compare (5.12)]. Without any deflector, the surface would be a paraboloid, and there would be just one image at the minimum. Now, imagine that there are two mass concentrations, situated such that the original minimum lies somewhere between them. These two masses (or groups of masses) will introduce two maxima in the surface, and, according to the theorem discussed in Sect. 5.4, in addition, there will be two saddle points. Thus, in the situation under consideration, there will be five images of the source (compare also the two point-mass lens discussed in Sect. 8.3). Depending on the details of the distribution, the two maxima can be very sharp, such that the magnification of the corresponding images is very small. In that case, there will be three bright images, and the two deflector components are not lying between them. Thus, it is conceivable that situations like that can account for candidate GL systems when no lens can be detected near the images.

For this idea to be attractive we must show that such shear-induced lens cases are not too rare to be considered a plausible explanation for 'dark matter' lens cases. Thus, the problem is necessarily of a statistical nature; it was studied with Monte-Carlo methods in [KA87.1] and [PA89.1], by randomly distributing deflectors in a given area of the lens plane and then finding all the solutions of the lens equation for this particular lens model, using the method described in [PA86.1] and at the beginning of Sect. 11.2.5. Some aspects of these simulations are described next.

Lens system. A realistic study of the question above would require a generated three-dimensional distribution of galaxies (or, nearly equivalently, a distribution of galaxies on several lens planes, as described in Chap. 9), and to assign to each galaxy a set of physical parameters (e.g., velocity dispersion, cut-off radius, core-radius, or even ellipticity, and so on). Since galaxies are not randomly distributed, but clustered, this would also have to be taken into account. We see that such a simulation would be extremely difficult and, worse, would require many free parameters that are ill-constrained by observations. Moreover, the simulation would be enormously time-consuming. To understand the essential properties of lensing statistics, it is desirable to have a simpler model, with fewer free parameters, which can be studied in detail. That is the approach of [KA87.1] and [PA89.1]. There, the lenses are distributed in a single lens plane, with a mean surface mass density of $\bar{\kappa} = 0.2$. This value was chosen because it is not too small for lensing to occur nearly exclusively by a single lens, and not so high as to be unrealistic. According to (5.25), the mean magnification of sources seen through such a lens plane is $(1-0.2)^{-2} \simeq 1.56$; the same mean magnification, relative to that in an empty cone in a Friedmann–Lemaître universe with $\Omega = 1$, $\tilde{\alpha} = 0$ would occur at redshift $z \simeq 1.631$. Hence, this model can be considered to be an approximate description for multiple imaging of sources in such a universe at a redshift of about 2, the typical redshift of lensed QSOs. Three different types of lenses were considered: point masses, singular isothermal spheres with cut-off radii, and Gaussian matter distributions, all with the same mass. For a length-scale in the lens plane equal to the Einstein radius ξ_0 of the mass of a single lens, the last two lens models are described by a single parameter: for the truncated isothermal spheres, the cut-off radius r, and for the Gaussian lenses, the 'one-sigma' width. Thus, in the notation of Sect. 8.1, for the isothermal sphere,

$$\kappa(x) = \frac{1}{2xr} \quad \text{for} \quad x \leq r$$

and is zero otherwise, and for the Gaussian model,

$$\kappa(x) = \frac{1}{2\sigma_r^2} e^{-\frac{x^2}{2\sigma_r^2}} .$$

The lenses were placed randomly in circular fields with radius $\sqrt{150}$, so that there are on average 30 lenses in each field.

Results. A very large number of statistical lens fields were generated and analyzed, in some cases up to 10^6 fields. Solving the lens equation for that many cases is one thing; 'real' work begins with the extraction of meaningful results from such a huge amount of data. The most straightforward quantity is the image multiplicity, shown in Fig. 12.19 for several models. The Gaussian lenses have a finite surface mass density everywhere, and thus, the odd-image theorem applies, whereas the isothermal sphere and the point-mass models can also produce an even number of images due to the singularity of their matter distribution. It is easily seen that, as expected, the probability for multiple images decreases sharply as the lenses become less compact.

Another convenient way to analyze the data is to define the image circle, the smallest circle containing all the images. The important result of the simulations is that in about 95% of all cases of multiple imaging, the center of at least one of the lenses is within the image circle. That means, referring to the original motivation of this study, that *shear-induced multiple imaging is rare*; in nearly all cases, one lens is near the images. We discuss this further below.

However, a more detailed look at the triple-image cases shows that most of them have two images at a minimum of ϕ, and one at a saddle point, quite in contrast to expectations – for a single Gaussian lens, multiple images are always of the minimum-saddle point-maximum type. This property shows that the multiple images are mainly due to the combined action of several lenses. In fact, for lenses with $\sigma_r \gtrsim 1.5$, practically no triple image geometry

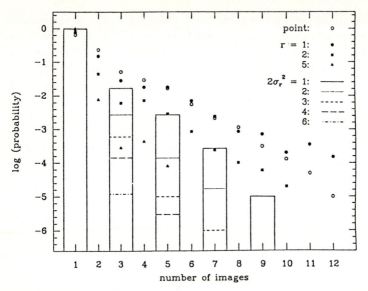

Fig. 12.19. The distribution of image multiplicity for Gaussian lenses (histograms) and truncated isothermal spheres (symbols); point masses are just a special case ($r = 0$) of the latter. Since for point mass lenses, owing to their compactness, about as many images as lenses occur, most of them with very small magnification, only images with magnification ≥ 0.1 are counted in this case

is of the 'standard' type. This result is rather surprising and is not yet easily understood, but it shows that shear *is* important.

Consider next the image separations, conveniently measured by the maximum separation for all images. In Fig. 12.20 the largest image magnification in multiple image systems is plotted as a function of this maximum separation $d_{\max}$. For a single point-mass lens, this separation is always larger than 2, whereas for a single truncated isothermal sphere with $r = 2$, the separation is always larger than 1. As the figure shows, this is roughly true even if one considers an ensemble of such lenses; however, rather surprisingly, there are many cases with a much larger separation than expected from a single lens. This is also seen for the Gaussian lenses, which show a broad distribution of image separations.

Finally, we note that, for the Gaussian model, many high-magnification cases occur that lack multiple images, quite in contrast to the isothermal sphere model, where most high magnification events occur in multiple image situations. This is because of the smoothness of the matter distribution, which allows light bundles to traverse an appreciable surface mass density without having strong shear. Note that this is quite in contrast to the discussion in Sect. 12.2, where the isothermal sphere models were predicted to lead to high magnifications only in multiple-image cases. It is clear that such a modification should be included in more realistic considerations of the expected frequency of lensed QSOs, in particular, if the amplification bias turns out to be important.

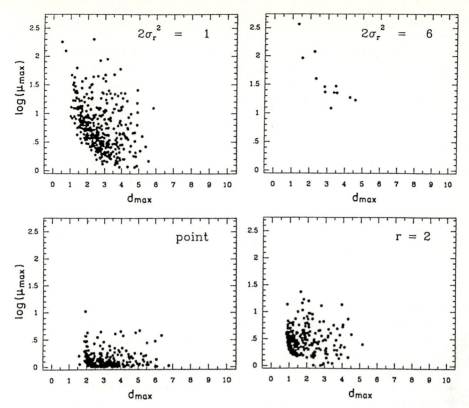

Fig. 12.20. The magnification of the brightest image as a function of the maximum separation for multiple image systems, for two Gaussian models, one isothermal sphere model, and the point-mass case. Note the broad distribution of d_{max} in all cases

The missing lens problem. Coming back to the question posed at the beginning of this subsection, the numerical simulations discussed above strongly indicate that the missing lens problem cannot be resolved by postulating shear-induced multiple imaging; the probability for this to occur is too small. However, the surprisingly large image separations seen in Fig. 12.20 somewhat ease the constraint on the mass-to-light ratio of the 'dark lens', because then, the image separations are frequently much larger than predicted by single-mass models. Or, for fixed (observed) angular separation, less mass is needed than would be expected from considering a single galaxy lens. Since the image separation scales with the square of the lens mass, a splitting twice as large as predicted from single lens theory would reduce the expected lens mass by a factor of four. Whether or not this 'solves' the missing lens problem cannot be decided on the basis of such a simple model.

Can the simplified assumptions made for this model significantly influence these results? This is difficult to say, without actually having carried out more realistic simulations. But probably all assumptions are noncritical, except for the randomness of the lens distribution. It may be that the results will quantitatively change if the clustering of galaxies is taken into account.

The missing lens can of course be a compact object with a mass distribution like a galaxy, which has failed to become luminous. Such dark galactic halos are predicted in cosmological models with biased galaxy formation, which have been suggested to account for the apparent discrepancy between the distribution of luminous galaxies and the large-scale distribution of dark matter as predicted from cosmological models (see [SU87.1] for references). Such dark galactic halos can act as lenses [NA88.1], although, as argued in [SU87.1], they may be not sufficiently compact. However, it was suggested in [SU87.1] that two such halos on the line-of-sight to a distant QSO can cause an observable image splitting, causing multiple QSOs without an observable lens (see also [NA89.2]). Such an explanation for the missing lens problem requires the density of dark galactic halos to be a significant fraction of the closure density of the universe. We want to point out that in such a scenario, the amplification bias of these objects on QSOs must be strong.

Another possible solution of the missing-lens problem must be mentioned [NA84.3]: what if the lens responsible for the splitting is not a galaxy, but a cluster of galaxies? Then, much larger separations than a few arcseconds are expected, typically about one arcminute, but it can happen that two images are very close to the (radial) critical curve for the cluster, and thus are highly magnified. A third image which would be produced can then be an arcminute away from this image pair, and can be much fainter. Thus, in this case, the observed images are a pair of radially-merging images. This suggestion can be tested when the source is resolved, i.e., mainly for compact radio sources. For the double QSO 0957+561, milliarcsecond radio cores were detected, with accompanying jets that are nearly parallel to each other, rather than being oriented towards each other, as would be expected for radially-merging images. Hence, at least for this case, the cluster hypothesis can be ruled out, but, at any rate, in this system the lens is observed. For other cases, the cluster hypothesis cannot be ruled out, and arguments related to the small probability of occurrence neglect the amplification bias introduced in this case. On the other hand, we know that a large part of the deflection in the GL system 0957+561 is done by the cluster in which the main lensing galaxy is hosted. As pointed out in [NA84.3], a weak galaxy, supported strongly by a compact cluster (which itself does not need to be critical, i.e., able to split images) can lead to large image splittings. Therefore, faint galaxies embedded in a compact cluster can cause strong lensing (see also Sect. 12.2) and may provide a way to solve the missing lens problem.

13. Gravitational lenses as astrophysical tools

Observations of GL systems can provide a wealth of astrophysical information. As F. Zwicky noted as early as one-half century ago [ZW37.1], GLs (1.) furnish an additional test of General Relativity; (2.) allow us to see fainter or more distant sources; and (3.) provide a means to estimate the mass of galaxies, of clusters of galaxies, and of unseen, massive aggregates of dark matter.

In this final chapter, we discuss applications of the GL effect, starting in Sect. 13.1 with the problem of determination of the parameters of a lens model, given an observed GL system. In particular, we investigate how one determines the lens mass and the Hubble constant, H_0, using a multiply-imaged QSO. The time delays for image pairs play a crucial role in this discussion. In Sect. 13.2, we investigate the luminous arcs produced by the GL action of clusters of galaxies on background galaxies. Simple models for observed arcs are discussed, as well as a simple analytic method for the construction of images of extended sources produced by nearly spherical lenses. As is the case for multiply-imaged QSOs, the mass of the lens cluster is the most robust model parameter. The recently discovered 'arclets' also have the potential of becoming a powerful tool to determine the mass distribution in clusters of galaxies.

In Sect. 13.3, we discuss several additional applications. Spectroscopy of multiply-imaged QSOs can be used to estimate the size of the objects responsible for their absorption-line systems. If the amplification bias affects the selection of such QSOs, a similar bias may affect the statistics of absorption-line systems. The light curve for a compact source which crosses a caustic can be used in principle to obtain information about the brightness distribution of the source; therefore, ML can be used as an 'occultation telescope' with very high resolution. If cosmic strings existed, they would reveal themselves as particularly simple GLs; the GL effect is therefore a unique means to discover these exotic objects.

We conclude in Sect. 13.4, with several topics that are usually excluded from discussions of gravitational lensing, but which are related to gravitational light deflection. To wit, we briefly discuss three examples, covering various extremes of the scale of gravitational deflection of radiation. First, the cosmic microwave background radiation, which presumably has traversed the universe from a redshift of $z \sim 1500$ until today, is deflected by the inhomogeneities of the universe, a process which has been proposed as the origin of its high isotropy. Second, the propagation of light rays in the Solar System is affected by the gravitational field of the Sun (and to a lesser degree, by the planets). Very accurate positional measurements must take this effect into account. Third, the strong gradients of gravitational fields in the vicinity of compact objects, such as neutron stars and black holes, affect the radiation emanating from these objects. The resulting effects are potentially observable, due to their singular character.

13.1 Estimation of model parameters

Importance of the time delay. Early in the 'modern history' of GLs, it was suggested ([RE64.2], [RE66.1]) that observations of a multiply-imaged source can be used to estimate physical parameters for the lens, in particular, its mass and H_0. Estimates of mass appear to be straightforward, since the image splitting increases in proportion to the lens mass. As to estimates of H_0, we note that nearly all observables in a GL system are dimensionless: the angular separations for image pairs, their angular positions relative to the lens, their flux ratios (image fluxes do not constrain the lens system, because the intrinsic luminosity of the source is unknown), the axial ratio and position angle of each (sufficiently small) elliptical image, and the redshifts of the source and the lens. The only dimensional observable is the relative time delay for image pairs. Using only the dimensionless observables, it is not possible to determine the cosmological distance scale, which is proportional to H_0^{-1}; however, the time delays for image pairs are proportional to this scale. In fact, from purely dimensional analysis we see that, since H_0 has dimensions of inverse time, the time delays for image pairs are $\propto H_0^{-1}$. Note that, since the lens mass is also a dimensional parameter, it cannot be determined from the dimensionless observables alone: the value of the time delay (or knowledge of H_0) is also necessary. Since H_0 at present is known only up to about a factor of two, any mass estimate made without measuring time delays is uncertain by at least the same factor. In this section, we study the prospects of determining these two quantities by considering a series of idealized situations. A time delay for pairs of images in a GL system is not well known at the moment; for the 0957+561 system, several measurements exist, claiming values for the time delay which differ by about 50% ([FL84.1], [SC86.1], [VA89.1], [SC90.3], [PR91.1], [RO91.1]). As discussed in [FA90.1] (see also Sect. 2.5.1), it is not clear yet whether these estimates should be considered reliable (however, see the discussion at the end of this section). In Sect. 13.1.3, we estimate the mass of the main lens galaxy in 0957+561, and the corresponding product $H_0\,\Delta t$.

The observables. The number and type of observables depend on the configuration of each GL system. We consider the lensing of a sufficiently small source, so that its images are distinguishable from each other; in some cases they may be resolved. The first observables that come to mind are (a) the source redshift z_s, determined from the spectra of the images; (b) the image positions $\boldsymbol{\theta}_i$, $i = 1, ...N$, where N is the total number of observed images; and (c) the fluxes S_i of the images. When the lens is observed, we also have (d) the lens redshift z_d and (e) the light distribution for the lens. For example, if the main contributor to the gravitational deflection is a galaxy whose position is measured, the image positions can also be measured relative to its center. In certain cases it is possible to investigate the structure of resolved images. That is the case for radio-bright QSO images such as those in the 0957+561 system, which have been studied with VLBI techniques (Sect. 2.5.1). In some cases it may also be possible to obtain additional observations of the lens itself. For example, the velocity dispersion of the lens galaxy, or that of its parent cluster, and its morphology, color and other parameters may be observable. These observations are useful in constraining a lens model: for instance, the velocity dispersion yields a corresponding estimate for the mass of a galaxy. Furthermore, in cases where the source is intrinsically variable on sufficiently short time-scales, the relative time delay for image pairs can be determined, at least in principle. Note that ML can cause flux variations in gravitational images. It may be difficult to separate intrinsic variations from ML; however, measurements over a sufficiently long time should allow one to distinguish between these two effects.

Relations between observables and the lens equation. The observables we have discussed may be used to estimate several quantities which appear in the ray-trace equation for GLs. Recall that this equation may be written as

$$\boldsymbol{\beta} = \boldsymbol{\theta} - \boldsymbol{\alpha}(\boldsymbol{\theta}) \quad , \tag{13.1a}$$

where $\boldsymbol{\beta}$ is the angular position of the source, and the scaled deflection angle $\boldsymbol{\alpha}$ is related to the "physical" deflection angle $\hat{\boldsymbol{\alpha}}$ through

$$\boldsymbol{\alpha} = \frac{D_{ds}}{D_s} \hat{\boldsymbol{\alpha}} \quad . \tag{13.1b}$$

Since (13.1a) must be satisfied for all image positions $\boldsymbol{\theta}_i$, for a fixed source position $\boldsymbol{\beta}$, we can subtract these relations in pairs, to eliminate the unobservable source position:

$$\boldsymbol{\theta}_i - \boldsymbol{\alpha}(\boldsymbol{\theta}_i) = \boldsymbol{\theta}_j - \boldsymbol{\alpha}(\boldsymbol{\theta}_j) \quad . \tag{13.2}$$

Since the intrinsic luminosity of the source is unknown, flux measurements do not yield the magnifications μ_i of the images, but with them we can calculate the ratio of magnifications for different images,

$$\frac{\mu_i}{\mu_j} = \frac{S_i}{S_j} \quad ; \tag{13.3}$$

also, recall that the magnification theorem (see Sect. 5.4) implies $\mu \geq 1$ for at least one image. More refined information can be obtained if the structure of the images can be measured. If a point source at β has images at $\boldsymbol{\theta}_i$, and a nearby point source at $\beta + \delta\beta$ has images at $\boldsymbol{\theta}_i + \delta\boldsymbol{\theta}_i$, then, for sufficiently small $|\delta\boldsymbol{\theta}_i|$,

$$\delta\beta \approx A(\boldsymbol{\theta}_i)\,\delta\boldsymbol{\theta}_i \equiv A_i\,\delta\boldsymbol{\theta}_i \quad , \quad \text{where } A(\boldsymbol{\theta}) = \frac{\partial \beta}{\partial \boldsymbol{\theta}}$$

is the Jacobian matrix of the lens equation. Although $\delta\beta$ is of course unobservable, it can be eliminated from the above equations by considering pairs of images once again:

$$\delta\boldsymbol{\theta}_i \approx A_i^{-1} A_j\, \delta\boldsymbol{\theta}_j \equiv A_{ij}\,\delta\boldsymbol{\theta}_j \quad . \tag{13.4}$$

The matrices A_{ij}, which represent the transformation of the structure from one image to another, can be obtained by observing three distinct points in each image that are not colinear (e.g., a triple source); from three such points, two independent difference vectors in each image can be obtained, and applying (13.4) yields the four components of the matrices A_{ij}. Note that, although the A_i are symmetric, the A_{ij} in general are not.

The time delay for two images was given in (5.46):

$$c\,\Delta t_{ij} = \frac{D_d\,D_s}{D_{ds}}(1+z_d)\left[\phi(\boldsymbol{\theta}_i,\beta) - \phi(\boldsymbol{\theta}_j,\beta)\right] \quad ,$$

where the Fermat potential is $\phi(\boldsymbol{\theta},\beta) = (\boldsymbol{\theta}-\beta)^2/2 - \psi(\boldsymbol{\theta})$, and $\psi(\boldsymbol{\theta})$ is the two-dimensional deflection potential. Using the ray-trace equation (13.1a), we can write $\phi(\boldsymbol{\theta},\beta) = [\boldsymbol{\alpha}(\boldsymbol{\theta})]^2/2 - \psi(\boldsymbol{\theta})$; thus

$$c\,\Delta t_{ij} = \frac{D_d\,D_s}{D_{ds}}(1+z_d) \\ \left\{\frac{1}{2}[\alpha^2(\boldsymbol{\theta}_i) - \alpha^2(\boldsymbol{\theta}_j)] - [\psi(\boldsymbol{\theta}_i) - \psi(\boldsymbol{\theta}_j)]\right\} \quad . \tag{13.5a}$$

Since the deflection angle is the gradient of ψ, we can express the last term in (13.5a) as an integral over $\boldsymbol{\alpha}$; the first term in (13.5a) can be simplified using (13.2):

$$c\,\Delta t_{ij} = \frac{D_d\,D_s}{D_{ds}}(1+z_d) \\ \left\{\frac{1}{2}(\boldsymbol{\theta}_i - \boldsymbol{\theta}_j)[\boldsymbol{\alpha}(\boldsymbol{\theta}_i) + \boldsymbol{\alpha}(\boldsymbol{\theta}_j)] - \int_{\boldsymbol{\theta}_j}^{\boldsymbol{\theta}_i} d\boldsymbol{\theta}'\,\boldsymbol{\alpha}(\boldsymbol{\theta}')\right\} \quad , \tag{13.5b}$$

where the line integral is computed along an arbitrary path in the image plane, from $\boldsymbol{\theta}_j$ to $\boldsymbol{\theta}_i$.

Number of independent constraints. Let us assume that we have observed a GL system with N images, where the lens is a galaxy near the images. Let us also assume that we have measured the redshift of this galaxy. We desire a model for the mass distribution of the galaxy that accounts for all the observations at hand. How many free parameters can a realistic, useful model have? To answer this question, we must count the number of independent observables. The image positions $\boldsymbol{\theta}_i$ yield $N - 1$ independent vector equations (13.2), i.e., $2(N - 1)$ independent scalar equations. The time delays yield $(N - 1)$ independent relations. If the image structure cannot be determined, the observed fluxes yield $(N - 1)$ relations between the image magnifications; hence, in this case, the total number of equations is $4(N - 1)$. If the image structure can be determined, and if at least three corresponding points in each image can be identified, the $(N - 1)$ independent matrices A_{ij} yield $4(N-1)$ relations, bringing the total number of equations to $7(N - 1)$.[1] Hence, a lens model containing this number of free parameters can be used to try to fit the observations. The parameters may be further constrained, e.g., by the absence of additional images that may be predicted by the models.

In subsection 13.1.2, we present an example for the type of parameter estimation we have discussed, including an estimation of H_0. But first, one might ask whether there might be variations of the lens parameters that leave all observables unchanged.

13.1.1 Invariance transformations

If a transformation of GL model parameters leaves all available observables invariant, these parameters cannot all be determined; thus, all lens models can be determined only up to such invariance transformations. Below, we discuss three transformations which leave all observables, save for the time delay, invariant, and two that also lead to an invariance of the time delay [GO88.1].

Note that the deflection angle $\hat{\boldsymbol{\alpha}}$ is obtained from (5.2) as a function of the surface mass density Σ in the lens plane. In some cases, we can observe the light distribution of the lens, i.e., its flux per unit solid angle. If we assume, for instance, a constant mass-to-light ratio for the lens, the measurement of its light distribution yields the mass per unit solid angle, which is proportional to ΣD_d^2. Therefore, let us rewrite $\hat{\boldsymbol{\alpha}}$ as

$$\hat{\boldsymbol{\alpha}}(\boldsymbol{\theta}) = \frac{4G}{c^2 D_\mathrm{d}} \int_{\mathbb{R}^2} d^2\theta' \left(D_\mathrm{d}^2 \Sigma\right)(\boldsymbol{\theta}') \frac{\boldsymbol{\theta} - \boldsymbol{\theta}'}{|\boldsymbol{\theta} - \boldsymbol{\theta}'|^2} \equiv \frac{4G}{c^2 D_\mathrm{d}} \mathbf{a}(\boldsymbol{\theta}) \quad , \quad (13.6a)$$

where in the last step we defined the function $\mathbf{a}(\boldsymbol{\theta})$, which has the dimension of a mass. The scaled deflection angle $\boldsymbol{\alpha}(\boldsymbol{\theta})$ is then

$$\boldsymbol{\alpha}(\boldsymbol{\theta}) = \frac{D_\mathrm{ds}}{D_\mathrm{s}} \hat{\boldsymbol{\alpha}}(\boldsymbol{\theta}) = \frac{4G}{c^2} \frac{D_\mathrm{ds}}{D_\mathrm{d} D_\mathrm{s}} \mathbf{a}(\boldsymbol{\theta}) \quad . \tag{13.6b}$$

If this expression is inserted into equation (13.5b), the cosmological distance

---

[1] The magnification ratios are contained in the matrices A_{ij}, and thus provide no independent constraints.

factors cancel out, as does H_0. Hence, if the deflection angle is described by the function $\mathbf{a}(\boldsymbol{\theta})$, the mass scale for the galaxy can be determined from the relative time delay without any knowledge of H_0.

Let us now express H_0 as $H_0 = h_{50} H_{50}$, where $H_{50} = 50 \,\mathrm{km\,s^{-1}\,Mpc^{-1}}$, and define

$$\mathcal{R} = \frac{r(z_\mathrm{d}, z_\mathrm{s})}{r(z_\mathrm{d})\, r(z_\mathrm{s})} \quad ; \tag{13.7a}$$

thus,

$$\boldsymbol{\alpha}(\boldsymbol{\theta}) = \frac{4G\,H_{50}}{c^3} h_{50}\, \mathcal{R}(z_\mathrm{d}, z_\mathrm{s})\, \mathbf{a}(\boldsymbol{\theta}) \quad , \tag{13.7b}$$

where $\mathcal{R}$ is independent of H_0.

Modification of the lens model. Suppose we have found a model for the lens which accounts well for all the observables. Let us assume that the scaled deflection angle for this lens model is $\boldsymbol{\alpha}_\mathrm{g}(\boldsymbol{\theta})$; let us also assume that the model, with $h_{50} = 1$, predicts the same value for the time delay as is determined from monitoring observations. Thus, for this model, the ray-trace equation is

$$\boldsymbol{\beta} = \boldsymbol{\theta} - \boldsymbol{\alpha}_\mathrm{g}(\boldsymbol{\theta}) \quad .$$

Let us now modify the lens model in four ways. First, we add a constant deflection angle $\boldsymbol{\alpha}_\mathrm{s}$, as could be caused by a straight cosmic string. Second, we multiply the surface mass density by a factor m. Third, we allow h_{50} to take on values different from unity, and fourth, we add a sheet of matter with constant (dimensionless) surface mass density κ_c. These changes result in a deflection angle

$$\boldsymbol{\alpha}'(\boldsymbol{\theta}) = h_{50}\, m\, \boldsymbol{\alpha}_\mathrm{g}(\boldsymbol{\theta}) + \kappa_\mathrm{c}\boldsymbol{\theta} + \boldsymbol{\alpha}_\mathrm{s} \quad ,$$

and the new lens equation reads

$$\boldsymbol{\beta} = (1 - \kappa_\mathrm{c})\boldsymbol{\theta} - h_{50}\, m\, \boldsymbol{\alpha}_\mathrm{g}(\boldsymbol{\theta}) - \boldsymbol{\alpha}_\mathrm{s} \quad . \tag{13.8}$$

Sufficient condition for invariance of observables. From the observed image positions $\boldsymbol{\theta}_i$, we calculate a source position which depends on $\boldsymbol{\alpha}_\mathrm{s}$, κ_c, m, and h_{50}. However, the source position $\boldsymbol{\beta}$ is unobservable, and the differences $\boldsymbol{\theta}_i - \boldsymbol{\theta}_j$ which are observable are unchanged if – see (13.2) –

$$\frac{h_{50}\, m}{1 - \kappa_\mathrm{c}} = 1 \quad . \tag{13.9}$$

The Jacobian matrix of the ray-trace equation (13.8) is

$$A = \frac{\partial \boldsymbol{\beta}}{\partial \boldsymbol{\theta}} = (1 - \kappa_\mathrm{c}) \left[1 - \frac{h_{50}\, m}{1 - \kappa_\mathrm{c}} \frac{\partial \boldsymbol{\alpha}_\mathrm{g}}{\partial \boldsymbol{\theta}} \right] \quad ,$$

which, however, is unobservable. The image transformation matrices A_{ij} [see (13.4)] are unchanged if (13.9) holds, since the factor $(1 - \kappa_c)$ in A cancels out in A_{ij}. Thus, condition (13.9) suffices to leave the relative image positions and relative magnification matrices invariant.

Finally, let us consider the change in the time delays Δt_{ij}. First, we note that (13.5b) is linear in $\boldsymbol{\alpha}(\boldsymbol{\theta})$ for fixed angular separations $(\boldsymbol{\theta}_i - \boldsymbol{\theta}_j)$. Second, any contribution to $\boldsymbol{\alpha}(\boldsymbol{\theta})$ which is linear in $\boldsymbol{\theta}$ does not contribute to Δt_{ij}, i.e., a term in $\boldsymbol{\alpha}$ of the form $\boldsymbol{\alpha}_0 + \Lambda\boldsymbol{\theta}$, where Λ is a 2×2 matrix, cancels out of the time delay calculation, as is easily verified by direct calculation.[2] We see from (13.5b) and (13.8) that

$$\Delta t_{ij} = m\, \Delta t_{ij}^{(0)} \quad ,$$

where $\Delta t_{ij}^{(0)}$ is the time delay for two images as predicted by the original lens model. Thus, we see that (13.9) is insufficient to keep the time delay invariant: we must require

$$m = 1 \quad ; \quad h_{50} = 1 - \kappa_c \quad , \qquad (13.10)$$

for all observables to be invariant (the combined 'similarity' and 'magnification' transformations of [GO88.1]).

The invariance with respect to simultaneous changes of H_0 and of the surface mass density κ_c, in accordance with (13.10), can be understood by noting that a uniform mass sheet acts as a Gaussian thin lens. It is thus equivalent to changing the distance to the source. Since the latter enters the lens equations only through the combination $D_{ds}/(D_d D_s)$, which is proportional to H_0, it is clear that the addition of a uniform mass sheet is equivalent to scaling H_0 with $(1 - \kappa_c)$.

From the preceding analysis, we see that a determination of H_0 is not possible unless the value of κ_c is known. However, since $\kappa_c \geq 0$, an upper limit for H_0 can always be obtained from observations of simple GL systems with measurable time delays. On the other hand, the first equality in (13.10) shows that the determination of the lens mass is independent of κ_c; therefore, an estimate of the mass of a GL should be less uncertain than one of H_0.

13.1.2 Determination of lens mass and Hubble constant

In this subsection, we discuss in detail a simplified procedure for the estimation of GL model parameters. A typical application would be the case of the 0957+561 system, but our discussion is sufficiently general to be useful in the analysis of similar GL systems. Our treatment is similar to that in [BO84.1], [FA85.2], and [BO86.1].

---

[2] This does not mean, however, that an external shear and convergence are unimportant for the analysis of the time delay, since including a finite γ and κ_c in a GL model usually means that the models parameters will change as a function of γ and κ_c [KA90.1], [FA91.2].

Parametrization of the lens model. Suppose we have observed two distinct objects, assumed to be gravitational images of a single source, and a galaxy, which we assume is the lens. The angular positions of the images relative to the center of this galaxy will be denoted by $\boldsymbol{\theta}_1, \boldsymbol{\theta}_2$. If the galaxy shows no sign of ellipticity, we may assume that its mass distribution is circularly symmetric; for simplicity, we make this assumption here. We choose a simple parametrization for the mass distribution; let $M(\theta)$ be the mass of the galaxy enclosed within angle θ,

$$M(\theta) = M_1 + M_2 \left(\frac{\theta}{\theta_2}\right)^\delta, \quad \text{for } \theta_1 \leq \theta \leq \theta_2, \quad (13.11a)$$

where $\theta_i \equiv |\boldsymbol{\theta}_i|$; only the matter distribution within this annulus enters our discussion from here on. For $\delta = 1$, (13.11a) describes the matter distribution of an isothermal sphere with a point mass (or simply a compact object) at its center. Let

$$M \equiv M(\theta_2) = M_1 + M_2 \quad (13.11b)$$

be the total mass within the radial angular position of the outer image, and

$$p = M_1/M \quad (13.11c)$$

be the fraction of this mass apportioned to the point mass.

In addition to the galaxy, we assume that the lens contains a component with characteristic length-scale much larger than the image separation (e.g., the galaxy is embedded in a cluster). This second lens component can be represented by a constant deflection $\boldsymbol{\alpha}_0$, a uniform surface mass density κ_c, and a shear, with magnitude γ and position angle ϕ. Thus, the lens equation is

$$\boldsymbol{\beta} = \left[1 - \kappa_c - \gamma \begin{pmatrix} \cos 2\phi & \sin 2\phi \\ \sin 2\phi & -\cos 2\phi \end{pmatrix}\right] \boldsymbol{\theta} \quad (13.12)$$
$$- \frac{4G}{c^2} \frac{D_{ds}}{D_d D_s} \frac{M(\theta)}{\theta} \frac{\boldsymbol{\theta}}{\theta} - \boldsymbol{\alpha}_0.$$

Determining equations. Since the source position is unobservable, we subtract the ray-trace equations for the two images to obtain

$$\left[1 - \gamma' \begin{pmatrix} \cos 2\phi & \sin 2\phi \\ \sin 2\phi & -\cos 2\phi \end{pmatrix}\right] (\boldsymbol{\theta}_1 - \boldsymbol{\theta}_2)$$
$$= U \left\{\left[p + (1-p)\left(\frac{\theta_1}{\theta_2}\right)^\delta\right] \frac{\boldsymbol{\theta}_1}{\theta_1^2} - \frac{\boldsymbol{\theta}_2}{\theta_2^2}\right\}, \quad (13.13a)$$

where we have divided the equation by $(1 - \kappa_c)$, and we let

$$\gamma' = \frac{\gamma}{1-\kappa_c} \quad ; \quad U = \frac{4GM}{c^2} \frac{D_{\mathrm{ds}}}{D_{\mathrm{d}} D_{\mathrm{s}}} \frac{1}{1-\kappa_c} \ . \qquad (13.13b)$$

If we measure all the angular positions in arcseconds, U can be written as

$$U \simeq 1.39 \, \frac{M_{12} \, h_{50}}{1-\kappa_c} \, \mathcal{R}(z_{\mathrm{d}}, z_{\mathrm{s}}) \ , \qquad (13.13c)$$

where $M_{12} = M/(10^{12} M_\odot)$, and $\mathcal{R}(z_{\mathrm{d}}, z_{\mathrm{s}})$ was defined in (13.7a). For a given galaxy model (p, δ), the vector equation (13.13a) provides two independent relations between the lens parameters γ', ϕ and U.

Time delays and magnifications. As mentioned in the previous subsection, the linear deflection caused by the larger mass distribution does not enter Δt_{12}, which depends only on the mass distribution in the galaxy. We evaluate the integral in (13.5b) along a radial path from $\boldsymbol{\theta}_2$ to $(\theta_1/\theta_2)\boldsymbol{\theta}_2$, and then on a circular arc centered on the galaxy to $\boldsymbol{\theta}_1$. For the second part of the path, $d\boldsymbol{\theta}$ is perpendicular to $\boldsymbol{\alpha}(\boldsymbol{\theta})$ and thus does not contribute to the integral. Hence, only the radial path contributes:

$$c \, \Delta t_{12} = \frac{4G}{c^2} (1 + z_{\mathrm{d}}) \left\{ \frac{1}{2} (\boldsymbol{\theta}_1 - \boldsymbol{\theta}_2) \left(\frac{M(\theta_1)}{\theta_1^2} \boldsymbol{\theta}_1 + \frac{M(\theta_2)}{\theta_2^2} \boldsymbol{\theta}_2 \right) \right.$$
$$\left. + M_1 \ln \left(\frac{\theta_2}{\theta_1}\right) + \frac{M_2}{\delta} \left[1 - \left(\frac{\theta_1}{\theta_2}\right)^\delta \right] \right\} \ . \qquad (13.14a)$$

Using (13.11), we rewrite this equation as

$$\frac{M_{12}}{T} = \frac{1.60}{1 + z_{\mathrm{d}}} \frac{1}{f(p, \delta)} \ , \qquad (13.14b)$$

where T is the time delay measured in years, and

$$f(p, \delta) = f_1(\delta) + p \, f_2(\delta) \ , \qquad (13.14c)$$

$$f_1(\delta) = \frac{1}{2} (\boldsymbol{\theta}_1 - \boldsymbol{\theta}_2) \left[\left(\frac{\theta_1}{\theta_2}\right)^\delta \frac{\boldsymbol{\theta}_1}{\theta_1^2} + \frac{\boldsymbol{\theta}_2}{\theta_2^2} \right] + \frac{1}{\delta} \left[1 - \left(\frac{\theta_1}{\theta_2}\right)^\delta \right] \ , \qquad (13.14d)$$

$$f_2(\delta) = \frac{1}{2} \left[1 - \left(\frac{\theta_1}{\theta_2}\right)^\delta \right] (\boldsymbol{\theta}_1 - \boldsymbol{\theta}_2) \frac{\boldsymbol{\theta}_1}{\theta_1^2} + \ln \frac{\theta_2}{\theta_1} - \frac{1}{\delta} \left[1 - \left(\frac{\theta_1}{\theta_2}\right)^\delta \right] \ . \qquad (13.14e)$$

Note that the mass is determined by (13.14b) for any chosen galaxy model parametrized by (p, δ). The mass depends neither on H_0, nor on the cosmological model, nor on the parameters for the linear perturbation, κ_c, γ and ϕ [BO86.1] [but note again, the galaxy parameters (p, δ) *will* depend on κ_c, γ, ϕ].

Let us also consider the Jacobian matrix for the ray-trace equation,

$$A(\boldsymbol{\theta}) = (1 - \kappa_c)$$
$$\times \left\{ \begin{pmatrix} 1 - \gamma' \cos 2\phi & -\gamma' \sin 2\phi \\ -\gamma' \sin 2\phi & 1 + \gamma' \cos 2\phi \end{pmatrix} - U \left[g(\theta) \mathcal{I} + \frac{g'(\theta)}{\theta} \begin{pmatrix} \theta_x^2 & \theta_x \theta_y \\ \theta_x \theta_y & \theta_y^2 \end{pmatrix} \right] \right\} \ , \qquad (13.15a)$$

where $\mathcal{I}$ is the identity matrix, and

$$g(\theta) = \frac{1}{\theta^2} \left[p + (1-p) \left(\frac{\theta}{\theta_2} \right)^\delta \right] \quad . \tag{13.15b}$$

The ratio of magnifications for the two images provides another independent relation for the lens parameters. In the case of the 0957+561 system, the image structure has been determined with VLBI techniques, which therefore yielded an estimate of the relative magnification matrix A_{12} [GO88.2]; such an estimate is correct, because structural changes in the images can occur only on time-scales much longer than the expected time delay, due to the relatively large size of the radio-emitting region of the source. Thus, four relations follow from (13.4) instead of one. However, our simple model is too unrealistic for us to attempt to account for such details of the image geometry [FA91.2]. In addition, since the magnification of images is much more sensitive to changes in the lens model (e.g., introduction of a non-zero ellipticity for the lens galaxy) than the image positions and the corresponding time delay, we will not use the relative magnification of the images as an additional constraint for our example.

13.1.3 Application to the 0957+561 system

Now, let us apply the equations above to the 0957+561 system, for which we assume (see, e.g., [YO80.1], [YO81.3], [ST80.1], [RO85.1], [GO88.2])

$$\begin{aligned} z_\mathrm{d} &= 0.36 \quad ; & z_\mathrm{s} &= 1.41 \quad ; \\ \boldsymbol{\theta}_1 &= (0.19, -0.97) \quad ; & \boldsymbol{\theta}_2 &= (1.43, 5.10) \quad , \end{aligned} \tag{13.16}$$

with the center of the mass distribution (13.11a) (representing the galaxy labeled G1 in [YO80.1]) at the center of coordinates.

Mass determination. From these quantities, we can compute the ratio M_{12}/T, according to (13.14), as a function of the galaxy parameters p and δ. In Fig. 13.1, this ratio is shown as a function of p, for various values of δ. We see that the ratio varies rapidly as both parameters vary, a property that provides a hint on the reliability of the mass determination. Without additional information, these estimates would vary by more than a factor of 3. This is not in conflict with [BO86.1], where the relative error of the mass estimated from observations of a GL system was quoted as less than 20%. In that paper, the mass contained within a circle with angular radius $(\theta_1 + \theta_2)/2$ was considered, and this mass is indeed much less sensitive than the mass contained within the radial position of the outer image. Furthermore, the study of low-redshift galaxies [DR83.1] shows that 'isothermal' distributions ($\delta \approx 1$, p small) fit the light distribution of galaxies well, and therefore we suspect that the same holds for their mass distributions.

Determination of H_0. Using the subtracted lens equation (13.13a), we can compute the values of U and γ' as a function of ϕ, as plotted in Fig. 13.2. Note that the factor of 2 in the arguments of the trigonometric functions shows that both quantities have period π; in addition, the transformation $\phi \to \phi + \pi/2$ changes the sign of all trigonometric functions and is thus equivalent to $\gamma' \to -\gamma'$. These periodicities are clearly visible in Fig. 13.2.

It is seen that the curves $U(\phi)$ all intersect at $U = 0$. Such behavior can be easily understood from the lens equation (13.13a). The matrix on

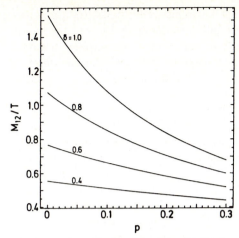

Fig. 13.1. The ratio M_{12}/T, where M_{12} is the mass of the galaxy within the outer image in units of $10^{12} M_\odot$ and T is the time delay in years, as a function of the lens parameter p, the relative contribution to the total mass of a point mass at the center of the galaxy. The curves are labeled with the parameter δ. The parameter values used in this figure are those for the 0957+561 GL system

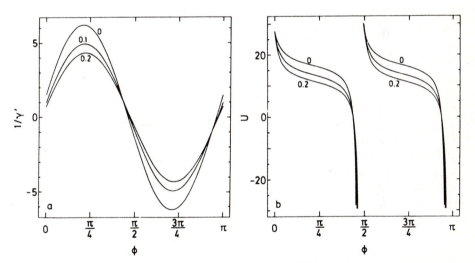

Fig. 13.2. (a) The inverse of the scaled shear γ', and (b) the parameter U as a function of ϕ, for the 0957+561 GL system. Note that U can become negative. The period of U is $\pi/2$. The curves are drawn for $\delta = 1$, and are labeled with the value of p

the left-hand side of this equation has eigenvalues $1 \pm \gamma'$, i.e., the eigenvalue zero occurs for $\gamma' = \pm 1$. If $\boldsymbol{\theta}_1 - \boldsymbol{\theta}_2$ is an eigenvector corresponding to the eigenvalue zero, U must vanish, independently of p and δ. This explains why the $1/\gamma'$ curves all intersect at $\gamma' = \pm 1$, at the same ϕ where $U = 0$. The fact that, for a given value of ϕ, both U and γ' can be determined, can be seen as follows: since the images are not colinear with the center of the galaxy, a non-zero value of the shear is needed. On the other hand, the parameter U describes the strength of the lens and is thus mainly responsible for the image splitting. Note that, for certain values of ϕ, U becomes negative. That is, for these values of ϕ it is not possible to account for the observed image positions without requiring the surface mass density κ_c to be larger than 1.

The variation of U with ϕ makes it difficult to determine U from the observed properties of the lens system. We could use additional information, e.g., the observed flux ratio of the images. However, as we remarked before, the magnifications are much more sensitive to variations of the lens model than the image positions and the time delays, and if we used the magnification ratio as a constraint, we would be reaching beyond the domain of applicability of our simple lens model. It should be noted in this context that the main lens galaxy in the 0957+561 system is not exactly spherical, which means that our model may be simpler than required to account for all the image properties. Knowledge of the approximate cluster center would yield ϕ. However, observations can only determine the center of brightness of the cluster, and, therefore, we must rely on assumed relations between visible and dark matter to attempt to find the center of mass of the cluster. Another key question concerns which galaxies in the field surrounding the QSO images belong to the cluster. No answer is yet possible, because redshifts have been published for only two galaxies other than the main lens galaxy (G1).

For illustration, we proceed as follows: the cluster is likely to be subcritical, i.e., $\kappa_c + \gamma < 1$; otherwise, additional images of the QSO are expected, but none are observed. Let us also assume that the cluster can be described as an isothermal sphere; therefore, we obtain a relation between κ_c and γ, $\kappa_c = \gamma$, or $\kappa_c = \gamma'/(1+\gamma')$. Our first requirement then implies $\gamma' < 1$. From Fig. 13.2a, we find that ϕ satisfies $0 \lesssim \phi \lesssim 3\pi/8$.

For the 0957+561 system, the cosmological factor $\mathcal{R}(z_d, z_s)$ is fairly well determined, because the redshifts in this system are not large. A plot of this factor is shown in Fig. 13.3a, as a function of the smoothness parameter $\tilde{\alpha}$, for several values of the density parameter Ω. The variation of $\mathcal{R}(z_d, z_s)$ between the most extreme parameters is less than 20%.

If we rewrite (13.13c) as

$$T h_{50} = \frac{U(1-\kappa_c)}{1.39(M_{12}/T)\mathcal{R}(z_d, z_s)} \quad , \tag{13.17}$$

with the aforementioned assumption about the cluster, and for a given choice of the cosmological parameter $\mathcal{R}(z_d, z_s)$, we can calculate the combination Th_{50} as a function of ϕ, as shown in Fig. 13.3b. We see that the curves are flat for $0 \lesssim \phi \lesssim \pi/4$, but become steeper for larger values of ϕ. We can interpret this property using Fig. 13.2 and equation (13.17): for $\phi \to 0$, γ' rises, thus the factor $1 - \kappa_c$ decreases; this decrease, however, is counteracted by a sharp increase of U as $\phi \to 0$; no such counteraction occurs for larger ϕ. The model parameters for the galaxy were varied for

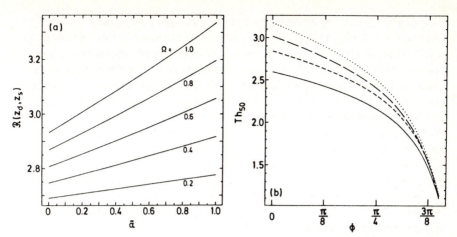

Fig. 13.3. (a) The cosmological factor $\mathcal{R}(z_d, z_s)$ (13.7a) for the redshifts found in the 0957+561 system, $z_d = 0.36$ and $z_s = 1.41$, as a function of the smoothness parameter $\tilde{\alpha}$. The curves are labeled with the value of Ω. (b) The product Th_{50} as a function of the ϕ, under the assumption that the cluster is a singular isothermal sphere, i.e., $\kappa_c = \gamma$. The value of $\mathcal{R}(z_d, z_s)$ was assumed to be 3.2, which corresponds to a flat cosmological model with some clumpiness ($\tilde{\alpha} \approx 0.65$). The model parameters for the galaxy are: $p = 0$, $\delta = 1$ (solid line); $p = 0$, $\delta = 0.9$ (short dashes); $p = 0.1$, $\delta = 1$ (long-dashes); $p = 0.1$, $\delta = 0.9$ (dotted)

Fig. 13.3b over a range that corresponds to current knowledge of elliptical galaxies, i.e., we assume the galaxies are nearly isothermal. If we use the observed cluster's center of brightness (as defined in [YO81.3]) to estimate ϕ, we see that ϕ is small. For an isothermal sphere model, we find that Th_{50} is ~ 2.5. Thus, from our simple model, with all its requisite assumptions, we would estimate H_0 to be $h_{50} \approx 2.5/T$.

Discussion of the method. One should keep in mind all the assumptions that were needed to estimate H_0, as we have discussed. First, our lens model is much too simple to be realistic; in particular, it is impossible to account for the ratio of magnifications with our model, unless the cluster parameters take on extreme values. Second, the assumption that the cluster is isothermal may be questioned; without it, the large variations of U as a function of small ϕ show that it would be impossible to determine h_{50} to within a factor below 2. Third, restricting the lensing galaxy to be nearly isothermal results in additional uncertainty in the estimate of h_{50}. This is in contrast to the determination of the lens mass, where the uncertainties are reasonably small. In particular, although the 'measurement' of H_0 would be even more uncertain for a higher-redshift GL system, the mass determination would remain independent of the cosmological parameters, which means this method is also useful for high-redshift galaxies. We will see in the following section that a similar method may be used to determine the mass of high-redshift clusters of galaxies.

A further criticism of the determination of H_0 from observations of GL systems was discussed in the literature ([AL85.1], [AL86.1]), based on the possibility that inhomogeneities in the universe occupy larger volumes than was previously thought. Observations of the distribution of galaxies [GE89.1] show that luminous matter is distributed in intersecting sheet-like structures, and that large voids exist. It is of course likely that a lens galaxy is part of a large-scale feature, so that its line-of-sight and a random line-of-sight would be affected differently. Therefore, the use of 'standard' distance-redshift relations may be misleading. We have partly accounted for this problem by leaving the smoothness $\tilde{\alpha}$ as a free parameter in our considerations. However, such an approach neglects the possible influence of large-scale shear in the universe, which so far is not properly accounted for; some model calculations may be found in [AL86.1]. But we note that recent simulations of light propagation through an inhomogeneous universe ([JA90.1], [BA91.2]) have revealed no indication for the large-scale distribution of (dark) matter to affect light bundles strongly.

As was pointed out in [KA90.1], there is another kind of invariance transformation. Suppose we found a set of GL models to reproduce an observed image configuration. We characterize these N lens models by their deflection potentials $\psi_i(\mathbf{x})$. Clearly, any superposition $\psi(\mathbf{x}) = \sum_{i=1}^{N} p_i \psi_i(\mathbf{x})$ with $\sum_{i=1}^{N} p_i = 1$ will also reproduce the image configuration, since the deflection angle is linear in the ψ_i. However, since the magnification is not a linear function of ψ, the predicted flux ratios will depend on the p_i. If, however, the number of observational constraints on the flux ratios is smaller than N, there will be a family of superpositions that reproduce the observed magnifications, showing again that a successful GL model is unlikely to be unique, unless further observational constraints about the lens are available. As we show in the following paragraph, significant progress has been achieved in modeling efforts on the 0957+561 GL system.

Despite the uncertainties in the determination of H_0 from observations of GL systems, such a method remains the only one we know of that might be able to yield an estimate for H_0 from high-redshift sources, whereas all other methods use low-redshift sources. If the 'GL method' and others yielded comparable results, we would have a consistent, significant argument in favor of Friedmann-Lemaître models. The standard, low-redshift methods have not yet reduced the factor of 2 uncertainty in H_0. Estimates with smaller uncertainties might well be derived from observations of the GL effect, as more observations are made, and in particular of simpler lens systems, i.e., where a single galaxy is responsible for the image splitting, without assistance from a cluster. If such estimates became available for a large number of GL systems, we could hope that statistical arguments would eventually whittle down the uncertainties. Finally, if large-scale inhomogeneities in fact resulted in a strong anisotropy of the redshift-distance relations, it would be unreasonable to speak of 'a Hubble constant', without specifying its definition in terms of measurable quantities.

From the difficulties to determine H_0 from gravitational lensing, it is clear that the determination of other cosmological parameters, such as Ω, will be even more difficult [LA82.1].

Model refinements. Several groups have developed different models for the 0957+561 system. These were intercompared, and a new, simple model was introduced, in [FA91.2]. The particular model introduced in [FA91.2] accounted for the known properties of the GL system, but assumed little about the galaxies that constitute the lens. The paucity of parameters achieved with this model allowed the first detailed study of uniqueness of the model parameters, and also served to demonstrate that, with two types of additional observations, a point estimate of H_0 could be made, with significantly lower uncertainty than that of 'classical estimates'. We now summarize the results of [FA91.2], and compare them with those derived from our simple model.

The results presented in [FA91.2] were derived from VLBI observations of the 0957+561 system [GO88.2], but they were also consistent with other observations, in the radio (VLA), as well as in the optical spectrum. The main lens galaxy, G1, was represented as a circularly symmetric, smooth, isothermal surface density profile, with core radius θ_c and velocity dispersion σ_v, accompanied by a compact nucleus with mass M_c. The compact nucleus was required to account for the relative magnification for the images A and B. The measured ellipticity of the light distribution of G1, if it is indicative of the ellipticity of its matter distribution, turned out to be unimportant for the results of [FA91.2]. Other galaxies (the 'cluster' in preceding subsections) that contribute to the light deflection were assumed to be embedded in a smoothly-distributed dark component with surface mass density κ_c. The light deflection attributable to these galaxies was represented purely as a shear, parametrized by γ and ϕ, as in the simple model which we have discussed. Such a simple representation sufficed to account for the observations, because of the invariance transformations: κ_c and σ_v may be scaled so that the observables, save for the time delay for A and B, Δt_{BA}, remain invariant. This invariance is analogous to that described for our simple model, where a simple relation for variation of the parameters m, κ_c and γ implied invariant observables. Thus, although the images yield no constraints on the amount of mass on the scale of the cluster, a measurement of σ_v would determine this amount. This relationship was exploited in [FA91.2] to derive the expression

$$H_0 = (90 \pm 10) \left(\frac{\sigma_v}{390 \, \text{km/s}} \right)^2 \left(\frac{\Delta t_{BA}}{1 \, \text{yr}} \right)^{-1} \text{km s}^{-1} \, \text{Mpc}^{-1} \quad .$$

The expression was left in terms of Δt_{BA}, because as we explained above, the claimed values are deemed unreliable, and to show that an accurate determination of Δt_{BA} and σ_v would indeed yield a useful estimate of H_0. We also see that an explicit relationship between H_0, τ_{BA}, and σ_v does exist, with an implicit dependence on the shear due to galaxies other than G1.

The estimate of θ_c presented in [FA91.2] was significantly larger than that of the light distribution of G1, perhaps indicating the presence of dark matter in G1. A significant point to note is that the model of [FA91.2] shows that the core of G1 must include a compact object. Although the outer part of the mass distribution (between the radii corresponding to A and B), and the total mass of G1 are accounted for well by an isothermal profile, a central density 'spike' is unavoidable, within the framework of this model. The observations available constrained σ_v, M_c and κ_c only to within factors of 2 to 3, but within reasonable ranges. These constraints would be pinned down by a measurement of σ_v, thus yielding not only cosmological information, but also local mass estimates for all the matter in the lens. One of the more tightly constrained parameters turned out to be ϕ, due to the near alignment of the radio jets in each image.

Time-delay and Hubble constant from 0957+561. The observed flux variations in both images of the 0957+561 GL system can be due either to intrinsic variability of the source (in which case they should be correlated), or due to ML, or a combination of both. Recently, we have investigated [FA91.1] whether the optical flux variations can be due to ML alone, using ML simulations of the type discussed in Sect. 12.4, adapted to the parameters of the 0957+561 system as derived from the GL model of [FA91.2]. The result of this study can be summarized as follows: lightcurves from ML can yield cross-correlation functions with height and width comparable to that corresponding to the observed light curves, provided the source is sufficiently compact and the effective transverse velocity (12.39a) is sufficiently large. These cross-correlations occur due to chance correlations between features in the ML lightcurves of both images. However, the corresponding flux variations are much larger than those in the observed light curves. On the other hand, if the source size is increased to that point where the ML variations are of the same order as that observed, the corresponding cross-correlation becomes extremely weak. From this study we can conclude that ML alone is not responsible for the flux variations of the images of 0957+561, but that at least part of it must be due to intrinsic variations of the source.

Several different methods for cross-correlating the light curves of both images are tried in [FA91.1], applied to the data sets of [SC90.3] and [VA89.1]. It turns out that these different methods, applied to two different data sets, yield consistent values for the time delay, however with much larger uncertainty than, e.g., claimed in [SC90.3]. We find that the time delay is about 460 days, with an uncertainty of about 10%. Since it appears that the "noise" in the cross-correlation function is larger than the error bars of individual flux measurements, one can conclude that some fraction of the flux variability is due to ML (see also [SC91.3]; for further discussions on the determination of Δt, see Sect. 2.5.1).

Recently, an attempt was made to determine the velocity dispersion of the galaxy G1, the main lens galaxy in the 0957+561 system [RH91.1]. The result, $\sigma_v = 303 \pm 50$ km/s, has to be considered as preliminary, due to the extremely difficult nature of the observations, but taking this value for σ_v and the estimate of the time delay as discussed above, one can obtain an estimate for H_0 from the equation given above, based on the detailed model in [FA91.2], yielding $H_0 = 50 \pm 17 \,\mathrm{km\,s^{-1}\,Mpc^{-1}}$. Whereas we want to stress all the uncertainties with which this estimate is burdened, it is nevertheless the first attempt to determine the Hubble constant from a gravitational lens system, and it measures the cosmic expansion over a good fraction of the Hubble length, in contrast to more conventional, local methods. It remains to be seen whether this estimate stands further checks.

13.2 Arcs in clusters of galaxies

13.2.1 Introduction

So far, we have considered the imaging of point-like sources by GLs, and discussed the location and magnification of the images. For historical reasons, QSOs and other compact extragalactic objects have been regarded as the most important sources in GL systems. However, the latter are not the only objects at high redshift which are subject to gravitational light deflection by foreground mass concentrations. It is natural to ask what the images of lensed galaxies look like and how they can be identified.

In Sect. 12.6, this problem was treated in a statistical way, because the inhomogeneity of the matter in the universe should distort the images of distant galaxies, thereby inducing additional ellipticity in their images. Since galaxies are not circular in general, this cannot be used to single out those sources which have been distorted considerably by lensing. Only when the distortion is extreme, one might hope to 'see' the lensing effect directly.

Historically, these considerations were made after the detection of highly distorted images of distant galaxies by clusters of galaxies. Several interpretations for these 'arcs' have been proposed (see Sect. 2.5.5), but observations, in particular spectroscopy of the arcs, have confirmed the GL interpretation.[3] In this section, we first investigate the conditions which can lead to arcs owing to gravitational lensing, and describe additional observations and interpretations. We then turn to the statistics of arcs and their less prominent, but more abundant versions, the arclets.

Arcs as images of sources close to a caustic. The image of an extended source is highly distorted if it occurs close to a critical curve, or, correspondingly, if the source is close to a caustic. An example is provided in Fig. 8.9 where a circular source is situated on a cusp and a very elongated image occurs tangentially to the critical curve. Note that in this case, the lens is highly asymmetric, i.e., axi-symmetry is not required to yield arc-like features. This is in contrast to the simple picture one might have from the detailed

---

[3] In fact, it is somewhat puzzling that these features were not predicted explicitly based on GL models. In an unpublished thesis, K. Chang investigated the imaging of extended sources by the Chang–Refsdal lens (see Sect. 8.2.2) and obtained very elongated images. Also, the detailed study of the two point-mass lens [SC86.4] led to arc-like images (see Fig. 8.9). Thus, the occurrence of arcs should not have come as a surprise.

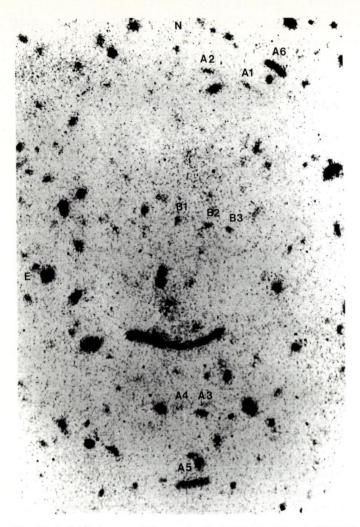

Fig. 13.4. CCD close-up of the arcs in Abell 370, labeled as in [FO88.1]. Note the main, giant arc, and the various arclets, labeled A1–A6 (courtesy of G. Soucail and B. Fort)

studies of symmetric lenses. For there, if an extended source is situated on the point caustic, a ring-shaped image ('Einstein ring') occurs. If the source is shifted slightly away from this point of symmetry, the ring splits into two arcs, such that the main arc is longer than, and further from the lens center than the counter arc. If a slight asymmetry is introduced, the point caustic unfolds into a caustic curve. The structure of the images remains essentially unchanged if the size of the caustic is smaller than the displacement of the source from the original point caustic, but will be very perturbed if the source lies inside the caustic. We consider this situation below; here we only note that, given a symmetric lens, arcs are easily obtained if the source is sufficiently aligned with the lens. However, as Fig. 8.9 shows, near symmetry is by no means a necessary condition for the occurrence of arcs.

If a small source is close to a fold, but not on a cusp, it has two images close to, and on opposite sides of, the corresponding critical curve in the lens plane (see Sect. 6.2). The two images are elongated in the direction of their separation. Moving the source onto the caustic leads to a fusion of the two images, as can be seen in Fig. 8.9. Hence, a fold caustic is able to produce highly elongated single images of an extended source, but, in a first approximation, this image will be fairly straight, and will not show the curvature observed in most of the arcs in clusters of galaxies. A curved elongated image can be produced if the source lies close to a cusp, as in Fig. 8.9 (or close to the caustic point in symmetric lens models). To understand the formation of arcs in this situation, we consider the simple model of a nearly symmetric lens, following in part the treatment of [BL88.1].

13.2.2 The nearly spherical lens

Consider a lens model where the deflection angle $\boldsymbol{\alpha}(\mathbf{x})$ is a sum of an axisymmetric deflection $\boldsymbol{\alpha}_s$ and a perturbation of the form $\epsilon \mathbf{a}(\mathbf{x})$. Thus, we consider the ray-trace equation

$$\mathbf{y} = \mathbf{x} - \boldsymbol{\alpha}_s(\mathbf{x}) - \epsilon \mathbf{a}(\mathbf{x}) \quad .$$

Owing to the symmetry of $\boldsymbol{\alpha}_s$, we can write $\boldsymbol{\alpha}_s(\mathbf{x}) = \alpha_s(x)\mathbf{n}$, where $\mathbf{n} = (\cos\varphi, \sin\varphi)$, and we have introduced polar coordinates, $\mathbf{x} = x(\cos\varphi, \sin\varphi)$. The asymmetric part of the deflection is to be considered as a perturbation in the sense that α_s and $|\mathbf{a}|$ are of the same order of magnitude, so that $\epsilon \ll 1$ is a perturbation parameter.

Without the perturbation ($\epsilon = 0$), the lens is assumed to have a single tangential critical curve at $x = x_0$, and the corresponding caustic is the point $\mathbf{y} = 0$. That is, the symmetric deflection angle satisfies $\alpha_s(x_0) = x_0$. For non-zero ϵ, this critical circle is distorted and the caustic point unfolds into a caustic curve with cusps; later this unfolding will be shown explicitly.

Critical curve and caustic. Let us first compute the position of the new critical curve. To that end, we introduce the new variable $\Delta = x - x_0$. Close to the 'old' critical circle, the lens equation can be rewritten, using the approximation $x - \alpha_s(x) \approx [1 - \alpha'_s(x_0)]\Delta$,

$$\mathbf{y}(x,\varphi) = [1 - \alpha'_s(x_0)]\Delta\, \mathbf{n}(\varphi) - \epsilon \begin{pmatrix} a_1(x_0,\varphi) \\ a_2(x_0,\varphi) \end{pmatrix} \quad . \tag{13.18}$$

The Jacobian $\det A$ for this equation is

$$\det A \equiv \det\left[\frac{\partial(y_1,y_2)}{\partial(x_1,x_2)}\right] = \frac{1}{x}\det\left[\frac{\partial(y_1,y_2)}{\partial(x,\varphi)}\right] \quad ,$$

so that

$$\det A = [1 - \alpha'_s(x_0)]^2 \Delta - \epsilon \left\{ [1 - \alpha'_s(x_0)] \left[\frac{\partial a_2}{\partial \varphi} \cos \varphi - \frac{\partial a_1}{\partial \varphi} \sin \varphi \right] \right\}$$
$$+ \mathcal{O}(\epsilon^2, \epsilon\Delta, \Delta^2) \quad . \tag{13.19}$$

Thus, the condition $\det A = 0$ leads to a deviation Δ_c of the new critical curve from the old critical circle at x_0 which is of order ϵ. Before proceeding, we note that the perturbing deflection can be obtained from a potential, $\mathbf{a}(\mathbf{x}) = \nabla \psi_p(\mathbf{x})$. Therefore,

$$\begin{aligned} a_1(\mathbf{x}) &= \frac{\partial \psi_p}{\partial x} \cos \varphi - \frac{1}{x} \frac{\partial \psi_p}{\partial \varphi} \sin \varphi \;, \\ a_2(\mathbf{x}) &= \frac{\partial \psi_p}{\partial x} \sin \varphi + \frac{1}{x} \frac{\partial \psi_p}{\partial \varphi} \cos \varphi \;. \end{aligned} \tag{13.20}$$

With the help of these last two expressions, we find that the second bracket in the second term of the right-hand side of (13.19) can be simplified, yielding up to first order in the small parameters ϵ and Δ

$$\det A = [1 - \alpha'_s(x_0)]^2 \Delta - \epsilon \left\{ [1 - \alpha'_s(x_0)] \left[\frac{\partial \psi_p}{\partial x} + \frac{1}{x} \frac{\partial^2 \psi_p}{\partial \varphi^2} \right] \right\} \;, \tag{13.19'}$$

where the derivatives are evaluated at (x_0, φ). The critical curve is then described by $x_c(\varphi) = x_0 + \Delta_c(\varphi)$, where Δ_c is the value of Δ where the determinant vanishes,

$$\Delta_c(\varphi) = \frac{\epsilon}{[1 - \alpha'_s(x_0)]} \left[\frac{\partial \psi_p}{\partial x} + \frac{1}{x_0} \frac{\partial^2 \psi_p}{\partial \varphi^2} \right]_{x=x_0} . \tag{13.21}$$

Inserting this expression into the ray-trace equation (13.18), and using (13.20), yields a parametrized description of the caustic,

$$\mathbf{y}_c(\varphi) = \frac{\epsilon}{x_0} \left[\frac{\partial^2 \psi_p}{\partial \varphi^2} \mathbf{n} - \frac{\partial \psi_p}{\partial \varphi} \mathbf{t} \right] \;, \tag{13.22}$$

where $\mathbf{t} = (-\sin \varphi, \cos \varphi)$ is a unit vector perpendicular to $\mathbf{n}$. Thus, the caustic depends solely on the φ-derivative of the potential at the location of the original critical circle $x = x_0$; the spatial extent of the caustic is directly proportional to the perturbation parameter ϵ.

The caustic has at least four cusps. A cusp occurs when the vector tangent to the critical curve is an eigenvector corresponding to the eigenvalue zero; the corresponding image of this critical point is then a cusp. Equivalently, at a cusp the vector tangent to the caustic changes sign. Since $d\mathbf{n}/d\varphi = \mathbf{t}$ and $d\mathbf{t}/d\varphi = -\mathbf{n}$, we have

$$\frac{d\mathbf{y}_c}{d\varphi} = \frac{\epsilon}{x_0} \left[\frac{\partial^3 \psi_p}{\partial \varphi^3} + \frac{\partial \psi_p}{\partial \varphi} \right] \mathbf{n} \equiv \frac{\epsilon}{x_0} D(\varphi) \mathbf{n} \;, \tag{13.23}$$

where again the derivatives of ψ_p are calculated at $x = x_0$. We then find:

$$\int_0^{2\pi} d\varphi\, D(\varphi) = 0 \quad;\quad \int_0^{2\pi} d\varphi\, D(\varphi)\, \sin(\varphi - \varphi_0) = 0 \quad,$$

for any φ_0. The first relation is due to the fact that $D(\varphi)$ is a sum of derivatives of a periodic function, so that its integral over one period vanishes; the second can be shown by integrating twice by parts the term with the third derivative, and noting that the boundary terms vanish due to the periodicity. We can now show that the last two integral relations require that there be at least four roots of $D(\varphi)$, by considering the Fourier expansion of $D(\varphi)$. The two integral relations then require the zero and first order Fourier coefficients to vanish. Thus,

$$D(\varphi) = \mathcal{R}e\left\{\sum_{k=m}^{\infty} c_k\, e^{ik\varphi}\right\} \equiv \mathcal{R}e\left[z(\varphi)\right] \quad,$$

where $\mathcal{R}e(z)$ denotes the real part of z, and c_m is the lowest-order coefficient which does not vanish, $m \geq 2$. Consider the curve Γ parametrized by $z(\varphi)$ in the complex plane; roots of $D(\varphi)$ are those points where $z(\varphi)$ intersects the imaginary axis. We show below that Γ encircles the origin at least twice for $0 \leq \varphi < 2\pi$; therefore, the curve has at least four intersections with the imaginary axis. We will assume that $z(\varphi) \neq 0$ for $0 \leq \varphi < 2\pi$; this is not a restriction, since an infinitesimal variation of the coefficients can always be applied to satisfy our assumption.

To represent explicitly the number of times Γ encircles the origin, we use the winding number $n(\Gamma, z_0)$ for a closed curve Γ around a point z_0 which does not lie on Γ,

$$n(\Gamma, z_0) = \frac{1}{2\pi i} \oint_\Gamma \frac{dz}{z - z_0} \quad.$$

Here, we are interested in the winding number of Γ around the origin, $z_0 = 0$. Let $u = e^{i\varphi}$; then,

$$n(\Gamma, 0) = \frac{1}{2\pi i} \oint_1 du\, \frac{dz/du}{z} \quad,$$

where now the integral is taken along the unit circle. Expressing z and its derivative in terms of u,

$$z = u^m \sum_{\ell=0}^{\infty} c_{\ell+m}\, u^\ell \quad;\quad \frac{dz}{du} = u^{m-1} \sum_{\ell=0}^{\infty} (\ell + m)\, c_{\ell+m}\, u^\ell \quad,$$

the winding number becomes

$$n(\Gamma, 0) = \frac{1}{2\pi i} \oint_1 du\, \frac{\sum_{\ell=0}^{\infty}(\ell + m) c_{\ell+m}\, u^\ell}{u \sum_{\ell=0}^{\infty} c_{\ell+m}\, u^\ell} \quad.$$

This integral can now be evaluated with the residue theorem,

$$n(\Gamma, 0) = \sum_i \mathrm{Res}\,(u_i) \quad,$$

where the u_i are the poles of the integrand within the unit circle. One pole is at $u_0 = 0$, for which we find $\mathrm{Res}\,(u_0) = m$. Now, consider a pole at $u_i \neq 0$, i.e.,

$$\sum_{\ell=0}^{\infty} c_{\ell+m}\, u_i^\ell = 0 \ .$$

Then,

$$\text{Res}\,(u_i) = \lim_{u \to u_i}\, (u - u_i)\, \frac{\sum\limits_{\ell=0}^{\infty}(\ell + m)\, c_{\ell+m}\, u^\ell}{u \sum\limits_{\ell=0}^{m} c_{\ell+m}\, u^\ell} \ .$$

Writing $u = u_i + \Delta$, we have in the limit $\Delta \to 0$, $u^\ell = u_i^\ell + \ell\, \Delta\, u_i^{\ell-1}$; inserting this value into the equation above, we see immediately that $\text{Res}\,(u_i) = 1$ for $u_i \neq 0$. Hence, the winding number is

$$n(\Gamma, 0) = m + N \ ,$$

where N is the number of roots of $\sum c_{\ell+m}\, u^\ell$ in the unit circle. Since $m \geq 2$, this proves that $D(\varphi)$ has at least four roots, and thus, the caustic has at least four cusps.

Geometrical construction of arcs. From the foregoing discussion, it is a straightforward matter to find a simple geometrical construction for the images of a source close to the caustic. Consider first the image of the radial line $\mathbf{x} = (x_0 + \Delta)\mathbf{n}(\varphi)$. According to the lens equation (13.18), which is valid for $\Delta \ll x_0$, the corresponding image curve is a straight line with the direction of $\mathbf{n}$, i.e., it is parallel to the original curve. Furthermore, at $\Delta = \Delta_\text{c}(\varphi)$, the line in the lens plane crosses the critical curve, and the corresponding image curve intersects the caustic at this point $\mathbf{y}_\text{c}(\varphi)$. But, as can be seen from (13.23), the vector tangent to the caustic at that point also has the direction of $\mathbf{n}$; thus, the image line is tangent to the caustic.

Our construction can now be reversed (see Fig. 13.5): first, compute the caustic from (13.22). Then, consider a point source close to the caustic. Draw all the lines from this point, tangent to the caustic; the polar angles of these lines, φ_i, can then be measured. Next, draw the radial lines $\varphi = \varphi_i$ from the center of the symmetric lens. The image points will then be close to the intersection of these radial lines with the Einstein circle $x = x_0$. To find the radial coordinates of the images, note that the lens equation (13.18) yields a radial magnification of $[1 - \alpha'_\text{s}(x_0)]$; hence, if the source point is at a distance dy_i from the point where the tangent touches the caustic, the corresponding distance in the lens plane is $dx_i = dy_i / [1 - \alpha'_\text{s}(x_0)]$, yielding for the distance Δ_i from the Einstein circle $\Delta_i = \Delta_\text{c}(\varphi_i) + dx_i$.

Our geometrical construction can be formulated algebraically in the following way. Let $\mathbf{y} = (y_1, y_2)$ be a source point close to the caustic. We can write its coordinates in the form

$$\mathbf{y} = y_n(\varphi)\mathbf{n}(\varphi) + y_t(\varphi)\mathbf{t}(\varphi) \ , \tag{13.24a}$$

where

$$y_n(\varphi) = y_1 \cos\varphi + y_2 \sin\varphi \ ; \quad y_t(\varphi) = -y_1 \sin\varphi + y_2 \cos\varphi \ . \tag{13.24b}$$

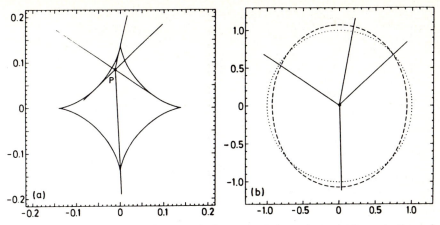

Fig. 13.5. (a) The caustic of a perturbed symmetric lens, chosen to be an isothermal sphere perturbed by a quadrupole term, with shear $\epsilon = 0.07$. Units are chosen such that $x_0 = 1$. The four tangents of the caustic on which the source point P lies are drawn; these determine the polar angles φ_i where images of P occur. (b) For the same lens, the Einstein circle (dotted) and the critical curve of the perturbed lens (dashed) are drawn, together with the four radial lines $\varphi = \varphi_i$, on which the images of the point source P are found

A parametric description of the caustic is given by (13.22), and from (13.23), we see that the tangent is proportional to $\mathbf{n}(\varphi)$. Hence, the tangent to the caustic at the point $y_c(\varphi)$ can be parametrized as

$$\mathbf{T}(\varphi, \lambda) = \lambda\, \mathbf{n}(\varphi) - \frac{\epsilon}{x_0} \frac{\partial \psi_\mathrm{p}}{\partial \varphi}\, \mathbf{t}(\varphi) \quad. \tag{13.25}$$

The angles φ_i are obtained as those where the point $\mathbf{y}$ lies on the tangent $\mathbf{T}(\varphi, \lambda)$. Since λ enters only the $\mathbf{n}$-component of $\mathbf{T}$, we must solve for the tangential component, i.e.,

$$y_1 \sin \varphi - y_2 \cos \varphi = \frac{\epsilon}{x_0} \frac{\partial \psi_\mathrm{p}}{\partial \varphi} \quad.$$

The solutions φ_i of this equation determine the directions in which images of the source point are found close to the critical curve. We could now determine λ from the $\mathbf{n}$-component of the above equation, but it is easier to write down the radial component of the ray-trace equation (13.18) for any φ_i, and solve for Δ_i:

$$\Delta_i = \frac{1}{[1 - \alpha'_\mathrm{s}(x_0)]} \left(y_n(\varphi_i) + \epsilon \frac{\partial \psi_\mathrm{p}}{\partial x}\bigg|_{x = x_0, \varphi = \varphi_i} \right) \quad, \tag{13.26}$$

where $y_n(\varphi)$ is given in (13.24b).

This construction for point sources can now be easily generalized for extended sources (see Fig. 13.6), by drawing the common tangents of the source and the caustic. For angles φ within these tangent pairs, image points close to the Einstein circle exist. This geometrical construction can be expressed in an algebraic form, which can be easily used for explicit calculation of the images of a source close to the caustic. For this, it is sufficient to find the images of the boundary line of the source. Let the source be circular, with center $\mathbf{y}$ and radius r. We can then check whether

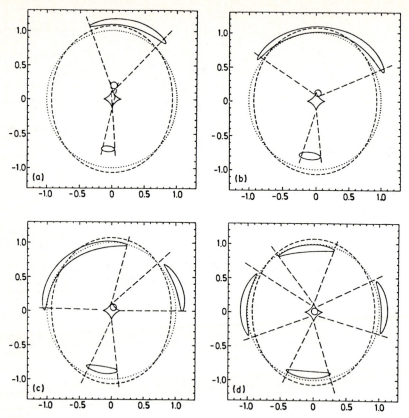

Fig. 13.6a–d. Construction of the images of an extended source close to the caustic of a perturbed symmetric lens. In this example, an isothermal sphere with a quadrupole perturbation is considered; the parameters are chosen such that $x_0 = 1$, $\epsilon = 0.07$. The dotted curve is the Einstein circle, the dashed closed curve is the critical curve of the perturbed lens. The astroid at the center of each figure is the caustic with its four cusps. The dashed lines are the common tangents of the source and the caustic; the source is circular and has radius $r = 0.05$ in all cases. The other solid curves are the boundary curves of the images. It is seen in all cases that the common tangents are very close to the edges of the images; the error is of order ϵx_0. (a) The source is outside the caustic; in this case two images occur. (b) If the source lies on a cusp, again two extended images are formed, but those source points within the caustic have three images in the main arc. (c) If the source lies on the caustic but not on a cusp, three images are formed. Those source points which are inside the caustic have two image points inside the most elongated image. (d) A source within the caustic has four separate images close to the Einstein circle

the tangents (13.25) intersect the source boundary. Such is the case if the equation

$$|\mathbf{T}(\varphi, \lambda) - \mathbf{y}| = r \tag{13.27a}$$

has, for given φ, real solutions λ. The equation can be solved using (13.24) and (13.25):

$$\lambda_{1,2} = y_n \pm \sqrt{r^2 - \left(\frac{\epsilon}{x_0}\frac{\partial \psi_\mathrm{p}}{\partial \varphi} + y_t\right)^2} \quad . \tag{13.27b}$$

Thus, real-valued solutions exist for a given φ if the term under the root in (13.27b) is non-negative. For those values of φ for which λ is real, we can also compute the two points $\hat{\mathbf{y}}_1(\varphi)$, $\hat{\mathbf{y}}_2(\varphi)$ of intersection of the tangent with the source boundary curve:

$$\hat{\mathbf{y}}_i(\varphi) = \mathbf{T}(\varphi, \lambda_i) \quad . \tag{13.28}$$

The **n**-component of $\hat{\mathbf{y}}_i(\varphi)$ is, according to (13.25), simply $\lambda_i(\varphi)$; hence, according to (13.26) we can calculate the boundary curve of the images of the source by

$$x_i(\varphi) = x_0 + \Delta_i(\varphi) = x_0 + \frac{1}{[1 - \alpha'_\mathrm{s}(x_0)]} \left(\lambda_i + \epsilon \frac{\partial \psi_\mathrm{p}}{\partial x}_{(x=x_0,\varphi)}\right) \quad . \tag{13.29}$$

The construction of images as we have described is easily performed on a computer; we apply this method to a simple example below (Figs. 13.5 and 13.6 were drawn for this example). As can be seen from Fig. 13.6, the whole arc can be constructed directly from the common tangents. The thickness of the arc is mainly given by the radial magnification, i.e., it is $\sim r/[1 - \alpha'_\mathrm{s}(x_0)]$.

Example: Isothermal sphere with small quadrupole term. In this case, we can choose the length scale such that the radius of the tangential critical curve is $x_0 = 1$; the deflection angle α_s is constant, thus $\alpha'_\mathrm{s}(x_0) = 0$. The perturbing deflection in this case is $\mathbf{a} = (x_1, -x_2)$, and the shear takes the role of the perturbation parameter ϵ. The perturbing deflection potential is $\psi_\mathrm{p} = (x_1^2 - x_2^2)/2 = (x^2 \cos 2\varphi)/2$. After differentiation, the equation for the caustic can be calculated from (13.22) as

$$\mathbf{y}_\mathrm{c}(\varphi) = 2\epsilon \begin{pmatrix} -\cos^3 \varphi \\ \sin^3 \varphi \end{pmatrix} \quad .$$

The caustic has four cusps, as can also be easily seen from the corresponding function $D(\varphi)$. The tangents to the caustic are

$$\mathbf{T}(\varphi, \lambda) = \lambda\,\mathbf{n}(\varphi) + \epsilon \sin(2\varphi)\,\mathbf{t}(\varphi) \quad .$$

A point source has images in those directions φ_i where the equation $y_t(\varphi) = \epsilon \sin(2\varphi)$ is satisfied; the corresponding radial distances from the Einstein circle at $x = x_0 = 1$ are $\Delta_i = y_n + \epsilon \cos(2\varphi)$. The specialization of (13.27–29) to this case is easily obtained; we drew Fig. 13.6 based on these equations. This figure demonstrates a few possible geometries for arc formation, but it should be noted that more complex arrangements are possible; for example, the source can have a size comparable to that of the caustic, and thus lie on more than one cusp; this is probably the case for the ring radio galaxy

MG1131+0456 (Sect. 2.5.6). Also, the source itself can well be asymmetric, so other shapes of images are possible.

13.2.3 Analysis of the observations; arcs as astronomical tools

Counterarcs and additional images. The analysis of Sect. 13.2.2 clearly answers the question of whether an arc must be accompanied by a counterarc. In general, this is not necessary, in agreement with the observations: of the arcs found so far, only that in A963 [LA88.1] has a possible counterarc. The counterarcs expected from symmetric lens models are strongly weakened if the symmetry is perturbed.

Whereas no prominent counterarc is expected in general, it is clear that multiple images must occur.[4] Owing to their faintness, they will be difficult to detect. Since arcs occur in rich, compact clusters, one must separate the gravitational images of background objects from the cluster galaxies. The best signature of a background object is its color, since the arcs found so far are very blue compared with normal, lower-redshift galaxies; note, however, that galaxies in the foreground cluster (e.g., those with strong star formation) can also be blue. Hence, blue color is not a unique signature for a background galaxy population, and it may be impossible to detect companion images of the source that gave rise to the arc.

If a cluster is to form arcs, its central density must approach the critical surface mass density (5.5) for multiple imaging. Although asymmetric matter distributions can form multiple images with a surface mass density smaller than $\Sigma_{\rm cr}$, it is still safe to assume that a cluster which forms arcs has a central surface mass density of the order of, or in excess of, $\Sigma_{\rm cr}$. This critical surface mass density is a function of the redshift of the source. Only for some of the arcs have redshifts been determined; they are about twice the redshift of the cluster (see Sect. 2.5.5). Of course, if a cluster is capable of forming an arc from a source at redshift $z_{\rm s}$, it can lead to multiple images of sources with $z > z_{\rm s}$. The angular size of the caustic is an increasing function of the source redshift, so that the probability for a source to be close to the caustic increases with redshift. At the same time, it is more probable for a source to lie on the caustic away from a cusp rather than on (or very close to) one. In such a case, we expect the source to have multiple images, two of which are merging tangentially, and strongly elongated in the direction of their separation. For higher-redshift sources, the size of the critical curve

---

[4] This is not always the case. One situation where a highly distorted image can occur, without additional images, is a 'naked cusp'. If we perturb a symmetric GL by external shear, say, the caustic corresponding to the radial critical curve will remain a smooth closed curve, but the caustic corresponding to the tangential critical curve has four cusps (an 'astroid'). For small shear, this astroid will be smaller than the 'radial caustic' but for larger shear, two cusps can lie outside the radial caustic (an example for this is given in Fig. 8.7b). An extended source located on such a cusp will have have only one highly distorted image, whereas if the cusp lies inside the radial caustic, (two) additional images must occur. A cusp not enclosed by another closed caustic curve ('naked cusp') can thus produce a single arc.

is also increased, so their images should occur at a larger distance from the center of the cluster. Indeed, observations of A370 have revealed a number of blue images, some of which occur in pairs, elongated in the direction of their separation (pairs A1, A2, and A3, A4 in Fig. 13.4, which were used to estimate the center of mass of the cluster [FO88.1]). Furthermore, the colors of the members of the pairs are equal within the measurement error. These observations are compatible with the picture outlined above, of sources at redshifts larger than that of the main arc. Recently, the redshift of arclet A5 was determined to be $z \simeq 1.3$ (Sect. 2.5.5), a value that confirms our expectations. In addition, three very blue images occur between the two massive elliptical galaxies (B1-B3); they can also be images of high-redshift sources. We should note that in the two point-mass lens model, there is always one image between the two masses; such an image is not very elongated. On the other hand, we must consider the possibility that the pairs of images are in fact interacting galaxies. In this case, their color could be due to enhanced star formation triggered by the tidal interaction between the galaxies. However, this hypothesis would not yield an natural answer to the question of why the separation vectors and elongations are roughly perpendicular to the direction to the center. For the cluster Cl2244−02, additional possible images have been reported [HA89.1].

Estimates of the lens mass. If one observes an Einstein ring, and if the redshifts of the source and the lens can be measured, the mass of the lens within the Einstein circle can be determined: the mean mass density in this circle is just the critical surface mass density $\Sigma_{\rm cr}$. Thus, up to the Hubble constant, which enters $\Sigma_{\rm cr}$, the lens mass could be determined. However, nature is more complex, and one usually must deal with asymmetric lenses (or a source offset from exact alignment for a symmetric lens) and, thus, 'incomplete Einstein rings'. Therefore, the estimate of the lens mass is more difficult. There are several possibilities. If one observes an arc and a counterarc, as in A963, where the two arcs are on opposite sides of the central cD galaxy, their separation can be used as a reasonable guess for the Einstein radius, although the lens might well be asymmetric. This 'radius', together with the redshifts of the source (which is unknown) and the lensing cluster, would then yield an estimate of the lens mass for the disk within the two arcs. Such an analysis was performed in [LA88.1] for A963, where the cD galaxy seems to define the center of the cluster fairly well; in addition, the velocity dispersion of the cluster can be measured, and its core radius can be estimated. These observations, together with the location of the arc, allow an estimate of the mass of the cD galaxy, which turns out to be of the order of $5 \times 10^{12} M_\odot$; note, however, that this estimate depends critically on the core radius of the cluster and on the (unknown) redshift of the source.

If only a single arc is observed, the mass determination is more uncertain. In those cases where the center of the cluster is well determined, one could in a first approximation use the distance of the arc from the center as an estimate of the corresponding 'Einstein radius', which then yields the mass

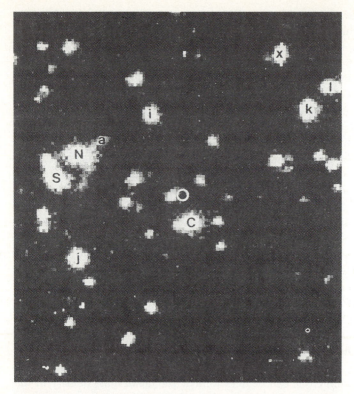

Fig. 13.7. (a) CCD image in the B band of Cl0500−24. C is a massive galaxy in the cluster. The redshifts of the galaxies S and N are slightly larger than that of the cluster; these appear to form a pair of interacting galaxies. An elongated feature is clearly seen to the west (right) of the two galaxies S and N; a is a galaxy in the cluster, which is not expected to be very massive, but marks the location of the elongated feature (termed 'arc' in the following). From redshift measurements of the galaxies in the cluster, its velocity dispersion is estimated to be 1300 ± 300 km/s. The circle to the north of C is the estimated center of brightness of the cluster. (b) Result of a model for the arc; the circle marks the position of the assumed center of gravity of the cluster, assumed to agree with the center of light. The two squares mark the location of galaxies S and N, to which a mass of $0.5\times10^{12}M_\odot$ (S) and $2\times10^{12}M_\odot$ (N) was assigned, as estimated from the color and brightness of these objects. The grey features are the images of a circular source. The small dot at the tip of the highly elongated image marks the location of galaxy a (from [WA89.1])

within this circle. If this is done for the arc in A370, we obtain a mass in agreement with estimates from dynamical observations. If the center of the cluster is ill-determined, one might be tempted to use the radius of curvature as a guess for the 'Einstein radius'; the robustness of such estimates remains to be demonstrated. At least in two cases, such an approach would be misleading: the curvature of the arcs in Cl0500−24 and A2390 is fairly small.

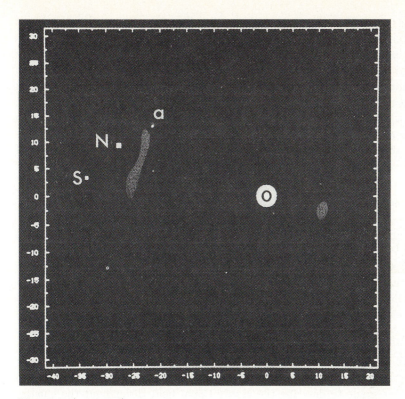

Fig. 13.7. (continued)

Model for the arc in Cl0500−24. In Fig. 13.7a, a CCD image of the cluster Cl0500−24 is shown, together with the position of the center of brightness for the cluster. To the west of the two galaxies S and N, a highly elongated feature is seen [GI88.1]. This feature, which is more easily seen on the difference image between B and R frames, is fairly straight; from the image, no curvature can be obtained. Although it is feasible that such a feature is a tidal arm produced by the interaction of the two galaxies, we would then have no natural explanation for the orientation of the feature (which will be called an 'arc' henceforth). If one assumes that this arc is due to gravitational light deflection, one can try to account for it with a model [WA89.1] (for a more detailed model, see [GI89.1]). The simplest assumption (without X-ray data for the cluster) is that the cluster is spherical, and that its center of mass agrees with the center of brightness. The mass of the two galaxies S and N can be estimated from their luminosity and color. The lens model, whose effects are shown in Fig. 13.7b, consists of an isothermal sphere centered at the center of light, and the three galaxies S, N, and a; the latter was only taken into account to mark the position of the arc, its deflection being unimportant due to its small expected mass. The only free parameter of the model is the velocity dispersion of the cluster. For a chosen value of σ_v, the source position is adjusted until an image occurs at the position of the arc. For a very wide range of velocity dispersion, this image is always elongated in the observed direction. Changes of the velocity dispersion change the length of the arc for fixed source radius. For the arc to have approximately the observed length and thickness, the value of the lens strength (which is a combination of velocity dispersion and unknown source redshift) is fairly well determined. The strength parameter can be taken to be the radius of the Einstein circle of the cluster: to obtain a reasonable image shape, this radius must be $\theta_c \approx 22''$. Then, the velocity dispersion of the cluster can be estimated to be

$$\sigma_v \approx 870\,\text{km/s} \left(\frac{D_s}{D_{ds}}\right)^{1/2} \left(\frac{\theta_c}{22''}\right)^{1/2}$$

For source redshifts in excess of $z_s \simeq 0.7$, this yields a velocity dispersion of about 1400 km/s, in agreement with the direct observational estimate.

A second model for this arc was tried, with the cluster represented as a point mass; in this case, it was very difficult to find acceptable images, in particular, the source size had to be chosen larger than $2''.6$, large compared to observed sizes of high-redshift galaxies. The isothermal sphere model needs a source size of about $1''.1$. The second model-predicted image seen in Fig. 13.7b is sufficiently bright to be observable. However, introducing a small shear in the lens model hardly changes the main arc at all, but can decrease the brightness of the second image considerably. Since no effort was made to describe the gravitational field throughout the cluster, but only that close to the arc, this second image does not lead to serious constraints on the lens model. Refinements of the model can be applied once observations improve.

A detailed model for the lens that produces the arc in Cl2244−02 was able to reproduce the location of the arc, and also the position of other images of the source, that may have been detected [HA89.1]. The free parameters in this model were the mass-to-light ratios of the galaxies and galaxy groups in the cluster. Fitting the image configuration to the observations then yielded estimates for these parameters, which, once more, were in good agreement with estimates obtained with other methods. Thus, this arc may also be understood under the lens hypothesis. Again, the essential check is the redshift measurement for the arcs.

Mass of individual galaxies. Close to the A370 arc (Fig. 13.4), several cluster galaxies are seen. One might wonder whether these galaxies perturb the arc. Although the structure of the arc is perturbed locally [NA88.2], this is very difficult to observe since arcs are usually unresolved in width. Thus, a local perturbation would escape the observer's attention, in particular because the galaxy outshines the arc locally. However, if the mass of the galaxy is sufficiently large, the arc could be deformed over its total width; this will occur if the additional deflection caused by the galaxy is of the order of the arc-width, or the resolution of the observation, whichever is larger. This case is very similar to the distortion of straight jets caused by a galaxy, as discussed in Sect. 12.6. Together with a good model for the generation of the arc, observed perturbations could then be used to estimate the mass of the perturbing galaxy.

Supernovae in arcs. The blue color of the arcs suggests that if the lensed sources are galaxies, considerable star formation must be going on within the latter. A large population of early-type stars means that the supernova rate in these galaxies must be high. It thus seems possible that one might observe a supernova in an arc over a convenient time scale of several years. As we saw in the preceding subsection, it is likely that arcs are the images of sources close to a cusp. Therefore, if the supernova explodes in that part of the galaxy which lies within the caustic and close to the cusp, three highly-magnified images might occur [KO88.2]. Owing to the time delay for the different images, they will be seen in succession. A simple method to describe such an event was given in [KO88.2]:

Consider the Fermat potential ϕ (5.11) close to an arc-shaped image of a source close to a cusp. As was described in Sect. 6.2, a point source moving from the inside towards the cusp has three images, two of which are merging tangentially. That means that the Fermat potential must be very flat at that point where the merging occurs, in the tangential direction. The geometrical picture one should have in mind is that of a long, not necessarily straight valley. This intuitive picture can be formulated algebraically by introducing a parameter which measures the length along the valley floor (λ, say) and denoting the distance from the surrounding ridge by δ. Then, the function ϕ can be written locally as $\phi \approx a\delta^2/2 + P(\lambda)$, where a, the parameter describing the curvature of the valley perpendicular to its extension, is assumed to be independent of λ. We now make the ansatz that $P(\lambda)$ is a fourth-order polynomial; its constant term is unimportant, and its linear term can be made to vanish by choosing the origin of λ. Thus,

$$\phi \approx \frac{p_2}{2}\lambda^2 + \frac{p_3}{6}\lambda^3 + \frac{p_4}{24}\lambda^4 + \frac{a}{2}\delta^2 \quad .$$

The determinant of the corresponding Jacobian matrix is

$$\det A = a\frac{d^2 P(\lambda)}{d\lambda^2} \quad .$$

The lens equation becomes

$$\delta = 0; \quad p_2\lambda + \frac{p_3}{2}\lambda^2 + \frac{p_4}{6}\lambda^3 = 0 \quad ;$$

the latter equation has three solutions if the source lies inside the cusp. These solutions, as well as the corresponding magnifications, are easily derived [KO88.2]; note that the two independent ratios of magnification and the two independent time delays do not depend on a. Thus, from these four observables, one can in principle determine the coefficients p_i. In fact, numerical solutions of a realistic arc model confirm the applicability of the simple method outlined above. Thus, the observation of a multiply-imaged supernova in an arc would yield strong constraints on a GL model.

Arcs as telescopes. Since arcs are highly magnified images of distant galaxies, we have the unique opportunity to obtain spectra of objects which otherwise would be too faint. In addition to the magnification, it is important that an arc is essentially a one-dimensional source of radiation, which can be used by spectroscopists to fit a slit along the arc and integrate the spectrum in this direction. The observed arcs are magnified by factors of about 10 to 20, which means that they are three magnitudes brighter than the unlensed source. Needless to say, this makes an enormous difference for the integration time to obtain a spectrum. Thus, clusters of galaxies which happen to be sufficiently strong lenses and which are sufficiently well-aligned with a high-redshift background galaxy can be used as telescopes for observations of these background objects! In fact, the recently determined redshift of the arc in the cluster Cl2244 [ME91.1] of $z \sim 2.2$, if correct, would be the highest redshift ever obtained for a "normal" galaxy, and the spectroscopy of that object would not have been feasable without the large magnification provided by the cluster lens.

13.2.4 Statistics of arcs and arclets

Statistics of arcs. Long before the discovery of the first GL, the probability for the occurrence of multiple images of a distant compact source was estimated [ZW37.2]. No such estimate was available for the occurrence of arcs at the time they were discovered; indeed, they had not even been predicted. Estimating the probability of finding arcs is still a difficult business. Although the distribution of galaxies in the universe is fairly well known, the density of galaxy clusters and their properties (e.g., ellipticity, velocity dispersion, core radius) are less well known. In particular, the evolution time for clusters is estimated to be comparable to the Hubble time in some cases, and thus it is premature to extrapolate from the observations of low-redshift clusters to higher redshifts. In fact, as we have argued above, gravitational lensing might be an important tool to investigate high-redshift clusters.

Still, we can ask for the probability that a given model cluster leads to the occurrence of highly elongated images, and the distribution of the properties of these images. Such an approach was taken in [GR88.2], where a number of clusters was randomly generated, according to a probability distribution for their core radius, shear parameter, and their distributions of galaxies. 300 clusters were generated, together with a random distribution of background galaxies. The study is too limited to yield statistical estimates for arc length, occurrence of counterarcs, etc., but it demonstrates that arcs should occur frequently: 37 arcs were found with length greater than $10''$, and 5 with length greater than $20''$. Thus, these simulations confirm the expectation that shorter arcs should occur more frequently. However, observations are clearly biased against detecting short arcs, as they are less luminous and less prominent, and thus can easily escape detection. Of the long arcs, the large majority were found to be oriented perpendicularly to the radius vector to the center of the corresponding cluster, confirming the idea that most highly-elongated images are due to tangentially-merging images. The occurrence of double images was rare, and for the longest arcs, not a single counterarc was found. Also, the addition of galaxies to the smooth cluster potential did not lead to the occurrence of strongly distorted arcs; usually, they remained very smooth, as we discussed above.

Arcs in these simulations occurred frequently in clusters where the central surface mass density was below Σ_{cr}; this is possible either because of the ellipticity of the cluster potential, which decreases the minimum surface mass density needed for multiple imaging compared to the axi-symmetric case, or because the deflection due to a single galaxy is strongly enhanced if the surface mass density is close to critical (more generally, if the lensing galaxy is at a location where the magnification of the cluster potential is large).

A further statistical study for the formation of arcs is described in [KO89.3], which presents a study of the angle subtended by arcs, measured from the centers of their corresponding lens clusters. In addition, the distribution of this angle for arcs and counterarcs was studied. The distribution clearly shows that all arcs observed so far can easily be explained by a slightly perturbed symmetric lens, although, of course, these may not be the most realistic models.

As we have seen, arcs are formed when an extended source lies close to a cusp (or a higher-order catastrophe of the lens mapping). Hence, the caustic structure of the lens is essential for the occurrence of arcs. Statistics of arcs based on symmetric lens models [NE89.1] are therefore not realistic.

Arclets. Every cluster of galaxies sufficiently compact to form long, highly distorted arcs will distort the images of other background sources, depending on their angular impact parameters. In the preceding discussion of the cluster A370, we saw that, in addition to the main arc, there are six additional fairly elongated and blue images, readily identified as arclets; one of these arclets and the main, 'giant' arc, have redshifts larger than the redshift of the putative lens, in accordance with the GL hypothesis. As was mentioned before, for every prominent arc, there must be numerous images which are less distorted, and which cannot be confirmed to be lensed by imaging alone, since in general galaxies are not 'round'. As discussed in Sect. 12.6, a separation of intrinsic asymmetric shape and gravitational distortion is not possible, unless the distortions are extreme.

On the other hand, image distortion by clusters proceeds in such a way that the major axis of the images is mainly oriented tangentially to the radius vector to the cluster center. If an ensemble of presumably background galaxies is observed near a cluster, one can check whether the images have a preferred alignment in the tangential direction. In Sect. 12.6, we discussed a search for gravitationally distorted background galaxies close to foreground galaxies, which yielded no positive signature [TY84.1]. The success of such a search will depend on the faintness of the images investigated (as fainter images belong to more distant sources, thus increasing the probability for appreciable distortion). Furthermore, since fainter images are more numerous, one can hope to discover 'coherent' distortion of several images around the same deflector.

Let us now discuss a recent study of very faint galaxy images close to clusters ([TY90.1], [TY90.2]; see also [GR90.1]). Deep CCD imaging surveys up to a limiting surface brightness of 29 B mag/arcsec^2 reveal a population of very blue galaxies; there are over 200000 of these objects per square degree. As argued in [TY88.1], the redshifts of these objects seem to cover the range between 1 and 3. Selecting those galaxies with anomalously blue color, one selects a background population with about 30 objects per square arcminute. This population thus has a density which is sufficient to map the distortion distribution of a compact foreground cluster.

Following this search strategy, systematic alignments orthogonal to the direction to the cluster center were found in two clusters [TY90.1]. Several faint arclets have been seen, but more important is the coherent alignment of the more numerous faint blue galaxy ellipticities. The histograms in [TY90.1] clearly show a preferred alignment tangent to the cluster center of blue galaxies, but not of red galaxies (which are presumably in the cluster or foreground objects). The observed distortion distribution was compared with simulations, where the background source distribution (with an assumed redshift distribution) was imaged through a model cluster, using the same software as was used to analyze the observations. With such a procedure, the velocity dispersions of the two observed clusters were determined in [TY90.1].

A more detailed study of the extraction of information from an observed distribution of arclets in a cluster is presented in [KO90.4] and [MI91.1]. Assuming that the intrinsic shape distribution of the background sources is known (e.g., it can be obtained from the observed distribution in directions not containing clusters) and that the sources are oriented randomly, the center of the cluster, as well as its radial density profile can be determined, depending on the number density of observed arclets. In principle, the ellipticity and orientation of the cluster can be determined as well. Equally important, the observations of arclets may be the most powerful method to determine the redshift distribution of these faint galaxies, if a sufficiently large sample of clusters with observed arclets becomes available. Furthermore, a search for arclets near multiply-imaged QSOs could yield further evidence for their GL origin. This is particularly true for the 'dark matter' lens candidates, where no sign of a lensing galaxy has been found. If these systems (e.g., 2345+007) are indeed lensed images, their angular separation may point towards a lensing galaxy embedded in a cluster. Observations of arclets may be used to detect such a mass concentration in the field of these multiple QSOs. Also, in the GL system 0957+561, the detection of arclets could in principle constrain the mass distribution of the cluster hosting the main lensing galaxy.

We leave this most timely subject here with the expectation that there will be rapid progress, in both observations and theoretical analysis.

13.3 Additional applications

In this section, we briefly describe a number of topics which have been suggested in the literature as potential applications of gravitational lensing. We start in Sect. 13.3.1 with estimates of the size of QSO absorption line systems from lensed QSOs (or close pairs of QSOs). A method is outlined in Sect. 13.3.2 for gaining information about the brightness distribution of QSOs. In Sect. 13.3.3 we consider the parallax effect, which can in principle be used to detect ML in individual objects; together with the method of Sect. 13.3.2, the absolute size and relative transverse velocity of the source can be determined. Cosmic strings as lenses are briefly discussed in Sect. 13.3.4. We then show in Sect. 13.3.5 how an upper limit of the mass of at least one QSO can be obtained, and discuss in Sect. 13.3.6 whether gravitational lensing can account for superluminal motion.

13.3.1 The size of QSO absorption line systems

Absorption lines in QSOs. In addition to the broad emission lines, the defining feature of QSOs, many QSO spectra also show absorption lines. QSO absorption lines are currently classified into three main categories:

(a) Broad absorption lines (BAL) with a width comparable to that of the emission lines. About 5% of all QSOs show these BAL; the redshift difference of the BAL and the emission lines, if interpreted as a velocity difference in the rest frame of the QSO, can be up to 0.2 times the velocity of light.

(b) Narrow lines of heavy elements, which can be identified only if more than one line per system is seen; e.g., doublets of CIV or MgII. These lines can have redshifts very close to that of the emission-line redshift of the QSO, but can also have a much lower redshift.

(c) In all QSOs where Lyα can be observed in emission, there are a large number of narrow absorption lines at shorter wavelengths than the emission line. Most of these lines are interpreted as Lyα absorption lines ("Lyα forest").

The BAL are usually interpreted as being connected to the QSO phenomenon, mainly because very few broad systems have been found with velocities $\gtrsim 0.2c$ relative to that corresponding to the emission line redshift of the QSO. The large relative velocities point toward a massive outflow phenomenon. Also, the narrow metal-line systems with redshifts close to the QSO's are believed to be closely related to the QSO. However, narrow metal-line systems with much smaller redshift than the QSO's, as well as the Lyα forest, are interpreted as being due to intervening material (for reviews, see [WE81.1], [SA87.2]). The nature of the absorbers is less clear. One can measure the column density of the absorbing species, and if several ionization species are observed at the same redshift, one can infer the ionization state of the absorber or the relative abundance of two or more elements. A direct measurement of the size of the absorbers is not possible; however, lensing can lead to estimates of their sizes in some cases.

The size of absorbers from lens systems and QSO pairs. Suppose one measures the spectra of the two images of a lensed QSO or of a close pair of QSOs. If both spectra have an absorption line at (nearly) the same redshift, which itself is much smaller than the emission redshift of the QSO (so that the absorber is taken to be intervening material), one can conclude that the absorber must be at least as large as the spatial separation of the two light beams at the redshift where the absorption is observed. In both cases – the lens system and the close pair of QSOs – the separation of the light beams varies with redshift. Hence, a single spectrum of such a pair of images can probe a range of physical sizes in the absorption systems.

The spectrum of the double QSO 0957+561 has two absorption line systems, one at $z_{abs} = 1.39$ ([WA79.1], [WE79.1]) and another one at $z_{abs} = 1.12$ ([YO81.4]; note that the emission redshift of the images is $z_{em} = 1.41$). The first system can be associated with the QSO, but if the lower redshift system is due to material ejected from the QSO, it has a relative velocity of $\sim 0.29c$. Alternatively, this system can simply be an intervening clump of matter not physically associated with the QSO. In that case, the system must have an extent in excess of 7 kpc, which is the separation of the two light beams at the absorption redshift. What is puzzling, however, is the fact that the redshift difference of the absorption lines in the two images is less than about 10 km/s. This leads to difficulties in interpreting the origin of this absorption system. Suppose the absorber is a galaxy. The typical large-scale velocities in galaxies are of the order of 200 km/s, due either to galactic rotation or to a velocity dispersion. It is thus unlikely that such a physical system would give rise to the same absorption redshift in both images. Another possibility is that the absorber is a large-scale intergalactic cloud, but there again the small redshift difference does not occur naturally. Thus, the nature of this absorption system is currently unclear.

The spectra of the images of Q2345+007 have several Lyα absorption lines [FO84.1], most of which are common to both images. Since for this system no lens has been observed (and it is not clear whether this is a lensing system at all), the separation of the light beams as a function of redshift is not known well. Nevertheless, it is possible to obtain a lower limit on the size of the Lyα clouds; since some absorption lines are not matched in both spectra, one can also estimate a probable value for the extent of the absorbers, which again depends on the unknown redshift of the lens; a typical value is about 8 kpc [FO84.1]

A more recent study of the absorption lines in this GL candidate system (S. Bajtlik, personal communication) has taken the correlation of the equivalent width of the lines into account, which provides additional information: if the correlation of the equivalent widths is strong, the separation of the two light rays at the redshift of the absorber must be much smaller than the size of the absorber. Assuming that 2345+007 is a GL system (see the discussion on this in Sect. 2.5.1) and that the redshift of the lens is 1.49, as suggested by the strong iron line absorption corresponding to that redshift, they obtained an estimate on the cloud size of between 10 and 20 kpc.

A similar study was conducted for the GL system UM673 ([SM90.1], [SM91.1]). All 68 Lyα lines seen in image B are also present in image A, and there are only two anti-coincidences (i.e., lines present in image A and not in B); these two anti-coincidences could be a MgII doublet. Cross-correlation of high-resolution spectra of both images yields a conservative lower limit on the size of Lyα clouds of about 12 kpc (for $h_{50} = 1$, $\Omega = 0$); if the two anti-coincidences are indeed MgII, this lower limit increases to 23 kpc. In addition, a strong correlation between the equivalent widths of corresponding lines indicates that the two light rays traverse the same clouds, proving that the absorption lines are caused by individual large clouds, instead of an ensemble of many extended clouds at the same redshift.

Information on the larger-scale structure of absorption systems can be obtained from QSO pairs with somewhat larger separation. For example, in the QSO pair Q0307−195A,B [SH83.1], in the spectrum of one QSO one absorption system is seen at a redshift corresponding to the emission redshift of the other QSO. Since the separation of these two QSOs (with redshifts z_1=2.144, z_2=2.122) corresponds to several hundred kpc, one infers that there is an absorption system physically associated with the nearer QSO, but at a distance comparable to the size of a cluster. Two additional common absorption systems are observed in this QSO pair, with a redshift difference corresponding to 297 km/s and 172 km/s, respectively. Recently, correlated absorption in a QSO pair was detected, implying very large-scale clustering of absorbers [SA87.3].

Lensing and absorption system statistics. If one assumes that all narrow metal absorption systems are galaxies, one can obtain their average cross-section from the frequency with which these systems occur in high-redshift QSOs and the assumed density of galaxies. However, these statistics might be influenced by lensing: suppose the counts of high-redshift QSOs are influenced by the amplification bias. In such a case, flux-limited samples contain a higher fraction of QSOs with a deflecting mass in the line-of-sight. If the deflectors which cause the magnification are physically related to absorbers, one should then obtain a larger mean number of absorption systems in the sample than would be expected just from chance superposition. This effect, and its relation to the cosmic evolution of absorbing system, was recently investigated [TH90.1]; no evidence was found for evolution of low-equivalent-width absorption systems, but that evolution is required for high-equivalent-width systems. Whereas the former ones occur frequently in high-redshift QSOs (typically more than one per object), they cannot be associated with a population of objects which can cause significant amplification bias. In contrast, the data on high-equivalent-width systems are compatible with the idea that the corresponding QSOs are biased into the sample by lensing. If the apparent evolution of high-equivalent-width metal absorption systems is to be attributed solely to this effect, nearly every bright QSO in the sample must be biased.

Absorption can also be viewed from the opposite point of view: if absorbers are related to matter concentrations, one expects a higher fraction of lensed QSOs for those with rich absorption line spectra [KR79.1]. However, as we saw in the preceding chapter, the most efficient lenses have redshift $\lesssim 1$ for QSOs at redshift 2 to 3; thus, the absorption lines corresponding to high-redshift absorption systems should not be strongly correlated with lens properties. But it may be that lower-redshift absorption systems show a positive correlation with lensing probability. In this respect, detailed spectral investigations of those QSOs with a nearby galaxy (as discussed in Sect. 12.3) should yield more information about the relation between absorption-line systems and galaxies.

Amplification bias and the proximity effect. The comoving number density of Lyα clouds increases with redshift, and can be described by a power law, $dN/dz \propto (1+z)^\gamma$ with $\gamma \approx 2$, although γ is not very well determined. However, it has been found that, for an individual QSO, this law ceases to be valid as one approaches the emission-line redshift of the QSO [MU86.1]. A study of this effect for a large sample of Lyα absorption lines (roughly 470 lines in 19 QSOs) [BA88.1] indicates that such a decrease ('proximity effect') can be accounted for by the ionizing radiation of the QSO. The idea is that the Lyα clouds are in ionization balance with their surrounding radiation field. Since the recombination time-scale is much longer than the typical ionization time, nearly all hydrogen atoms in these clouds are ionized. The fraction of neutral hydrogen atoms is roughly inversely proportional to the density of the ionizing photons. Assuming that the radiation field is a superposition of an UV background and the radiation from the QSO, a cloud closer to a QSO sees a higher density of these photons, so that the density of neutral atoms decreases, and fewer systems (for a given equivalent width of the absorption lines) are observed close to the QSO. If this explanation is correct, it has two important implications: first, one can measure the density of ionizing photons at high redshift, and second, the proximity effect yields a measure of the UV luminosity of the QSOs, as it should be more pronounced for high-luminosity objects. If this effect could be confirmed, it would allow measurements of the luminosity in two different ways, by the proximity effect and by the usual measurement of the QSO flux. Since the former measurement is not influenced by gravitational flux magnification, a comparison between these two values of the luminosity could in principle be used to obtain the magnification factor of lensed QSOs, or even to detect ML.

13.3.2 Scanning of the source by caustics

Point source near caustic. It was shown in Chap. 6 that the magnification of a point source close to a caustic varies as the inverse square root of its distance to the caustic.[5] This result was obtained by a series expansion of the lens equation about a point on the corresponding critical curve in the lens plane. Two bright images of the source are formed next to the critical curve when the source changes its position from the 'negative' side of the caustic to the 'positive' side. Consider a caustic which can be approximated locally as a straight line, and let it be described by $y_1 = 0$. Let $y_1 > 0$ be the positive side of the caustic, i.e., if the point source is at $y_1 > 0$ there are two bright images close to the corresponding critical curve. Then, the magnification of the point source is

$$\mu_p = \mu_0 + C y_1^{-1/2} H(y_1) \quad , \tag{13.30}$$

where $H(x)$ is the Heaviside step function. The first term is the magnification of all the other images (together) of the source, which is assumed to vary much less with the position of the source than that of the two critical images; C is a constant of proportionality which characterizes the caustic locally. Although C may vary along the caustic, we assume it does so slowly, so that C can be considered to be constant (compare also the discussion in Sect. 6.4).

Extended source. Let us consider an extended source, and let $\mathbf{y}_c$ be a fixed point on the source (e.g., center of brightness). The magnification of this source is then

$$\mu_e = \mu_0 + C\, S_0^{-1} \int d^2 y\, I(\mathbf{y})\, y_1^{-1/2} H(y_1) \quad , \tag{13.31}$$

where $I(\mathbf{y})$ is the brightness profile of the source at position $\mathbf{y}$, and S_0 is the integral of the brightness over the source. If the source's brightness profile is constant over time, one has $I(\mathbf{y}) = J(\mathbf{y} - \mathbf{y}_c)$, so (13.31) becomes

$$\mu_e(y_{1c}) = \mu_0 + \frac{C}{S_0} \int_{-y_{1c}}^{\infty} dz\, j(z)(z + y_{1c})^{-1/2} \quad , \tag{13.32a}$$

where we have defined

$$j(y_1) = \int dy_2\, J(\mathbf{y}) \quad , \tag{13.32b}$$

i.e., $j(y_1)$ is the scaled brightness profile integrated along the direction parallel to the caustic.

The motion of the source across the caustic leads to a change of y_{1c} with time. We can approximate locally the velocity as a constant, so that y_{1c} is a linear function of time t, $y_{1c} = u + vt$, where v is the effective transverse velocity perpendicular to the caustic. If one measures the flux at

---

[5] In this section, we only consider folds.

several times during the source's crossing of the caustic, one obtains several integrals of the one-dimensional brightness profile $j(y_1)$. This is very similar to occultation measurements of a source during a lunar eclipse, where the different points on the light curve are integrals of the brightness of that part of the source which is not yet covered by the Moon. This light curve can then be used to obtain the one-dimensional brightness profile of the source.

The brightness profile. Let us assume that the source is finite, and let us further assume that one can identify the instant when the first source point crosses the caustic; without loss of generality we take the source velocity to be in the positive y_1 direction. That source point which first crosses the caustic can be used as the 'origin' of the source, i.e., this point can be identified with $\mathbf{y}_c$. Then, the brightness profile $j(z)$ vanishes for all $z \geq 0$. We can then rewrite (13.32), after multiplication by S_0,

$$S(t) = S_* + \int_0^t \frac{d\tau}{\sqrt{\tau}} j^*(\tau - t) \quad , \tag{13.33a}$$

where $S_* = \mu_0 S_0$ is the flux of all images but the two critical ones, and

$$j^*(t) = \sqrt{v} C j(vt) \quad ; \tag{13.33b}$$

hence, the dependence of the light curve on the velocity v has been scaled into j^*. Note that $j^*(t)$ is proportional to the one-dimensional brightness profile at the position of the caustic at observer's time t. The constant u vanished due to the identification of $t = 0$ with the time the point $\mathbf{y}_c$ crosses the caustic.

The problem is how to invert the above relation, i.e., to obtain a scaled brightness profile $j^*(\tau)$ from an observed light curve $S(t)$. Two different methods were given in [GR88.1]. The first is the direct inversion of the integral equation (13.33a). S_* is assumed to be measurable – it is the flux of the source before it has reached the caustic. Rewriting (13.33a) by substituting $y = t - \tau$ as

$$S(t) - S_* = \int_0^t \frac{dy}{\sqrt{t-y}} j^*(-y) \quad , \tag{13.34a}$$

we can invert this Abel integral equation [TR85.1]:

$$j^*(-\tau) = -\frac{1}{\pi} \frac{d}{d\tau} \int_0^\tau dt \, \frac{S(t) - S_*}{\sqrt{\tau - t}} \quad . \tag{13.34b}$$

The disadvantage of this method of inversion is the differential operator; the integral must then be well-determined as a function of τ before (13.34b) can be applied. However, the measurements of $S(t)$ can only be obtained at a finite number of points, and each data point has an observational uncertainty. It turns out [GR88.1] that this method is very sensitive to these errors and thus should not be used.

Inversion with the polygon approximation. A second method proposed in [GR88.1] is the representation of the brightness profile as a polygon. Suppose one measures the flux $S(t)$ at times t_i, $i = 0, 1, 2...$, such that $t_0 = 0$. Define $S(t_i) = S_i$; then, from (13.34a) we obtain

$$S_i = S_* + \int_0^{t_i} \frac{dy}{\sqrt{t_i - y}} j^*(-y) = S_* + \sum_{k=0}^{i-1} \int_{t_k}^{t_{k+1}} \frac{dy}{\sqrt{t_i - y}} j^*(-y) \quad .$$

If we approximate $j^*(y)$ piecewise with linear functions of the form

$$j^*(-y) = \frac{t_{k+1} - y}{t_{k+1} - t_k} j_k + \frac{y - t_k}{t_{k+1} - t_k} j_{k+1} \quad \text{for} \quad t_k \leq y \leq t_{k+1} \quad ,$$

where $j_k = j^*(-t_k)$, the above equation becomes

$$S_i = S_* + \sum_{k=0}^{i-1} (A_{ik} j_k + B_{ik} j_{k+1}) \quad ,$$

with

$$A_{ik} = \int_{t_k}^{t_{k+1}} \frac{dy}{\sqrt{t_i - y}} = \frac{2}{3} \frac{\left[(t_i - t_k)^{\frac{3}{2}} - (t_i - t_{k+1})^{\frac{3}{2}}\right]}{t_{k+1} - t_k}$$

$$- 2 \frac{t_i - t_{k+1}}{t_{k+1} - t_k} \left[\sqrt{t_i - t_k} - \sqrt{t_i - t_{k+1}}\right] \quad ,$$

$$B_{ik} = \int_{t_k}^{t_{k+1}} \frac{dy}{\sqrt{t_i - y}} \frac{y - t_k}{t_{k+1} - t_k} = 2 \frac{t_i - t_k}{t_{k+1} - t_k} \left[\sqrt{t_i - t_k} - \sqrt{t_i - t_{k+1}}\right]$$

$$- \frac{2}{3} \frac{\left[(t_i - t_k)^{\frac{3}{2}} - (t_i - t_{k+1})^{\frac{3}{2}}\right]}{t_{k+1} - t_k} \quad .$$

From the equations above,

$$S_{i+1} - S_i = \sum_{k=0}^{i-1} \left[\left(A_{i+1,k} - A_{ik}\right) j_k + \left(B_{i+1,k} - B_{ik}\right) j_{k+1}\right] \quad (13.35)$$
$$+ A_{i+1,i} j_i + B_{i+1,i} j_{i+1} \quad .$$

Equation (13.35) can readily be solved for j_{i+1} if all j_k, $k \leq i$ are known. Thus, the j_i can be obtained iteratively from (13.35). We therefore obtain the one-dimensional brightness profile at discrete points. This method was applied in [GR88.1] to simulated light curves of sources generated by the ray-shooting method (compare Sect. 12.4), for two different assumed brightness profiles of the source. These profiles could be recovered from the generated light curve in a satisfactory approximation. Thus, it seems that the approximations made in this section, i.e., local straightness of the caustic, as well as constancy of C along the caustic and the approximate constancy of S_*, are fairly well satisfied for this situation.

A practical application of the polygon method to observed light curves will be fairly difficult, for several reasons. First of all, the inversion of the integral equation (13.34a) is based on the assumption that the time t_0 at which the first source point crosses the caustic can be identified from the data with sufficient accuracy. Whether or not this will be possible depends critically on the density of available data points, and also on the accuracy of the photometric observations. Second, whereas the test cases considered by [GR88.1] used very smooth light curves (due to good ray statistics in the ray-shooting program), observational data are noisy. Since the inversion scheme discussed above solves (13.34) iteratively, an error in the measurement will propagate throughout the brightness profile. For this reason, the observed data points should be smoothed before using the above scheme, rather than using the scheme and smoothing the resulting brightness profile. A better way for reconstructing the brightness profile was given in [GR90.2], where a regularization method is used to reconstruct the source profile from noisy data; this method yields fairly satisfactory results, even when applied to noisy data.

The method of eclipsing a source with GL caustics may yield some insight into the brightness structure of sources much too small to be resolved with other methods; it is, therefore, potentially very useful. Note that the method can at best yield the shape of the one-dimensional brightness distribution; its scale cannot be obtained, since the transverse velocity is not known. A method to obtain this velocity and the absolute scale of the source is discussed below.

13.3.3 The parallax effect

In Sect. 13.3.2, we considered the possibility of investigating the brightness structure of QSOs from lensing as the source crosses a caustic; the light curve is a convolution of the universal point-source magnification and the one-dimensional brightness distribution of the source, from which the latter can be obtained, in principle, by deconvolution. Related to this effect is the observation of the same source from two or more different, well-separated positions. In the derivation of the time delay in Sect. 5.3, we showed that a displacement of the observer is equivalent to a displacement of the source [see (5.37)]. Now, suppose that two telescopes with separation ζ, measured in the observer's plane, observe a source simultaneously. This is equivalent to observing the source at two instances, between which the source position has changed by $\eta = \left[\frac{D_{ds}}{D_d(1+z_d)}\right]\zeta$; the only difference is that during this time interval the source may change intrinsically, whereas in the case where two telescopes observe the source simultaneously, any difference in the flux of the images of the source is due to a propagation effect. Such observations can be used to find out whether a given source is indeed microlensed [GR86.1]. If the apparent magnitudes measured by two separate observers disagree, then either the source radiates highly anisotropically (which is very unlikely, as the corresponding angle as seen by the source is of the order of microarcseconds), or this difference is due to intervening material. Such material can be an absorption column, but if the source is a lensing candidate, it seems reasonable to attribute this difference to the slightly different relative positions of source, lens and observers. In fact, apart from intrinsic variability, the two observers should measure the same light curve of the source, but with a finite time delay.

In order to see how widely the telescopes should be separated for this effect to be measurable, we first note that the flux of a source changes drastically if it crosses a caustic. The shortest time-scale of flux variation is thus a fraction of the time it takes the source to move a distance equal to its own radius (or, equivalently, the time it takes the caustic to sweep over the source). If ρ is the size of the source, a significant flux variation occurs if the source position changes by $q\rho$ in the direction perpendicular to a caustic, where q is a factor which depends on the structure of the caustic and on the accuracy of the flux measurements. From the above equation, we see that a distance $q\rho$ in the source plane is equivalent to a distance

$$d = \frac{D_d(1+z_d)}{D_{ds}} q\rho$$

in the observer plane. Hence, the necessary separation depends on ρ and the relative redshifts of the source and the lens; for fixed source redshift, it is smaller for smaller lens redshifts. For the GL system 2237+0305, $D_{\rm d} \ll D_{\rm ds}$, which makes it the best candidate for such a measurement. If the size of the source is about 10^{15} cm, and if we assume $q \sim 0.1$, the necessary separation of telescopes is of the order of 10^{13} cm, or one Astronomical Unit. That means that for such a favorable lens candidate, the parallax effect could in principle be observed from planetary spacecraft. A space project has been proposed for a telescope to be brought to a distance of one Thousand Astronomical Units (TAU); this would then allow to measure the parallax effect for GL systems where the redshifts of the source and the lens are comparable.

The time delay for the light curves of an image measured with two or more telescopes can be used to measure the effective transverse velocity of the caustic [GR88.1]. Thus, such a measurement would fix the physical scale of the lensing geometry, and the absolute size of the source could be determined – in contrast to the discussion in Sect. 13.3.2, where only the ratio of source size and effective velocity was probed.

13.3.4 Cosmic strings

The 1980's have brought about a marriage of two apparently diametrically-opposed disciplines in physics, particle physics and cosmology (for a review, see, e.g., [DO81.1], [BA83.2]). To name just two applications of this amalgam, the primordial nucleosynthesis depends on the life-time of the neutron, as well as on the cosmic density in baryons and the density parameter of the universe, and a finite rest mass of the neutrino would lead to the conclusion that most of the mass in the universe could be in such very light particles. On the other hand, primordial nucleosynthesis in connection with observational determination of the abundance of He^3 and Li^7 yields constraints on the number of different species of neutrinos.

A dramatic evolution of cosmology was triggered by realizing that the universe is the ideal laboratory for high energy physics, as the temperature in the very early phase of the cosmos was sufficiently high for the strong interaction to be symmetric with respect to the electro-weak interaction. The threshold for symmetry breaking as predicted by Grand Unification Theories (GUTs) is about 10^{15} GeV. GUTs predict that if matter cools through this temperature, a phase transition must occur, which breaks the symmetry, leading to the idea of the 'inflationary universe' model [GU81.1] (for reviews see, e.g., [BR85.1], [BO88.1]). A discussion of these highly interesting matters is beyond the scope of this book; however, there is one possible consequence of inflation which may have an application in gravitational lensing. Namely, the phase transition can give rise to topological defects, monopoles, strings, and domain walls (see the review [VI85.2]). Here, we will only briefly discuss the possible influence of cosmic strings on the light propagation in the universe, although strings have a number of other attractive properties, e.g., they can help to form galaxies [ZE80.1], [VI81.1].

Locally, a string is described by a single parameter μ_s, its mass per unit length. It has a pressure along its length which is the negative value of its mass density, and thus a tension. $\mu_s = 1$ corresponds to one Planck mass per Planck length, $\approx 1.35 \times 10^{28}$ g/cm. The value of μ_s can be predicted in principle from GUTs, and is supposed to be $\approx 10^{-6}$. Strings have no end, they are either infinite or form closed loops. The tension of the strings will straighten out any kinks, so that infinite strings are straight.[6] Closed loops will oscillate, and it can be shown that their motion is relativistic, which leads to the emission of gravitational waves. Hence, a scenario where cosmic strings play a role in cosmology predicts a background of gravitational waves. Using the timing observations of the millisecond pulsar 1937+21, or the binary pulsar 1913+16, the cosmological density of strings can be constrained (e.g., [HO84.2]).

The exterior metric of a vacuum string was calculated in the weak field limit ($\mu_s \ll 1$) in [VI81.2] and for the general case in [GO85.1]. Locally, the metric is flat, and can best be visualized as a cone, whose tip is the location of the string. If one considers such a cone, it is easy to see that, for some pairs of points lying on 'opposite' sides of the tip, there are two geodesics which connect them; similarly, if a string lies between a source and the observer, there can be two images of the source, one on either side of the string. Since the metric is locally flat, there is no 'local' deflection, i.e., there is no local distortion of the images of the source; in particular, the magnification of the images is 1. Distortions of source images are expected for extended sources, if only part of them are multiply imaged; this could lead to images with sharp edges. The angular separation between two images formed by a string is

$$\Delta\theta = \frac{D_{ds}}{D_s} 8\pi\mu_s \sin\vartheta \quad,$$

where ϑ is the angle between the string and the line-of-sight to the source. For $\mu_s = 10^{-6}$, $8\pi\mu_s = 5.2$ arcseconds, so strings can lead to double images with observable angular separations. The time delay for the images is of the order $\Delta t \sim 4\pi\mu_s \Delta\theta D_d/c$, which for $\mu_s = 10^{-6}$ is of the order of 10 years. Note that all numerical estimates given here are for static strings, and they have to be modified for relativistically moving strings.

Thus, sources lying behind a straight (or nearly straight) string in a strip of angular width $\sim \Delta\theta$ have two unmagnified images. Since all the sources in that strip have double images, one should expect a 'chain' of image pairs. If the source is a QSO, the two spectra need not be identical, since QSO spectra can vary on time-scales shorter than the expected time delay. The same is true for the redshifts, if these are determined from the broad

---

[6] although, in a recent paper [VI90.1], the existence of kinky cosmic strings were discussed

emission lines, whose profiles can also vary, as well as their mean redshifts. There are a number of QSO pairs which have a separation of a few arcminutes and very similar redshifts [PA86.3]. The pair with the smallest redshift difference is 1146+111, first detected in [AR80.1], mentioned in [PA86.3] as a possible case for lensing by a string, and further investigated in [TU86.1] (for more references, see the discussion of this source in Sect. 2.5.8). The image separation is 2.6 arcminutes, and the redshift difference about 10^{-3}. The spectral differences are compatible with the lensing hypothesis, because of the large expected time delay; however, the lack of any other obvious pair of sources in this field gives rise to serious doubts as to whether this can be a double image caused by a string (see the discussion in Sect. 2.5.8). The optical depth for lensing by strings is about $4\mu_s$ [GO85.1]. Recently, four very close pairs of galaxies were observed [CO87.1]; the redshifts of the pairs are quite similar. The image separation is about 2 arcminutes. However, the geometry of these galaxies is complex, so the required shape of the string is quite odd. Radio observations of this group of galaxies [HE90.1] (see also [HU90.1]) do not support the lensing hypothesis for this system; in particular, one galaxy in one of the four pairs shows radio emission, whereas the others are radio quiet.

The foregoing discussion has implicitly assumed that the strings are at rest; however, they are actually thought to move relativistically. This yields a modification of the preceding results; e.g., the angular separation is modified, depending on the angle between the line-of-sight to the source and the velocity of the string [VI86.1]. In addition, a temperature jump of the cosmic microwave background across the string is expected in this case ([KA84.1], [VI86.1], [TR86.1]).

Finally, string loops are more complex. For them, the metric is no longer locally flat, so the image magnification is no longer unity. More complex image geometries are possible in this case, and, due to loop oscillations, image positions and magnifications are expected to change in time [HO84.1].

13.3.5 Upper limits to the mass of some QSOs

QSOs are not only the most natural sources for GL systems; they can also act as lenses. In a pair of QSOs with vastly different redshift but very small angular separation, we expect the foreground QSO to influence the image of the background QSO. In particular, the background QSO can be magnified, and its image can be distorted. If the mass of the foreground QSO is sufficiently large, it may generate a second image of the background QSO. Observational constraints on such a secondary image can then yield an upper limit to the mass of the foreground QSO.

The distortion of a single image of the background QSO will be very difficult to determine, as the intrinsic source geometry is not known; hence, any observed ellipticity can be due either to intrinsic source shape, or to lensing. Therefore, it seems that this method [DY82.1] is difficult to apply. However, the lack of any additional images of the background QSO can be easily applied whenever a very close pair of QSOs is found. The best case for this method is the QSO pair 1548+114A,B, with components separated by 4.8 arcseconds, and with redshifts $z_A = 0.44$, $z_B = 1.90$. A very crude upper limit on the mass of QSO A can be obtained from the requirement that the image of B is the primary image, which implies that this image lies outside the critical curve of the lens QSO A; this yields an upper limit of $7 \times 10^{12} M_\odot$ [GO74.1]. A more refined analysis, which uses the fact that strict limits are available on the flux of any additional image of QSO B, yields an upper limit of $1.8 \times 10^{12} M_\odot$, within one arcsecond of the QSO A [IO86.1]. This analysis also allows for three galaxies seen close to the two QSOs, and for a finite surface mass density of the cluster with which QSO A is associated. As can be expected, a finite smooth surface mass density of the cluster lowers this limit on the mass of A. Note that this upper limit is the tightest observational constraint available on the masses of QSOs.

13.3.6 Gravitational lensing and superluminal motion

The discovery that compact extragalactic radio sources have components which separate with an apparent velocity in excess of the velocity of light is one of the crucial results of Very Long Baseline Interferometry (VLBI). Multi-component sources observed at different times show that the difference in angular separation, multiplied by the angular distance to the source as evaluated from its redshift, and divided by the time between the observations, yields a speed larger than c. Although this phenomenon was predicted as early as in 1966 [RE66.3], it nevertheless puzzled the astronomers when it was first discovered [CO71.1]. Of the several possible explanations listed in [KE81.1], one is presently preferred by most, namely that a relativistic source component is moving almost along the line-of-sight; this, in combination with the finite speed of light can cause apparent superluminal velocities (for a review of superluminal sources, see [ZE87.1]).

Gravitational lensing can also influence the apparent angular velocity of a source. The Jacobian matrix of the ray-trace equation yields the transformation of the velocity vector in the source plane to that in the image plane. If the source is close to a caustic, its velocity can be greatly magnified due to the lens action. This was recognized and put forward as a possible explanation for superluminal effects in [BA71.1], [CH79.2], [CH80.1], and [SH86.3]. However, there are several difficulties associated with this hypothesis. First of all, most compact radio sources which are sufficiently strong to be observable with VLBI show superluminal motion, i.e., this effect is common among compact sources. That means that, if lensing is responsible for this effect, it must also be the reason why these sources are observed in the first place; in other words, lensing must cause a very strong amplification bias for these compact radio sources. Second, whenever a superluminal source has a kiloparsec-scale radio jet, it is usually well aligned with the VLBI structure. Lensing does not provide any explanation for this, as the direction of the velocity, as well as its amplitude, is modified by light deflection. Third, the best-studied superluminal sources show a remarkable constancy of the separation velocity of components, whereas lensing would predict $v \propto r^{-1/2}$, where r is the distance of the moving source component from the caustic. In addition, one always observes superluminal expansion, and never contraction, whereas lensing could yield either behavior. Finally, the two components (in the simplest case, a 'core' and a 'blob') have different radio spectra, so they cannot be images of the same source. Thus, we conclude that lensing is not a likely explanation for the origin of superluminal motion. Nevertheless, there may be superluminal sources where the detailed behavior of the movement of the source components can be affected by lensing. For example, the trajectories of some of these components (e.g., in 3C345) are curved, and it is possible that this can be accounted for by lensing; in this connection, we note that this source has an optical light curve that reminds us strongly of simulated light curves due to ML, as discussed in Sect. 12.4.

13.4 Miscellaneous topics

This final section lists a number of possible applications of gravitational light deflection, which have not been discussed in great detail in the literature, or which are fairly distinct from the main topic of this book: the influence of light bending on observable anisotropy of the cosmic microwave background, the light deflection in the Solar System, and light deflection in strong gravitational fields, such as near neutron stars.

13.4.1 Lensing and the microwave background

As we saw in the previous section, lensing by moving cosmic strings can induce an anisotropy of the cosmic microwave background (CMB). On the other hand, one can think that lensing can 'smear out' anisotropies in the CMB, since the optical depth up to the surface of last scattering $z \approx 1500$ is non-negligible. The current upper limits on the CMB anisotropy are very stringent, as the latest results from the COBE satellite constrain the total anisotropy to be less than four parts in 10^5 at the 95% confidence level (Mather, J. 1990, personal communication). Such numbers provide tight constraints on the formation of galaxies, so that a possibility to have a smoothing due to lensing appears to be attractive. This question is fairly difficult to answer, since the corresponding equation for light propagation is highly non-linear. Furthermore, the lens population at low redshifts consists of galaxies, clusters, and other non-linear density perturbations of the universe, whereas for high redshifts, the density perturbations are probably linear. How and when the transition took place depends on the formation of structure, which is poorly known.

The propagation of light rays through a universe with non-linear density perturbations was studied in the framework of CMB radiation in [KA88.5] and [LI88.3]. Several effects are of potential importance: light rays sampled in a radio telescope with a beam of angular width $\Delta\theta$ come from a larger angular region of the last scattering surface, if the light bundle propagated through an empty cone in a clumpy universe. If there is a small number of lenses within the beam, some rays in the beam will be scattered appreciably, and thus the beam contains rays from various regions of the last scattering surface. Both effects influence the anisotropy correlation function. Both of the aforementioned papers agree that anisotropies cannot be smoothed out on scales larger than about 1 arcminute, but the situation is less clear on smaller scales. We feel that this problem deserves more detailed consideration, and that the present status is fairly unclear. Related work can be found in [DY76.1], [GU86.1], [CO89.1], [SA89.1], and [TO89.1].

13.4.2 Light deflection in the Solar System

The detection of the light bending close to the Sun was one of the major tests of General Relativity; the maximum deflection of 1.75 arcseconds was just barely measurable in 1919, but still large enough to distinguish between values predicted by GR and the 'classical' value as calculated by Soldner [SO04.1]. Currently, much more accurate angular measurements are possible, for example with radio telescopes and, in particular, with VLBI. Planned space missions will increase the accuracy of angular measurements. This implies, on the one hand, that the angular deflection can be measured with much higher accuracy, but on the other hand, that the positions of all observed sources must be corrected for this deflection. For example, a source located at right angles to the Sun will suffer an apparent shift of position of about 4 milliarcseconds. For a discussion of other effects due to deflection in the Solar System, see [CO84.1] and [RO84.1].

13.4.3 Light deflection in strong fields

Throughout this book we have considered the deflection of light in a weak gravitational field. In particular, a consequence there of is that the bending angles of light rays are very small. This assumption is essential for the lensing formalism used throughout this book. There are, however, situations of astrophysical relevance in which the gravitational fields are strong and where light bending leads to new effects. For example, the radii of neutron stars are only two to four times their Schwarzschild radii, so photons coming close to a neutron star follow very curved paths. Of course, the same is true for black holes. Light emitted deep inside such gravitational potentials is influenced strongly by the field, i.e., the radiation seen by an observer at large distance is significantly modified by gravitational effects. One is the gravitational redshift. Second, if the source is not spherically symmetric, its angular radiation pattern is also changed. This is of potential importance for accretion discs around neutron stars and black holes [CU75.1], and for the radiation from accreting neutron stars whose strong magnetic fields lead to channeling of the accreted matter onto the magnetic poles; hence, in this case, the source is highly anisotropic (for a discussion of these effects, as well as for the GL effect in close binaries, see [FT86.1], [RI88.1], [GO86.2], [ME87.1], [SC89.1], [WO88.1], [NO89.2]).

References

[AB65.1] Abramowitz, M. & Stegun, I.A. 1965, *Handbook of mathematical functions*, Dover, New York.

[AD89.1] Adam, G., Bacon, R., Courtes, G., Georgelin, Y., Monnet, G. & Pecontal, E. 1989, "Observations of the Einstein cross 2237+030 with the TIGER integral field spectrograph," *Astr. Ap.*, **208**, L15.

[AL73.1] Allen, C.W. 1973, *Astrophysical quantities*, Athlone Press, London.

[AL85.1] Alcock, C. & Anderson, N. 1985, "On the use of measured time delays in gravitational lenses to determine the Hubble constant," *Ap. J.*, **291**, L29.

[AL86.1] Alcock, C. & Anderson, N. 1986, "On the distance measure for gravitational lenses," *Ap. J.*, **302**, 43.

[AN86.1] Anderson, N. & Alcock, C. 1986, "The effect of clustering of galaxies on the statistics of gravitational lenses," *Ap. J.*, **300**, 56.

[AN90.1] Angonin, M.C., Vanderriest, C. & Surdej, J. 1990, "Bidimensional spectrography of the 'Clover Leaf' H1413+117 at sub-arcsec spatial resolution," in: [ME90.2]

[AN90.2] Angonin, M.C., Remy, M., Surdej, J. & Vanderriest, C. 1990, "First spectroscopic evidence of microlensing on a BAL quasar? The case of H 1413+117," *Astr. Ap.*, **233**, L5.

[AR80.1] Arp, H.C. & Hazard, C. 1980, "Peculiar configurations of quasars in two adjacent areas of the sky," *Ap. J.*, **240**, 726.

[AR84.1] Arp, H. & Gavazzi, G. 1984, "A third quasar close to NGC3842," *Astr. Ap.*, **139**, 240.

[AR84.2] Arnold, V.J. 1984, *Catastrophe theory*, Springer-Verlag, New York.

[AR85.1] Arfken, G. 1985, *Mathematical methods for physicists*, Academic Press, Orlando.

[AR85.2] Arnold, V.J., Varchenko, A.N. & Gasein-Zade, S.M. 1985 (Vol. I), 1988 (Vol. II), *Singularities of differentiable maps*, Birkhäuser Publ. Comp.

[AR86.1] Arp, H. 1986, "Recent discoveries of a supermass in the universe," *Nature*, **322**, 316.

[AR87.1] Arp, H. 1987, *Quasars, redshifts, and controversies*, Interstellar Media, Berkeley.

[AR89.1] Arnold, V.I. 1989, *Mathematical methods of classical mechanics*, 2nd ed., Springer-Verlag, New York.

[AR90.1] Aragón Salamanca, A. & Ellis, R.S. 1990, "Optical-infrared studies of arcs in Abell 370," in: [ME90.2]

[AV81.1] Avni, Y. 1981, "On gravitational lenses and the cosmological evolution of quasars," *Ap. J.*, **248**, L95.

[AV88.1] Avni, Y. & Shulami, I. 1988, "'Flux conservation' by a Schwarzschild gravitational lens," *Ap. J.*, **332**, 113.

[BA65.1] Barnothy, J.M. 1965, "Quasars and the gravitational image intensifier," *Astron. J.*, **70**, 666.

[BA68.1] Barnothy, J. & Barnothy, M.F. 1968, "Galaxies as gravitational lenses," *Science*, **162**, 348.

[BA71.1] Barnothy, J.M. & Barnothy, M.F. 1971, "Rapid differential proper motion in the radio fine structure of 3C279," *Bull. AAS*, **3**, 472.

[BA73.1] Bardeen, J.M. 1973, "Timelike and null geodesics in the Kerr metric," in: *Black holes*, DeWitt, C. & DeWitt, B.S. (eds.), Gordon and Breach (New York), p. 215.

[BA77.1] Baldwin, J.A. 1977, "Luminosity indicators in the spectra of quasi-stellar objects," *Ap. J.*, **214**, 679.

[BA83.1] Bacon, R. & Nieto, J.-L. 1983, "Optical variability of QSOs and gravitational lenses in galactic haloes," *Nature*, **301**, 400.

[BA83.2] Barrow, J.D. 1983, "Cosmology and elementary particles," *Fundamentals Cosm. Phys.*, **8**, 83.

[BA84.1] Bahcall, J.N. 1984, "K giants and the total amount of matter near the sun," *Ap. J.*, **287**, 926.

[BA85.1] Bahcall, J.N., Hut, P. & Tremaine, S. 1985, "Maximum mass of objects that constitute unseen disk material," *Ap. J.*, **290**, 15.

[BA86.1] Bahcall, J.N., Bahcall, N.A. & Schneider, D.P. 1986, "Multiple quasars for multiple images," *Nature*, **323**, 515.

[BA86.2] Barnothy, J.M. & Barnothy, M.F. 1986, "Morphological aberration in gravitational-lens images," *Astron. J.*, **91**, 755.

[BA87.1] Babul, A., Paczyński, B. & Spergel, D. 1987, "Gamma-ray bursts from superconducting cosmic strings at large redshifts," *Ap. J.*, **316**, L49.

[BA88.1] Bajtlik, S., Duncan, R.C. & Ostriker, J.P. 1988, "Quasar ionization of Lyman-alpha clouds: the proximity effect, a probe of the ultraviolet background at high redshift," *Ap. J.*, **327**, 570.

[BA88.2] Barthel, P.D. & Miley, G.K. 1988, "Evolution of radio structure in quasars: a new probe of protogalaxies?," *Nature*, **333**, 319.

[BA89.1] Barnothy, J.M. 1989, "History of gravitational lenses and the phenomena they produce," in: [MO89.1]

[BA90.1] Bartelmann, M. & Schneider, P. 1990, "Influence of microlensing on quasar statistics," *Astr. Ap.*, **239**, 113.

[BA91.1] Babul, A. & Lee, M.H. 1991, "Gravitational lensing by smooth inhomogeneities in the universe," *M.N.R.A.S.*, **250**, 407.

[BA91.2] Bartelmann, M. & Schneider, P. 1991, "Gravitational lensing by large-scale structures," *Astr. Ap.*, **248**, 353.

[BE79.1] Benson, J.R. & Cooke, J.H. 1979, "High-intensification regions of gravitational lenses," *Ap. J.*, **227**, 360.

[BE87.1] Begelman, M.C. & Blandford, R.D. 1987, "Luminous arcs from galactic bow shocks," *Nature*, **330**, 46.

[BE90.1] Bergmann, A.G., Petrosian, V. & Lynds, R. 1990, "Gravitational lens models of arcs in clusters," *Ap. J.*, **350**, 23.

[BI87.1] Binney, J. & Tremaine, S. 1987, *Galactic dynamics*, Princeton University Press, Princeton.

[BL75.1] Bliokh, P.V. & Minakov, A.A. 1975, "Diffraction of light and lens effect of the stellar gravitation field," *Astrophys. and Space Sci.*, **34**, L7.

[BL81.1] Blandford, R.D. & Jaroszyński, M. 1981, "Gravitational distortion of the images of distant radio sources in an inhomogeneous universe," *Ap. J.*, **246**, 1.

[BL86.1] Blandford, R. & Narayan, R. 1986, "Fermat's principle, caustics, and the classification of gravitational lens images," *Ap. J.*, **310**, 568.

[BL87.1] Blandford, R.D., Phinney, E.S. & Narayan, R. 1987, "1146+111B,C: A giant gravitational lens?," *Ap. J.*, **313**, 28.

[BL87.2] Blandford, R.D. & Kochanek, C.S. 1987, "Gravitational imaging by isolated elliptical potential wells. I. Cross sections," *Ap. J.*, **321**, 658.

[BL87.3] Blandford, R.D. & Kochanek, C.S. 1987, "Gravitational lenses," in: *Dark matter in the universe*, Bahcall, J., Piran, T. & Weinberg, S. (eds.), World Scientific, Singapore.

[BL88.1] Blandford, R.D. & Kovner, I. 1988, "Formation of arcs by nearly circular gravitational lenses," *Phys. Rev.*, **38A**, 4028.

[BL89.1] Blandford, R.D., Kochanek, C.S., Kovner, I. & Narayan, R. 1989, "Gravitational lens optics," *Science*, **245**, 824.

[BL89.2] Bliokh, P.V. & Minakov, A.A. 1989, *Gravitational lensing*, (in Russian).

[BL90.1] Blandford, R.D. 1990, "Concluding remarks," in: [ME90.2]

[BO73.1] Bourassa, R.R., Kantowski, R. & Norton, T.D. 1973, "The spheroidal gravitational lens," *Ap. J.*, **185**, 747.

[BO75.1] Bourassa, R.R. & Kantowski, R. 1975, "The theory of transparent gravitational lenses," *Ap. J.*, **195**, 13.

[BO76.1] Bourassa, R.R. & Kantowski,R. 1976, "Multiple image probabilities for a spheroidal gravitational lens," *Ap. J.*, **205**, 674.

[BO79.1] Bontz, R.J. 1979, "The gravitational lens effect and pregalactic halo objects," *Ap. J.*, **233**, 402.

[BO80.1] Born, M. & Wolf, E. 1970, *Principles of optics*, Pergamon Press, Oxford

[BO83.1] Borgeest, U. 1983, "The difference in light travel time between gravitational lens images," *Astr. Ap.*, **128**, 162.

[BO84.1] Borgeest, U. & Refsdal, S. 1984, "The Hubble parameter: An upper limit from QSO 0957+561A,B," *Astr. Ap.*, **141**, 318.

[BO86.1] Borgeest, U. 1986, "Determination of galaxy masses by the gravitational lens effect," *Ap. J.*, **309**, 467.

[BO88.1] Börner, G. 1988, *The early universe*, Springer, Heidelberg.

[BO88.2] Bouchet, F.R., Bennet, D.P. & Stebbins, A. 1988, "Patterns of the cosmic microwave background from evolving string networks," *Nature*, **335**, 410.

[BO88.3] Boyle, B.J., Shanks, T. & Peterson, B.A. 1988, "The evolution of optically selected QSO's," in: [OS88.1]

[BO88.4] Börner, G. & Ehlers, J. 1988, "Was there a big bang?" *Astr. Ap.*, **204**, 1.

[BR73.1] Brill, D. 1973, "Observational contacts of general relativity," in: *Relativity, Astrophysics and Cosmology*, W. Israel (ed.), Reidel Publ. Comp., Dordrecht, p. 134

[BR75.1] Bröcker, Th. & Lander, L. 1975, *Differentiable germs and catastrophes*, Cambridge University Press, Cambridge.

[BR84.1] Bray, I. 1984, "Spheroidal gravitational lenses," *M.N.R.A.S.*, **208**, 511.

[BR85.1] Brandenberger, R.H. 1985, "Quantum field theory methods and inflationary universe models," *Rev. Mod. Phys.*, **57**, 1.

[BR89.1] Braun, E. & Milgrom, M. 1989, "Light reprocessing in the echo model of luminous arcs," *Ap. J.*, **337**, 644.

[BU71.1] Burbidge, E.M., Burbidge, G.R., Solomon, P.M. & Strittmatter, P.A. 1971, "Apparent associations between bright galaxies and quasi-stellar objects," *Ap. J.*, **170**, 233.

[BU79.1] Burbidge, G. 1979, "Redshifts and distances," *Nature*, **282**, 451.

[BU81.1] Burke, W.L. 1981, "Multiple gravitational imaging by distributed masses," *Ap. J.*, **244**, L1.

[BU85.1] Burbidge, G. 1985, "A comment on the discovery of the QSO and related galaxy 2237+0305" *Astron. J.*, **90**, 1399.

[BU86.1] Burke, B.F. 1986, "Gravitational lenses: observations," in: [SW86.1]

[BU89.1] Buchert, T. 1989, "A class of solutions in Newtonian cosmology and the pancake theory," *Astr. Ap.*, **223**, 9.

[BU90.1] Burke, B.F. 1990, "The MIT search program for gravitational lenses," in: [ME90.2]

[CA81.1] Canizares, C.R. 1981, "Gravitational focusing and the association of distant quasars with foreground galaxies," *Nature*, **291**, 620.

[CA82.1] Canizares, C.R. 1982, "Manifestations of a cosmological density of compact objects in quasar light," *Ap. J.*, **263**, 508.

[CA87.1] Canizares, C.R. 1987, "Gravitational lenses as tools in observational cosmology," in: [HE87.2]

[CA88.1] Canizares, C.R. 1988, "Quasar distribution: not too close for comfort," *Nature*, **336**, 309.

[CH24.1] Chwolson, O. 1924, "Über eine mögliche Form fiktiver Doppelsterne," *Astr. Nachrichten*, **221**,329.

[CH43.1] Chandrasekhar, S. 1943, "Stochastic Problems in Physics and Astronomy," *Rev. Mod. Phys.*, **15**, 1.

[CH76.1] Chang, K. & Refsdal, S. 1976, "Time delay effects for measuring cosmological distances," Colloques internationaux du C.N.R.S. No.263.

[CH79.1] Chang, K. & Refsdal, S. 1979, "Flux variations of QSO 0957+561A,B and image splitting by stars near the light path," *Nature*, **282**, 561.

[CH79.2] Chitre, S.M. & Narlikar, J.V. 1979, "On the apparent superluminal separation of radio source components," *M.N.R.A.S.*, **187**, 655.

[CH80.1] Chitre, S.M. & Narlikar, J.V. 1980, "Gravitational screens and superluminal separation in quasars," *Ap. J.*, **235**, 335.

[CH84.1] Chang, K. 1984, "Maximum flux amplification by star disturbances in gravitational lens galaxies," *Astr. Ap.*, **130**, 157.

[CH84.2] Chang, K. & Refsdal, S. 1984, "Star disturbances in gravitational lens galaxies," *Astr. Ap.*, **132**, 168.

[CH84.3] Chu, Y., Zhu, X., Burbidge, G. & Hewitt, A. 1984, "Statistical evidence for possible association between QSOs and bright galaxies," *Astr. Ap.*, **138**, 408.

[CH87.1] Christian, C.A., Crabtree, D. & Waddell, P. 1987, "Detection of the lensing galaxy in PG 1115+080" *Ap. J.*, **312**, 45.

[CH87.2] Chandrasekhar, S. 1987, *Ellipsoidal figures of equilibrium*, Dover, Mineola.

[CO62.1] Courant, R. & Hilbert, D. 1962, *Methods of mathematical physics*, vol. II, Interscience Publ., John Wiley & Sons, New York.

[CO71.1] Cohen, M.H., Cannon, W., Purcell, G.H., Shaffer, D.B., Broderick, J.J., Kellermann, K.I. & Jauncey, D.L. 1971, "The small-scale structure of radio galaxies and quasi-stellar sources at 3.8 centimeters," *Ap. J.*, **170**, 207.

[CO75.1] Cooke, J.H. & Kantowski, R. 1975, "Time delays for multiply imaged quasars," *Ap. J.*, **195**, L11.

[CO84.1] Cowling, S.A. 1984, "Gravitational light deflection in the Solar System," *M.N.R.A.S.*, **209**, 415.

[CO87.1] Cowie, L.L. & Hu, E.M. 1987, "The formation of families of twin galaxies by string loops," *Ap. J.*, **318**, L33.

[CO89.1] Cole, S. & Efstathiou, G. 1989, "Gravitational lensing of fluctuations in the microwave background radiation," *M.N.R.A.S.*, **239**, 195.

[CR86.1] Crawford, C.S., Fabian, A.C. & Rees, M.J. 1986, "Double clusters and gravitational lenses," *Nature*, **323**, 514.

[CR88.1] Crampton, D., Cowley, A.P., Hickson,P., Kindl, E., Wagner, R.M., Tyson, J.A. & Guillixson, C. 1988, "1343.4+2640: A close quasar pair," *Ap. J.*, **330**, 184.

[CR89.1] Crampton, D., McClure, R.D., Fletcher, J.M. & Hutchings, J.B. 1989, "A search for closely spaced gravitational lenses," *Astron. J.*, **98**, 1188.

[CU75.1] Cunningham, C.T. 1975, "The effects of redshifts and focusing on the spectrum of an accretion disk around a Kerr black hole," *Ap. J.*, **202**, 788.

[DA65.1] Dashevskii, V.M. & Zel'dovich, Ya.B. 1965, "The propagation of light in a nonhomogeneous nonflat universe II.," *Sov. Astr.*, **8**, 854.

[DA66.1] Dashevskii, V.M. & Slysh, V.I. 1966, "On the propagation of light in a nonhomogeneous universe," *Sov. Astr.*, **9**, 671.

[DA87.1] Damour, T. 1987, "The problem of motion in Newtonian and Einsteinian gravity," Ch. 6, in: *300 Years of Gravitation*, S.W. Hawking and W. Israel (eds.), Cambridge University Press.

[DE86.1] Deguchi, S. & Watson, W.D. 1986, "Wave effects in gravitational lensing of electromagnetic radiation," *Phys. Rev. D*, **34**, 1708.

[DE86.2] Deguchi, S. & Watson, W.D. 1986, "Diffraction in gravitational lensing for compact objects of low mass," *Ap. J.*, **307**, 30.

[DE87.1] Deguchi, S. & Watson, W.D. 1987, "Statistical treatment of fluctuations in the gravitational focusing of light due to stellar masses within a gravitational lens," *Phys. Rev. Lett.*, **59**, 2814.

[DE87.2] Deguchi, S. & Watson, W.D. 1987, "Electron scintillation in gravitationally lensed images of astrophysical radio sources," *Ap. J.*, **315**, 440.

[DE88.1] De Robertis, M.M. & Yee, H.K.C. 1988, "Spatially resolved spectroscopy of the gravitational lens system 2237+030," *Ap. J.*, **332**, L49.

[DE88.2] Deguchi, S. & Watson, W.D. 1988, "An analytic treatment of gravitational microlensing for sources of finite size at large optical depths," *Ap. J.*, **335**, 67.

[DJ84.1] Djorgovski, S. & Spinrad, H. 1984, "Discovery of a new gravitational lens," *Ap. J.*, **282**, L1.

[DJ87.1] Djorgovski, S., Perley, R., Meylan, G. & McCarthy, P. 1987, "Discovery of a probable binary quasar," *Ap. J.*, **321**, L17.

[DO81.1] Dolgov, A.D. & Zeldovich, Ya.B. 1981, "Cosmology and elementary particles," *Rev. Mod. Phys.*, **53**, 1.

[DR83.1] Dressler, A. & Gunn, G.E. 1983, "Spectroscopy of galaxies in distant clusters. II. The population of the 3C295 cluster," *Ap. J.*, **270**, 7.

[DY72.1] Dyer, C.C. & Roder, R.C. 1972, "The distance-redshift relation for universes with no intergalactic medium," *Ap. J.*, **174**, L115.

[DY73.1] Dyer, C.C. & Roder, R.C. 1973, "Distance-redshift relations for universes with some intergalactic medium," *Ap. J.*, **180**, L31.

[DY74.1] Dyer, C.C. & Roder, R.C. 1974, "Observations in locally inhomogeneous cosmological models," *Ap. J.*, **189**, 167.

[DY76.1] Dyer, C.C. 1976, "The gravitational perturbation of the cosmic background radiation by density concentrations," *M.N.R.A.S.*, **175**, 429.

[DY77.1] Dyer, C.C. 1977, "Optical scalars and the spherical gravitational lens," *M.N.R.A.S.*, **180**, 231.

[DY80.1] Dyer, C.C. & Roder, R.C. 1980, "A range of time delays for the double quasar 0957+561 A,B," *Ap. J.*, **241**, L133.

[DY80.2] Dyer, C.C. & Roder, R.C. 1980, "Possible multiple imaging by spherical galaxies," *Ap. J.*, **238**, L67.

[DY81.1] Dyer, C.C. & Roder, R.C. 1981, "Toward a realistic nebular gravitational lens," *Ap. J.*, **249**, 290.

[DY81.2] Dyer, C.C. & Roder, R.C. 1981, "On the transition from Weyl to Ricci focusing," *Gen. Rel. Grav.*, **13**, 1157.

[DY82.1] Dyer, C.C. & Roder, R.C. 1982, "A method for estimating the masses of some quasars," *Ap. J.*, **256**, 386.

[DY84.1] Dyer, C.C. 1984, "Image separation statistics for multiply images quasars," *Ap. J.*, **287**, 26.

[DY88.1] Dyer, C.C. & Oattes, L.M. 1988, "On the effects of inhomogeneity on the luminosity structure of the past null cone," *Ap. J.*, **326**, 50.

[ED20.1] Eddington, A.S. 1920, *Space, time and gravitation*, Cambridge University Press, Cambridge.

[EH59.1] Ehlers, J. & Sachs, R.K. 1959, "Erhaltungssätze für die Wirkung in elektromagnetischen und gravischen Strahlungsfeldern," *Z. Physik*, **155**, 498.

[EH61.1] Ehlers, J. 1961, "Beiträge zur relativistischen Mechanik kontinuierlicher Medien," *Akad. Wiss. Mainz Abh., Math. Nat. Kl.*, **1**, 1.

[EH67.1] Ehlers, J. 1967, "Zum Übergang von der Wellenoptik zur geometrischen Optik in der allgemeinen Relativitätstheorie," *Zeitschrift für Naturforschung*, **22a**, 1328.

[EH73.1] Ehlers, J. 1973, "Survey of general relativity theory," in: *Relativity, Astrophysics and Cosmology*, W. Israel (ed.), Reidel Publ. Comp., Dordrecht, p. 1

[EH81.1] Ehlers, J. 1981, "Über den Newtonschen Grenzwert der Einsteinschen Gravitationstheorie," in: *Grundlagenprobleme der modernen Physik*, J. Nitsch, J. Pfarr and E.W. Stachow (eds.), Bibliographisches Institut, Mannheim, p. 65.

[EH86.1] Ehlers, J. & Schneider, P. 1986, "Self-consistent probabilities for gravitational lensing in inhomogeneous universes," *Astr. Ap.*, **168**, 57.

[EH86.2] Ehlers, J. 1986, "On limit relations between, and approximative explanations of, physical theories," in: *Logic, Methodology and Philosophy of Science VII*, R.B. Marcus, G.J.W. Dorn, P. Weingartner (eds.), Dordrecht, North-Holland, p. 387.

[EI11.1] Einstein, A. 1911, "Über den Einfluß der Schwerkraft auf die Ausbreitung des Lichtes," *Annalen der Physik*, **35**, 898.

[EI15.1] Einstein, A. 1915, "Erklärung der Perihelbewegung des Merkur aus der allgemeinen Relativitätstheorie," *Sitzungber. Preuß . Akad. Wissensch.*, erster Halbband, p.831.

[EI36.1] Einstein, A. 1936, "Lens-like action of a star by the deviation of light in the gravitational field," *Science*, **84**, 506.

[EL72.1] Ellis, G.F.R. & Sciama, D.W. 1972, "Global and non-global problems in cosmology," in: *General Relativity*, O'Raifeartaigh, L. (ed.), Clarendon Press (Oxford), p. 35.

[EL86.1] Ellis, G.F.R. & Schreiber, G. 1986, "Observational and dynamic properties of small universes," *Phys. Lett. A*, **115**, 97.

[EL88.1] Ellis, G.F.R. 1988, "Fitting and averaging in cosmology," *Proc. 2nd. Canadian Conf. on General Relativity and Astrophysics*, A. Coley et al. (Eds.), World Scientific, p. 1.

[EL91.1] Ellis, R., Allington-Smith, J. & Smail, I. 1991, "Spectroscopy of arcs in the rich cluster Abell 963," *M.N.R.A.S.*, **249**, 184.

[ER90.1] Erdl, H. 1990, "Klassifizierung der kritischen Linien von Zwei-Punktmassen-Gravitationslinsen," Diploma-thesis, TU München.

[ET33.1] Etherington, I.M.H. 1933, "On the definition of distance in general relativity," *Phil. Mag.*, **15**, 761.

[FA76.1] Faber, S.M. & Jackson, R.E. 1976, "Velocity dispersions and mass-to-light ratios for elliptical galaxies," *Ap. J.*, **204**, 668.

[FA83.1] Falco, E.E. 1983, "Gravitational lenses: a graphical representation," Master thesis, MIT, Cambridge.

[FA85.1] Falco, E.E. 1985, "The ultimate cosmological tool: does it work?" Ph. D. thesis, CfA, Cambridge.

[FA88.1] Falco, E.E., Gorenstein, M.V. & Shapiro, I.I. 1988, "Can we measure H_0 with VLBI observations of gravitational images?" in: *IAU Symp. 129*, Reid, M.J. & Moran, J.M. (eds.), Reidel, p.208.

[FA90.1] Falco, E.E., Shapiro, I.I. & Krolik, J.H. 1990, "0957+561: the time delay revisited," in: [ME90.2]

[FA91.1] Falco, E.E., Wambsganss, J. & Schneider, P. 1991, "The role of microlensing in estimates of the relative time delay for the gravitational images of Q0957+561," *M.N.R.A.S.*, **251**, 698.

[FA91.2] Falco, E.E., Gorenstein, M.V. & Shapiro, I.I. 1990, "New model for the 0957+561 gravitational lens system: bounds on masses of possible black hole and dark matter and prospects for estimation of H_0," *Ap. J.*, **372**, 364.

[FI73.1] Field, G.B., Arp, H. & Bahcall, J.N. 1973, *The redshift controversy*, Benjamin, Reading.

[FI89.1] Filippenko, A.V. 1989, "Evidence for Mg II $\lambda 2798$ profile differences in the gravitationally lensed QSO 2237+0305A,B," *Ap. J.*, **338**, L49.

[FL84.1] Florentin-Nielsen, R. 1984, "Determination of difference in light travel time for QSO 0957-561A,B," *Astr. Ap.*, **138**, L19.

[FO75.1] Fomalont, E.B. & Sramek, R.A. 1975, "A confirmation of Einstein's general theory of relativity by measuring the bending of microwave radiation in the gravitational field of the sun," *Ap. J.*, **199**, 749.

[FO76.1] Fomalont, E.B. & Sramek, R.A. 1976, "Measurements of the solar gravitational deflection of radio waves in agreement with general relativity," *Phys. Rev. Lett.*, **36**, 1475.

[FO84.1] Foltz, C.B., Weymann, R.J., Röser, H.-J. & Chaffee Jr., F.H. 1984, "Improved lower limits on Lyman-alpha forest cloud dimensions and additional evidence supporting the gravitational lens nature of 2345+007A,B," *Ap. J.*, **281**, L1.

[FO85.1] Foy, R., Bonneau, D. & Blazit, A. 1985, "The multiple QSO PG1115+08: A fifth component?" *Astr. Ap.*, **149**, L13.

[FO88.1] Fort, B., Prieur, J.L., Mathez, G., Mellier, Y. & Soucail, G. 1988, "Faint distorted structures in the core of A370: Are they gravitationally lensed galaxies at $z \simeq 1$?" *Astr. Ap.*, **200**, L17.

[FO90.1] Fort, B. 1990, "Clusters of galaxies: a new observable class of gravitational lenses," in: [ME90.2]

[FR83.1] Friedrichs, H. & Stewart, J.M. 1983, "Characteristic initial data and wavefront singularities in general relativity," *Proc. Roy. Soc. London*, **A385**, 345.

[FR90.1] Freyberg, M. 1990, "Lichtausbreitung in inhomogenen Universen," diploma thesis, TU München.

[FT86.1] Ftaclas, C., Kearney, M.W. & Pechenick, K. 1986, "Hot spots on neutron stars. II The observer's sky," *Ap. J.*, **300**, 203.

[FU88.1] Fugmann, W. 1988, "Galaxies near distant quasars: observational evidence for statistical gravitational lensing," *Astr. Ap.*, **204**, 73.

[FU89.1] Fugmann, W. 1989, "Galaxies near distant quasars: observational evidence for statistical gravitational lensing (Part II)," *Astr. Ap.*, **222**, 45.

[FU89.2] Futamase, T. & Sasaki, M. 1989, "Light propagation and the distance-redshift relation in a realistic inhomogeneous universe," *Phys. Rev. D*, **40**, 2502.

[FU90.1] Fugmann, W. 1990, "Statistical gravitational lensing and the Lick catalogue of galaxies," *Astr. Ap.*, **240**, 11.

[GE89.1] Geller, M.J. & Huchra, J.P. 1989, "Mapping the universe," *Science*, **246**, 897.

[GI81.1] Gilmore, R. 1981, *Catastrophe theory for scientists and engineers*, Wiley, New York.

[GI86.1] Gioia, I.M., Maccacaro, T, Schild, R.E., Giommi, P. & Stocke, J.T. 1986, "New X-ray and optical observations of the X-ray discovered QSO-galaxy pair 1E 0104.2+3153," *Ap. J.*, **307**, 497.

[GI88.1] Giraud, E. 1988, "A possible candidate of arclike structure in a southern cluster of galaxies," *Ap. J.*, **334**, L69.

[GI89.1] Giraud, E., Schneider, P. & Wambsganss, J. 1989, "The arc in Cl 0500-24," *ESO Messenger*, **56**, 62.

[GO73.1] Golubitsky, M. & Guillemin, V. 1973, *Stable mappings and their singularities*, Springer-Verlag, New York.

[GO74.1] Gott III, J.R. & Gunn, J.E. 1974, "The double quasar 1548+115a,b as a gravitational lens," *Ap. J.*, **190**, L105.

[GO80.1] Gorenstein, M.V., Shapiro, I.I., Cohen, N.L., Falco, E.E., Kassim, N., Rogers, A.E.E., Whitney, A.R., Preston, R.A. & Rius, A. 1980, "VLBI observations of the "twin quasar" 0957+56 A,B," *Bull. AAS*, **12**, 498.

[GO81.1] Gott III, J.R. 1981, "Are heavy halos made of low mass stars? A gravitational lens test," *Ap. J.*, **243**, 140.

[GO82.1] Gondhalekar, P.M. & Wilson, R. 1982, "IUE observations of variability and differences in the UV spectra of double quasar 0957+561 A,B," *Nature*, **296**, 415.

[GO83.1] Gorenstein, M.V., Shapiro, I.I., Cohen, N.L., Corey, B.E., Falco, E.E., Marcaide, J.M., Rogers, A.E.E., Whitney, A.R., Porcas, R.W., Preston, R.A. &

Rius, A. 1983, "Detection of a compact radio source near the center of a gravitational lens: Quasar image or galactic core?" *Science*, **219**, 54.

[GO84.1] Gorenstein, M.V., Shapiro, I.I., Rogers, A.E.E., Cohen, N.L., Corey, B.E., Porcas, R.W., Falco, E.E., Bonometti, R.J., Preston, R.A., Rius, A. & Whitney, A.R. 1984, "The milli-arcsecond images of Q 0957+561," *Ap. J.*, **287**, 538.

[GO85.1] Gott III, J.R. 1985, "Gravitational lensing effects of vacuum strings: Exact solutions," *Ap. J.*, **288**, 422.

[GO86.1] Gott III, J.R. 1986, "Is QSO 1146+111B,C due to lensing by a cosmic string?" *Nature*, **321**, 420.

[GO86.2] Gorham, P.W. 1986, "Gravitational light deflection effects in eclipsing binary pulsars," *Ap. J.*, **303**, 601.

[GO86.3] Goodman, J. 1986, "Are gamma-ray bursts optically thick?" *Ap. J.*, **308**, L47.

[GO87.1] Goodman, J.J., Romani, R.W., Blandford, R.D. & Narayan, R. 1987, "The effects of caustics on scintillating radio sources," *M.N.R.A.S.*, **229**, 73.

[GO88.1] Gorenstein, M.V., Falco, E.E. & Shapiro, I.I. 1988, "Degeneracies in parameter estimates for models of gravitational lens systems," *Ap. J.*, **327**, 693.

[GO88.2] Gorenstein, M.V., Cohen, N.L., Shapiro, I.I., Rogers, A.E.E., Bonometti, R.J., Falco, E.E., Bartel, N. & Marcaide, J.M. 1988, "VLBI observations of the gravitational lens system 0957+561: Structure and relative magnification of the A and B images," *Ap. J.*, **334**, 42.

[GR80.1] Greenfield, P.E., Burke, B.F. & Roberts, D.H. 1980, "The double quasar 0957+561 as a gravitational lens: Further VLA observations," *Nature*, **286**, 865.

[GR80.2] Greenfield, P.E., Roberts, D.H. & Burke, B.F. 1980, "The double quasar 0957+561: Examination of the gravitational lens hypothesis using the Very Large Array," *Science*, **208**, 495.

[GR80.3] Gradshteyn, I.S. & Ryzhik, I.M. 1980, *Table of integrals, series and products*, Academic Press.

[GR85.1] Greenfield, P.E., Roberts, D.H. & Burke, B.F. 1985, "The gravitationally lensed quasar 0957+561: VLA observations and mass models," *Ap. J.*, **293**, 370.

[GR86.1] Grieger, B., Kayser, R. & Refsdal, S. 1986, "A parallax effect due to gravitational micro-lensing," *Nature*, **324**, 126.

[GR86.2] Green, R.F. 1986, "Evolution of the luminosity function," in: [SW86.1]

[GR88.1] Grieger, B., Kayser, R. & Refsdal, S. 1988, "Gravitational micro-lensing as a clue to quasar structure," *Astr. Ap.*, **194**, 54.

[GR88.2] Grossman, S.A. & Narayan, R. 1988, "Arcs from gravitational lensing," *Ap. J.*, **324**, L37.

[GR89.1] Grossman, S.A. & Narayan, R. 1989, "Gravitationally lensed images in Abell 370," *Ap. J.*, **344**, 637.

[GR89.2] Grieger, B., Kayser, R., Refsdal, S. & Stabell, R. 1989, "The two point mass gravitational lens," *Abhandlungen aus der Hamburger Sternwarte*, **X,4**, Hamburg-Bergedorf.

[GR89.3] Grossman, S.A. & Narayan, R. 1989, "The versatile elliptical gravitational lens," in: [MO89.1]

[GR90.1] Grossman, S.A. 1990, "Probing rich galaxy clusters with mini-arcs," in: [ME90.2]

[GR90.2] Grieger, B. 1990, "The deconvolution of the quasar structure from the lightcurve of a high amplification event with the regularisation method," in: [ME90.2]

[GR91.1] Griest, K. 1991, "Galactic microlensing as a method of detecting massive compact halo objects," *Ap. J.*, **366**, 412.

[GU67.1] Gunn, J.E. 1967, "A fundamental limitation on the accuracy of angular measurements in observational cosmology," *Ap. J.*, **147**, 61.

[GU67.2] Gunn, J.E. 1967, "On the propagation of light in inhomogeneous cosmologies. I. Mean effects," *Ap. J.*, **150**, 737.

[GU81.1] Guth, A.H. 1981, "Inflationary universe: a possible solution to the horizon and flatness problem," *Phys. Rev.*, **23D**, 347.

[GU86.1] Gurvits, L.I. & Mitrofanov I.G. 1986, "Perturbation of the background radiation by a moving gravitational lens," *Nature*, **324**, 349.

[HA64.1] Hammersley, J.M. & Handscomb, D.C. 1964, *Monte Carlo methods*, Chapman and Hall, London.

[HA73.1] Hawking, S.W. & Ellis, G.F.R. 1973, *The large scale structure of space-time*, Cambridge University Press, Cambridge

[HA85.1] Hammer, F. 1985, "Gravitational amplification of light: General solution in the vacuole model," *Astr. Ap.*, **152**, 262.

[HA85.2] Hammer, F. 1985, "Gravitational amplification of light by a non-static and thick lens," *Astr. Ap.*, **144**, 408.

[HA86.1] Hammer, F. & Nottale, L. 1986, "Gravitational amplification of brightest cluster galaxies by foreground clusters," *Astr. Ap.*, **167**, 1.

[HA86.2] Hammer, F. & Nottale, L. 1986, "VV 172: An effect of gravitational amplification by a massive halo?" *Astr. Ap.*, **155**, 420.

[HA86.3] Hammer, F., Nottale, L. & Le Fèvre, O. 1986, "Overluminosity of distant radiogalaxies: Evolution or gravitational amplification?" *Astr. Ap.*, **169**, L1.

[HA87.1] Hammer, F., 1987, "A gravitational lensing model of the strange ring-like structure in A370," in: *High redshift and primeval galaxies*, Bergeron, J., Kunth, D., Rocca-Volmerange, B. & Tran Thran Van, J. (eds.), Proceedings of the Third IAP Workshop.

[HA89.1] Hammer, F., Le Fèvre, O., Jones, J., Rigaut, F. & Soucail, G. 1989, "Probable additional gravitational images related to the Cl2244-02 arc and B, V, R photometry of the cluster core," *Astr. Ap.*, **208**, L7.

[HA89.2] Hammer, F. & Rigaut, F. 1989, "Giant luminous arcs from lensing: determination of the mass distribution inside distant cluster cores," *Astr. Ap.*, **226**, 45.

[HA89.3] Hansen, L., Jørgensen, H.E., Nørgaard-Nielsen, H.U., Ellis, R.S. & Couch, W.J. 1989, "A supernova at $z = 0.28$ and the rate of distant supernovae," *Astr. Ap.*, **211**, L9.

[HA90.1] Hammer, F. & Le Fèvre, O. 1990, "High spatial resolution imaging of 10 3CR galaxies with $z \geq 1$ and statistical evidence for selection effects from gravitational amplification," *Ap. J.*, **357**, 38.

[HA91.1] Hammer, F. & Le Fèvre, O. 1991, "New NTT discoveries on distant galaxies and gravitational lensing," *ESO Messenger*, **63**, 59.

[HE59.1] Heckmann, O.H.L. & Schücking, E. 1959, "Newtonsche und Einsteinsche Kosmologie," in: *Encyclopedia of Physics*, **53**, 489, ed. by Flügge, S., Springer, Berlin.

[HE80.1] Hege, E.K., Angel, J.R.P., Weymann, R.J. & Hubbard, E.N. 1980, "Morphology of the triple QSO PG1115+08," *Nature*, **287**, 416.

[HE80.2] Hewitt, A. & Burbidge, G. 1980, "A revised catalog of quasi-stellar objects," *Ap. J. Suppl.*, **43**, 57.

[HE81.1] Hege, E.K., Hubbard, E.N., Strittmatter, P.A. & Worden, S.P. 1981, "Speckle interferometry observations of the triple QSO PG 1115+08," *Ap. J.*, **248**, L1.

[HE86.1] Henry, J.P. & Heasley, J.N. 1986, "High-resolution imaging from Mauna Kea: The triple quasar in 0.3-arc s seeing," *Nature*, **321**, 139.

[HE86.2] Hewitt, J.N. 1986, "A search for gravitational lensing," Ph. D. thesis, MIT, Cambridge.

[HE86.3] Hegyi, D.J. & Olive, K.A. 1986, "A case against baryons in galactic halos," *Ap. J.*, **303**, 56.

[HE87.1] Hewitt, J.N., Turner, E.L., Lawrence, C.R., Schneider, D.P., Gunn, J.E., Bennett, C.L., Burke, B.F., Mahoney, J.H., Langston, G.I., Schmidt, M., Oke, J.B. & Hoessel, J.G. 1987, "The triple radio source 0023+171: A candidate for a dark gravitational lens," *Ap. J.*, **321**, 706.

[HE87.2] Hewitt, A., Burbidge, G. & Fang, L.-Z. (eds.) 1987, *Observational Cosmology*, IAU Symposium 124 (Reidel, Dordrecht).

[HE88.1] Hewitt, J.N., Turner, E.L., Schneider, D.P., Burke, B.F., Langston, G.I. & Lawrence, C.R. 1988, "Unusual radio source MG 1131+0456: a possible Einstein ring," *Nature*, **333**, 537.

[HE88.2] Heisler, J. & Ostriker, J.P. 1988, "Models of the quasar population. II. The effects of dust obscuration," *Ap. J.*, **332**, 543.

[HE89.1] Hewitt, J.N., Burke, B.F., Turner, E.L., Schneider, D.P., Lawrence, C.R., Langston, G.I. & Brody, J.P. 1989, "Results of the VLA gravitational lens survey," in: [MO89.1]

[HE89.2] Hewett, P.C., Webster, R.L., Harding, M.E., Jedrzejewski, R.I., Foltz, C.B., Chaffee, F.H., Irwin, M.J. & Le Fevre, O. 1989, "A new wide-separation gravitational lens candidate," *Ap. J.*, **346**, L61.

[HE90.1] Hewitt, J.N., Perley, R.A., Turner, E.L. & Hu, E.M. 1990, "Radio observations of a candidate cosmic string gravitational lens," *Ap. J.*, **356**, 57.

[HE90.2] Heflin, M.B., Gorenstein, M.V., Lawrence, C.R., Burke, B.F. & Shapiro, I.I. 1990, "First epoch VLBI observations of the gravitational lens system 2016+112," in: [ME90.2]

[HI86.1] Hills, J.G. 1986, "Limitations on the masses of objects constituting the missing mass in the galactic disk and the galactic halo," *Astron. J.*, **92**, 595.

[HI87.1] Hinshaw, G. & Krauss, L.M. 1987, "Gravitational lensing by isothermal spheres with finite core radii: Galaxies and dark matter," *Ap. J.*, **320**, 468.

[HI90.1] Hintzen, P., Maran, S.P., Michalitsianos, A.G., Foltz, C.B., Chaffee, Jr., F.H. & Kafatos, M. 1990, "Moderate-resolution spectroscopy of the lensed quasar 2237+0305: a search for Ca II absorption due to the interstellar medium in the foreground lensing galaxy," *Astron. J.*, **99**, 45.

[HO84.1] Hogan, C. & Narayan, R. 1984, "Gravitational lensing by cosmic strings," *M.N.R.A.S.*, **211**, 575.

[HO84.2] Hogan, C.J. & Rees, M.J. 1984, "Gravitational interactions of cosmic strings," *Nature*, **311**, 109.

[HO89.1] Hogan, C.J., Narayan, R. & White, S.D.M. 1989, "Quasar lensing by galaxies?" *Nature*, **339**, 106.

[HU85.1] Huchra, J., Gorenstein, M., Kent, S., Shapiro, I., Smith, G., Horine, E. & Perley, R. 1985, "2237+0305: A new and unusual gravitational lens," *Astron. J.*, **90**, 691.

[HU86.1] Huchra, J.P. 1986, "Is 1146+111B,C a lensed quasar or a quasar pair?" *Nature*, **323**, 784.

[HU90.1] Hu, E.M. 1990, "Investigation of a candidate string-lensing field," *Ap. J.*, **360**, L7.

[IB83.1] Ibanez, J. 1983, "Gravitational lenses with angular momentum," *Astr. Ap.*, **124**, 175.

[IO86.1] Iovino, A. & Shaver, P.A. 1986, "Gravitational lensing in the QSO pair Q 1548+114A,B," *Astr. Ap.*, **166**, 119.

[IR89.1] Irwin, M.J., Webster, R.L., Hewett, P.C., Corrigan, R.T. & Jedrzejewski, R.I. 1989, "Photometric variations in the Q2237+0305 system: first detection of a microlensing event," *Astron. J.*, **98**, 1989.

[IS89.1] Isaacson, J.A. & Canizares, C.R. 1989, "Probabilities for gravitational lensing by point masses in a locally inhomogeneous universe," *Ap. J.*, **336**, 544.

[JA75.1] Jackson, J.D. 1975, *Classical electrodynamics*, 2nd ed., John Wiley & Sons, New York.

[JA89.1] Jaroszynski, M. 1989, "Multi-plane lensing by isothermal spheres," *Acta Astr.*, **39**, 301.

[JA90.1] Jaroszynski, M., Park, C., Paczynski, B. & Gott, J.R. 1990, "Weak gravitational lensing due to large-scale structure of the universe," *Ap. J.*, **365**, 22.

[JA91.1] Jaroszynski, M. 1991, "Gravitational lensing in a universe model with realistic mass distribution," *M.N.R.A.S.*, **249**, 430.

[JA91.2] Jauncey, D.L. *et al.* 1991, "An unusually strong Einstein ring in the source PKS1830-211," *Nature*, **352**, 132.

[KA69.1] Kantowski, R. 1969, "Corrections in the luminosity-redshift relations of the homogeneous Friedmann models," *Ap. J.*, **155**, 89.

[KA83.1] Kayser, R. & Refsdal, S. 1983, "The difference in light travel time between gravitational lens images," *Astr. Ap.*, **128**, 156.

[KA84.1] Kaiser, N. & Stebbins, A. 1984, "Microwave anisotropy due to cosmic strings," *Nature*, **310**, 391.

[KA86.1] Katz, N., Balbus, S. & Paczyński, B. 1986, "Random scattering approach to gravitational microlensing," *Ap. J.*, **306**, 2.

[KA86.2] Kayser, R., Refsdal, S. & Stabell, R. 1986, "Astrophysical applications of gravitational micro-lensing," *Astr. Ap.*, **166**, 36.

[KA87.1] Katz, N. & Paczyński, B. 1987, "Gravitational lensing by an ensemble of isothermal galxies," *Ap. J.*, **317**, 11.

[KA88.1] Katz, J.I. & Jackson, S. 1988, "Predictions for arc polarization," *Astr. Ap.*, **197**, 29.

[KA88.2] Kayser, R. & Schramm, T. 1988, "Imaging procedures for gravitational lenses," *Astr. Ap.*, **191**, 39.

[KA88.3] Kayser, R. & Refsdal, S. 1988, "Gravitational lenses and the brightness distribution of distant sources," *Astr. Ap.*, **197**, 63.

[KA88.4] Kayser, R. 1988, "The micro-lensing scenario for the BL Lac object AO 235+164," *Astr. Ap.*, **206**, L8.

[KA88.5] Kashlinsky, A. 1988, "Small-scale fluctuations in the microwave background radiation and multiple gravitational lensing," *Ap. J.*, **331**, L1.

[KA89.1] Kayser, R., Weiss, A., Refsdal, S. & Schneider, P. 1989, "Gravitational microlensing due to an ensemble of compact objects with different masses," *Astr. Ap.*, **214**, 4.

[KA89.2] Kayser, R. & Refsdal, S. 1989, "Detectability of gravitational microlensing in the quasar QSO 2237+0305," *Nature*, **338**, 745.

[KA89.3] Kayser, R. & Witt, H.J. 1989, "Amplification near gravitational lens caustics," *Astr. Ap.*, **221**, 1.

[KA90.1] Kayser, R. 1990, "Macro- and micromodels for gravitational lenses," *Ap. J.*, **357**, 309.

[KA90.2] Kayser, R., Surdej, J., Condon, J.J., Kellermann, K.I., Magain, P., Remy, M. & Smette, A. 1990, "New observations and gravitational lens models of the cloverleaf quasar H1413+117," *Ap. J.*, **364**, 15.

[KA90.3] Kasai, M., Futamase, T. & Takahara, F. 1990, "Angular diameter distance in a clumpy universe," *Phys. Lett. A*, **147**, 97.

[KA91.1] Kassiola, A., Kovner, I. & Blandford, R.D. 1991, "Bounds on intergalactic compact objects from observations of compact radio sources," (preprint).

[KE32.1] Kermack, W.O., McCrea, W.H. & Whittaker, E.T. 1932, "On properties of null geodesics and their application to the theory of radiation," *Proc. Roy. Soc. Edinb.*, **53**, 31.

[KE81.1] Kellermann, K.I. & Pauliny-Toth, I.I.K. 1981, "Compact radio sources," *Ann. Rev. Astr. Ap.*, **19**, 373.

[KE82.1] Keel, W.C. 1982, "The nature of the light variations in the double QSO Q0957+561," *Ap. J.*, **255**, 20.

[KE82.2] Keel, W.C. 1982, "Observational tests of QSO amplification by condensed objects in galactic halos," *Ap. J.*, **259**, L1.

[KE88.1] Kent, S.M. & Falco, E.E. 1988, "A model for the gravitational lens system 2237+0305," *Astron. J.*, **96**, 1570.

[KE90.1] Kellog, E., Falco, E.E., Forman, W., Jones, C. & Slane, P. 1990, "X-ray observations of gravitatoinal lenses," in: [ME90.2]

[KI66.1] King, I.R. 1966, "The structure of star clusters: III. Some simple dynamical models," *Astron. J.*, **71**, 64.

[KI74.1] Kippenhahn, R. & de Vries, H.L. 1974, "Statistics of apparent distances between QSO's and bright galaxies," *Astrophys. and Space Sci.*, **26**, 131.

[KI83.1] Kirshner, R.P., Oemler, A., Schechter, P.L. & Shectman, S.A. 1983, "A deep survey of galaxies," *Astron. J.*, **88**, 1285.

[KI89.1] Kidger, M.R. 1989, "The optical variability of 3C 345," *Astr. Ap.*, **226**, 9.

[KI90.1] Kidger, M.R. & de Diego, J.A. 1990, "Rapid variations and a very deep minimum in the light curve of 3C 345," *Astr. Ap.*, **227**, L25.

[KL63.1] Klimov, Yu.G. 1963, "The deflection of light rays in the gravitational fields of galaxies," *Sov. Phys. Doklady*, **8**, 119.

[KO87.1] Kovner, I. 1987, "The marginal gravitational lens," *Ap. J.*, **321**, 686.

[KO87.2] Kochanek, C.S. & Blandford, R.D. 1987, "Gravitational imaging by isolated elliptical potential wells. II. Probability distributions," *Ap. J.*, **321**, 676.

[KO87.3] Kovner, I. 1987, "The quadrupole gravitational lens," *Ap. J.*, **312**, 22.

[KO87.4] Kovner, I. 1987, "Giant luminous arcs from gravitational lensing," *Nature*, **327**, 193.

[KO87.5] Kovner, I. 1987, "The thick gravitational lens: A lens composed of many elements at different distances," *Ap. J.*, **316**, 52.

[KO87.6] Kovner, I. & Milgrom, M. 1987, "Concerning the limit on the mean mass distribution of galaxies from their gravitational lens effect," *Ap. J.*, **321**, L113.

[KO88.1] Kochanek, C.S. & Apostolakis, J. 1988, "The two-screen gravitational lens," *M.N.R.A.S.*, **235**, 1073.

[KO88.2] Kovner, I. & Paczyński, B. 1988, "Supernovae in luminous arcs," *Ap. J.*, **335**, L9.

[KO89.1] Kochanek, C.S., Blandford, R.D., Lawrence, C.R. & Narayan, R. 1989, "The ring cycle: an iterative lens reconstruction technique applied to MG1131+0456," *M.N.R.A.S.*, **238**, 43.

[KO89.2] Kovner, I. 1989, "Maximal lens bounds on QSO-galaxy association," *Ap. J.*, **341**, L1.

[KO89.3] Kovner, I. 1989, "Diagnostics of compact clusters of galaxies by giant luminous arcs," *Ap. J.*, **337**, 621.

[KO90.1] Kovner, I. 1990, "Fermat principle in arbitrary gravitational fields," *Ap. J.*, **351**, 114.

[KO90.2] Kochanek, C.S. & Lawrence, C.R. 1990, "Where have all the lenses gone?" *Astron. J.*, **99**, 1700.

[KO90.3] Kochanek, C.S. 1990, "The theory and practice of radio ring lenses," in: [ME90.2]

[KO90.4] Kochanek, C.S. 1990, "Inversions of cluster gravitational lenses and the determination of source redshifts," *M.N.R.A.S.*, **247**, 135.

[KO90.5] Kolb, E.W. & Turner, M.G. 1990, *The early universe*, Addison-Wesley Publ. Comp., Redwood.

[KO90.6] Kovner, I. 1990, "Cosmic gravitational diagnostics," in: [ME90.2]

[KO91.1] Kochanek, C.S. 1991, "Selection effects in optical surveys for gravitational lenses," *Ap. J.*, **379**, 517.

[KR66.1] Kristian, J. & Sachs, R.K. 1966, "Observations in cosmology," *Ap. J.*, **143**, 379.

[KR79.1] Krolik, J.H. & Kwan, J. 1979, "Quasar image doubling associated with absorption systems," *Nature*, **281**, 550.

[KR80.1] Kramer, D., Stephani, H., MacCallum, M. & Herlt, E. 1980, *Exact solutions of Einstein field equations*, VEB-Verlag, Berlin.

[LA71.1] Lawrence, J.K. 1971, "Focusing of gravitational radiation by interior gravitational fields," *Nuovo Cimento*, **6B**, 225.

[LA82.1] Lacroix, G. & Schneider, J. 1982, "The precision on the measure of q_0 using the gravitational lensing effect," *Astr. Ap.*, **115**, 54.

[LA84.1] Lawrence, C.R., Schneider, D.P., Schmidt, M., Bennett, C.L., Hewitt, J.N., Burke, B.F., Turner, E.L. & Gunn, J.E. 1984, "Discovery of a new gravitational lens system," *Science*, **223**, 46.

[LA86.1] Lawrence, C.R., Readhead, A.C.S., Moffet, A.T. & Birkinshaw, M. 1986, "Isotropy of the microwave background radiation near 1146+111 B and C," *Astron. J.*, **92**, 1235.

[LA88.1] Lavery, R.J. & Henry, J.P. 1988, "Two blue arcs associated with the cD galaxy of Abell 963," *Ap. J.*, **329**, L21.

[LA89.1] Langston, G.I., Schneider, D.P., Conner, S., Carilli, C.L., Lehar, J., Burke, B.F., Turner, E.L., Gunn, J.E., Hewitt, J.N. & Schmidt, M. 1989, "MG

1654+1346: an Einstein ring image of a quasar radio lobe," *Astron. J.*, **97**, 1283.

[LA90.1] Langston, G.I., Conner, S.R., Lehar, J., Burke, B.F. & Weiler, K.W. 1990, "Galaxy mass deduced from the structure of Einstein ring MG 1654+1346," *Nature*, **344**, 43.

[LA90.2] Langston, G.I., Conner, S.R., Heflin, M.B., Lehár, J. & Burke, B.F. 1990, "Faint radio sources and gravitational lensing," *Ap. J.*, **353**, 34.

[LA95.1] Laplace, P.S. 1795, "Exposition du système du monde".

[LE21.1] Lenard, P. 1921, *Annalen der Physik*, **65**, 593.

[LE87.1] Le Fèvre, O., Hammer, F., Nottale, L. & Mathez, G. 1987, "Is 3C324 the first gravitationally lensed giant galaxy?" *Nature*, **326**, 268.

[LE88.1] Le Fèvre, O., Hammer, F., Nottale, L., Mazure, A. & Christian, C. 1988, "Peculiar morphology of the high-redshift radio galaxies 3C 13 and 3C 256 in subsecond seeing," *Ap. J.*, **324**, L1.

[LE88.2] Le Févre, O., Hammer, F. & Jones, J. 1988, "High spatial resolution of some of the most distant 3CR galaxies," *Ap. J.*, **331**, L73.

[LE88.3] Le Fèvre, O. & Hammer, F. 1988, "Imaging of very distant 3CR galaxies: high spatial resolution data for seven galaxies with $1.176 \leq z \leq 1.841$," *Ap. J.*, **333**, L37.

[LE90.1] Le Fèvre, O. & Hammer, F. 1990, "3CR 208.1: a radio-loud quasar at $z = 1.02$ gravitationally amplified by a foreground Seyfert galaxy at $z = 0.159$," *Ap. J.*, **350**, L1.

[LE90.2] Lee, M.H. & Paczyński, B. 1990, "Gravitational lensing by a smoothly variable three-dimensional mass distribution," *Ap. J.*, **357**, 32.

[LE90.3] Lee, M.H. & Spergel, D.N. 1990, "An analytical approach to gravitational lensing by an ensemble of axisymmetric lenses," *Ap. J.*, **357**, 23.

[LI64.1] Liebes Jr., S. 1964, "Gravitational lenses," *Phys. Rev.*, **133**, B835.

[LI66.1] Lindquist, R.W. 1966, "Relativistic transport theory," *Ann. Phys. (USA)* **37**, 487.

[LI88.1] Linder, E.V., Schneider, P. & Wagoner, R.V. 1988, "Gravitational lensing statistics of amplified supernovae," *Ap. J.*, **324**, 786.

[LI88.2] Linder, E.V. & Schneider, P. 1988, "Quasar-galaxy associations by gravitational lensing revisited," *Astr. Ap.*, **204**, L8.

[LI88.3] Linder, E.V. 1988, "Isotropy of the microwave background by gravitational lensing," *Astr. Ap.*, **206**, 199.

[LL81.1] Lloyd, C. 1981, "Intrinsic variations of the double quasar 0957+56 AB," *Nature*, **294**, 727.

[LO19.1] Lodge, O.J. 1919, "Gravitation and light," *Nature*, **104**, 354.

[LO88.1] Lottermoser, M. 1988, "Über den Newtonschen Grenzwert der Allgemeinen Relativitätstheorie und die relativistische Erweiterung Newtonscher Anfangsdaten," Ph. D. thesis, University of Munich.

[LY86.1] Lynds, R. & Petrosian, V. 1986, "Giant luminous arcs in galaxy clusters," *Bull. AAS*, **18**, 1014.

[LY89.1] Lynds, R. & Petrosian, V. 1989, "Luminous arcs in clusters of galaxies," *Ap. J.*, **336**, 1.

[MA58.1] Mattig, W. 1958, "Über den Zusammenhang zwischen Rotverschiebung und Scheinbarer Helligkeit," *Astron. Nachr.*, **284**, 109.

[MA81.1] Maslov, V.P. & Fedoriuk, M.V. 1981, *Semi-classical approximation in quantum mechanics*, Reidel Publ. Comp., Dordrecht.

[MA82.1] Mandzhos, A.V. 1982, "The mutual coherence of gravitational-lens images," *Sov. Astr. Let.*, **7**, 213.

[MA82.3] Maccacaro, T., Feigelson, E.D., Fener, M., Giacconi, R., Gioio, I.M., Griffiths, R.E., Murray, S.S., Zamorani, G., Stocke, J. & Liebert, J. 1982, "A medium sensitivity X-ray survey using the *Einstein* Observatory: the Log N − Log S relation for extragalactic X-ray sources," *Ap. J.*, **253**, 504.

[MA84.1] Marshall, H.L., Avni, Y., Braccesi, A., Huchra, J.P., Tananbaum, H., Zamorani, G. & Zitelli, V. 1984, "A complete sample of quasars at $B = 19.80$," *Ap. J.*, **283**, 50.

[MA85.1] Marshall, H.L. 1985, "The evolution of optically selected quasars with $z < 2.2$ and $B < 20$," *Ap. J.*, **299**, 109.

[MA88.1] Magain, P., Surdej, J., Swings, J.-P., Borgeest, U., Kayser, R., Kühr, H., Refsdal, S. & Remy, M. 1988, "Discovery of a quadruply lensed quasar: the 'clover leaf' H1413+117," *Nature*, **334**, 327.

[MA90.1] Mather, J.C., Cheng, E.S., Eplee, R.E. Jr., Isaacman, R.B., Meyer, S.S., Shafer, R.A., Weiss, R., Wright, E.L., Bennett, C.L., Boggess, N.W., Dwek, E., Gulkis, S., Hauser, M.G., Janssen, M., Kelsall, T., Lubin, P.M., Moseley, S.H. Jr., Murdock, T.L., Silverberg, R.F., Smoot, G.F. & Wilkinson, D.T. 1990, "A preliminary measurement of the cosmic microwave background spectrum by the cosmic background explorer (COBE) satellite," *Ap. J.*, **354**, L37.

[MC85.1] McKenzie, R.H. 1985, "A gravitational lens produces on odd number of images," *J. Math. Phys.*, **26**, 1592.

[ME63.1] Metzner, A.W.K. 1963, "Observable properties of large relativistic masses," *J. Math. Phys.*, **4**, 1194.

[ME87.1] Mészáros, P. & Riffert, H. 1987, "On the radius of neutron stars," *Ap. J.*, **323**, L127.

[ME88.1] Mellier, Y., Soucail, G., Fort, B. & Mathez, G. 1988, "Photometry, spectroscopy and content of the distant cluster of galaxies Abell 370," *Astr. Ap.*, **199**, 13.

[ME88.2] Mészáros, P. & Riffert, H. 1988, "Gravitational light bending near neutron stars. II. Accreting pulsar spectra as a function of phase," *Ap. J.*, **327**, 712.

[ME89.1] Meylan, G. & Djorgovski, S. 1989, "UM 425: a new gravitational lens candidate," *Ap. J.*, **338**, L1.

[ME90.1] Mellier, Y., Soucail, G., Fort, B., Le Borgne, J.-F. & Pello, R. 1990, "Modeling the giant arcs in A370 and A2390," in: [ME90.2]

[ME90.2] Mellier, Y., Fort, B. & Soucail, G. 1990, "Gravitational lensing," *Lecture Notes in Physics*, **360**, Springer-Verlag.

[ME91.1] Mellier, Y., 1991, "Determination of arc redshift in Cl2244" (preprint).

[MI73.1] Misner, C.W., Thorne, K.S. & Wheeler, J.A. 1973, *Gravitation*, Freeman, New York.

[MI81.1] Miller, J.S., Antonucci, R.R.J. & Keel, W.C. 1981, "Variability of the double quasar 0957+561 A,B," *Nature*, **289**, 153.

[MI84.1] Michell, J., 1784, "On the means of discovering the distance, magnitude, &c. of the fixed stars...," *Trans. R. Soc. London*, **74**, 35, reprinted in: *Black holes: Selected reprints*, S. Detweiler (ed.), Am. Assoc. of Physics Teachers, Stony Brook, NY, 1982.

[MI88.1] Miller, J.S. & Goodrich, R.W. 1988, "Spectrophotometry and polarimetry of the giant luminous arcs in the clusters 2244-02 and Abell 370," *Nature*, **331**, 685.

[MI91.1] Miralda-Escude, J. 1991, "Gravitational lensing by clusters of galaxies: constraining the mass distribution," *Ap. J.*, **370**, 1.

[MO89.1] Moran, J.M., Hewitt, J.N. & Lo, K. Y. 1989, "Gravitational lenses," *Lecture Notes in Physics*, **330**, Springer-Verlag.

[MU86.1] Murdoch, H.S., Hunstead, R.W., Pettini, M. & Blades, J.C. 1986, "Absorption spectrum of the $z = 3.78$ QSO 2000-330. II. The redshift and equivalent width distributions of primordial hydrogen clouds," *Ap. J.*, **309**, 19.

[NA82.1] Narasimha, D., Subramanian, K. & Chitre, S.M. 1982, "Spheroidal gravitational lenses and the triple quasar," *M.N.R.A.S.*, **200**, 941.

[NA84.1] Narasimha, D., Subramanian, K. & Chitre, S.M. 1984, "Gravitational lens model of the double QSO 0957+561 A,B incorporating *VLBI* features," *M.N.R.A.S.*, **210**, 79.

[NA84.2] Narasimha, D., Subramanian, K. & Chitre, S.M. 1984, "Gravitational lens models for the triple radio source MG 2016+112," *Ap. J.*, **283**, 512.

[NA84.3] Narayan, R., Blandford, R. & Nityananda, R. 1984, "Multiple imaging of quasars by galaxies and clusters," *Nature*, **310**, 112.

[NA86.1] Narasimha, D., Subramanian, K. & Chitre, S.M. 1986, "'Missing image' in gravitational lens systems?" *Nature*, **321**, 45.

[NA86.2] Narayan, R. 1986, "Gravitational lensing – models," in: [SW86.1]

[NA87.1] Narasimha, D., Subramanian, K. & Chitre, S.M. 1987, "The gravitational lens system 2016+112 revisited," *Ap. J.*, **315**, 434.

[NA88.1] Narayan, R. & White, S.D.M. 1988, "Gravitational lensing in a cold dark matter universe," *M.N.R.A.S.*, **231**, 97p.

[NA88.2] Narasimha, D. & Chitre, S.M. 1988, "Giant luminous arcs in galaxy clusters," *Ap. J.*, **332**, 75.

[NA89.1] Narayan, R. 1989, "Gravitational lensing and quasar-galaxy correlations," *Ap. J.*, **339**, L53.

[NA89.2] Narasimha, D. & Chitre, S.M. 1989, "Lensing of extended sources by dark galactic halos," *Astron. J.*, **97**, 327.

[NA90.1] Narayan, R. & Schneider, P. 1989, "Why are the Bl Lac objects AO0235+164 and PKS0537-441 not multiply imaged by gravitational lensing?" *M.N.R.A.S.*, **243**, 192.

[NE86.1] Nemiroff, R.J. 1986, "Random gravitational lensing," *Astrophys. and Space Sci.*, **123**, 381.

[NE88.1] Nemiroff, R.J. 1988, "AGN broad emission line amplification from gravitational microlensing," *Ap. J.*, **335**, 593.

[NE89.1] Nemiroff, R.J. & Dekel, A. 1989, "The probability of arc lensing by clusters of galaxies," *Ap. J.*, **344**, 51.

[NE90.1] Nemiroff, R.J. 1990, "Gravitationally lensed arcs as galaxy redshift indicators," in: [ME90.2]

[NI77.1] Nieto, J.-L. 1977, "Quasar-galaxy pairs and surface density of quasars," *Nature*, **270**, 411.

[NI78.1] Nieto, J.-L. 1978, "Statistical studies of the quasar-galaxy associations," *Astr. Ap.*, **70**, 219.

[NI79.1] Nieto, J.-L. 1979, "Optically variable quasars and bright galaxies," *Astr. Ap.*, **74**, 152.

[NI82.1] Nieto, J.-L. & Seldner, M. 1982, "Further investigations on possible correlations between QSOs and the Lick Catalogue of galaxies," *Astr. Ap.*, **112**, 321.

[NI84.1] Nityananda, R. & Ostriker, J.P. 1984, "Gravitational lensing by stars in a galaxy halo: Theory of combined weak and strong scattering," *J. Astrophys. Astr.*, **5**, 235.

[NI88.1] Nieto, J.-L., Roques, S., Llebaria, A., Vanderriest, Ch., Lelièvre, G., Serego Alighieri di, S., Macchetto, F.D. & Perryman, M.A.C. 1988, "High-resolution imaging of the double QSO 2345+007: Evidence for subcomponents," *Ap. J.*, **325**, 644.

[NO82.1] Nottale, L. 1982, "Perturbation of the magnitude-redshift relation in an inhomogeneous relativistic model: The redshift equations," *Astr. Ap.*, **110**, 9.

[NO82.2] Nottale, L. 1982, "Perturbation of the magnitude-redshift relation in an inhomogeneous relativistic model: II. Correction to the Hubble law behind clusters," *Astr. Ap.*, **114**, 261.

[NO83.1] Nottale, L. 1983, "Perturbation of the magnitude-redshift relation in an inhomogeneous relativistic model," *Astr. Ap.*, **118**, 85.

[NO86.1] Nottale, L. 1986, "The eruptive BL Lac object 0846+51W1: A gravitationally lensed QSO?" *Astr. Ap.*, **157**, 383.

[NO86.2] Nottale, L. & Chauvineau, B. 1986, "Multiple gravitational lensing," *Astr. Ap.*, **162**, 1.

[NO89.1] Nørgaard-Nielsen, H.U., Hansen, L., Jørgensen, H.E., Salamanca, A.A., Ellis, R.S. & Couch, W.J. 1989, "The discovery of a type Ia supernova at a redshift of 0.31," *Nature*, **339**, 523.

[NO89.2] Nollert, H.-P., Ruder, H., Herold, H. & Kraus, U. 1989, "The relativistic 'looks' of a neutron star," *Astr. Ap.*, **208**, 153.

[NO90.1] Nottale, L. 1990, "Gravitational redshifts and lensing by large scale structures," in: [ME90.2]

[OH74.1] Ohanian, H.C. 1974, "On the focusing of gravitational radiation," *Int. J. Theoret. Phys.*, **9**, 425.

[OH83.1] Ohanian, H.C. 1983, "The caustics of gravitational 'lenses'," *Ap. J.*, **271**, 551.

[OL90.1] Olive, K., Schramm, D., Steigman, G. & Walker, T. 1990, "Big-bang nucleosynthesis revisited," *Phys. Lett. B*, **236**, 454.

[OS85.1] Ostriker, J.P. & Vietri, M. 1985, "Are some BL Lac objects artefacts of gravitational lensing?" *Nature*, **318**, 446.

[OS86.1] Ostriker, J.P. & Vietri, M. 1986, "The statistics of gravitational lensing. III. Astrophysical consequences of quasar lensing," *Ap. J.*, **300**, 68.

[OS86.2] Ostriker, J.P. & Vishniac, E.T. 1986, "Effect of gravitational lenses on the microwave background, and 1146+111B,C," *Nature*, **322**, 804.

[OS88.1] Osmer, P.S., Porter, A.C, Green, R.F. & Foltz, C.B. (eds.) 1988, *Proceedings of a workshop on optical surveys for quasars*, Astronomical Society of the Pacific Conference Series, Vol. 2.

[OS90.1] Ostriker, J.P. & Vietri, M. 1990, "Are some BL Lacs artefacts of gravitational lensing?" *Nature*, **344**, 45.

[PA86.1] Paczyński, B. 1986, "Gravitational microlensing at large optical depth," *Ap. J.*, **301**, 503.

[PA86.2] Paczyński, B. 1986, "Gamma-ray bursters at cosmological distances," *Ap. J.*, **308**, L43.

[PA86.3] Paczyński, B. 1986, "Will cosmic strings be discoverd using the Space Telescope?" *Nature*, **319**, 567.

[PA86.4] Paczyński, B. 1986, "Gravitational microlensing by the galactic halo," *Ap. J.*, **304**, 1.

[PA86.5] Paczyński, B. 1986, "Is there a black hole in the sky?" *Nature*, **321**, 419.

[PA87.1] Paczyński, B. 1987, "Giant luminous arcs discovered in two clusters of galaxies," *Nature*, **325**, 572.

[PA87.2] Paczyński, B. 1987, "Gravitational microlensing and gamma-ray bursts," *Ap. J.*, **317**, L51.

[PA88.1] Padmanabhan, T. & Subramanian, K. 1988, "The focussing equation, caustics and the condition for multiple imaging by thick gravitational lenses," *M.N.R.A.S.*, **233**, 265.

[PA89.1] Paczyński, B. & Wambsganss, J. 1989, "Gravitational lensing by a smoothly variable surface mass density," *Ap. J.*, **337**, 581.

[PE66.1] Penrose, R. 1966, "General-relativistic energy flux and elementary optics," in: *Essays in Honor of Václav Hlavatý*, Hoffmann, B. (ed.), Indiana Univ. Press, Bloomington, p. 259

[PE82.1] Peacock, J.A. 1982, "Gravitational lenses and cosmological evolution, *M.N.R.A.S.*, **199**, 987."

[PE86.1] Peacock, J.A. 1986, "Flux conservation and random gravitational lensing," *M.N.R.A.S.*, **223**, 113.

[PE88.1] Pello-Descayre, R., Soucail, G., Sanahuja, B., Mathez, G. & Ojero, E. 1988, "Detection and photometry of a complex system in the center of the A 2218 cluster of galaxies," *Astr. Ap.*, **190**, L11.

[PE89.1] Petrosian, V. 1989, "Arc in clusters of galaxies as gravitational lens images," in: [MO89.1]

[PE90.1] Pello, R., Le Borgne, J.F., Mathez, G., Mellier, Y., Sanahuja, B. & Soucail, G. 1990, "First results on the spectroscopy of the arc-like object in Abell 2390," in: [ME90.2]

[PE90.2] Perlick, V. 1990, "On Fermat's principle in general relativity: I. The general case," *Class. Quantum Grav.*, **7**, 1319.

[PE91.1] Pello, R., Le Borgne, J.-F., Soucail, G., Mellier, Y. & Sanahuja, B. 1991, "A straight gravitational image in Abell 2390: a striking case of lensing by a cluster of galaxies," *Ap. J.*, **366**, 405.

[PH62.1] Pham Mau Quan 1962, "Le principle de Fermat en relativité generale," in: *Les Theories Relativistes de la Gravitation*, Proceeding Royarmont Conf., CNRS, Paris, p. 165

[PH86.1] Phinney, E.S. & Blandford, R.D. 1986, "Q1146+111B,C quasar pair: Illusion or delusion?" *Nature*, **321**, 569.

[PO78.1] Poston, T. & Steward, I. 1978, *Catastrophe theory and its applications*, Pitman, London.

[PO79.1] Pooley, G.G., Browne, I.W.A., Daintree, E.J., Moore, P.K., Noble, R.G. & Walsh, D. 1979, "Radio studies of the double QSO, 0957+561A,B," *Nature*, **280**, 461.

[PO79.2] Porcas, R.W., Booth, R.S., Browne, I.W.A., Walsh, D. & Wilkinson, P.N. 1979, "VLBI observations of the double QSO, 0957+561 A,B," *Nature*, **282**, 385.

[PO81.1] Porcas, R.W., Booth, R.S., Browne, I.W.A., Walsh, D. & Wilkinson, P.N. 1981, "VLBI structures of the images of the double QSO 0957+561," *Nature*, **289**, 758.

[PR73.1] Press, W.H. & Gunn, J.E. 1973, "Method for detecting a cosmological density of condensed objects," *Ap. J.*, **185**, 397.

[PR86.1] Press, W.H., Flannery, B.P., Teukolsky, S.A. & Vetterling, W.T. 1986, *Numerical Recipes*, Cambridge University Press, Cambridge

[PR91.1] Press, W.H., Rybicki, G.B. & Hewitt, J.N. 1991, "The time delay of gravitational lens 0957+561. I. Methodology, and analysis of optical photom etric data," *Ap. J.*, (in press)

[QU89.1] Quirrenbach, A., Witzel, A., Qian, S.J., Krichbaum, T., Hummel, C.A. & Alberdi, A. 1989, "Rapid radio polarization variability in the quasar 0917+624," *Astr. Ap.*, **226**, L1.

[RA88.1] Rao, A.P. & Subrahmanyan, R. 1988, "1830−211 − a flat-spectrum radio source with double structure," *M.N.R.A.S.*, **231**, 229.

[RA91.1] Rauch, K.P. & Blandford, R.D. 1991, "Microlensing and the structure of AGN accretion disks," (preprint).

[RE64.1] Refsdal, S. 1964, "The gravitational lens effect," *M.N.R.A.S.*, **128**, 295.

[RE64.2] Refsdal, S. 1964, "On the possibility of determining Hubble's parameter and the masses of galaxies from the gravitational lens effect," *M.N.R.A.S.*, **128**, 307.

[RE66.1] Refsdal, S. 1966, "On the possibility of determining the distances and masses of stars from the gravitational lens effect," *M.N.R.A.S.*, **134**, 315.

[RE66.2] Refsdal, S. 1966, "On the possibility of testing cosmological theories from the gravitational lens effect," *M.N.R.A.S.*, **132**, 101.

[RE66.3] Rees, M.J. 1966, "Appearance of relativistically expanding radio sources," *Nature*, **211**, 468.

[RE70.1] Refsdal, S. 1970, "On the propagation of light in universes with inhomogeneous mass distribution," *Ap. J.*, **159**, 357.

[RE79.1] Reasenberg, R.D., Shapiro, I.I., MacNeil, P.E., Goldstein, R.B., Breidenthal, J.C., Brenkle, J.P., Cain, D.L., Kaufman, T.L., Komarek, T.A. & Zygielbaum, A.I. 1979, "Viking relativity experiment: Verification of signal retardation by solar gravity," *Ap. J.*, **234**, L219.

[RE82.1] Rees, M.J. & Kashlinsky, A. 1982, "Galaxy formation, clustering and the 'hidden mass'," in: *Progress in cosmology*, A.W. Wolfendale (ed.), Reidel, p.259.

[RH91.1] Rhee, G. 1991, "An estimate of the Hubble constant from the gravitational lensing of quasar Q 0957+561," *Nature*, **350**, 211.

[RI79.1] Rindler, W. 1979, *Essential Relativity*, Springer-Verlag, New York.

[RI88.1] Riffert, H. & Mészáros, P. 1988, "Gravitational light bending near neutron stars. I. Emission from columns and hot spots," *Ap. J.*, **325**, 207.

[RI88.2] Rix, H.-W. & Hogan, C.J. 1988, "Quasar microlensing and dark matter," *Ap. J.*, **332**, 108.

[RO35.1] Robertson, H.P. 1935, "Kinematics and world-structure," *Ap. J.*, **82**, 284.

[RO72.1] Rozyczka, M. 1972, "Counts of galaxies in the fields surrounding quasars," *Acta Astr.*, **22**, 93.

[RO79.1] Roberts, D.H., Greenfield, P.E. & Burke, B.F. 1979, "The double quasar 0957+561: A radio study at 6-centimeters wavelength," *Science*, **205**, 894.

[RO84.1] Robertson, D.S. & Carter, W.E. 1984, "Relativistic deflection of radio signals in the solar gravitational field measured with VLBI," *Nature*, **310**, 572.

[RO85.1] Roberts, D.H., Greenfield, P.E., Hewitt, J.N., Burke, B.F. & Dupree, A.K. 1985, "The multiple images of the quasar 0957+561," *Ap. J.*, **293**, 356.

[RO91.1] Roberts, D.H., Lehar, J., Hewitt, J.N. & Burke, B.F. 1991, "The Hubble constant from VLA measurement of the time delay in the double quasar 0957+561," *Nature*, **352**, 43.

[SA55.1]   Salpeter, E.E. 1955, "The luminosity function and stellar evolution," *Ap. J.*, **121**, 161.

[SA61.1]   Sachs, R.K. 1961, "Gravitational waves in general relativity, VI: The outgoing radiation condition," *Proc. Roy. Soc. London*, **A264**, 309.

[SA67.1]   Sachs, R.K. & Wolfe, A.M. 1967, "Perturbations of a cosmological model and angular variations of the microwave background," *Ap. J.*, **147**, 73.

[SA71.1]   Sanitt, N. 1971, "Quasi-stellar objects and gravitational lenses," *Nature*, **234**, 199.

[SA76.1]   Sanitt, N. 1976, "3C268.4 – Evidence for the presence of gravitationally-lensed secondary image," *M.N.R.A.S.*, **174**, 91.

[SA82.1]   Saunders, P.T. 1982, *An introduction to catastrophe theory*, Cambridge University Press, Cambridge

[SA85.1]   Saslaw, W.C., Narasimha, D. & Chitre, S.M. 1985, "The gravitational lens as an astronomical diagnostic," *Ap. J.*, **292**, 348.

[SA87.1]   Sasaki, M. 1987, "The magnitude-redshift relation in a perturbed Friedmann universe," *M.N.R.A.S.*, **228**, 653.

[SA87.2]   Sargent, W.L.W. 1987, "QSO absorption lines and cosmology," in: [HE87.2]

[SA87.3]   Sargent, W.L.W. & Steidel, C.C. 1987, "Absorption in the wide QSO pair Tololo 1037-2704 and Tololo 1038-2712: evidence for a specially aligned supercluster at $z = 2$?" *Ap. J.*, **322**, 142.

[SA89.1]   Sasaki, M. 1989, "Gravitational lens effect on anisotropies of the cosmic microwave background," *M.N.R.A.S.*, **240**, 415.

[SC21.1]   Schouten, J.A. 1921, "Über die konforme Abbildung n-dimensionaler Mannigfaltigkeiten mit quadratischer Maßbestimmung auf eine Mannigfaltigkeit mit euklidischer Maßbestimmung," *Math. Z.*, **11**, 58.

[SC63.1]   Schmidt, M. 1963, "3C273: A star-like object with large red-shift," *Nature*, **197**, 1040.

[SC80.1]   Schechter, P.L. 1980, titMass-to-light ratios for elliptical galaxies, *Astron. J.*, **85**, 801.

[SC83.1]   Schmidt, M. & Green, R.F. 1983, "Quasar evolution derived from the Palomar Bright Quasar Survey and other complete quasar surveys," *Ap. J.*, **269**, 352.

[SC84.1]   Schild, R.E. & Weekes, T. 1984, "CCD brightness monitoring of the twin QSO 0957+561," *Ap. J.*, **277**, 481.

[SC84.2]   Schneider, P. 1984, "The amplification caused by gravitational bending of light," *Astr. Ap.*, **140**, 119.

[SC85.1]   Schneider, P. & and Schmid-Burgk, J. 1985, "Mutual coherence of gravitationally lensed images," *Astr. Ap.*, **148**, 369.

[SC85.2]   Schneider, D.P., Lawrence, C.R., Schmidt, M., Gunn, J.E., Turner, E.L., Burke, B.F. & Dhawan, V. 1985, "Deep optical and radio observations of the gravitational lens system 2016+112," *Ap. J.*, **294**, 66.

[SC85.3]   Schneider, P. 1985, "A new formulation of gravitational lens theory, time-delay, and Fermat's principle," *Astr. Ap.*, **143**, 413.

[SC86.1]   Schild, R.E. & Cholfin, B. 1986, "CCD camera brightness monitoring of Q 0957+561A,B," *Ap. J.*, **300**, 209.

[SC86.2]   Schneider, D.P., Gunn, J.E., Turner, E.L., Lawrence, C.R., Hewitt, J.N., Schmidt, M. & Burke, B.F. 1986, "The third image, the redshift of the lens, and new components of the gravitational lens 2016+112," *Astron. J.*, **91**, 991.

[SC86.3]   Schneider, P. 1986, "Statistical gravitational lensing and quasar-galaxy associations," *Ap. J.*, **300**, L31.

[SC86.4] Schneider, P. & Weiss, A. 1986, "The two point mass lens: detailed investigation of a special asymmetric lens," *Astr. Ap.*, **164**, 237.

[SC86.5] Schramm, T. 1986, "Properties of gravitational lens mappings," in: *The post-recombination universe*, N. Kaiser & A.N. Lasenby (eds.), Kluwer.

[SC87.1] Schneider, P. 1987, "The amplification bias by gravitational lensing on observed number counts of quasars: Effects of a finite source size," *Ap. J.*, **316**, L7.

[SC87.2] Schramm, T. & Kayser, R. 1987, "A simple imaging procedure for gravitational lenses," *Astr. Ap.*, **174**, 361.

[SC87.3] Schneider, P. & Weiss, A. 1987, "A gravitational lens origin of AGN-variability? Consequences of micro-lensing," *Astr. Ap.*, **171**, 49.

[SC87.4] Schneider, P. 1987, "Apparent number density enhancement of quasars near foreground galaxies due to gravitational lensing. I. Amplification cross sections," *Astr. Ap.*, **179**, 71.

[SC87.5] Schneider, P. 1987, "Apparent number density enhancement of quasars near foreground galaxies due to gravitational lensing. II. The amplification probability distribution and results," *Astr. Ap.*, **179**, 80.

[SC87.6] Schneider, P. & Wagoner, R.V. 1987, "Amplification and polarization of supernovae by gravitational lenses," *Ap. J.*, **314**, 154.

[SC87.7] Schneider, P. 1987, "Statistical gravitational lensing: influence of compact objects on the number counts of quasars," *Astr. Ap.*, **183**, 189.

[SC87.8] Schneider, P. 1987, "An analytically soluble problem in fully nonlinear statistical gravitational lensing," *Ap. J.*, **319**, 9.

[SC87.9] Schmidt, M. 1987, "Cosmic distribution of optically selected quasars," in: [HE87.2]

[SC88.1] Schneider, D.P., Turner, E.L., Gunn, J.E., Hewitt, J.N., Schmidt, M. & Lawrence, C.R. 1988, "High-resolution CCD imaging and derived gravitational lens models of 2237+0305," *Astron. J.*, **95**, 1619.

[SC88.2] Schneider, P. & Weiss, A. 1988, "Light propagation in inhomogeneous universes," *Ap. J.*, **327**, 526.

[SC88.3] Schneider, P. & Weiss, A. 1988, "Light propagation in inhomogeneous universes: the ray shooting method," *Ap. J.*, **330**, 1.

[SC89.1] Schneider, J. 1989, "Possible gravitational amplification in the binary pulsar 1957+20," *Astr. Ap.*, **214**, 1.

[SC89.2] Schneider, P. 1989, "The number excess of galaxies around high redshift quasars," *Astr. Ap.*, **221**, 221.

[SC89.3] Schneider, P. 1989, "Gravitational lensing and light propagation in an inhomogeneous universe," in: *Cosmology and Gravitational Lensing*, G. Börner, T. Buchert, and P. Schneider (eds.), MPA-P3 (proceedings, Garching 1989).

[SC90.1] Schneider, P. 1989, "Microlensing," in: [ME90.?]

[SC90.2] Schneider, P. & Wambsganss, J. 1990, "Are the broad emission lines of quasars affected by gravitational microlensing?" *Astr. Ap.*, **237**, 42.

[SC90.3] Schild, R. 1990, "The time delay of Q0957+561 A,B from ten years of optical monitoring," in: [ME90.2]

[SC90.4] Schramm, T. 1990, "Realistic elliptical potential wells for gravitational lens models," *Astr. Ap.*, **231**, 19.

[SC90.5] Schild, R.E. 1990, "The time delay in the twin QSO Q0957+561," *Astron. J.*, **100**, 1771.

[SC91.1] Schneider, P. & Weiss, A. 1991, "A practical approach to (nearly) elliptical gravitational lens models," *Astr. Ap.*, **247**, 269.

[SC91.2] Schneider, P. 1991, "Inversion of the amplification bias and the number excess of foreground galaxies around high-redshift QSOs," *Astr. Ap.* (in press).

[SC91.3] Schild, R. & Smith, R.C. 1991, "Microlensing in the Q0957+561 gravitational mirage," *Astron. J.*, **101**, 813.

[SC91.4] Schneider, P. & Weiss, A. 1991, "The gravitational lens equation near cusps," (preprint).

[SE83.1] Setti, G. & Zamorani, G. 1983, "Can all quasars be gravitationally lensed Seyfert's nuclei?" *Astr. Ap.*, **118**, L1.

[SE84.1] Setti, G. 1984, "Active galactic nuclei and their cosmological evolution," in: *X-ray and UV emission from active galactic nuclei*, W. Brinkmann & J. Trümper (eds.), Garching, 1984, p. 243.

[SH83.1] Shaver, P.A. & Robertson, J.G. 1983, "Common absorption systems in the spectra of the QSO pair Q 0307-195A,B," *Ap. J.*, **268**, L57.

[SH86.1] Shaklan, S.B. & Hege, E.K. 1986, "Detection of the lensing galaxy in PG 1115+08," *Ap. J.*, **303**, 605.

[SH86.2] Shaver, P.A. & Cristiani, S. 1986, "Test of the gravitational lens hypothesis for the quasar pair 1146+111B,C," *Nature*, **321**, 585.

[SH86.3] Shouping, X. & Ruffini, R. 1986, "Superluminal separation caused by gravitational lens effect of neutrino objects," *Astrophys. and Space Sci.*, **123**, 1.

[SI74.1] Silva de, L.N.K. 1974, "On gravitational-lens quasars," *Ap. J.*, **189**, 177.

[SM90.1] Smette, A., Surdej, J., Shaver, P.A., Foltz, C.B., Chaffee, F.H. & Magain, P. 1990, "Preliminary analysis of high-resolution spectra for UM673 A & B," in: [ME90.2]

[SM91.1] Smette, A., Surdej, J., Shaver, P.A., Foltz, C.B., Chaffee, F.H., Weymann, R.J., Williams, R.E. & Magain, P. 1991, "A spectroscopic study of UM673 A& B: on the size of Lyα clouds," (preprint).

[SO04.1] Soldner, J. 1804, "Über die Ablenkung eines Lichtstrahls von seiner geradlinigen Bewegung durch die Attraktion eines Weltkörpers, an welchem er nahe vorbeigeht," *Berliner Astron. Jahrb. 1804*, p. 161.

[SO80.1] Soifer, B.T., Neugebauer, G., Matthews, K., Becklin, E.E., Wynn-Williams, C.G. & Capps, R. 1980, "IR observations of the double quasar 0957+561 A,B and the intervening galaxy," *Nature*, **285**, 91.

[SO84.1] Sol, H., Vanderriest, C., Lelievre, G., Pedersen, H. & Schneider, J. 1984, "UBVr photometry of 2345+007 A,B, and the gravitational lens hypothesis," *Astr. Ap.*, **132**, 105.

[SO87.1] Soucail, G., Fort, B., Mellier, Y. & Picat, J.P. 1987, "A blue ring-like structure in the center of the A 370 cluster of galaxies," *Astr. Ap.*, **172**, L14.

[SO87.2] Soucail, G., Mellier, Y., Fort, B., Hammer, F. & Mathez, G. 1987, "Further data on the blue ring-like structure in A 370," *Astr. Ap.*, **184**, L7.

[SO88.1] Soucail, G., Mellier, Y., Fort, B., Mathez, G. & Cailloux, M. 1988, "The giant arc in A370: Spectroscopic evidence for gravitational lensing from a source at $z = 0.724$," *Astr. Ap.*, **191**, L19.

[SO90.1] Soucail, G., Mellier, Y., Fort, B., Mathez, G. & Cailloux, M. 1990, "Spectroscopy of arcs in Cl 2244-02 and A370 (A5)," in: [ME90.2]

[SR78.1] Sramek, R.A. & Weedman, D.W. 1978, "An optical and radio study of quasars," *Ap. J.*, **221**, 468.

[ST64.1] Sternberg, S. 1964, *Lectures on differential geometry*, Prentice-Hall, Englewood Cliffs, New Jersey.

[ST76.1]   Stephani, H. & Herlt, E. 1976, "Wave optics of the spherical gravitational lens – I: Diffraction of a plane electromagnetic wave," *Intern. J. Theoret. Physics*, **15**, 45.

[ST78.1]   Stephani, H. & Herlt, E. 1978, "Wave optics of the spherical gravitational lens – II: Diffraction of a plane electromagnetic wave by a black hole," *Intern. J. Theoret. Physics*, **17**, 189.

[ST80.1]   Stockton, A. 1980, "The lens galaxy of the twin QSO 0957+561," *Ap. J.*, **242**, L141.

[ST83.1]   Stocke, J.T., Liebert, J., Gioia, I.M., Griffiths, R.E., Maccacaro, T., Danziger, I.J., Kunth, D. & Lub, J. 1983, "The *Einstein Observatory* medium sensitivity survey: optical identifications for a complete sample of X-ray sources," *Ap. J.*, **273**, 458.

[ST84.1]   Straumann, N. 1984, *General relativity and relativistic astrophysics*, Springer, Berlin.

[ST86.1]   Stark, A.A., Dragovan, M., Wilson, R.W. & Gott III, J.R. 1986, "Observations of the cosmic background radiation near the double quasar 1146+111B,C," *Nature*, **322**, 805.

[ST87.1]   Stocke, J.T., Schneider, P., Morris, S.L., Gioia, I.M., Maccacaro, T. & Schild, R.E. 1987, "X-ray-selected AGNs near bright galaxies," *Ap. J.*, **315**, L11.

[ST88.1]   Stickel, M., Fried, J.W. & Kühr H. 1988, "AO 0235+164: the case of a gravitationally microlensed BL Lac object?" *Astr. Ap.*, **198**, L13.

[ST88.2]   Stickel, M., Fried, J.W. & Kühr H. 1988, "PKS0537-441: A gravitationally lensed blazar?" *Astr. Ap.*, **206**, L30.

[ST89.1]   Stickel, M., Fried, J.W. & Kühr, H. 1989, "The gravitational lens hypothesis for 0846+51 W1 supported by new observations," *Astr. Ap.*, **224**, L27.

[ST90.1]   Steidel, C.C. & Sargent, W.L.W. 1990, "A reexamination of the gravitational lens interpretation of the close QSO pair Q2345+007A,B: implications for the nature of the Lyman alpha forest clouds," *Astron. J.*, **99**, 1693.

[SU84.1]   Subramanian, K. & Chitre, S.M. 1984, "The quasar Q 2345+007 A,B: a case for the double gravitational lens?" *Ap. J.*, **276**, 440.

[SU85.1]   Subramanian, K. & Chitre, S.M. 1985, "Minilensing of multiply imaged quasars: Flux variations and vanishing of images," *Ap. J.*, **289**, 37.

[SU86.1]   Subramanian, K. & Cowling, S.A. 1986, "On local conditions for multiple imaging by bounded, smooth gravitational lenses," *M.N.R.A.S.*, **219**, 333.

[SU87.1]   Subramanian, K., Rees, M.J. & Chitre, S.M. 1987, "Gravitational lensing by dark galactic haloes," *M.N.R.A.S.*, **224**, 283.

[SU87.2]   Subramanian, K. & Chitre, S.M. 1987, "The intensity ratio in different wave bands between images of gravitationally lensed quasars as a probe of dark matter," *Ap. J.*, **313**, 13.

[SU87.3]   Surdej, J. Magain, P., Swings, J.-P., Borgeest, U., Courvoisier, T.J.-L., Kayser, R., Kellermann, K.I., Kühr, H. & Refsdal, S. 1987, "A new case of gravitational lensing," *Nature*, **329**, 695.

[SU88.1]   Surdej, J., Magain, P., Swings, J.-P., Borgeest, U., Courvoisier, T.J.-L., Kayser, R., Kellermann, K.I., Kühr, H. & Refsdal, S. 1988, "Observations of the new gravitational lens system UM673=Q0142-100," *Astr. Ap.*, **198**, 49.

[SU88.2]   Surdej, J., Swings, J.-P., Magain, P., Borgeest, U., Kayser, R., Refsdal, S., Courvoisier, T.J.-L., Kellermann, K.I. & Kühr, H. 1988, "Search for gravitational lensing from a survey of highly luminous quasars," in: [OS88.1]

[SU90.1]   Surdej, J. 1990, "Observational aspect of gravitational lensing," in: [ME90.2]

[SU90.2] Subrahmanyan, R., Narasimha, D., Pramesh Rao, A. & Swarup, G. 1990, "1830-211: gravitationally lensed images of a flat-spectrum radio core?" *M.N.R.A.S.*, **246**, 263.

[SW83.1] Swings, J.P. 1983, *Quasars and gravitational lensing*, Proceedings of the 24th Liège International Astrophysical Colloquium, University of Liège.

[SW86.1] Swarup, G. & Kapahi, V.K. 1986, *Quasars*, IAU Symp. 119, Reidel, Dordrecht.

[SY64.1] Synge, J.L. 1964, *Relativity: The general theory*, North-Holland Publ. Comp., Amsterdam.

[TH90.1] Thomas, P.A. & Webster, R.L. 1990, "Gravitational lensing and evolution in quasar absorption systems," *Ap. J.*, **349**, 437.

[TO89.1] Tomita, K. & Watanabe, K. 1989, "Gravitational lens effect on the cosmic background radiation due to nonlinear inhomogeneities," *Prog. Theor. Phys.*, **82**, 563.

[TO90.1] Tomita, K. & Watanabe, K. 1990, "Gravitational lens effect on the images of high redshift objects," *Prog. Theor. Phys.*, **83**, 467.

[TR81.1] Treder, H.-J. & Jackish, G. 1981, "On Soldner's value of Newtonian deflection of light," *Astron. Nachr.*, **302**, 275.

[TR85.1] Tricomi, F.G. 1985, *Integral equations*, Dover, Mineola.

[TR86.1] Traschen, J., Turok, N. & Brandenberger, R. 1986, "Microwave anisotropies from cosmic strings," *Phys. Rev.*, **34D**, 919.

[TU80.1] Turner, E.L. 1980, "The effect of undetected gravitational lenses on statistical measures of quasar evolution," *Ap. J.*, **242**, L135.

[TU84.1] Turner, E.L., Ostriker, J.P. & Gott, J.R. 1984, "The statistics of gravitational lenses: the distributions of image angular separations and lens redshifts," *Ap. J.*, **284**, 1.

[TU86.1] Turner, E.L., Schneider, D.P., Burke, B.F., Hewitt, J.N., Langston, G.I., Gunn, J.E., Lawrence, C.R. & Schmidt, M. 1986, "An apparent gravitational lens with an image separation of 2.6 arc min," *Nature*, **321**, 142.

[TU86.2] Tucker, W.H. & Schwartz, D.A. 1986, "The X-ray background and the evolution of quasars," *Ap. J.*, **308**, 53.

[TU88.1] Turner, E.L., Hillenbrand, L.A., Schneider, D.P., Hewitt, J.N. & Burke, B.F. 1988, "Spectroscopic evidence supporting the gravitational lens hypothesis for 1635+267 A,B," *Astron. J.*, **96**, 1682.

[TU90.1] Turner, E.L., Wardle, M.J. & Schneider, D.P. 1990, "A technique for detecting massive collapsed objects in the dark halo of the galaxy," *Astron. J.*, **100**, 146.

[TY81.1] Tyson, J.A. 1981, "Gravitational lensing and the relation between QSO and galaxy magnitude-number counts," *Ap. J.*, **248**, L89.

[TY83.1] Tyson, J.A. 1983, "Expected number of multiple QSOs from galaxy and QSO surface density data," *Ap. J.*, **272**, L41.

[TY84.1] Tyson, J.A., Valdes, F., Jarvis, J.F. & Mills Jr., A.P. 1984, "Galaxy mass distribution from gravitational light deflection," *Ap. J.*, **281**, L59.

[TY85.1] Tyson, J.A. 1985, "Image distortion near galaxies: Dwarfs or lensing?" *Nature*, **316**, 799.

[TY85.2] Tyson, J.A. & Gorenstein, M.V. 1985, "Resolving the nearest gravitational lens," *Sky & Tel.*, **70**, 319.

[TY86.1] Tyson, J.A. & Gullixson, C.A. 1986, "Colors of objects in the field of the double quasi-stellar object 1146+111B,C," *Science*, **233**, 1183.

[TY86.2] Tyson, J.A., Seitzer, P., Weymann, R.J. & Foltz, C. 1986, "Deep CCD images of 2345+007: Lensing by dark matter," *Astron. J.*, **91**, 1274.

[TY86.3] Tyson, J.A. 1986, "Galaxy clustering in QSO fields at $z = 1 - 1.5$: evidence for galaxy luminosity evolution," *Astron. J.*, **92**, 691.

[TY88.1] Tyson, J.A. 1988, "Deep CCD survey: galaxy luminosity and color evolution," *Astron. J.*, **96**, 1.

[TY90.1] Tyson, J.A., Valdes, F. & Wenk, R.A. 1990, "Detection of systematic gravitational lens galaxy image alignments: mapping dark matter in galaxy clusters," *Ap. J.*, **349**, L1.

[TY90.2] Tyson, J.A. 1990, "Lensing the background population of galaxies," in: [ME90.2]

[VA48.1] de Vaucouleurs, G. 1952, *Ann. d'Ap.*, **11**, 247.

[VA82.1] Vanderriest, C., Bijaoui, A., Félenbok, P., Lelièvre, G., Schneider, J. & Wlérick, G. 1982, "Photometrie de 0957+561; detection de variations a courte periode," *Astr. Ap.*, **110**, L11.

[VA83.1] Valdes, F., Tyson, J.A. & Jarvis, J.F. 1983, "Alignment of faint galaxy images: Cosmological distortion and rotation," *Ap. J.*, **271**, 431.

[VA86.1] Vanderriest, C., Wlérick, G., Lelièvre, G., Schneider, J., Sol, H., Horville, D., Renard, L. & Servan, B. 1986, "La variabilité du mirage gravitationnel P.G. 1115+080," *Astr. Ap.*, **158**, L5.

[VA89.1] Vanderriest, C., Schneider, J., Herpe, G., Chevreton, M., Moles, M. & Wlérick, G. 1989, "The value of the time delay $\Delta T(A, B)$ for the "double" quasar 0957+561 from optical photometric monitoring," *Astr. Ap.*, **215**, 1.

[VE83.1] Véron, P. 1983, "Quasar surveys and cosmic evolution," in: [SW83.1]

[VI81.1] Vilenkin, A. 1981, "Cosmological density fluctuations produced by vacuum strings," *Phys. Rev. Lett.*, **46**, 1169.

[VI81.2] Vilenkin, A. 1981, "Gravitational field of vacuum domain walls and strings," *Phys. Rev.*, **23D**, 852.

[VI83.1] Vietri, M. & Ostriker, J.P. 1983, "The statistics of gravitational lenses: Apparent changes in the luminosity function of distant sources due to passage of light through a single galaxy," *Ap. J.*, **267**, 488.

[VI85.1] Vietri, M. 1985, "The statistics of gravitational lenses. II. Apparent evolution in the quasars' luminosity function," *Ap. J.*, **293**, 343.

[VI85.2] Vilenkin, A. 1985, "Cosmic strings and domain walls," *Phys. Rep.*, **121**, 263.

[VI86.1] Vilenkin, A. 1986, "Looking for cosmic strings" *Nature*, **322**, 613.

[VI90.1] Vilenkin, A. 1990, "Kinky cosmic strings," *Nature*, **343**, 591.

[WA35.1] Walker, A.G. 1935, "On Riemannian spaces with spherical symmetry about a line and the conditions for isotropy in general relativity," *Quant. Journ. Math. Oxford Sci.* **6**, 81.

[WA79.1] Walsh, D., Carswell, R.F. & Weymann, R.J. 1979, *Nature*, **279**, 381.

[WA84.1] Wampler, E.J., Gaskell, C.M., Burke, W.L. & Baldwin, J.A. 1984, "Spectrophotometry of two complete samples of flat radio spectrum quasars," *Ap. J.*, **276**, 403.

[WA84.2] Wald, R.M. 1984, *General relativity*, University of Chicago Press, Chicago.

[WA85.1] Wampler, E.J. & Ponz, D. 1985, "Optical selection effects that bias quasar evolution studies," *Ap. J.*, **198**, 448.

[WA86.1] Wagoner, R.V. 1986, in: *Physical Cosmology*, R. Balian, J. Audouze & D.N. Schramm (eds.), p.179, North Holland.

[WA89.1] Wambsganss, J., Giraud, E., Schneider, P. & Weiss, A. 1989, "A gravitational lens explanation for the arc-like structure in Cl0500-24," *Ap. J.*, **337**, L73.

[WA89.2] Walsh, D. 1989, "0957+561: the unpublished story," in: [MO89.1]

[WA89.3] Wambsganss, J., Paczyński, B. & Katz, N. 1989, "Microlensing model for QSO 2237+0305," in: [MO89.1]

[WA90.1] Wambsganss, J., Paczynski, B. & Schneider, P. 1990, "Interpretation of the micro-lensing event in QSO 2237+0305," *Ap. J. (Letters)*, **358**, L33.

[WA90.2] Wambsganss, J., Paczyński, B. & Katz, N. 1990, "A microlensing model for QSO 2237+0305," *Ap. J.*, **352**, 407.

[WA90.3] Watanabe, K. & Tomita, K. 1990, "Cosmological observations in an inhomogeneous universe: distance-redshift relation," *Ap. J.*, **355**, 1.

[WA90.4] Wambsganss, J. 1990, "Gravitational microlensing," report MPA 550, Garching.

[WA90.5] Watanabe K. & Sasaki, M. 1990, "The role of shear in the cosmologial distance," *Hiroshima University preprint* RRK 90-8, March.

[WE72.1] Weinberg, S. 1972, *Gravitation and cosmology*, Wiley, New York.

[WE76.1] Weinberg, S. 1976, "Apparent luminosities in a locally inhomogeneous universe," *Ap. J.*, **208**, L1.

[WE79.1] Weymann, R.J., Chaffee Jr., F.H., Davis, M., Carleton, N.P., Walsh, D. & Carswell, R.F. 1979, "Multiple-mirror telescope observations of the twin QSOs 0957+561 A,B," *Ap. J.*, **233**, L43.

[WE80.1] Weymann, R.J., Latham, D., Angel, J.R.P., Green, R.F., Liebert, J.W., Turnshek, D.A., Turnshek, D.E. & Tyson, J.A. 1980, "The triple QSO PG1115+08: Another probable gravitational lens," *Nature*, **285**, 641.

[WE81.1] Weymann, R.J., Carswell, R.F. & Smith, M.G. 1981, "Absorption lines in the spectra of quasistellar objects," *Ann. Rev. Astr. Ap.*, **19**, 41.

[WE82.1] Weedman, D.W., Weymann, R.J., Green, R.F. & Heckman, T.M. 1982, "Discovery of a third gravitational lens," *Ap. J.*, **255**, L5.

[WE85.1] Webster, R.L. 1985, "Gravitational lensing and galaxy shape," *M.N.R.A.S.*, **213**, 871.

[WE86.1] Webster, R. & Fitchett, M. 1986, "Views with a gravitational lens," *Nature*, **324**, 617.

[WE86.2] Weedman, D.W. 1986, *Quasar astronomy*, Cambridge University Press, Cambridge.

[WE88.1] Webster, R.L., Hewett, P.C. & Irwin, M.J. 1988, "An automated survey for gravitational lenses," *Astron. J.*, **95**, 19.

[WE88.2] Webster, R.L., Hewett, P.C., Harding, M.E. & Wegner, G.A. 1988, "Detection of statistical gravitational lensing by foreground mass distributions," *Nature*, **336**, 358.

[WE90.1] Webster, R.L. & Hewett, P.C. 1990, "Quasar-galaxy associations," in: [ME90.2]

[WE91.1] Weir, N. & Djorgovski, S. 1991, "High-resolution imaging of the double QSO 2345+007," *Astron. J.*, **101**, 66.

[WH55.1] Whitney, H. 1955, "Mappings of the plane into the plane," *Ann. Math.*, **62**, 374.

[WI80.1] Wills, B.J. & Wills, D. 1980, "Spectrophotometry of the double QSO, 0957+561," *Ap. J.*, **238**, 1.

[WI88.1] Will, C.M. 1988, "Henry Cavendish, Johann von Soldner, and the deflection of light," *Am. J. Phys.*, **56**, 413.

[WI90.1] Witt, H.J. 1990, "Investigation of high amplification events in light curves of gravitationally lensed quasars," *Astr. Ap.*, **236**, 311.

[WO88.1] Wood, K.S., Ftaclas, C. & Kearney, M. 1988, "The effect of gravitational lensing on pulsed emission from neutron stars exhibiting quasi-periodic oscillation," *Ap. J.*, **324**, L63.

[WU90.1] Wu, X. 1990, "Multiple gravitational lensing and flux conservation," *Astr. Ap.*, **232**, 3.

[WU90.2] Wu, X. 1990, "The magnitude-redshift relation in a locally inhomogeneous universe," *Astr. Ap.*, **239**, 29.

[YA89.1] Yanny, B., York, D.G. & Gallagher, J.S. 1988, "The extent of nebular emission associated with the $z = 0.525$ absorber near AO 0235+164," *Ap. J.*, **338**, 735.

[YE87.1] Yee, H.K.C. & Green, R.F. 1987, "Surveys of fields around quasars. IV. Luminosity of galaxies at $z \approx 0.6$ and preliminary evidence for the evolution of the environment of radio-loud quasars," *Ap. J.*, **319**, 28.

[YE88.1] Yee, H.K.C. 1988, "High-resolution imaging of the gravitational lens system candidate 2237+030," *Astron. J.*, **95**, 1331.

[YO80.1] Young, P., Gunn, J.E., Kristian, J., Oke, J.B. & Westphal, J.A. 1980, "The double quasar Q0957+561 A,B: A gravitational lens image formed by a galaxy at $z = 0.39$," *Ap. J.*, **241**, 507.

[YO81.1] Young, P. 1981, "Q 0957+561: Effects of random stars on the gravitational lens," *Ap. J.*, **244**, 756.

[YO81.2] Young, P., Deverill, R.S., Gunn, J.E., Westphal, J.A. & Kristian, J. 1981, "The triple quasar Q1115+080A,B,C: A quintuple gravitational lens image?" *Ap. J.*, **244**, 723.

[YO81.3] Young, P., Gunn, J.E., Kristian, J., Oke, J.B. & Westphal, J.A. 1981, "Q 0957+561: Detailed models of the gravitational lens effect," *Ap. J.*, **244**, 736.

[YO81.4] Young, P., Sargent, W.L.W., Boksenberg, A. & Oke, J.B. 1981, "The origin of a new absorption system discovered in both components of the double QSO Q 0957+561," *Ap. J.*, **249**, 415.

[ZE64.1] Zel'dovich, Ya.B. 1964, "Observations in a universe homogeneous in the mean," *Sov. Astr.*, **8**, 13.

[ZE70.1] Zel'dovich, Ya.B. 1970, "Gravitational instability: an approximate theory for large density perturbations," *Astr. Ap.*, **5**, 84.

[ZE80.1] Zeldovich, Ya.B. 1980, "Cosmological fluctuations produced near a singularity," *M.N.R.A.S.*, **192**, 663.

[ZE87.1] Zensus, J.A. & Pearson, T.J. 1987, *Superluminal radio sources*, Cambridge University Press, Cambridge.

[ZU85.1] Zuiderwijk, E.J. 1985, "The gravitational lens effect and the surface density of quasars near foreground galaxies," *M.N.R.A.S.*, **215**, 639.

[ZW37.1] Zwicky, F. 1937, "Nebulae as gravitational lenses," *Phys. Rev.*, **51**, 290.

[ZW37.2] Zwicky, F. 1937, "On the probability on detecting nebulae which act as gravitational lenses," *Phys. Rev.*, **51**, 679.

Index of Individual Objects

0023+171 49, 63–64, 280
0142−100 (UM673) 49, 60, 404, 503
0218+357 49
0235+164 425, 434
0307−195 503
0414+0534 49, 71, 279
0846+51W1 88, 425
0957+561 10, 48–58, 66, 90, 500
− absorption lines 48, 53, 502
− lens galaxy and cluster 50, 466, 478, 482
− microlensing 10, 54, 57, 421, 482
− "missing" third image 58, 466
− models 52, 55–57, 279, 481–482
− time delay measurements and Hubble constant 45, 56–58, 468, 476–483
− VLBI structure 51–54, 307, 469, 476, 481
1115+080 49, 64–66, 279, 445
− merging images 65, 191
− variability 66
1120+019 (UM425) 71, 279, 445
1131+0456 49, 77–82, 90
− image reconstruction 79–82, 307, 492
1145−071 87
1146+111 84, 86–87, 511
1156+295 88, 425
1343+164 87
1413+117 49, 70–71, 279, 404, 445
1548+114A,B 512
− upper limit of QSO mass 512
1549+3047 49
1635+267 59–60, 84, 86, 280, 404
1654+1346 49, 82–83
− determination of lens mass 83
1830−211 49, 83

2016+112 10, 49, 60–63, 402
− lens models 62, 279–280
2237+0305 49, 66–70, 445, 509
− emission line profile differences 431–432
− microlensing 69–70, 88, 344, 425–429, 431–432
− models 68–69, 279, 425
− statistical considerations 67
2345+007 58–59, 86, 280, 500, 502–503
3C238 85
3C324 85
3C345 88, 425, 513
Abell 370 49, 72–74, 484, 494, 496, 499
− arclets 74, 484, 493
− dark matter in cluster 74
− photometry and spectroscopy 73–74, 493
Abell 545 49, 84
Abell 949 85
Abell 963 49, 75–76, 492–493
Abell 1525 84
Abell 1689 84
Abell 2218 49, 72, 77, 84
Abell 2390 49, 76–77, 494
− straight arc and beak-to-beak singularity 77
Cl0024+16 49, 84
Cl0302+17 49
Cl0500−24 49, 84, 494–496
− straight arc and beak-to-beak singularity 77
Cl1409+52 89
Cl2244−02 49, 72, 74, 278, 493, 496, 498
VV172 447

545

Subject Index

Abel integral equation 506
Abell catalog 72
Aberration 110–111
Absorption lines *see* QSO absorption lines
Accretion discs 428, 515
Achromaticity of GLs 44, 123, 161
Action principle for light rays 100
Active galactic nuclei 23
– size of emission regions 379
– X–ray selected 380, 406, 446
– *see also* QSOs, Seyfert galaxies, BL Lac objects
Adiabatic invariants of electromagnetic field packets 100
Affine parameter of light rays 95, 104, 112, 132, 134–137
Age of the universe 130
Amplification bias 5, 37, 39, 143, 309, 366, 373–378, 408, 419, 435–448, 503
– deconvolution 442–444
– dependence on source size 362, 377, 435–437
– due to microlensing 377, 386, 394–395, 436–442, 446
– effects on source counts 88, 374–377, 438, 440
– – of high–redshift radio galaxies 85–86, 446
– – of QSOs 8, 378, 418
– influence on GL statistics 84, 89, 279, 380, 385, 389–394, 403–404
– redshift dependence 373, 377, 446
– signature 444–447
– *see also* QSO–galaxy associations
Anastigmatic focusing 109, 234, 286
– *see also* Ricci focusing
Angular diameter distance 22, 26, 110, 114, 116, 132, 134–143, 146
– anisotropy 138
– definition 110
– in clumpy universe 137–143, 345, 349
– in lens equation 127, 147, 157, 283

– in smooth Friedmann–Lemaître universe 137, 142, 145, 345
– *see also* Dyer–Roeder equation
Angular resolution bias 380, 402–403
Angular separation of images 4, 27, 49, 312, 393, 396–398, 401, 464–466
APM survey 403
Arclets 74, 77, 484, 493, 499–500
Arcs 10, 46–47, 72–77, 483–500
– and counterarcs 73, 75, 484, 492–493, 498
– curvature 485
– formation by nearly symmetric lenses 488–491
– observations and models *see* individual objects
– statistics 498–500
– straight arcs 77, 178, 494
– supernovae in arcs 497
Area distortion of lens mapping 33, 161–162, 165, 319
Arnold, V.J. 184
Arp, H. 404
Arrival time difference *see* Time delay
Arrival time surface *see* Wavefront
Astigmatic focusing 109, 234
– *see also* Weyl focusing
Astroid 490
Average flux in inhomogeneous universes 133
Axi-symmetric GLs 229–249, 296
– critical curves and caustics 213, 233–234
– deflection angle and potential 230–232
– Fermat potential and lens equation 231–232
– general properties 230–238
– Jacobian matrix 232–233
– multiple image diagram 237
– wave optics for 220–222

Background galaxy population 500

547

Background radiation *see* Cosmic microwave background, X-ray background
Baldwin effect 433
Barnothy, J.M. 7
Beak-to-beak catastrophe *see* Catastrophes
Beam of photons 104–117
- expansion and contraction 107
- propagation 108
- shear 108
Bifurcation, bifurcation surfaces 198, 288
BL Lac objects 23
- associations with foreground galaxies 88, 407, 434
- GL relation to QSOs 7, 88, 433–434
- microlensing hypothesis 433–434
- redshifts 433–434
- variability, and microlensing 88, 433
Black holes 2, 123, 445, 457, 514
Blandford, R.D. 8, 452
Bolometric luminosity 116
Bourassa, R.R. 9
Bright Quasar Survey 64
Brightness profile of sources 215, 238
- one dimensional 506–508
- reconstruction from microlensing light curve 504–508
Brightness ratio of lensed images 33, 301, 380–394, 399
Broad absorption line QSOs 71, 501
Broad emission lines 378, 429–433
- broad line region 420, 430–434
- effects of microlensing 44, 71, 429–433
- model profiles 431–432
Burbidge, G. 67
Burke, B.F. 23
Butterfly catastrophe *see* Catastrophes

C points, cusp point 252–255, 257–258
Canonical forms of singularities 187
Carswell, R.F. 9, 48
Cassini ovals 257
Catastrophes
- beak-to-beak 207–211, 213, 262, 264, 288, 339
-- models for straight arcs 77
- butterfly 194, 265, 342
- catastrophe tree 213–214
- cusp 192–198
-- even number of cusps theorem 213
-- generic arc models 264, 278, 485

-- local solution of lens equation 195–196
-- naked cusp 278, 492
-- relation between image magnifications 196, 319
- elliptic umbilic 199–203, 213, 288, 331, 344
- fold 186–192, 194, 197, 234
-- angular separation of images 190
-- magnification distribution 190
-- time delay of images 191
- hyperbolic umbilic 199, 201–203, 213
-- in generic 'elliptical' lenses 278
- lips 207–211, 213, 274–276, 400
-- in generic 'elliptical' lenses 278
- stability and genericity 197–198, 202, 207, 211–212
- swallowtail 194, 203–207, 213, 265
-- in microlensing 339
- Whitney's theorem 197
- wigwam 194
- *see also* Singularities
Catastrophe theory 184, 198, 212, 288
Caustic 34–35, 112, 115, 134, 163–165, 188, 193, 277–278
- convexity 192, 344, 365
- degenerate caustic 200, 212, 233, 237, 400, 484
- different notions of 198
- for special lens models 252–254, 258–260, 262, 266, 485–488
- inside and outside 183
- large caustic 198, 201–202, 206, 211–212, 242
- magnification of extended source 238
- radial and tangential caustic 237, 278, 400
- tangent vector 186, 206, 486
Cavendish, H. 1–2
cD galaxy 42, 75, 493
Center for Astrophysics redshift survey 66–67
Central limit theorem 325
Centrally condensed lens 236–237, 244–247, 270–271, 273–274, 396, 461
Chang, K. 9, 483
Chang–Refsdal lens 213, 255–261, 357, 420, 436
- caustic and critical curve 257–260, 316, 344
- magnification cross-section 315–320
- scaling of cross-sections 316–318
Change of redshift along ray 135

548

Characteristic angular and length scale 28
Characteristic function 321–323
Characteristics of Maxwell's equations 95
Christoffel symbols *see* Connection coefficients
Chwolson, O. 4
Classification of active galactic nuclei 433
Classification of ordinary images 163, 285
– *see also* Type of images
Classification of singularities *see* Catastrophes
Closure density of the universe 22, 129
Clumpiness parameter *see* Smoothness parameter
Clumpy universe 131, 136–140, 176, 281
– definition of magnification 351–352
– flux conservation 132–134, 374
– matching to Friedmann–Lemaître background 132–134, 349
– optical depth 347–348
– self-consistent probabilities 345–348, 357–358
– *see also* Dyer–Roeder equation
Clusters of galaxies 50, 364, 402, 466, 492–496
– dark matter distribution 89–90, 131, 312, 388, 500
– gravitational potential 449
– intracluster gas 312
COBE 513
Coherence of light 150, 155, 217, 219
Coherence time 219
Coherent alignment and distortion 452, 499–500
Comoving density 345, 373, 381
Compact objects as lenses 88, 381–385
– compact halo objects (MACHOS) 6, 40, 456–459
– – detectability and time-scales 39, 458–459
– cosmologically distributed 381–385, 436–442, 453
– cosmologically significant density 385
– dark matter candidate 8, 445, 453
– in lens galaxies 9, 320, 394–395, 407, 411–415, 419
– – *see also* Chang–Refsdal lens
– *see also* Microlensing, Schwarzschild lens
Compact radio sources 42, 512
– variability 447
Complex amplitude of light bundles 97, 152
Complex scattering function 160, 267
Complex signal 154
Confirmation of lensing events 43, 46–47, 88
Conformal curvature tensor *see* Weyl tensor
Conformal factor 92, 95, 103, 128, 151
Conformal invariance 91–92, 98, 101, 109, 150–152
Conformal time 144
Conformally static spacetimes 104, 149, 151
Conformally stationary spacetimes 96, 103
Conjugate points 103, 115, 150, 163, 234, 243
– multiplicity 115, 163, 234
Connection coefficients 91, 119–120
Connection vectors 105
– transport equation 106
Conservation law
– covariant conservation law 92
– of energy–momentum tensor 92–93, 123
– of photon number 100, 115, 132, 161
– *see also* Flux conservation
Constant of gravity 21
Construction of images 298
– geometrically 488
– numerically 298, 488–491
Continuum emission region 39, 373, 428
Control parameter 184, 198, 211–212
Convergence and Ricci focusing 162–163
Convexity of caustics 192, 344, 365
Cooke, J.H. 9
Core size of lenses 244, 395, 434
– influence on lensing properties 395–400
Corrected luminosity distance 110, 113–116, 132, 141
Cosmic microwave background 131, 342, 511
– anisotropy limits 131, 513
– isotropization by lensing 514
Cosmic redshift 128
Cosmic scale function and time 128–131, 144
Cosmic strings 86, 472, 509–511
Cosmological constant 22, 120, 129–131
Cosmological density 509

- density parameter 22
- linear and nonlinear perturbations 364

Cosmological distances 22
- see also Angular-diameter distance, Luminosity distance

Cosmological evolution 309
- of AGNs 379–380

Cosmological Fermat potential and lens mapping 146–147
- see also Fermat potential

Cosmological models 22
- fine-grained models 131–132
- Friedmann–Lemaître models 127–132

Counterarcs 74–76

Critical density 22, 129, 436

Critical curve 34–35, 185, 193
- critical images 164–165, 318
- critical points 162, 184
- definition 165
- for special lens models 244–248, 254, 257–260, 262, 266, 269, 485–486
- merging images 35, 66, 165, 183, 189–190, 246
- radial and tangential critical curve 233–235, 237, 244–248, 251, 396, 399
- tangent vector 186, 188, 192, 204
- see also Catastrophes

Critical surface mass density 158, 177, 397, 492

Cross-hairs imaging 299

Cross-section 23, 310–320, 345–347
- general definition 311
- for magnification 313–320, 361, 400
-- asymptotic behavior 318–320, 332, 394
-- of extended sources 313–318
-- of point sources 318–320, 386
-- of Schwarzschild lens 310, 381
- for multiple images 311–313, 381, 386, 396, 399–401
- no overlap condition 332, 346, 375, 391

Curvature tensor 91
- Weyl and Ricci part 109

Cusp singularity see Catastrophes

Cuspoids 194

Dark lenses 42, 59, 461, 465–466, 500

Dark matter 131, 364, 388, 395, 453
- constraints from nucleosynthesis 445
- dynamics of galaxy clusters 89, 131
- in our Galaxy 456–459

de Vaucouleurs profile 68, 398

Deceleration parameter 130

Deconvolution of microlensing light curve 504–508

Deflection angle 32, 124
- asymptotic behavior 160, 172
- at the Sun 2–3, 21, 514
- general expression 125, 158
- mathematical properties 160
- scaled deflection angle 22, 158–159, 230–231, 284

Deflection potential 22, 127, 159–160, 184
- asymptotic behavior 160
- for symmetric lenses 231–232
- mathematical properties 160

Deflection probability see Microlensing

Deflector sphere 30

Degenerate caustic 200, 233–234, 237–239, 252

Demagnification, in core of lens galaxies 175, 278

Density enhancement see QSO–galaxy associations

Density parameter 22, 130

Determination of parameters from lensing 5, 55, 79, 147–148, 468–483
- Hubble constant 7, 56–57, 148, 468, 473–483
- lens mass 5, 7, 83, 473–476, 493–494
- mass distribution in clusters 58, 74, 492–497

Deviation vector 111–113
- see also Jacobi vector

Differential deflection 152, 161

Differential magnification 63, 430

Differential reddening 64

Diffraction pattern due to source near caustic 224

Dimensionless observables 166, 468

Disks as lenses 240–243

Distances in curved spacetimes 26, 110–115, 132–134
- concepts 110
- distance–redshift relation 132–143
- see also Affine parameter, Angular-diameter distance, Luminosity distance

Distortion of images 165, 234–235, 286, 448–452, 500
- see also Arclets, Arcs

Distortion of light bundles 33–34, 161, 448

Dragging of inertial frames 124

Dyer, C.C. 8, 138, 398

Dyer–Roeder equation 137–143, 348
- angular-diameter distance 140, 146, 158, 345, 349
- as integral equation 352–353
- comparison of solutions 141, 353
- explicit solutions 141–143
- prescription for light propagation 8, 345
- see also Clumpy universe
Dynamic range 311, 380, 402

Eddington, A.S. 4
Effective energy tensor
- of locally plane waves 99
- of the vacuum 120
Effective index of refraction 104, 123
Effective transverse velocity 421
Eikonal equation 94–96, 152
Einstein, A. 1, 3–4
- Einstein curvature tensor 91, 120
- Einstein deflection angle 3, 21, 25, 124, 231
- Einstein radius 240
- Einstein ring 4–6, 10–11, 29, 69, 77–83, 242, 280, 402, 493
- Einstein's field equation 119
-- linearized 121–123
Electromagnetic field 91, 150, 154
Elementary catastrophes see Catastrophes
Ellipsoidal mass distribution see Elliptical lenses
Elliptical umbilic singularity see Catastrophes
Elliptical galaxies 266–267, 449–451
Elliptical lenses 277–280
- generic properties 277–280
- nearly elliptical lenses 271–274
- with elliptical mass distribution 266–268
- with elliptical potential 180–181, 266–271, 274–277, 296, 399–401
Ellipticity
- of galaxies 449–451
- of images 449–451
- of radio galaxies 451
Elongated images 451, 485
- see also Arcs, Arclets
Emission lines see Broad emission lines
Empty cone 134, 139, 149–150, 175, 349, 351–352, 358–359, 374, 463
- see also Clumpy universe, Dyer–Roeder equation
Energy–momentum tensor 120

- conservation law 92
- of electromagnetic field 92
- of locally plane wave 99
Enhancement factor see QSO–galaxy associations
Equivalent width 53, 58, 62, 433, 502–504
Escape velocity 1
Etherington's reciprocity law 111, 116
- for Dyer–Roeder distance 141
Euler–Lagrange equation for light rays 124
Event horizon 2, 176
Evolution
- of absorption line systems 504
- of galaxies 131, 405
- of QSOs 378–380, 435
Expansion of photon beam 108
Extended source 435
- image distortion 165, 234, 263, 448–452
-- see also Arclets, Arcs
- maximum magnification 215–216, 314, 435
- seen through star field
  see Microlensing
Extinction
Extragalactic nebulae 5, 309, 380

F points, fold points 252–255, 257–258
Faber–Jackson relation 387–388, 398, 417
Family of centrally condensed lenses 244–247
- with elliptical potential 270–271, 273–274
Faraday rotation 78, 80, 99
Fermat potential 22, 126–127, 152, 159, 163, 177–179, 184, 299, 462
- asymptotic behavior 161
- for Schwarzschild lens 220
- for symmetric lenses 231–232
- in cosmology 146–149
- interpretation 170–171
- local expansion 173, 186–188, 193, 204
- mathematical properties 160–161
- relation to time delay 171
Fermat's principle 100–104, 125, 143, 147
- in conformally stationary spacetimes 103, 123
- in GL theory 159, 166, 171, 288–290

551

Flux conservation 35–36, 132–134, 150, 352, 374, 413, 435
- in microlensing 327, 412
Flux ratio of images 33, 301, 380–394, 399
Flux-limited samples 37, 85, 309, 390, 406
Flux–redshift relation 135, 143
- see also Distances in curved spacetimes
Focusing 109, 113–114, 143
- focusing equation 109, 134, 137, 163
- focusing properties of ray bundles see Optical scalars
- focusing theorem 132–134, 176
Fold singularity see Catastrophes
Four–momentum and velocity 121
Fraction of lensed QSOs 388–393
Frequency shift 96–97, 149
Frequency vector of locally plane waves 94, 135
Fresnel–Kirchhoff diffraction integral 153
Freundlich, E. 3
Friedmann equation 129–131, 136
Friedmann–Lemaître model of the universe 22, 127–132, 480
Fringes 222, 224–225
Fugmann, W. 406
Fundamental observers 128

Galaxies
- clustering 418, 465
- in lens clusters 73, 493, 496–497
- mass distribution 267
Gamma-ray bursts 460
- lensing effects 460–461
Gauge transformation of phase function 98
Gaussian brightness profile 216, 427
Gaussian thin lens 473
General lensing situation 30–32
General relativity 1
- tests 3
Generalized quadrupole lens 291–293
Generic mappings 197
Genericity of singularities 197, 212–213
- see also Catastrophes
Geodesic deviation 111, 113
- see also Jacobi equation
Geodesics 95
- geodesic equation 95, 100
Geometrical optics approximation 6, 150

- validity 95, 162, 217, 227
Geometrical time delay see Time delay
Geometrically thin lens 32, 125, 164, 172, 281
GL (=gravitational lens) candidate systems
- confirmation 43, 46–47
- expectations 42, 46
- see also individual objects
Grand Unification Theories 509–510
Granularity of lenses 394–395
- see also Microlensing
Gravitational field equation see Einstein's field equation
Gravitational field strength 120
Gravitational potential 120, 122, 149–151, 166
- vector potential 123
Gravitational redshift 136, 514
Gravitational waves 121, 510
Gravitomagnetic field 124
Grid search method 300–303
Gunn, J.E. 8–9, 384

Halos of galaxies 386
- our Galaxy 456–459
- see also Microlensing, Compact objects as lenses
Hamilton's equations 95, 100
- Hamiltonian 95
Hammer, F. 81
Handedness of images 164–165
Helmholtz equation 152
Hierarchical tree method 305–307
High magnification
- cross-section and probability 318–320
- events 391
- see also Microlensing
High redshift objects 446, 483
HLQ survey 403–404
Hubble atlas of galaxies 405
Hubble constant 22, 129–130
- determination from lensing 7, 148, 468
- – general method 473–476
- – in 0957+561 56–57, 476–483
- Hubble law 129, 140
- Hubble length, time, and volume 22, 144, 387
Huyghens' principle 153
Hyperbolic umbilic singularity see Catastrophes

Image configuration 279–280

- allied and opposed images 399
- topography 177–181

Images formed by lensing
- image distortion 54, 165, 448–452
-- near critical curve 234, 483
-- see also Arclets, Arcs
- multiplicity 253–255, 260, 463
- observables 468–469
- of extended source 263, 298, 448–452
-- see also Arcs

Implicit function theorem 185, 192, 196, 203, 211
Index of a vector field 173, 285
- index theorem 172–174
- of ordinary images 173–174
Index of refraction 104, 123, 152
Inflationary universe 132, 509
Infrared divergence 325, 334
Intensity see Specific flux and intensity
Interference of light 217, 219
- near fold catastrophe 222–227
Intrinsic variability versus microlensing 88, 424–425, 433–434
Invariance transformations 471–473, 480–481
Inversion of the lens equation 158, 194, 229, 295
Isochrones 178–181
Isothermal sphere 245, 359–362
- core radius 273, 386, 395
- lens model 245, 273, 395–397, 481
- quadrupole perturbation 491
- see also Singular isothermal sphere

Jacobi vector, map, equation 111–115
Jacobian determinant of lens equation 162, 185
Jacobian matrix of lens equation 23, 33, 162
- for multiple deflection 285
-- asymmetry 285–286
- for quadrupole lenses 251, 257
- for random star field 328, 330
- for symmetric lenses 232–233
- geometrical meaning of eigenvalues 165, 235
- kernel 192, 194, 206
- rank at critical points 185, 203, 234
Jaroszyński, M. 452
Jodrell Bank 48

k-correction 454
Kantowski, R. 8–9
King profile 52, 398

Kirchhoff integral 151–153
Kitt Peak National Observatory 72
Klimov, Yu.G. 6
Kovner, I. 100

Lagrange equation 100
Laplace, P.S. 1–2
Large caustic see Caustic
Large-scale matter distribution 342, 364, 466, 480
- image distortion 451–452
- pancake model 210
Last scattering surface 513
Le Fevre, O. 81
Lemaître, G. 129
Lemniscate 178
Lenard, P. 3
Lens equation (lens mapping) 31, 127, 157–161
- as gradient mapping 127, 159
- containing observer position 167–169
- for random star field 304, 327
- general properties 157–181
- in cosmology 135, 143–150
- singularities see Catastrophes
- surjectivity 161, 183
Lens models
- construction from resolved images 307–308
- for clumpy universe 349–350
- see also Axi-symmetric GLs, Two point-mass lens, Elliptical lenses, Quadrupole lenses
Lens plane 31, 125, 151, 157
Lens redshift distribution 383–384, 397
Lens statistics see Statistical lensing
Lens surveys see Surveys for GLs
Lens-induced variability 39, 88, 420–425
- see also Microlensing
Lensing cross-section see Cross-section
Lensing within our Galaxy 40, 456–459
Levi-Civita connection see Connection coefficients
Lick catalogue of galaxies 418–419
Liebes, S. 6
Light cone
- caustic 112, 115, 198, 242
- future and past (half-) light cone 91, 111, 115, 132
Light deflection at the Sun 2–3, 25, 514
Light propagation in inhomogeneous universe 7, 132–143, 348–366, 480, 514
- as lens problem 349–350

- magnification probabilities 347–366
- numerical simulations 140, 349–351, 355–366
- weakly affected light bundles 353–355
- with galaxies 359–363
- with point-mass lenses 350–359, 364–366
- see also Clumpy universe, Dyer–Roeder equation, Optical scalars

Light rays 94–96, 100
Light travel time 166
Limaçon 179
Line-of-sight velocity dispersion see Velocity dispersion
Linearized field equations 121–123
Lips singularity see Catastrophes
Locally inhomogeneous universe 7–8, 131
- see also Clumpy universe

Locally plane wave 93–100
- amplitude and phase 93, 97
- circular frequency 93
- effective energy–momentum tensor 99
- frequency and wave vector 94

Lodge, O.J. 3
Luminosity distance 22, 110, 116, 132–133, 150
Luminosity function
- of galaxies 387
- of QSOs 373, 375, 378–380, 408–409
Luminous arcs see Arcs
Lynds, R. 10, 72

Macrolensing 256
Magnification 5, 23, 33–36, 149–150, 161–163, 351, 445
- as image distortion 33–34, 161–165
- cross-sections 313–320
-- for isothermal sphere 312–313
-- for point-mass lens 310, 313–315
- events see Microlensing
- in wave optics 218–227
- matrix see Jacobian matrix of lens equation
- maximum value in wave optics 35, 220, 222–224
- near fold catastrophes 190–192, 215–216, 222–227, 246, 318–320
- near isolated caustic points 220–222, 237–239, 246
- of extended source 35, 38, 162, 215–216, 238, 240, 303, 313–318, 505
- pattern see Microlensing

- variability see Microlensing
Magnification theorem 35, 132, 172–175
- generalization 176
- proof 175
Magnification-multiplicity conjecture 445
Magnitude–redshift relation 143
Mandl, R.W. 4–5
Marginal lenses 274–277
- with elliptical potentials 274–277, 399–400
Markov's method 321–322
Mass estimate of GLs 74, 83, 451, 468, 473–477, 479, 493
Mass-to-light ratio 68, 74, 85, 89, 425, 461, 471
Matter content of the universe 129
Matter tensor 120
- for perfect fluid 120, 122
- of the universe 129
Maximum probabilities 366–369, 408–411
Maxwell's equations 91–92, 150, 152
Maxwell's stress tensor 92
McGraw, J. 455
Medium Sensitivity Survey 380, 406
Merging of images 35, 183, 190, 195, 466
Metamorphoses 198–215
- see also Catastrophes
Metric of spacetime 91, 119
- Friedmann–Lemaître metric 128–129
- weak-field approximation 150
Michell, J. 1–2
Microlensing (ML) 43, 62, 87–88, 215, 255, 394, 413–415, 419–434
- collective effects 339, 343, 423
- diagnostics and detection 420–421, 433
- hierarchical tree method 305–307
- high magnification events 88, 427–428
- magnification pattern 338–344, 429
- magnification probability
-- analytical estimates 332
-- asymptotic expression 333–334
- microimages 327, 333, 338, 419
- morphology of macroimages 326–328
- observed cases
-- in 0957+561 421, 482
-- in 2237+0305 88, 420, 425–429
- optical depth 338, 342–344, 423, 434
- ray shooting simulations 304–307, 338–344
- synthetic light curves 88, 422–428

- typical scales 39, 419, 423
- *see also* Numerical methods, Random star field

Microwave background radiation *see* Cosmic microwave background

Minkowski metric 91, 121

Missing image problem 175, 278

Missing lens problem 58–59, 461–466

MIT–Green Bank survey 71, 400

Model fitting 89–90, 277, 299–300
- *see also* Numerical methods

Model light curves *see* Microlensing

Monopoles and domain walls 132, 509

Monte Carlo simulations 337, 385, 451, 462

Morse index 218

Morse point 184

Multiple images
- cross-sections 311–313
- necessary and sufficient conditions 177, 236
- statistics *see* Statistical lensing
- topography 177–181

Multiple lens plane theory 281–293, 350
- conditions for validity 281–282
- deflection potential 284
- generalized quadrupole lens 282, 291–293
- Jacobian matrix 285
- lens equation 282–284
- – derived from Fermat's principle 288–290
- rotation of images 282, 286

Mutual coherence of lensed images 217

N point-mass lens 265–266

Naked cusp 278, 492

Nearly axisymmetric lens 485–492

Neutrino 509

Neutron stars 123, 460, 514

New Technology Telescope (NTT) 81

Newton, I. 1

Newton's approximation scheme 266, 302

Newtonian cosmology 129

Newtonian potential 122, 149

Nieto, J.-L. 405

Non-linearity of statistical lensing 321, 342, 355

Notation 21–24, 91

Nucleosynthesis 445, 509

Null geodesics *see* Light rays

Number of cusps theorem 213

Number of images 189, 236, 266, 295, 338
- changes of 165, 183, 189
- theorem 172–175

Numerical methods 295–308
- construction of lens models from extended images 307–308
- grid search method 300–303
- images of extended sources 298
- in microlensing simulations 303–307, 338–344
- interactive model fitting 299–300

Observables in GL systems 468–469
- dimensional vs. dimensionless observables 166, 468
- number of constraint for models 470–471

Observational biases 311, 379, 393, 396
- *see also* Amplification bias, Angular resolution bias

Observational cosmology 130

Observer sphere 30

Occultation 506

Odd number of images theorem 31, 35, 172–175, 237
- generalization 176, 284–285
- proof 174

On-average Friedmann–Lemaître universe *see* Clumpy universe

Optical axis 30, 146, 157, 167, 229, 421

Optical depth 23, 360, 447–448
- for compact halo objects 40, 457–458
- for cosmologically distributed point masses 347–348, 356–358, 442, 447
- for Galactic objects 39–40
- in microlensing 342, 423

Optical focusing equation *see* Focusing equation

Optical imaging surveys *see* Surveys for GLs

Optical scalars 106–109, 282
- geometrical interpretation 107
- transport equation 108, 140
- *see also* Focusing

Optically selected AGNs 378–380, 406

Optically violently variable QSOs 434, 447

Ordinary image 163–165, 183–185

Orientation of GL images 34, 164–165, 451–452
- radial and tangent direction 449–451

Overdensity of galaxies
- large-scale overdensities 418–419

- maximum overdensities 408–411
- near 3CR sources 86, 446
- near BL Lac objects 407, 434, 445
- near flat-spectrum radio sources 418–419
- near QSOs 405–406, 416–418, 445
- – see also QSO–galaxy associations
- statistical significance 415–418
- theoretical estimates 407–418
- – including microlensing 413–415
- – with smooth lenses 411–413

Paczyński, B. 86
Pair annihilation 460
Pairs of foreground–background objects 451
Palomar Observatory Sky Survey 48, 64
Pancake model 210
Parallax effect in microlensing 420, 508–509
Parameter estimates for GLs see Determination of parameters from lensing
Parametrized lens models 184, 299, 471, 474
Parity of images 34, 54, 162–165, 189
Pattern recognition 89
Pello, R. 76
Pennypacker, C. 445
Perfect fluid 120, 129
Perlick, V. 100
Perturbed axisymmetric lenses see Nearly axisymmetric lens, Quadrupole lens
Perturbed Plummer model 252–255
Petrosian, V. 10, 72
Phase function 92, 95, 98, 152
Photon number
- current density 99
- conservation law 100, 132, 161
Photons 95, 99
- four momentum 99–100
Physical associations see QSO–galaxy associations
Planck length and mass 510
Plane wave see Locally plane wave
Plummer model 245
- see also Perturbed Plummer model
Point-mass lens see Schwarzschild lens
Poisson's equation 160, 192, 272
Polarization vector 80, 98, 152
- transport equation 98

Polarization of light from supernovae 456
Polynomial normal form of mappings 187
Post-Minkowskian metric 122
Potential time delay see Time delay
Poynting vector 92
Press, W.H. 8–9, 384
Primordial abundance of elements 445, 509
Principle of equivalence 3, 119
Principle of stationary arrival time 125
- see also Fermat's principle
Probability distribution
- for magnification 329–334, 365–366, 391, 407, 436–440
- for shear 329, 350–351
- from Markov's method 321–322
Probability of lensing 5
- for arc formation 498–499
- for multiple images 380–401
- – for separations and flux ratios 381–398
- in clumpy universe 344–348
- self-consistent probabilities 346
- see also Statistical lensing, Cross-section, Optical depth
Proper distance 137
Proper length and time 128, 135
Proximity effect 504

QSO absorption lines 501–504
- broad absorption lines 71, 501
- classification 501
- Lyα forest 501
- – GL size estimate of Lyα clouds 58, 502–503
- – proximity effect 504
- metal absorption lines 59, 71, 501
- – association with foreground objects 70, 503
- – in QSO0957+561 48, 53, 502
- statistics 503–504
QSO broad lines see Broad emission lines
QSO catalog of Hewitt and Burbidge 405
QSO counts and surveys 41, 378–380
QSO masses, constraints from lensing 511–512
QSO pairs 86–87
QSO variability 377, 379–380
QSO–galaxy associations 404–419, 447–448

- models 407–417
- observations 404–407, 416–419
- *see also* Overdensity of galaxies

QSOs 6, 23, 39, 373
- as lensed nuclei of Seyfert galaxies 7, 433, 435
- cosmological evolution 379–380
- high redshift 379
- luminosity function 373
-- *see also* Source counts
- radio loud vs. radio quiet 42, 405, 419

Quadrupole lens 249–261, 277, 296
- caustics and critical curves 251–252
- in multiple lens plane theory 291–293
- two length-scales problem 249, 291
- *see also* Chang–Refsdal lens, Perturbed Plummer model

Quasars *see* QSOs
Quasiperiodic behavior 427
Quasi-stellar objects, radio sources *see* QSOs

Radial caustic and critical curve *see* Critical curve
Radio galaxies 23
- at high redshift 85–86, 446
Radio interferometry 42–43, 380
Radio rings *see* Einstein ring
Radio sources 385
- GL distortions 452
Radio surveys for GL candidates 42, 60, 71, 401–403
Random star field 57, 320–344
- correlated deflection probability 334–337
- deflection probability 322–328
- magnification fluctuation 343
- magnification probability 329–334
- scaling relation 330–331, 335, 344
- shear distribution 328–330
- statistical homogeneity 320, 420
- *see also* Microlensing

Ray shooting 303–307, 316, 338–344, 364–366
- *see also* Microlensing, Numerical methods
Ray system 104
Ray–trace equation *see* Lens equation
Recurrence of γ-ray bursters 460
Reciprocity relation *see* Etherington's reciprocity law
Reconstruction technique 79
Reddening of images 48, 50, 67
Redshift 128

Redshift controversy 66–67, 90, 404
Redshift difference of GL images 43, 50, 53, 58–59, 431
Redshift–distance relation 8, 132–143
- *see also* Distances in curved spacetimes
Redshift–magnitude relation 131
Refractive index 104, 123, 152
Refsdal, S. 6–7, 9, 56
Regularization method 507
Relative magnification 60, 469
- relative magnification matrix 54, 299, 470, 473
Representative mapping 187, 193, 200
Resolution limit 42, 311
- *see also* Angular resolution bias
Retarded potential 122
- retarded solution of linearized field equation 122
Ricci curvature tensor 91, 120
Ricci focusing 134, 137, 163, 360
Riemann curvature tensor 91, 120
Ring images *see* Einstein ring
Robertson, H.P. 128
Robertson–Walker metrics and spacetimes 128, 150, 152
Roeder, R.C. 8, 138
Roots of one-dimensional equations 296–297
ROSAT 55, 84
Rotation curves of galaxies 131, 243, 386, 445, 456

Sachs, R.K. 7
Salpeter mass function 426
Sanitt, N. 8, 50, 85
Sard's theorem 185
Scalar amplitude 98
Scalar curvature tensor 91
Scaling relations
- for random star field 330–331, 335, 344
- of cross-sections 316–318
Schechter function 387, 417, 451
Schwarzschild lens 6, 25–29, 148, 239–240
- angular separation 27, 37
- critical curve and caustic 36
- cross-sections
-- for flux ratio 311
-- for magnification 310, 313–315
-- for multiple images 312, 381
- length and angular scale 27, 159
- magnification 36–38, 220, 240, 422

557

- time delay 148, 220, 240
- wave optics effects 29, 220–221

Schwarzschild metric 25, 133–134
Schwarzschild radius 2, 21–22, 25
Second Reference Catalogue of Bright Galaxies 405
Selection function 402–403
Separation of images *see* Angular separation of images
Seyfert galaxies 7, 23
Shape of ordinary images 165
Shapiro, I.I. 6
Shapiro time delay 125
Shear 22, 162–163, 233, 328
- for multiple deflection 353–355
- of beams of photons 108, 134, 139
- tensor 108

Shear-induced lensing 461–464
Sheet of matter as GL 240–243
Signature of microlensing 420–421
Similarity of QSO spectra 45, 60, 86–87
Simple lens models 229–280
Simulated light curves *see* Microlensing
Singular isothermal sphere 243–244, 412–415
- image positions and magnification 244
- lensing cross-sections 312, 386–387
- plus smooth matter sheet 312–313, 386–395
- surface mass distribution 243
- time delay 244
- *see also* Isothermal sphere

Singular potentials 399
Singularities 211–213
- of differential maps 184
- of generic lens maps 197–198
- of wavefront 115
- stability and genericity 192, 196–197, 202, 207, 211
- *see also* Catastrophes

Smoothness parameter 22, 138, 143
Solar mass and radius 21
Soldner, J. 2, 514
Source counts 374
- lens-dominated 378, 391–392, 438, 440–441, 447
- observed 378–380, 409

Source plane 31, 151
Source size 373, 428–430, 435
Space curvature 130
Spacetime metric 91
Spatial light paths 124
Spatial metric of the universe 128

- spatial Riemannian metric 103

Specific flux and intensity 33, 116, 154–155, 217
- transport equation 116

Specific luminosity and surface brightness 115, 155, 217
Speckle interferometry 65
Spectral differences due to lensing 44–45
- *see also* Microlensing

Speculative GL candidates 84–86
Spherically symmetric lenses *see* Axi-symmetric GLs
Stability of ordinary images 185
Stability of singularities *see* Catastrophes
Standard candle 372, 448, 453
Standard cosmological model 127, 131, 406
State variable (in catastrophe theory) 184
Stationary phase 153, 218, 221
Stationary points 172–173
Statistical lensing 88–89, 309–466
- arc statistics 498–500
- distortion of images 448–452
- in clumpy universe 345–348
- multiply-image statistics 380–404
- nonlinear effects 321, 342, 355
- *see also* Compact objects as lenses, Cross-section, Light propagation in inhomogeneous universe, Microlensing, Overdensity of galaxies, QSO-galaxy associations, Supernovae

Statistical symmetry of microlensing light curves 425, 446
Stickel, M. 407
Stocke, J.T. 406
Stokes polarization parameters 154
Stress–energy–momentum tensor *see* Energy–momentum tensor
Strings *see* Cosmic strings
Sunyaev–Zeldovich effect 87
Superluminal motion 512–513
Supernovae 7, 453–456
- in arcs 497

Surdej, J. 71
Surface brightness 33
- constancy through light deflection 33, 161, 307

Surface mass density 32, 125, 158
- critical value 158, 199

- distribution for elliptical potentials 269–271
- mathematical properties 160, 235–236

Surveys for GLs 42, 71, 84, 401–404
Swallowtail singularity *see* Catastrophes
Swiss-cheese model 8, 136–138, 363
Symmetry breaking 131
Systematic alignments of images 451, 499–500

Tangential caustic and critical curve *see* Critical curve
Tensor field of electromagnetic field strengths 91, 98, 150
- *see also* Energy–momentum tensor

Test of General Relativity 3, 5
Third Cambridge Radio Sample (3CR) 71, 85, 379, 446
Tidal forces 250
- tidal gravitational field 29, 120, 139, 320
- tidal matrix 250

Time delay 6, 9, 43, 56, 143–147, 166–171, 240, 288–289
- and Hubble constant 7, 56–57, 468, 473–483
- derivation from wavefront method 169
- for image pairs 127, 147, 159, 167, 170, 470
- general expression 126, 146, 170
- geometrical term 9, 125–127, 144–145, 166
- of critical image pairs 66, 191
- potential term 9, 125–127, 145, 166
- relation to Fermat potential 170–171

Time-dependent magnification *see* Microlensing
Time-ordering of images 179
Topography of GL images 177–181
Topological defects 509
Totally geodesic submanifold 230
Transmission factor 154, 217–218, 223
Transport equation
- for amplitude and polarization 97–98
- for connection vectors 106
- for optical scalars 108–109
- for shear 109, 139

Transport of images 302–303
Transverse velocity 39, 69, 421
Truncation of Taylor expansion, significant terms 187–188
Turner, E.L. 392

Two length-scales problem 249, 255, 291, 312, 320, 420
Two lens plane theory *see* Multiple lens plane theory
Two point-mass lens 261–265
- arbitrary mass ratio 264
- as two lens plane problem 288
- equal masses 261–264
- with external shear 264–265

Type of images 163, 172, 185, 218, 244, 246, 255, 330
Tyson, J.A. 89, 405–406

Umbilic *see* Catastrophes
Unbiased sample 89
Uniform mass sheet 240, 412
Uniform ring as GL model 247–249

Vacuole model 138
Vacuum energy density 120, 129–130
Variability of flux 43, 421–425, 446
- time-scale 39
- *see also* Microlensing

Variational principle for light paths 124
- *see also* Fermat's priniple
Velocity dispersion 52, 243, 338, 387, 392, 451, 469, 482
Velocity of light 21
Very Large Array (VLA) 50, 384, 402
Very Long Baseline Interferometry (VLBI) 8, 51–54, 56, 384–385, 512
Volterra integral equation 353
Véron, P. 24

Walker, A.G. 128
Walsh, D. 9, 48, 50
Wave effects 35
Wave optics 150–155
- diffraction near folds 190, 222–227
- in GL theory 217–227

Wave vector of locally plane waves 94
Wavefronts 94, 132, 163, 176
- geometry 95, 166–170
- singularities 95, 115, 163

Weak field condition 122
Webster, R.L. 406, 416
Weyl conformal curvature tensor 91, 109, 137
Weymann, R.J. 9, 48, 64
Whitney, H. 197
Whitney's theorem 197
Wigwam singularity *see* Catastrophes
Winding number 487
WKB approximation 93–99, 117

X-ray background 378–379, 457
X-ray selected AGNs 380, 406, 446

Young, P. 10, 52, 57

Zwicky, F. 4–5, 309, 380
Zworykin, V.K. 5

3 5282 00491 6097

Druck: Strauss Offsetdruck, Mörlenbach
Verarbeitung: Schäffer, Grünstadt